Categorification and Higher Representation Theory

CONTEMPORARY MATHEMATICS

683

Categorification and Higher Representation Theory

Anna Beliakova
Aaron D. Lauda
Editors

American Mathematical Society
Providence, Rhode Island

2010 *Mathematics Subject Classification.* Primary 81R50, 17B10, 20C08, 14F05, 18D10, 17B50, 17B55, 17B67.

Library of Congress Cataloging-in-Publication Data

Library of Congress Cataloging-in-Publication Data
Names: Beliakova, Anna, 1968– editor. | Lauda, Aaron, 1981– editor.
Title: Categorification and higher representation theory / Anna Beliakova, Aaron Lauda, editors.
Description: Providence, Rhode Island : American Mathematical Society, [2017] | Series: Contemporary mathematics ; volume 683 | Includes bibliographical references.
Identifiers: LCCN 2016041983 | ISBN 9781470424602 (alk. paper)
Subjects: LCSH: Categories (Mathematics) | Mathematical analysis. | Algebra. | AMS: Quantum theory – Groups and algebras in quantum theory – Quantum groups and related algebraic methods. msc | Nonassociative rings and algebras – Lie algebras and Lie superalgebras – Representations, algebraic theory (weights). msc | Group theory and generalizations – Representation theory of groups – Hecke algebras and their representations. msc | Algebraic geometry – (Co)homology theory – Sheaves, derived categories of sheaves and related constructions. msc | Category theory; homological algebra – Categories with structure – Monoidal categories (multiplicative categories), symmetric monoidal categories, braided categories. msc | Nonassociative rings and algebras – Lie algebras and Lie superalgebras – Modular Lie (super)algebras. msc | Nonassociative rings and algebras – Lie algebras and Lie superalgebras – Homological methods in Lie (super)algebras. msc | Nonassociative rings and algebras – Lie algebras and Lie superalgebras – Kac-Moody (super)algebras; extended affine Lie algebras; toroidal Lie algebras. msc
Classification: LCC QA169 .C3744 2017 | DDC 512/.62–dc23 LC record available at https://lccn.loc.gov/2016041983
Contemporary Mathematics ISSN: 0271-4132 (print); ISSN: 1098-3627 (online)

DOI: http://dx.doi.org/10.1090/conm/683

♾ The paper used in this book is acid-free and falls within the guidelines
established to ensure permanence and durability.
Visit the AMS home page at http://www.ams.org/

10 9 8 7 6 5 4 3 2 1 22 21 20 19 18 17

Dedicated to Mikhail,
who taught us to look higher

Contents

Preface

The emergent mathematical philosophy of categorification is reshaping our view of modern mathematics by uncovering a hidden layer of structure in mathematics, revealing richer and more robust structures capable of describing more complex phenomena. This philosophy has led to a number of shocking new results and has ushered in a new area of interaction between algebra, geometry, and theoretical physics. Categorification is a powerful tool for relating various branches of mathematics and exploiting the commonalities between fields. It provides a language emphasizing essential features and allowing precise relationships between vastly different fields.

Categorified representation theory, or higher representation theory, aims to understand a new level of structure present in representation theory. Rather than studying actions of algebras on vector spaces where algebra elements act by linear endomorphisms of the vector space, higher representation theory describes the structure present when algebras act on categories, with algebra elements acting by functors. The new level of structure in higher representation theory arises by studying the natural transformations between functors. Often these natural transformations can be systematically described as representations of some monoidal category that categorifies the original algebra.

Representation theory has proven to be an especially fertile ground for categorification. This is in large part due to the geometric methods which pervade the subject. Geometric representation theory provides tools not only for classifying and studying representations but also for constructing natural bases with positivity and integrality properties. Through 'geometrization', various representation theoretic objects are realized via categories of sheaves on algebraic varieties. Passing from geometrization to categorification, these categories of sheaves are reinterpreted as categorical representations and the structure of functors between these categories, along with the higher structure of natural transformations, are abstracted and combinatorially encoded into higher categorical structures. Under categorification, basis elements are reinterpreted as indecomposable objects in a category, while structure constants become dimensions of Hom spaces. The positivity and integrality are then manifest. This enhanced perspective brings into play a powerful new set of tools that deepens our understanding of traditional representation theory.

This volume exhibits some of the current trends in higher representation theory and the diverse techniques that are being employed. These articles illustrate many important trends for the field including categorifications of quantum groups, Hecke algebras, and Heisenberg algebras, as well as related geometries. Some of the most pressing current directions are explored including roots of unity, the role of Cherednik algebras, higher analogs of crystal bases, interactions with modular representation theory, geometric representation theory, and topology. This volume also showcases the many applications of higher representation theory to classical representation theory.

This volume was inspired by the conference Categorification in Algebra, Geometry, and Physics (a conference in honor of the 60th Birthday of Christian Blanchet), which took place at IESC, Menasina, Cargese, France, from May 4th to May 8th, 2015. This conference made clear that categorification is a rapidly emerging area of intense study. It also elucidated the need for a reference for newcommers to the field to learn the types of tools used in categorification, the problems where these tools have been successfully applied, and the future directions in which the field is moving. Our aim with this volume is to address this need. To this end, we have solicited articles from experts in categorification from around the world who were invited to share their unique perspective.

The organizers wish to thank the John Templeton Foundation and the University of Zurich for their generous support in making this conference possible.

Anna Beliakova
Aaron Lauda

Contemporary Mathematics
Volume **683**, 2017
http://dx.doi.org/10.1090/conm/683/13722

Rational Cherednik algebras and categorification

Ivan Losev

ABSTRACT. In this survey article we review Kac-Moody and Heisenberg algebra actions on the categories $\mathcal{O}$ of the rational Cherednik algebras associated to groups $G(\ell,1,n)$. Using these actions we solve basic representation theoretic problems for these categories such as the classification of finite dimensional irreducible representations and computation of characters of the irreducibles.

1. Introduction

In this survey we study categorical actions of Lie algebras on categories $\mathcal{O}$ for rational Cherednik algebras. Rational Cherednik algebras were introduced by Etingof and Ginzburg in [**EG**] (in fact, the definition is already contained in [**D**], though in a somewhat implicit and sketchy form). These are associative algebras $H_{\mathbf{c}}(W)$ constructed from complex reflection groups W, where $\mathbf{c}$ is a parameter (see Section 2.1). They share many common features with the universal enveloping algebras of semisimple Lie algebras. In particular, one can consider the category $\mathcal{O}_{\mathbf{c}}(W)$ for $H_{\mathbf{c}}(W)$ that is similar in many aspects to the BGG categories $\mathcal{O}$, see Section 2.2.

For the purposes of this paper, the most important family of complex reflection groups is the infinite series $G(\ell,1,n)\cong \mathfrak{S}_n \ltimes (\mathbb{Z}/\ell\mathbb{Z})^n$. This is because the category $\mathcal{O}_{\mathbf{c}} := \bigoplus_{n\geqslant 0}\mathcal{O}_{\mathbf{c}}(G(\ell,1,n))$ (for some nice, in a sense, most interesting choices of the parameter $\mathbf{c}$) carries categorical actions of the Kac-Moody algebras $\hat{\mathfrak{sl}}_e$ ([**Sh**]) and of the Heisenberg algebra $\mathfrak{Heis}$ ([**SV**]), see Sections 3.2 and 3.5. These actions provide a categorification of the classical actions of the aforementioned algebras on the level ℓ Fock space $\mathcal{F}^\ell$ that are of importance as they provide an explicit combinatorial model for integrable representations of the algebras $\hat{\mathfrak{sl}}_e$ and $\mathfrak{Heis}$. The categorification of the $\hat{\mathfrak{sl}}_e$-action is one of the most interesting examples of categorical Kac-Moody actions on highest weight categories, while the categorical action of $\mathfrak{Heis}$ we consider is, perhaps, the most general known representation theoretic example of the categorical actions of this algebra.

In the other direction, our categorical actions play a crucial role in the study of the categories $\mathcal{O}_{\mathbf{c}}$. For example, they are of great help in determining the multiplicity formulas, [**GL, RSVV, Lo4, W**], this is studied in Section 5. In more detail, one can establish various (technical) uniqueness results for categorifications of Fock

2010 *Mathematics Subject Classification.* Primary 05E10, 16G99, 17B67, 20C08.

spaces and hence prove equivalences of the categories $\mathcal{O}_{\mathbf{c}}$ with categories, where the multiplicities are known.

Another application is to the computation of supports of simple objects in $\mathcal{O}_{\mathbf{c}}$ (and, in particular, in classifying the finite dimensional irreducible $H_{\mathbf{c}}(G(\ell,1,n))$-modules, that is a very basic representation theoretic problem), [**Lo2**, **SV**, **Lo8**], see Section 4. The paper [**SV**] counts the number of simples with given support proving a conjecture of Etingof from [**Et**]. The papers [**Lo2**, **Lo8**] provide combinatorial recipes for computing the support of a given simple object.

Another important object studied in this paper is the wall-crossing functors from [**Lo6**] that are derived equivalences between Cherednik categories $\mathcal{O}$ with different parameters.

This paper consists of four sections. It is mostly a survey paper, the only new results appear in Section 5.7.

In Section 2 we deal with the general rational Cherednik algebras. We recall the definitions of rational Cherednik algebras, Section 2.1, and of their categories $\mathcal{O}$, Section 2.2, recall a connection between the categories $\mathcal{O}_{\mathbf{c}}(W)$ and the Hecke algebra $\mathcal{H}_{\mathbf{q}}(W)$, Section 2.3. Then we recall induction and restriction functors for categories $\mathcal{O}$ constructed in [**BE**], Section 2.4. We finish by producing abelian, Section 2.5, and derived equivalences, Section 2.6, between categories $\mathcal{O}_{\mathbf{c}}(W), \mathcal{O}_{\mathbf{c}'}(W)$ for different parameters $\mathbf{c}, \mathbf{c}'$. Here we follow [**R2**, **Lo6**].

In Section 3 we concentrate on the categories $\mathcal{O}_{\mathbf{c}}(G(\ell,1,n))$. We introduce a Kac-Moody categorical action on $\mathcal{O}_{\mathbf{c}}$ in Section 3.2 following [**Sh**]. Section 3.3 deals with crystals. There we define crystals for categorical actions and explain results of [**Lo2**] that compute crystals for actions on highest weight categories. In Section 3.4 we explain an equivalence of $\mathcal{O}_c(\mathfrak{S}_n)$ and the category of modules over a q-Schur algebra $S_\epsilon(n,n)$, [**R2**, **Lo4**]. Using this equivalence we construct a categorical Heisenberg action on $D^b(\mathcal{O}_{\mathbf{c}})$ following [**SV**] in Section 3.5.

In Section 4 we will apply categorical actions and wall-crossing bijections from Section 2.6 to study the supports of simple modules. We will see that the supports are described by two integers that are computed using crystals associated to Kac-Moody and Heisenberg actions. We will compute the filtration by support on K_0 following [**SV**]. We will also explain combinatorial rules to compute supports that were found in [**Lo8**].

In Section 5 we explain a proof of a conjecture of Varagnolo and Vasserot, [**VV**], on an equivalence of the categories $\mathcal{O}_{\mathbf{c}}(G(\ell,1,n))$ and certain truncations of affine parabolic categories of type A. We follow [**Lo4**] – the conjecture was first proved in [**RSVV**] but the approach from [**Lo4**] relies on the categorical actions much more than that of [**RSVV**]. The category equivalence allows to compute the multiplicities in $\mathcal{O}_{\mathbf{c}}(G(\ell,1,n))$ proving a conjecture of Rouquier, [**R2**].

To finish the introduction we would like to briefly mention a different potential connection of Rational Cherednik algebras to categorification, namely to Khovanov-Rozansky homology of torus knots. The homology is expected to be related to bi-graded characters of the minimally supported modules over the algebras $H_c(\mathfrak{S}_n)$ that appear in Section 3.4. This relation is not completely understood and is a bit far from the topics of the present survey so we just refer the reader to [**GORS**] for details.

Acknowledgements. I would like to thank Sasha Kleshchev for stimulating discussions. This work was partially supported by the NSF under grants DMS-1161584, DMS-1501558.

2. Rational Cherednik algebras and categories $\mathcal{O}$

2.1. Rational Cherednik algebras. In this section we will introduce Rational Cherednik algebras following the seminal paper of Etingof and Ginzburg, [**EG**].

2.1.1. *Complex reflection groups.* Let $\mathfrak{h}$ be a finite dimensional vector space over $\mathbb{C}$ and let W be a finite subgroup of $\mathrm{GL}(\mathfrak{h})$. By a (complex) reflection in W we mean an element s such that $\mathrm{rk}(s-1)=1$. We write S for the set of all reflections in W. We say that W is a *complex reflection group* if W is generated by S. For example, every Coxeter group is a complex reflection group. Here is another family of examples that is of importance for this paper.

EXAMPLE 2.1. Let ℓ and n be positive integers. We form the group $G(\ell,1,n):=\mathfrak{S}_n\ltimes\mu_\ell^n$, where $\mathfrak{S}_n$ stands for the symmetric group on n letters and $\mu_\ell\subset\mathbb{C}^\times$ is the group of ℓth roots of 1. The group $\mathfrak{S}_n$ acts on μ_ℓ^n by permuting the factors and we use this action to form the semi-direct product. The group $G(\ell,1,n)$ acts on $\mathfrak{h}:=\mathbb{C}^n$ as follows. The group $\mathfrak{S}_n$ permutes the coordinates. Each factor μ_ℓ acts on its own copy of $\mathbb{C}$ by $\eta.z=\eta z,\eta\in\mu_\ell,z\in\mathbb{C}$.

Let us describe the subset S of complex reflections. For $\eta\in\mu_\ell$, let $\eta_{(k)}$ denote η in the kth copy of $\mu_\ell\subset G(\ell,1,n)$. All these elements are reflections. The other reflections are $(ij)\eta_{(i)}\eta_{(j)}^{-1}$, where $i<j\in\{1,\ldots,n\}$ and $\eta\in\mu_\ell$. Here we write (ij) for the transposition in $\mathfrak{S}_n$ that permutes i and j. Clearly, S generates W and so W is a complex reflection group.

We note that, for $\ell=1$, we get $G(1,1,n)=\mathfrak{S}_n$, while, for $\ell=2$, we get the Weyl group of type B. All other groups $G(\ell,1,n)$ are not Coxeter groups.

The classification of complex reflection groups is known thanks to [**ST**]. There is one infinite series, $G(\ell,r,n)$, where r divides ℓ. The group $G(\ell,r,n)$ is the normal subgroup of $G(\ell,1,n)$ consisting of all elements of the form $\sigma\prod_{i=1}^n\eta_{j,(i)}$, where $\sigma\in\mathfrak{S}_n$ and $\prod_{i=1}^n\eta_j^r=1$. For example, $G(2,2,n)$ is the Weyl group of type D_n. And then there are several exceptional groups.

2.1.2. *Definition of RCA.* A Rational Cherednik algebra (RCA) depends on a parameter $\mathbf{c}$, where $\mathbf{c}:S\to\mathbb{C}$ is a conjugation invariant function. We write $\mathfrak{P}$ for the space of all possible parameters $\mathbf{c}$. Clearly, $\mathfrak{P}$ is a vector space whose dimension equals $|S/W|$, the number of W-conjugacy classes in S.

For $s\in S$, let $\alpha_s\in\mathfrak{h}^*$ and $\alpha_s^\vee\in\mathfrak{h}$ be eigenvectors for s with eigenvalues different from 1 (analogs of roots and coroots for Weyl groups). We partially normalize them by requiring $\langle\alpha_s,\alpha_s^\vee\rangle=2$.

Recall that for an algebra A and a finite group Γ acting on A by automorphisms we can form the *smash-product* algebra $A\#\Gamma$. As a vector space, this algebra is $A\otimes\mathbb{C}\Gamma$, while the product is given by $(a_1\otimes\gamma_1)\cdot(a_2\otimes\gamma_2)=a_1\gamma_1(a_2)\otimes\gamma_1\gamma_2$, where $\gamma_1(a_2)$ denotes the image of a_2 under the action of γ_1. The definition is given in such a way that an $A\#\Gamma$-module is the same thing as a Γ-equivariant A-module.

Now we define the RCA $H_{\mathbf{c}}$ as the quotient of $T(\mathfrak{h}\oplus\mathfrak{h}^*)\#W$ by the following relations:
(2.1)
$$[x,x']=[y,y']=0,\quad [y,x]=\langle y,x\rangle-\sum_{s\in S}\mathbf{c}(s)\langle\alpha_s,x\rangle\langle\alpha_s^\vee,y\rangle s,\quad x,x'\in\mathfrak{h}^*,y,y'\in\mathfrak{h}.$$

Note that we get $H_0=D(\mathfrak{h})\#W$.

When we need to indicate the dependence of $H_{\mathbf{c}}$ on W or on $\mathfrak{h}$ and W, we write $H_{\mathbf{c}}(W)$ or $H_{\mathbf{c}}(W,\mathfrak{h})$.

Note that we can define the algebra $H_{R,\mathbf{c}}$ over a $\mathbb{C}$-algebra R. For $\mathbf{c}$ we take a W-invariant map $S\to R$.

2.1.3. *Examples.* Let us give two concrete examples: for groups μ_ℓ and $\mathfrak{S}_n$.

EXAMPLE 2.2. Let $W=\mathbb{Z}/\ell\mathbb{Z}$ and $\dim\mathfrak{h}=1$. Set $C:=1-2\sum_{s\in\mu_\ell\setminus\{1\}}\mathbf{c}(s)s$. Then $H_{\mathbf{c}}=\mathbb{C}\langle x,y\rangle\#W/([y,x]=C)$.

EXAMPLE 2.3. Let $W=\mathfrak{S}_n$ and $\mathfrak{h}$ be its reflection representation,
$$\mathfrak{h}=\{(x_1,\ldots,x_n)\in\mathbb{C}^n|x_1+\ldots+x_n=0\}.$$
We have one class of complex reflections, and so $\mathbf{c}$ is a single complex number (to be denoted by c). Then H_c is the quotient of $\mathbb{C}\langle x_1,\ldots,x_n,y_1,\ldots,y_n\rangle\#\mathfrak{S}_n$ by the following relations:
$$\sum_{i=1}^n x_i=\sum_{i=1}^n y_i=0,$$
$$[x_i,x_j]=[y_i,y_j]=0,\forall i,j\in\{1,\ldots,n\},$$
$$[y_i,x_j]=c(ij),i\neq j,$$
$$[y_i,x_i]=1-c\sum_{j\neq i}(ij).$$

2.1.4. *PBW property and triangular decomposition.* The algebra $H_{\mathbf{c}}$ is filtered with $\deg\mathfrak{h}^*=\deg W=0,\deg\mathfrak{h}=1$ (the filtration by the "order of a differential operator"). We have a natural epimorphism $S(\mathfrak{h}\oplus\mathfrak{h}^*)\#W\twoheadrightarrow\operatorname{gr}H_{\mathbf{c}}$.

The following fundamental result (the PBW property for Rational Cherednik algebras) is due to Etingof and Ginzburg, see [**EG**, Theorem 1.3].

THEOREM 2.4. *The epimorphism $S(\mathfrak{h}\oplus\mathfrak{h}^*)\#W\twoheadrightarrow\operatorname{gr}H_{\mathbf{c}}$ is an isomorphism.*

The following corollary of Theorem 2.4 was observed in [**GGOR**, Section 3].

COROLLARY 2.5. *The multiplication map $S(\mathfrak{h}^*)\otimes\mathbb{C}W\otimes S(\mathfrak{h})\to H_{\mathbf{c}}$ is an isomorphism of vector spaces.*

The isomorphism $S(\mathfrak{h}^*)\otimes\mathbb{C}W\otimes S(\mathfrak{h})\xrightarrow{\sim}H_{\mathbf{c}}$ is often called the *triangular decomposition.* The reader should compare it to the triangular decomposition $U(\mathfrak{n}^-)\otimes U(\mathfrak{t})\otimes U(\mathfrak{n})\xrightarrow{\sim}U(\mathfrak{g})$ for a complex semisimple Lie algebra $\mathfrak{g}$.

2.1.5. *Spherical subalgebras.* Let $e=|W|^{-1}\sum_{w\in W}w$ be the trivial idempotent in $\mathbb{C}W\subset H_{\mathbf{c}}$. Consider the subspace $eH_{\mathbf{c}}e\subset H_{\mathbf{c}}$. It is closed under multiplication, and e is a unit. We call $eH_{\mathbf{c}}e$ the *spherical subalgebra* of $H_{\mathbf{c}}$. Note that $\operatorname{gr}eH_{\mathbf{c}}e=e(S(\mathfrak{h}\oplus\mathfrak{h}^*)\#W)e\cong S(\mathfrak{h}\oplus\mathfrak{h}^*)^W$.

We have a functor $H_{\mathbf{c}}$-mod $\to eH_{\mathbf{c}}e$-mod, $M\mapsto eM(=M^W)$. This functor is an equivalence if and only if $H_{\mathbf{c}}=H_{\mathbf{c}}eH_{\mathbf{c}}$. In this case, we say that the parameter $\mathbf{c}$ is *spherical.*

It was shown in [**Lo6**, Corollary 3.5] that there is a non-empty Zariski open subset of spherical parameters in $\mathfrak{P}$.

2.1.6. *Generalizations and ramifications.* Rational Cherednik algebras are a special case of Symplectic reflection algebras also introduced in [**EG**]. Those are filtered deformations of $S(V)\#\Gamma$, where V is a symplectic vector space and Γ is a finite group of linear symplectomorphisms of V.

On the other hand, there are trigonometric and elliptic deformations of $H_{\mathbf{c}}(W)$ that make sense in the case when W is a Weyl group, see an explanation in [**EG**, Section 1]. The elliptic deformation is known as the double affine Hecke algebra (DAHA) and was introduced earlier by Cherednik (hence the name "Rational Cherednik algebra").

2.2. Categories $\mathcal{O}$. Here we will define the category $\mathcal{O}_{\mathbf{c}}(W)$ for the RCA $H_{\mathbf{c}}$ following [**GGOR**].

2.2.1. *Definition of the category and Verma modules.* By definition, the category $\mathcal{O}_{\mathbf{c}}(W)$ consists of all modules with locally nilpotent action of $\mathfrak{h} \subset H_{\mathbf{c}}$ that are finitely generated over $H_{\mathbf{c}}$ or, equivalently (under the condition that $\mathfrak{h}$ acts locally nilpotently), over $S(\mathfrak{h}^*) \subset H_{\mathbf{c}}$.

The triangular decomposition allows one to define Verma modules over $H_{\mathbf{c}}$ that give examples of objects in $\mathcal{O}_{\mathbf{c}}$. The subalgebra $S(\mathfrak{h})\#W \subset H_{\mathbf{c}}$ plays the role of $U(\mathfrak{b}) \subset U(\mathfrak{g})$ and $\mathbb{C}W \subset S(\mathfrak{h})\#W$ plays the role of $U(\mathfrak{t})$. So the Verma modules are parameterized by the irreducible representations of W: given $\tau \in \operatorname{Irr}(W)$, we set $\Delta_{\mathbf{c}}(\tau) := H_{\mathbf{c}} \otimes_{S(\mathfrak{h})\#W} \tau$, where $\mathfrak{h}$ acts on τ by 0. Thanks to the triangular decomposition, the Verma module $\Delta_{\mathbf{c}}(\tau)$ is naturally isomorphic to $S(\mathfrak{h}^*)\otimes\tau$ as an $S(\mathfrak{h}^*)\#W$-module (where W acts diagonally and $S(\mathfrak{h}^*)$ acts by multiplications on the left).

We identify $K_0(\mathcal{O}_{\mathbf{c}}(W))$ with $K_0(W\text{-mod})$ by sending the class $[\Delta_{\mathbf{c}}(\tau)]$ to $[\tau]$.

Let us now give a combinatorial description of the irreducible representations of $G(\ell, 1, n)$.

EXAMPLE 2.6. The set $\operatorname{Irr}(G(\ell,1,n))$ is in a natural bijection with the set $\mathcal{P}_\ell(n)$ of ℓ-multipartitions λ of n, i.e., the set of ℓ partitions $\lambda^{(0)}, \ldots, \lambda^{(\ell-1)}$ with $\sum_{i=0}^{\ell-1} |\lambda^{(i)}| = n$. An irreducible module V_λ corresponding to $\lambda \in \mathcal{P}_\ell(n)$ is constructed as follows.

The product $G(\ell,1,\lambda) := \prod_{i=0}^{\ell-1} G(\ell,1,|\lambda^{(i)}|)$ naturally embeds into $G(\ell,1,n)$. Let $V_{\lambda^{(i)}}$ denote the irreducible $\mathfrak{S}_{|\lambda^{(i)}|}$-module labeled by the partition $\lambda^{(i)}$. We equip $V_{\lambda^{(i)}}$ with the structure of a $G(\ell,1,|\lambda^{(i)}|)$-module by making all $\eta_{(j)}$ act by η^i. Denote the resulting $G(\ell,1,|\lambda^{(i)}|)$-module by $V^{(i)}_{\lambda^{(i)}}$. Let V_λ denote the $G(\ell,1,n)$-module induced from the $G(\ell,1,\lambda)$-module $V^{(0)}_{\lambda^{(0)}} \boxtimes V^{(1)}_{\lambda^{(1)}} \boxtimes \ldots \boxtimes V^{(\ell)}_{\lambda^{(\ell-1)}}$. The modules V_λ form a complete collection of the irreducible $G(\ell,1,n)$-modules.

2.2.2. *Euler element and c-function.* There is a so called *Euler element* $h \in H_{\mathbf{c}}$ satisfying $[h,x] = x, [h,y] = -y, [h,w] = 0$. It is constructed as follows. Pick a basis $y_1, \ldots, y_n \in \mathfrak{h}$ and let $x_1, \ldots, x_n \in \mathfrak{h}^*$ be the dual basis. For $s \in S$, let λ_s denote the eigenvalue of s in $\mathfrak{h}^*$ different from 1. Then

$$h = \sum_{i=1}^{n} x_i y_i + \frac{n}{2} - \sum_{s\in S} \frac{2c(s)}{1 \quad \lambda_s} s. \tag{2.2}$$

Using h, we can see that every finite dimensional module M lies in $\mathcal{O}_{\mathbf{c}}(W)$. Indeed, the number of h-eigenvalues in M is finite and so $\mathfrak{h}$ acts nilpotently on M.

The Euler element acts on $\tau \subset \Delta_{\mathbf{c}}(\tau)$ by a scalar denoted by c_τ (and called the c-function of τ).

We will need to compute the function c_τ for the groups $G(\ell,1,n)$ (up to a summand independent of τ). For this, it is convenient to introduce new parameters.

Let H denote a hyperplane of the form $\mathfrak{h}^s$, $s\in S$. We can find elements $h_{H,j}\in\mathbb{C}$ with $j=0,\ldots,\ell_H-1$ and $h_{H,j}=h_{H',j}$ for $H'\in WH$ such that

$$c(s)=\sum_{j=1}^{\ell-1}\frac{1-\lambda_s^j}{2}(h_{\mathfrak{h}^s,j}-h_{\mathfrak{h}^s,j-1}). \tag{2.3}$$

Clearly, for fixed H, the numbers $h_{H,0},\ldots,h_{H,\ell_H-1}$ are defined up to a common summand.

We can recover the elements $h_{H,i}$ by the formula

$$h_{H,i}=\frac{1}{\ell_H}\sum_{s\in W_H\setminus\{1\}}\frac{2c(s)}{\lambda_s-1}\lambda_s^{-i}. \tag{2.4}$$

Note that $\sum_{i=0}^{\ell_H-1}h_{H,i}=0$ in this case.

Note that, for $W=G(\ell,1,n)$ with $\ell,n>1$, we have two conjugacy classes of hyperplanes, their representatives are given by the equations $x_1=0$ and $x_1=x_2$. Set $\kappa:=-c(s)$, where s is a reflection in $\mathfrak{S}_n$, and $h_i:=h_{H,i}$, where H is the hyperplane $x_1=0$.

Let us get back to computing the functions c_λ, where $\lambda\in\mathcal{P}_\ell(n)\cong\operatorname{Irr}(G(\ell,1,n))$. We view elements of $\mathcal{P}_\ell(n)$ as ℓ-tuples of Young diagrams. Let b be a box of λ. It can be characterized by three numbers x,y,i, where x is the number of column, y is the number of row, and i is the number of the diagram $\lambda^{(i)}$ containing b. Further, we set $c_b:=\kappa\ell(x-y)+\ell h_i$ and $c_\lambda:=\sum_{b\in\lambda}c_b$. Up to a summand independent of λ, the number c_λ coincides with c_{V^λ} defined above. This is proved in [**R2**, Section 6.1].

2.2.3. *Simple objects and finiteness properties.* Using the Euler element one can establish many basic structural results about the categories $\mathcal{O}_{\mathbf{c}}(W)$. For example, let us describe the simple objects in $\mathcal{O}_{\mathbf{c}}(W)$.

PROPOSITION 2.7. *Every Verma module $\Delta_{\mathbf{c}}(\tau)$ has a unique simple quotient, to be denoted by $L_{\mathbf{c}}(\tau)$. The map $\tau\mapsto L_{\mathbf{c}}(\tau)$ is a bijection between* $\operatorname{Irr}(W)$ *and* $\operatorname{Irr}(\mathcal{O}_{\mathbf{c}}(W))$.

Also we can establish the following result about the structure of $\mathcal{O}_{\mathbf{c}}(W)$. The proof is in [**GGOR**, Section 2], see, in particular, [**GGOR**, Corollaries 2.8,2.16].

PROPOSITION 2.8. *The category $\mathcal{O}_{\mathbf{c}}(W)$ has enough projectives and all objects there have finite length.*

The two propositions imply that $\mathcal{O}_{\mathbf{c}}(W)$ is equivalent to the category of modules over the finite dimensional algebra $\operatorname{End}_{\mathcal{O}_{\mathbf{c}}(W)}(P)^{opp}$, where $P:=\bigoplus_{\tau\in\operatorname{Irr}(W)}P_{\mathbf{c}}(\tau)$, and $P_{\mathbf{c}}(\tau)$ denotes the projective cover of $L_{\mathbf{c}}(\tau)$.

2.2.4. *Highest weight structure.* The classical BGG category $\mathcal{O}$ has certain upper triangularity properties that are formalized in the notion of a highest weight category. Here we will see that the categories $\mathcal{O}_{\mathbf{c}}(W)$ are highest weight as well, this result was established in [**GGOR**].

Let us start by recalling the general notion of a highest weight category (over a field). Let $\mathcal{C}$ be a $\mathbb{C}$-linear abelian category equivalent to A-mod for some finite dimensional algebra A. For $L \in \mathrm{Irr}(\mathcal{C})$, let P_L denote the projective cover of L. Equip $\mathrm{Irr}(\mathcal{C})$ with a partial order $\leqslant$. For $L \in \mathrm{Irr}(\mathcal{C})$, consider the subcategory $\mathcal{C}_{\leqslant L}$, the Serre span of $L' \in \mathrm{Irr}(\mathcal{C})$ with $L' \leqslant L$. Let Δ_L denote the projective cover of L in $\mathcal{C}_{\leqslant L}$. The object Δ_L coincides with the maximal quotient of P_L lying in $\mathcal{C}_{\leqslant L}$.

DEFINITION 2.9. We say that $\mathcal{C}$ is a highest weight category (with respect to the order $\leqslant$) if, for every L, the kernel of $P_L \twoheadrightarrow \Delta_L$ is filtered by $\Delta_{L'}$ with $L' > L$.

The objects Δ_L are called *standard.*

Let us get back to the categories $\mathcal{O}_{\mathbf{c}}(W)$. Fix a parameter $\mathbf{c}$. Define a partial order $\leqslant_{\mathbf{c}}$ on $\mathrm{Irr}(W)$ as follows. Recall that to $\tau \in \mathrm{Irr}(W)$ we can assign the complex number c_τ depending on $\mathbf{c}$. We set $\tau \leqslant_{\mathbf{c}} \xi$ if $\tau = \xi$ or $c_\tau - c_\xi \in \mathbb{Z}_{>0}$.

The following result is established in [**GGOR**, Theorem 2.19].

PROPOSITION 2.10. *The category $\mathcal{O}_{\mathbf{c}}(W)$ is highest weight with respect to the order $\leqslant_{\mathbf{c}}$. The standard $\Delta_{L(\tau)}$ coincides with the Verma module $\Delta_{\mathbf{c}}(\tau)$.*

2.2.5. *Costandard and tilting objects.* Recall that in any highest weight category $\mathcal{C}$ one has costandard objects $\nabla_L, L \in \mathrm{Irr}(\mathcal{C})$, with $\dim \mathrm{Ext}^i(\Delta_L, \nabla_{L'}) = \delta_{i,0}\delta_{L,L'}$.

In the case of the category $\mathcal{O}_{\mathbf{c}}(W)$ one can construct the costandard objects $\nabla_{\mathbf{c}}(\tau)$ as follows. Consider the parameter $\mathbf{c}^*$ defined by $c^*(s) := -c(s^{-1})$. There is an isomorphism $H_{\mathbf{c}}(W,\mathfrak{h}) \xrightarrow{\sim} H_{\mathbf{c}^*}(W,\mathfrak{h}^*)^{opp}$ that is the identity on $\mathfrak{h}^*, \mathfrak{h}$ and is the inversion on W. This isomorphism allows to define a contravariant equivalence (the *naive duality*) $\mathcal{O}_{\mathbf{c}^*}(W,\mathfrak{h}^*) \xrightarrow{\sim} \mathcal{O}_{\mathbf{c}}(W,\mathfrak{h})$ that maps M to its restricted dual $M^\vee := \bigoplus_{a\in\mathbb{C}} M_a^*$, where we write M_a for the generalized eigenspace with eigenvalue a of h in M. We set $\nabla_{\mathbf{c}}(\tau) := \Delta_{\mathbf{c}^*}(\tau^*)^\vee$, where τ^* is the dual to τ.

Recall that by a tilting object in a highest weight category one means an object that is both standardly filtered (i.e., admits a filtration with standard quotients) and is costandardly filtered. The indecomposable tilting objects are in bijection with $\mathrm{Irr}(\mathcal{C})$: for any $L \in \mathcal{C}$, there is a unique indecomposable tilting object T_L that admits an inclusion $\Delta_L \hookrightarrow T_L$ with standardly filtered cokernel. We write $T_{\mathbf{c}}(\tau)$ instead of $T_{L_{\mathbf{c}}(\tau)}$.

2.2.6. *Generic semisimplicity.* The c-function and the highest weight structure give a sufficient (but not necessary) criterium for $\mathcal{O}_{\mathbf{c}}(W)$ to be semisimple. Let us write $\varpi_{\xi\tau}$ for the element of $\mathfrak{P}^*$ given by $\mathbf{c} \mapsto c_\xi - c_\tau$.

LEMMA 2.11. *Suppose that $\varpi_{\xi\tau}(\mathbf{c}) \notin \mathbb{Z}\setminus\{0\}$ for all $\xi, \tau \in \mathrm{Irr}(W)$. Then the category $\mathcal{O}_{\mathbf{c}}(W)$ is semisimple.*

2.2.7. *Supports.* Every object $M \in \mathcal{O}_{\mathbf{c}}(W)$ is finitely generated over $S(\mathfrak{h}^*) = \mathbb{C}[\mathfrak{h}]$. Thanks to this we can define the support $\mathrm{Supp}(M)$ of M in $\mathfrak{h}$, this will be the support of M viewed as a coherent sheaf on $\mathfrak{h}$. By definition, this is a closed subvariety in $\mathfrak{h}$.

It turns out that $\mathrm{Supp}(M)$ is the union of strata for the *stabilizer stratification* of $\mathfrak{h}$. The strata are numbered by the conjugacy classes of possible stabilizers for the W-action on $\mathfrak{h}$ (these stabilizers are called *parabolic subgroups* of W). Namely, to a parabolic subgroup $\underline{W} \subset W$ we assign the locally closed subvariety $X(\underline{W}) := \{b \in \mathfrak{h} | W_b = \underline{W}\}$. Note that $\overline{X(\underline{W})} := \bigsqcup_{W'} X(W')$, where the union is taken over the conjugacy classes of all parabolic subgroups W' containing a conjugate of $\underline{W}$.

Clearly, for an exact sequence $0 \to M' \to M \to M'' \to 0$, we have $\mathrm{Supp}(M) = \mathrm{Supp}(M') \cup \mathrm{Supp}(M'')$. This, in principle, reduces the computation of supports to the case of simple modules. The following result was proved in [**BE**, Section 3.8].

LEMMA 2.12. *Let* $L \in \mathrm{Irr}(\mathcal{O}_{\mathbf{c}}(W))$. *Then there is a parabolic subgroup* $\underline{W} \subset W$ *such that* $\mathrm{Supp}(M) = \overline{X(\underline{W})}$.

It turns out that $\dim \mathrm{Supp}(L)$ can be recovered from the highest weight structure on $\mathcal{O}_{\mathbf{c}}(W)$. Let T denote the sum of all indecomposable tilting objects in $\mathcal{O}_{\mathbf{c}}(W)$.

LEMMA 2.13. *The number* $\dim \mathfrak{h} - \dim \mathrm{Supp}(L)$ *coincides with the minimal number* i *such that* $\mathrm{Ext}^i_{\mathcal{O}_{\mathbf{c}}(W)}(L, T) \neq 0$. *It also coincides with the minimal number* j *such that* $\mathrm{Ext}^j_{\mathcal{O}_{\mathbf{c}}(W)}(T, L) \neq 0$.

This is [**RSVV**, Lemma 6.2], a closely related result was obtained earlier in [**GGOR**, Corollary 4.14].

2.2.8. *Example.* As an example, consider the category $\mathcal{O}_{\mathbf{c}}(W)$ for the group $W = \mu_\ell$. First of all, note that all Verma modules are isomorphic to $\mathbb{C}[x]$ as $S(\mathfrak{h}^*) = \mathbb{C}[x]$-modules.

For $j = 0, 1, \ldots, \ell-1$, let Δ_j be the Verma module defined by the representation $\tau : \eta \mapsto \eta^j$. We are going to compute $y \cdot x^n$ for $x^n \in \Delta_j$. Let $\eta \in \mu_\ell$ is a generator of μ_ℓ. We get

$$
\begin{aligned}
y \cdot x^n =& [y, x] \cdot x^{n-1} + x(y \cdot x^{n-1}) = (1 - 2\sum_{i=1}^{\ell-1} c(\eta^i)\eta^i) \cdot x^{n-1} + x(y \cdot x^{n-1}) = \\
& (1 - 2\sum_{i=1}^{\ell-1} c(\eta^i)\eta^{i(1-n)}\eta^{ij})x^{n-1} + x(y \cdot x^{n-1}) = \\
& (n - 2\sum_{i=1}^{\ell-1} c(\eta^i)(\eta^{i(1-n)} + \eta^{i(2-n)} + \ldots + 1)\eta^{ij})x^{n-1} = \\
& (n + \sum_{i=1}^{\ell-1} \frac{2c(\eta^i)}{1-\eta^{-i}}\eta^{i(j-n)} - \sum_{i=1}^{\ell-1} \frac{2c(\eta^i)}{1-\eta^{-i}}\eta^{ij})x^{n-1} = \\
& (n + \ell h_{j-n} - \ell h_j)x^{n-1}.
\end{aligned}
$$

Here we write h_j for $h_{H,j}$ and we assume that h_j is numbered mod ℓ. In particular, we see that $\mathrm{Hom}_{\mathcal{O}_{\mathbf{c}}(W)}(\Delta_k, \Delta_j)$ is one dimensional provided there is $n \geqslant 0$ congruent to $j - k$ modulo ℓ such that $h_j = h_k + n/\ell$, and is zero else. In particular, either $\Delta_j = L_j$ or L_j is finite dimensional.

2.3. KZ functor. Here we will recall the KZ functor introduced in [**GGOR**]. It connects the category $\mathcal{O}_{\mathbf{c}}(W)$ to the category of modules over the Hecke algebra of W and is a crucial tool to study $\mathcal{O}_{\mathbf{c}}(W)$.

2.3.1. *Localization lemma.* Let $\mathfrak{h}^{reg}$ denote the open subset of $\mathfrak{h}$ consisting of all v with $W_v = \{1\}$, equivalently $\mathfrak{h}^{reg} = \mathfrak{h} \setminus \bigcup_{s \in S} \ker \alpha_s$. Consider an element $\delta \in \mathbb{C}[\mathfrak{h}]^W$ whose set of zeroes in $\mathfrak{h}$ coincides with $\mathfrak{h} \setminus \mathfrak{h}^{reg}$. We can take $\delta = \left(\prod_{s \in S} \alpha_s\right)^k$, where k is a suitable integer so that $\delta \in \mathbb{C}[\mathfrak{h}]^W$.

Note that $[\delta, x] = [\delta, w] = 0$ for all $x \in \mathfrak{h}^*, w \in W$. Also note that $[\delta, y] \in S(\mathfrak{h}^*)\#W$ and hence $[\delta, [\delta, y]] = 0$. It follows that the endomorphism $[\delta, \cdot]$ of $H_{\mathbf{c}}$ is locally nilpotent. So the set $\{\delta^k, k \geqslant 0\}$ satisfies the Ore conditions and we have the localization $H_{\mathbf{c}}[\delta^{-1}]$ consisting of right fractions.

We have an algebra homomorphism $H_{\mathbf{c}} \to D(\mathfrak{h}^{reg})\#W$ (the Dunkl homomorphism) defined on generators $x \in \mathfrak{h}^*, w \in W, y \in \mathfrak{h}$ as follows:

$$x \mapsto x,\ w \mapsto w,\ y \mapsto y + \sum_{s \in S} \frac{2c(s)\langle \alpha_s, y\rangle}{(1-\lambda_s)\alpha_s}(s-1).$$

This homomorphism factors through $H_{\mathbf{c}}[\delta^{-1}] \to D(\mathfrak{h}^{reg})\#W$ because δ is invertible in $D(\mathfrak{h}^{reg})\#W$. The following lemma is easy.

LEMMA 2.14. *The homomorphism $H_{\mathbf{c}}[\delta^{-1}] \to D(\mathfrak{h}^{reg})\#W$ is an isomorphism.*

2.3.2. *Localization functor.* Let $M \in \mathcal{O}_{\mathbf{c}}(W)$. Thanks to Lemma 2.14, we can view $M[\delta^{-1}]$ as a module over $D(\mathfrak{h}^{reg})\#W$. This module is finitely generated over $\mathbb{C}[\mathfrak{h}^{reg}]\#W$. So $M[\delta^{-1}]$ is a W-equivariant local system over $\mathfrak{h}^{reg}$. Computing $\Delta_{\mathbf{c}}(\tau)[\delta^{-1}]$ explicitly, we see that the latter has regular singuularities, see [**GGOR**, Proposition 5.7]. Therefore the same is true for any $M \in \mathcal{O}_c(W)$. So we get an exact functor $M \mapsto M[\delta^{-1}]$ from $\mathcal{O}_{\mathbf{c}}(W)$ to the category $\mathrm{Loc}_{rs}^W(\mathfrak{h}^{reg})$ of W-equivariant regular local systems on $\mathfrak{h}^{reg}$.

Pick a point $p \in \mathfrak{h}^{reg}/W$ and let π denote the quotient morphism $\mathfrak{h}^{reg} \to \mathfrak{h}^{reg}/W$. According to Deligne, the category $\mathrm{Loc}_{rs}^W(\mathfrak{h}^{reg})$ is equivalent to $\pi_1(\mathfrak{h}^{reg}/W, p)\text{-mod}_{fin}$ (the category of finite dimensional modules) via $N \mapsto [\pi_*(N)^W]_p$.

The group $\pi_1(\mathfrak{h}^{reg}/W, p)$ is known as the braid group of W and is denoted by B_W. When W is a Coxeter group, we get the classical braid group. In general, the structure of B_W was studied in [**BMR**]. The group B_W is generated by elements T_H, where H runs over the set of reflection hyperplanes for W. Roughly speaking, T_H is a curve obtained by rotating a point in $\mathfrak{h}^{reg}$ close to H by the angle of $2\pi/|W_H|$ (here W_H is the pointwise stabilizer of H). Note that if H, H' are W-conjugate, then $T_H, T_{H'}$ are conjugate in B_W.

EXAMPLE 2.15. Let us consider the case when $W = G(\ell, 1, n)$. When $\ell = 1$, we get the usual type A braid group. When $\ell > 1$, we get the affine braid group of type A. It is given by generators $T_0, \ldots, T_{n-1}$ subject to the following relations: $T_iT_j = T_jT_i$ when $|i-j| > 1$, $T_iT_{i+1}T_i = T_{i+1}T_iT_{i+1}$ for $i > 1$, and $T_0T_1T_0T_1 = T_1T_0T_1T_0$.

2.3.3. *Hecke algebras.* We want to determine the essential image of the functor $\mathcal{O}_{\mathbf{c}}(W) \to B_W\text{-mod}_{fin}$. It turns out that this image coincides with $\mathcal{H}_{\mathbf{q}}(W)\text{-mod}_{fin} \hookrightarrow B_W\text{-mod}_{fin}$, where $\mathcal{H}_{\mathbf{q}}(W)$ is a quotient of $\mathbb{C}B_W$ called the Hecke algebra of W, and $\mathbf{q}$ is a parameter recovered from $\mathbf{c}$.

First, let us explain how the parameter $\mathbf{q}$ is computed. It is a collection $q_{H,i}$ of nonzero complex numbers, where H is a reflection hyperplane for W, $i = 0, 1, \ldots, |W_H| - 1$, and $q_{H,i} = q_{H',i}$ if H and H' are W-conjugate. We set

$$q_{H,j} := \exp(2\pi\sqrt{-1}(h_{H,j} + j/\ell_H)), \tag{2.5}$$

where $h_{H,j}$ is recovered from (2.3). Note that the parameters $q_{H,0}, \ldots, q_{H,|W_H|-1}$ are defined up to a common multiple.

Following [**BMR**, 4C], define the Hecke algebra $\mathcal{H}_{\mathbf{q}}(W)$ as the quotient of $\mathbb{C}B_W$ by the following relations

$$\prod_{i=1}^{|W_H|}(T_H - q_{H,i}) = 0, \tag{2.6}$$

where H runs over the set of reflection hyperplanes for W.

When W is a Coxeter group, we get the usual Iwahori-Hecke algebra.

EXAMPLE 2.16. Let us consider the case of $W = G(\ell, 1, n)$. Then $\mathcal{H}_{\mathbf{q}}(W)$ is the cyclotomic Hecke algebra (a.k.a. Ariki-Koike algebra). This algebra is the quotient of $\mathbb{C}B_W$ by the following relations: $(T_i+1)(T_i-q) = 0, i = 1, \ldots, n-1$, and $\prod_{i=0}^{\ell-1}(T_0 - Q_i)$, where $q = \exp(2\pi\sqrt{-1}\kappa)$ and $Q_i = \exp(2\pi\sqrt{-1}(h_j + j/\ell))$. Recall that the parameters $\kappa, h_0, \ldots, h_{\ell-1}$ were introduced in 2.2.2. Note that $\dim \mathcal{H}_{\mathbf{q}}(W) = |W|$ in this case, [**AK**].

It was shown in [**GGOR**, Theorem 5.13] that the functor $\mathcal{O}_{\mathbf{c}}(W) \to B_W\text{-mod}_{fin}$ decomposes as the composition of $\mathrm{KZ}_{\mathbf{c}} : \mathcal{O}_{\mathbf{c}}(W) \to \mathcal{H}_{\mathbf{q}}(W)\text{-mod}_{fin}$ (the KZ functor) and the inclusion of $\mathcal{H}_{\mathbf{q}}(W)\text{-mod}_{fin} \hookrightarrow B_W\text{-mod}_{fin}$.

PROPOSITION 2.17. *The functor* $\mathrm{KZ}_{\mathbf{c}} : \mathcal{O}_{\mathbf{c}}(W) \to \mathcal{H}_{\mathbf{q}}(W)\text{-mod}_{fin}$ *is essentially surjective.*

It is expected (and is shown in many cases) that $\dim \mathcal{H}_{\mathbf{q}}(W) = |W|$. Proposition 2.17 was proved in [**GGOR**, Theorem 5.15] under that assumption. The assumption was removed in [**Lo7**].

2.3.4. *Properties of KZ functor.* Let us list some properties of the KZ functor obtained in [**GGOR**, Section 5].

PROPOSITION 2.18. *The following is true.*

(1) *The KZ functor* $\mathcal{O}_{\mathbf{c}}(W) \twoheadrightarrow \mathcal{H}_{\mathbf{q}}(W)\text{-mod}_{fin}$ *is a quotient functor. Its kernel is the subcategory* $\mathcal{O}_{\mathbf{c},tor}(W) \subset \mathcal{O}_{\mathbf{c}}(W)$ *consisting of all modules in* $\mathcal{O}_{\mathbf{c}}(W)$ *that are torsion over* $\mathbb{C}[\mathfrak{h}]$ *(equivalently, whose support is a proper subvariety in* $\mathfrak{h}$*).*
(2) *The functor* $\mathrm{KZ}_{\mathbf{c}}$ *is defined by a projective object* P_{KZ} *in* $\mathcal{O}_{\mathbf{c}}(W)$ *that is also injective. The multiplicity of* $\Delta_{\mathbf{c}}(\tau)$ *in* P_{KZ} *equals* $\dim \tau$.
(3) $\mathrm{KZ}_{\mathbf{c}}$ *is fully faithful on the projective objects in* $\mathcal{O}_{\mathbf{c}}(W)$. *It is also fully faithful on the tilting objects in* $\mathcal{O}_{\mathbf{c}}(W)$.
(4) *Suppose that the parameter* $\mathbf{q}$ *satisfies the following condition: for any reflection hyperplane* H, *we have* $q_{H,i} \neq q_{H,j}$ *for* $i \neq j$. *Then* $\mathrm{KZ}_{\mathbf{c}}$ *is fully faithful on all standardly filtered objects.*

For $W = G(\ell, 1, n)$, the condition in 4) means that $q \neq -1$ and $Q_i \neq Q_j$ for $i \neq j$.

2.4. Induction and restriction functors. In this section we review another major tool to study the categories $\mathcal{O}_{\mathbf{c}}(W)$, the induction and restriction functors introduced by Bezrukavnikov and Etingof in [**BE**]. These functors relate categories $\mathcal{O}_{\mathbf{c}}(W)$ and $\mathcal{O}_{\mathbf{c}}(\underline{W})$, where $\underline{W}$ is a parabolic subgroup in W. More precisely, we have functors $\mathrm{Res}_{\underline{W}}^W : \mathcal{O}_{\mathbf{c}}(W) \to \mathcal{O}_{\mathbf{c}}(\underline{W})$ and $\mathrm{Ind}_{\underline{W}}^W : \mathcal{O}_{\mathbf{c}}(\underline{W}) \to \mathcal{O}_{\mathbf{c}}(W)$. When we write $\mathcal{O}_{\mathbf{c}}(\underline{W})$, by $\mathbf{c}$ we mean the restriction of $\mathbf{c}$ to $\underline{W} \cap S$. Since we also consider restriction/induction functors for other categories, we will sometimes write ${}^{\mathcal{O}}\mathrm{Res}_{\underline{W}}^W$, ${}^{\mathcal{O}}\mathrm{Ind}_{\underline{W}}^W$.

The construction of the functors is technical. We are not going to explain the construction, we will just explain the properties.

2.4.1. *Exactness and adjointness properties.* We have the following properties of Res and Ind.

PROPOSITION 2.19. *The functors* $\mathrm{Res}^W_{\underline{W}}$ *and* $\mathrm{Ind}^W_{\underline{W}}$ *are biadjoint. Hence they are exact.*

The claims that $\mathrm{Res}^W_{\underline{W}}$ and $\mathrm{Ind}^W_{\underline{W}}$ are exact was checked in [**BE**, Section 3.5]. The claim that $\mathrm{Ind}^W_{\underline{W}}$ is right adjoint to $\mathrm{Res}^W_{\underline{W}}$ is a consequence of the construction in *loc.cit.*. The other adjointness was established in [**Sh**, Section 2.4] under some restrictions on W and in [**Lo1**], in general.

In fact, [**Lo1**] proved a stronger statement. Recall the parameter $\mathbf{c}^*$, contravariant equivalences $M \mapsto M^\vee$ (taking the restricted duals) $\mathcal{O}_{\mathbf{c}}(W,\mathfrak{h}) \xrightarrow{\sim} \mathcal{O}_{\mathbf{c}^*}(W,\mathfrak{h}^*)$ and $\mathcal{O}_{\mathbf{c}}(\underline{W},\underline{\mathfrak{h}}) \xrightarrow{\sim} \mathcal{O}_{\mathbf{c}^*}(\underline{W},\underline{\mathfrak{h}}^*)$ from 2.2.5. Here $\underline{\mathfrak{h}}(=\mathfrak{h}/\mathfrak{h}^{\underline{W}})$ is the reflection representation of $\underline{W}$.

PROPOSITION 2.20. *The functors* $\mathrm{Res}^W_{\underline{W}}, \mathrm{Ind}^W_{\underline{W}}$ *intertwine the duality functors, i.e., we have functorial isomorphisms* $\mathrm{Res}^W_{\underline{W}}(M^\vee) \cong \mathrm{Res}^W_{\underline{W}}(M)^\vee$ *and* $\mathrm{Ind}^W_{\underline{W}}(N^\vee) \cong \mathrm{Ind}^W_{\underline{W}}(N)^\vee$.

This is the main result of [**Lo1**].

2.4.2. *Relation to KZ.* We have a natural homomorphism $\mathcal{H}_{\mathbf{q}}(\underline{W}) \to \mathcal{H}_{\mathbf{q}}(W)$. This gives rise to an exact restriction functor ${}^{\mathcal{H}}\mathrm{Res}^W_{\underline{W}} : \mathcal{H}_{\mathbf{q}}(W)\text{-mod}_{fin} \to \mathcal{H}_{\mathbf{q}}(\underline{W})\text{-mod}_{fin}$.

The following proposition is [**Sh**, Theorem 2.1].

PROPOSITION 2.21. *The KZ functors intertwine the restriction functors:* $\underline{\mathrm{KZ}}_{\mathbf{c}} \circ {}^{\mathcal{O}}\mathrm{Res}^W_{\underline{W}} \cong {}^{\mathcal{H}}\mathrm{Res}^W_{\underline{W}} \circ \mathrm{KZ}_{\mathbf{c}}$. *Here we write* $\underline{\mathrm{KZ}}_{\mathbf{c}}$ *for the KZ functor* $\mathcal{O}_{\mathbf{c}}(\underline{W}) \to \mathcal{H}_{\mathbf{q}}(\underline{W})\text{-mod}_{fin}$.

As was explained in [**Lo7**], the algebra $\mathcal{H}_{\mathbf{q}}(W)$ has the maximal finite dimensional quotient to be denoted by $\mathcal{H}^f_{\mathbf{q}}(W)$ here. The same, of course, applies to $\mathcal{H}_{\mathbf{q}}(\underline{W})$. We get induction, ${}^{\mathcal{H}}\mathrm{Ind}^W_{\underline{W}}$, and coinduction, ${}^{\mathcal{H}}\mathrm{Coind}^W_{\underline{W}}$, functors $\mathcal{H}_{\mathbf{q}}(\underline{W})\text{-mod}_{fin} \to \mathcal{H}_{\mathbf{q}}(W)\text{-mod}_{fin}$ associated to the homomorphism $\mathcal{H}^f_{\mathbf{q}}(\underline{W}) \to \mathcal{H}^f_{\mathbf{q}}(W)$. As explained in [**Lo6**, Section 2.8], Proposition 2.21 has the following corollary.

COROLLARY 2.22. *We have an isomorphism of functors* ${}^{\mathcal{H}}\mathrm{Ind}^W_{\underline{W}} \cong {}^{\mathcal{H}}\mathrm{Coind}^W_{\underline{W}}$. *The KZ functors intertwine the induction functors.*

We have one more corollary of Proposition 2.21, see [**Sh**, Corollary 2.5].

COROLLARY 2.23. *Let* $\underline{W}' \subset \underline{W}$ *be a parabolic subgroup. Then we have an isomorphism of functors* $\mathrm{Res}^W_{\underline{W}'} \cong \mathrm{Res}^{\underline{W}}_{\underline{W}'} \circ \mathrm{Res}^W_{\underline{W}}$. *An analogous claim holds for the induction functors.*

2.4.3. *Behavior on* K_0 *and on (co)standardly filtered objects.*

LEMMA 2.24. *The object* $\mathrm{Res}^W_{\underline{W}}(\Delta_{\mathbf{c}}(\tau))$ *(resp.,* $\mathrm{Res}^W_{\underline{W}}(\nabla_{\mathbf{c}}(\tau))$*) is filtered with quotients* $\Delta_{\mathbf{c}}(\tau')$ *(resp.,* $\nabla_{\mathbf{c}}(\tau')$*) for* $\tau' \in \mathrm{Irr}(\underline{W})$. *The multiplicity of* $\Delta_{\mathbf{c}}(\tau')$ *(or of* $\nabla_{\mathbf{c}}(\tau')$*) in the filtration equals* $\dim \mathrm{Hom}_{\underline{W}}(\tau',\tau)$. *The similar claims are true for* $\mathrm{Ind}^W_{\underline{W}}(\Delta_{\mathbf{c}}(\tau'))$ *and* $\mathrm{Ind}^W_{\underline{W}}(\nabla_{\mathbf{c}}(\tau'))$.

PROOF. The claim about $\mathrm{Res}^W_{\underline{W}}(\Delta_{\mathbf{c}}(\tau))$ is quite straightforward from the construction of the functor, see [**Sh**, Proposition 1.9]. The claim about $\mathrm{Res}^W_{\underline{W}}(\nabla_{\mathbf{c}}(\tau))$ follows from Proposition 2.20. The claims about the induction functors follow from the biadjointness. □

In particular, the map $[\mathrm{Res}^W_{\underline{W}}] : K_0(\mathcal{O}_{\mathbf{c}}(W)) = K_0(W\text{-mod}) \to K_0(\mathcal{O}_{\mathbf{c}}(\underline{W})) = K_0(\underline{W}\text{-mod})$ coincides with the usual restriction $K_0(W\text{-mod}) \to K_0(\underline{W}\text{-mod})$. The similar claim holds for the induction. This was observed already in [**BE**, Section 3.6].

2.4.4. *Behavior on supports.* Recall that the support of a module from category $\mathcal{O}_{\mathbf{c}}(W)$ was defined in 2.2.7. Here we will investigate the interplay between the supports and the induction and restriction functors.

Let us start with the restriction functor. The following result was established in [**SV**, Proposition 2.2].

LEMMA 2.25. *Let $M \in \mathcal{O}_{\mathbf{c}}(W)$ be such that* $\mathrm{Supp}(L) = \overline{X(W')}$. *Let $\underline{W}$ be a parabolic subgroup in W. Then*

$$\mathrm{Supp}(\mathrm{Res}^W_{\underline{W}}(M)) = \bigcup_{\underline{W}'} \underline{X}(\underline{W}').$$

Here the union is taken over all ($\underline{W}$-conjugacy classes of) parabolic subgroups $\underline{W}' \subset \underline{W}$ that are conjugate to W' in W. By $\underline{X}(\underline{W}')$, we denote the stratum of $\underline{\mathfrak{h}}$ (the reflection representation of $\underline{W}$) corresponding to the stabilizer $\underline{W}'$.

In particular, this lemma implies that $\mathrm{Res}^W_{\underline{W}}(M) = 0$ if and only if W' is not conjugate to a subgroup of $\underline{W}$, while $\mathrm{Res}^W_{\underline{W}}(M)$ is finite dimensional if and only if W' is conjugate to $\underline{W}$.

Let us proceed to the induction functors. The following result is an easy consequence of [**SV**, Proposition 2.7].

LEMMA 2.26. *Let $\underline{L}$ be a simple in $\mathcal{O}_{\mathbf{c}}(\underline{W})$ and $\underline{W}' \subset \underline{W}$ be such that* $\mathrm{Supp}(\underline{L}) = \underline{X}(\underline{W}')$. *Then, for every quotient/sub M of* $\mathrm{Ind}^W_{\underline{W}}(\underline{L})$, *we have* $\mathrm{Supp}(M) = X(\underline{W}')$.

2.5. Abelian equivalences. Here we will establish some equivalences between categories $\mathcal{O}_{\mathbf{c}}(W)$ and $\mathcal{O}_{\mathbf{c}'}(W)$ with different parameters $\mathbf{c}, \mathbf{c}'$.

2.5.1. *Twist by a one-dimensional character.* Now let χ be a one-dimensional character of W. Given $\mathbf{c} \in \mathfrak{P}$, define $\mathbf{c}^\chi \in \mathfrak{P}$ by $c^\chi(s) = \chi(s)^{-1}c(s)$. We have an isomorphism $\psi_\chi : H_{\mathbf{c}} \xrightarrow{\sim} H_{\mathbf{c}^\chi}$ given on the generators by $x \mapsto x, y \mapsto y, w \mapsto \chi(w)w$. This gives rise to an equivalence $\psi_{\chi*} : \mathcal{O}_{\mathbf{c}}(W) \xrightarrow{\sim} \mathcal{O}_{\mathbf{c}^\chi}(W)$ that maps $\Delta_{\mathbf{c}}(\tau)$ to $\Delta_{\mathbf{c}^\chi}(\chi \otimes \tau)$.

2.5.2. *Chambers and walls.* From now on, in this section we will consider parameters $\mathbf{c}, \mathbf{c}'$ giving the same parameter $\mathbf{q}$ for the Hecke algebra. By (2.3) and (2.5) there is a $\mathbb{Z}$-lattice $\mathfrak{P}_{\mathbb{Z}} \subset \mathfrak{P}$ such that parameters $\mathbf{c}, \mathbf{c}' \in \mathfrak{P}$ give the same parameter $\mathbf{q}$ if and only if $\mathbf{c}' - \mathbf{c} \in \mathfrak{P}_{\mathbb{Z}}$. For example, when W is a real group, $\mathfrak{P}_{\mathbb{Z}} = \mathbb{Z}^{S/W}$.

We will see that $\mathcal{O}_{\mathbf{c}}(W)$ and $\mathcal{O}_{\mathbf{c}'}(W)$ are equivalent provided $\mathbf{c}, \mathbf{c}'$ lie in the same "chamber". Let us explain what kind of chambers we consider.

Consider the shifted lattice $\mathbf{c}+\mathfrak{P}_{\mathbb{Z}}$. Set $\Pi_{\xi\tau} := \ker \varpi_{\xi\tau}$ (recall that the functions $\varpi_{\xi\tau} \in \mathfrak{P}^*$ were defined in 2.2.6). Consider the hyperplanes $\Pi_{\xi\tau}$ such that $\varpi_{\xi\tau}(\mathbf{c}) \in \mathbb{Q}$. We call them *walls* for $\mathbf{c} + \mathfrak{P}_{\mathbb{Z}}$ (note that the walls do not need to intersect

$\mathbf{c}+\mathfrak{P}_{\mathbb{Z}}$). In any case, the hyperplanes $\Pi_{\xi\tau}$ split $\mathbf{c}+\mathfrak{P}_{\mathbb{Z}}$ into the union of chambers that are polyhedral cones.

2.5.3. *Main results.* We have the following.

PROPOSITION 2.27. *Suppose that $\mathbf{c}' \in \mathbf{c}+\mathfrak{P}_{\mathbb{Z}}$ lies in the closure of the chamber containing* $\mathbf{c}$. *Then there is an abelian equivalence $\Phi_{\mathbf{c}'\leftarrow\mathbf{c}} : \mathcal{O}_{\mathbf{c}}(W) \xrightarrow{\sim} \mathcal{O}_{\mathbf{c}'}(W)$ intertwining the KZ functors.*

In fact, under an additional condition one can also ensure that an equivalence from Proposition 2.27 maps $\Delta_{\mathbf{c}}(\tau)$ to $\Delta_{\mathbf{c}'}(\tau)$. Namely, according to [**BC**, Section 7.2], there is a group homomorphism $\mathfrak{P}_{\mathbb{Z}} \to \operatorname{Bij}(\operatorname{Irr}(W))$ called the KZ twist and denoted by tw. Set $\underline{\mathfrak{P}}_{\mathbb{Z}} := \ker \mathsf{tw}$. For example, for $W = G(\ell, 1, n)$, the KZ twist is trivial, see [**GL**, 6.4.7] for explanation and references, while for some other groups, such as G_2, it is not.

DEFINITION 2.28. For two highest weight categories $\mathcal{O}^1, \mathcal{O}^2$ with identified sets of irreducibles $\operatorname{Irr}(\mathcal{O}^1) \cong \operatorname{Irr}(\mathcal{O}^2) \cong \Lambda$, by a *highest weight* equivalence $\mathcal{O}^1 \xrightarrow{\sim} \mathcal{O}^2$, we mean an equivalence $\mathcal{O}^1 \xrightarrow{\sim} \mathcal{O}^2$ of abelian categories that maps Δ^1_L to Δ^2_L, for every $L \in \Lambda$.

PROPOSITION 2.29. *Under the assumptions of Proposition* 2.27, *suppose, in addition, that $\mathbf{c}' - \mathbf{c} \in \underline{\mathfrak{P}}_{\mathbb{Z}}$. Then one can take a highest weight equivalence $\Phi_{\mathbf{c}'\leftarrow\mathbf{c}}$ in Proposition* 2.27.

Proposition 2.27 is a consequence of Proposition 2.29. In fact, for $\psi \in \mathfrak{P}_{\mathbb{Z}}$, we have *shift functors* $\mathsf{Sh}_{\psi} : \mathcal{O}_{\mathbf{c}}(W) \to \mathcal{O}_{\mathbf{c}+\psi}(W)$, they are given by taking tensor products with shift bimodules, see, e.g., [**Lo6**, Section 3.1]. For a fixed ψ, this functor is an equivalence for a Zariski generic $\mathbf{c}$, this can be deduced from [**Lo6**, Corollary 3.5].

2.5.4. *Highest weight covers.* A general technique of proving results like Proposition 2.29 is due to Rouquier, [**R2**, Section 4.2], who proved Proposition 2.29 under the assumption that $\mathrm{KZ}_{\mathbf{c}}, \mathrm{KZ}_{\mathbf{c}'}$ are fully faithful on all standardly filtered objects, see (4) of Proposition 2.18. An extension of Rouquier's technique found in [**Lo4**, Section 7] allows to remove the additional restriction on $\mathbf{c}$.

One obvious obstruction to an existence of an equivalence $\mathcal{O}_{\mathbf{c}}(W) \to \mathcal{O}_{\mathbf{c}'}(W)$ with $\Delta_{\mathbf{c}}(\lambda) \mapsto \Delta_{\mathbf{c}'}(\lambda)$ is that highest weight orders may be different. The condition that $\mathbf{c}'$ lies in the closure of the chamber containing $\mathbf{c}$ guarantees that $\leqslant^{\mathbf{c}}$ refines $\leqslant^{\mathbf{c}'}$ (strictly speaking, here we need to modify the order and require $\tau <^{\mathbf{c}} \xi$ if $c_\tau - c_\xi \in \mathbb{Q}_{>0}$, not in $\mathbb{Z}_{>0}$, but this does not matter). So there is a common order on $\operatorname{Irr}(\mathcal{O}_{\mathbf{c}}(W)) \cong \operatorname{Irr}(\mathcal{O}_{\mathbf{c}'}(W))$ making $\mathcal{O}_{\mathbf{c}}(W), \mathcal{O}_{\mathbf{c}'}(W)$ into highest weight categories.

Now let us consider a more general setting. Let $\mathcal{C}$ be a $\mathbb{C}$-linear abelian category equivalent to A-mod, where A is a finite dimensional associative algebra. Let $\mathcal{O}^1, \mathcal{O}^2$ be two highest weight categories with identified posets $\operatorname{Irr}(\mathcal{O}^1), \operatorname{Irr}(\mathcal{O}^2)$, let us denote this common poset by Λ. Assume that there are quotient functors $\pi^i : \mathcal{O}^i \twoheadrightarrow \mathcal{C}$.

General speculation: If $\mathcal{C}$ is "large enough", then there is a highest weight equivalence $\varphi : \mathcal{O}^1 \to \mathcal{O}^2$ and $\pi^2 \circ \varphi \cong \pi^1$.

The condition that π^1, π^2 are fully faithful on standardly filtered objects (below we will say that π^i is 0-*faithful*), morally, says that $\mathcal{C}$ is "large enough", but is not sufficient to establish the existence of φ. On the other hand, what is sufficient is the following two conditions.

(L1) $\pi^1(\Delta^1_L) = \pi^2(\Delta^2_L)$ for all $L \in \Lambda$.

(L2) For any standardly filtered objects $M, N \in \mathcal{O}^i, i = 1,2$, we have $\operatorname{Ext}^j_{\mathcal{O}^i}(M,N) = \operatorname{Ext}^j_{\mathcal{C}}(\pi M, \pi N)$ for $j = 0,1$. In this situation we say that π^i is 1-*faithful.*

Indeed, (L2) guarantees that π^i restricts to an equivalence $\mathcal{O}^{i\Delta} \xrightarrow{\sim} \mathcal{C}^{\pi^i(\Delta)}$, where $\mathcal{O}^{i\Delta}$ is the full subcategory of standardly filtered objects in $\mathcal{O}^i$ and $\mathcal{C}^{\pi^i(\Delta)}$ denotes the full subcategory of all objects in $\mathcal{C}$ filtered by $\pi^i(\Delta_L), L \in \Lambda$. (L1) guarantees that $\mathcal{C}^{\pi^1(\Delta)} = \mathcal{C}^{\pi^2(\Delta)}$. This gives an equivalence $\mathcal{O}^{1\Delta} \xrightarrow{\sim} \mathcal{O}^{2\Delta}$ that uniquely extends to an equivalence $\mathcal{O}^1 \xrightarrow{\sim} \mathcal{O}^2$ with desired properties. Unfortunately, it is almost never possible to establish (L1) and (L2) directly.

2.5.5. *Deformations.* A solution found by Rouquier was to use deformations. Let R be a formal power series ring over $\mathbb{C}$. Suppose that we have R-linear categories $\mathcal{O}^i_R, \mathcal{C}_R$ (for example, if we have an R-algebra A_R that is a free R-module specializing to A, then we take $\mathcal{C}_R := A_R$-mod). Standard objects in $\mathcal{O}^i$ do not have higher self-extensions and so uniquely deform to $\mathcal{O}^i_R$. So do the projective objects. The deformation of the projective object defining π^i gives the quotient functor $\pi^i_R : \mathcal{O}^i_R \twoheadrightarrow \mathcal{C}_R$. Suppose that the base change $\mathcal{C}_{\operatorname{Frac}(R)}$ is semisimple and that $\pi^i_{\operatorname{Frac}(R)}$ is an equivalence. This gives rise to identifications $\operatorname{Irr}(\mathcal{O}^i) = \operatorname{Irr}(\mathcal{O}^i_{\operatorname{Frac}(R)}) = \operatorname{Irr}(\mathcal{C}_{\operatorname{Frac}(R)})$. We further assume that the resulting identification $\operatorname{Irr}(\mathcal{O}^1) \cong \operatorname{Irr}(\mathcal{O}^2)$ coincides with our initial identification.

As Rouquier checked in [**R2**, Lemma 4.48], an analog of (L1) holds for $\mathcal{O}^1_R, \mathcal{O}^2_R$. Further, in [**R2**, Proposition 4.42] he has checked that if π^i is 0-faithful, then π^i_R is 1-faithful. So (L1) and (L2) hold for the deformed categories, and we get an equivalence $\varphi_R : \mathcal{O}^1_R \xrightarrow{\sim} \mathcal{O}^2_R$ that then specializes to an equivalence $\varphi : \mathcal{O}^1 \to \mathcal{O}^2$ with the desired properties.

2.5.6. *Cherednik categories $\mathcal{O}$.* Let us return to the situation when the categories of interest are $\mathcal{O}_{\mathbf{c}}(W), \mathcal{O}_{\mathbf{c}'}(W)$. Let us explain how to deform them. We can form the $\mathbb{C}[\mathfrak{P}]$-algebra $H_{\mathfrak{P}}$ that is the quotient of $T(\mathfrak{h} \oplus \mathfrak{h}^*)\#W[\mathfrak{P}]$ by the relations similar to above but where we replace $c(s) \in \mathbb{C}$ with the basis element $c(s) \in \mathfrak{P}^*$ corresponding to the conjugacy class of s in S. The specialization of $H_{\mathfrak{P}}$ to $\mathbf{c} \in \mathfrak{P}$ coincides with $H_{\mathbf{c}}$. We can define the category $\mathcal{O}_{\mathfrak{P}}(W)$ and the Verma modules there as before.

Now pick a sufficiently generic line ℓ through $\mathbf{c}$. Let R denote the completion $\mathbb{C}[\ell]^{\wedge_{\mathbf{c}}}$, this a formal power series algebra in one variable that is also an algebra over $\mathbb{C}[\mathfrak{P}]$. We can consider the specialization $H_{R,\mathbf{c}} := R \otimes_{\mathbb{C}[\mathfrak{P}]} H_{\mathfrak{P}}$ and consider the correspoding category $\mathcal{O}_{R,\mathbf{c}}$. The algebra R is naturally identified with the completion of $\mathbb{C}[\ell + \mathbf{c}' - \mathbf{c}]$ at $\mathbf{c}$. So we can also form the algebra $H_{R,\mathbf{c}'}$ and the category $\mathcal{O}_{R,\mathbf{c}'}$.

On the other hand, we also have the R-algebra $\mathcal{H}_{R,\mathbf{q}}$. It still has the maximal quotient of finite rank over R, the algebra that we denote by $\mathcal{H}^f_{R,\mathbf{q}}$. The KZ functors extend to quotient functors $\mathcal{O}_{R,\mathbf{c}} \twoheadrightarrow \mathcal{H}^f_{R,\mathbf{q}}$-mod, $\mathcal{O}_{R,\mathbf{c}'} \twoheadrightarrow \mathcal{H}^f_{R,\mathbf{q}}$-mod. We take $\mathcal{C}_R := \mathcal{H}^f_{R,\mathbf{q}}$-mod, $\mathcal{O}^1_R := \mathcal{O}_{R,\mathbf{c}}, \mathcal{O}^2_R := \mathcal{O}_{R,\mathbf{c}'}$.

We need to check that the categories $\mathcal{O}^i_{\operatorname{Frac}(R),\mathbf{c}}$ are semisimple. This has to do with the choice of ℓ – it is generic. Then the categories $\mathcal{O}_{\operatorname{Frac} R,\mathbf{c}}(W), \mathcal{O}_{\operatorname{Frac} R,\mathbf{c}'}(W)$ are semisimple for the reasons explained in 2.2.6. The KZ functor does not kill $\Delta_{R,\mathbf{c}}(\tau)$ because this module is not torsion. So $\pi^i_{\operatorname{Frac}(R)}$ is a category equivalence.

We need to show that the chain of identifications

$$\mathrm{Irr}(W) \xrightarrow{\sim} \mathrm{Irr}(\mathcal{O}^1_{\mathrm{Frac}(R)}) \xrightarrow{\sim} \mathrm{Irr}(\mathcal{C}_{\mathrm{Frac}(R)}) \xrightarrow{\sim} \mathrm{Irr}(\mathcal{O}^2_{\mathrm{Frac}(R)}) \xrightarrow{\sim} \mathrm{Irr}(W)$$

gives the identity. It is a general fact that it gives the bijection $\mathsf{tw}(\mathbf{c}'-\mathbf{c})$, more or less by the definition of tw. By our choice of $\mathbf{c}', \mathbf{c}$, this bijection is the identity, [**BC**, Theorem 7.2].

This completes the proof of Proposition 2.29 in the case when the functors $\mathsf{KZ}_{\mathbf{c}}, \mathsf{KZ}_{\mathbf{c}'}$ are 0-faithful.

2.5.7. *Extended quotients.* Let us explain how to modify the argument above when the functors $\mathsf{KZ}_{\mathbf{c}}, \mathsf{KZ}_{\mathbf{c}'}$ are not 0-faithful. The idea is to consider some intermediate quotient between $\mathcal{O}_{\mathbf{c}}(W)$ and $\mathcal{H}^f_{\mathbf{q}}(W)$-mod. This idea has appeared in [**Lo4**, Section 7] in a different situation and was used in [**Lo6**, Section 4.2] to prove Proposition 2.29.

Recall that $\mathsf{KZ}_{\mathbf{c}}$ is the quotient functor defined by the sum of the projectives $P_{\mathbf{c}}(\tau)$ such that $\mathrm{Supp}(L_{\mathbf{c}}(\tau)) = \mathfrak{h}$. Now consider the sum P^1 of all projectives $P_{\mathbf{c}}(\tau)$ such that $\mathrm{codim}_{\mathfrak{h}} \mathrm{Supp}(L_{\mathbf{c}}(\tau)) = 1$. Set $\tilde{P} := P_{KZ} \oplus P^1$. Let $\tilde{\mathcal{C}}_{\mathbf{c}}$ denote the quotient category of $\mathcal{O}_{\mathbf{c}}(W)$ defined by $\tilde{P}$. The functor $\mathsf{KZ}_{\mathbf{c}} : \mathcal{O}_{\mathbf{c}}(W) \twoheadrightarrow \mathcal{H}^f_{\mathbf{q}}(W)$-mod factorizes as $\mathsf{KZ}_{\mathbf{c}} = \underline{\pi}_{\mathbf{c}} \circ \overline{\pi}_{\mathbf{c}}$, where $\overline{\pi}_{\mathbf{c}}$ is the quotient functor $\mathcal{O}_{\mathbf{c}}(W) \twoheadrightarrow \tilde{\mathcal{C}}_{\mathbf{c}}$ defined by $\tilde{P}$ and $\underline{\pi}_{\mathbf{c}} : \tilde{\mathcal{C}}_{\mathbf{c}} \twoheadrightarrow \mathcal{H}^f_{\mathbf{q}}(W)$-mod is the quotient functor defined by the direct summand P_{KZ} of $\tilde{P}$. We will write $\tilde{P}_{\mathbf{c}}, \tilde{P}_{\mathbf{c}'}$ when we need to indicate the dependence on the parameter.

The following lemma was proved in [**Lo4**] (the proof is contained in that of Proposition 8.1 of *loc.cit.*). See also [**RSVV**, Lemma 2.8].

LEMMA 2.30. *The quotient functors $\overline{\pi}_{\mathbf{c}}, \overline{\pi}_{\mathbf{c}'}$ are 0-faithful.*

Note that the condition in 4) of Proposition 2.18 precisely means that the category $\mathcal{O}_{\mathbf{c}}(W_H)$ is semisimple for every reflection hyperplane H, equivalently, there are no finite dimensional modules in this category. Using Lemma 2.25, we see that $\mathrm{codim}_{\mathfrak{h}} \mathrm{Supp}(L_{\mathbf{c}}(\tau)) \leqslant 1$ implies $\mathrm{Supp}\, L_{\mathbf{c}}(\tau) = \mathfrak{h}$. So Lemma 2.30 can be regarded as a generalization of 4) of Proposition 2.18.

Now let us explain how to modify the deformation argument in order to establish a desired equivalence $\mathcal{O}_{\mathbf{c}}(W) \xrightarrow{\sim} \mathcal{O}_{\mathbf{c}'}(W)$. Let R have the same meaning as before. We can consider the deformation $P^1_R \in \mathcal{O}_{R,\mathbf{c}}(W)$ of the projective object P^1. So we get the deformation $\tilde{\mathcal{C}}_{R,\mathbf{c}}$ of $\tilde{\mathcal{C}}_{\mathbf{c}}$ that comes with quotient functors $\overline{\pi}_{R,\mathbf{c}} : \mathcal{O}_{R,\mathbf{c}}(W) \twoheadrightarrow \tilde{\mathcal{C}}_{R,\mathbf{c}}, \underline{\pi}_{R,\mathbf{c}} : \tilde{\mathcal{C}}_{R,\mathbf{c}} \twoheadrightarrow \mathcal{H}^f_{R,\mathbf{q}}(W)$.

PROPOSITION 2.31. *We have an equivalence $\tilde{\mathcal{C}}_{R,\mathbf{c}} \xrightarrow{\sim} \tilde{\mathcal{C}}_{R,\mathbf{c}'}$ that intertwines the quotient functors $\underline{\pi}_{R,\mathbf{c}}$ and $\underline{\pi}_{R,\mathbf{c}'}$.*

SKETCH OF PROOF. The first case to consider is when $\dim \mathfrak{h} = 1$ so that $\tilde{\mathcal{C}}_? = \mathcal{O}_?(W)$. This is a very easy case that can be done by hand, compare to [**Lo4**, Lemma 7.3] or [**RSVV**, 2.4.4].

In the general case, we can argue as follows. It is sufficient to show that

(1) $\mathsf{KZ}_{R,?}$ is fully faithful on $\tilde{P}_{R,?}$ (the deformation of $\tilde{P}_?$), where $? = \mathbf{c}, \mathbf{c}'$.
(2) $\mathsf{KZ}_{R,\mathbf{c}}(\tilde{P}_{R,\mathbf{c}}) \cong \mathsf{KZ}_{R,\mathbf{c}'}(\tilde{P}_{R,\mathbf{c}'})$.

(1) follows from (3) in Proposition 2.18. In the proof of (2) we can replace $\tilde{P}_{R,\mathbf{c}}$ with $\bigoplus_H \mathrm{Ind}_W^{W_H}(P_{R,H})$, where $P_{R,H}$ is a projective generator of $\mathcal{O}_{R,\mathbf{c}}(W_H)$, these two projectives have the same indecomposable summands (perhaps, with different

multiplicities). We modify $\tilde{P}_{R,\mathbf{c}'}$ similarly. Then we use the case of $\dim\mathfrak{h}=1$ and Corollary 2.22 to prove (2). Note that a similar argument combined with biadjointness of Res and Ind deduces (1) from the case of $\dim\mathfrak{h}=1$ (we just need to know that $\mathsf{KZ}_{R,\mathbf{c}}$ is fully faithful on the projective objects when $\dim\mathfrak{h}=1$). □

Now we can complete the proof of Proposition 2.29 as in 2.5.5.

2.6. Derived equivalences. Again, we pick $\mathbf{c}\in\mathfrak{P}$ and consider the shifted lattice $\mathbf{c}+\mathfrak{P}_{\mathbb{Z}}$. As we have seen in the previous section, the categories $\mathcal{O}_{\mathbf{c}}(W)$ and $\mathcal{O}_{\mathbf{c}'}(W)$ are equivalent provided $\mathbf{c},\mathbf{c}'$ lie in the same chamber and the equivalence intertwines the KZ functors. A natural question is what happens when $\mathbf{c}$ and $\mathbf{c}'$ lie in different chambers. Here is the most basic version of an answer.

THEOREM 2.32. *Let $\mathbf{c},\mathbf{c}'\in\mathfrak{P}$ satisfy $\mathbf{c}'-\mathbf{c}\in\mathfrak{P}_{\mathbb{Z}}$. Then there is a derived equivalence $D^b(\mathcal{O}_{\mathbf{c}}(W))\xrightarrow{\sim}D^b(\mathcal{O}_{\mathbf{c}'}(W))$ that intertwine the KZ functors.*

This result was conjectured in [**R2**] and proved in [**GL**, Section 5] for $W=G(\ell,1,n)$ and in [**Lo6**] for an arbitrary W. The crucial role in the latter is played by so called wall-crossing functors.

2.6.1. *Wall-crossing functors.* Let F be a face in an open chamber C with $\mathbf{c}\in C$. Let C' be a chamber that is opposite to C with respect to F (for example, if F is a codimension one face, then we take the unique other chamber adjacent to F). Pick $\mathbf{c}'\in(\mathbf{c}+\underline{\mathfrak{P}}_{\mathbb{Z}})\cap C'$. Let $\mathfrak{P}_0:=\mathbf{c}+\operatorname{Span}_{\mathbb{C}}(F)$, this is an affine subspace in $\mathfrak{P}$.

We have the following important remark that follows from Proposition 2.29.

REMARK 2.33. For every $\mathbf{c}^0\in F\cap\underline{\mathfrak{P}}_{\mathbb{Z}}$, we can replace $\mathbf{c},\mathbf{c}'$ with $\mathbf{c}+\mathbf{c}^0,\mathbf{c}'+\mathbf{c}^0$ without changing the categories $\mathcal{O}$. In particular, and this is going to be very important in what follows, in studying an interplay between $\mathcal{O}_{\mathbf{c}},\mathcal{O}_{\mathbf{c}'}$, we may assume that that $\mathbf{c}$ is Zariski generic in $\mathfrak{P}$.

The following is [**Lo6**, Theorem 1.1]. It implies Theorem 2.32.

PROPOSITION 2.34. *Suppose $\mathbf{c}'-\mathbf{c}\in\underline{\mathfrak{P}}_{\mathbb{Z}}$. Then there is a derived equivalence $\mathfrak{WC}_{\mathbf{c}'\leftarrow\mathbf{c}}:D^b(\mathcal{O}_{\mathbf{c}}(W))\to D^b(\mathcal{O}_{\mathbf{c}'}(W))$ intertwining the KZ functors.*

SKETCH OF PROOF. Set $\psi:=\mathbf{c}'-\mathbf{c}$. Consider the specializations $H_{\mathfrak{P}_0+\psi},H_{\mathfrak{P}_0}$ of $H_{\mathfrak{P}}$. In [**Lo6**, Section 5.2], we have produced a $H_{\mathfrak{P}_0+\psi}$-$H_{\mathfrak{P}_0}$-bimodule to be denoted by $\mathcal{B}_{\mathfrak{P}_0}(\psi)$. It is *Harish-Chandra* in the sense of [**BEG1**, Section 3] meaning that it is finitely generated as a bimodule, and the operators $[a,\cdot]:\mathcal{B}_{\mathfrak{P}_0}(\psi)\to\mathcal{B}_{\mathfrak{P}_0}(\psi)$ are locally nilpotent for all $a\in S(\mathfrak{h})^W\cup S(\mathfrak{h}^*)^W$ (note that $S(\mathfrak{h})^W,S(\mathfrak{h}^*)^W$ are included into both $H_{\mathfrak{P}_0+\psi},H_{\mathfrak{P}_0}$ and so it makes sense to consider the adjoint operators above). Now pick $\hat{\mathbf{c}}\in\mathfrak{P}_0$ and consider the specialization $\mathcal{B}_{\hat{\mathbf{c}}}(\psi)$ of $\mathcal{B}_{\mathfrak{P}_0}(\psi)$. It defines a functor

$$\mathcal{B}_{\hat{\mathbf{c}}}(\psi)\otimes^L_{H_{\hat{\mathbf{c}}}}\bullet:D^b(\mathcal{O}_{\hat{\mathbf{c}}}(W))\to D^b(\mathcal{O}_{\hat{\mathbf{c}}+\psi}(W)). \tag{2.7}$$

A crucial property of this functor is that if $\hat{\mathbf{c}}$ is Weil generic (=lies outside of countably many algebraic subvarieties in $\mathfrak{P}_0$), then (2.7) is a derived equivalence. Using generic flatness kind arguments, see Step 3 in the proof of [**Lo6**, Proposition 5.3], we then show that the functor $\mathfrak{WC}_{\mathbf{c}'\leftarrow\mathbf{c}}$ is an equivalence for a Zariski generic $\hat{\mathbf{c}}\in\mathfrak{P}_0$. Since we can achieve that $\mathbf{c}$ is Zariski generic by replacing $\mathbf{c}$ with $\mathbf{c}+\mathbf{c}^0$, this proves Proposition 2.34. □

2.6.2. *Behavior of WC on K_0.* Here is an easy but useful property of $\mathfrak{WC}_{\mathbf{c}'\leftarrow\mathbf{c}}$, see [**Lo8**, 3.1.1].

LEMMA 2.35. *The following is true.*

(1) *If $\hat{\mathbf{c}} \in \mathfrak{P}_0$ is Weil generic, then*

$$\mathfrak{WC}_{\hat{\mathbf{c}}'\leftarrow\hat{\mathbf{c}}}(\Delta_{\hat{\mathbf{c}}}(\tau)) \cong \nabla_{\hat{\mathbf{c}}'}(\tau)$$

for any $\tau \in \operatorname{Irr}(W)$.

(2) *For a Zariski generic $\mathbf{c} \in \mathfrak{P}_0$, the object $\mathfrak{WC}_{\mathbf{c}'\leftarrow\mathbf{c}}(\Delta_{\mathbf{c}}(\tau)) \in D^b(\mathcal{O}_{\mathbf{c}'}(W))$ lies in $\mathcal{O}_{\mathbf{c}'}(W)$ and its class in K_0 coincides with that of $\Delta_{\mathbf{c}'}(\tau)$.*

(3) *In particular, $\mathfrak{WC}_{\mathbf{c}'\leftarrow\mathbf{c}}$ induces the identity map on the K_0 groups.*

2.6.3. *Perverse equivalences.* The equivalence $\mathfrak{WC}_{\mathbf{c}'\leftarrow\mathbf{c}}$ introduced above has an important property, it is *perverse*. Let us explain what this means, following [**R1**, Section 2.6].

Suppose that we have two abelian categories $\mathcal{C}^1, \mathcal{C}^2$ that are equipped with finite filtrations by Serre subcategories: $\mathcal{C}^j = \mathcal{C}^j_0 \supset \mathcal{C}^j_1 \supset \ldots \supset \mathcal{C}^j_n \supset \mathcal{C}^j_{n+1} = 0$. A derived equivalence $\varphi : D^b(\mathcal{C}^1) \to D^b(\mathcal{C}^2)$ is said to be *perverse* with respect to these filtrations if the following three conditions hold:

(P1) φ restricts to an equivalence between $D^b_{\mathcal{C}^1_i}(\mathcal{C}^1)$ and $D^b_{\mathcal{C}^2_i}(\mathcal{C}^2)$. Here we write $D^b_{\mathcal{C}^j_i}(\mathcal{C}^j)$ for the full subcategory of $D^b(\mathcal{C}^j)$ of all objects with homology in $\mathcal{C}^j_i$.

(P2) For $M \in \mathcal{C}^1_i$, we have $H_k(\varphi M) = 0$ for $k < i$ and $H_k(\varphi M) \in \mathcal{C}^2_{i+1}$ for $k > i$.

(P3) The functor $M \mapsto H_i(\varphi M)$ defines an equivalence $\mathcal{C}^1_i/\mathcal{C}^1_{i+1} \xrightarrow{\sim} \mathcal{C}^2_i/\mathcal{C}^2_{i+1}$.

Let us explain how the filtrations on $\mathcal{C}^1 = \mathcal{O}_{\mathbf{c}}(W), \mathcal{C}^2 = \mathcal{O}_{\mathbf{c}'}(W)$ look like, see [**Lo6**, Section 6.2]. We have chains of two-sided ideals $H_{\mathfrak{P}_0} = I^{\mathfrak{P}_0}_{n+1} \supset I^{\mathfrak{P}_0}_n \supset \ldots \supset I^{\mathfrak{P}_0}_1 \supset I^{\mathfrak{P}_0}_0 = \{0\}, H_{\mathfrak{P}_0+\psi} = I^{\mathfrak{P}_0+\psi}_{n+1} \supset I^{\mathfrak{P}_0+\psi}_n \supset \ldots \supset I^{\mathfrak{P}_0+\psi}_1 \supset I^{\mathfrak{P}_0+\psi}_0 = \{0\}$ with the following properties. For a Weil generic $\hat{\mathbf{c}} \in \mathfrak{P}_0$, the specialization $I^{\hat{\mathbf{c}}}_i$ (resp., $I^{\hat{\mathbf{c}}+\psi}_i$) is the intersection of the annihilators of all modules $M \in \mathcal{O}_{\hat{\mathbf{c}}}(W)$ (resp., $M \in \mathcal{O}_{\hat{\mathbf{c}}+\psi}(W)$) with $\dim \operatorname{Supp}(M) < i$. In particular, this property implies that

$$(I^{\hat{\mathbf{c}}}_i)^2 = I^{\hat{\mathbf{c}}}_i, (I^{\hat{\mathbf{c}}+\psi}_i)^2 = I^{\hat{\mathbf{c}}+\psi}_i. \tag{2.8}$$

By generic flatness reasons, (2.8) holds for a Zariski generic $\hat{\mathbf{c}}$ as well. In particular, as before, we may assume that it holds for $\hat{\mathbf{c}} = \mathbf{c}$. Let $\mathcal{C}^1_i$ consist of all modules in $\mathcal{O}_{\mathbf{c}}(W)$ annihilated by $I^{\mathbf{c}}_i$, by (2.8) this subcategory is closed under extensions. Define $\mathcal{C}^2_i \subset \mathcal{O}_{\mathbf{c}'}(W)$ similarly.

The following claim is [**Lo6**, Theorem 6.1].

PROPOSITION 2.36. *The equivalence $\mathfrak{WC}_{\mathbf{c}'\leftarrow\mathbf{c}} : D^b(\mathcal{O}_{\mathbf{c}}) \xrightarrow{\sim} D^b(\mathcal{O}_{\mathbf{c}'})$ is perverse with respect to the filtrations introduced above.*

SKETCH OF PROOF. First, one proves this claim for a Weil generic $\hat{\mathbf{c}}$. Then one translates (P1)-(P3) to statements involving Tor's and Ext's between $\mathcal{B}_{\mathbf{c}}(\psi), H_{\mathbf{c}}/I^{\mathbf{c}}_i$, $H^{\mathbf{c}'}/I^{\mathbf{c}'}_i$. One uses generic flatness arguments to establish these statements for a Zariski generic $\mathbf{c}$. This establishes (P1)-(P3) for such $\mathbf{c}$. □

2.6.4. *Essential walls.* Proposition 2.36 allows to show that some wall-crossing functors are actually highest weight equivalences. Recall hyperplanes $\Pi_{\xi_T} \subset \mathfrak{P}$ introduced in 2.5.2. We say that such a hyperplane is *essential* for a parameter $\mathbf{c}$ if, for a Weil generic parameter $\hat{\mathbf{c}} \in \mathbf{c}+\Pi_{\xi_T}$, the category $\mathcal{O}_{\hat{\mathbf{c}}}(W)$ is not semisimple.

COROLLARY 2.37. *Let F be a codimension 1 face spanning a non-essential wall for $\mathbf{c}$. Then the corresponding functor $\mathfrak{WC}_{\mathbf{c}'\leftarrow\mathbf{c}}$ restricts to a highest weight equivalence $\mathcal{O}_{\mathbf{c}}(W) \xrightarrow{\sim} \mathcal{O}_{\mathbf{c}'}(W)$.*

See [**Lo8**, Lemma 3.7] for the proof.

2.6.5. *Wall-crossing and restriction functors.* Morally, wall-crossing functors intertwine the restriction and induction functors. The following result is proved in [**Lo8**, Proposition 3.5].

PROPOSITION 2.38. *Let $\mathbf{c}, F, \mathbf{c}', \psi, \mathfrak{P}_0$ be as before. We assume that*

(*) *for a Weil generic $\hat{\mathbf{c}} \in \mathfrak{P}_0$, the parameters $\hat{\mathbf{c}}, \hat{\mathbf{c}}+\psi$ lie in the opposite chambers for both $W, \underline{W}$.*

Then there are isomorphisms of functors

$$\underline{\mathfrak{WC}}_{\mathbf{c}'\leftarrow\mathbf{c}} \circ \operatorname{Res}^W_{\underline{W}} \cong \operatorname{Res}^W_{\underline{W}} \circ \mathfrak{WC}_{\mathbf{c}'\leftarrow\mathbf{c}},$$
$$\mathfrak{WC}_{\mathbf{c}'\leftarrow\mathbf{c}} \circ \operatorname{Ind}^W_{\underline{W}} \cong \operatorname{Ind}^W_{\underline{W}} \circ \underline{\mathfrak{WC}}_{\mathbf{c}'\leftarrow\mathbf{c}},$$

where we write $\underline{\mathfrak{WC}}_{\mathbf{c}'\leftarrow\mathbf{c}}$ for the wall-crossing functor associated to $\underline{W}$.

See [**Lo8**, 3.2.3] for the proof and the discussion of the assumption (*).

2.6.6. *Wall-crossing bijections.* An important feature of a perverse equivalence $\varphi : D^b(\mathcal{C}^1) \xrightarrow{\sim} D^b(\mathcal{C}^2)$ is that it induces a bijection $\operatorname{Irr}(\mathcal{C}^1) \xrightarrow{\sim} \operatorname{Irr}(\mathcal{C}^2)$. We have the bijection $\operatorname{Irr}(\mathcal{C}^1_i/\mathcal{C}^1_{i+1}) \xrightarrow{\sim} \operatorname{Irr}(\mathcal{C}^2_i/\mathcal{C}^2_{i+1})$ given by the equivalence in (P3) for any i. These bijections constitute a desired bijection $\operatorname{Irr}(\mathcal{C}^1) \xrightarrow{\sim} \operatorname{Irr}(\mathcal{C}^2)$. For the wall-crossing functor $\mathfrak{WC}_{\mathbf{c}'\leftarrow\mathbf{c}}$ the corresponding bijection $\operatorname{Irr}(W) \to \operatorname{Irr}(W)$ will be called the *wall-crossing* bijection and denoted by $\mathfrak{wc}_{\mathbf{c}'\to\mathbf{c}}$.

We have the following two important properties of wall-crossing bijections established in [**Lo8**]. The first property is a consequence of the construction of a wall-crossing functor as the derived tensor product with a Harish-Chandra bimodule.

LEMMA 2.39. *The wall-crossing bijections preserve supports.*

This is [**Lo8**, Corollary 2.13].

PROPOSITION 2.40. *The wall-crossing bijection $\mathfrak{wc}_{\mathbf{c}+\psi\leftarrow\mathbf{c}}$ is independent of the choice of a Zariski generic $\mathbf{c} \in \mathfrak{P}_0$.*

This is [**Lo8**, Proposition 3.1].

The second property allows to reduce the computation of the wall-crossing bijection to the case when $\mathbf{c}$ is Weil generic in $\mathfrak{P}_0$. For such a $\mathbf{c}$, the categories $\mathcal{O}_{\mathbf{c}}(W), \mathcal{O}_{\mathbf{c}+\psi}(W)$ simplify and it is easier to compute the wall-crossing bijections there. We will do some explicit computations in the case of $W = G(\ell,1,n)$ in Section 4.4 following [**Lo8**, Section 5.4].

3. Cyclotomic categories $\mathcal{O}$ and categorification

3.1. Cyclotomic categories $\mathcal{O}$ and Fock spaces.

3.1.1. *Recap.* From now on we are interested in the groups $W = G(\ell, 1, n)$. Recall that the set $\operatorname{Irr}(W)$ is identified with the set $\mathcal{P}_\ell(n)$ of the ℓ-multipartitions of n, see Example 2.6.

We will write $\mathcal{O}_{\mathbf{c}}(n)$ for $\mathcal{O}_{\mathbf{c}}(G(\ell, 1, n))$.

Recall the parameters $\kappa, h_0, \ldots h_{\ell-1}$, 2.2.2. In the case when $\kappa = 0$, we get $H_{\mathbf{c}} = H_{\mathbf{c}}(\mu_\ell)^{\otimes n} \# \mathfrak{S}_n$ and so the category $\mathcal{O}_{\mathbf{c}}(n)$ coincides with the category of $\mathfrak{S}_n$-equivariant objects in $\mathcal{O}_{\mathbf{c}}(1)^{\otimes n}$. This is an easy case and we are not going to consider it below. So we assume that $\kappa \neq 0$. In this case we introduce another set of parameters, $s_0, \ldots, s_{\ell-1}$, by $h_i = \kappa s_i - i/\ell$. We write $\mathbf{s}$ for the collection $(s_0, \ldots, s_{\ell-1})$ (note that $s_0, \ldots, s_{\ell-1}$ are defined up to a common summand). We often write $\mathcal{O}_{\kappa,\mathbf{s}}(n)$ instead of $\mathcal{O}_{\mathbf{c}}(n)$.

Also recall that the Hecke algebra $\mathcal{H}_{\mathbf{q}}(W)$ coincides with the cyclotomic Hecke algebra, $\mathcal{H}_{q,\mathbf{s}}(n)$ with parameters $q = \exp(2\pi\sqrt{-1}\kappa)$ and $Q_i := \exp(2\pi\sqrt{-1}\kappa s_i)$, Example 2.16.

3.1.2. *Order on* $\mathcal{O}_{\mathbf{c}}$. Recall, 2.2.2, that to a box b with coordinates (x, y) in the partition $\lambda^{(i)}$ we assign the number $c_b = \kappa\ell(x - y) + \ell h_i = \kappa\ell(x - y + s_i) - i$. We will write $\operatorname{cont}^{\mathbf{s}}(b)$ for $x - y + s_i$. We can take the order $\leqslant_{\mathbf{c}}$ defined by $\lambda \leqslant_{\mathbf{c}} \mu$ if $\lambda = \mu$ or $\sum_{b\in\lambda} c_b - \sum_{b\in\mu} c_b \in \mathbb{Z}_{>0}$ for a highest weight order on $\mathcal{O}_{\mathbf{c}}(W)$. It turns out that a rougher order will also work.

Define an equivalence relation on boxes by $b \sim b'$ if $\kappa(\operatorname{cont}^{\mathbf{s}}(b) - \operatorname{cont}^{\mathbf{s}}(b')) \in \mathbb{Z}$. We write $b \leqslant b'$ if $b \sim b'$ and $c_b - c_{b'} \in \mathbb{Z}_{\geqslant 0}$. Define the order $\preceq_{\mathbf{c}}$ on $\mathcal{P}_\ell(n)$ as follows: we set $\lambda \preceq_{\mathbf{c}} \lambda'$ if one can order boxes $b_1, \ldots, b_n$ of λ and $b'_1, \ldots, b'_n$ of λ' so that $b_i \leqslant b'_i$ for any i. Clearly, $\lambda \preceq_{\mathbf{c}} \lambda'$ implies $\lambda \leqslant_{\mathbf{c}} \lambda'$.

The following result is due to Dunkl and Griffeth, [**DG**, Theorem 1.2].

PROPOSITION 3.1. *One can take $\preceq_{\mathbf{c}}$ for a highest weight order for $\mathcal{O}_{\mathbf{c}}(n)$.*

3.1.3. *Fock space.* We set $\mathcal{O}_{\kappa,\mathbf{s}} := \bigoplus_{n\geqslant 0} \mathcal{O}_{\kappa,\mathbf{s}}(n)$. We have the basis $|\lambda\rangle := [\Delta_{\kappa,\mathbf{s}}(\lambda)]$ in $K_0^{\mathbb{C}}(\mathcal{O}_{\kappa,\mathbf{s}})$ indexed by $\lambda \in \mathcal{P}_\ell := \bigsqcup_{n\geqslant 0} \mathcal{P}_\ell(n)$. In other words, $K_0^{\mathbb{C}}(\mathcal{O}_{\kappa,\mathbf{s}})$ is the level ℓ Fock space.

3.1.4. *Decomposition.* For certain values of $\mathbf{s}$, the category $\mathcal{O}_{\kappa,\mathbf{s}}$ can be decomposed into the product of categories $\mathcal{O}_{\kappa,?}$ for smaller ℓ.

We define an equivalence relation $\sim_{\mathbf{c}}$ on $\{0, \ldots, \ell - 1\}$ by setting $i \sim_{\mathbf{c}} j$ if the ith and jth partitions can contain equivalent boxes, i.e., $s_i - s_j \in \kappa^{-1}\mathbb{Z} + \mathbb{Z}$. For an equivalence class α, we write $\mathbf{s}(\alpha)$ for $(s_i)_{i\in\alpha}$ and $\mathcal{P}_\alpha$ for the subset of all $\lambda \in \mathcal{P}_\ell$ with $\lambda^{(j)} = \varnothing$ for $j \notin \alpha$. Form the category $\bigotimes_\alpha \mathcal{O}_{\kappa,\mathbf{s}(\alpha)}$. The simples in this category are labelled by the set $\prod_\alpha \mathcal{P}_\alpha$ that is naturally identified with $\mathcal{P}_\ell$.

PROPOSITION 3.2. *There is a highest weight equivalence $\mathcal{O}_{\kappa,\mathbf{s}} \xrightarrow{\sim} \bigotimes_\alpha \mathcal{O}_{\kappa,\mathbf{s}(\alpha)}$.*

In the case when the assumptions of (4) of Proposition 2.18 are satisfied, this result was established in [**R2**, Section 6]. In general, it can be proved using the extension of Rouquier's approach explained in 2.5.7.

Using Proposition 3.2, we can reduce the study of categories $\mathcal{O}_{\kappa,\mathbf{s}}$ to the case when we have just one equivalence class in $\{0, 1, \ldots, \ell - 1\}$.

3.1.5. *Essential walls.* The definition of an essential wall was given in 2.6.4. Here we are going to describe the essential walls for the groups $G(\ell, 1, n)$. According to [**Lo8**, Lemma 5.5], we have the following essential walls

(1) $\kappa = 0$ for the parameters $\mathbf{c}$, where the κ-component of $\mathbf{c}$ is a rational number with denominator between 2 and n.
(2) $h_i - h_j = \kappa m$ with $i \neq j$ and $|m| < n$ – for the parameters $\mathbf{c}$ satisfying $s_i - s_j - m \in \kappa^{-1}\mathbb{Z}$.

These walls split $\mathbf{c} + \mathfrak{P}_{\mathbb{Z}}$ into the union of polyhedral chambers to be called *essential chambers*. Recall, Corollary 2.37, that the categories $\mathcal{O}$ with parameters in one essential chamber are highest weight equivalent.

3.2. Categorical Kac-Moody action.

3.2.1. *Kac-Moody action on the Fock space.* Let $\kappa, \mathbf{s}$ be the same as in 3.1.1.

Define the Kac-Moody algebra $\mathfrak{g}_\kappa$ as $\hat{\mathfrak{sl}}_e$ if κ is rational with denominator e and as $\mathfrak{sl}_\infty$ if κ is irrational. Here $\mathfrak{sl}_\infty$ stands for the Kac-Moody algebra of infinite rank associated to the type A Dynkin diagram that is infinite in both direction. When $e = 1$, we assume that $\mathfrak{g}_\kappa = \{0\}$.

Define the Kac-Moody algebra $\mathfrak{g}_{\kappa,\mathbf{s}}$ (to be denoted also as $\mathfrak{g}_{\mathbf{c}}$) as the product of several copies of $\mathfrak{g}_\kappa$, one per equivalence class for $\sim_{\mathbf{c}}$ in $\{0, 1, \ldots, \ell-1\}$. This algebra has generators e_z, f_z, where z runs over the subset in $\mathbb{C}/\kappa^{-1}\mathbb{Z}$ of the elements of the form $s_i + m$, where m is an integer. We are going to define an action of $\mathfrak{g}_{\kappa,\mathbf{s}}$ on $\mathcal{F}^\ell$, the Fock space of level ℓ.

The action is defined as follows. We say that a box b is a z-box if $\mathrm{cont}^{\mathbf{s}}(b)$ equals to z in $\mathbb{C}/\kappa^{-1}\mathbb{Z}$. We set $f_z|\lambda\rangle := \sum_\mu |\mu\rangle$, where the sum is taken over all ℓ-partitions μ that are obtained from λ by adding a z-box. Similarly, we set $e_z|\lambda\rangle = \sum_\nu |\nu\rangle$, where the sum is taken over all ν obtained from λ by removing a z-box.

We write $\mathcal{F}_{\kappa,\mathbf{s}}$ for the space $\mathcal{F}^\ell$ equipped with this $\mathfrak{g}_\kappa$-action. We note that we have a natural isomorphism of $\mathfrak{g}_\kappa$-modules, $\mathcal{F}_{\kappa,\mathbf{s}} = \bigotimes_{i=0}^{\ell-1} \mathcal{F}_{\kappa,s_i}$.

3.2.2. *Type A categorical Kac-Moody action.* Let $\mathcal{C}$ be an abelian $\mathbb{C}$-linear category, where all objects have finite length. A *type A categorical Kac-Moody action* on $\mathcal{C}$ as defined in [**R3**, 5.3.7, 5.3.8] consists of the following data:

(1) exact endo-functors E, F of $\mathcal{C}$ and a number $q \in \mathbb{C} \setminus \{0, 1\}$,
(2) adjointness morphisms $1 \to EF, FE \to 1$,
(3) endomorphisms $X \in \mathrm{End}(E), T \in \mathrm{End}(E^2)$.

These data are supposed to satisfy the axioms to be provided below. We will need the following notation. Let I be a subset in $\mathbb{C}^\times$. Define a Kac-Moody algebra $\mathfrak{g}_I$ as follows. Define an unoriented graph structure on I by connecting z and z' if $z'z^{-1} = q^{\pm 1}$. Then $\mathfrak{g}_I$ is the Kac-Moody algebra defined from I, it is the product of several copies of $\hat{\mathfrak{sl}}_e$ if q is a primitive root of unity of order e, and is the product of several copies of $\mathfrak{sl}_\infty$ if q is not a root of unity. For example, taking $q = \exp(2\pi\sqrt{-1}\kappa)$ and $I = \{Q_0, \ldots, Q_{\ell-1}\}$, we get $\mathfrak{g}_I = \mathfrak{g}_{\kappa,\mathbf{s}}$.

Here are the axioms of a categorical action.

(i) F is isomorphic to the left adjoint of E.
(ii) For any d, the map $X_i \mapsto \mathbf{1}^{i-1}X\mathbf{1}^{d-i}, T_i \mapsto \mathbf{1}^{i-1}T\mathbf{1}^{d-1-i}$ extends to a homomorphism $\mathcal{H}_q^{aff}(d) \to \mathrm{End}(E^d)$.
(iii) Let $E = \bigoplus_{z\in\mathbb{C}} E_z$ be the decomposition into eigen-functors according to X, and $F = \bigoplus_{z\in\mathbb{C}} F_z$ be the decomposition coming from (2). The operators $[E_z], [F_z]$ give rise to an integrable representation of $\mathfrak{g}_I$ on $K_0^{\mathbb{C}}(\mathcal{C})$, where $I := \{z \in \mathbb{C}^\times | E_z \neq 0\}$.

(iv) Let $\mathcal{C}_\nu$ denote the Serre subcategory of $\mathcal{C}$ spanned by the simples L with $[L] \in K_0^{\mathbb{C}}(\mathcal{C}_\nu)$, where $K_0^{\mathbb{C}}(\mathcal{C}_\nu)$ is the ν weight space for the $\mathfrak{g}_I$-module $K_0^{\mathbb{C}}(\mathcal{C})$. Then $\mathcal{C} = \bigoplus_\nu \mathcal{C}_\nu$.

3.2.3. *Example: cyclotomic Hecke algebras.* Let us provide an example that first appeared in [**CR**, Section 7.2]. Let us write $\mathcal{H}_{q,\mathbf{s}}(n)$ for the cyclotomic Hecke algebra as in 3.1.1. Set $\mathcal{C} := \bigoplus_{n\geqslant 0} \mathcal{H}_{q,\mathbf{s}}(n)\text{-mod}$.

Let us explain the categorification data in this case. We take the same q as the eponymous parameter for our Hecke algebra and $I := \{Q_0, \ldots, Q_{\ell-1}\}$. Let ${}^{\mathcal{H}}\operatorname{Res}^n_{n-1}$ denote the restriction functor $\mathcal{H}_{q,\mathbf{s}}(n)\text{-mod} \to \mathcal{H}_{q,\mathbf{s}}(n-1)\text{-mod}$ (we set ${}^{\mathcal{H}}\operatorname{Res}^0_{-1} = 0$) and let ${}^{\mathcal{H}}\operatorname{Coind}^{n-1}_n$ denote the coinduction functor, the right adjoint of the restriction functor. We set $E := \bigoplus_{n=0}^\infty {}^{\mathcal{H}}\operatorname{Res}^n_{n-1}$ and $F := \bigoplus_{n=0}^\infty {}^{\mathcal{H}}\operatorname{CoInd}^n_{n+1}$ so that F is naturally identified with the right adjoint of E.

The endomorphism X on the summand ${}^{\mathcal{H}}\operatorname{Res}^n_{n-1}$ is given by the multiplication by $X_n \in \mathcal{H}_{q,\mathbf{s}}(n)$. Similarly, on the summand ${}^{\mathcal{H}}\operatorname{Res}^n_{n-2}$ of $E^2 = \bigoplus_{n\geqslant 0} {}^{\mathcal{H}}\operatorname{Res}^n_{n-2}$ we define T as the multiplication by T_{n-1}.

It is shown in [**A**, Section 13.6] that $K_0^{\mathbb{C}}(\mathcal{C})$ is the irreducible highest weight $\mathfrak{g}_I$-module $L(\omega_{\mathbf{s}})$ whose highest weight equals $\omega_{\mathbf{s}} := \sum_{z\in I} n_z\omega_z$, where ω_z is the fundamental weight in the vertex z and n_z is the number s_i such that $\exp(2\pi\sqrt{-1}\kappa s_i) = z$. The category $\mathcal{C}$ admits a weight decomposition as in axiom (iv). And so it indeed carries a categorical $\mathfrak{g}_I$-action.

3.2.4. *Categorical Kac-Moody action on $\mathcal{O}_{\mathbf{c}}$.* Now let us proceed to defining a categorical $\mathfrak{g}_{\mathbf{c}}$-action on $\mathcal{O}_{\mathbf{c}}$ that categorifies the $\mathfrak{g}_{\mathbf{c}}$-action on $\mathcal{F}^\ell$ from 3.2.1. Here we follow [**Sh**].

Consider the functors ${}^{\mathcal{O}}\operatorname{Res}^n_{n-1} := {}^{\mathcal{O}}\operatorname{Res}^{G(\ell,1,n)}_{G(\ell,1,n-1)}$ and

$$
{}^{\mathcal{O}}\operatorname{Ind}^n_{n+1} := {}^{\mathcal{O}}\operatorname{Ind}^{G(\ell,1,n)}_{G(\ell,1,n+1)}.
$$

We set $E := \bigoplus_{n=0}^\infty {}^{\mathcal{O}}\operatorname{Res}^n_{n-1}$ and $F := \bigoplus_{n=0}^\infty {}^{\mathcal{O}}\operatorname{Ind}^n_{n+1}$. This functor F is naturally the right adjoint of E but it is also isomorphic to the left adjoint of E (by Proposition 2.19) so we have (i).

Now let us explain how to construct the endomorphisms $X \in \operatorname{End}(E), T \in \operatorname{End}(E^2)$. We have the KZ functor $\mathsf{KZ}_{\mathbf{c}} : \mathcal{O}_{\mathbf{c}} \twoheadrightarrow \bigoplus_{n=0}^\infty \mathcal{H}_{q,\mathbf{s}}(n)\text{-mod}$, the sum of KZ functors $\mathcal{O}_{\mathbf{c}}(n) \twoheadrightarrow \mathcal{H}_{q,\mathbf{s}}(n)\text{-mod}$. By Proposition 2.21,

$$
{}^{\mathcal{H}}E \circ \mathrm{KZ}_{\mathbf{c}} \cong \mathrm{KZ}_{\mathbf{c}} \circ {}^{\mathcal{O}}E. \tag{3.1}
$$

Since $\mathsf{KZ}_{\mathbf{c}}$ is fully faithful on the projectives objects, see (3) of Proposition 2.18, we deduce that isomorphism (3.1) induces an isomorphism $\operatorname{End}({}^{\mathcal{H}}E) \cong \operatorname{End}({}^{\mathcal{O}}E)$ that gives as an element X in the right hand side. Similarly, we have an isomorphism $\operatorname{End}({}^{\mathcal{H}}E^2) \cong \operatorname{End}({}^{\mathcal{O}}E^2)$ that gives us $T \in \operatorname{End}({}^{\mathcal{O}}E^2)$. (ii) follows.

Let us explain how to prove (iii) and (iv) (the same techniques work for cyclotomic Hecke algebras). First, consider the case when the parameter $\hat{\mathbf{c}}$ is Weil generic. Here the category $\mathcal{O}_{\hat{\mathbf{c}}}$ is semisimple and $\mathrm{KZ}_{\hat{\mathbf{c}}}$ is an equivalence. We have $E\Delta_{\hat{\mathbf{c}}}(\lambda) \cong \bigoplus_\nu \Delta_{\hat{\mathbf{c}}}(\nu)$, where the sum is taken over all ν obtained from λ by removing a box. The endomorphism X acts on $\Delta_{\hat{\mathbf{c}}}(\nu)$ by $z(\hat{\mathbf{c}}, \lambda\setminus\nu) := q^{x-y}Q_i$, where $\lambda \setminus \nu = (x, y, i)$, see [**A**, Section 13.6]. Now take an arbitrary $\mathbf{c}$ and pick a Weil generic line $\mathcal{L}$ through $\mathbf{c}$ so that $\mathcal{O}_{\hat{\mathbf{c}}}$ is semisimple for any $\hat{\mathbf{c}} \in \ell \setminus \{\mathbf{c}\}$. The class $[E_z\Delta_{\mathbf{c}}(\lambda)]$ coincides with $\sum_\nu'[\Delta_{\hat{\mathbf{c}}}(\nu)]$, where sum is taken over all ν

such that the number $z(\hat{\mathbf{c}}, \lambda \setminus \nu)$ approaches z as $\hat{\mathbf{c}} \in \mathcal{L}$ approaches $\mathbf{c}$. It follows that $[E_z\Delta_{\mathbf{c}}(\lambda)] = \sum_\nu [\Delta_{\mathbf{c}}(\nu)]$, where the summation is taken over all ν such that $\nu \subset \lambda, \lambda \setminus \nu$ is a z-box. Repeating this argument for F and the object $\nabla_{\mathbf{c}}(\lambda)$, we get $[F_z\nabla_{\mathbf{c}}(\lambda)] = \sum_\mu [\nabla_{\mathbf{c}}(\mu)]$, where the sum is taken over all μ such that $\lambda \subset \mu$ and $\mu \setminus \lambda$ is a z-box. Since $[\nabla_{\mathbf{c}}(\lambda)] = [\Delta_{\mathbf{c}}(\lambda)]$, we see that the operators $[E_z], [F_z]$ define the action of $\mathfrak{g}_{\mathbf{c}}$ on $\mathcal{F}^\ell$ considered in 3.2.1. This establishes (iii).

(iv) is shown in a similar fashion. For a multi-subset A of $\mathbb{C}/\kappa^{-1}\mathbb{Z}$ define the subspace $\mathcal{F}^\ell(A) \subset \mathcal{F}^\ell$ (resp., the subcategory $\mathcal{O}_{\mathbf{c}}(A)$) to be the linear span of $|\lambda\rangle$ (resp., the Serre span of $\Delta_{\mathbf{c}}(\lambda)$) with $\{\mathrm{cont}^{\mathbf{s}}(\lambda)\} = A$. We consider the central subalgebras $\mathbb{C}[X_1^{\pm 1}, \ldots, X_n^{\pm n}]^{\mathfrak{S}_n} \subset \mathcal{H}_q^{aff}(n)$. Since $\mathrm{KZ}_{\mathbf{c}}$ is fully faithful on the projective objects, we see that $\mathbb{C}[X_1^{\pm 1}, \ldots, X_n^{\pm n}]^{\mathfrak{S}_n}$ maps into the endomorphism algebra of the identity functor of $\mathcal{O}_{\mathbf{c}}(n)$ and so functorially acts on objects of $\mathcal{O}_{\mathbf{c}}(n)$. A multiset A of cardinality n defines a character $\chi_A : \mathbb{C}[X_1^{\pm 1}, \ldots, X_n^{\pm n}]^{\mathfrak{S}_n} \to \mathbb{C}^\times$ that sends f to $f(\exp(2\pi\sqrt{-1}a_1), \ldots, \exp(2\pi\sqrt{-1}a_n))$, where $A = \{a_1, \ldots, a_n\}$. A degeneration argument similar to the previous paragraph shows that $\mathcal{O}_{\mathbf{c}}(A)$ consists precisely of the objects $M \in \mathcal{O}_{\mathbf{c}}(n)$ (where $n = |A|$) such that $\mathbb{C}[X_1^{\pm 1}, \ldots, X_n^{\pm 1}]^{\mathfrak{S}_n}$ acts on M with generalized eigen-character χ_A. From here it follows that $K_0^{\mathbb{C}}(\mathcal{O}_{\mathbf{c}}(A)) = \mathcal{F}^\ell(A)$.

3.2.5. *Compatibility with highest weight structure.* The categorical $\mathfrak{g}_{\mathbf{c}}$-action on $\mathcal{O}_{\mathbf{c}}$ is highest weight in the sense of [**Lo2, Lo3**]. The definition from [**Lo3**] is quite technical, let us explain what it boils down to in our case. The following result is a direct consequence of 3.2.4 and Proposition 2.20.

LEMMA 3.3. *The object $E_z\Delta_{\mathbf{c}}(\lambda)$ has a filtration by $\Delta_{\mathbf{c}}(\nu)$, where ν runs over all ℓ-partitions contained in λ such that $\lambda \setminus \nu$ is a z-box, each $\Delta_{\mathbf{c}}(\nu)$ occurs with multiplicity* 1. *Similarly, $F_z\Delta_{\mathbf{c}}(\lambda)$ has a filtration by $\Delta_{\mathbf{c}}(\mu)$, where μ runs over all ℓ-partitions contained in λ such that $\mu \setminus \lambda$ is a z-box, each $\Delta_{\mathbf{c}}(\mu)$ occurs with multiplicity* 1.

3.3. Crystal.

3.3.1. *Crystal for a categorical action.* Let us recall the definition of a crystal corresponding to a categorical $\mathfrak{g}_I$-action. The crystal structure will be defined for each $z \in I$ separately so we will get I copies of an $\mathfrak{sl}_2$-crystal.

In this paper, by an $\mathfrak{sl}_2$-crystal we mean a set C with maps $\tilde{e}, \tilde{f} : C \to C \sqcup \{0\}$ such that the following holds.

- For any $v \in C$ there are m and n such that $\tilde{e}^n v = \tilde{f}^m v = 0$.
- Moreover, for $u, v \in C$, the equalities $\tilde{e}u = v$ and $\tilde{f}v = u$ are equivalent.

Now let $\mathcal{C}$ be a $\mathfrak{g}_I$-categorification. We will introduce a $\mathfrak{g}_I$-crystal structure on the set $\mathrm{Irr}(\mathcal{C})$. Namely, pick $L \in \mathrm{Irr}(\mathcal{C})$ and consider the object E_zL. If it is nonzero, it has simple head (=maximal semisimple quotient) and simple socle (=maximal semisimple sub) and those two are isomorphic, [**CR**, Proposition 5.20]. We take that simple object for $\tilde{e}_zL$ if $E_zL \neq 0$. We set $\tilde{e}_zL = 0$ if $E_zL = 0$. We define $\tilde{f}_zL$ similarly. That we get a $\mathfrak{g}_I$-crystal follows from [**CR**, Proposition 5.20] combined with [**BK**, Section 5].

3.3.2. *Computation for $\mathcal{O}_{\mathbf{c}}$.* Now let us explain how to compute the crystal on $\mathcal{P}_\ell = \mathrm{Irr}(\mathcal{O}_{\mathbf{c}})$, which was done in [**Lo2**]. In order to compute $\tilde{e}_z\lambda, \tilde{f}_z\lambda$, we first record the *$z$-signature* of λ that is a sequence of $+$'s and $-$'s. Then we perform a certain reduction procedure getting what we call the *reduced z-signature.* Based on that signature, we can then compute $\tilde{e}_z\lambda, \tilde{f}_z\lambda$. What we get is a crystal on $\mathcal{P}_\ell$ very similar to what was discovered by Uglov in [**U**].

Let us start by constructing the z-signature of λ. Let us take the addable and removable z-boxes of λ and order them in a decreasing way according to the order $\prec_{\mathbf{c}}$ introduced above. Here is an example when $\ell = 2, \kappa = -1/2, s_0 = 0, s_1 = -1$, we take z to be 0 modulo 2 and $\lambda^{(0)} = (2,2), \lambda^{(1)} = (3, 1^3)$. We get boxes $b_1, \ldots, b_5$.

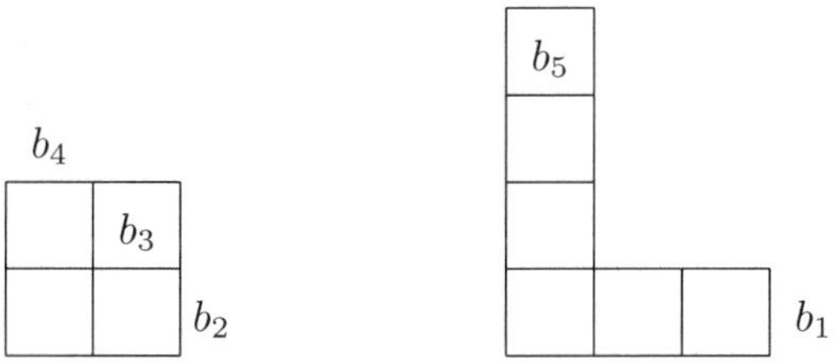

To get the signature we write a $+$ for each addable box and a $-$ for each removable box. In the example above, we get $++-+-$.

Now we are going to reduce the signature using the following steps. Initially, we have a sequence of $+$'s and $-$'s. We erase consecutive $-+$ leaving empty spaces. On each next step, if we have $-,+$ in this order separated by empty spaces, we erase them. We continue until there is no $-$'s to the left of a $+$. What we get is the *reduced signature*. It is easy to see that it does not depend on the order in which we perform our steps. In the example above, we get $++-$.

To define $\tilde{e}_z\lambda$ we pick the leftmost $-$ in the reduced z-signature of λ and remove the box in the corresponding position of λ (in the reduced signature, we replace this $-$ with a $+$). If the reduced z-signature of λ consists only of $+$'s, we set $\tilde{e}_z\lambda = 0$. To define $\tilde{f}_z\lambda$ we pick the rightmost $+$ in the reduced signature of λ and add the corresponding box to λ. If there are no $+$'s in the reduced signature we set $\tilde{f}_z\lambda = 0$.

In the example above we remove the box b_5 to get $\tilde{e}_z\lambda = ((2,2),(3,1^2))$ and add the box b_2 to get $\tilde{f}_z\lambda = ((3,2),(3,1^3))$.

3.3.3. *Special cases.* Here we are going to give a partial description of several crystals that are going to appear below.

EXAMPLE 3.4. Consider the case $\ell = 1, \kappa < 0$ and let $s_0 = 0$. Then in the κ^i-signature of $\lambda \in \mathcal{P}_1$ we include all addable/removable boxes (x, y) with $x - y$ congruent to i modulo e, where e stands for the denominator of κ (if κ is irrational, then we take $e = +\infty$ and our condition becomes $x = y$).

We note that in this case all connected components are isomorphic to the component of $\varnothing$ via an explicit isomorphism. Namely, the singular vertices are precisely the partitions divisible by e (meaning that all parts are divisible by e). For such a partition μ, an isomorphism $\mathcal{P}_1(\varnothing) \xrightarrow{\sim} \mathcal{P}_1(\mu)$ is given by $\lambda \to \lambda + \mu$ (the sum is also taken component-wise).

The component of $\varnothing$ consists of all partitions, where each column occurs less than e times.

EXAMPLE 3.5. Now consider the case when $\ell = 2$, and κ is irrational. Suppose that $s_1 = 0, s_2 = m$, where m is an integer. A box $b = (x, y, 1)$ is a κ^i-box if and only if $x - y = i$. A box $b = (x, y, 2)$ is a κ^i-box if and only if $x - y = i - m$. Note that a κ^i-box from the first partition is always smaller than an i-box from the second partition. The κ^i-signature of any λ consists of at most two elements.

Let us describe the singular bi-partitions λ in this case. Those are the partitions, where all signatures consist of $+$'s only or are equal to $-+$. This implies, in particular, that $\lambda^{(1)} = \varnothing$. The only removable box in $\lambda^{(2)}$ is a κ^0-box so it has content m. So the singular bi-partitions are the bipartitions of the form $(\varnothing, ((k+m)^k))$, where k is a positive integer. Here and below, in the notation for a partition, a superscript means the multiplicity.

EXAMPLE 3.6. We still consider the case $\ell = 2, \kappa$ is irrational. But now take $s_1 = 0, s_2 = m - \kappa^{-1}$. The description of the κ^i-boxes is the same but the order is reversed: a κ^i-box from the first partition is always bigger than a κ^i-box from the second partition. So the singular bi-partitions are precisely the bi-partitions of the form $((k^{m+k}), \varnothing)$.

3.4. Type A category $\mathcal{O}$ and q-Schur algebra. In order to proceed to a categorical Heisenberg action, we need to establish an equivalence of categories $\mathcal{O}_c(\mathfrak{S}_n)$ and the categories of modules over q-Schur algebras. This equivalence was established by Rouquier in [**R2**] when $c \notin 1/2 + \mathbb{Z}$ using the approach described in Section 2.5. An alternative approach based on induction and restriction functors to be explained in this section was found in [**Lo4**, Appendix].

3.4.1. *q-Schur algebras.* Fix n and consider $N \geqslant n$. We can form Lusztig's form (i.e., the form with divided powers) of the quantum group $U_\epsilon(\mathfrak{gl}_N)$, where $\epsilon := \exp(\pi\sqrt{-1}\kappa)$. It makes sense to speak about polynomial representations of $U_\epsilon(\mathfrak{gl}_N)$ of degree n, those are the modules where all weights $(w_1, \ldots, w_N)$ satisfy

$$w_N \geqslant 0, w_1 + \ldots + w_N = n. \tag{3.2}$$

The action of $U_\epsilon(\mathfrak{gl}_N)$ on any such module factors through the *ϵ-Schur algebra* $\mathcal{S}_\epsilon(N, n)$ that, by definition, is the image of $U_\epsilon(\mathfrak{gl}_N)$ in $\operatorname{End}(V^{\otimes n})$, where V stands for the tautological $U_\epsilon(\mathfrak{gl}_N)$-module $\mathbb{C}^N$.

The category of finite dimensional $U_\epsilon(\mathfrak{gl}_N)$-modules is known to be highest weight, where the standard objects are the Weyl modules, and the order is, for example, the dominance ordering on highest weights (one can also take rougher orders). Recall that the dominance ordering is defined by $w \leqslant w'$ if $\sum_{i=1}^N w_i = \sum_{i=1}^N w_i'$ and $\sum_{i=1}^j w_i \leqslant \sum_{i=1}^j w_i'$.

It follows that the weights satisfying (3.2) form a poset ideal. So $\mathcal{S}_\epsilon(N, n)$ is a highest weight subcategory of the category of finite dimensional $U_\epsilon(\mathfrak{gl}_N)$-modules. Also it is easy to see that, for $N \geqslant n$, we have a highest weight equivalence $\mathcal{S}_\epsilon(N+1, n)\text{-mod} \xrightarrow{\sim} \mathcal{S}_\epsilon(N, n)\text{-mod}$. The standard object labelled by a partition λ in $\mathcal{S}_\epsilon(N, n)$-mod will be denoted by $\Delta^S(\lambda)$.

There is an alternative definition of the q-Schur algebras (up to a Morita equivalence), see [**DJ**]. An equivalence with the previous definition is explained in [**M**, Section 6].

Namely, consider the one-dimensional $\mathcal{H}_q(n)$-module, where the generator T_i of $\mathcal{H}_q(n)$ acts by q. We denote this module by by Triv_n. For $\lambda \in \mathcal{P}(n)$, set

$$IT(\lambda) := {}^{\mathcal{H}}\operatorname{Ind}_{\mathfrak{S}_n}^{\mathfrak{S}_\lambda}(\operatorname{Triv}_{\lambda_1} \boxtimes \ldots \boxtimes \operatorname{Triv}_{\lambda_k}).$$

Here $\lambda = (\lambda_1, \ldots, \lambda_k)$ and we write $\mathfrak{S}_\lambda$ for the parabolic subgroup $\mathfrak{S}_{\lambda_1} \times \ldots \times \mathfrak{S}_{\lambda_k} \subset \mathfrak{S}_n$. Then the $S_\epsilon(N, n)$-mod is equivalent to the category of right modules over

$$\operatorname{End}\Big(\bigoplus_{\lambda \in \mathcal{P}(n)} IT(\lambda)\Big).$$

In other words, the category of indecomposable projectives in $S_\epsilon(N, n)$-mod is identified with the category of *Young modules* $Y(\lambda)$. By definition, $Y(\lambda)$ is the unique indecomposable summand of $IT(\lambda)$ that does not occur in $IT(\lambda')$ for $\lambda' > \lambda$. The projective cover $P^S(\lambda)$ of $\Delta^S(\lambda)$ corresponds to $Y(\lambda)$.

Note that $\mathcal{H}_q(n) = IT((1^n))$. It follows that we get a quotient functor Sch : $S_\epsilon(N,n)$-mod $\twoheadrightarrow \mathcal{H}_q(n)$-mod called the Schur functor. The reason for the name is that, in the quantum group realization of $S_\epsilon(N,n)$-mod, this functor is given by $\operatorname{Hom}_{S_\epsilon(N,n)}(V^{\otimes n}, \bullet)$, where V is the tautological $U_\epsilon(\mathfrak{gl}_N)$-module. By the second construction of $S_\epsilon(N,n)$-mod, the functor Sch is fully faithful on the projectives.

3.4.2. *Equivalence theorem.* Here we are going to prove the following theorem.

PROPOSITION 3.7. *Let* $c > 0$. *There is a highest weight equivalence* φ : $\mathcal{O}_c(n) \xrightarrow{\sim} S_\epsilon(N,n)$-mod *that intertwines the quotients* KZ_c *and* Sch *to* $\mathcal{H}_q(n)$-mod.

PROOF. Since both KZ_c, Sch are fully faithful on the projectives, it is enough to check that $\mathrm{KZ}_c(P_c(\lambda)) \cong Y(\lambda)$. Let us define the object $I\Delta_c(\lambda) \in \mathcal{O}_c(n)$ as follows:

$$I\Delta_c(\lambda) = {}^{\mathcal{O}}\operatorname{Ind}_{\mathfrak{S}_n}^{\mathfrak{S}_\lambda} \left(\Delta_{|\lambda_1|}((\lambda_1)) \boxtimes \Delta_{|\lambda_2|}((\lambda_2)) \boxtimes \ldots \boxtimes \Delta_{|\lambda_k|}((\lambda_k))\right).$$

Note that the object $I\Delta_c(\lambda)$ is projective in $\mathcal{O}_c(n)$. Indeed, each $\Delta_c((\lambda_i))$ is projective in $\mathcal{O}_c(|\lambda_i|)$ because the label (λ_i) is maximal in the highest weight order. Since the induction functor maps projective objects to projective objects, we see that $I\Delta_c(\lambda)$ is projective. Further, from Lemma 2.24 we deduce that λ is the minimal label of a standard that appears in the filtration of $I\Delta_c(\lambda)$ and it occurs once. From here we deduce that $I\Delta_c(\lambda) \cong P_c(\lambda) \oplus \bigoplus_{\mu>\lambda} P_c(\mu)^{\oplus ?}$. By [**GGOR**, 6.2.1], $\mathrm{KZ}_c(\Delta_c((n))) \cong \mathrm{Triv}_n$. By Corollary 2.22,

$$\mathrm{KZ}_c(I\Delta_c(\lambda)) \cong IT(\lambda).$$

We conclude that $\mathrm{KZ}_c(P_c(\lambda)) = Y(\lambda)$. □

We finish this section with the following lemma proved in [**SV**, Section B.3].

LEMMA 3.8. *The equivalence* φ *intertwines* $\operatorname{Ind}_{\mathfrak{S}_n}^{\mathfrak{S}_{n_1} \otimes \mathfrak{S}_{n_2}}$ *with* $\bullet \otimes \bullet$.

3.4.3. *Modules* $L_c(e\mu)$. Suppose that n is divisible by e, the denominator of c. The quantum Frobenius homomorphism $U_\epsilon(\mathfrak{gl}_N) \twoheadrightarrow U_{(-1)^e}(\mathfrak{gl}_N)$ (see [**Lu**, 35.1]), gives rise to the epimorphism $S_\epsilon(N,n) \twoheadrightarrow S_{(-1)^e}(N, n/e)$. This gives rise to an exact functor $S_{-1}(N,n/e)$-mod $\to S_\epsilon(N,n)$-mod to be denoted by Fr^*. The objects in the essential image of Fr^* are the sums of the simple objects $L^S(e\mu)$.

Thanks to Proposition 3.7, we can determine the class of $L_c(e\mu)$ in $K_0(\mathcal{O}_c(n))$. Recall that the latter is identified with $K_0(\mathfrak{S}_n$-mod) that in its turn is identified with the space of degree n symmetric polynomials in $\mathbb{Z}[z_1, z_2, \ldots]$ so that the class $[\lambda]$ becomes the Schur polynomial s_λ.

The following result is implicit in [**SV**, Proposition 5.13] and is more explicit in [**EGL**, Theorem 1.4].

LEMMA 3.9. *The class of* $[L_c(e\mu)]$ *in* $K_0(\mathcal{O}_c(n))$ *corresponds to* $s_\mu(z_1^e, z_2^e, \ldots)$.

One of the reasons why we are interested in these modules is that they have the minimal possible support, see Section 4.

3.5. Categorical Heisenberg action. Let $\kappa < 0$ be a rational number with denominator e. Suppose that $\kappa e s_0, \ldots, \kappa e s_{\ell-1}$ be integers. These data define an action of $\hat{\mathfrak{sl}}_e$ on the level ℓ Fock space $\mathcal{F}^\ell$. Recall that we write $\mathcal{F}_{\kappa,\mathbf{s}}$ for $\mathcal{F}^\ell$ equipped with this action.

3.5.1. *Heisenberg action on the Fock space.* There is an action of the Heisenberg algebra $\mathfrak{Heis}$ on $\mathcal{F}_{\kappa,\mathbf{s}}$ commuting with the action of $\hat{\mathfrak{sl}}_e$. Recall that $\mathfrak{Heis}$ has generators $b_i, i \in \mathbb{Z}$, where $[b_i, b_j] = i\delta_{i,j} b_0$. We are only interested in the representations, where b_0 acts by 1.

So let us start by defining a Heisenberg action on the level one Fock $\mathcal{F}_{\kappa,a}, a \in \mathbb{Z}$ (it will be independent of the choice of a). Let $i > 0$. Define an operator b_i on $\mathcal{F}_{\kappa,a} = \bigoplus_{n\geqslant 0} K_0^{\mathbb{C}}(\mathfrak{S}_n)$ by

$$[V_\lambda] \mapsto \sum_{j=0}^{ie-1} [\operatorname{Ind}_{\mathfrak{S}_{n+ie}}^{\mathfrak{S}_n \times \mathfrak{S}_{ie}} V_\lambda \boxtimes V_{(ie-j,1^j)}].$$

Define the operator b_{-i} as the adjoint of b_i with respect to the symmetric bilinear form on $\mathcal{F}_{\kappa,a}$ given by $(|\lambda\rangle, |\mu\rangle) = \delta_{\lambda\mu}$. One can check that this indeed defines a representation of $\mathfrak{Heis}$ in $\mathcal{F}_{\kappa,a}$.

In the realization of $\bigoplus_{a\in\mathbb{Z}} \mathcal{F}_{\kappa,a}$ as the semi-infinite wedge $\Lambda^{\infty/2}\mathbb{C}^e[z^{\pm 1}]$, the operator $b_i, i > 0$, is given by

$$v_{i_1} \wedge v_{i_2} \wedge \ldots \mapsto \sum_{j=1}^{\infty} v_{i_1} \wedge \ldots \wedge v_{i_{j-1}} \wedge v_{i_j+ei} \wedge v_{i_{j+1}} \wedge \ldots$$

This realization shows that the actions of $\mathfrak{Heis}$ and $\hat{\mathfrak{sl}}_e$ commute.

We take the tensor product representation of $\mathfrak{Heis}$ in $\mathcal{F}_{\kappa,\mathbf{s}} = \bigotimes_{i=0}^{\ell-1} \mathcal{F}_{\kappa,s_i}$. It commutes with the representation of $\hat{\mathfrak{sl}}_e$.

3.5.2. *Functors A_μ.* Let μ be a partition of some positive integer d. Following [**SV**, Section 5.1], we consider functors $A_\mu : \mathcal{O}_{\mathbf{c}} \to \mathcal{O}_{\mathbf{c}}$ mapping $\mathcal{O}_{\mathbf{c}}(n)$ to $\mathcal{O}_{\mathbf{c}}(n+de)$. These functors are defined by

$$A_\mu(M) := \operatorname{Ind}_{G(\ell,1,n+de)}^{G(\ell,1,n)\times\mathfrak{S}_{de}} M \boxtimes L_c(e\mu).$$

The functors A_μ have derived right adjoint functors

$$RA_\mu^*(N) := R\operatorname{Hom}_{\mathcal{O}_c(de)}(L_c(e\mu), \operatorname{Res}_{G(\ell,1,n)\times\mathfrak{S}_{de}}^{G(\ell,1,n+de)} N).$$

The following results were established in [**SV**]. The first part follows from Proposition 5.4 there and the second part is Proposition 5.15.

PROPOSITION 3.10. *The following is true.*

- $A_\mu \circ A_{\mu'} \cong A_{\mu'} \circ A_\mu$ *for any partitions* μ, μ'.
- $A_\mu \circ E_i \cong E_i \circ A_\mu, A_\mu \circ F_i = F_i \circ A_\mu$.

By adjointness, we get the following claim.

COROLLARY 3.11. *We have* $RA_\mu^* \circ RA_{\mu'}^* \cong RA_{\mu'}^* \circ RA_\mu^*$ *and* $RA_\mu^* \circ E_i \cong E_i \circ RA_\mu^*, RA_\mu^* \circ F_i \cong F_i \circ RA_\mu^*$.

3.5.3. *From functors A_μ to a Heisenberg action.* Let h_d denote the power symmetric polynomial of degree d, i.e., $h_d = \sum_{i=1}^{\infty} z_i^d$. We have the following identity for symmetric functions:

$$h_d = \sum_{i=0}^{d-1} (-1)^i s_{(d-i,1^i)}. \tag{3.3}$$

From Lemma 3.9 combined with (3.3) it follows that

$$\sum_{i=0}^{ed-1}(-1)^i[\Delta_{\mathbf{c}}((d-i,1^i))] = \sum_{i=0}^{d-1}(-1)^i[L_{\mathbf{c}}((d-i)e,e^{i-1})]. \tag{3.4}$$

COROLLARY 3.12. *The operator* $\sum_{i=0}^{d-1}(-1)^i[A_{(d-i,1^i)}]$ *on* $K_0(\mathcal{O}_{\mathbf{c}}) = \mathcal{F}_{\kappa,\mathbf{s}}$ *coincides with the Heisenberg operator* b_d. *The operator* $\sum_{i=0}^{d-1}(-1)^i[RA^*_{(d-i,1^i)}]$ *coincides with* b_{-d}.

PROOF. The claim about the operators b_d follows from (3.4) and the definition of b_d. To prove the claim about the operators b_{-d} we observe that the symmetric form on $\mathcal{F}_{\kappa,\mathbf{s}}$ is the Euler form on K_0: $([M],[N]) = \sum_{i\geqslant 0}(-1)^i \dim \operatorname{Ext}^i(M,N)$. To see the coincidence of the forms, we recall that $[\nabla_{\mathbf{c}}(\mu)] = [\Delta_{\mathbf{c}}(\mu)]$ and

$$\dim \operatorname{Ext}^i(\Delta_{\mathbf{c}}(\lambda),\nabla_{\mathbf{c}}(\mu)) = \delta_{\lambda\mu}\delta_{i0}. \quad \square$$

3.5.4. *Application to finite dimensional modules in* $\mathcal{O}_{\mathbf{c}}(n)$. The Heisenberg categorical action (meaning the functors A_μ, RA^*_ν) was used in [**SV**] to determine the filtration by supports on $K_0(\mathcal{O}_{\mathbf{c}}(n))$. We will discuss this in more detail in Section 4. For now we will just explain the first step in this direction.

PROPOSITION 3.13. *Let* $L \in \operatorname{Irr}(\mathcal{O}_{\mathbf{c}}(n))$ *be finite dimensional. Then* $e_i[L] = b_{-d}[L] = 0$.

PROOF. The condition that L is finite dimensional is equivalent to $\operatorname{Res}^W_{\underline{W}} L = 0$ for any proper parabolic subgroup $\underline{W} \subset W := G(\ell,1,n)$. In particular, we see that $EL = 0$ and $RA^*_\mu L = 0$ for any μ. The former equality implies $E_i L = 0$ for any i, while the latter yields $b_d[L] = 0$ for any d, thanks to Corollary 3.12. $\square$

4. Supports of simple modules

In this section we study the supports of simple modules in the category $\mathcal{O}_{\kappa,\mathbf{s}}$. We determine all theoretically possible supports in Section 4.1. Then we relate the supports to crystal structures on $\mathcal{P}_\ell$. Then we introduce and compute the filtration by supports on $K_0^{\mathbb{C}}(\mathcal{O}_{\kappa,\mathbf{s}})$ following [**SV**]. Finally, we explain a combinatorial recipe to compute the support of $L_{\kappa,\mathbf{s}}(\lambda)$ following [**Lo8**].

4.1. Possible supports. Let e denote the denominator of κ. Set $W = G(\ell,1,n)$. For non-negative integers p,q satisfying $p+eq \leqslant n$ (if κ is irrational, we assume that $e=+\infty$, and so q is automatically 0), we set $W_{p,q} = G(n-p-eq,1,\ell)\times \mathfrak{S}_e^q$, this is the stabilizer of the point of the form $(x_1,\ldots,x_p,y_1,\ldots,y_1,\ldots,y_q,\ldots,y_q, 0,\ldots,0)$, where $x_1,\ldots,x_p,y_1,\ldots,y_q$ are pairwise different complex numbers and each $y_1,\ldots,y_q$ occurs e times.

When $e=1$, we assume that $p=0$.

Recall the subvariety $\overline{X(W_{p,q})} := W\mathfrak{h}^{W_{p,q}} \subset \mathfrak{h}$.

LEMMA 4.1. *Let* $L \in \operatorname{Irr}(\mathcal{O}_{\kappa,\mathbf{s}}(n))$. *Then there are* p,q *with* $p+eq\leqslant n$ *such that* $\operatorname{Supp}(L) = \overline{X(W_{p,q})}$. *If* $\ell=1$, *then* $p=0$.

PROOF. We know, Lemma 2.12, that $\operatorname{Supp}(L) = \overline{X(\underline{W})}$ for some parabolic subgroup $\underline{W}\subset W$. Also we know that $\operatorname{Res}^W_{\underline{W}}(L)$ is nonzero finite dimensional. By [**BEG2**], there is a finite dimensional representation in $\mathcal{O}_c(m)$ if and only if m is the denominator of c. The subgroup $\underline{W}$ is conjugate to $G(k,1,\ell)\times\mathfrak{S}_{m_1}\times\ldots\mathfrak{S}_{m_s}$

for some $k, m_1, \ldots, m_s$ (when $\ell = 1$ we assume that $k = 0$). We must have $m_1 = \ldots = m_s = e$. This implies the claim of the lemma. □

4.2. Supports vs crystals. For $\lambda \in \mathcal{P}_\ell(n)$, we write $p_{\kappa,\mathbf{s}}(\lambda), q_{\kappa,\mathbf{s}}(\lambda)$ for the numbers p, q such that $\mathrm{Supp}(L_{\kappa,\mathbf{s}}(\lambda)) = \overline{X}(W_{p,q})$.

4.2.1. *$p_{\kappa,\mathbf{s}}(\lambda)$ and $\mathfrak{g}_{\kappa,\mathbf{s}}$-crystal.* Let $\mathfrak{g}_I$ be a type A Kac-Moody algebra and let C be a highest weight $\mathfrak{g}_I$-crystal, where "highest weight" means that for every $v \in C$ there is $k \in \mathbb{Z}_{\geqslant 0}$ such that $\tilde{e}_{i_1} \ldots \tilde{e}_{i_k} v = 0$ for any $i_1, \ldots, i_k \in I$. We define the *depth* of v as $k - 1$ for the minimal such k. For example, v has depth 0 if and only if $\tilde{e}_i v = 0$ for all $i \in I$.

The $\mathfrak{g}_{\kappa,\mathbf{s}}$-crystal $\mathcal{P}_{\kappa,\mathbf{s}}$ (i.e., the set $\mathcal{P}_\ell$ equipped with the crystal structure determined by $\kappa, \mathbf{s}$) is highest weight. The following result was obtained in [**Lo2**, Section 5.5].

PROPOSITION 4.2. *The number $p_{\kappa,\mathbf{s}}(\lambda)$ coincides with the depth of λ in $\mathcal{P}_{\kappa,\mathbf{s}}$.*

4.2.2. *The map $\tilde{a}_\mu$.* Now we start to explain how to compute $q_{\kappa,\mathbf{s}}$. Assume for the time being that κ is a negative rational number with denominator e and $\kappa e s_1, \ldots, \kappa e s_\ell$ are integers. The general case can be reduced to this one, we will explain how to do this below.

Let $\mathcal{P}^{sing}_{\kappa,\mathbf{s}}$ denote the set of depth 0 elements in the $\mathfrak{g}_{\kappa,\mathbf{s}}$-crystal $\mathcal{P}_{\kappa,\mathbf{s}}$. We write $\mathcal{P}^{sing,q}_{\kappa,\mathbf{s}}$ for the subset of all $\lambda \in \mathcal{P}^{sing,q}_{\kappa,\mathbf{s}}$ such that $q_{\kappa,\mathbf{s}}(\lambda) = q$.

We are going to define a map $(\mu, \lambda) \mapsto \tilde{a}_\mu \lambda$ from $\mathcal{P}_1(q) \times \mathcal{P}^{sing,0}_{\kappa,\mathbf{s}}$ to $\mathcal{P}^{sing,q}_{\kappa,\mathbf{s}}$ following [**SV**, Section 5] that happens to be a bijection.

Pick $\lambda \in \mathcal{P}^{sing,0}_{\kappa,\mathbf{s}}(n)$ (i.e., $\lambda \in \mathcal{P}_\ell(n)$ such that $L_{\kappa,\mathbf{s}}(\lambda)$ is finite dimensional) and consider the object

$$A_\mu L_{\kappa,\mathbf{s}}(\lambda) = \mathrm{Ind}^{G(n,1,\ell)\times\mathfrak{S}_{e|\mu|}}_{G(n+e|\mu|,1,\ell)} \left(L_{\kappa,\mathbf{s}}(\lambda) \boxtimes L_\kappa(e\mu)\right).$$

The following is a key result about the structure of $A_\mu L_{\kappa,\mathbf{s}}(\lambda)$, see [**SV**, Sections 5.4-5.6].

PROPOSITION 4.3. *The following is true:*

(1) *The head and the socle of $A_\mu L_{\kappa,\mathbf{s}}(\lambda)$ are isomorphic simple objects with label to be denoted by $\tilde{a}_\mu\lambda$. We have $\tilde{a}_\mu\lambda \in \mathcal{P}^{sing,|\mu|}_{\kappa,\mathbf{s}}$.*
(2) *The multiplicity of $L_{\kappa,\mathbf{s}}(\tilde{a}_\mu\lambda)$ in $A_\mu L_{\kappa,\mathbf{s}}(\lambda)$ equals $\ell^{|\mu|}\dim V_\mu$, where we write V_μ for the irreducible $\mathfrak{S}_{|\mu|}$-module labelled by μ.*
(3) *Let $\lambda' \neq \tilde{a}_\mu\lambda$ be such that $L_{\kappa,\mathbf{s}}(\lambda')$ is a simple constituent of $A_\mu L_{\kappa,\mathbf{s}}(\lambda)$. Then $p_{\kappa,\mathbf{s}}(\lambda') = 0$ and $q_{\kappa,\mathbf{s}}(\lambda') < |\mu|$.*

So we indeed get a map $\mathcal{P}_1(q) \times \mathcal{P}^{sing,0}_{\kappa,\mathbf{s}} \to \mathcal{P}^{sing,q}_{\kappa,\mathbf{s}}$. The following is [**SV**, Proposition 5.33].

PROPOSITION 4.4. *The map $(\mu, \lambda) \mapsto \tilde{a}_\mu\lambda$ is a bijection*

$$\mathcal{P}_1(q) \times \mathcal{P}^{sing,0}_{\kappa,\mathbf{s}} \xrightarrow{\sim} \mathcal{P}^{sing,q}_{\kappa,\mathbf{s}}.$$

4.2.3. *Heisenberg crystal.* The set $\mathcal{P}_1$ comes with a natural $\mathfrak{sl}_\infty$-crystal structure that is the same as on $\mathcal{P}_\kappa$ for irrational κ. The bijection $\mathcal{P}_1 \times \mathcal{P}^{sing,0}_{\kappa,\mathbf{s}} \xrightarrow{\sim} \mathcal{P}^{sing}_{\kappa,\mathbf{s}}$ allows to carry this structure over to $\mathcal{P}^{sing}_{\kappa,\mathbf{s}}$. Then we have the following result proved in [**Lo8**, 5.1.2].

PROPOSITION 4.5. *There is a unique $\mathfrak{sl}_\infty$-crystal structure on $\mathcal{P}_{\kappa,\mathbf{s}}$ that commutes with the $\hat{\mathfrak{sl}}_e$-crystal structure and extends the $\mathfrak{sl}_\infty$-crystal structure on $\mathcal{P}^{sing}_{\kappa,\mathbf{s}}$ introduced in the previous paragraph. The number $q_{\kappa,\mathbf{s}}(\lambda)$ coincides with the depth of λ in this crystal.*

Since this $\mathfrak{sl}_\infty$-crystal is a crystal analog of a Heisenberg algebra action, we call it the *Heisenberg crystal.*

Comparing to the $\hat{\mathfrak{sl}}_e$-crystal operators, the crystal operators for the $\hat{\mathfrak{sl}}_\infty$-crystal are hard to compute. A relatively easy case is when one of the parameters s_i is much less than the others. In order to compute the crystal operators in this case, let us recall the division with remainder for Young diagrams. For a Young diagram $\nu = (\nu_1, \ldots, \nu_k)$ and an integer $e > 1$ we get the partial quotient $\nu' = (\nu'_1, \ldots, \nu'_k)$ and the remainder $\nu'' = (\nu''_1, \ldots, \nu''_k)$, both are Young diagrams. The rule to determine ν', ν'' is that $\nu_i = e\nu'_i + \nu''_i$ for all i and that $|\nu'|$ is maximal possible. For example, if $\nu = (7, 3, 1)$ and $e = 3$, then we get $\nu' = (1), \nu'' = (4, 3, 1)$. Note that the partition ν lies in the $\hat{\mathfrak{sl}}_e$-crystal component of $e\nu'$ and goes to ν'' under the isomorphism of the components of $\varnothing$ and of $e\nu'$, see Example 3.4.

PROPOSITION 4.6. *Assume that $j \in \{0, 1, \ldots, \ell - 1\}$ is such that $s_j < s_i - n$ for any other i. Let $\lambda \in \mathcal{P}_{\kappa,\mathbf{s}}(n)$. Let $\lambda^{(j)} = e\nu' + \nu''$ (division with remainder). Then $q_{\kappa,\mathbf{s}}(\lambda) = |\nu'|$. Moreover, the crystal operator $\tilde{e}_j, j \in \mathbb{Z}$, for $\mathfrak{sl}_\infty$ sends λ to*

$$\underline{\lambda} := (\lambda^{(1)}, \ldots, \lambda^{(j-1)}, e(\tilde{e}_j\nu') + \nu'', \lambda^{(j+1)}, \ldots, \lambda^{(\ell)}),$$

if $\tilde{e}_j\nu' \neq 0$ and to 0 else. Here $\tilde{e}_j\nu'$ is computed in the $\mathfrak{sl}_\infty$-crystal on the set of partitions.

This is proved in [**Lo8**, Section 5.2]. Below, we will explain how to compute the crystal operators for $\mathfrak{sl}_\infty$, in general, using the wall-crossing bijections.

4.2.4. *Reduction.* Let us explain how to reduce the computation of $q_{\kappa,\mathbf{s}}$ to the case when $\kappa < 0$ and $\kappa e s_i \in \mathbb{Z}$ for all i.

First of all, assume that $\kappa > 0$. Let χ be the one-dimensional character of $G(\ell, 1, n)$ that is the identity of μ_ℓ and is the sign on $\mathfrak{S}_n$. Then we have an equivalence $\mathcal{O}_{\kappa,\mathbf{s}} \xrightarrow{\sim} \mathcal{O}_{-\kappa,-\mathbf{s}}$ that sends $\Delta(\lambda)$ to $\Delta(\lambda^t)$ (and hence $L(\lambda)$ to $L(\lambda^t)$), where $\bullet^t$ means the component wise transpose. This reduces the computation of $q_{\kappa,\mathbf{s}}$ to the case when $\kappa < 0$.

Now let us explain what to do if not all $\kappa e s_0, \ldots \kappa e s_{\ell-1}$ are integers (up to adding the same summand to $s_0, \ldots, s_{\ell-1}$). This is precisely the case when there is more than one equivalence class for $\sim_{\mathbf{c}}$ in $\{0, 1, \ldots, \ell-1\}$ (see 3.1.4 for the definition of $\sim_{\mathbf{c}}$). The following result follows easily from Proposition 3.2, see [**Lo8**, Corollary 4.3].

LEMMA 4.7. *We have $q_{\kappa,\mathbf{s}}(\lambda) = \sum_\alpha q_{\kappa,\mathbf{s}^\alpha}(\lambda^\alpha)$.*

4.3. Filtration by support on K_0. Propositions 4.3, 4.4 together with properties of $\hat{\mathfrak{sl}}_e$-crystals were used in [**SV**] to describe the filtration by supports on $K_0^{\mathbb{C}}(\mathcal{O}_{\kappa,\mathbf{s}}) = \mathcal{F}_{\kappa,\mathbf{s}}$. Namely, for $p, q \in \mathbb{Z}_{\geqslant 0}$, let $\mathcal{O}^{p;q}_{\kappa,\mathbf{s}}$ denote the span of $[L_{\kappa,\mathbf{s}}(\lambda)]$ with $p_{\kappa,\mathbf{s}}(\lambda) \leqslant p, q_{\kappa,\mathbf{s}}(\lambda) \leqslant q$. Further, we let $\mathcal{F}^{p;q}_{\kappa,\mathbf{s}}$ to denote the span of elements of the form

$$f_{i_1} \ldots f_{i_p} b_{j_1} \ldots b_{j_k} v,$$

where $p' \leqslant p, j_1 + \ldots + j_k \leqslant q$ and $v \in \mathcal{F}_{\kappa,\mathbf{s}}$ is a vector annihilated by all e_i and all b_{-j}.

The following result was obtained in [**SV**], see the proof of Proposition 6.5 there.

PROPOSITION 4.8. *We have* $K_0^{\mathbb{C}}(\mathcal{O}^{p,q}_{\kappa,\mathbf{s}}) = \mathcal{F}^{p,q}_{\kappa,\mathbf{s}}$.

4.4. Combinatorial computation of supports via wall-crossing bijections. We assume that κ is a negative rational number with denominator e and the numbers $\kappa e s_0, \ldots, \kappa e s_{\ell-1}$ are integers.

Thanks to Proposition 4.6 we can compute the numbers $q_{\kappa,\mathbf{s}}(\lambda)$ in the case when one of the numbers s_j is much less than the others. We call (essential) chambers, see 3.1.5, satisfying these conditions *asymptotic*. In order to compute $q_{\kappa,\mathbf{s}}(\lambda)$ (and the crystal operators for the Heisenberg crystal) we will "move" a general parameter $(\kappa, \mathbf{s})$ to an asymptotic chamber by applying wall-crossing bijections through the essential walls $s_i - s_j = m$, where m is an integer (with fixed residue modulo e). We will fix j and use different i.

Two crucial properties that allow us to compute this wall-crossing bijection are as follows.

PROPOSITION 4.9. *Let* $\mathbf{c} = (\kappa, \mathbf{s})$ *and* $\mathbf{c}' = (\kappa', \mathbf{s}')$ *be two parameters with integral difference whose chambers are separated by the wall* Π *given by* $h_i - h_j = \kappa m$ *for* $m \in \mathbb{Z}$. *Let* $\mathfrak{wc}_{\mathbf{c}' \leftarrow \mathbf{c}} : \mathcal{P}_\ell \to \mathcal{P}_\ell$. *Then the following is true.*

(1) *For a Zariski generic* $\tilde{\mathbf{c}} \in \mathbf{c} + \Pi, \tilde{\mathbf{c}}' := \tilde{\mathbf{c}} + (\mathbf{c}' - \mathbf{c})$, *the bijection* $\mathfrak{wc}_{\tilde{\mathbf{c}}' \leftarrow \tilde{\mathbf{c}}}$ *is independent of the choice of* $\tilde{\mathbf{c}}$.
(2) *The bijection* $\mathfrak{wc}_{\tilde{\mathbf{c}}' \leftarrow \tilde{\mathbf{c}}}$ *commutes with the* $\mathfrak{g}_{\tilde{\mathbf{c}}}$*-crystal.*

The first part is a special case of Proposition 2.40. The second part is an easy consequence of 2.38.

The bijection $\mathfrak{wc}_{\mathbf{c}' \leftarrow \mathbf{c}}$ is then computed as follows. We choose a Weil generic parameter $\tilde{\mathbf{c}} \in \mathbf{c} + \Pi$. In this case there are $\ell - 1$ equivalence classes for $\sim_{\tilde{\mathbf{c}}}$, and $i \sim_{\tilde{\mathbf{c}}} j$. Furthermore, $\mathfrak{g}_{\tilde{\mathbf{c}}} = \mathfrak{sl}_\infty^{\ell-1}$ and the crystal structure on $\mathcal{P}_{\tilde{\mathbf{c}}}$ is the product of $\ell - 2$ of the level 1 crystal for $\mathfrak{sl}_\infty$ and one copy of a level 2 crystal for $\mathfrak{sl}_\infty$, the latter affects partitions number i and j. In fact, there is a unique isomorphism of crystals $\mathcal{P}_{\tilde{\mathbf{c}}} \to \mathcal{P}_{\tilde{\mathbf{c}}'}$ that maps $\mathcal{P}_{\tilde{\mathbf{c}}}(n)$ to $\mathcal{P}_{\tilde{\mathbf{c}}'}(n)$, this can be deduced from Examples 3.5,3.6. This uniqueness allows to recover $\mathfrak{wc}_{\mathbf{c}' \leftarrow \mathbf{c}}$. For details the reader is referred to [**Lo8**, Section 5.4].

5. Category equivalences and multiplicities

5.1. Kazhdan-Lusztig category. Recall that we have a highest weight category equivalence $\mathcal{O}_c(n) \xrightarrow{\sim} \mathcal{S}_\epsilon(m,n)$-mod, where $\mathcal{S}_\epsilon(m,n)$-mod is the Serre subcategory in $U_\epsilon(\mathfrak{gl}_m)$-mod$_{fin}$ generated by simples $L(\lambda)$, where λ is a highest weight of the form $(\lambda_1, \ldots, \lambda_m)$, where $\lambda_1 \geqslant \lambda_2 \geqslant \ldots \geqslant \lambda_m \geqslant 0$.

The category $U_\epsilon(\mathfrak{gl}_m)$ has an alternative realization, the so called *Kazhdan-Lusztig category* of representations of the affine Lie algebra $\hat{\mathfrak{g}} := \hat{\mathfrak{gl}}_m$. Namely, let $\hat{G}$ denote the corresponding Kac-Moody group, a central extension of $\mathrm{GL}_m(\mathbb{C}((t)))$, $I \subset \hat{G}$ be the Iwahori subgroup. We consider the category $\mathcal{O}^{\mathfrak{b}}_\kappa$ consisting of all I-integrable $\hat{\mathfrak{g}}$-modules with weights bounded from above and such that the standard central element $C \in \hat{\mathfrak{g}}$ acts by $\kappa^{-1} - m$ (here we assume that κ is negative). Set $G_+ := \mathrm{GL}_m(\mathbb{C}[[t]]) \subset \hat{G}$. By the Kazhdan-Lusztig category (to be denoted by $\mathcal{O}^{\mathfrak{g}}_\kappa$) one means the full subcategory of $\mathcal{O}^{\mathfrak{b}}_\kappa$ consisting of G_+-integrable modules. This is a highest weight category whose standard objects are the so called Weyl modules

$$W(\lambda) := U_{\kappa^{-1}-m}(\hat{\mathfrak{g}}) \otimes_{U(\mathfrak{g}_+)} V(\lambda),$$

where we write $V(\lambda)$ for the finite dimensional irreducible $\mathfrak{g}$-module with highest weight λ.

A deep result of Kazhdan and Lusztig (valid for any reductive group) is that there is a highest weight category equivalence $\mathcal{O}^{\mathfrak{g}}_\kappa \cong U_\epsilon(\mathfrak{g})\text{-mod}_{fin}$. In order to produce this equivalence, Kazhdan and Lusztig introduce a braided monoidal structure on $\mathcal{O}^{\mathfrak{g}}_\kappa$, the so called *fusion product* $\dot{\otimes} : \mathcal{O}^{\mathfrak{g}}_\kappa \times \mathcal{O}^{\mathfrak{g}}_\kappa \to \mathcal{O}^{\mathfrak{g}}_\kappa$. To compute the fusion product of modules $V_1, \ldots, V_k \in \mathcal{O}^{\mathfrak{g}}_\kappa$ one needs to fix different points $z_1, \ldots, z_k$; the different choices of points lead to isomorphic products $V_1 \dot{\otimes} V_2 \dot{\otimes} \ldots \dot{\otimes} V_k$ that form an $\mathfrak{S}_k$-equivariant local system on $\{(z_1, \ldots, z_k) \in \mathbb{C}^k | z_i \neq z_j\}$ (the $\mathfrak{S}_k$ action permutes both coordinates and the tensor factors). The monodromy of this local system gives the braided structure.

5.2. Truncated affine parabolic category $\mathcal{O}$. Now let $m \geqslant n$ and let $\mathfrak{g} = \mathfrak{gl}_m$. Let $\mathcal{O}^{\mathfrak{g}}_\kappa(n)$ denote the Serre subcategory of $\mathcal{O}^{\mathfrak{g}}_\kappa$ spanned by the Weyl modules $W(\lambda)$ with $\lambda_m \geqslant 0, \sum_{i=1}^m \lambda_i = n$. We have a highest weight equivalence $\mathcal{S}_\epsilon(m,n)\text{-mod} \xrightarrow{\sim} \mathcal{O}^{\mathfrak{g}}_\kappa(n)$ and hence a highest weight equivalence $\mathcal{O}_\kappa(n) \xrightarrow{\sim} \mathcal{O}^{\mathfrak{g}}_\kappa(n)$. The point of emphasizing the latter equivalence is that it can be generalized to $\ell > 1$ as was conjectured by Varagnolo-Vasserot in **[VV]** and proved in **[RSVV, Lo4]**.

Let us explain the conjecture by introducing "higher level" analogs of the categories $\mathcal{O}^{\mathfrak{g}}_\kappa(n) \subset \mathcal{O}^{\mathfrak{g}}_\kappa$. We assume that $\kappa = -\frac{1}{e}, s_0, \ldots, s_{\ell-1}$ are integers, in the end of the section we will explain what to do in the general case.

We modify $s_0, \ldots, s_{\ell-1}$ by adding a common summand so that $s_i > n$ for all i. Set $m := s_0 + \ldots + s_{\ell-1}$. Let $\mathfrak{p} \subset \mathfrak{g}$ be the parabolic subalgebra of block upper triangular matrices, where the blocks have sizes $s_0, s_1, \ldots, s_{\ell-1}$ (in this order). We have the corresponding parahoric subgroup $\hat{P} \subset \hat{G}$. Consider the full subcategory $\mathcal{O}^{\mathfrak{p}}_\kappa$ inside the affine category $\mathcal{O}$ consisting of all $\hat{P}$-integrable modules. This is "almost" a highest weight category in the following sense.

Consider the set

$$\mathbb{Z}^{\mathbf{s}} = \{(a_1, \ldots, a_m) | a_1 > \ldots > a_{s_0}, a_{s_0+1} > \ldots > a_{s_0+s_1}, \ldots, \\ a_{s_0+\ldots+s_{\ell-2}+1} > \ldots > a_m\}.$$

For $A \in \mathbb{Z}^{\mathbf{s}}$, let $\Delta^{\mathfrak{p}}(A)$ denote the parabolic Verma module with ρ-shifted highest weight A so that $\mathcal{O}^{\mathfrak{p}}_\kappa$ is the Serre span of $\Delta^{\mathfrak{p}}(A)$. There is a so called *linkage order* on $\mathbb{Z}^{\mathbf{s}}$ (that we are not going to define). It has the property that for all $A \in \mathbb{Z}^{\mathbf{s}}$ the poset ideals $\{A' \in \mathbb{Z}^{\mathbf{s}} | A' \leqslant A\}$ are finite. It turns out that, for any finite poset ideal $\Lambda \subset \mathbb{Z}^{\mathbf{s}}$, the Serre span of $\Delta^{\mathfrak{p}}(A), A \in \Lambda$, is a highest weight category with standard objects $\Delta^{\mathfrak{p}}(A)$. The category $\mathcal{O}^{\mathfrak{p}}_\kappa$ itself does not have enough projective objects (but has enough projective pro-objects) and so it is not highest weight in the sense introduced in 2.2.4.

Now recall the poset $\mathcal{P}_\ell(n)$ with order $\preceq_{\mathbf{c}}$, see 3.1.2. We can embed $\mathcal{P}_\ell(n)$ into $\mathbb{Z}^{\mathbf{s}}$ by $(\lambda^{(0)}, \ldots, \lambda^{(\ell-1)}) \mapsto A_\lambda$, where A_λ is given by

$$(s_0 + \lambda^{(0)}_1, s_0 - 1 + \lambda^{(0)}_2, s_0 - 2 + \lambda^{(0)}_3, \ldots, 1, s_1 + \lambda^{(1)}_1, s_1 + \lambda^{(1)}_2 - 1, \ldots, \\ 1, \ldots, s_{\ell-1} + \lambda^{(\ell-1)}_1, \ldots, 1).$$

Here we use the inequalities $n \leqslant s_i$ for all i.

One can show that $\mathcal{P}_\ell(n) \subset \mathbb{Z}^{\mathbf{s}}$ is a poset ideal with respect to the linkage order and, moreover, the order $\preceq_{\mathbf{c}}$ on $\mathcal{P}_\ell(n)$ refines the restriction of the linkage order, see, e.g., [**Lo4**, Section 2.3].

It is convenient for us to view elements of $\mathbb{Z}^{\mathbf{s}}$ as some infinite to the left analogs of Young diagrams. Namely, we view an element of $\mathbb{Z}^{\mathbf{s}}$ as a sequence of ℓ collections of boxes on the plane, where ith collection is infinite to the left, consists of s_i rows, and the rightmost positions of boxes in each row (non-strictly) decrease from top to bottom. The embedding $\mathcal{P}_\ell(n) \hookrightarrow \mathbb{Z}^{\mathbf{s}}$ takes an ℓ-tuple of diagrams and adjoins the columns with nonpositive numbers of height s_i to $\lambda^{(i)}$.

For $\mathcal{O}^{\mathfrak{p}}_\kappa(n)$ we take the Serre span of $\Delta^{\mathfrak{p}}(A_\lambda), \lambda \in \mathcal{P}_\ell(n)$.

The following conjecture was made in [**VV**, 8.8] (under unnecessary restrictions on parameters).

CONJECTURE 5.1. There is a highest weight equivalence $\mathcal{O}_{\kappa,\mathbf{s}}(n) \xrightarrow{\sim} \mathcal{O}^{\mathfrak{p}}_\kappa(n)$.

In particular, this conjecture allows to prove that the category $\mathcal{O}_{\kappa,\mathbf{s}}(n)$ is standard Koszul and describe its Koszul dual, [**RSVV**, Section 7.3], see also [**W**, Section 6].

5.3. Categorical action on affine parabolic category. In order to prove Conjecture 5.1 we will use techniques similar to what was used in Section 2.5. First of all, we need to produce a quotient functor $\pi : \mathcal{O}^{\mathfrak{p}}_\kappa(n) \twoheadrightarrow \mathcal{H}_{q,\mathbf{s}}(n)$-mod. This is one of the places in the proof of Conjecture 5.1, where a categorical $\hat{\mathfrak{sl}}_e$-action is used.

5.3.1. *Cartan component functor.* Let $\omega_i, i \in \mathbb{Z}/e\mathbb{Z}$, denote the fundamental weight for $\hat{\mathfrak{sl}}_e$. Set $\omega_{\mathbf{s}} := \sum_{i=0}^{\ell-1} \omega_{s_i}$. Now let $\mathcal{C}$ be an $\hat{\mathfrak{sl}}_e$-categorification. Assume that $\mathcal{C}_{\omega_{\mathbf{s}}} \cong \mathrm{Vect}$, while $\mathcal{C}_\eta \neq 0$ implies $\eta \leqslant \omega_{\mathbf{s}}$. In particular, $[\mathcal{C}]$ is a highest weight integrable representation with highest weight $\omega_{\mathbf{s}}$. It follows that the irreducible $\hat{\mathfrak{sl}}_e$-module $V(\omega_{\mathbf{s}})$ with highest weight $\omega_{\mathbf{s}}$ is a direct summand of $[\mathcal{C}]$ (we call it the *Cartan component*, by analogy with tensor product modules). The following result of Rouquier, [**R3**, 5.1.2] should be thought as a categorical analog of the existence of the Cartan component.

PROPOSITION 5.2. *Let* $\mathbf{1}$ *denote the indecomposable projective object in* $\mathcal{C}_{\omega_{\mathbf{s}}}$. *Then* $\mathrm{End}_{\mathcal{C}}(F^n\mathbf{1})^{opp} \cong \mathcal{H}^{\mathbf{s}}_q(n)$. *In particular, the projective object* $\bigoplus_{n\geqslant 0} F^n\mathbf{1}$ *defines a quotient functor* $\mathcal{C} \twoheadrightarrow \bigoplus_{n\geqslant 0}\mathcal{H}^{\mathbf{s}}_q(n)$ *that is a morphism of* $\hat{\mathfrak{sl}}_e$*-categorifications.*

5.3.2. *Categorical action on* $\mathcal{O}^{\mathfrak{p}}_\kappa$. So we would like to realize $\mathcal{O}^{\mathfrak{p}}_\kappa(n)$ as the degree n component in a categorification $\mathcal{C}$ like in 5.3.1. We will do something weaker to be explained in 5.3.3. Now we are going to produce a categorical $\hat{\mathfrak{sl}}_e$-action on the whole category $\mathcal{O}^{\mathfrak{p}}_\kappa$.

For $M \in \mathcal{O}^{\mathfrak{g}}_\kappa, N \in \mathcal{O}^{\mathfrak{p}}_\kappa$, we can still form the fusion product $M\dot{\otimes}N$ and this will be an object of $\mathcal{O}^{\mathfrak{p}}_\kappa$. If M is standardly filtered, then the endo-functor $M\dot{\otimes}\bullet$ is exact. Moreover, there is a contravariant duality D on $(\mathcal{O}^{\mathfrak{p}}_\kappa)^\Delta$ that maps $W(\mu)$ to $W(\mu^*)$ (where μ^* denotes the highest weight of $V(\mu)^*$), see [**Lo4**, Sections 2.6,2.9]. The functor $DM\dot{\otimes}\bullet$ is biadjoint to $M\dot{\otimes}\bullet$. See [**VV**, Corollary 7.3] for the proofs.

Moreover, we have an $\mathfrak{S}_n$-equivariant local system $M^{\dot{\otimes}n}\dot{\otimes}N$ on $(\mathbb{C}^\times \setminus \{0\})^n$ (we place the n copies of M to different nonzero points $z_1, \ldots, z_n \in \mathbb{C}$ and N to 0). So the affine braid group B^{aff}_n acts by automorphisms on the functor $M^{\dot{\otimes}n}\dot{\otimes}\bullet$. The action of the subgroup $B_n \subset B^{aff}_n$ comes from the B_n-action on $M^{\dot{\otimes}n}$.

Now let us define the categorical $\hat{\mathfrak{sl}}_e$-action on $\mathcal{O}^{\mathfrak{p}}_\kappa$. Let μ be the highest weight of $\mathbb{C}^m$. Set $F := W(\mu)\dot{\otimes}\bullet, E := W(\mu^*)\dot{\otimes}\bullet$. We have already mentioned that these functors are biadjoint so we get (i). The B_n^{aff}-action on F^n factors through $\mathcal{H}_q^{aff}(n)$. This is because the B_n-action on $W(\mu)^{\dot{\otimes} n}$ factors through $\mathcal{H}_q(n)$, which follows from results of Kazhdan and Lusztig recalled in Section 5.1. So we get (ii).

In order to establish (iii) and (iv), we need to understand the behavior of F, E on the standard objects $\Delta^{\mathfrak{p}}(A)$. There is a general result of Varagnolo and Vasserot, [**VV**, Proposition A2.6], that $W(\mu)\dot{\otimes}\Delta^{\mathfrak{p}}(A)$ is standardly filtered, and the multiplicity of $\Delta^{\mathfrak{p}}(A')$ in this filtration coincides with the multiplicity of $\Delta^{\mathfrak{p}}_{fin}(A')$ in $V(\lambda)\otimes\Delta^{\mathfrak{p}}_{fin}(A)$, where we write $\Delta^{\mathfrak{p}}_{fin}(A)$ for the parabolic Verma module for $(\mathfrak{g},\mathfrak{p})$ with ρ-shifted highest weight A. It follows that $F\Delta^{\mathfrak{p}}(A)$ has a filtration with all possible quotients $\Delta^{\mathfrak{p}}(A+\epsilon_i)$, where ϵ_i is the ith coordinate vector, each quotient occurs once. A similar result holds for $E\Delta^{\mathfrak{p}}(A)$ (we need to use $A-\epsilon_i$ instead of $A+\epsilon_i$). In other words, for a virtual ℓ-partition λ, the object $F\Delta^{\mathfrak{p}}(\lambda)$ is filtered with $\Delta^{\mathfrak{p}}(\lambda')$, where λ' is obtained from λ by adding a box. One can show that the action of X on the subquotient $\Delta^{\mathfrak{p}}(\lambda')$ is by q^d, where d is the content of the box $\lambda'\setminus\lambda$.

In particular, we have the following result. Consider the level 0 $\hat{\mathfrak{sl}}_e$-module $\mathbb{C}^{\mathbb{Z}}$ with basis $v_i, i\in\mathbb{Z}$, and the action given by $f_i v_j = \delta_{i,j \bmod e} v_{j+1}, e_i v_{j+1} = \delta_{i,j \bmod e} v_j$.

COROLLARY 5.3. *We have an isomorphism* $K_0^{\mathbb{C}}(\mathcal{O}^{\mathfrak{p}}_\kappa)\cong\bigwedge^{s_0}\mathbb{C}^{\mathbb{Z}}\otimes\bigwedge^{s_1}\mathbb{C}^{\mathbb{Z}}\otimes\ldots\otimes\bigwedge^{s_{\ell-1}}\mathbb{C}^{\mathbb{Z}}$ *of* $\hat{\mathfrak{sl}}_e$*-modules that maps* $[\Delta^{\mathfrak{p}}(A)]$ *to the monomial vector*

$$v_{a_1}\wedge v_{a_2}\ldots\wedge v_{a_{s_0}}\otimes v_{a_{s_0}+1}\wedge\ldots\wedge v_{a_{s_0+s_1}}\otimes\ldots$$

In particular, (iii) follows.

Let us explain why (iv) is true. The category $\mathcal{O}^{\mathfrak{p}}_\kappa$ splits into the direct sum of blocks. If two parabolic Verma modules $\Delta^{\mathfrak{p}}(A)$ and $\Delta^{\mathfrak{p}}(A')$ lie in the same block, then $A \bmod e$ and $A' \bmod e$ are $\mathfrak{S}_m$-conjugate. On the other hand, $[\Delta^{\mathfrak{p}}(A)]$, $[\Delta^{\mathfrak{p}}(A')]$ lie in the same weight space if and only if $A \bmod e, A' \bmod e$ are $\mathfrak{S}_m$-conjugate. So for the weight subcategories we can take suitable sums of blocks.

5.3.3. *Categorical truncation.* Consider the subcategory

$$\mathcal{O}^{\mathfrak{p}}_\kappa(\leqslant n) := \bigoplus_{j=0}^{n}\mathcal{O}^{\mathfrak{p}}_\kappa(j),$$

where, recall, we assume that $n < s_i$. The functors F_i map $\mathcal{O}^{\mathfrak{p}}_\kappa(j)$ to $\mathcal{O}^{\mathfrak{p}}_\kappa(j+1)$ for $j<n$. The functors $E_i, i\neq 0$, send $\mathcal{O}^{\mathfrak{p}}_\kappa(j)$ to $\mathcal{O}^{\mathfrak{p}}_\kappa(j-1)$. This is because all removable i-boxes in the diagram representing A_λ are the same as in λ. However, the diagram of A_λ contains ℓ removable zero boxes that are not in λ, those are the boxes $(0, s_i, i)$. So E_0 does not map $\mathcal{O}^{\mathfrak{p}}_\kappa(j)$ to $\mathcal{O}^{\mathfrak{p}}_\kappa(j-1)$.

Fortunately, this can be fixed because the ℓ-boxes $(0,s_i,i)$ are the minimal removable 0-boxes in the diagram of A_λ. It follows from results of [**Lo3**, Section 5] that the functor $F_0:\mathcal{O}^{\mathfrak{p}}_\kappa(j-1)\to\mathcal{O}^{\mathfrak{p}}_\kappa(j)$ admits a biadjoint functor to be denoted by $\underline{E}_0$. The functors $F:\mathcal{O}^{\mathfrak{p}}_\kappa(\leqslant n-1)\to\mathcal{O}^{\mathfrak{p}}_\kappa(\leqslant n), \underline{E} := \underline{E}_0\oplus\bigoplus_{i\neq 0}E_i : \mathcal{O}^{\mathfrak{p}}_\kappa(\leqslant n)\to\mathcal{O}^{\mathfrak{p}}_\kappa(\leqslant n-1)$ give $\mathcal{O}^{\mathfrak{p}}_\kappa(\leqslant n)$ the structure of a *restricted* categorical $\hat{\mathfrak{sl}}_e$-action. This categorical action categorifies the restriction of the $\hat{\mathfrak{sl}}_e$-action to $\mathcal{F}^\ell(\leqslant n) := \bigoplus_{i=0}^n\mathcal{F}^\ell(n)$.

Proposition 5.2 is still true for restricted categorifications. In particular, we get a quotient functor $\mathcal{O}^{\mathfrak{p}}_{\kappa}(n) \twoheadrightarrow \mathcal{H}_{q,\mathbf{s}}(n)$-mod.

5.4. Deformations. So now we have quotient functors

$$\pi^1 : \mathcal{O}^{\mathfrak{p}}_{\kappa}(n) \twoheadrightarrow \mathcal{H}^{\mathbf{s}}_{\kappa}(n)\text{-mod}, \pi^2 : \mathcal{O}_{\kappa,\mathbf{s}}(n) \twoheadrightarrow \mathcal{H}^{\mathbf{s}}_{\kappa}(n)\text{-mod}.$$

Similarly to Section 2.5, in order to produce a highest weight equivalence $\mathcal{O}^{\mathfrak{p}}_{\kappa}(n) \xrightarrow{\sim} \mathcal{O}_{\kappa,\mathbf{s}}(n)$ we will need to consider deformations of $\mathcal{O}^{\mathfrak{p}}_{\kappa}(n), \mathcal{O}_{\kappa,\mathbf{s}}(n), \mathcal{H}_{q,\mathbf{s}}(n)$-mod. Unlike in *loc.cit.*, it does not seem that one-parameter deformations are sufficient. Following a key idea of [**RSVV**], we will use two parameter deformations.

Deformations of $\mathcal{O}_{\kappa,\mathbf{s}}(n), \mathcal{H}_{q,\mathbf{s}}(n)$-mod are basically the same as in 2.5.6, but instead of a generic line through $\mathbf{c}$ we need to take a generic plane. Let us explain how to deform $\mathcal{O}^{\mathfrak{p}}_{\kappa}(n)$ together with the quotient functor $\pi^2 : \mathcal{O}^{\mathfrak{p}}_{\kappa}(n) \twoheadrightarrow \mathcal{H}_{q,\mathbf{s}}(n)$-mod.

Pick formal variables $x_0, \ldots, x_\ell$. Set $\tilde{R} = \mathbb{C}[[x_0, \ldots, x_\ell]]/(x_1 + \ldots + x_\ell)$. We will deform our categories over $\tilde{R}$ (and then we can specialize to a suitable algebra of formal power series in two variables). First, let us define the deformed category $\mathcal{O}^{\mathfrak{p}}_{\kappa,\tilde{R}}$. It consists of all $\hat{\mathfrak{g}} \otimes \tilde{R}$-modules M subject to the following conditions:

- The action of $\hat{\mathfrak{g}} \otimes \tilde{R}$ on M is $\tilde{R}$-linear. Moreover, M is finitely generated over $U(\hat{\mathfrak{g}}) \otimes \tilde{R}$.
- The level of M is $(x_0 - \frac{1}{e})^{-1}$.
- The action of the parabolic subalgebra $\hat{\mathfrak{p}} \subset \hat{\mathfrak{g}}$ is locally finite, meaning that every element of M is contained in a finitely generated R-submodule stable under $\hat{\mathfrak{p}}$.
- For any i, the element $\mathrm{id}_i \in \mathfrak{gl}_{s_i} \subset \mathfrak{p}$ acts on M diagonalizably with eigenvalues in $\mathbb{Z} + x_i$. Moreover, after specializing the elements x_i to 0, the action of $\hat{\mathfrak{p}}$ on $M/(x_0, \ldots, x_\ell)M$ integrates to $\hat{P}$.

This category still has parabolic Verma modules $\Delta^{\mathfrak{p}}(A)$ labelled by elements of $\mathbb{Z}^{\mathbf{s}}$. So we can form the highest weight subcategory $\mathcal{O}^{\mathfrak{p}}_{\kappa,\tilde{R}}(n) \subset \mathcal{O}^{\mathfrak{p}}_{\kappa,\tilde{R}}$ whose poset is $\mathcal{P}_\ell(n)$. We note that this category is equivalent to $A_{\tilde{R}}$-mod for some $\tilde{R}$-algebra $A_{\tilde{R}}$ that is a free $\tilde{R}$-module of finite rank. So we can consider the base change $\mathcal{O}^{\mathfrak{p}}_{\kappa,S}(n)$ for any $\tilde{R}$-algebra S.

We still have a fusion product functor $\bullet\dot{\otimes}\bullet : \mathcal{O}^{\mathfrak{g}}_{\kappa,\tilde{R}} \boxtimes \mathcal{O}^{\mathfrak{p}}_{\kappa,\tilde{R}} \to \mathcal{O}^{\mathfrak{p}}_{\kappa,\tilde{R}}$. Using this we can define the functors E, F and endomorphisms $X \in \mathrm{End}(F), T \in \mathrm{End}(F^2)$. This gives rise to a quotient functor $\mathcal{O}^{\mathfrak{p}}_{\kappa,\tilde{R}}(n) \twoheadrightarrow \mathcal{H}_{q,\mathbf{s},\tilde{R}}(n)$-mod. Here $\mathcal{H}_{q,\mathbf{s},\tilde{R}}(n)$ is the cyclotomic Hecke algebra over $\tilde{R}$ with parameters $q := \exp(2\pi\sqrt{-1}(\kappa^{-1} + x_0)), Q_i := \exp(2\pi\sqrt{-1}(\kappa^{-1} + x_0)(s_i + x_i))$.

Using the formulas $x_0 \mapsto x_0 - 1/e, x_i \mapsto (x_0 - 1/e)(s_i + x_i) - i/\ell$ we define an embedding $\mathrm{Spec}(\tilde{R}) \hookrightarrow \mathfrak{P}$. Let $\mathfrak{P}_0$ be a generic plane through $\mathbf{c} \in \mathfrak{P}$. Let R be the quotient of $\tilde{R}$ that is the algebra of functions on the preimage of $\mathfrak{P}_0$ in $\mathrm{Spec}(\tilde{R})$. So R is isomorphic to the algebra of formal power series in two variables.

We have two highest weight categories $\mathcal{O}^1_R := \mathcal{O}^{\mathfrak{p}}_{\kappa,R}(n), \mathcal{O}^2_R := \mathcal{O}_{\kappa,\mathbf{s},R}(n)$ with the same posets of simples and with quotient functors

$$\pi^i_R : \mathcal{O}^i_R \twoheadrightarrow \mathcal{C}_R := \mathcal{H}_{q,\mathbf{s},R}(n)\text{-mod}.$$

It is not difficult to see that both π^i_R become equivalences after changing the base ring from R to the fraction field $\mathrm{Frac}(R)$. However, establishing faithfulness properties of π^1_R that are needed to apply techniques described in 2.5.5 is difficult.

The main reason why we need a two-parametric deformation is that one can show that the functor π_R^1 is 0-faithful (=fully faithful on standardly filtered objects). The following result is an extension of [**R2**, Proposition 4.42].

PROPOSITION 5.4. *Assume that for every point* $\mathbf{p} \in \operatorname{Spec}(R)$ *of codimension* 1, *the functor* $\pi_{\mathbf{p}}^1 : \mathcal{O}_{\mathbf{p}}^1 \twoheadrightarrow \mathcal{C}_{\mathbf{p}}$ *is* (-1)*-faithful, i.e., faithful on standardly filtered objects. Then* π_R^1 *is* 0*-faithful.*

This is proved in [**Lo4**, Proposition 3.1], related but weaker results can be found in [**RSVV**, 2.4.2].

5.5. Category equivalence. In this section, we will explain how to prove a slightly weaker ("asymptotic") version of Conjecture 5.1.

5.5.1. *Equivalence theorem.* Here we are going to provide a general result on an equivalence of two highest weight covers of the same category.

Let R be an algebra of formal power series over $\mathbb{C}$ and let p denote the closed point in $\operatorname{Spec}(R)$. Suppose that we have two highest weight categories $\mathcal{O}_R^i, i = 1, 2$, over R such that the base changes $\mathcal{O}_{\operatorname{Frac}(R)}^i$ are split semisimple.

Let $\bar{P}_R^i \in \mathcal{O}_R^i$ be projective objects with a fixed isomorphism $\operatorname{End}(\bar{P}_R^1) \xrightarrow{\sim} \operatorname{End}(\bar{P}_R^2)$. Let $\mathcal{C}_R$ denote the category of right $\operatorname{End}(\bar{P}_R^i)$-modules. So we have quotient functors $\bar{\pi}_R^i : \mathcal{O}_R^i \to \mathcal{C}_R$. Further, let $P_R^i \subset \bar{P}_R^i, i = 1, 2$, be a direct summand with the following properties.

- P_R^i deforms an injective object in $\mathcal{O}_p^i$.
- We have $\bar{\pi}_R^1(P_R^1) = \bar{\pi}_R^2(P_R^2)$.
- The base change $P_{\operatorname{Frac}(R)}^i$ generates $\mathcal{O}_{\operatorname{Frac}(R)}^i$.
- Under the identification $\operatorname{Irr}(\mathcal{O}_{\operatorname{Frac}(R)}^1) \cong \operatorname{Irr}(\mathcal{C}_{\operatorname{Frac}(R)}) \cong \operatorname{Irr}(\mathcal{O}_{\operatorname{Frac}(R)}^2)$, there is a common highest weight order on the labeling sets $\operatorname{Irr}(\mathcal{O}_{\operatorname{Frac}(R)}^i)$.

The following claim is [**Lo4**, Theorem 3.4].

PROPOSITION 5.5. *Suppose, in addition, that the following is true.*

- *The functor* π_R^1 *is* 0*-faithful.*
- *The functor* π_R^2 *is* 1*-faithful.*
- *For any projective* $\hat{P}_R^2 \in \mathcal{O}_R^2$, *there is an embedding* $\hat{P}_R^2 \hookrightarrow (P_R^2)^{\oplus m}$ *(for some* m*) with standardly filtered cokernel.*

Then there is a highest weight equivalence $\mathcal{O}_R^1 \xrightarrow{\sim} \mathcal{O}_R^2$ *intertwining the quotient functors* $\bar{\pi}_R^i$.

5.5.2. *Categories* $\mathcal{C}_R^i$. We will take $P_R^i := F^n \Delta_R^i(\varnothing)$. The choice of the objects $\bar{P}_R^i$ is more subtle. Let p denote the closed point of R. It follows from [**R2**, Proposition 4.42] that the functor $\bar{\pi}_R^2$ is 1-faithful provided the functor $\bar{\pi}_p^2$ is 0-faithful. Take

$$\bar{P}_R^2 := P_R^2 \oplus \bigoplus_\lambda P_R^2(\lambda),$$

where the summation is taken over all $\lambda \in \mathcal{P}_\ell(n)$ such that $\operatorname{codim}_{\mathfrak{h}} \operatorname{Supp} L_p(\lambda) = 1$. Using results from 4.2.1, 4.2.3, it is not difficult to describe the set of λ with this property explicitly. First of all, note that $P_R^2(\lambda)$ appears as a summand of P_R^2 if and only if $\operatorname{Supp} L_p^2(\lambda) = \mathfrak{h}$ if and only if λ lies in the crystal component of $\varnothing$ in $\mathcal{P}_{\kappa,\mathbf{s}}$. Further, $\operatorname{Supp} L_p(\lambda)$ has codimension 1 if and only if one of the following conditions holds:

- λ lies in the connected component of the crystal $\mathcal{P}_{\kappa,\mathbf{s}}$ that contains a singular multipartition of 1.
- for $e=2$, λ lies in the crystal component of ν, which is a minimal multipartition of 2 that is not a column.

One can show that $\bar{\pi}^2_p$ is 0-faithful, see the proof of [**Lo4**, Proposition 8.1] or [**RSVV**, Lemma 2.8].

Now set $\bar{P}^1_R := P^1_R \oplus \bigoplus_\lambda P^1_R(\lambda)$, where the summation is taken over the same set of λ. What we need to prove is an isomorphism $\operatorname{End}(\bar{P}^1_R) = \operatorname{End}(\bar{P}^2_R)$. As was shown in [**Lo4**, Section 7], this reduces to proving the following four claims

(a1) $\pi^1_R(P^1_R(\lambda)) \cong \pi^2_R(P^2_R(\lambda))$ for any singular λ with $|\lambda|=1$.
(b1) $\pi^1_R(P^2_R(\nu)) \cong \pi^2_R(P^2_R(\nu))$ for ν as above.
(a2) $\operatorname{Hom}_{\mathcal{O}^1_R(1)}(P^1_R(\lambda), P^1_R(\lambda')) \xrightarrow{\sim} \operatorname{Hom}_{\mathcal{H}_{q,\mathbf{s},R}(1)}(\pi^1_R P^1_R(\lambda), \pi^1_R P^1_R(\lambda'))$ for any singular λ, λ' with $|\lambda|=|\lambda'|=1$.
(b2) $\operatorname{End}_{\mathcal{O}^1_R(2)}(P^1_R(\nu)) \xrightarrow{\sim} \operatorname{Hom}_{\mathcal{H}_{q,\mathbf{s},R}(2)}(\pi^1_R P^1_R(\nu))$.

Recall that "singular" means "annihilated by all $\tilde{e}_i$".

The proofs of (a1),(a2) are easy. In order to prove (b1) and (b2) we need an explicit construction of $P^1_R(\nu), P^2_R(\nu)$. It is not difficult to see that $P^2_R(\nu)$ is the projection to the block corresponding to ν of $\operatorname{Ind}^{\mathfrak{S}_2}_{G(\ell,1,2)} \Delta^2_{A,R}((2))$, where we write $\Delta^2_{A,R}((2))$ for the Verma module over $H_R(\mathfrak{S}_2)$ corresponding to the trivial $\mathfrak{S}_2$-module. This motivates us to consider the object $\Delta^1_{A,R}((2)) \dot{\otimes} \Delta^1_R(\varnothing)$ in $\mathcal{O}^1_R(2)$ and define $Q^1_R(\nu)$ as its projection to the block corresponding to ν. Clearly, (b1) holds if we replace $P^1_R(\nu)$ with $Q^1_R(\nu)$. It turns out that $Q^1_R(\nu) \cong P^1_R(\nu)$, [**Lo4**, Proposition 7.5].

5.5.3. *Checking* (-1)*-faithfulness.* Now let us explain how to check (-1)-faithfulness of the functor $\pi^1_{\mathbf{p}}$ at codimension 1 points of $\operatorname{Spec}(R)$. This is done using categorical actions and their crystals. This approach only allows to establish a weaker version of Conjecture 5.1: we need to assume that $s_i \gg n$ for all i.

Namely, we still have functors $\underline{E}, F$ on $\mathcal{O}^1_{\mathbf{p}}(\leqslant N)$ (for some $N \gg n$) and the natural transformations $X \in \operatorname{End}(F), T \in \operatorname{End}(F^2)$. These data give rise to a (restricted) categorical Kac-Moody action on $\mathcal{O}^1_{\mathbf{p}}(\leqslant N)$. The algebra acting is $\mathfrak{g}_{\tilde{\mathbf{c}}}$ (this Kac-Moody algebra was defined in 3.2.1), where $\tilde{\mathbf{c}} \in \mathfrak{P} \otimes \mathbf{k}_{\mathbf{p}}$ is the element corresponding to the homomorphism $\mathbb{C}[\mathfrak{P}] \to R \to \mathbf{k}_{\mathbf{p}}$. Equivalently, if $\mathbf{p}$ is an intersection of a hyperplane in $\mathfrak{P}$ with $\operatorname{Spec}(R)$, then we can take a Weil generic point $\tilde{\mathbf{c}}$ in this hyperplane. So we have three options.

(i) $\mathfrak{g}_{\tilde{\mathbf{c}}} = \mathfrak{sl}_\infty^{\oplus \ell}$. In this case, $\pi^1_{\mathbf{p}}$ is an equivalence.
(ii) $\mathbf{p}$ is the intersection of the hyperplane $y_0 = -\frac{1}{e}$ with $\operatorname{Spec}(R)$. In this case $\mathfrak{g}_{\tilde{\mathbf{c}}} = \hat{\mathfrak{sl}}_e^{\oplus \ell}$.
(iii) $\mathbf{p}$ is the intersection of $\operatorname{Spec}(R)$ with the hyperplane of the form $s_i - s_j = m$ (where m is an integer with $|m| < N$ and i,j are different elements of $\{0,1,\ldots,\ell-1\}$). In this case $\mathfrak{g}_{\tilde{\mathbf{c}}} = \hat{\mathfrak{sl}}_\infty^{\oplus \ell - 1}$.

In (i), there is nothing to prove. To deal with (ii) and (iii) we use the following result.

PROPOSITION 5.6. *Let $\mathcal{O}(\leqslant N)$ be a restricted highest weight $\mathfrak{g}_{\tilde{\mathbf{c}}}$-categorification (over an arbitrary field $\mathbf{k}$) of the level ℓ Fock space $\mathcal{F}_{\tilde{\mathbf{c}}}$ as in (ii) or (iii). Pick*

$n \ll N$. Then the quotient functor π defined by the projective object $F^n\mathbf{1}$ is (-1)-faithful.

PROOF. We are going to sketch the proof (a complete proof can be found in [**Lo4**, Section 8.2]). The proof is in several steps, all but the last one work for all $\mathbf{c}$.

Step 1. Let $\mathcal{P}_\ell^0$ denote the connected component of $\varnothing$ in the $\mathfrak{g}_{\mathbf{c}}$-crystal. What we need to show is that if $\lambda \notin \mathcal{P}_\ell^0$, then $\operatorname{Hom}_{\mathcal{O}}(L(\lambda), T(\mu)) = 0$. It is easy to reduce to the case when λ is singular (i.e., annaihilated by all operators $\tilde{e}_z$). Note that in this case $\operatorname{Hom}_{\mathcal{O}}(L(\lambda), T(\mu)) = 0$ provided $T(\mu)$ is a direct summand of a tilting object of the form $F_iT(\mu')$. Using results from [**Lo3**], we can determine one summand of $F_iT(\mu')$ combinatorially starting from μ'. For this, we need a so called *dual* crystal structure on $\mathcal{P}_\ell$. It is defined as follows. We write down the i-signature of μ' but now cancel $+-$ rather than $-+$. In this way we get crystal operators $\tilde{e}_i^*, \tilde{f}_i^*$. By [**Lo4**, Lemma 4.2], $T(\tilde{f}_i^*\mu')$ is a direct summand of $F_iT(\mu')$. We conclude that, for singular λ, we have $\operatorname{Hom}_{\mathcal{O}}(L(\lambda), T(\mu)) = 0$ unless $\tilde{e}_i^*\mu = 0$ for all i. In this case, we say that μ is *cosingular.* So we need to show that $\operatorname{Hom}_{\mathcal{O}}(L(\lambda), T(\mu)) = 0$ when $\lambda \neq \varnothing$ is singular, and μ is cosingular.

Step 2. For λ with $\tilde{e}_i\lambda = 0$ we can define the reflection $\sigma_i\lambda$ of λ by $\sigma_i\lambda := \tilde{f}_i^k\lambda$, where k is minimal with $\tilde{f}_i^{k+1}\lambda = 0$. Similarly, for μ with $\tilde{e}_i^*\mu = 0$, we can define $\tilde{\sigma}_i^*\mu$. When λ is singular, and w is in the Weyl group of $\mathfrak{g}_{\mathbf{c}}$ (such that $|w\lambda| \leqslant N$), then $w\lambda$ is well-defined, in particular, is independent of the choice of a reduced expression for w. Similarly, $w^*\mu$ is well-defined for a cosingular μ. A key observation for us is that, if $\tilde{e}_i\lambda = 0, \tilde{e}_i^*\mu = 0$, then $\dim \operatorname{Hom}_{\mathcal{O}}(L(\lambda), T(\mu)) = \dim \operatorname{Hom}_{\mathcal{O}}(L(\sigma_i\lambda), T(\sigma_i^*\mu))$. It follows that for a singular λ and a cosingular μ we have

$$\dim \operatorname{Hom}_{\mathcal{O}}(L(\lambda), T(\mu)) = \dim \operatorname{Hom}_{\mathcal{O}}(L(w\lambda), T(w^*\mu)). \tag{5.1}$$

Step 3. If we are in situation (iii), then actually $\lambda \not\leqslant \mu$ for any singular λ and any cosingular μ, this can be deduced from Examples 3.5,3.6. So we are done by Step 1. In case (ii), there is w such that $w\lambda \not\leqslant w\mu$, see [**Lo4**, Section 6.3]. So we are done by (5.1). It is here that we use that N is large enough. □

This completes the proof of the asymptotic version of Conjecture 5.1.

5.6. Multiplicities. Conjecture 5.1 allows to compute the multiplicities in $\mathcal{O}_{\kappa,\mathbf{s}}(n)$ because the multiplicities in $\mathcal{O}_\kappa^{\mathfrak{p}}$ are known, they are given by (the values at $q = 1$ of) suitable parabolic Kazhdan-Lusztig polynomials.

There is an alternative way to present this result. Namely, in [**U**], Uglov defined a $U_v(\hat{\mathfrak{sl}}_e)$-action on $\mathcal{F}_{\kappa,\mathbf{s}}(v)$, where $U_v(\hat{\mathfrak{sl}}_e)$ is the quantum group over $\mathbb{C}(v)$. He introduced a $\mathbb{C}[v^{-1}]$-lattice in $\mathcal{F}_{\kappa,\mathbf{s}}(v)$ and a bar-involution of $\mathcal{F}_{\kappa,\mathbf{s}}(v)$. Using these data, one can define a so called dual canonical basis, $b_{\mathbf{s}}(\lambda) \in \mathcal{F}_{\kappa,\mathbf{s}}[v^{-1}]$ indexed by $\lambda \in \mathcal{P}_\ell$. As was checked in [**VV**, Section A4], the specialization of $b_{\mathbf{s}}(\lambda)$ to $v = 1$ coincides $[L_\kappa^{\mathfrak{p}}(\lambda)] \in K_0^{\mathbb{C}}(\mathcal{O}_\kappa^{\mathfrak{p}}(n))$. This was conjectured by Rouquier in [**R2**, Section 6].

5.7. Complements. Above in this section we were dealing with the case when $\kappa = -\frac{1}{e}$, $s_0, \ldots, s_{\ell-1} \in \mathbb{Z}$. Let us explain how to describe the category $\mathcal{O}_{\kappa,\mathbf{s}}(n)$ in general. We can assume that there is a single equivalence class in $\{0, 1, \ldots, \ell-1\}$ with respect to $\sim_{\kappa,\mathbf{s}}$. We can always reduce to this case using Proposition 3.2.

Let us consider the case when κ is irrational. In this case the category $\mathcal{O}_{\kappa,\mathbf{s}}(n)$ is equivalent to the sum of suitable blocks in a suitable parabolic category $\mathcal{O}$ for $\mathfrak{gl}_m$ (for some m). This was checked in [**GL**, Section 6] under the faithfulness restrictions on the parameters $s_0, \ldots, s_{\ell-1}$. These restrictions can be removed by considering extended quotients, similarly to 2.5.7.

Now consider the case when κ is rational. Using the equivalence relating positive and negative κ, we can assume that $\kappa = -\frac{r}{e}$, where $r > 0$ is coprime to e. The condition that $\{0, 1, \ldots, \ell-1\}$ is a single equivalence class means that $rs_i \in \mathbb{Z}$ (up to a common shift of the s_i's, as usual). We are going to prove the following result.

PROPOSITION 5.7. *The category $\mathcal{O}_{\kappa,\mathbf{s}}(n)$ is equivalent to the category $\mathcal{O}_{-1/e,\mathbf{s}'}(n)$ for a suitable collection $\mathbf{s}' \in \mathbb{Z}^\ell$. The equivalence sends $\Delta_{\kappa,\mathbf{s}}(\lambda)$ to $\Delta_{-1/e,\mathbf{s}'}(\sigma\lambda)$, where $\sigma \in \mathfrak{S}_\ell$ and $\sigma\lambda$ is the ℓ-partition obtained from λ by permuting $\lambda^{(0)}, \ldots, \lambda^{(\ell-1)}$ according to σ.*

Since the proof has not appeared elsewhere, we are going to provide it here in full.

We start with a lemma that was proved in [**Lo5**, Section 3.4] (the proof is an elementary combinatorics).

LEMMA 5.8. *Let $\mathbf{c} = (\kappa, \mathbf{s})$ be a parameter with $\kappa = -\frac{r}{e}, \mathbf{s} \in (\frac{1}{r}\mathbb{Z})^\ell$. Pick $M \gg 0$ and integers $0 < m_0 < \ldots < m_{\ell-1} \ll M$, and set $\mathbf{c}' := (M\kappa, s_0 + m_0/\kappa, s_1 + m_1/\kappa, \ldots, s_{\ell-1} + m_{\ell-1}/\kappa)$ Then the orders on the boxes coming from $\mathbf{c}, \mathbf{c}'$ coincide and hence the orders $\preceq_{\mathbf{c}}, \preceq_{\mathbf{c}'}$ on the multipartitions are equivalent.*

Next, we have the following result.

LEMMA 5.9. *Let $\mathbf{c}$ be a parameter given by $(\kappa, \mathbf{s})$. Suppose that $\mathbf{c}$ is spherical. Then, for any $\sigma \in \mathfrak{S}_\ell$, there is an equivalence $\mathcal{O}_{\mathbf{c}} \xrightarrow{\sim} \mathcal{O}_{\sigma.\mathbf{c}}$ that maps $\Delta_{\mathbf{c}}(\lambda)$ to $\Delta_{\sigma.\mathbf{c}}(\sigma\lambda)$, and preserves the supports.*

This lemma was proved in [**GL**, 4.1.4]. Here $\sigma.\mathbf{c}$ denotes the parameter obtained from $\mathbf{c}$ by permuting the s_i's according to σ.

PROOF OF PROPOSITION 5.7. We will produce $\mathbf{c}' = (-1/e, \mathbf{s}')$ in several steps.

Lemma 5.8 and Proposition 2.29 allow us to pass from a parameter $\mathbf{c}$ to a Zariski generic (and, in particular, spherical) parameter $\mathbf{c}^1$ (since the spherical parameters are known explicitly, [**DG**, Theorem 1.1], one can write $\mathbf{c}^1$ explicitly as well). Fix an integer r with $0 \leqslant r < e$. In each of the ℓ diagrams we have precisely one diagonal of boxes with shifted content ($s_i^1 + x - y$, for the box (x, y, i)) congruent r modulo e and lying in $[r, e + r - 1)$. Let $(i_0, \ldots, i_{\ell-1})$ be the indexes of the diagrams such that the diagonals are in the increasing order (with respect to the preorder on boxes from 3.1.2), note that the ordering of indexes is independent of the choice of r. Let $\sigma \in \mathfrak{S}_\ell$ be defined by $\sigma^{-1}(j) = i_j$. So the analogously defined sequence of indexes for the parameter $\mathbf{c}^2 := \sigma.\mathbf{c}^1$ is $(0, 1, \ldots, \ell-1)$. We get an equivalence $\mathcal{O}_{\mathbf{c}^1} \xrightarrow{\sim} \mathcal{O}_{\mathbf{c}^2}$ by using Lemma 5.9. Set $\mathbf{c}^3 := (\kappa^2, s_0^3, \ldots, s_{\ell-1}^3)$, where $\mathbf{s}^3$ is such that all differences $s_i^3 - s_j^3$ are integral and the order on the diagonals is the same as before, it is clear that such $\mathbf{s}^3$ exists. So $\mathbf{c}^2, \mathbf{c}^3$ are equivalent and we get a highest weight equivalence $\mathcal{O}_{\mathbf{c}^2} \xrightarrow{\sim} \mathcal{O}_{\mathbf{c}^3}$ by Proposition 2.29. Now set $\mathbf{c}' := (-\frac{1}{e}, s_0^3, \ldots, s_{\ell-1}^3)$. ☐

References

[A] S. Ariki, *Representations of quantum algebras and combinatorics of Young tableaux*, University Lecture Series, vol. 26, American Mathematical Society, Providence, RI, 2002. Translated from the 2000 Japanese edition and revised by the author. MR1911030

[AK] S. Ariki and K. Koike, *A Hecke algebra of* $(\mathbf{Z}/r\mathbf{Z})\wr\mathfrak{S}_n$ *and construction of its irreducible representations*, Adv. Math. **106** (1994), no. 2, 216–243, DOI 10.1006/aima.1994.1057. MR1279219

[BK] A. Berenstein and D. Kazhdan, *Geometric and unipotent crystals. II. From unipotent bicrystals to crystal bases*, Quantum groups, Contemp. Math., vol. 433, Amer. Math. Soc., Providence, RI, 2007, pp. 13–88, DOI 10.1090/conm/433/08321. MR2349617

[BEG1] Y. Berest, P. Etingof, and V. Ginzburg, *Cherednik algebras and differential operators on quasi-invariants*, Duke Math. J. **118** (2003), no. 2, 279–337, DOI 10.1215/S0012-7094-03-11824-4. MR1980996

[BEG2] Y. Berest, P. Etingof, and V. Ginzburg, *Finite-dimensional representations of rational Cherednik algebras*, Int. Math. Res. Not. **19** (2003), 1053–1088, DOI 10.1155/S1073792803210205. MR1961261

[BC] Y. Berest and O. Chalykh, *Quasi-invariants of complex reflection groups*, Compos. Math. **147** (2011), no. 3, 965–1002, DOI 10.1112/S0010437X10005063. MR2801407

[BE] R. Bezrukavnikov and P. Etingof, *Parabolic induction and restriction functors for rational Cherednik algebras*, Selecta Math. (N.S.) **14** (2009), no. 3-4, 397–425, DOI 10.1007/s00029-009-0507-z. MR2511190

[BMR] M. Broué, G. Malle, and R. Rouquier, *Complex reflection groups, braid groups, Hecke algebras*, J. Reine Angew. Math. **500** (1998), 127–190. MR1637497

[CR] J. Chuang and R. Rouquier, *Derived equivalences for symmetric groups and* $\mathfrak{sl}_2$*-categorification*, Ann. of Math. (2) **167** (2008), no. 1, 245–298, DOI 10.4007/annals.2008.167.245. MR2373155

[D] V. G. Drinfel′d, *Degenerate affine Hecke algebras and Yangians* (Russian), Funktsional. Anal. i Prilozhen. **20** (1986), no. 1, 69–70. MR831053

[DG] C. Dunkl and S. Griffeth, *Generalized Jack polynomials and the representation theory of rational Cherednik algebras*, Selecta Math. (N.S.) **16** (2010), no. 4, 791–818, DOI 10.1007/s00029-010-0032-0. MR2734331

[DJ] R. Dipper and G. James, *The q-Schur algebra*, Proc. London Math. Soc. (3) **59** (1989), no. 1, 23–50, DOI 10.1112/plms/s3-59.1.23. MR997250

[Et] P. Etingof, *Symplectic reflection algebras and affine Lie algebras* (English, with English and Russian summaries), Mosc. Math. J. **12** (2012), no. 3, 543–565, 668–669. MR3024822

[EG] P. Etingof and V. Ginzburg, *Symplectic reflection algebras, Calogero-Moser space, and deformed Harish-Chandra homomorphism*, Invent. Math. **147** (2002), no. 2, 243–348, DOI 10.1007/s002220100171. MR1881922

[EGL] P. Etingof, E. Gorsky, and I. Losev, *Representations of rational Cherednik algebras with minimal support and torus knots*, Adv. Math. **277** (2015), 124–180, DOI 10.1016/j.aim.2015.03.003. MR3336085

[GGOR] V. Ginzburg, N. Guay, E. Opdam, and R. Rouquier, *On the category* $\mathcal{O}$ *for rational Cherednik algebras*, Invent. Math. **154** (2003), no. 3, 617–651, DOI 10.1007/s00222-003-0313-8. MR2018786

[GL] I. G. Gordon and I. Losev, *On category* $\mathcal{O}$ *for cyclotomic rational Cherednik algebras*, J. Eur. Math. Soc. (JEMS) **16** (2014), no. 5, 1017–1079, DOI 10.4171/JEMS/454. MR3210960

[GORS] E. Gorsky, A. Oblomkov, J. Rasmussen, and V. Shende, *Torus knots and the rational DAHA*, Duke Math. J. **163** (2014), no. 14, 2709–2794, DOI 10.1215/00127094-2827126. MR3273582

[Lo1] I. Losev, *On isomorphisms of certain functors for Cherednik algebras*, Represent. Theory **17** (2013), 247–262, DOI 10.1090/S1088-4165-2013-00437-5. MR3054265

[Lo2] I. Losev, *Highest weight* $\mathfrak{sl}_2$*-categorifications I: crystals*, Math. Z. **274** (2013), no. 3-4, 1231–1247, DOI 10.1007/s00209-012-1114-9. MR3078265

[Lo3] I. Losev, *Highest weight* $\mathfrak{sl}_2$*-categorifications II: Structure theory*, Trans. Amer. Math. Soc. **367** (2015), no. 12, 8383–8419, DOI 10.1090/tran/6311. MR3403059

[Lo4] I. Losev, *Proof of Varagnolo-Vasserot conjecture on cyclotomic categories $\mathcal{O}$*. arXiv:1305.4894. Accepted by Selecta Math.

[Lo5] I. Losev, *Abelian localization for cyclotomic Cherednik algebras*, Int. Math. Res. Not. IMRN **18** (2015), 8860–8873, DOI 10.1093/imrn/rnu216. MR3417695

[Lo6] I. Losev. *Derived equivalences for Rational Cherednik algebras*. arXiv:1406.7502.

[Lo7] I. Losev, *Finite-dimensional quotients of Hecke algebras*, Algebra Number Theory **9** (2015), no. 2, 493–502, DOI 10.2140/ant.2015.9.493. MR3320850

[Lo8] I. Losev. *Supports of simple modules in cyclotomic Cherednik categories O* . arXiv:1509.00526.

[Lu] G. Lusztig, *Introduction to quantum groups*, Progress in Mathematics, vol. 110, Birkhäuser Boston, Inc., Boston, MA, 1993. MR1227098

[M] S. Martin, *Schur algebras and representation theory*, Cambridge Tracts in Mathematics, vol. 112, Cambridge University Press, Cambridge, 1993. MR1268640

[R1] R. Rouquier, *Derived equivalences and finite dimensional algebras*, International Congress of Mathematicians. Vol. II, Eur. Math. Soc., Zürich, 2006, pp. 191–221. MR2275594

[R2] R. Rouquier, *q-Schur algebras and complex reflection groups* (English, with English and Russian summaries), Mosc. Math. J. **8** (2008), no. 1, 119–158, 184. MR2422270

[R3] R. Rouquier, *2-Kac-Moody algebras*. arXiv:0812.5023.

[RSVV] R. Rouquier, P. Shan, M. Varagnolo, and E. Vasserot, *Categorifications and cyclotomic rational double affine Hecke algebras*, Invent. Math. **204** (2016), no. 3, 671–786, DOI 10.1007/s00222-015-0623-7. MR3502064

[Sh] P. Shan, *Crystals of Fock spaces and cyclotomic rational double affine Hecke algebras* (English, with English and French summaries), Ann. Sci. Éc. Norm. Supér. (4) **44** (2011), no. 1, 147–182. MR2760196

[SV] P. Shan and E. Vasserot, *Heisenberg algebras and rational double affine Hecke algebras*, J. Amer. Math. Soc. **25** (2012), no. 4, 959–1031, DOI 10.1090/S0894-0347-2012-00738-3. MR2947944

[ST] G. C. Shephard and J. A. Todd, *Finite unitary reflection groups*, Canadian J. Math. **6** (1954), 274–304. MR0059914

[U] D. Uglov, *Canonical bases of higher-level q-deformed Fock spaces and Kazhdan-Lusztig polynomials*, Physical combinatorics (Kyoto, 1999), Progr. Math., vol. 191, Birkhäuser Boston, Boston, MA, 2000, pp. 249–299. MR1768086

[VV] M. Varagnolo and E. Vasserot, *Cyclotomic double affine Hecke algebras and affine parabolic category $\mathcal{O}$*, Adv. Math. **225** (2010), no. 3, 1523–1588, DOI 10.1016/j.aim.2010.03.028. MR2673739

[W] B. Webster. *Rouquier's conjecture and diagrammatic algebra*. arXiv:1306.0074.

Department of Mathematics, Northeastern University, Boston, Massachusetts 02115

E-mail address: i.loseu@neu.edu

Contemporary Mathematics
Volume **683**, 2017
http://dx.doi.org/10.1090/conm/683/13720

Categorical actions on unipotent representations of finite classical groups

Olivier Dudas, Michela Varagnolo, and Eric Vasserot

Abstract. We review the categorical representation of a Kac-Moody algebra on unipotent representations of finite unitary groups in non-defining characteristic, using Harish-Chandra induction and restriction. Then, we extend this construction to finite reductive groups of types B or C in non-defining characteristic. We show that the decategorified representation is isomorphic to a direct sum of level 2 Fock spaces. We deduce that the Harish-Chandra branching graph coincides with the crystal graph of these Fock spaces. We also obtain derived equivalences between blocks, yielding Broué's abelian defect group conjecture for unipotent ℓ-blocks at linear primes ℓ.

Contents

2010 *Mathematics Subject Classification.* Primary 06B15.

This research was partially supported by the ANR grant number ANR-13-BS01-0001-01.

Introduction

Let $\mathbf{G}$ be a connected algebraic group defined over a finite field $\mathbb{F}_q$. The finite group $G = \mathbf{G}(\mathbb{F}_q)$ of its $\mathbb{F}_q$-rational points is a *finite reductive group.* The irreducible representations of G over fields of characteristic ℓ prime to q fall into Harish-Chandra series, which are defined in terms of Harish-Chandra induction R_L^G and restriction ${}^*R_L^G$ from proper Levi subgroups L of G. The isomorphism classes in each series are parametrized by the simple modules of the *ramified Hecke algebras*, which are realized as the endomorphism algebras of the Harish-Chandra induction of cuspidal representations. Therefore, the classification of isomorphism classes of irreducible representations of G can be reduced to the following two problems

(a) classification of the cuspidal irreducible representations,
(b) determination of the ramified Hecke algebras.

This was achieved by Lusztig in [**54**] when $\ell = 0$ but it remains open for representations in positive characteristic for most of the finite reductive groups. By results of Geck-Hiss-Malle [**32**], we know however that the ramified Hecke algebras are indeed Hecke algebras of finite type, only the parameters of the deformation are unknown in general.

When G is a classical group, it turns out that most of the structure of the ramified Hecke algebras does not depend on the corresponding unipotent cuspidal representations. This suggests to rather study the endomorphism algebra of the Harish-Chandra induction functor R_L^G rather than the endomorphism algebra of the induced representation. This was already achieved in [**10**] for $G = \mathrm{GL}_n(q)$. Our goal is to extend Chuang-Rouquier's approach to other classical groups.

In the first part of this paper we focus on the case of finite unitary groups $\mathrm{GU}_n(q)$, reviewing our previous work [**18**], whereas in the last part we deal with the case of groups of types B or C. We will work with both ordinary representations (characteristic zero) and modular representations in non-defining characteristic (characteristic ℓ prime to q). More precisely, the field of coefficients R of the representations will be an extension of either $\mathbb{Q}_\ell$ or $\mathbb{F}_\ell$.

Let $\mathrm{G}_n(q)$ be one of the families of finite classical groups among $\mathrm{GU}_{2n}(q)$, $\mathrm{GU}_{2n+1}(q)$, $\mathrm{Sp}_{2n}(q)$ and $\mathrm{SO}_{2n+1}(q)$. Using the tower of inclusion of groups $\cdots \subset \mathrm{G}_n(q) \subset \mathrm{G}_{n+1}(q) \subset \cdots$ one can form the abelian category of unipotent representations

$$\mathscr{U}_R := \bigoplus_{n \geqslant 0} \mathrm{G}_n(q)\text{-umod}.$$

Furthermore, under mild assumption on ℓ, we can modify the Harish-Chandra induction and restriction functors to obtain an adjoint pair (E, F) of functors on $\mathscr{U}_R$. The functor F corresponds to a Harish-Chandra induction from $\mathrm{G}_n(q)$ to $\mathrm{G}_{n+1}(q)$ whereas E corresponds to the restriction. Note that only specific Levi subgroups are considered, and we must work with a variation of the usual Harish-Chandra theory (the *weak Harish-Chandra theory*) introduced in [**36**].

In this framework, problem (a) amounts to finding the modules V such that $EV = 0$ and problem (b) is about the structure of $\mathrm{End}_G(F^m V)$ for such cuspidal modules V. As mentioned before, most of the structure of this endomorphism algebra is already contained in $\mathrm{End}(F^m)$. In §4.2 and §6.2, we construct natural transformations X of F and T of F^2 which endow $\mathrm{End}(F^m)$ with a morphism from an affine Hecke algebra $\mathbf{H}_m^{q^\delta}$ of type $A_{m-1}^{(1)}$ with parameter q^δ where the integer δ is determined as follows

- $\delta = 2$ if $\mathrm{G}_n(q)$ is $\mathrm{GU}_{2n}(q)$ or $\mathrm{GU}_{2n+1}(q)$,
- $\delta = 1$ if $\mathrm{G}_n(q)$ is $\mathrm{Sp}_{2n}(q)$ or $\mathrm{SO}_{2n+1}(q)$.

Back to our original problem, the evaluation at a cuspidal module V provides a natural map $\mathbf{H}_m^{q^\delta} \to \mathrm{End}_G(F^m V)$. Then, we prove that this map induces a natural isomorphism between $\mathrm{End}_G(F^m V)$ and a level 2 cyclotomic quotient of $\mathbf{H}_m^{q^\delta}$ whose parameters are naturally attached to V, see Theorems 4.7 and 6.5.

Next, we prove that the eigenvalues of X belong to

- $I_2 = (-q)^{\mathbb{Z}}$ if $\mathrm{G}_n(q)$ is $\mathrm{GU}_{2n}(q)$ or $\mathrm{GU}_{2n+1}(q)$,
- $I_1 = q^{\mathbb{Z}} \sqcup (-q^{\mathbb{Z}})$ if $\mathrm{G}_n(q)$ is $\mathrm{Sp}_{2n}(q)$ or $\mathrm{SO}_{2n+1}(q)$.

Then, we can form a Lie algebra $\mathfrak{g}$ corresponding to the quiver with vertices I_δ and arrows given by multiplication by q^δ. When working in characteristic zero, the Lie algebra $\mathfrak{g}$ is isomorphic to two copies of $\mathfrak{sl}_{\mathbb{Z}}$. In positive characteristic ℓ prime to q it will depend on whether $-q$ is a power of q^δ. This reflects the difference of behaviour for unipotent representations when working at linear or unitary primes. To explain this we denote by d, e, f the order of q^2, $-q$ and q modulo ℓ. Two situations occur:

- ℓ is a *linear prime* if $\mathrm{G}_n(q)$ is $\mathrm{GU}_{2n}(q)$ or $\mathrm{GU}_{2n+1}(q)$ and e is even (then $e = 2d$ and $-q \notin q^{2\mathbb{Z}} \bmod \ell$), or if $\mathrm{G}_n(q)$ is $\mathrm{Sp}_{2n}(q)$ or $\mathrm{SO}_{2n+1}(q)$ and f is odd (then $f = d$ and $-q \notin q^{\mathbb{Z}} \bmod \ell$). In that case $\mathfrak{g}$ is a subalgebra of $(\widehat{\mathfrak{sl}}_d)^{\oplus 2}$;

- ℓ is a *unitary prime* if $\mathrm{G}_n(q)$ is $\mathrm{GU}_{2n}(q)$ or $\mathrm{GU}_{2n+1}(q)$ and e is odd (then $e = d$ and $-q \equiv (q^2)^{(e+1)/2} \bmod \ell$), or if $\mathrm{G}_n(q)$ is $\mathrm{Sp}_{2n}(q)$ or $\mathrm{SO}_{2n+1}(q)$ and f is even (then $f = 2d$ and $-q \equiv q^{d+1} \bmod \ell$). In that case $\mathfrak{g}$ is isomorphic to $\widehat{\mathfrak{sl}}_{f/\delta}$.

In each case we prove that the representation datum (E, F, X, T) induces a categorical action of $\mathfrak{g}$ on $\mathscr{U}_R$. See Theorems 4.10 and 4.16 for finite unitary groups, and Theorems 6.11, 6.12 and 6.18 for groups of types B and C.

In particular, let $E = \bigoplus E_i$ and $F = \bigoplus F_i$ be the decomposition of the functors into generalized i-eigenspaces for X. Then $[E_i]$, $[F_i]$ act as the Chevalley generators of $\mathfrak{g}$ on the Grothendieck group $[\mathscr{U}_R]$ of $\mathscr{U}_R$ and many problems on $\mathscr{U}_R$ have a Lie-theoretic counterpart. For example,

- weakly cuspidal modules correspond to highest weight vectors,
- the decomposition of $\mathscr{U}_R$ into weak Harish-Chandra series corresponds to the decomposition of the $\mathfrak{g}$-module $[\mathscr{U}_R]$ into a direct sum of irreducible highest weight modules,
- the parameters of the ramified Hecke algebra attached to a weakly cuspidal unipotent module V are given by the weight of $[V]$,
- the blocks of $\mathscr{U}_R$, or equivalently the unipotent ℓ-blocks, correspond to the weight spaces for the action of $\mathfrak{g}$ (inside a Harish-Chandra series if ℓ is a linear prime).

Such observations were already used in other situations, e.g., for cyclotomic rational double affine Hecke algebras.

For this dictionary to be efficient one needs to determine the $\mathfrak{g}$-module structure on $[\mathscr{U}_R]$. This is done in §4.5 and §6.5 by looking at the action of $[E_i]$ and $[F_i]$ on the basis of $[\mathscr{U}_R]$ formed by unipotent characters and their ℓ-reduction, which play a role similar to the role of the *standard modules* in the categorifications of cyclotomic rational double affine Hecke algebras mentioned above. On this basis the action can be made explicit, and we prove that there is a natural $\mathfrak{g}$-module isomorphism

$$\mathbb{C} \otimes_{\mathbb{Z}} [\mathscr{U}_R] \xrightarrow{\sim} \bigoplus_{t \geqslant 0} \mathbf{F}(Q_t)$$

between the complexified Grothendieck group of $\mathscr{U}_R$ and a direct sum of level 2 Fock spaces $\mathbf{F}(Q_t)$, each of which corresponds to an ordinary Harish-Chandra series. Through this isomorphism, the basis of unipotent characters (or their ℓ-reduction) is sent to the standard monomial basis.

Our original motivation for constructing a categorical action of $\mathfrak{g}$ on $\mathscr{U}_R$ comes from a conjecture of Gerber-Hiss-Jacon [**36**] for finite unitary groups, which predicts an explicit relation between the Harish-Chandra branching graph and the crystal graph of the Fock spaces $\bigoplus_{t \geqslant 0} \mathbf{F}(Q_t)$ when e is odd. See also [**37**]. Using our categorical methods and the unitriangularity of the decomposition matrix we obtained in [**18**] a complete proof of the conjecture, see Theorem 4.20. We extend here this result to groups of type B and C (see Theorem 6.22). Note however that in order to do so, we must rely on a conjecture of Geck concerning the unitriangular shape of the decomposition matrices that we have recalled in Conjecture 6.21.

A similar result can be deduced when ℓ is a linear prime. However, in that case, the situation is already well-understood by the work of Gruber-Hiss [**39**] on classical groups. The case where ℓ is unitary is considered as more challenging and our categorification techniques give the first major result in that direction since the case of $\mathrm{GL}_n(q)$ was solved by Dipper-Du [**16**]. This solves completely the problem of classification of irreducible unipotent modules for unitary groups and groups of types B or C mentioned at the beginning of the introduction, and yields a combinatorial description of the weak Harish-Chandra series.

For finite unitary groups, another categorical construction can be used in order to get the usual (non weak) Harish-Chandra series, by adapting some techniques from [**62**]. It relies on a categorification of the Heisenberg representation of level 2 on the Fock spaces $\mathbf{F}(Q_t)$ mentioned above. We explain this in §5. This yields a complete classification of the cuspidal unipotent modules, as explained in Theorem 5.10 and §5.5.3 (and a computation of the parameters of all ramified Hecke algebras).

By the work of Chuang-Rouquier, categorical actions also provide derived equivalences between weight spaces. In our situation, these weight spaces are exactly the unipotent ℓ-blocks and we obtain many derived equivalences between blocks with the same local structure. Together with Livesey's construction of good blocks in the linear prime case, we deduce a proof of Broué's abelian defect group conjecture, see Theorems 4.18 (or [**18**]) and 6.20 (which does not rely on Conjecture 6.21).

The paper is organized as follows. In Section 1 we set our notations and recall basic facts on categorical actions, perfect bases and derived equivalences. In Section 2 we introduce the Fock spaces, which are certain integrable representations of Kac-Moody algebras. They have a crystal graph which can be defined combinatorially. In Section 3 we recall standard results on unipotent representations of finite reductive groups in non-defining characteristic. In Section 4 we recall the categorical representation on the unipotent modules of unitary groups given in [**18**], and the main applications proved there. Sections 1 to 4 do not contain new material and are a review of the main results of [**18**]. In Section 5 we explain the role of the Heisenberg categorical action and its relation with Harish-Chandra series. The results here are new and were announced in [**18**]. Section 6 deals with the categorical representation on the unipotent modules of groups of type B,C and its applications to weak Harish-Chandra series and derived equivalences. The results in this final section are new.

1. Categorical representations

Throughout this section, R will denote a noetherian commutative domain (with unit).

1.1. Rings and categories.

An R-*category* $\mathscr{C}$ is an additive category enriched over the tensor category of R-modules. All the functors F on $\mathscr{C}$ will be assumed to be R-linear. Given such a functor, we denote by 1_F or sometimes F the identity element in the endomorphism ring $\mathrm{End}(F)$. The identity functor on $\mathscr{C}$ will be denoted by by $1_{\mathscr{C}}$. A composition of functors E and F is written as EF, while a composition of morphisms of functors (or natural transformations) ψ and ϕ is written as $\psi \circ \phi$. We say that $\mathscr{C}$ is *Hom-finite* if the Hom spaces are finitely generated over R. If the category $\mathscr{C}$ is abelian or exact, we denote by $[\mathscr{C}]$ the complexified Grothendieck group and by $\mathrm{Irr}(\mathscr{C})$ the set of isomorphism classes of simple objects of $\mathscr{C}$. The class of an object M of $\mathscr{C}$ in the Grothendieck group is denoted by $[M]$. An exact endofunctor F of $\mathscr{C}$ induces a linear map on $[\mathscr{C}]$ which we will denote by $[F]$.

Assume that $\mathscr{C}$ is Hom-finite. Given an object $M \in \mathscr{C}$ we set $\mathscr{H}(M) = \mathrm{End}_{\mathscr{C}}(M)^{\mathrm{op}}$. It is an R-algebra which is finitely generated as an R-module. We denote by $\mathfrak{F}_M$ the functor $\mathfrak{F}_M = \mathrm{Hom}_{\mathscr{C}}(M, -) : \mathscr{C} \longrightarrow \mathscr{H}(M)\text{-mod}$.

Assume now that $\mathscr{C} = H\text{-mod}$, where H is an R-algebra with 1 which is finitely generated and free over R. We abbreviate $\mathrm{Irr}(H) = \mathrm{Irr}(\mathscr{C})$. Given an homomorphism $R \to S$, we can form the S-category $S\mathscr{C} = SH\text{-mod}$ where $SH = S \otimes_R H$. Given another R-category $\mathscr{C}'$ as above and an exact functor $F : \mathscr{C} \to \mathscr{C}'$, then F is represented by a projective object $P \in \mathscr{C}$. We set $SF = \mathrm{Hom}_{S\mathscr{C}}(SP, -) : S\mathscr{C} \to S\mathscr{C}'$.

Let K be the field of fractions of R, $A \subset R$ be a subring which is integrally closed in K and $\theta : R \to \mathrm{k}$ be a ring homomorphism into a field k such that k is the field of fractions of $\theta(A)$. If $\mathrm{k}H$ is split, then there is a *decomposition map* $d_\theta : [KH\text{-mod}] \longrightarrow [\mathrm{k}H\text{-mod}]$, see e.g. [**33**, sec. 3.1] for more details.

1.2. Kac-Moody algebras of type A and their representations.

The Lie algebras which will act on the categories we will study will always be finite sums of Kac-Moody algebras of type A_∞ or $A_{e-1}^{(1)}$. They will arise from quivers of the same type.

1.2.1. *Lie algebra associated with a quiver.* Let $v \in R^\times$ and $I \subset R^\times$. We assume that $v \neq 1$ and that I is stable by multiplication by v and v^{-1} with finitely many orbits. To the pair (I, v) we associate a quiver $I(v)$ (also denoted by I) as follows:

- the vertices of $I(v)$ are the elements of I;
- the arrows of $I(v)$ are $i \to iv$ for $i \in I$.

Since I is assumed to be stable by multiplication by v and v^{-1}, such a quiver is the disjoint union of quivers of type A_∞ if v is not a root of unity, or of cyclic quivers of type $A_{e-1}^{(1)}$ if v is a primitive e-th root of 1.

The quiver $I(v)$ defines a symmetric generalized Cartan matrix $A = (a_{ij})_{i,j \in I}$ with $a_{ii} = 2$, $a_{ij} = -1$ when $i \to j$ or $j \to i$ and $a_{ij} = 0$ otherwise. To this Cartan matrix one can associate the (derived) Kac-Moody algebra $\mathfrak{g}'_I$ over $\mathbb{C}$, which has Chevalley generators e_i, f_i for $i \in I$, subject to the usual relations.

More generally, let $(\mathrm{X}_I, \mathrm{X}_I^\vee, \langle \bullet, \bullet \rangle_I, \{\alpha_i\}_{i \in I}, \{\alpha_i^\vee\}_{i \in I})$ be a *Cartan datum* associated with A, *i.e.*, we assume that

- X_I and $\mathrm{X}_I^\vee$ are free abelian groups,
- the simple coroots $\{\alpha_i^\vee\}$ are linearly independent in $\mathrm{X}_I^\vee$,
- for each $i \in I$ there exists a fundamental weight $\Lambda_i \in \mathrm{X}_I$ satisfying $\langle \alpha_j^\vee, \Lambda_i \rangle_I = \delta_{ij}$ for all $j \in I$,
- $\langle \bullet, \bullet \rangle_I : \mathrm{X}_I^\vee \times \mathrm{X}_I \longrightarrow \mathbb{Z}$ is a perfect pairing such that $\langle \alpha_j^\vee, \alpha_i \rangle_I = a_{ij}$.

Let $\mathrm{Q}_I^\vee = \bigoplus \mathbb{Z}\alpha_i^\vee$ be the coroot lattice and $\mathrm{P}_I = \bigoplus \mathbb{Z}\Lambda_i$ be the weight lattice. Then, the Kac-Moody algebra $\mathfrak{g}_I$ corresponding to this datum is the Lie algebra generated by the Chevalley generators e_i, f_i for $i \in I$ and the Cartan algebra $\mathfrak{h} = \mathbb{C} \otimes \mathrm{X}_I^\vee$. An element $h \in \mathfrak{h}$ acts by $[h, e_i] = \langle h, \alpha_i \rangle e_i$. The Lie algebra $\mathfrak{g}'_I$ is the derived subalgebra $[\mathfrak{g}_I, \mathfrak{g}_I]$.

EXAMPLE 1.1. When $I = v^{\mathbb{Z}}$ two cases arise.

(a) If I is infinite, then $\mathfrak{g}'_I$ is isomorphic to $\mathfrak{sl}_{\mathbb{Z}}$, the Lie algebra of traceless matrices with finitely many non-zero entries.

(b) If v has finite order e, then I is isomorphic to a cyclic quiver of type $A^{(1)}_{e-1}$. We can form $\mathrm{X}^\vee = \mathrm{Q}^\vee \oplus \mathbb{Z}\partial$ and $\mathrm{X} = \mathrm{P} \oplus \mathbb{Z}\delta$ with $\langle \partial, \Lambda_i \rangle = 0$, $\langle \partial, \alpha_i \rangle = \delta_{i1}$ and $\delta = \sum_{i \in I} \alpha_i$. The pairing is non-degenerate, and $\mathfrak{g}_I$ is isomorphic to the Kac-Moody algebra

$$\widehat{\mathfrak{sl}}_e = \mathfrak{sl}_e(\mathbb{C}) \otimes \mathbb{C}[t, t^{-1}] \oplus \mathbb{C}c \oplus \mathbb{C}\partial.$$

An explicit isomorphism sends e_{v^i} (resp. f_{v^i}) to the matrix $E_{i,i+1} \otimes 1$ (resp. $E_{i+1,i} \otimes 1$) if $i \neq e$ and e_1 (resp. f_1) to $E_{e,1} \otimes t$ (resp. $E_{1,e} \otimes t^{-1}$). Via this isomophism the central element c corresponds to $\sum_{i \in I} \alpha_i^\vee$, and the derived algebra $\mathfrak{g}'_I$ to $\widetilde{\mathfrak{sl}}_e = \mathfrak{sl}_e(\mathbb{C}) \otimes \mathbb{C}[t, t^{-1}] \oplus \mathbb{C}c$.

To avoid cumbersome notation, we may write $\mathfrak{g} = \mathfrak{g}_I$, $\mathrm{P} = \mathrm{P}_I$, $\mathrm{Q}^\vee = \mathrm{Q}^\vee_I$, etc. when there is no risk of confusion.

1.2.2. *Integrable representations.* Let V be a $\mathfrak{g}$-module. Given $\omega \in \mathrm{X}$, the *ω-weight space* of V is $V_\omega = \{v \in V \mid \alpha^\vee \cdot v = \langle \alpha^\vee, \omega \rangle\, v,\ \forall \alpha^\vee \in \mathrm{Q}^\vee\}$. We denote by $\mathcal{O}^{\mathrm{int}}$ the category of *integrable highest weight* modules, *i.e.* $\mathfrak{g}$-modules V satisfying

- $V = \bigoplus_{\omega \in \mathrm{X}} V_\omega$ and $\dim V_\omega < \infty$ for all $\omega \in \mathrm{X}$,
- the action of e_i and f_i is locally nilpotent for all $i \in I$,
- there exists a finite set $F \subset \mathrm{X}$ such that $\mathrm{wt}(V) \subset F + \sum_{i \in I} \mathbb{Z}_{\leqslant 0}\alpha_i$.

Let $\mathrm{X}^+ = \{\omega \in \mathrm{X} \mid \langle \alpha_i^\vee, \omega \rangle \in \mathbb{N} \text{ for all } i \in I\}$ be the set of *integral dominant weights.* Given $\Lambda \in \mathrm{X}^+$, we denote by $\mathbf{L}(\Lambda)$ the simple integrable highest weight module with highest weight Λ.

1.2.3. *Quantized enveloping algebras.* Let u be a formal variable and $A = \mathbb{C}[u, u^{-1}]$. Let $U_u(\mathfrak{g})$ be the quantized enveloping algebra over $\mathbb{C}(u)$. Let $U_A(\mathfrak{g}) \subset U_u(\mathfrak{g})$ be Lusztig's divided power version of $U_u(\mathfrak{g})$. For each integral weight Λ the module $\mathbf{L}(\Lambda)$ admits a deformed version $\mathbf{L}_u(\Lambda)$ over $U_u(\mathfrak{g})$ and an integral form $\mathbf{L}_A(\Lambda)$ which is the $U_A(\mathfrak{g})$-submodule of $\mathbf{L}_u(\Lambda)$ generated by the highest weight vector $|\Lambda\rangle$. Let $\mathcal{O}_u^{\mathrm{int}}$ be the category consisting of the $U_u(\mathfrak{g})$-modules which are (possibly infinite) direct sums of $\mathbf{L}_u(\Lambda)$'s. If $V_u \in \mathcal{O}_u^{\mathrm{int}}$, then its integral form V_A is the corresponding sum of the modules $\mathbf{L}_A(\Lambda)$. It depends on the choice of a family of highest weight vectors of the constituents of V_u.

1.3. Categorical representations on abelian categories.

In this section we recall from [**10**, **59**] the notion of a categorical action of $\mathfrak{g}$. It consists of the data of functors E_i, F_i lifting the Chevalley generators e_i, f_i of $\mathfrak{g}$, together with an action of an affine Hecke algebra on $(\bigoplus_{i \in I} F_i)^m$.

1.3.1. *Affine Hecke algebras and representation data.* Let $\mathscr{C}$ be an abelian R-category and $v \in R^\times$.

DEFINITION 1.2. A *representation datum* on $\mathscr{C}$ with parameter v is a tuple (E, F, X, T) where E, F are bi-adjoint functors $\mathscr{C} \to \mathscr{C}$ and $X \in \mathrm{End}(F)^\times$, $T \in \mathrm{End}(F^2)$ are endomorphisms of functors satisfying the following conditions:

(a) $1_F T \circ T 1_F \circ 1_F T = T 1_F \circ 1_F T \circ T 1_F$,

(b) $(T + 1_{F^2}) \circ (T - v 1_{F^2}) = 0$,

(c) $T \circ (1_F X) \circ T = vX1_F$.

This definition can also be formulated in terms of actions of affine Hecke algebras. For $m \geqslant 1$, let $\mathbf{H}^v_{R,m}$ be the *affine Hecke algebra* of type A_{m-1} over R. It is generated by $T_1, \ldots, T_{m-1}, X_1^{\pm 1}, \ldots, X_m^{\pm 1}$ subject to the well-known relations. We will also set $\mathbf{H}^v_{R,0} = R$.

Given (E, F) a pair of biadjoint functors, and $X \in \mathrm{End}(F)$, $T \in \mathrm{End}(F^2)$, the tuple (E, F, X, T) is a representation datum if and only if for each $m \in \mathbb{N}$, the map

$$\begin{array}{rcl} \phi_{F^m} : \mathbf{H}^v_{R,m} & \longrightarrow & \mathrm{End}(F^m) \\ X_k & \longmapsto & 1_{F^{m-k}} X 1_{F^{k-1}} \\ T_l & \longmapsto & 1_{F^{m-l-1}} T 1_{F^{l-1}} \end{array}$$

is a well-defined R-algebra homomorphism.

1.3.2. *Categorical representations.* We assume now that R is a field and that $\mathscr{C}$ is Hom-finite. We fix a pair (I, v) as in §1.2 and we denote by $\mathfrak{g} = \mathfrak{g}_I$ the Lie algebra associated to that pair.

DEFINITION 1.3 ([**59**]). A $\mathfrak{g}$*-representation* on $\mathscr{C}$ consists of a representation datum (E, F, X, T) on $\mathscr{C}$ and of a decomposition $\mathscr{C} = \bigoplus_{\omega \in \mathrm{X}} \mathscr{C}_\omega$. For each $i \in I$, let F_i, E_i be the generalized i-eigenspaces of X acting on F, E respectively. We assume in addition that

(a) $F = \bigoplus_{i \in I} F_i$ and $E = \bigoplus_{i \in I} E_i$,
(b) the action of $[E_i]$ and $[F_i]$ for $i \in I$ endow $[\mathscr{C}]$ with a structure of integrable $\mathfrak{g}$-module such that $[\mathscr{C}]_\omega = [\mathscr{C}_\omega]$,
(c) $E_i(\mathscr{C}_\omega) \subset \mathscr{C}_{\omega + \alpha_i}$ and $F_i(\mathscr{C}_\omega) \subset \mathscr{C}_{\omega - \alpha_i}$.

We say that the tuple (E, F, X, T) and the decomposition $\mathscr{C} = \bigoplus_{\omega \in \mathrm{X}} \mathscr{C}_\omega$ is a $\mathfrak{g}$*-categorification* of the integrable $\mathfrak{g}$-module $[\mathscr{C}]$.

1.4. Minimal categorical representations.

For most of the results in the rest of Section 1 we will assume that R is a field and that I is finite. In particular $v \in R^\times$ will be a root of unity.

Let $m \geqslant 0$, $v \in R^\times$ and $\mathbf{H}^v_{R,m}$ be the affine Hecke algebra as defined in §1.3.1. We fix a tuple $Q = (Q_1, \ldots, Q_l)$ in $(R^\times)^l$. The *cyclotomic Hecke algebra* $\mathbf{H}^{Q;\,v}_{R,m}$ is the quotient of $\mathbf{H}^v_{R,m}$ by the two-sided ideal generated by $\prod_{i=1}^l (X_1 - Q_i)$.

Assume now that R is a field. Any finite dimensional $\mathbf{H}^{Q;\,v}_{R,m}$-module M is the direct sum of the weight subspaces

$$M_\nu = \{v \in M \mid (X_r - i_r)^d v = 0,\ r \in [1, m],\ d \gg 0\}, \quad \nu = (i_1, \ldots, i_m) \in R^m.$$

Decomposing the regular module, we get a system of orthogonal idempotents $\{e_\nu; \nu \in R^m\}$ in $\mathbf{H}^{Q;\,v}_{R,m}$ such that $e_\nu M = M_\nu$ for each M. The eigenvalues of X_r are always of the form $Q_i v^j$ for some $i \in \{1, \ldots, l\}$ and $j \in \mathbb{Z}$. As a consequence, if we set $I = \bigcup Q_i v^{\mathbb{Z}}$, then $e_\nu = 0$ unless $\nu \in I$. The pair (I, v) satisfies the assumptions of §1.2 and we can consider a corresponding Kac-Moody algebra $\mathfrak{g}_I$ and its root lattice Q_I. Given $\alpha \in \mathrm{Q}_I^+$ of height m, let $e_\alpha = \sum_\nu e_\nu$ where the sum runs over the set of all tuples such that $\sum_{r=1}^m \alpha_{i_r} = \alpha$. The nonzero e_α's are the primitive central idempotents in $\mathbf{H}^{Q;\,v}_{R,m}$.

To the dominant weight $\Lambda_Q = \sum_{i=1}^{l} \Lambda_{Q_i}$ of $\mathfrak{g}_I$ and to any $\alpha \in \mathrm{Q}_I^+$ we associate the following abelian categories:

$$\mathscr{L}(\Lambda_Q) = \bigoplus_{m \in \mathbb{N}} \mathbf{H}_{R,m}^{Q;\,v}\text{-mod} \quad \text{and} \quad \mathscr{L}(\Lambda_Q)_{\Lambda_Q - \alpha} = e_\alpha \mathbf{H}_{R,m}^{Q;\,v}\text{-mod}\,.$$

For any $m < n$, the R-algebra embedding of the affine Hecke algebras $\mathbf{H}_{R,m}^{v} \hookrightarrow \mathbf{H}_{R,n}^{v}$ given by $T_i \mapsto T_i$ and $X_j \mapsto X_j$ induces an embedding $\mathbf{H}_{R,m}^{Q;\,v} \hookrightarrow \mathbf{H}_{R,n}^{Q;\,v}$. The R-algebra $\mathbf{H}_{R,n}^{Q;\,v}$ is free as a left and as a right $\mathbf{H}_{R,m}^{Q;\,v}$-module. This yields a pair of exact adjoint functors $(\mathrm{Ind}_m^n, \mathrm{Res}_m^n)$ between $\mathbf{H}_{R,n}^{Q;\,v}$-mod and $\mathbf{H}_{R,m}^{Q;\,v}$-mod. They induce endofunctors E and F of $\mathscr{L}(\Lambda_Q)$ by $E = \bigoplus_{m \in \mathbb{N}} \mathrm{Res}_m^{m+1}$ and $F = \bigoplus_{m \in \mathbb{N}} \mathrm{Ind}_m^{m+1}$. The right multiplication on $\mathbf{H}_{R,m+1}^{Q;\,v}$ by X_{m+1} yields an endomorphism of the functor Ind_m^{m+1}. The right multiplication by T_{m+1} yields an endomorphism of Ind_m^{m+2}. We define $X \in \mathrm{End}(F)$ and $T \in \mathrm{End}(F^2)$ by $X = \bigoplus_m X_{m+1}$ and $T = \bigoplus_m T_{m+1}$.

This construction yields a categorification of the simple highest module $\mathbf{L}(\Lambda_Q)$ of $\mathfrak{g}_I$. Indeed, a theorem of Kang and Kashiwara implies that this holds in the more general setting of cyclotomic quiver Hecke algebras of arbitrary type.

THEOREM 1.4 ([**46**], [**49**]).

(a) *The endofunctors E and F of $\mathscr{L}(\Lambda_Q)$ are biadjoint.*
(b) *The tuple (E, F, X, T) and the decomposition $\mathscr{L}(\Lambda_Q) = \bigoplus_{\omega \in \mathrm{X}} \mathscr{L}(\Lambda_Q)_\omega$ is a $\mathfrak{g}_I$-categorification of $\mathbf{L}(\Lambda_Q)$.* □

This categorical representation is called the *minimal categorical $\mathfrak{g}_I$-representation* of highest weight Λ_Q.

The $\mathfrak{g}_I$-modules we are interested in are direct sums of various irreducible highest weight modules $\mathbf{L}(\Lambda_Q)$. Let (I, v) as in §1.2, and $\mathfrak{g} = \mathfrak{g}_I$ be a corresponding Kac-Moody algebra. Let (E, F, X, T) be a $\mathfrak{g}$-representation on an abelian R-category $\mathscr{C}$. We want to relate $\mathscr{C}$ to minimal categorical $\mathfrak{g}_I$-representations. To do that, recall that for any $m \geqslant 0$ we have an R-algebra homomorphism $\phi_{F^m} : \mathbf{H}_{R,m}^{Q;\,v} \longrightarrow \mathrm{End}(F^m)^{\mathrm{op}}$. Given an object M in $\mathscr{C}$, it specializes to an R-algebra homomorphism

$$\mathbf{H}_{R,m}^{Q;\,v} \longrightarrow \mathrm{End}(F^m M)^{\mathrm{op}} =: \mathscr{H}(F^m M).$$

PROPOSITION 1.5 ([**59**]). *Assume that the simple roots are linearly independent in* X. *Let (E, F, X, T) be a representation of $\mathfrak{g}$ in a abelian R-category $\mathscr{C}$, and $M \in \mathscr{C}_\omega$. Assume that $EM = 0$ and $\mathrm{End}_{\mathscr{C}}(M) = R$. Then*

(a) *$\omega \in \mathrm{X}^+$ is an integral dominant weight,*
(b) *if we write $\Lambda_Q = \sum_{i \in I} \langle \alpha_i^\vee, \omega \rangle \Lambda_i = \sum_{p=1}^{l} \Lambda_{Q_p}$ for some $Q = (Q_1, \dots, Q_l) \in I^l$ and $l \geqslant 1$, then for all $m \geq 0$ the map ϕ_{F^m} factors to an R-algebra isomorphism*

$$\mathbf{H}_{R,m}^{Q;\,v} \xrightarrow{\sim} \mathscr{H}(F^m M).$$

□

1.5. Crystals.

We start by a review of Kashiwara's theory of perfect bases and crystals. We will be working with the Kac-Moody algebra $\mathfrak{g}$ coming from a pair (I, v) as in §1.2.

DEFINITION 1.6. An *abstract crystal* is a set B together with maps $\mathrm{wt} : B \to \mathrm{P}$, $\varepsilon_i, \varphi_i : B \to \mathbb{Z} \sqcup \{-\infty\}$ and $\widetilde{e}_i, \widetilde{f}_i : B \to B \sqcup \{0\}$ for all $i \in I$ satisfying the following properties:

(a) $\varphi_i(b) = \varepsilon_i(b) + \langle \alpha_i^\vee, \mathrm{wt}(b) \rangle$,
(b) $\mathrm{wt}(\widetilde{e}_i b) = \mathrm{wt}(b) + \alpha_i$ and $\mathrm{wt}(\widetilde{f}_i b) = \mathrm{wt}(b) - \alpha_i$,
(c) $b = \widetilde{e}_i b'$ if and only if $\widetilde{f}_i b = b'$, where $b, b' \in B$, $i \in I$,
(d) if $\varphi_i(b) = -\infty$, then $\widetilde{e}_i b = \widetilde{f}_i b = 0$,
(e) if $b \in B$ and $\widetilde{e}_i b \in B$, then $\varepsilon_i(\widetilde{e}_i b) = \varepsilon_i(b) - 1$ and $\varphi_i(\widetilde{e}_i b) = \varphi_i(b) + 1$,
(f) if $b \in B$ and $\widetilde{f}_i b \in B$, then $\varepsilon_i(\widetilde{f}_i b) = \varepsilon_i(b) + 1$ and $\varphi_i(\widetilde{f}_i b) = \varphi_i(b) - 1$.

Note that by (a), the map φ_i is entirely determined by ε_i and wt. We may therefore omit φ_i in the data of an abstract crystal and denote it by $(B, \widetilde{e}_i, \widetilde{f}_i, \varepsilon_i, \mathrm{wt})$.

An isomorphism between crystals B_1, B_2 is a bijection $\psi : B_1 \sqcup \{0\} \longrightarrow B_2 \sqcup \{0\}$ such that $\psi(0) = 0$ which commutes with wt, ε_i, φ_i $\widetilde{f}_i$,, $\widetilde{e}_i$.

Let V_u be an integrable $U_u(\mathfrak{g})$-module in $\mathcal{O}_u^{\mathrm{int}}$. Let V_A be an integral form of V_u. A *lower crystal lattice* in V_u is a free $\mathbb{C}[u]$-submodule $\mathcal{L}$ of V_A such that $V_A = A\mathcal{L}$, $\mathcal{L} = \bigoplus_{\lambda \in \mathrm{X}} \mathcal{L}_\lambda$ with $\mathcal{L}_\lambda = \mathcal{L} \cap (V_A)_\lambda$ and $\mathcal{L}$ is preserved by the *lower Kashiwara crystal operators* $\widetilde{e}_i^{\,\mathrm{lw}}$, $\widetilde{f}_i^{\,\mathrm{lw}}$ on V_u. A *lower crystal basis* of V_u is a pair $(\mathcal{L}, B)$ where $\mathcal{L}$ is a lower crystal lattice of V_u and B is a basis of $\mathcal{L}/u\mathcal{L}$ such that we have $B = \bigsqcup_{\lambda \in \mathrm{X}} B_\lambda$ where $B_\lambda = B \cap (\mathcal{L}_\lambda / u\,\mathcal{L}_\lambda)$, $\widetilde{e}_i^{\,\mathrm{lw}}(B)$, $\widetilde{f}_i^{\,\mathrm{lw}}(B) \subset B \sqcup \{0\}$ and $b' = \widetilde{f}_i^{\,\mathrm{lw}} b$ if and only if $b = \widetilde{e}_i^{\,\mathrm{lw}} b'$ for each $b, b' \in B$. A *lower global basis* of V_u (or a *canonical basis*) is an A-basis $\mathbf{B}$ of V_A such that the lattice $\mathcal{L} = \bigoplus_{b \in \mathbf{B}} \mathbb{C}[u]\, b$ and the basis $B = \{b \bmod u\,\mathcal{L} \,|\, b \in \mathbf{B}\}$ of $\mathcal{L}/u\,\mathcal{L}$ form a lower crystal basis. One defines in a similar way an *upper crystal lattice*, an *upper crystal basis* and an *upper global basis* (or a *dual canonical basis*) using the *upper Kashiwara crystal operators* $\widetilde{e}_i^{\,\mathrm{up}}$, $\widetilde{f}_i^{\,\mathrm{up}}$ on V_u, see, e.g., [**47**, def. 4.1,4.2]. Any $U_u(\mathfrak{g})$-module in $\mathcal{O}_u^{\mathrm{int}}$ admits a lower crystal, an upper crystal and a global basis.

If $(\mathcal{L}, B)$, $(\mathcal{L}^\vee, B^\vee)$ are lower, upper crystal bases, then $(B, \widetilde{e}_i^{\,\mathrm{lw}}, \widetilde{f}_i^{\,\mathrm{lw}})$, $(B^\vee, \widetilde{e}_i^{\,\mathrm{up}}, \widetilde{f}_i^{\,\mathrm{up}})$ are abstract crystals.

Let E_i, F_i, u^h with $i \in I$, $h \in \mathrm{X}^\vee$, be the standard generators of $U_u(\mathfrak{g})$. There exists a unique non-degenerate symmetric bilinear form $(\bullet, \bullet)$ on the module $\mathbf{L}_u(\Lambda)$ with highest weight vector $|\Lambda\rangle$ satisfying

- $(|\Lambda\rangle, |\Lambda\rangle) = 1$,
- $(E_i x, y) = (x, F_i y)$, $(F_i x, y) = (x, E_i y)$, $(u^h x, y) = (x, u^h y)$,
- $(\mathbf{L}_u(\Lambda)_\lambda, \mathbf{L}_u(\Lambda)_\mu) = 0$ if $\lambda \neq \mu$.

If $(\mathcal{L}, B)$ is a lower crystal basis of $\mathbf{L}_u(\Lambda)$ then the pair $(\mathcal{L}^\vee, B^\vee)$ such that $\mathcal{L}^\vee = \{x \in \mathbf{L}_u(\Lambda) \,|\, (x, \mathcal{L}) \subset \mathbb{C}[u]\}$ and $B^\vee$ is the basis of $\mathcal{L}^\vee / u\,\mathcal{L}^\vee$ which is dual to B with respect to the non-degenerate bilinear form $\mathcal{L}^\vee / u\,\mathcal{L}^\vee \times \mathcal{L}/u\,\mathcal{L} \to \mathbb{C}$ induced by $(\bullet, \bullet)$, is an upper crystal basis. Further, taking a basis element in B to the dual basis element in $B^\vee$ is a crystal isomorphism $(B, \widetilde{e}_i^{\,\mathrm{lw}}, \widetilde{f}_i^{\,\mathrm{lw}}) \to (B^\vee, \widetilde{e}_i^{\,\mathrm{up}}, \widetilde{f}_i^{\,\mathrm{up}})$. Therefore, if $\mathbf{B}$ is a lower global basis of $\mathbf{L}_u(\Lambda)$ then the dual basis $\mathbf{B}^\vee$ with respect to the non-degenerate bilinear form $(\bullet, \bullet)$ is an upper global basis and the corresponding abstract crystals $(B, \widetilde{e}_i^{\,\mathrm{lw}}, \widetilde{f}_i^{\,\mathrm{lw}})$ and $(B^\vee, \widetilde{e}_i^{\,\mathrm{up}}, \widetilde{f}_i^{\,\mathrm{up}})$ are canonically isomorphic.

1.6. Perfect bases.

The crystals that we will consider in this paper all come from particular bases of $\mathfrak{g}_I$-modules called *perfect bases*. Let us define them. Let $V \in \mathcal{O}^{\mathrm{int}}$ be an integrable highest weight $\mathfrak{g}_I$-module. For each $i \in I$ and $x \in V$ we define

$$\ell_i(x) = \max\{k \in \mathbb{N} \mid e_i^k x \neq 0\} = \min\{k \in \mathbb{N} \mid e_i^{k+1} x = 0\}$$

with the convention that $\ell_i(0) = -\infty$. For each integer k, we also consider the vector spaces

$$V_i^{\leqslant k} = \{x \in V \mid \ell_i(x) \leqslant k\}, \quad V^{\leqslant k} = \bigcap_{i \in I} V_i^{\leqslant k}, \quad V_i^k = V_i^{\leqslant k} / V_i^{< k}.$$

DEFINITION 1.7. A basis B of V is *perfect* if

(a) $B = \bigsqcup_{\mu \in X_I} B_\mu$ where $B_\mu = B \cap V_\mu$,
(b) for any $i \in I$, there is a map $\mathbf{e}_i : B \to B \sqcup \{0\}$ such that for any $b \in B$, we have
  (i) if $\ell_i(b) = 0$, then $\mathbf{e}_i b = 0$,
  (ii) if $\ell_i(b) > 0$, then $\mathbf{e}_i b \in B$ and $e_i b \in \mathbb{C}^\times \, \mathbf{e}_i b + V_i^{< \ell_i(b) - 1}$,
(c) if $\mathbf{e}_i b = \mathbf{e}_i b' \neq 0$ for $b, b' \in B$, then $b = b'$.

Any $\mathfrak{g}$-module in $\mathcal{O}^{\mathrm{int}}$ admits a perfect basis. More precisely, we have the following.

PROPOSITION 1.8. *If V is an integrable $\mathfrak{g}$-module in $\mathcal{O}^{int}$ with a quantum deformation V_u, then the specialization at $u = 1$ of an upper global basis of V_u is a perfect basis of V.* □

To any categorical representation we associate a perfect basis as in [**61**, prop. 6.2]. More precisely, let R be a field (of any characteristic) and consider a $\mathfrak{g}$-representation on an abelian artinian R-category $\mathscr{C}$. Then, for each $i \in I$ we define the maps

$$\widetilde{E}_i : \mathrm{Irr}(\mathscr{C}) \to \mathrm{Irr}(\mathscr{C}) \sqcup \{0\}, \quad [L] \mapsto [\mathrm{soc}(E_i(L))],$$
$$\widetilde{F}_i : \mathrm{Irr}(\mathscr{C}) \to \mathrm{Irr}(\mathscr{C}) \sqcup \{0\}, \quad [L] \mapsto [\mathrm{hd}(F_i(L))].$$

PROPOSITION 1.9. *The tuple $\big(\mathrm{Irr}(\mathscr{C}), \widetilde{E}_i, \widetilde{F}_i\big)$ defines a perfect basis of $[\mathscr{C}]$.* □

We now recall how to construct an abstract crystal from a perfect basis B. We set $\widetilde{e}_i = \mathbf{e}_i$. For all $b \in B$ we set $\widetilde{f}_i b = b'$ if $\mathbf{e}_i b' = b$ for some $b' \in B$, and 0 otherwise. Then it follows easily from the definition that $(B, \widetilde{e}_i, \widetilde{f}_i, \ell_i, \mathrm{wt})$ is an abstract crystal.

We finish this section with a result which will be important to identify the crystal obtained by the categorification with the crystal of some Fock space. For each $i \in I$ and $k \in \mathbb{N}$, we set $B^{\leqslant k} = V^{\leqslant k} \cap B$ and $B_i^{\leqslant k} = V_i^{\leqslant k} \cap B$.

PROPOSITION 1.10. *Let B and B' be perfect bases of $V \in \mathcal{O}^{\mathrm{int}}$.*

(a) *$B^{\leqslant k}$ and $B_i^{\leqslant k}$ are bases of $V^{\leqslant k}$ and $V_i^{\leqslant k}$.*
(b) *Assume that there is a bijection $\varphi : B \to B'$ and a partial order $\leqslant$ on B such that $\varphi(b) \in b + \sum_{c > b} \mathbb{C}\, c$ for each $b \in B$. Then the map φ is a crystal isomorphism $B \xrightarrow{\sim} B'$.* □

1.7. Derived equivalences.

Given V an integrable $\mathfrak{g}$-module and $i \in I$, one can consider the action of the simple reflection $s_i = \exp(-f_i)\exp(e_i)\exp(-f_i)$ on V. For each weight $\omega \in \mathrm{X}$, this action maps a weight space V_ω to $V_{s_i(\omega)}$ with $s_i(\omega) = \omega - \langle \alpha_i^\vee, \omega \rangle \alpha_i$. If $\mathscr{C}$ is a categorification of V, then it restricts to an $\mathfrak{sl}_2(\mathbb{C})$-categorification in the sense of Chuang-Rouquier. In particular, the simple objects are weight vectors for the categorical $\mathfrak{sl}_2(\mathbb{C})$-action. Thus, the theory of Chuang-Rouquier can be applied and [**10**, thm. 6.6] implies that s_i can be lifted to a derived equivalence Θ_i of $\mathscr{C}$.

THEOREM 1.11. *Assume that R is a field. Let (E, F, X, T) be a representation of $\mathfrak{g}$ in an abelian R-category $\mathscr{C}$, and $i \in I$. Then there exists a derived self-equivalence Θ_i of $\mathscr{C}$ which restricts to derived equivalences $\Theta_i : D^b(\mathscr{C}_\omega) \xrightarrow{\sim} D^b(\mathscr{C}_{s_i(\omega)})$ for all weights $\omega \in \mathrm{X}$. Furthermore, $[\Theta_i] = s_i$ as a linear map of $[\mathscr{C}]$.* □

2. Representations on Fock spaces

Let R be a noetherian commutative domain with unit. As in §1.2, we fix an element $v \in R^\times$ and a subset I of $R^\times$ which is stable by multiplication by v and v^{-1}. We explained in §1.2.1 how one can associate a Lie algebra $\mathfrak{g} = \mathfrak{g}_I$ to this data. In this section we recall the construction of (charged) Fock spaces which are particular integrable representations of $\mathfrak{g}$.

2.1. Combinatorics of l-partitions.

2.1.1. *Partitions and l-partitions.* A *partition* of n is a non-increasing sequence of non-negative integers $\lambda = (\lambda_1 \geqslant \lambda_2 \geqslant \cdots)$ whose terms add up to n. We denote by $\mathscr{P}_n$ the set of partitions of n and by $\mathscr{P} = \bigsqcup_n \mathscr{P}_n$ be the set of all partitions. Given a partition λ, we write $|\lambda|$ for the *weight* of λ. We associate to $\lambda = (\lambda_1, \lambda_2, \dots)$ the *Young diagram* $Y(\lambda)$ defined by $Y(\lambda) = \{(x, y) \in \mathbb{Z}_{>0} \times \mathbb{Z}_{>0} \mid y \leqslant \lambda_x\}$. It may be visualised by an array of boxes in left justified rows with λ_x boxes in the x-th row. If λ, μ are partitions of n then we write $\lambda \geqslant \mu$ if for all $n \geqslant i \geqslant 1$ we have $\sum_{j=1}^i \lambda_j \geqslant \sum_{j=1}^i \mu_j$. This relation defines a partial order on $\mathscr{P}$ called the *dominance order*. Let λ^* denote the partition dual (or conjugate) to λ.

An *l-partition* of n is an l-tuple of partitions whose weights add up to n. We denote by $\mathscr{P}_n^l$ be the set of l-partitions of n and by $\mathscr{P}^l = \bigsqcup_n \mathscr{P}_n^l$ the set of all l-partitions. The Young diagram of the l-partition $\lambda = (\lambda^1, \dots, \lambda^l)$ is the set $Y(\lambda) = \bigsqcup_{p=1}^l Y(\lambda^p) \times \{p\}$. Its weight is the integer $|\lambda| = \sum_p |\lambda^p|$.

2.1.2. *Residues and content.* We fix $Q = (Q_1, \dots, Q_l) \in I^l$. Let λ be an l-partition and $A = (x, y, p)$ be a node in $Y(\lambda)$. The *(Q, v)-shifted residue* of the node A is the element of I given by $\mathrm{res}(A, Q)_I = v^{y-x} Q_p$. Let $n_i(\lambda, Q)_I$ be the number of nodes of (Q, v)-shifted residue i in $Y(\lambda)$. If λ, μ are l-partitions such that $|\mu| = |\lambda| + 1$ we write $\mathrm{res}(\mu - \lambda, Q)_I = i$ if $Y(\mu)$ is obtained by adding a node of (Q, v)-shifted residue i to $Y(\lambda)$. A *charge* of the tuple $Q = (Q_1, \dots, Q_l)$ is an l-tuple of integers $s = (s_1, \dots, s_l)$ such that $Q_p = v^{s_p}$ for all $p = 1, \dots, l$. Conversely, given $I \subset R^\times$ and $v \in R^\times$ as in §1.2, any ℓ-tuple of integers $s \in \mathbb{Z}^l$ defines a tuple $Q = (v^{s_1}, \dots, v^{s_l})$ with charge s. The *s-shifted content* of the box $A = (x, y, p)$ is the integer $\mathrm{ct}^s(A) = s_p + y - x$. It is related to the residue of A by the formula $\mathrm{res}(A,Q)_I = v^{\mathrm{ct}^s(A)}$. We will also write $p(A) = p$. We will call *charged l-partition* a pair (μ, s) in $\mathscr{P}^l \times \mathbb{Z}^l$.

2.1.3. *l-cores and l-quotients.* We start with the case $l = 1$. The set of *β-numbers* of a charged partition $(\lambda, d) \in \mathscr{P} \times \mathbb{Z}$ is the set given by $\beta_d(\lambda) = \{\lambda_u + d + 1 - u \mid u \geqslant 1\}$. The charged partition (λ, d) is uniquely determined by the set $\beta_d(\lambda)$.

For any positive integer e, an *e-hook* of (λ, d) is a pair $(x, x+e)$ such that $x + e \in \beta_d(\lambda)$ and $x \notin \beta_d(\lambda)$. Removing the e-hook $(x, x+e)$ corresponds to replacing $x + e$ with x in $\beta_d(\lambda)$. We say that the charged partition (λ, d) is an *e-core* if it does not have any e-hook. This does not depend on d.

Next, we construct a bijection $\tau_l : \mathscr{P} \times \mathbb{Z} \to \mathscr{P}^l \times \mathbb{Z}^l$. It takes the pair (λ, d) to (μ, s), where $\mu = (\mu^1, \dots, \mu^l)$ is an l-partition and $s = (s_1, \dots, s_l)$ is a l-tuple in $\mathbb{Z}^l(d) = \{s \in \mathbb{Z}^l \mid s_1 + \cdots + s_l = d\}$. The bijection is uniquely determined by the relation $\beta_d(\lambda) = \bigsqcup_{p=1}^{l} \big(p - l + l\beta_{s_p}(\mu^p)\big)$.

The bijection τ_l takes the pair $(\lambda, 0)$ to $(\lambda^{[l]}, \lambda_{[l]})$, where $\lambda^{[l]}$ is the *l-quotient* of λ and $\lambda_{[l]}$ lies in $\mathbb{Z}^l(0)$. Since λ is an l-core if and only if $\lambda^{[l]} = \emptyset$, this bijection identifies the set of l-cores and $\mathbb{Z}^l(0)$. We define the *l-weight* $w_l(\lambda) := |\lambda^{[l]}|$ of the partition λ to be the weight of its l-quotient.

We will mostly consider the bijection τ_l for $l = 2$. In particular, a 2-core is either $\Delta_0 = \emptyset$ or a triangular partition $\Delta_t = (t, t-1, \dots, 1)$ with $t \in \mathbb{N}$. We abbreviate $\sigma_t = (\Delta_t)_{[2]}$, and we write $\sigma_t = (\sigma_1, \sigma_2)$. We have

$$\sigma_t = \begin{cases} (-t/2,\, t/2) & \text{if } t \text{ is even,} \\ ((1+t)/2, -(1+t)/2) & \text{if } t \text{ is odd.} \end{cases} \tag{2.1}$$

For each bipartition μ, let $\varpi_t(\mu)$ denote the unique partition with 2-quotient μ and 2-core Δ_t. Thus, the bijection τ_2 maps $(\varpi_t(\mu), 0)$ to the pair (μ, σ_t).

2.2. Fock spaces.

For a reference for the results presented in this section, see for example [**65**], [**66**]. Let $Q = (Q_1, \dots, Q_l) \in I^l$. It defines an integral dominant weight $\Lambda_Q = \sum_{p=1}^{l} \Lambda_{Q_p} \in \mathrm{P}^+$. The *Fock space* $\mathbf{F}(Q)_I$ is the $\mathbb{C}$-vector space with basis $\{|\lambda, Q\rangle_I \mid \lambda \in \mathscr{P}^l\}$ called the *standard monomial basis*, and action of e_i, f_i for all $i \in I$ given by

$$f_i(|\lambda, Q\rangle_I) = \sum_{\mu} |\mu, Q\rangle_I, \qquad e_i(|\mu, Q\rangle_I) = \sum_{\lambda} |\lambda, Q\rangle_I, \tag{2.2}$$

where the sums run over all partitions such that $\mathrm{res}(\mu - \lambda, Q)_I = i$. This endows $\mathbf{F}(Q)_I$ with a structure of $\mathfrak{g}'$-module. The Fock space $\mathbf{F}(Q)_I$ can also be equipped with a symmetric non-degenerate bilinear form $\langle \bullet, \bullet \rangle_I$ for which the standard monomial basis is orthonormal. To avoid cumbersome notation, we shall omit the subscript I when not necessary. It is easy to see that every element of the standard monomial basis is a weight vector.

The $\mathfrak{g}'$-submodule of $\mathbf{F}(Q)$ generated by $|\emptyset, Q\rangle$ is isomorphic to $\mathbf{L}(\Lambda_Q)$. Furthermore, if $I = A_\infty$, then $\mathbf{F}(Q) = \mathbf{L}(\Lambda_Q)$. Using the minimal categorification $\mathscr{L}(\Lambda_Q)$ of $\mathbf{L}(\Lambda_Q)$, the map $\mathbf{L}(\Lambda_Q) \to \mathbf{F}(Q)$ can be made more explicit.

To explain this, let us first recall briefly the definition of the *Specht modules*. Assume that R has characteristic 0 and contains a primitive l-th root ζ of 1, so R is a splitting field of the complex reflection group $G(l,1,m)$. Let $\mathrm{Irr}(R\mathfrak{S}_m) = \{\phi_\lambda \,|\, \lambda \in \mathscr{P}_m\}$ be the standard labelling of the characters of the symmetric group. Then

$$\mathrm{Irr}(RG(l,1,m)) = \{\mathscr{X}_\lambda \,|\, \lambda \in \mathscr{P}^l_m\}$$

is the labelling of the simple modules such that $\mathscr{X}_\lambda$ is induced from the $G(l,1,|\lambda_1|)\times \ldots \times G(l,1,|\lambda_l|)$-module

$$\phi_{\lambda^{(1)}}\chi^0 \otimes \phi_{\lambda^{(2)}}\chi^1 \otimes \cdots \otimes \phi_{\lambda^{(l)}}\chi^{l-1}.$$

Here, we denote by χ^p the one-dimensional module of the $|\lambda_p|$-th cartesian power of the cyclic group $G(l,1,1)$ given by the p-th power of the determinant, see, e.g., [**33**, sec. 5.1.3]. Recall that (for every field R) the R-algebra $\mathbf{H}^{Q;\,v}_{R,m}$ is split and that it is semisimple if and only if we have, see, e.g., [**58**, sec. 3.2],

$$\prod_{i=1}^m (1+v+\cdots+v^{i-1}) \prod_{a<b} \prod_{-m<r<m} (v^r\,Q_a - Q_b) \neq 0.$$

Thus, by Tits' deformation theorem, under the specialization $v \mapsto 1$ and $Q_p \mapsto \zeta^{p-1}$, the labelling of $\mathrm{Irr}(RG(l,1,m))$ yields a canonical labelling

$$\mathrm{Irr}(\mathbf{H}^{Q;\,v}_{R,m}) = \{S(\lambda)^{Q;\,v}_R \,|\, \lambda \in \mathscr{P}^l_m\}.$$

Now, if R is a commutative domain with fraction field K of characteristic 0 as above, we define the $\mathbf{H}^{Q;\,v}_{R,m}$-module $S(\lambda)^{Q,\,v}_R$ as in [**60**, sec. 2.4.3] or [**33**, sec. 5.3], using $S(\lambda)^{Q;\,v}_K$ and the dominance order on $\mathscr{P}^l_m$, and if $\theta\,:\,R \to \mathrm{k}$ is a ring homomorphism such that k is the fraction field of $\theta(R)$ we set $S(\lambda)^{Q;\,v}_{\mathrm{k}} = \mathrm{k}S(\lambda)^{Q,\,v}_R$.

PROPOSITION 2.1. *Let R be a field of characteristic 0 which contains a primitive l-th root of 1. The composition $[\mathscr{L}(\Lambda_Q)] \xrightarrow{\sim} \mathbf{L}(\Lambda_Q) \to \mathbf{F}(Q)$ obtained from Theorem* 1.4 *sends the class of $S(\lambda)^{Q,\,v}_R$ to the standard monomial $|\lambda, Q\rangle$.* □

For each $p = 1,\ldots,l$, let I_p be the subquiver of I corresponding to the subset $v^{\mathbb{Z}}Q_p$ of I. We define a relation on $\{1,\ldots,l\}$ by $i \sim j \iff I_i = I_j$. Let $\Omega = \{1,\ldots,l\}/\sim$ be the set of equivalence classes for this relation. Given $p \in \Omega$, we denote by Q_p the tuple of $(Q_{i_1},\ldots,Q_{i_r})$ where $(i_1,\ldots,i_r)$ is the ordered set of elements in p. The decomposition $I = \bigsqcup_{p\in\Omega} I_p$ yields a canonical decomposition of Lie algebras $\mathfrak{g}'_I = \bigoplus_{p\in\Omega}\mathfrak{g}'_{I_p}$. The corresponding decomposition of Fock spaces is given in the following proposition.

PROPOSITION 2.2. *The map $|\lambda, Q\rangle_I \longmapsto \otimes_{p\in\Omega}|\lambda^p, Q_p\rangle_{I_p}$ yields an isomorphism of $\mathfrak{g}'_I$-modules $\mathbf{F}(Q)_I \xrightarrow{\sim} \bigotimes_{p\in\Omega}\mathbf{F}(Q_p)_{I_p}$.* □

2.3. Charged Fock spaces.

A *charged Fock space* is a pair $\mathbf{F}(s) = (\mathbf{F}(Q), s)$ such that $s \in \mathbb{Z}^l$ is a charge of Q, that is $Q = (v^{s_1},\ldots,v^{s_l})$. Throughout this section, we will always assume that I is either of type A_∞ or a cyclic quiver. For more general quivers we can invoke Proposition 2.2 to reduce to that case.

2.3.1. *The $\mathfrak{g}$-action on the Fock space.* The action of $\mathfrak{g}'$ on $\mathbf{F}(Q)$ can be extended to an action of $\mathfrak{g}$ when Q admits a charge s. We describe this action in the case where v has finite order e, and $l = 1$. In that case $I = v^{\mathbb{Z}}$ is isomorphic to the cyclic quiver $A_{e-1}^{(1)}$ and the charge s is just an integer $d \in \mathbb{Z}$ such that $Q = v^d$. If we fix the affine simple root to be α_1, then $\mathrm{X} = \mathrm{P} \oplus \mathbb{Z}\delta$ and $\mathrm{X}^\vee = \mathrm{Q}^\vee \oplus \mathbb{Z}\partial$ with $\delta = \sum_{i \in I} \alpha_i$ and $\partial = \Lambda_1^\vee$ (see Example 1.1 for more details).

We define the integer

$$\Delta(d,e) = \big(\bar{d}(1-\bar{d}/e) + d(d/e-1)\big)/2, \tag{2.3}$$

where $\bar{d}$ is the residue of d modulo e in $[0, e-1]$. Then, the action of the derivation ∂ on $\mathbf{F}(d) = (\mathbf{F}(Q), d)$ is

$$\partial(|\lambda, Q\rangle) = -\big(n_1(\lambda, Q) + \Delta(d,e)\big)|\lambda, Q\rangle.$$

For this action the weight of a standard basis element is

$$\mathrm{wt}(|\lambda, Q\rangle) = \Lambda_Q - \sum_{i \in I} n_i(\lambda, Q)\, \alpha_i - \Delta(d,e)\, \delta. \tag{2.4}$$

We now describe the action of the affine Weyl group of $\mathfrak{g}$ on $\mathbf{F}(d)$. For $i \in I \smallsetminus \{1\}$, we denote by $\alpha_i^{\mathrm{cl}} = 2\Lambda_i - \Lambda_{iv} - \Lambda_{iv^{-1}}$ and $\Lambda_i^{\mathrm{cl}} = \Lambda_i - \Lambda_1$ the i-th simple root and fundamental weight of $\mathfrak{sl}_e$. These (classical) simple roots span the lattice of classical roots Q^{cl}. The affine Weyl group of $\mathfrak{g}$ is $W = \mathfrak{S}_I \ltimes \mathrm{Q}^{\mathrm{cl}}$. It acts linearly on X. We will denote by t_γ the action of an element $\gamma \in \mathrm{Q}^{\mathrm{cl}}$, i.e., for each $\alpha \in \mathrm{X}$ we set

$$t_\gamma(\alpha) = \alpha + (\alpha : \delta)\,\gamma - (\alpha : \gamma)\,\delta - \frac{1}{2}(\alpha : \delta)(\gamma : \gamma)\,\delta \tag{2.5}$$

where $(\bullet : \bullet)$ is the standard symmetric non-degenerate bilinear form on $\mathrm{X} \times \mathrm{X}$. Now, we consider the element $\pi_s = \sum_{i \in I}(s_i - s_{iv})\,\Lambda_i$. If $s \in \mathbb{Z}^I(d)$, then $\pi_s - \Lambda_Q^{\mathrm{cl}} \in \mathrm{Q}^{\mathrm{cl}}$ and we can consider the corresponding operator $t_{\pi_s - \Lambda_Q^{\mathrm{cl}}}$ on X.

PROPOSITION 2.3. *Let λ and ν be two partitions. Let $s \in \mathbb{Z}^I(d)$ be such that $(\emptyset, s) = \tau_e(\lambda_{[e]}, d)$, where $\lambda_{[e]}$ is the e-core of λ.*

(a) *The weight of $|\lambda, Q\rangle$ equals $\mathrm{wt}(|\lambda, Q\rangle) = t_{\pi_s - \Lambda_Q^{\mathrm{cl}}}(\Lambda_Q) - w_e(\lambda)\,\delta$.*

(b) *The weights of $|\lambda, Q\rangle$ and $|\nu, Q\rangle$ are W-conjugate if and only if $w_e(\lambda) = w_e(\nu)$.* □

In the particular case where the charge is zero, then $s = \lambda_{[e]}$, and

$$\mathrm{wt}(|\lambda, 1\rangle) = t_{\pi_{\lambda_{[e]}}}(\Lambda_1) - w_e(\lambda)\,\delta. \tag{2.6}$$

Therefore weight spaces are parametrized by pairs (ν, w) where ν is an e-core and w is a non-negative integer. The basis element $|\lambda, 1\rangle$ is in the weight space corresponding to $(\lambda_{[e]}, w_e(\lambda))$.

2.3.2. *The crystal of the Fock space.* We explain here how to associate an abstract crystal to a charged Fock space $\mathbf{F}(s)$. By Proposition 2.2, we can assume that I is either cyclic or of type A_∞. We assume that the reader is familiar with [**65**].

When I has type A_∞, Uglov's bases coincide with the standard monomial basis and the discussion is trivial in that case. We will therefore assume that v has finite order e and $I = v^{\mathbb{Z}}$, so that I has type $A_{e-1}^{(1)}$.

The $\mathfrak{g}$-module $\mathbf{F}(Q)$ admits a quantum deformation $\mathbf{F}_u(s)$ with an A-lattice $\mathbf{F}_A(s)$, which is a free A-module with basis $\{|\mu, s\rangle \mid \mu \in \mathscr{P}^l\}$. It is equipped with an integrable representation of $U_A(\mathfrak{g})$. Note that the action of the Chevalley generators e_i and f_i depends on the choice of the charge s. The representation of $\mathfrak{g}$ on $\mathbf{F}(s)$ given in §2.3.1 is recovered by specializing the parameter u to 1. Uglov has constructed a remarkable A-basis $\mathbf{B}_u^+(s) = \{\mathbf{b}_u^+(\mu, s) \mid \mu \in \mathscr{P}^l\}$ of $\mathbf{F}_A(s)$ which is a lower global basis for the representation of $U_u(\mathfrak{g})$ on $\mathbf{F}_u(s)$.

Next, we consider the pairing $(\bullet, \bullet)$ on $\mathbf{F}_u(s)$ defined in [**66**, sec. 4.3]. Then, let $\mathbf{B}_u^\vee(s) = \{\mathbf{b}_u^\vee(\mu, s) \mid \mu \in \mathscr{P}^l\}$ be the $\mathbb{C}(u)$-basis of $\mathbf{F}_u(s)$ dual to $\mathbf{B}_u^+(s)$ with respect to the bilinear form $(\bullet, \bullet)$. It is an upper global basis of $\mathbf{F}_u(s)$. Let $\mathbf{B}^\vee(s) = \{\mathbf{b}^\vee(\mu, s) \mid \mu \in \mathscr{P}^l\}$ be the specialization at $u = 1$ of $\mathbf{B}_u^\vee(s)$, with the obvious labeling of its elements. It is a perfect basis of the usual Fock space $\mathbf{F}(s) = \mathbf{F}_u(s)|_{u=1}$ by Proposition 1.8.

We equip the set of l-partitions with the abstract crystal structure $B(s) = (\mathscr{P}^l, \widetilde{e}_i, \widetilde{f}_i)$ defined in [**44**]. Let $B(s) = \{b(\mu, s) \mid \mu \in \mathscr{P}^l\}$ be the obvious labeling. The map $\mathbf{b}^\vee(\mu, s) \in \mathbf{B}^\vee(s) \longmapsto b(\mu, s) \in B(s)$ is a crystal isomorphism.

3. Unipotent representations

In this section we record standard results on unipotent representations of finite reductive groups in non-defining characteristic. A good reference is [**7**].

3.1. Basics.

By an *ℓ-modular system* we will mean a triple $(K, \mathscr{O}, \mathrm{k})$ where K is a field of characteristic zero, $\mathscr{O}$ is a complete discrete valuation ring with fraction field K, and k is the residue field of $\mathscr{O}$ with $\mathrm{char}(\mathrm{k}) = \ell$. When working with representations of a finite group Γ, we will always assume that $(K, \mathscr{O}, \mathrm{k})$ is a *splitting ℓ-modular system for* Γ, which means that K and k are splitting fields for all subgroups of Γ. When Γ comes from an algebraic group in characteristic p, we will in addition assume that $\ell \neq p$. This case is usually referred to as the *non-defining characteristic* case.

Let R be any commutative domain (with 1) and Γ be a finite group. We will assume that p is invertible in R and $\mathscr{O}$. Let $R\Gamma$ denote the group ring of Γ over R. For any subset $S \subseteq \Gamma$ such that $|S|$ is invertible in R, let e_S be the element $e_S = |S|^{-1} \sum_{g \in S} g$ in $R\Gamma$. When S is a subgroup, e_S is an idempotent. If R is not a field, an $R\Gamma$-module which is free as an R-module will be called an *$R\Gamma$-lattice.*

We shall identify isomorphism classes of irreducible representations over K with their character. We denote by $R\,\mathrm{Irr}(\Gamma)$ the R-module of class functions $\Gamma \to R$. When R is a field containing K, $R\,\mathrm{Irr}(\Gamma)$ is endowed with the canonical scalar product $\langle \bullet, \bullet \rangle_\Gamma$, for which the set of irreducible characters $\mathrm{Irr}(K\Gamma)$ of $K\Gamma$ is an orthonormal basis.

3.2. Unipotent KG-modules.

By a *rational group* we will mean a pair $(\mathbf{G}, F)$ consisting of an algebraic group $\mathbf{G}$ over $\overline{\mathbb{F}}_q$ and a Frobenius endomorphism F of $\mathbf{G}$. Assume that $\mathbf{G}$ is a connected reductive group. Fix a parabolic subgroup $\mathbf{P}$ of $\mathbf{G}$ and an F-stable Levi complement $\mathbf{L}$ of $\mathbf{P}$. We *do not* assume $\mathbf{P}$ to be F-stable. Write $L = \mathbf{L}^F$ and $G = \mathbf{G}^F$.

Let $R^{\mathbf{G}}_{\mathbf{L}\subset\mathbf{P}}$ and ${}^*R^{\mathbf{G}}_{\mathbf{L}\subset\mathbf{P}}$ denote respectively the Lusztig induction and restriction maps from $\mathbb{Z}\operatorname{Irr}(L)$ to $\mathbb{Z}\operatorname{Irr}(G)$. We will assume that the Mackey formula holds for $R^{\mathbf{G}}_{\mathbf{L}\subset\mathbf{P}}$ and ${}^*R^{\mathbf{G}}_{\mathbf{L}\subset\mathbf{P}}$, which we know for the groups we will focus on later (see [**2**] for more details). Under this condition, the Lusztig induction and restriction do not depend on the choice of the parabolic subgroup $\mathbf{P}$. We abbreviate $R^{\mathbf{G}}_{\mathbf{L}\subset\mathbf{P}} = R^{\mathbf{G}}_{\mathbf{L}}$ and ${}^*R^{\mathbf{G}}_{\mathbf{L}\subset\mathbf{P}} = {}^*R^{\mathbf{G}}_{\mathbf{L}}$.

Let $\mathbf{B}$ be an F-stable Borel subgroup of $\mathbf{G}$ and $\mathbf{T}$ be an F-stable maximal torus of $\mathbf{G}$ with $\mathbf{T} \subset \mathbf{B}$. Let $\mathbf{N}$ be the normalizer of $\mathbf{T}$ in $\mathbf{G}$. Write $B = \mathbf{B}^F$, $T = \mathbf{T}^F$ and $N = \mathbf{N}^F$. The groups $\mathbf{B}$, $\mathbf{N}$ form a reductive BN-pair of $\mathbf{G}$ with Weyl group $\mathbf{W} = \mathbf{W}_{\mathbf{G}}$ given by $\mathbf{W} = \mathbf{N}/\mathbf{T}$. Since $\mathbf{B}$, $\mathbf{N}$ are stable by F and $\mathbf{G}$ is connected, the finite groups B, N form a split BN-pair of G whose Weyl group $W = W(T)$ is given by $W = \mathbf{W}^F = N/T$.

The G-conjugacy classes of F-stable maximal tori of $\mathbf{G}$ are parametrized by the F-conjugacy classes in $\mathbf{W}$. For each $w \in \mathbf{W}$ let $\mathbf{T}_w$ be an F-stable maximal torus in the G-conjugacy class parametrized by the F-conjugacy class of w. Under conjugation by some element of $\mathbf{G}$, the pair $(\mathbf{T}_w, F)$ is identified with the pair $(\mathbf{T}, wF)$. In particular, we have $T_w \simeq \mathbf{T}^{wF}$ and $W(T_w) \simeq \mathbf{W}^{wF}$. The virtual characters $R^{\mathbf{G}}_{\mathbf{T}_w}(1)$ obtained by induction of the trivial representation of the tori T_w are called the *virtual Deligne-Lusztig characters*.

Definition 3.1. An irreducible KG-module is *unipotent* if its character, say χ, occurs as a constituent of a virtual Deligne-Lusztig character $R^{\mathbf{G}}_{\mathbf{T}_w}(1)$ for some element $w \in W$, *i.e.*, if we have $\langle \chi, R^{\mathbf{G}}_{\mathbf{T}_w}(1)\rangle_G \neq 0$.

We denote by KG-umod the full subcategory of KG-mod consisting of the modules which are sums of irreducible unipotent modules. The objects of this category are the *unipotent KG-modules.*

Remark 3.2. Fix an element σ in the normalizer of $\mathbf{L}$ in $\mathbf{G}$ such that the conjugation by σ yields an element in the normalizer of $\mathbf{W}_{\mathbf{L}}$ in $\mathbf{W}$. Let us denote it again by the symbol σ. Then, we can equip the subgroup $\mathbf{L} \subseteq \mathbf{G}$ with the rational structure with the Frobenius endomorphism σF and we have Lusztig induction and restriction maps $R^{\mathbf{G},F}_{\mathbf{L},\sigma F}$ and ${}^*R^{\mathbf{G},F}_{\mathbf{L},\sigma F}$ between $\mathbb{Z}\operatorname{Irr}(\mathbf{L}^{\sigma F})$ and $\mathbb{Z}\operatorname{Irr}(G)$. In the particular case $\sigma = 1$ considered above we have $R^{\mathbf{G},F}_{\mathbf{L},F} = R^{\mathbf{G}}_{\mathbf{L}}$ and ${}^*R^{\mathbf{G},F}_{\mathbf{L},F} = {}^*R^{\mathbf{G}}_{\mathbf{L}}$. If $\mathbf{T}_w$ is as above then we have also $R^{\mathbf{G},F}_{\mathbf{T},wF} = R^{\mathbf{G}}_{\mathbf{T}_w}$ and ${}^*R^{\mathbf{G},F}_{\mathbf{T},wF} = {}^*R^{\mathbf{G}}_{\mathbf{T}_w}$.

3.3. Unipotent kG-modules and ℓ-blocks.

As a result of the lifting of idempotents, the blocks of $\mathcal{O}G$ and kG correspond by reduction. Both are usually called the *ℓ-blocks* of G. For $R = \mathcal{O}$ or k, any block B of RG is of the form $B = RG \cdot b$, where b is a central primitive idempotent of RG. The unit b of B is called the *block idempotent of B*. We will also call *block of RG-mod associated with B* the Serre subcategory generated by the simple modules on which b acts non-trivially. The ℓ-blocks of G induce a partition of $\operatorname{Irr}(KG)$ such that the piece associated with B is the set of all irreducible characters χ of KG with $\chi(b) = \chi(1)$. If $\chi \in \operatorname{Irr}(KG)$, we will write $B(\chi) \subseteq \operatorname{Irr}(KG)$ for the piece containing χ. When there is no risk of confusion, we will also call $B(\chi)$ an ℓ-block of G.

DEFINITION 3.3. An ℓ-block of $\mathscr{O}G$ is *unipotent* if it contains at least one unipotent KG-module. A simple kG-module is *unipotent* if it lies in a unipotent block of kG.

We denote by kG-umod the Serre subcategory of kG-mod generated by the simple unipotent kG-modules. It corresponds to the sum of unipotent blocks of kG-mod. The *unipotent* kG*-modules* are by definition the objects of this category.

Recall that $(K, \mathscr{O}, \mathrm{k})$ is a splitting ℓ-modular system for G. To this system one can associate a decomposition map $d_{\mathscr{O}G} : [KG\text{-mod}] \to [\mathrm{k}G\text{-mod}]$. By [**40**, thm. 3.1], a simple kG-module is unipotent if and only if it is a constituent of the ℓ-reduction of a unipotent KG-module, see also [**5**]. In other words, the classes of unipotent modules are exactly the image of unipotent characters through the decomposition map. We will denote by $d_{\mathscr{U}} : [KG\text{-umod}] \to [\mathrm{k}G\text{-umod}]$ the restriction of this map to unipotent characters.

THEOREM 3.4 (Geck-Hiss [**29**], Geck [**27**]). *Assume ℓ is good for* $\mathbf{G}$ *and that* $\ell \nmid |Z(\mathbf{G})/Z(\mathbf{G})^\circ|$. *Then the map $d_{\mathscr{U}}$ is a linear isomorphism* $[KG\text{-umod}] \xrightarrow{\sim} [\mathrm{k}G\text{-umod}]$. □

Note that if $\mathbf{G}$ is semisimple with no component of type A, then the condition $\ell \nmid |Z(\mathbf{G})/Z(\mathbf{G})^\circ|$ is superfluous since every prime divisor of $|Z(\mathbf{G})/Z(\mathbf{G})^\circ|$ is a bad prime.

Given a positive integer f, let Φ_f be the fth cyclotomic polynomial. If $\mathbf{T} \subset \mathbf{G}$ is an F-stable torus, then the order of $T := \mathbf{T}^F$ is a polynomial in q with coefficients in $\mathbb{Z}$. We say that $\mathbf{T}$ (and by extension T) is a Φ_f*-torus* if this polynomial is a power of Φ_f. An F-stable Levi subgroup $\mathbf{L} \subseteq \mathbf{G}$ is f*-split* if it is the centralizer in $\mathbf{G}$ of a Φ_f-torus. A *unipotent f-pair* is a pair $(\mathbf{L}, \chi)$ where $\mathbf{L}$ is an f-split Levi subgroup and χ is an irreducible unipotent KL-module. The pair $(\mathbf{L}, \chi)$ is f*-cuspidal* if for every proper f-split Levi subgroup $\mathbf{M} \subseteq \mathbf{L}$ we have ${}^*R_{\mathbf{M}}^{\mathbf{L}}(\chi) = 0$.

Now assume that f is the smallest positive integer such that ℓ divides $q^f - 1$. Under the assumption that ℓ is good, unipotent ℓ-blocks correspond to f-cuspidal f-pairs.

THEOREM 3.5 (Broué-Malle-Michel [**4**], Cabanes-Enguehard [**7**]). *Assume ℓ is good for* $\mathbf{G}$, *and $\ell \neq 3$ if* $\mathbf{G}$ *has a component of type* ${}^3D_4$. *Then there is a bijection between the G-conjugacy classes of unipotent f-cuspidal f-pairs and the set of unipotent ℓ-blocks of G which takes the class of $(\mathbf{L}, \chi)$ to the ℓ-block $B_{\mathbf{L},\chi}$ such that the irreducible unipotent characters in $B_{\mathbf{L},\chi}$ are exactly the irreducible constituents of $R_{\mathbf{L}}^{\mathbf{G}}(\chi)$.* □

3.4. Harish-Chandra series.

Assume now that the parabolic subgroup $\mathbf{P} \subseteq \mathbf{G}$ considered in §3.2 is F-stable. In that case the group L is G-conjugate to a standard Levi subgroup of G. Let R_L^G and ${}^*R_L^G$ be the corresponding Harish-Chandra induction and restriction functors between RL-mod and RG-mod. Let $P = \mathbf{P}^F$ and $U = \mathbf{U}^F$, where $\mathbf{U} \subset \mathbf{P}$ is the unipotent radical of $\mathbf{P}$. Notice that the Harish-Chandra induction is the special case of Lusztig induction for 1-split Levi subgroups.

The order of U is a power of q, hence it is invertible in R. Thus, for all $M \in RL$-mod, $N \in RG$-mod we have

$$R_L^G(M) = RG \cdot e_U \otimes_{RL} M \quad \text{and} \quad {}^*R_L^G(N) = e_U \cdot RG \otimes_{RG} N.$$

We will say that the functors R_L^G and ${}^*R_L^G$ are *represented* by the (RG, RL)-bimodule $RG \cdot e_U$ and the (RL, RG)-bimodule $e_U \cdot RG$ respectively.

Here are some well-known basic properties of the functors R_L^G, ${}^*R_L^G$.

(a) R_L^G, ${}^*R_L^G$ do not depend on P,
(b) R_L^G, ${}^*R_L^G$ are exact and left and right adjoint to one another,
(c) if $L \subseteq M \subseteq G$ are Levi subgroups there are isomorphisms of functors $R_L^G = R_M^G R_L^M$ and ${}^*R_L^G = {}^*R_L^M {}^*R_M^G$.

Let $R = K$ or k. An irreducible RG-module E is *cuspidal* if ${}^*R_L^G(E) = 0$ for all proper standard Levi subgroups $L \subsetneq G$. A *cuspidal pair of RG* is a pair (L, E) where L is as above and $E \in \mathrm{Irr}(RL)$ is cuspidal. Since the group L is uniquely recovered from E, from now on we may omit it from the notation. Then, the set $\mathrm{Irr}(RG, E) \subseteq \mathrm{Irr}(RG)$ consisting of the constituents of the head of $R_L^G(E)$ is equal to the set of constituents of the socle of $R_L^G(E)$ and is called the *Harish-Chandra series* of (L, E). The R-algebra $\mathscr{H}(RG, E) = \mathrm{End}_{RG}(R_L^G(E))^{\mathrm{op}}$ is its *ramified Hecke algebra*. We have the following facts:

(d) the Harish-Chandra series form a partition of $\mathrm{Irr}(RG)$,
(e) the functor $\mathfrak{F}_{R_L^G(E)}$ as defined in §1.1 yields a bijection

$$\mathrm{Irr}(RG, E) \stackrel{1:1}{\longleftrightarrow} \mathrm{Irr}(\mathscr{H}(RG, E)).$$

The following result is well-known. It follows from the fact that the ℓ-rational series are stable by Harish-Chandra induction (see for example [**3**, thm. 10.3] for a more general statement).

PROPOSITION 3.6 ([**3**]). *If $R = K$ or* k, *then the Harish-Chandra induction and restriction functors preserve the category of unipotent RG-modules.* □

4. Finite unitary groups

This section is devoted to the construction of categorical actions on the category of unipotent representations of finite unitary groups $\mathrm{GU}_n(q)$.

Let R be a commutative domain with unit. Under mild assumptions on R, we construct in §4.2 a representation datum on the abelian category

$$RG\text{-mod} := \bigoplus_{n \geqslant 0} \mathrm{GU}_n(q)\text{-mod}$$

given by Harish-Chandra induction and restriction. When R is an extension of $\mathbb{Q}_\ell$ or $\mathbb{F}_\ell$, the categorical datum restricts to the category $\mathscr{U}_R$ of unipotent representations of RG-mod. On this smaller category, the eigenvalues of X are powers of $-q$, which hints that the Lie algebra $\mathfrak{g}$ that should act on $[\mathscr{U}_R]$ corresponds to the quiver with vertices $(-q)^{\mathbb{Z}}$ and arrows given by multiplication by q^2.

We prove that the representation datum lifts indeed to a categorical action of $\mathfrak{g}$ on $\mathscr{U}_R$. The situation depends on the parity of e, the order of $-q$ modulo ℓ. When

e is even, i.e., ℓ is a linear prime, $\mathfrak{g}$ is a subalgebra of $(\widehat{\mathfrak{sl}_{e/2}})^{\oplus 2}$ and each ordinary Harish-Chandra series categorifies a level 2 Fock space for $\mathfrak{g}$. When e is odd (unitary prime case), $\mathfrak{g} \simeq \widehat{\mathfrak{sl}_e}$ and weight spaces correspond to unipotent ℓ-blocks, which are now transverse to the ordinary Harish-Chandra series.

For studying the weight space decomposition of $[\mathscr{U}_R]$ as well as the action of the affine Weyl group we use the action of a bigger Lie algebra $\mathfrak{g}_\circ$, which comes from Harish-Chandra induction and restriction for general linear groups. Going from linear to unitary groups by Ennola duality introduces signs for this action (see Lemma 4.14). It is to be expected that the action of $\mathfrak{g}_\circ$ on $[\mathscr{U}_R]$ lifts to an action by triangulated functors on $D^b(\mathscr{U}_R)$ coming from Lusztig induction, although we will not use it. See [**19**] for more details.

4.1. Definition.

Fix a positive integer n. We equip the reductive algebraic group $\mathbf{GL}_n = \mathrm{GL}_n(\overline{\mathbb{F}}_q)$ with the standard Frobenius map $F_q : \mathbf{GL}_n \to \mathbf{GL}_n$, $(a_{ij}) \mapsto (a_{ij}^q)$ given by raising every coefficient to the qth power. The finite general linear group $\mathrm{GL_n}(q)$ is given by the fixed points of $\mathbf{GL}_n$ under F_q. In this section we will work with a twisted version of this group obtained by twisting the Frobenius map. Let J_n be the $n \times n$ matrix with entry 1 in $(i, n-i+1)$ and zero elsewhere. We will often write $J = J_n$ when there is no risk of confusion on the size of the matrices. We define a new Frobenius map F on $\mathbf{GL}_n$, called the *twisted* Frobenius map, by setting $F = F_q \circ \alpha$ where $\alpha(g) = J \cdot {}^t g^{-1} \cdot J$ for each $g \in \mathbf{GL}_n$. The *finite unitary group* $G_n = \mathrm{GU}_n(q)$ is then given by

$$G_n = (\mathbf{GL}_n)^F = \{g \in \mathbf{GL}_n\,;\, F(g) = g\}.$$

Since $F^2 = F_{q^2}$ we have $G_n \subset GL_n$, where we abbreviate

$$GL_n = \mathbf{GL}_n(q^2) := (\mathbf{GL}_n)^{F^2}.$$

By convention we also define $G_0 = \{1\}$ to be the trivial group.

We equip $\mathbf{GL}_n$ with the standard split BN-pair such that $\mathbf{B}$ is the subgroup of upper triangular matrices and $\mathbf{N}$ is the subgroup of all monomial matrices. Since $\mathbf{B}$, $\mathbf{N}$ are stable by F and $\mathbf{GL}_n$ is connected, the groups $B = \mathbf{B}^F$, $N = \mathbf{N}^F$ form a split BN-pair of the finite group G_n. Let $\mathbf{T}$ be the diagonal torus in $\mathbf{GL}_n$ and $T = \mathbf{T}^F$. Let $\mathbf{W} = \mathbf{W}_n$ be the Weyl group of $\mathbf{GL}_n$ and $W = W_n$ be the Weyl group of G_n. We have $\mathbf{W} \simeq \mathfrak{S}_n$, and F induces on $\mathbf{W}$ the automorphism given by conjugation with the longest element w_0. We will embed $\mathbf{W}$ in G_n using permutation matrices, so that w_0 corresponds to J. We have $W = \mathbf{W}^F = C_{\mathbf{W}}(w_0)$. It is a Weyl group of type B_m if $n = 2m$ or $2m+1$.

Let $\varepsilon_1, \ldots, \varepsilon_n$ be the characters of $\mathbf{T}$ such that $t = \mathrm{diag}(\varepsilon_1(t), \varepsilon_2(t), \ldots, \varepsilon_n(t))$ for all $t \in \mathbf{T}$. The roots (resp. simple roots) of $\mathbf{GL}_n$ are given by $\{\varepsilon_i - \varepsilon_j \,|\, i \neq j\}$ (resp. $\{\varepsilon_r - \varepsilon_{r+1}\}$). Let $s_r = (r, r+1)$ be the simple reflection in $\mathbf{W}$ associated with the simple root $\alpha_r = \varepsilon_r - \varepsilon_{r+1}$. The action of F on the roots induces an automorphism σ of the Dynkin diagram of $\mathbf{GL}_n$ such that

$$F \circ \alpha_r^\vee = q\sigma(\alpha_r)^\vee \quad \text{with} \quad \sigma(\alpha_r) = \alpha_{n-r}.$$

For each root α, let $\mathbf{U}_\alpha$ and $\alpha^\vee \in \mathrm{Hom}(\mathbb{G}_m, \mathbf{T})$ be the corresponding root subgroup and cocharacter. We also choose an isomorphism $u_\alpha : \mathbb{G}_a \xrightarrow{\sim} \mathbf{U}_\alpha$ such that

$F(u_{\sigma(\alpha)}(t)) = u_\alpha(-t^q)$. Note that a one-parameter subgroup of T has either (q^2-1) or $(q+1)$ elements, and a root subgroup of G_n has either q or q^2 elements.

The standard Levi subgroups $L_{r,\lambda}$ of G_n are parametrized by pairs (r,λ) where r is a non-negative integer and $\lambda = (\lambda_1, \lambda_2, \dots, \lambda_s)$ is a tuple of positive integers such that $n = r + 2\sum_{u=1}^{s} \lambda_u$. The group $L_{r,\lambda}$ consists of all matrices of $\mathbf{GL}_n$ which belong to G_n and are of block-diagonal type

$$\prod_{u=1}^{s} \mathbf{GL}_{\lambda_u} \times \mathbf{GL}_r \times \prod_{u=s}^{1} \mathbf{GL}_{\lambda_u}.$$

Consequently we have a group isomorphism $L_{r,\lambda} \simeq G_r \times \prod_u GL_{\lambda_u}$. If m is a positive integer we abbreviate $L_{r,1^m} = L_{r,(1^m)}$ and $L_{r,m} = L_{r,(m)}$.

4.2. The representation datum on RG-mod.

Let R be a commutative domain with unit. We assume that $q(q^2-1)$ in invertible in R. Using parabolic induction and restriction, we show in this section how to construct a representation datum on

$$RG\text{-mod} := \bigoplus_{n \in \mathbb{N}} RG_n\text{-mod}.$$

Fix a positive integer n. Parabolic (or Harish-Chandra) induction provides functors between L-mod and G_n-mod for any standard Levi subgroup $L = L_{r,m} \subset G_n$. Since we want functors between G_r-mod and G_n-mod we will consider a slight variation of the usual parabolic induction.

Let $0 \leqslant r < n$. We denote by V_r the unipotent radical of the standard parabolic subgroup $P_{r,1} \subset G_{r+2}$ with Levi complement $L_{r,1}$. Let $U_r \subset G_{r+2}$ be the subgroup given by

$$U_r = V_r \rtimes \mathbb{F}_{q^2}^\times = \begin{pmatrix} * & * & \cdots & \cdots & * \\ & 1 & & (0) & \vdots \\ & & \ddots & & \vdots \\ & (0) & & 1 & * \\ & & & & * \end{pmatrix} \tag{4.1}$$

so that $P_{r,1} = V_r \rtimes L_{r,1} \simeq U_r \rtimes G_r$. If $n-r$ is even, we set $U_{n,r} = U_{n-2} \rtimes \cdots \rtimes U_r$ and we define $e_{n,r} = e_{U_{n,r}}$ to be the corresponding idempotent. In particular, we have $e_{r+2,r} = e_{U_r}$. We embed G_r into the Levi subgroup $L_{r,1^m} = G_r \times GL_1^m$ in the obvious way. This yields an embedding $G_r \subset G_n$ and functors

$$\begin{aligned} F_{n,r} &= RG_n \cdot e_{n,r} \otimes_{RG_r} \bullet \; : \; RG_r\text{-mod} \longrightarrow RG_n\text{-mod}, \\ E_{r,n} &= e_{n,r} \cdot RG_n \otimes_{RG_n} \bullet \; : \; RG_n\text{-mod} \longrightarrow RG_r\text{-mod}. \end{aligned}$$

Note that $F_{n,r}$ can be seen as the composition of the inflation $G_r\text{-mod} \longrightarrow L_{r,1^m}\text{-mod}$ with the parabolic induction from $L_{r,1^m}$ to G_n.

An endomorphism of the functor $F_{n,r}$ can be represented by an (RG_n, RG_r)-bimodule endomorphism of $RG_n \cdot e_{n,r}$, or equivalently by an element of $e_{n,r} \cdot RG_n \cdot e_{n,r}$ centralizing RG_r. Thus, the elements

$$X_{r+2,r} = (-q)^r e_{r+2,r}(1, r+2)\, e_{r+2,r}, \qquad T_{r+4,r} = q^2 e_{r+4,r}(1,2)(r+3, r+4)\, e_{r+4,r}$$

define respectively natural transformations of the functors $F_{r+2,r}$ and $F_{r+4,r}$. We set

$$F = \bigoplus_{r\geqslant 0} F_{r+2,r}, \qquad X = \bigoplus_{r\geqslant 0} X_{r+2,r}, \qquad T = \bigoplus_{r\geqslant 0} T_{r+4,r}.$$

PROPOSITION 4.1. *The tuple (E,F,X,T) defines a representation datum with parameter q^2 on RG-mod $= \bigoplus_{n\in\mathbb{N}} RG_n$-mod.* □

4.3. The categories of unipotent modules $\mathscr{U}_K$ and $\mathscr{U}_{\mathrm{k}}$.

From now on, we fix a prime number ℓ such that $\ell \nmid q\,(q^2-1)$, and an ℓ-modular system $(K,\mathscr{O},\mathrm{k})$ with $\mathbb{Q}_\ell \subset K$ and $\overline{\mathbb{F}}_\ell \subset \mathrm{k}$. Although this modular system is not large enough for all the finite groups encountered, it will be enough for our purpose since we will be working with unipotent characters only, which are defined over $\mathbb{Q}_\ell$ (see below).

Throughout the following sections, we will denote by d, e and f the order of q^2, $-q$ and q in $\mathrm{k}^\times$. In particular $e \neq 1,2$. If e is odd, then $d = e$ and $f = 2e$; if e is even, then $d = e/2$ and either $f = e/2$ if $e \equiv 2 \bmod 4$ or $f = e$ if $e \equiv 0 \bmod 4$.

4.3.1. *The category $\mathscr{U}_K$.* Fix a positive integer n. By [**56**], the irreducible unipotent KG_n-modules are labelled by partitions of n. Their character can be directly constructed from the virtual Deligne-Lusztig characters. Namely, for each $w \in \mathfrak{S}_n$, fix an F-stable maximal torus $\mathbf{T}_w \subseteq \mathbf{GL}_n$ in the G_n-conjugacy class parametrized by the $\mathfrak{S}_n$-conjugacy class of wF in the coset $\mathfrak{S}_nF$, with the convention that $\mathbf{T}_1 = \mathbf{T}$. Then the class function

$$\chi_\lambda = |\mathfrak{S}_n|^{-1} \sum_{w\in\mathfrak{S}_n} \phi_\lambda(ww_0) R_{\mathbf{T}_w}^{\mathbf{GL}_n}(1) \in \mathbb{Z}\mathrm{Irr}(G_n) \tag{4.2}$$

is, up to a sign, an irreducible unipotent character. Since the cohomology groups are defined over $\mathbb{Q}_\ell \subset K$ and since any character of $\mathfrak{S}_n$ has rational values, either χ_λ or $-\chi_\lambda$ is the character of some irreducible KG_n-module which we denote by E_λ. By abuse of notation we will still denote by E_λ its isomorphism class. Let ε_λ be the sign determined by the relation $\chi_\lambda = \varepsilon_\lambda\, E_\lambda$.

Recall that $G_0 = \{1\}$. We call the category of unipotent KG-modules the category

$$\mathscr{U}_K := \bigoplus_{n\in\mathbb{N}} KG_n\text{-umod}.$$

This category is abelian semisimple. From the previous discussion we have

$$\mathrm{Irr}(\mathscr{U}_K) = \{E_\lambda \,|\, \lambda \in \mathscr{P}\},$$

where by convention $\mathrm{Irr}(KG_0) = \{E_\emptyset\}$.

4.3.2. *The category $\mathscr{U}_{\mathrm{k}}$.* Using the ℓ-modular system we have decomposition maps $d_{\mathscr{O}G_n}$ which by Theorem 3.4 restrict to linear isomorphisms

$$d_{\mathscr{O}G_n} : [KG_n\text{-umod}] \xrightarrow{\sim} [\mathrm{k}G_n\text{-umod}].$$

Note that we do not need to assume that KG_n is a split algebra since we are only interested in unipotent KG_n-modules, which are all absolutely irreducible.

For unitary groups, this map is actually unitriangular with respect to the basis of irreducible modules and the dominance order.

THEOREM 4.2 ([**26**]). *There is a unique labelling $\{D_\lambda\}_{\lambda\in\mathscr{P}_n}$ of the unipotent simple* $\mathrm{k}G_n$*-modules such that* $d_{\mathscr{O}G_n}([E_\lambda]) \in [D_\lambda] + \sum_{\nu>\lambda}\mathbb{Z}[D_\nu]$ *where $>$ is the dominance order on partitions of n.* □

We call the category of unipotent $\mathrm{k}G$-modules the category

$$\mathscr{U}_{\mathrm{k}} := \bigoplus_{n\in\mathbb{N}} \mathrm{k}G_n\text{-umod}\,.$$

This is an abelian category which is not semisimple. The isomorphism classes of simple objects are $\mathrm{Irr}(\mathscr{U}_{\mathrm{k}}) = \{D_\lambda\}_{\lambda\in\mathscr{P}}$ and the decomposition map yields a $\mathbb{Z}$-linear isomorphism $d_{\mathscr{U}} : [\mathscr{U}_K] \xrightarrow{\sim} [\mathscr{U}_{\mathrm{k}}]$. The following result is a consequence of the unitriangularity of this map.

PROPOSITION 4.3. *Given a partition $\lambda \vdash n$ there is a unique $\mathscr{O}G_n$-lattice $\widetilde{E}_\lambda$ such that*

(a) $K\widetilde{E}_\lambda = E_\lambda$ *as a* KG_n*-module,*
(b) $V_\lambda := \mathrm{k}\widetilde{E}_\lambda$ *is an indecomposable* $\mathrm{k}G_n$*-module with head isomorphic to* D_λ*,*
(c) $d_{\mathscr{U}}([E_\lambda]) = [V_\lambda]$. □

4.3.3. *Blocks of $\mathscr{U}_{\mathrm{k}}$.* A block of $\mathscr{U}_{\mathrm{k}}$ is an indecomposable summand of $\mathscr{U}_{\mathrm{k}}$. Therefore blocks of $\mathscr{U}_{\mathrm{k}}$ correspond to the unipotent blocks of $\mathrm{k}G_n$ where n runs over $\mathbb{N}$. Recall that e is the order of $-q$ modulo ℓ.

PROPOSITION 4.4 ([**22**]). *The map $E_\lambda \longmapsto (\lambda_{[e]}, w_e(\lambda))$ yields a bijection between unipotent ℓ-blocks and pairs (s,w) where $s\in\mathbb{Z}^e(0)$ and $w\in\mathbb{N}$.* □

Recall that $\nu\mapsto\nu_{[e]}$ induces a bijection between e-cores and $\mathbb{Z}^e(0)$. Given ν an e-core we will denote by $B_{\nu,w}$ or $B_{\nu_{[e]},w}$ the unipotent ℓ-block containing all the unipotent characters E_λ such that ν is the e-core of λ and $w_e(\lambda)=w$. It is a block of $\mathrm{k}G_n$ with $n=|\nu|+ew$.

It also follows from the classification of blocks that when $w<e$ and $e\neq 1$, the defect group of $B_{\nu,w}$ is an elementary abelian ℓ-group of rank w (see [**4**]). In particular, when $w=0$ the module D_λ is simple and projective, isomorphic to V_λ, and when $w=1$ the defect group of $B_{\nu,w}$ is a cyclic group. The structure of such blocks was determined in [**24**].

4.3.4. *The weak Harish-Chandra series.* For this section we assume that R is one of the fields K or k.

Let $r,m\geqslant 0$ and $n=r+2m$. The inflation from G_r to $L_{r,1^m}$ yields an equivalence between RG_r-umod and $RL_{r,1^m}$-umod, since the Deligne-Lusztig varieties depend only on the semisimple type of the reductive group. This equivalence intertwines the functors $E_{n,r}$, $F_{n,r}$ with the parabolic restriction and induction ${}^*R^{G_n}_{L_{r,1^m}}$, $R^{G_n}_{L_{r,1^m}}$. Therefore working with $\mathscr{U}_R$ and the functors E and F is the same as working in the usual framework of unipotent representations and Harish-Chandra induction/restriction from Levi subgroups. Note however that we do not consider all the standard Levi subgroups, but only the ones that are conjugate to $L_{r,1^m}$. Therefore we need to consider a slight variation of the usual Harish-Chandra theory. The following definitions and properties are taken from [**36**].

DEFINITION 4.5. Fix a non-negative integer n.

(a) An RG_n-module D is *weakly cuspidal* if ${}^*R^{G_n}_L(D)=0$ for any Levi subgroup $L\subsetneq G_n$ which is G_n-conjugate to a subgroup of the form $L_{r,1^m}$.

(b) A *weakly cuspidal pair* of RG_n is a pair which is G_n-conjugate to $(L_{r,1^m}, D)$ for some $n = r + 2m$ and some weakly cuspidal irreducible RG_r-module D which is viewed as a $\mathrm{k}L_{r,1^m}$-module by inflation.

(c) The *weak Harish-Chandra series* $\mathrm{WIrr}(RG_n, D)$ of RG_n determined by the weakly cuspidal pair (L, D) is the set of the constituents of the head of $R_L^{G_n}(D)$. By [**36**], it coincides with the set of the constituents of its socle.

If D is a unipotent RG_r-module, then D is weakly cuspidal if and only if $E(D) = 0$. Moreover, if D is irreducible, the weak Harish-Chandra series coincides with the set of irreducible constituents in the head of $F^m D$. Therefore it makes sense to define the weak Harish-Chandra series of D in $\mathscr{U}_R$ by

$$\mathrm{WIrr}(RG, D) = \bigsqcup_{n \geqslant r} \mathrm{WIrr}(RG_n, D).$$

It is the set of irreducible constituents in the head of $F^k D$ for some $k \geqslant 0$ (or equivalently in the socle). As in the case of the usual theory, the weak Harish-Chandra series of $\mathscr{U}_R$ form a partition of $\mathrm{Irr}(\mathscr{U}_R)$, i.e., we have $\mathrm{Irr}(\mathscr{U}_R) = \bigsqcup \mathrm{WIrr}(RG, D)$ where the sum runs over the set of isomorphism classes of weakly cuspidal unipotent irreducible modules D.

When $R = K$, weakly cuspidal unipotent modules coincide with cuspidal modules and were determined in [**53**]. One of the main results in this paper is the classification of the weakly cuspidal modules in the case where $R = \mathrm{k}$.

4.4. The $\mathfrak{g}_\infty$-representation on $\mathscr{U}_K$.

In this section we show how the categorical datum defined in §4.2 yields a categorical representation on $\mathscr{U}_R$ in the case where $R = K$. This is achieved by translating into our framework the theory of Lusztig and Howlett-Lehrer on endomorphism algebras of representations induced from a cuspidal module in [**53**] and [**42**].

4.4.1. *Action of E and F.* Since every unipotent character is a linear combination of (virtual) Deligne-Lusztig characters – we call such a class function a *uniform function* – it is well-known how to compute the action of E and F on the category $\mathscr{U}_K$.

LEMMA 4.6. *Let λ be a partition. Then $[E]([E_\lambda]) = \sum_\mu [E_\mu]$ and $[F]([E_\lambda]) = \sum_{\mu'} [E_{\mu'}]$ where μ (resp. μ') runs over the partitions that are obtained from λ by removing (resp. adding) a 2-hook.* □

We deduce that a unipotent KG_n module E_λ is weakly cuspidal if and only if there exists $t \geqslant 0$ such that $n = t(t+1)/2$ and $\lambda = \Delta_t = (t, t-1, \ldots, 1)$. We will set $E_t := E_{\Delta_t}$. Consequently, the weakly cuspidal pairs are, up to conjugation, the pairs $(L_{r,1^m}, E_t)$ with $r = t(t+1)/2$. The partition into series is thus $\mathrm{Irr}(\mathscr{U}_K) = \bigsqcup_{t \in \mathbb{N}} \mathrm{WIrr}(KG, E_t)$.

4.4.2. *The ramified Hecke algebra.* Let r, m be non-negative integers and $n = r + 2m$. As mentioned in §4.3.4, the inflation from G_r to $L_{r,1^m}$ yields an equivalence between KG_r-umod and $KL_{r,1^m}$-umod which intertwines the functor $F_{n,r}$ with the parabolic induction $R_{L_{r,1^m}}^{G_n}$. In particular, we have a canonical isomorphism

$$\mathscr{H}(KG_n, E_t) := \mathrm{End}_{KG_n}(F^m(E_t))^{\mathrm{op}} \xrightarrow{\sim} \mathrm{End}_{KG_n}(R_{L_{r,1^m}}^{G_n}(E_t))^{\mathrm{op}}.$$

Now recall from §1.3.1 that to the categorical datum (E, F, X, T) is attached a map $\phi_{F^m} : \mathbf{H}^{q^2}_{K,m} \to \mathrm{End}(F^m)$. The evaluation of this map at the module E_t yields a K-algebra homomorphism

$$\phi_{K,m} : \mathbf{H}^{q^2}_{K,m} \to \mathscr{H}(KG_n, E_t), \quad X_k \mapsto X_k(E_t), \quad T_l \mapsto T_l(E_t).$$

By [**53**], $\mathscr{H}(KG_n, E_t)$ is isomorphic to a Hecke algebra $\mathbf{H}^{Q_t;\,q^2}_{K,m}$ of type B_m with

$$Q_t = \begin{cases} \big((-q)^{-1-t}\,,\,(-q)^t\big) & \text{if } t \text{ is even,} \\ \big((-q)^t\,,\,(-q)^{-1-t}\big) & \text{if } t \text{ is odd.} \end{cases} \tag{4.3}$$

The previous map provides indeed such an isomorphism.

THEOREM 4.7. *Let $t, m \geqslant 0$ and $n = t(t+1)/2 + 2m$. Then the map $\phi_{K,m}$ factors through a K-algebra isomorphism $\mathbf{H}^{Q_t;\,q^2}_{K,m} \xrightarrow{\sim} \mathscr{H}(KG_n, E_t)$.* □

4.4.3. *Parametrization of the weak Harish-Chandra series of $\mathscr{U}_K$.* Let $W(B_m)$ be the Weyl group of type B_m, and $t_0, t_1, \ldots, t_{m-1}$ be the generators corresponding to the following Dynkin diagram

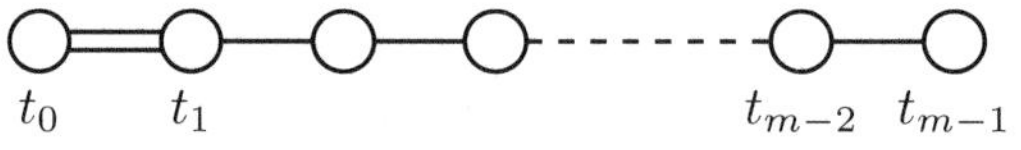

Let us first recall the construction of the irreducible characters of $W(B_m)$. We denote by σ_m the linear character of $W(B_m)$ such that $\sigma_m(t_0) = -1$ and $\sigma_m(t_i) = 1$ for all $i > 0$. Given $\lambda \vdash m$ a partition of m, we write $\widetilde{\phi}_\lambda$ for the inflation to $W(B_m)$ of the irreducible character of $\mathfrak{S}_m$ corresponding to λ. Given a, b such that $a + b = m$, one can consider the subgroups $W(B_a) \times \mathfrak{S}_b \subset W(B_a) \times W(B_b)$ of $W(B_m)$ where $W(B_a) \times \mathfrak{S}_b$ is a parabolic subgroup generated by $\{t_0, t_1, \ldots, t_{a-1}\} \cup \{t_{a+1}, \ldots, t_{m-1}\}$, and $W(B_a) \times W(B_b)$ is obtained by adding the reflection $t_a \cdots t_0 \cdots t_a$. The irreducible character of $W(B_m)$ associated to a bipartition (λ, μ) of m is

$$\mathscr{X}_{\lambda,\mu} = \mathrm{Ind}^{W(B_m)}_{W(B_{|\lambda|}) \times W(B_{|\mu|})} (\widetilde{\phi}_\lambda \boxtimes \sigma_{|\mu|} \widetilde{\phi}_\mu).$$

By Tits deformation theorem, the specialization $q \mapsto 1$ yields a bijection

$$\mathrm{Irr}(\mathbf{H}^{Q_t;\,q^2}_{K,m}) \xrightarrow{\sim} \mathrm{Irr}(W(B_m))$$

from which we obtain a canonical labelling of the irreducible representations of $\mathbf{H}^{Q_t;\,q^2}_{K,m}$. We write $\mathrm{Irr}(\mathbf{H}^{Q_t;\,q^2}_{K,m}) = \{S(\lambda, \mu)^{Q_t;\,q^2}_K\}_{(\lambda,\mu) \in \mathscr{P}^2_m}$. Setting $T_0 = (-1)^t q^{-1-t} X$, we have now the quadratic relation

$$(T_0 + 1)(T_0 - q^{2t+1}) = 0.$$

Using this generator instead of X, we obtain the usual presentation for a Hecke algebra of type B_m with parameters (q^{2t+1}, q^2). The endomorphism of $KW(B_m)$ which is obtained from the renormalization $\mathbf{H}^{Q_t;\,q^2}_{K,m} \xrightarrow{\sim} \mathbf{H}^{(q^{2t+1},1);\,q^2}_{K,m}$ at $q = 1$ is the identity on $\mathfrak{S}_m$ but sends t_0 to $(-1)^t t_0$. Therefore this renormalization sends $S(\lambda, \mu)$ to $S(\lambda, \mu)$ if t is even, and to $S(\mu, \lambda)$ if t is odd.

COROLLARY 4.8. *Let $t, m \geqslant 0$ and $n = t(t+1)/2 + 2m$. Then the map $\phi_{K,m}$ and the functor $\mathfrak{F}_{F^m(E_t)}$ induce a bijection $\mathrm{WIrr}(KG_n, E_t) \xleftrightarrow{1:1} \mathrm{Irr}(\mathbf{H}^{Q_t;\,q^2}_{K,m})$ sending E_λ to $S(\lambda^{[2]})^{Q_t,q^2}_K$ for all partitions $\lambda \vdash n$ with 2-core $\Delta_t = (t, t-1, \ldots, 1)$.* □

4.4.4. *The $\mathfrak{g}_\infty$-representation on $\mathscr{U}_K$.* The functors E, F preserve the subcategory $\mathscr{U}_K$ by Proposition 3.6, hence (E, F, X, T) yields a representation datum on $\mathscr{U}_K$. In order to extend it to a categorical representation on $\mathscr{U}_K$, one should consider the quiver $I(q^2)$ with vertices given by the various eigenvalues of X and arrows $i \longrightarrow q^2 i$. In this section we will view the integer q as an element of $K^\times$ in the obvious way.

DEFINITION 4.9. Let I_∞ denote the subset $(-q)^{\mathbb{Z}}$ of $K^\times$. We define $\mathfrak{g}_\infty$ to be the (derived) Kac-Moody algebra associated to the quiver $I_\infty(q^2)$.

To avoid cumbersome notation, we will write for short $I_\infty = I_\infty(q^2)$, and $(-)_\infty = (-)_{I_\infty}$. We denote by $\{\Lambda_i\}$, $\{\alpha_i\}$ and $\{\alpha_i^\vee\}$ the fundamental weights, simple roots and simple coroots of $\mathfrak{g}_\infty$. Here X_∞ coincides with $\mathrm{P}_\infty = \bigoplus \mathbb{Z}\Lambda_i$. Consequently, there is a Lie algebra isomorphism $(\mathfrak{sl}_{\mathbb{Z}})^{\oplus 2} \overset{\sim}{\to} \mathfrak{g}_\infty$ such that $(\alpha_d^\vee, 0) \longmapsto \alpha^\vee_{-q^{2d-1}}$ and $(0, \alpha_d^\vee) \longmapsto \alpha^\vee_{q^{2d}}$.

For any $t, m, n \in \mathbb{N}$, let (KG_n, E_t)-mod be the Serre subcategory of $\mathscr{U}_K$ generated by the modules $F^m(E_t)$ with $n = r + 2m$ and $r = t(t+1)/2$. We define

$$\mathscr{U}_{K,t} = \bigoplus_{n \geqslant 0} (KG_n, E_t)\text{-mod}.$$

Then $\mathscr{U}_K = \bigoplus_{t \geqslant 0} \mathscr{U}_{K,t}$. We obtain our first categorification result.

THEOREM 4.10. *Let $t \geqslant 0$ and Q_t be as in (4.3).*

(a) *The Harish-Chandra induction and restriction functors yield a representation of $\mathfrak{g}_\infty$ on $\mathscr{U}_{K,t}$ which is isomorphic to $\mathscr{L}(\Lambda_{Q_t})_\infty$.*
(b) *The map $|\mu, Q_t\rangle_\infty \longmapsto [E_{\varpi_t(\mu)}]$ induces an isomorphism of $\mathfrak{g}_\infty$-modules $\mathbf{F}(Q_t)_\infty \overset{\sim}{\to} [\mathscr{U}_{K,t}]$.* □

4.5. The $\mathfrak{g}_e$-representation on $\mathscr{U}_\mathrm{k}$.

By Proposition 3.6, the representation datum (E, F, X, T) on $\mathrm{k}G$-mod induces a representation datum on $\mathscr{U}_\mathrm{k}$. Since the abelian category $\mathscr{U}_\mathrm{k}$ is not semisimple, to extend the representation datum to a categorical $\mathfrak{g}$-representation one needs to prove that weight spaces of $\mathscr{U}_\mathrm{k}$ are sums of blocks. This will be done combinatorially by studying a representation of a bigger Lie algebra $\mathfrak{g}_{e,\circ}$.

4.5.1. *The Lie algebras $\mathfrak{g}_e$ and $\mathfrak{g}_{e,\circ}$.* The eigenvalues of X on E and F are all powers of $-q$. If we denote again by q the image of q under the canonical map $\mathscr{O} \twoheadrightarrow \mathrm{k}$, then the eigenvalues of X on $\mathrm{k}E$ and $\mathrm{k}F$ belong to the finite set $(-q)^{\mathbb{Z}} \subset \mathrm{k}^\times$. This set has exactly e elements, where e is the order of $-q$ in $\mathrm{k}^\times$.

DEFINITION 4.11. We define I_e to be the subset $(-q)^{\mathbb{Z}}$ of $\mathrm{k}^\times$. We denote by I_e and $I_{e,\circ}$ the finite quivers $I_e(q^2)$ and $I_e(-q)$.

The quivers I_e and $I_{e,\circ}$ have the same set of vertices, but the arrows in I_e are the composition of two consecutive arrows in $I_{e,\circ}$. The quiver $I_{e,\circ}$ is cyclic, whereas the quiver I_e is cyclic if e is odd, and is a union of two cyclic quivers if e is even. Therefore the corresponding Kac-Moody algebras are isomorphic to $\widehat{\mathfrak{sl}}_e$ or $(\widehat{\mathfrak{sl}}_{e/2})^{\oplus 2}$.

To avoid cumbersome notation, we will write $(\bullet)_{e,\circ} = (\bullet)_{I_{e,\circ}}$ and $(\bullet)_e = (\bullet)_{I_e}$. We must introduce Lie algebras $\mathfrak{g}_{e,\circ}$ and $\mathfrak{g}_e$ such that $\mathfrak{g}'_{e,\circ} = [\mathfrak{g}_{e,\circ}, \mathfrak{g}_{e,\circ}]$ and $\mathfrak{g}'_e =$

$[\mathfrak{g}_e, \mathfrak{g}_e]$. The Chevalley generators of $\mathfrak{g}'_{e,\circ}$ and $\mathfrak{g}'_e$ are $e_{i,\circ}, f_{i,/\circ}$ and e_i, f_i respectively, for $i \in (-q)^{\mathbb{Z}}$. There is a Lie algebra homomorphism $\kappa : \mathfrak{g}'_e \longrightarrow \mathfrak{g}'_{e,\circ}$ defined by

$$\kappa(e_i) = [e_{-qi,\circ}, e_{i,\circ}] \quad \text{and} \quad \kappa(f_i) = [f_{-qi,\circ}, f_{i,\circ}].$$

It restricts to a map between the coroot lattices sending $\alpha_i^\vee$ to $\alpha_{i,\circ}^\vee + \alpha_{-qi,\circ}^\vee$

We denote by $\mathfrak{g}_{e,\circ}$ the Kac-Moody algebra associated with the lattices $\mathrm{X}_{e,\circ} = \mathrm{P}_{e,\circ} \oplus \mathbb{Z}\delta_\circ$ and $\mathrm{X}_{e,\circ}^\vee = \mathrm{Q}_{e,\circ}^\vee \oplus \mathbb{Z}\partial_\circ$, where $\delta_\circ = \sum \alpha_{i,\circ}$, $\partial_\circ = \Lambda_{1,\circ}^\vee$ and the pairing $\mathrm{X}_{e,\circ}^\vee \times \mathrm{X}_{e,\circ} \longrightarrow \mathbb{Z}$ is given by

$$\langle \alpha_{j,\circ}^\vee, \Lambda_{i,\circ}\rangle_{e,\circ} = \delta_{ij}, \quad \langle \partial_\circ, \Lambda_{i,\circ}\rangle_{e,\circ} = \langle \alpha_{j,\circ}^\vee, \delta_\circ\rangle_{e,\circ} = 0, \quad \langle \partial_\circ, \delta_\circ\rangle_{e,\circ} = 1.$$

Then $\mathfrak{g}_{e,\circ}$, $\mathfrak{g}'_{e,\circ}$ are isomorphic to $\widehat{\mathfrak{sl}}_e$, $\widetilde{\mathfrak{sl}}_e$ respectively.

Let $\widetilde{\mathfrak{g}}_e$ be the usual Kac-Moody algebra associated with I_e. Its derived Lie subalgebra is equal to $\mathfrak{g}'_e$. Let $\widetilde{\mathrm{X}}_e$ and $\widetilde{\mathrm{X}}_e^\vee$ be the lattices corresponding to $\widetilde{\mathfrak{g}}_e$. The map $\kappa : \mathfrak{g}'_e \longrightarrow \mathfrak{g}'_{e,\circ}$ may not extend to a morphism of Lie algebras $\widetilde{\mathfrak{g}}_e \longrightarrow \mathfrak{g}_{e,\circ}$. For this reason we will define $\mathfrak{g}_e$ to be the Lie subalgebra $\mathfrak{g}'_e \oplus \mathbb{C}\partial$ of $\widetilde{\mathfrak{g}}_e$, where ∂ is the element given by $\partial = \Lambda_1^\vee + \Lambda_{-q^{-1}}^\vee$. We can view $\mathfrak{g}_e$ as the Kac-Moody algebra associated with the lattice $\mathrm{X}_e^\vee = \mathrm{Q}_e^\vee \oplus \mathbb{Z}\partial$ above and a lattice $\mathrm{X}_e = \mathrm{P}_e \oplus \mathbb{Z}\delta$ that we define below case-by-case.

If e is odd, then I_e is a cyclic quiver and $\mathfrak{g}_e = \widetilde{\mathfrak{g}}_e$ is isomorphic to $\widehat{\mathfrak{sl}}_e$. Let α_1 be the affine root, then we have $\widetilde{\mathrm{X}}_e = \mathrm{P}_e \oplus \mathbb{Z}\tilde{\delta}$ and $\widetilde{\mathrm{X}}_e^\vee = \mathrm{Q}_e^\vee \oplus \mathbb{Z}\widetilde{\partial}$ with $\tilde{\delta} = \sum \alpha_i$ and $\widetilde{\partial} = \Lambda_1^\vee$. Then, we set $\mathrm{X}_e = \mathrm{P}_e \oplus \mathbb{Z}\delta$ and $\delta = \tilde{\delta}/2$.

If e is even, then I_e is the disjoint union of two cyclic quivers and $\tilde{\mathfrak{g}}_e$ is isomorphic to $(\widehat{\mathfrak{sl}}_{e/2})^{\oplus 2}$. Let α_1 and $\alpha_{-q^{-1}}$ be the affine roots, then we have $\widetilde{\mathrm{X}}_e = \mathrm{P}_e \oplus \mathbb{Z}\delta_1 \oplus \mathbb{Z}\delta_2$ and $\widetilde{\mathrm{X}}_e^\vee = \mathrm{Q}_e^\vee \oplus \mathbb{Z}\partial_1 \oplus \mathbb{Z}\partial_2$ with $\delta_1 = \sum_{j \text{ odd}} \alpha_{-q^j}$, $\delta_2 = \sum_{j \text{ even}} \alpha_{q^j}$, $\partial_1 = \Lambda_{-q^{-1}}^\vee$ and $\partial_2 = \Lambda_1^\vee$. Then, we set $\mathrm{X}_e = \widetilde{\mathrm{X}}_e/(\delta_1 - \delta_2)$ and $\delta = (\delta_1 + \delta_2)/2$.

In both cases the perfect pairing $\mathrm{X}_e^\vee \times \mathrm{X}_e \longrightarrow \mathbb{Z}$ is induced in the obvious way by the pairing $\widetilde{\mathrm{X}}_e^\vee \times \widetilde{\mathrm{X}}_e \longrightarrow \mathbb{Z}$.

LEMMA 4.12. *There is a well-defined morphism of Lie algebras $\mathfrak{g}_e \longrightarrow \mathfrak{g}_{e,\circ}$ which extends κ and whose restriction to $\mathrm{X}_e^\vee$ is given by $\kappa(\alpha_i^\vee) = \alpha_{i,\circ}^\vee + \alpha_{-qi,\circ}^\vee$ and $\kappa(\partial) = \partial_\circ$. The restriction $\kappa : \mathrm{X}_e^\vee \longrightarrow \mathrm{X}_{e,\circ}^\vee$ has an adjoint $\kappa^* : \mathrm{X}_{e,\circ} \longrightarrow \mathrm{X}_e$ such that $\kappa^*(\Lambda_{i,\circ}) \equiv \Lambda_i + \Lambda_{-q^{-1}i}$ mod δ and $\kappa^*(\delta_\circ) = \delta$.* □

4.5.2. *The $\mathfrak{g}'_e$-action on $[\mathscr{U}_{\mathrm{k}}]$.* The quotient map $\mathscr{O} \twoheadrightarrow \mathrm{k}$ induces a morphism of quivers $\mathrm{sp} : I_\infty \longrightarrow I_e$ and a surjective morphism of abelian groups $\mathrm{sp} : \mathrm{P}_\infty \twoheadrightarrow \mathrm{P}_e$ sending Λ_i to $\Lambda_{\mathrm{sp}(i)}$. In addition any integrable representation V of $\mathfrak{g}_\infty$ can be "restricted" to an integrable representation of the derived algebra $\mathfrak{g}'_e$, where $e_i \in \mathfrak{g}'_e$ (resp. $f_i \in \mathfrak{g}'_e$) act as $\sum_{\mathrm{sp}(j)=i} e_j$ (resp. $\sum_{\mathrm{sp}(j)=i} f_j$). From the definition of the action of $\mathfrak{g}_\infty$ and $\mathfrak{g}_e$ on Fock spaces, see (2.2), we deduce that the map $|\mu, Q_t\rangle_\infty \mapsto |\mu, Q_t\rangle_e$ induces the following isomorphism of $\mathfrak{g}'_e$-modules

$$\mathrm{sp} : \mathrm{Res}_{\mathfrak{g}'_e}^{\mathfrak{g}_\infty} \mathbf{F}(Q_t)_\infty \xrightarrow{\sim} \mathbf{F}(Q_t)_e. \tag{4.4}$$

Under the decomposition map, this isomorphism endows $[\mathscr{U}_{\mathrm{k}}]$ with a structure of $\mathfrak{g}'_e$-module which is compatible with the one coming from the representation datum.

PROPOSITION 4.13. *For each $i \in I_e$, let $\mathrm{k}E_i$ (resp. $\mathrm{k}F_i$) be the generalized i-eigenspace of X on $\mathrm{k}E$ (resp. $\mathrm{k}F$).*

(a) $[\mathrm{k}E_i], [\mathrm{k}F_i]$ *endow* $[\mathscr{U}_\mathrm{k}]$ *with a structure of* $\mathfrak{g}'_e$*-module.*

(b) $d_{\mathscr{U}} : [\mathscr{U}_K] \longrightarrow [\mathscr{U}_\mathrm{k}]$ *induces a* $\mathfrak{g}'_e$*-module isomorphism* $\mathrm{Res}^{\mathfrak{g}_\infty}_{\mathfrak{g}'_e} [\mathscr{U}_K] \xrightarrow{\sim} [\mathscr{U}_\mathrm{k}]$.

(c) $|\mu, Q_t\rangle_e \longmapsto [V_{\varpi_t(\mu)}]$ *induces a* $\mathfrak{g}'_e$*-module isomorphism* $\bigoplus_{t\in\mathbb{N}} \mathbf{F}(Q_t)_e \xrightarrow{\sim} [\mathscr{U}_\mathrm{k}]$. □

4.5.3. *The* $\mathfrak{g}_e$*-action on* $[\mathscr{U}_\mathrm{k}]$. We now define an action of $\mathfrak{g}_e$ on $[\mathscr{U}_\mathrm{k}]$ by extending the action from $\mathfrak{g}'_e$ to $\mathfrak{g}_e$ on $\bigoplus_{t\in\mathbb{N}} \mathbf{F}(Q_t)_e$. This amounts to extending the grading from P_e to $\mathrm{X}_e = \mathrm{P}_e \oplus \mathbb{Z}\delta$.

Any integrable $\mathfrak{g}_{e,\circ}$-representation (resp. $\mathfrak{g}'_{e,\circ}$-representation) can be "restricted" to an integrable $\mathfrak{g}_e$-representation (resp. $\mathfrak{g}'_e$-representation) through κ. We denote by $\mathrm{Res}^{\mathfrak{g}_{e,\circ}}_{\mathfrak{g}_e}$ and $\mathrm{Res}^{\mathfrak{g}'_{e,\circ}}_{\mathfrak{g}'_e}$ the corresponding operations.

Let $a : \mathscr{P} \to \mathbb{N}$ be Lusztig's a-function, see [**54**, 4.4.2]. Recall that $a(\lambda) = \sum_i (i-1)\lambda_i$ if $\lambda = (\lambda_1, \lambda_2, \dots)$.

LEMMA 4.14. *The assignment* $|\mu, Q_t\rangle_e \longmapsto (-1)^{a(\varpi_t(\mu))}|\varpi_t(\mu), 1\rangle_{e,\circ}$ *induces an isomorphism of* $\mathfrak{g}'_e$*-modules* $\bigoplus_{t\in\mathbb{N}} \mathbf{F}(Q_t)_e \xrightarrow{\sim} \mathrm{Res}^{\mathfrak{g}'_{e,\circ}}_{\mathfrak{g}'_e} \mathbf{F}(1)_{e,\circ}$. □

Now, consider the Fock space $\mathbf{F}(1)_{e,\circ}$ as a charged Fock space for the charge $s = 0$. This endows $\mathbf{F}(1)_{e,\circ}$ with an integrable representation of $\mathfrak{g}_{e,\circ}$ as in §2.3.1. Consequently, we can endow $\bigoplus_{t\in\mathbb{N}} \mathbf{F}(Q_t)_e$ and $[\mathscr{U}_\mathrm{k}]$ with integrable representations of $\mathfrak{g}_e$ such that there are $\mathfrak{g}_e$-module isomorphisms

$$\begin{aligned} \bigoplus_{t\in\mathbb{N}} \mathbf{F}(Q_t)_e &\xrightarrow{\sim} [\mathscr{U}_\mathrm{k}] \xrightarrow{\sim} \mathrm{Res}^{\mathfrak{g}_{e,\circ}}_{\mathfrak{g}_e} \mathbf{F}(1)_{e,\circ} \\ |\mu, Q_t\rangle_e &\mapsto [V_{\varpi_t(\mu)}] \mapsto (-1)^{a(\varpi_t(\mu))}|\varpi_t(\mu), 1\rangle_{e,\circ}. \end{aligned} \tag{4.5}$$

4.5.4. *The* $\mathfrak{g}_e$*-action on* $\mathscr{U}_\mathrm{k}$. The following lemma holds.

LEMMA 4.15. *Let* λ *and* μ *be partitions of* n. *If* V_μ, V_λ *belong to the same block of* $\mathscr{U}_\mathrm{k}$ *then* $[V_\lambda]$, $[V_\mu]$ *have the same weight for the action of* $\mathfrak{g}_e$. □

We deduce that the classes of the simple modules $[D_\lambda]$ are also weight vectors. Given $\omega \in \mathrm{X}_e$, we define $\mathscr{U}_{\mathrm{k},\omega}$ to be the Serre subcategory of $\mathscr{U}_\mathrm{k}$ generated by the simple modules D_λ such that the element $[D_\lambda]$ in $[\mathscr{U}_\mathrm{k}]$ has weight ω. We have $\mathscr{U}_\mathrm{k} = \bigoplus_{\omega\in\mathrm{X}_e} \mathscr{U}_{\mathrm{k},\omega}$. We can now formulate the following.

THEOREM 4.16. *The representation datum associated with Harish-Chandra induction and restriction and the decomposition* $\mathscr{U}_\mathrm{k} = \bigoplus_{\omega\in\mathrm{X}_e} \mathscr{U}_{\mathrm{k},\omega}$ *yield a categorical representation of* $\mathfrak{g}_e$ *on* $\mathscr{U}_\mathrm{k}$. *Furthermore, the map* $|\mu, Q_t\rangle_e \longmapsto [V_{\varpi_t(\mu)}]$ *induces a* $\mathfrak{g}'_e$*-module isomorphism* $\bigoplus_{t\in\mathbb{N}} \mathbf{F}(Q_t)_e \xrightarrow{\sim} [\mathscr{U}_\mathrm{k}]$. □

4.6. Derived equivalences of blocks of $\mathscr{U}_\mathrm{k}$.

Recall that d, e and f denote respectively the order of q^2, $-q$ and q modulo ℓ.

4.6.1. *Characterization of the blocks of* $\mathscr{U}_\mathrm{k}$. Here we investigate which block can occur in a given weight space.

Assume e is even. Given $t \in \mathbb{N}$ and $\omega \in \mathrm{X}_e$, we define $\mathscr{U}_{\mathrm{k},t}$ (resp. $\mathscr{U}_{\mathrm{k},t,\omega}$) to be the Serre subcategory of $\mathscr{U}_\mathrm{k}$ generated by simple modules D_λ where λ has 2-core Δ_t (resp. with in addition $\omega_\lambda = \omega$). With e being even, any pair of partitions with the same e-core have the same 2-core, therefore $\mathscr{U}_{\mathrm{k},t}$ is a direct summand of $\mathscr{U}_\mathrm{k}$.

PROPOSITION 4.17. *Let* $\omega \in \mathrm{X}_e$.

(a) *If* e *is odd, then the category* $\mathscr{U}_{\mathrm{k},\omega}$ *is an indecomposable summand of* $\mathscr{U}_{\mathrm{k}}$.
(b) *If* e *is even, then the category* $\mathscr{U}_{\mathrm{k},t,\omega}$ *is an indecomposable summand of* $\mathscr{U}_{\mathrm{k},t}$ *and* $\mathscr{U}_{\mathrm{k}}$ *for all* $t \in \mathbb{N}$. □

4.6.2. *Derived equivalences of blocks of* $\mathscr{U}_{\mathrm{k}}$. It is not difficult to compute the orbits of the affine Weyl group on the weight spaces of $[\mathscr{U}_{\mathrm{k}}]$. Hence, we can apply Proposition 4.17 and the work of Chuang and Rouquier [**10**] to produce derived equivalences between blocks of $\mathscr{U}_{\mathrm{k}}$ in the same orbit. Using the results of Livesey [**51**] on the structure of good blocks for linear primes, we deduce that Broué's abelian defect group conjecture holds for unipotent blocks when e is even.

THEOREM 4.18. *Assume* e *is even. Let* B *be a unipotent block of* G_n *over* k *or* $\mathscr{O}$, *and* D *be a defect group of* B. *If the group* D *is abelian, then the block* B *is derived equivalent to its Brauer correspondent in* $N_{G_n}(D)$. □

4.7. The crystals of $\mathscr{U}_K$ and $\mathscr{U}_{\mathrm{k}}$.

Now, we want to compare the crystals of the categorical representations $\mathscr{U}_K$ and $\mathscr{U}_{\mathrm{k}}$ (which are related to Harish-Chandra induction and restriction) with the crystals of the Fock spaces related to $[\mathscr{U}_K]$ and $[\mathscr{U}_{\mathrm{k}}]$. This solves the main conjecture of Gerber-Hiss-Jacon [**36**] and gives a combinatorial way to compute the (weak) Harish-Chandra branching graph and the Hecke algebras associated to the weakly cuspidal unipotent modules.

4.7.1. *Crystals and Harish-Chandra series.* Recall that to any categorical representation one can associate a perfect basis, and hence an abstract crystal. In the previous sections we constructed a categorical action on the categories of unipotent representations over K (denoted by $\mathscr{U}_K$) and over k (denoted by $\mathscr{U}_{\mathrm{k}}$). From these two categorical representations we get the following abstract crystals

(a) $B(\mathscr{U}_K) = \big(\mathrm{Irr}(\mathscr{U}_K), \widetilde{E}_i, \widetilde{F}_i\big)$ where $\mathrm{Irr}(\mathscr{U}_K) = \{[E_\lambda] \,|\, \lambda \in \mathscr{P}\}$,
(b) $B(\mathscr{U}_{\mathrm{k}}) = \big(\mathrm{Irr}(\mathscr{U}_{\mathrm{k}}), \widetilde{E}_i, \widetilde{F}_i\big)$ where $\mathrm{Irr}(\mathscr{U}_{\mathrm{k}}) = \{[D_\lambda] \,|\, \lambda \in \mathscr{P}\}$.

The (uncolored) crystal graph associated with $B(\mathscr{U}_R)$ coincides with the weak Harish-Chandra branching graph, and its connected components with the weak Harish-Chandra series.

PROPOSITION 4.19. *Let* $R = K$ *or* k, *and* $I = (-q)^{\mathbb{Z}} \subset R^\times$. *Let* $D, M, N \in \mathrm{Irr}(\mathscr{U}_R)$.

(a) D *is weakly cuspidal if and only if* $\widetilde{E}_i(D) = 0$ *for all* $i \in I$.
(b) N *appears in* $\mathrm{hd}(F(M))$ *if and only if there exists* $i \in I$ *such that* $N \simeq \widetilde{F}_i M$.
(c) *If* D *is weakly cuspidal, then*

$$\mathrm{WIrr}(RG, D) = \{\widetilde{F}_{i_1} \cdots \widetilde{F}_{i_m}(D) \mid m \in \mathbb{N},\, i_1, \ldots, i_m \in I\}.$$

□

Note that by [**53**], the ordinary Harish-Chandra series and weak Harish-Chandra series on unipotent modules coincide when $R = K$.

4.7.2. *Comparison of the crystals.* In this section we will assume that e is odd. To any charged Fock space one can associate an abstract crystal, see §2.3.2. We now show how to choose the charge for each Fock space $\mathbf{F}(Q_t)_e$ so that the crystal will coincide with the Harish-Chandra branching graph.

Define $s_t = (s_1, s_2) = -\frac{1}{2}(e+1, 0) + \sigma_t$ where σ_t is as in (2.1). We have $Q_p = q^{2s_p}$ for each $p = 1, 2$. We denote by $B(s_t)_e$ the corresponding abstract crystal of $\mathbf{F}(Q_t)_e$, with the canonical labeling $B(s_t)_e = \{b(\mu, s_t) \mid \mu \in \mathscr{P}^2,\, t \in \mathbb{N}\}$. Finally, we set $B_e = \bigsqcup_{t\in\mathbb{N}} B(s_t)_e$.

THEOREM 4.20. *The map $b(\mu, s_t) \mapsto [D_{\varpi_t(\mu)}]$ is a crystal isomorphism $B_e \xrightarrow{\sim} B(\mathscr{U}_{\mathrm{k}})$.* □

We deduce the following corollaries.

COROLLARY 4.21. *The modules D_λ and D_ν lie in the same weak Harish-Chandra series if and only if the corresponding vertices of the abstract crystal B_e belong to the same connected component. In particular, if this holds then λ and ν have the same 2-core.* □

COROLLARY 4.22. *Let λ be a partition of $r > 0$ such that D_λ is weakly cuspidal.*

(a) *The e-core of λ is a 2-core $\Delta_t = (t, t-1, \ldots, 1)$,*
(b) *the weight of the class $[D_\lambda]$ with respect to the $\mathfrak{g}'_e$-action on $[\mathscr{U}_{\mathrm{k}}]$ is Λ_{Q_t},*
(c) *for each $m \geqslant 0$ and $n = r + 2m$, the map $\phi_{\mathrm{k},m} : \mathbf{H}^{q^2}_{\mathrm{k},m} \longrightarrow \mathrm{End}_{\mathrm{k}G_n}(F^m)^{\mathrm{op}}$ factors through an algebra isomorphism $\mathbf{H}^{Q_t;\, q^2}_{\mathrm{k},m} \xrightarrow{\sim} \mathscr{H}(\mathrm{k}G_n, D_\lambda)$.* □

5. The representation of the Heisenberg algebra on $\mathscr{U}_{\mathrm{k}}$

Recall that $G_n = \mathrm{GU}_n(q)$ is the finite unitary group and that $(K, \mathscr{O}, \mathrm{k})$ is an ℓ-modular system as in §3.1. Every cuspidal $\mathrm{k}G_n$-module is weakly cuspidal. Therefore, every Harish-Chandra series in $\mathrm{Irr}(\mathscr{U}_{\mathrm{k}})$ is partitioned into weak Harish-Chandra series. Proposition 4.19 and Theorem 4.20 yield a complete (combinatorial) description of the partition of $\mathrm{Irr}(\mathscr{U}_{\mathrm{k}})$ into weak Harish-Chandra series using the decomposition of $[\mathscr{U}_{\mathrm{k}}]$ for the action of $\mathfrak{g}_e$. In this section we generalize this result to usual Harish-Chandra series.

To do so, we introduce another categorical construction by adapting some techniques from [**62**]. It relies on a categorification of the Heisenberg representation of level 2 on the Fock spaces $\mathbf{F}(Q_t)$ mentioned above. This yields a complete classification of the cuspidal unipotent modules, as explained in Theorem 5.10 and §5.5.3 (and a computation of the parameters of all ramified Hecke algebras). The results here are new and were announced in [**18**]. In [**18**] we emphasized the relation between the Heisenberg action on $[\mathscr{U}_{\mathrm{k}}]$ and q-Schur algebras. To shorten the exposition, here we only express this new action in terms of Lusztig induction and we refer to [**18**] for more details.

Throughout this section we will assume that e, the order of $-q$ modulo ℓ is odd, and hence equal to d, the order of q^2 modulo ℓ.

5.1. The Heisenberg action on $[\mathscr{U}_\mathrm{k}]$.

5.1.1. *The Heisenberg algebra.* Set $\mathbf{\Lambda} = \bigoplus_{m\in\mathbb{N}} \mathbb{C}\mathrm{Irr}(\mathfrak{S}_m)$ where $\mathfrak{S}_0$ is the trivial group. We equip the vector space $\mathbf{\Lambda}$ with the bilinear form $\langle\bullet,\bullet\rangle_{\mathbf{\Lambda}} = \bigoplus_{m\in\mathbb{N}}\langle\bullet,\bullet\rangle_{\mathfrak{S}_m}$ and the vector space $\mathbf{\Lambda}\otimes\mathbf{\Lambda}$ with the tensor square of $\langle\bullet,\bullet\rangle_{\mathbf{\Lambda}}$. The induction and restriction yield a pair of linear maps $\mathrm{Ind}_{n,m}^{n+m}$ and $\mathrm{Res}_{n,m}^{n+m}$ between the $\mathbb{C}$-vector spaces $\mathbf{\Lambda}$ and $\mathbf{\Lambda}\otimes\mathbf{\Lambda}$ which are adjoint with respect to these bilinear forms.

Fix a positive integer a. A level a representation of the Heisenberg algebra, or an $\mathfrak{H}_a$-module, is a $\mathbb{C}$-vector space V with a family of endomorphisms b_n, b_n^* labelled by integers $n \in a\mathbb{Z}_{>0}$ satisfying the relations $[b_n, b_m] = [b_n^*, b_m^*] = 0$ and $[b_n, b_m^*] = -n\delta_{n,m}$. For each partition λ we write $a\lambda = (a\lambda_1, a\lambda_2, \dots)$ and $b_{a\lambda} = \prod_i b_{a\lambda_i}$, $b_{a\lambda}^* = \prod_i b_{a\lambda_i}^*$.

If λ is a partition of m and $w \in \mathfrak{S}_m$ is of *cycle-type* $\rho(w) = \lambda$, we denote by $c_\lambda = c_w$ the conjugacy class of w. The characteristic function of c_λ is a class function on $\mathfrak{S}_m$, and as such it can be viewed as an element in $\mathbf{\Lambda}$. We abbreviate $c_m = c_{(m)}$ for the conjugacy class of the m-cycles. Let z_λ be the number of elements of the centralizer of w in $\mathfrak{S}_m$. We have $\langle c_\lambda, c_\mu\rangle = z_\lambda^{-1}\delta_{\lambda,\mu}$.

There is a unique representation of $\mathfrak{H}_1$ on $\mathbf{\Lambda}$ such that for every $\phi \in \mathbb{C}\mathrm{Irr}(\mathfrak{S}_r)$ and $\psi \in \mathbb{C}\mathrm{Irr}(\mathfrak{S}_n)$ with $n = r + m$ we have

$$b_\lambda(\phi) = \mathrm{Ind}_{r,m}^n(\phi\otimes z_\lambda c_\lambda), \quad b_\lambda^*(\psi) = \langle \mathrm{Res}_{r,m}^n(\psi)\,,\, z_\lambda c_\lambda\rangle_{\mathbf{\Lambda}}. \tag{5.1}$$

5.1.2. *The Heisenberg action on* $\mathbf{F}(Q)$. Recall that e is the order of $-q$ modulo ℓ. We defined in §4.5.1 a Kac-Moody algebra $\mathfrak{g}_e$ corresponding to a quiver I_e. From now one we will assume that e is odd and we will abbreviate $\mathfrak{g} = \mathfrak{g}_e$. Then $\mathfrak{g}$ is isomorphic to $\widehat{\mathfrak{sl}}_e$. We can now define the representation of $\mathfrak{H}_{le}$ on $\mathbf{F}(Q)$, where $Q = (Q_1, \dots, Q_l) \in I_e^l$. The Fock space $\mathbf{F}(Q)$ admits a representation of $\mathfrak{H}_l$ which is identified with the l-th tensor power of the representation of $\mathfrak{H}_1$ on $\mathbf{\Lambda}$ in (5.1) under the $\mathbb{C}$-linear isomorphism $\mathbf{F}(Q) \xrightarrow{\sim} \mathbf{\Lambda}^{\otimes l}$ such that $|\mu, Q\rangle \mapsto \phi_{\mu^1}\otimes\cdots\otimes\phi_{\mu^l}$. Then, by (2.2), the representation of $\mathfrak{g}'$ on $\mathbf{F}(Q)$ commutes with the action of the subalgebra $\mathfrak{H}_{le}$ of $\mathfrak{H}_l$. We will apply this construction to $l = 1$ or 2.

5.1.3. *The Heisenberg action on* $[\mathscr{U}_\mathrm{k}]$. The previous section yields a representation of $\mathfrak{H}_{2e}$ in $\mathbf{F}(Q_t)_e$ for each $t \in \mathbb{N}$ and a representation of $\mathfrak{H}_e$ in $\mathbf{F}(1)_{e,\mathrm{o}}$. We want to compare them.

To do that, consider the $\mathbb{C}$-linear isomorphism $\bigoplus_{t\in\mathbb{N}}\mathbf{F}(Q_t)_e \xrightarrow{\sim} \mathbf{F}(1)_{e,\mathrm{o}}$ studied in Lemma 4.14, which is given by $|\mu, Q_t\rangle_e \mapsto (-1)^{a(\lambda)}|\lambda, 1\rangle_{e,\mathrm{o}}$ with $\lambda = \varpi_t(\mu)$.

LEMMA 5.1. *The $\mathfrak{g}_e$-linear isomorphism in Lemma 4.14 is an isomorphism of $\mathfrak{H}_{2e}$-modules $\bigoplus_{t\in\mathbb{N}}\mathbf{F}(Q_t)_e \xrightarrow{\sim} \mathrm{Res}_{\mathfrak{H}_{2e}}^{\mathfrak{H}_e}\mathbf{F}(1)_{e,\mathrm{o}}$.*

PROOF. Fix an integer $t \geqslant 0$. We must check that the linear map

$$\mathbf{F}(Q_t)_e \simeq \mathbf{\Lambda}\otimes\mathbf{\Lambda} \to \mathbf{\Lambda}$$

such that

$$|\mu, Q_t\rangle_e \mapsto \phi_{\mu^1}\otimes\phi_{\mu^2} \mapsto (-1)^{a(\lambda)}\phi_\lambda, \quad \forall \mu \in \mathscr{P}^2$$

intertwines the operator $b_n\otimes 1 + 1\otimes b_n$ on $\mathbf{\Lambda}\otimes\mathbf{\Lambda}$ with the operator b_{2n} on $\mathbf{\Lambda}$ if $n \in e\mathbb{Z}_{>0}$. Here, we have set $\lambda = \varpi_t(\mu)$.

To do this, note that we have

$$b_{2n}(\phi_\lambda) = \sum_\nu (-1)^{N(\lambda,x,2n)}\,\phi_\nu, \tag{5.2}$$

where the sum runs over all partitions ν obtained by adding a $2n$-hook $(x, x+2n)$ to the charged partition $(\lambda, 0)$ and $N(\lambda, x, 2n)$ is the number of elements in $\beta_0(\lambda) \cap (x, x+2n)$. Recall that $\sigma_t = (\sigma_1, \sigma_2)$. Since

$$\beta_0(\lambda) = (-1 + 2\beta_{\sigma_1}(\mu^1)) \sqcup (2\beta_{\sigma_2}(\mu^2)),$$

we have

$$N(\lambda, x, 2n) = \begin{cases} N(\mu^1, y - \sigma_1, n) + |(2\beta_{\sigma_2}(\mu^2)) \cap (x, x+2n)| & \text{if } x = -1 + 2y, \\ |(-1 + 2\beta_{\sigma_1}(\mu^1)) \cap (x, x+2n)| + N(\mu^2, y - \sigma_2, n) & \text{if } x = 2y. \end{cases}$$

Furthermore, adding a $2n$-hook $(x, x+2n)$ to λ is equivalent to adding an n-hook to μ^1 if x is odd or to μ^2 if x is even. So we are reduced to check that if λ, ν, x are as in the sum (5.2), then $a(\nu) - a(\lambda)$ has the same parity as the integer $b(\lambda, x)$ given by

- $|(2\beta_{\sigma_2}(\mu^2)) \cap (x, x+2n)|$ if x is odd,
- $|(-1 + 2\beta_{\sigma_1}(\mu^1)) \cap (x, x+2n)|$ if x is even.

Now a partition ν obtained by adding a $2n$-hook to λ can also be obtained by adding n successive 2-hooks to λ. Therefore it is enough to check that the parity of $a(\nu) - a(\lambda)$ and $b(\lambda, x)$ are the same when $n = 1$. Since $a(\lambda) = \sum_i (i-1)\lambda_i$, the integer $a(\nu) - a(\lambda)$ is even if $\nu \setminus \lambda$ is a vertical 2-hook (increasing λ_i by 2 for some i) or odd if $\nu \setminus \lambda$ is a horizontal 2-hook (increasing λ_i and $\lambda_{i+1} = \lambda_i$ by 1 for some i). On the other hand, we have $b(\lambda, x) = 0$ (resp. $b(\lambda, x) = 1$) for a vertical 2-hook (resp. for a horizontal 2-hook). □

We equip $[\mathscr{U}_k]$ with the unique representation of $\mathfrak{H}_{2e}$ such that the isomorphism $\bigoplus_{t \in \mathbb{N}} \mathbf{F}(Q_t)_e \xrightarrow{\sim} [\mathscr{U}_k]$ in Theorem 4.16 is $\mathfrak{H}_{2e}$-equivariant. From Lemmas 4.14 and 5.1 we deduce that there is a $\mathfrak{g}_e \times \mathfrak{H}_{2e}$-equivariant isomorphism given by

$$(5.3) \qquad [\mathscr{U}_k] \xrightarrow{\sim} \mathbf{F}(1)_{e,\mathrm{o}}, \quad [V_\lambda] \mapsto (-1)^{a(\lambda)} |\lambda, 1\rangle_{e,\mathrm{o}}.$$

REMARK 5.2. Set $A(\lambda) = m(m-1)/2 - a(\lambda^*)$ with $m = |\lambda|$. By [**12**, lem. 4.2] we have $(-1)^{A(\lambda)} = \varepsilon_\lambda$ where ε_λ was defined in §4.3.1. From this we can check that

$$\lambda_{[2]} = \nu_{[2]} \Rightarrow (-1)^{a(\lambda)} \varepsilon_\lambda = (-1)^{a(\nu)} \varepsilon_\nu.$$

Hence, the endomorphism of $[\mathscr{U}_k]$ such that $[V_\lambda] \mapsto (-1)^{a(\lambda)} \varepsilon_\lambda [V_\lambda]$ commutes with the action of $\mathfrak{g}_e$ and $\mathfrak{H}_{2e}$. This shows that we could have used the sign ε_λ instead of $(-1)^{a(\lambda)}$ in the isomorphism (5.3).

5.2. The modular Harish-Chandra series of GL_m.

Recall that $GL_n := \mathrm{GL}_n(q^2)$ denotes the finite general linear group over a finite field with q^2 elements.

5.2.1. *The unipotent modules of* GL_n. The set of unipotent characters of KGL_n is

$$(5.4) \qquad \mathrm{Unip}(KGL_n) = \{L_\lambda \mid \lambda \in \mathscr{P}_n\},$$

where

$$(5.5) \qquad L_\lambda = |\mathfrak{S}_n|^{-1} \sum_{w \in \mathfrak{S}_n} \phi_\lambda(w) R_{\mathbf{T}_w}^{\mathbf{GL}_n}(1)$$

and $\mathbf{T}_w$ is an F_{q^2}-stable maximal torus of $\mathbf{GL}_n$ of cycle-type $\rho(w)$, with the convention that $\mathbf{T}_1 = \mathbf{T}$ is the split torus of diagonal matrices. There is a unique labeling of the simple unipotent modules of $\mathrm{k}GL_n$ given by

$$\mathrm{Unip}(\mathrm{k}GL_n) = \{S_\lambda \mid \lambda \in \mathscr{P}_n\}$$

where

$$d_{\mathscr{O}GL_n}([L_\lambda]) = [S_\lambda] \text{ modulo } \bigoplus_{\mu > \lambda} \mathbb{Z}\,[S_\mu].$$

As in the case of unitary groups (see Proposition 4.3), there exists a unique indecomposable unipotent $\mathscr{O}GL_n$-lattice $\widetilde{W}_\lambda$ with character L_λ and such that $W_\lambda := k\widetilde{W}_\lambda$ is indecomposable with simple head S_λ.

5.2.2. *The modular Steinberg character and Harish-Chandra series.* First, let us introduce the following notation. Given non-negative integers m_j with $j \geqslant -1$, we will say that a partition λ of m is *cuspidal of type* (m_j) if λ is of the following form

$$\lambda = (1^{(m_{-1})}, e^{(m_0)}, (e\ell)^{(m_1)}, (e\ell^2)^{(m_2)}, \dots)$$

in exponential notation.

For a partition λ as above, we consider the Levi subgroup of $\mathbf{GL}_m$ given by

$$\mathbf{GL}_\lambda = (\mathbf{GL}_1)^{m_{-1}} \times \prod_{j \geqslant 0} (\mathbf{GL}_{e\ell^j})^{m_j}.$$

Recall that $GL_\lambda = \mathbf{GL}_\lambda(q^2)$. The unipotent module $S_{(1^m)}$ is the head of $W_{(1^n)}$. Since $W_{(1^n)}$ is a modular reduction of the Steinberg character $L_{(1^m)}$ of KGL_m, we call $S_{(1^m)}$ the *modular Steinberg representation* of $\mathrm{k}GL_m$ and we write $St_m = S_{(1^m)}$. Let St_λ be the $\mathrm{k}GL_\lambda$-module given by

$$St_\lambda = (St_1)^{\otimes m_{-1}} \otimes \bigotimes_{j \geqslant 0} (St_{e\ell^j})^{\otimes m_j}.$$

The following result is due to Dipper, James and Dipper-Du [**16**], see also [**8**, sec. 19].

PROPOSITION 5.3.

(a) *The cuspidal pairs of* $\mathrm{k}GL_m$ *are the* GL_m*-conjugates of the pairs* (GL_λ, St_λ) *where* λ *runs over the set of all cuspidal partitions of* m.

(c) *If* λ *is cuspidal of type* (m_j), *then we have a* k*-algebra isomorphism*

$$\mathscr{H}(\mathrm{k}GL_m, St_\lambda) \simeq \mathbf{H}^{1;\,q^2}_{\mathrm{k},m_{-1}} \otimes \bigotimes_{j \geqslant 0} \mathbf{H}^{1;1}_{\mathrm{k},m_j}.$$

□

Recall that here $\mathbf{H}^{1;\,u}_{\mathrm{k},m}$ denotes the Hecke algebra of type $\mathfrak{S}_m$ over k with parameter u. In particular, we have $\mathbf{H}^{1;\,1}_{\mathrm{k},m} = \mathrm{k}\mathfrak{S}_m$.

5.3. The Heisenberg functors.

For $R = K$ or k we call the category of unipotent RGL-modules the category

$$\mathscr{V}_R = \bigoplus_{m \in \mathbb{N}} RGL_m\text{-umod}.$$

Recall that the vector space $\mathbf{\Lambda}$ is equipped with the bilinear form $\langle \bullet, \bullet \rangle_{\mathbf{\Lambda}}$. It induces a bilinear form $\langle \bullet, \bullet \rangle_{\mathscr{V}}$ on $[\mathscr{V}_{\mathrm{k}}]$ via the isomorphism of vector spaces

$$[\mathscr{V}_{\mathrm{k}}] \xrightarrow{\sim} \mathbf{\Lambda}, \quad [W_\lambda] \mapsto \phi_\lambda. \tag{5.6}$$

Now, let us come back to the category $\mathscr{U}_{\mathrm{k}}$ of unipotent modules of the unitary group. Let $\langle \bullet, \bullet \rangle_{\mathscr{U}}$ be the unique bilinear form on $[\mathscr{U}_{\mathrm{k}}]$ such that the map (5.3) is an isometry.

Given $n, r, m > 0$ such that $n = r + 2m$, we write $L_{r,m} \simeq G_r \times GL_m$. Given a unipotent RGL_m-module X we consider the following functors

$$B_X : RG_r\text{-umod} \to RG_n\text{-umod}, \ M \mapsto R_{L_{r,m}}^{G_n}(M \otimes X),$$

$$B_X^* : RG_n\text{-umod} \to RG_r\text{-umod}, \ M \mapsto \mathrm{Hom}_{RGL_m}\big(X, {}^*R_{L_{r,m}}^{G_n}(M)\big).$$

The functor B_X^* is right adjoint to B_X. In the particular case where $m = 1$ and $X = R$ is the trivial module, we recover the functors F, E defined in §4.2. The functor B_X is exact, hence it induces a $\mathbb{C}$-linear endomorphism $[B_X]$ of $[\mathscr{U}_{\mathrm{k}}]$. However, the functor B_X^* is not exact and the module X may not have a finite projective dimension in $\mathrm{k}GL_m$-umod. Hence, the right derived functor of B_X^* may not yield any endomorphism of $[\mathscr{U}_{\mathrm{k}}]$. Nevertheless we can avoid this problem by defining the $\mathbb{C}$-linear map

$$[B_X^*] = \langle\, [X]\,,\, [{}^*R_{L_{r,m}}^{G_n}](\bullet)\,\rangle_{\mathscr{V}} : [\mathrm{k}G_n\text{-umod}] \to [\mathrm{k}G_r\text{-umod}].$$

The Harish-Chandra induction relative to the subgroups $GL_{n_1} \times GL_{n_2}$ of GL_n for $n_1 + n_2 = n$ yields a bifunctor $\bullet \odot \bullet : \mathscr{V}_R \otimes \mathscr{V}_R \to \mathscr{V}_R$.

PROPOSITION 5.4. *Let X, X' be objects in $\mathscr{V}_{\mathrm{k}}$.*

(a) *There is an isomorphism of functors $B_X B_{X'} \simeq B_{X \odot X'} \simeq B_{X'} B_X$.*
(b) *There is an equality of linear maps $[B_X^*] \circ [B_{X'}^*] = [B_{X \odot X'}^*] = [B_{X'}^*] \circ [B_X^*]$.*
(c) *The linear maps $[B_X]$, $[B_X^*]$ are adjoint relatively to the bilinear form on $[\mathscr{U}_{\mathrm{k}}]$.*
(d) *If $M \in \mathscr{U}_{\mathrm{k}}$ is cuspidal, then $[B_X^*]([M]) = 0$.* □

5.4. The categorification of the Heisenberg action on $[\mathscr{U}_{\mathrm{k}}]$.

Recall that e is odd. The representation of $\mathfrak{H}_{2e}$ on $[\mathscr{U}_{\mathrm{k}}]$ in §5.1.3 is determined by linear operators b_n, b_n^* with n a positive multiple of $2e$ satisfying the relations $[b_n, b_m] = [b_n^*, b_m^*] = 0$ and $[b_n, b_m^*] = -n\, \delta_{n,m}$ Id. In this section we will re-interpret *some* of these operators in representation theoretic terms using the category $\mathscr{U}_{\mathrm{k}}$.

Now, fix a positive integer m and a cuspidal partition λ of m. Consider the module

$$X_\lambda = R_{GL_\lambda}^{GL_m}(St_\lambda) = R_{GL_\lambda}^{GL_m}\Big((St_1)^{\otimes m_{-1}} \otimes \bigotimes_{j \geqslant 0} (St_{e\ell^j})^{\otimes m_j}\Big).$$

If we denote by $\mathbf{T}_\lambda$ an F_{q^2}-stable maximal torus of $\mathbf{GL}_m$ of cycle-type λ, the class of X_λ coincides with the Deligne-Lusztig character induced from $\mathbf{T}_\lambda$.

LEMMA 5.5. *If λ is a cuspidal partition of m then $[X_\lambda] = R_{\mathbf{T}_\lambda}^{\mathbf{GL}_m}(1)$ in $[\mathscr{V}_{\mathrm{k}}]$.*

PROOF. The finite group $T_\lambda = \mathbf{T}_\lambda(q^2)$ is isomorphic to

$$\mathbf{GL}_1(q^2)^{m_{-1}} \times \prod_{j\geqslant 0} \mathbf{GL}_1(q^{2e\ell^j})^{m_j}.$$

By [**8**, thm. 19.18], for each integer $j \geqslant 0$ there is a character θ of $K\mathbf{GL}_1(q^{2e\ell^j})$ of order a power of ℓ such that we have

$$R_{\mathbf{T}_{(e\ell^j)}}^{\mathbf{GL}_{e\ell^j}}(1) = d_{\mathscr{O}GL} R_{\mathbf{T}_{(e\ell^j)}}^{\mathbf{GL}_{e\ell^j}}([\theta]) = (-1)^{e\ell^j-1}[St_{e\ell^j}] = [St_{e\ell^j}].$$

We deduce that $R_{\mathbf{T}_\lambda}^{\mathbf{GL}_\lambda}(1) = [St_\lambda]$. By transitivity of the Lusztig induction we get

$$[X_\lambda] = R_{GL_\lambda}^{GL_m}([St_\lambda]) = R_{\mathbf{T}_\lambda}^{\mathbf{GL}_m}(1).$$

□

Given a cuspidal partition λ, we set $B_\lambda = B_{X_\lambda}$. Note that the endofunctor F of $\mathscr{U}_{\mathrm{k}}$ is equal to B_1. The functor B_λ^* is not well-defined, but we may still consider the linear endomorphism $[B_\lambda^*]$ of $[\mathscr{U}_{\mathrm{k}}]$. They are related to the Heisenberg operators considered earlier by the following proposition (which will not be used in the rest of the paper).

PROPOSITION 5.6. *Assume that $\ell \neq 2$. Let λ be a cuspidal partition of m of type (m_j) with $m_{-1} = 0$. Then, we have the equalities $b_{2\lambda} = [B_\lambda]$ and $b_{2\lambda}^* = [B_\lambda^*]$ in* $\mathrm{End}([\mathscr{U}_{\mathrm{k}}])$.

To prove the proposition we need more material. Fix a positive integer a. We abbreviate $GL_m^{(a)} = \mathbf{GL}_m(q^{2a})$ and we consider the category

$$\mathscr{V}_R^{(a)} := \bigoplus_{m\in\mathbb{N}} RGL_m^{(a)}\text{-umod}.$$

First, we view $GL_m^{(a)}$ as the group of $\mathbb{F}_{q^{2a}}$-rational points of the rational group $(\mathbf{GL}_m\,,\,F_{q^{2a}})$. Unipotent irreducible modules of KGL_m and $KGL_m^{(a)}$ are both parametrized by partitions of m. Therefore there is an isomorphism $[\mathscr{V}_{\mathrm{k}}] \overset{\sim}{\to} [\mathscr{V}_{\mathrm{k}}^{(a)}]$ sending $[W_\mu]$ to $[W_\mu^{(a)}]$ for any partition μ of m. More generally we will denote by $[V^{(a)}]$ the image of $[V]$ under this map. By composition we obtain $\mathbb{C}$-linear isomorphisms given by

$$(5.7)\qquad [\mathscr{V}_K^{(a)}] \overset{\sim}{\to} [\mathscr{V}_{\mathrm{k}}^{(a)}] \overset{\sim}{\to} \mathbf{\Lambda}, \quad [L_\mu^{(a)}] \mapsto [W_\mu^{(a)}] \mapsto \phi_\mu.$$

Next, we view $GL_m^{(a)}$ as the group of rational points of $\mathbf{GL}_m^{(a)} = ((\mathbf{GL}_m)^a\,,\,\sigma_a F_{q^2})$, where σ_a is the automorphism of $(\mathbf{GL}_m)^a$ such that $(g_1, g_2, \ldots, g_a) \mapsto (g_2, \ldots, g_a, g_1)$. Note that $\mathbf{GL}_m^{(a)}$ is a Levi subgroup of $\mathbf{GL}_{am} = (\mathbf{GL}_{am}\,,\,F_{q^2})$, but the rational structures are different. The Lusztig induction and restriction yield $\mathbb{C}$-linear maps

$$R_{\mathbf{GL}_m^{(a)}}^{\mathbf{GL}_{am}} : [\mathrm{k}GL_m^{(a)}\text{-umod}] \to [\mathrm{k}GL_{am}\text{-umod}],$$
$${}^*R_{\mathbf{GL}_m^{(a)}}^{\mathbf{GL}_{am}} : [\mathrm{k}GL_{am}\text{-umod}] \to [\mathrm{k}GL_m^{(a)}\text{-umod}].$$

Let $\mathscr{F}^a$ be the $\mathbb{C}$-linear map $\mathbf{\Lambda} \to \mathbf{\Lambda}$ such that $\mathscr{F}^a(z_\lambda c_\lambda) = z_{a\lambda} c_{a\lambda}$ for each partition λ. Let $\mathscr{F}_a$ be the transposed map relative to the bilinear form on $\mathbf{\Lambda}$. We have $\mathscr{F}_a(c_\lambda) = c_\mu$ if $\lambda = a\mu$ and $\mathscr{F}_a(c_\lambda) = 0$ if $\lambda \neq a\mu$ for all μ.

LEMMA 5.7.

(a) *The isomorphism* (5.7) *identifies the maps* $R^{\mathbf{GL}_{am}}_{\mathbf{GL}_m^{(a)}}$ *and* ${}^*R^{\mathbf{GL}_{am}}_{\mathbf{GL}_m^{(a)}}$ *with the endomorphisms* $\mathscr{F}^a$ *and* $\mathscr{F}_a$ *of* $\mathbf{\Lambda}$.

(b) *The isomorphism* (5.6) *identifies the maps* $R^{GL_{m+n}}_{GL_m \times GL_n}$ *and* ${}^*R^{GL_{m+n}}_{GL_m \times GL_n}$ *with the maps* $\mathrm{Ind}^{m+n}_{m,n}$ *and* $\mathrm{Res}^{m+n}_{m,n}$ *between* $\mathbf{\Lambda} \otimes \mathbf{\Lambda}$ *and* $\mathbf{\Lambda}$.

(c) *If* $n = r + 2m$, *the isomorphism* $[\mathscr{U}_\mathrm{k}] \xrightarrow{\sim} \mathbf{\Lambda}$ *sending* $\varepsilon_\lambda[V_\lambda]$ *to* ϕ_λ *and the isomorphism* $[\mathscr{V}_\mathrm{k}] \xrightarrow{\sim} \mathbf{\Lambda}$ *in* (5.7) *identify the maps* $R^{G_n}_{L_{r,m}}$ *and* ${}^*R^{G_n}_{L_{r,m}}$ *with the maps* $\mathrm{Ind}^n_{r,2m}(\bullet \otimes \mathscr{F}^2(\bullet))$ *and* $(1 \otimes \mathscr{F}_2)\,\mathrm{Res}^n_{r,2m}$ *between* $\mathbf{\Lambda} \otimes \mathbf{\Lambda}$ *and* $\mathbf{\Lambda}$.

PROOF. The proof of the lemma is standard, see, e.g., [**63**, prop. 6.1] or [**4**, sec. 3.A]. Let us give some details.

First, we prove part (a). We abbreviate $n = am$. Consider the rational groups $(\mathbf{GL}_n, F_{q^2})$ and $((\mathbf{GL}_m)^a, \sigma_a F_{q^2})$. We will denote them by $\mathbf{G}$ and $\mathbf{G}^{(a)}$ respectively. The Weyl groups of $\mathbf{G}$ and G are identified with $\mathfrak{S}_n$. The Weyl group of $\mathbf{G}^{(a)}$ is $(\mathfrak{S}_m)^a$. Let σ_a denote also the permutation in $\mathfrak{S}_n$ of cycle-type (a^m) which normalizes the subgroup $(\mathfrak{S}_m)^a$ and induces the automorphism such that $(w_1, w_2, \ldots, w_a) \mapsto (w_2, \ldots, w_a, w_1)$. The action of the Frobenius homomorphism on the Weyl group of $\mathbf{G}^{(a)}$ is given by the conjugation by σ_a. The Weyl group of $G^{(a)}$ is $\mathfrak{S}_m$.

By (5.5) we have

$$[L_\lambda] = \sum_{\nu \vdash n} \langle \phi_\lambda\,,\, c_\nu \rangle_{\mathbf{\Lambda}}\, R^{\mathbf{G}}_{\mathbf{T}_\nu}(1).$$

Hence, the isometry $[\mathscr{V}_\mathrm{k}] \xrightarrow{\sim} \mathbf{\Lambda}$ such that $[L_\lambda] \mapsto \phi_\lambda$ maps the virtual module $R^{\mathbf{G}}_{\mathbf{T}_\lambda}(1)$ to $z_\lambda\, c_\lambda$ for each partition λ of n. Similarly, the isometry $[\mathscr{V}_\mathrm{k}^{(a)}] \to \mathbf{\Lambda}$ in (5.7) maps $[L^{(a)}_\mu]$ to ϕ_μ for each partition μ of m, hence it maps the virtual module $\big(R^{\mathbf{GL}_m}_{\mathbf{T}_\mu}(1)\big)^{(a)}$ to $z_\mu\, c_\mu$.

With the notation from Remark 3.2 for Lusztig induction, we have

$$\begin{aligned}\big(R^{\mathbf{GL}_m}_{\mathbf{T}_\mu}(1)\big)^{(a)} &= \big(R^{\mathbf{GL}_m\,,\,F_{q^2}}_{\mathbf{T}_\mu\,,\,F_{q^2}}(1)\big)^{(a)} \\ &= R^{\mathbf{GL}_m\,,\,F_{q^{2a}}}_{\mathbf{T}_\mu\,,\,F_{q^{2a}}}(1) \\ &= R^{(\mathbf{GL}_m)^a\,,\,\sigma_a F_{q^2}}_{\mathbf{T}_w\,,\,\sigma_a F_{q^2}}(1),\end{aligned}$$

where $w = (w_1, w_2, \ldots, w_a)$ is any element in $(\mathfrak{S}_m)^a$ such that $w_1 w_2 \cdots w_a$ is of cycle-type μ. By transitivity of Lusztig induction we deduce that

$$R^{\mathbf{G}}_{\mathbf{G}^{(a)}}\big(R^{\mathbf{GL}_m}_{\mathbf{T}_\mu}(1)\big)^{(a)} = R^{\mathbf{G}}_{\mathbf{T}_{w\sigma_a}}(1) = R^{\mathbf{G}}_{\mathbf{T}_{a\mu}}(1),$$

because $w\sigma_a$ is of cycle-type $a\mu$. Therefore, up to the isometries above, the map

$$R^{\mathbf{G}}_{\mathbf{G}^{(a)}} : [\mathscr{V}_K^{(a)}] \to [\mathscr{V}_K]$$

is identified with the linear endomorphism $\mathscr{F}^a$ of $\mathbf{\Lambda}$. This proves the first identity in part (a). The second one follows by adjunction.

Now, we prove part (b). Consider the $\mathbb{F}_{q^2}$-rational group $(\mathbf{GL}_{m+n}, F_{q^2})$ with its F_{q^2}-stable Levi subgroup $(\mathbf{GL}_m \times \mathbf{GL}_n, F_{q^2})$. The isometry $[KGL_m\text{-umod}] \to$

$\mathbb{C}\mathrm{Irr}(\mathfrak{S}_m)$ in (5.6) maps the virtual module $R_{\mathbf{T}_\mu}^{\mathbf{GL}_m}(1)$ to $z_\mu c_\mu$ for each partition μ of m. By the transitivity of Lusztig induction we get

$$R_{GL_m\times GL_n}^{GL_{m+n}} R_{\mathbf{T}_\mu\times\mathbf{T}_\lambda}^{\mathbf{GL}_m\times\mathbf{GL}_n}(1) = R_{\mathbf{T}_\mu\times\mathbf{T}_\lambda}^{\mathbf{GL}_{m+n}}(1).$$

This proves the first identity in part (b). The second one follows by adjunction.

It remains to prove part (c). Consider the $\mathbb{F}_q$-rational group $(\mathbf{GL}_n, F)$ with its F-stable Levi subgroup $\mathbf{L}_{r,m}$. We will abbreviate $\mathbf{G}_n = (\mathbf{GL}_n, F)$ and we will identify the rational groups $(\mathbf{L}_{r,m}, F)$ and $(\mathbf{GL}_r \times (\mathbf{GL}_m)^2, (F, \sigma_2 F_q))$, where σ_2 is an element of $\mathfrak{S}_{2m}$ of cycle-type (2^m). Let w_n denote the longest element in $\mathfrak{S}_n$. The Weyl groups of $\mathbf{GL}_n$ and G_n are identified with $\mathfrak{S}_n$ and $W_n = C_{\mathfrak{S}_n}(w_n)$. The Weyl group of $\mathbf{L}_{r,m}$ is identified with $\mathfrak{S}_r \times (\mathfrak{S}_m)^2$ with the action of the Frobenius endomorphism F given by conjugation by (w_r, σ_2). Here, we embed $\mathfrak{S}_r \times (\mathfrak{S}_m)^2$ into $\mathfrak{S}_n$ via the map $(u, v) \mapsto (w_m v_1 w_m, u, v_2)$ if $v = (v_1, v_2)$ with $v_1, v_2 \in \mathfrak{S}_m$. The Weyl group of $L_{r,m}$ is identified with $W_r \times \mathfrak{S}_m$.

From (4.2) we deduce that the isometry $[\mathscr{U}_K] \xrightarrow{\sim} \mathbf{\Lambda}$ such that $\chi_\lambda \mapsto \phi_\lambda$ maps the virtual module $R_{\mathbf{T}_\lambda}^{\mathbf{G}_r}(1)$ to $z_\lambda c_\lambda$ for each partition λ of r. In the proof of (a) we already observed that the isometry $[\mathscr{V}_K] \xrightarrow{\sim} \mathbf{\Lambda}$ in (5.7) maps the virtual module $R_{\mathbf{T}_\mu\,,\,F_{q^2}}^{\mathbf{GL}_m\,,\,F_{q^2}}(1) = R_{\mathbf{T}_\mu\times\mathbf{T}_1\,,\,\sigma_2F_q}^{(\mathbf{GL}_m)^2\,,\,\sigma_2F_q}(1)$ to $z_\mu c_\mu$ for each partition μ of m. By transitivity of Lusztig induction we get

$$R_{\mathbf{L}_{r,m}\,,\,F}^{\mathbf{GL}_n\,,\,F}\big(R_{\mathbf{T}_\lambda\times\mathbf{T}_\mu\times\mathbf{T}_1\,,\,(F,\sigma_2F_q)}^{\mathbf{GL}_r\times(\mathbf{GL}_m)^2\,,\,(F,\sigma_2F_q)}(1)\big) = R_{\mathbf{T}_\mu\times\mathbf{T}_\lambda\times\mathbf{T}_1\,,\,F}^{\mathbf{GL}_n\,,\,F}(1).$$

We deduce that the map $R_{L_{r,m}}^{G_n}$ is identified with the map $\mathbb{C}\mathrm{Irr}(\mathfrak{S}_r) \otimes \mathbb{C}\mathrm{Irr}(\mathfrak{S}_m) \to \mathbb{C}\mathrm{Irr}(\mathfrak{S}_n)$ given by $\mathrm{Ind}_{r,2m}^n(\bullet \otimes \mathscr{F}^2(\bullet))$. This proves the first identity in part (c). The second one follows by adjunction. $\square$

PROOF OF PROPOSITION 5.6. Let $n = r + 2m$. Under the isomorphism (5.6) and the isomorphism $[\mathscr{U}_\Bbbk] \xrightarrow{\sim} \mathbf{\Lambda}$ given by $\varepsilon_\lambda[V_\lambda] \mapsto \phi_\lambda$, Lemma 5.7 identifies the maps $R_{L_{r,m}}^{G_n}$ and ${}^*R_{L_{r,m}}^{G_n}$ between $[\mathscr{U}_\Bbbk \otimes \mathscr{V}_\Bbbk]$ and $[\mathscr{V}_\Bbbk]$ with the $\mathbb{C}$-linear maps $\mathbf{\Lambda}\otimes\mathbf{\Lambda} \to \mathbf{\Lambda}$ and $\mathbf{\Lambda} \to \mathbf{\Lambda}\otimes\mathbf{\Lambda}$ given by

$$\phi_\nu \otimes \phi_\lambda \mapsto \mathrm{Ind}_{r,2m}^n(\phi_\nu \otimes \mathscr{F}^2(\phi_\lambda)), \qquad \phi_\nu \mapsto (1 \otimes \mathscr{F}_2)\,\mathrm{Res}_{r,2m}^n(\phi_\nu). \tag{5.8}$$

Now, assume that λ is cuspidal of type (m_j). By Lemmas 5.5, 5.7 the class of X_λ in $[\mathscr{V}_\Bbbk]$ is identified with the element $z_\lambda c_\lambda \in \mathbf{\Lambda}$. From (5.1) we deduce that the operator $[B_\lambda] = \bigoplus_r R_{L_{r,m}}^{G_{r+2m}}(\bullet \otimes [X_\lambda])$ on $[\mathscr{U}_\Bbbk]$ is identified with the operators $b_{2\lambda}$ on $\mathbf{\Lambda}$. Hence, the proposition follows from Lemma 5.1 and Remark 5.2. $\square$

5.5. Cuspidal modules and highest weight vectors.

By Proposition 4.19, the set of isomorphism classes of weakly cuspidal modules in $\mathscr{U}_\Bbbk$ is a basis of the space $[\mathscr{U}_\Bbbk]^0$ of all elements in $[\mathscr{U}_\Bbbk]$ which are killed by the map $[E_i]$ for each $i \in I_e$. We define

$$[\mathscr{U}_\Bbbk]^{\mathrm{hw}} = \{x \in [\mathscr{U}_\Bbbk]^0 \mid b_n^*(x) = 0,\ \forall n \in 2e\mathbb{Z}_{>0}\}.$$

Then, we have the following inclusion.

LEMMA 5.8. $\{[D] \mid D \in \mathrm{Irr}(\mathscr{U}_\Bbbk)$ *is cuspidal* $\} \subseteq [\mathscr{U}_\Bbbk]^{\mathrm{hw}}$.

PROOF. For any partition λ we set

$$a_{2\lambda} = \sum_{\nu} \chi_{\nu}^{\lambda}\, z_{\nu}^{-1}\, b_{2\nu}, \quad a_{2\lambda}^{*} = \sum_{\nu} \chi_{\nu}^{\lambda}\, z_{\nu}^{-1}\, b_{2\nu}^{*},$$

where $\chi_{\nu}^{\lambda} = \langle \phi_{\lambda}\,,\, z_{\nu}\, c_{\nu} \rangle_{\mathbf{\Lambda}}$ is the value of the irreducible character ϕ_{λ} on a permutation of cycle-type ν. Since the $\mathscr{P} \times \mathscr{P}$-matrix with entries χ_{ν}^{λ} is invertible, we have

$$[\mathscr{U}_{\mathrm{k}}]^{\mathrm{hw}} \supseteq \{x \in [\mathscr{U}_{\mathrm{k}}]^{0} \,|\, a_{2\lambda}^{*}(x) = 0,\, \forall \lambda \in \mathscr{P}\}.$$

Now, for any partition λ we set $A_{\lambda} = B_{W_{\lambda}}$ and $[A_{\lambda}^{*}] = [B_{W_{\lambda}}^{*}]$. Under the isomorphism $[\mathscr{U}_{\mathrm{k}}] \simeq \mathbf{\Lambda}$, the operator $[A_{\lambda}]$ on $[\mathscr{U}_{\mathrm{k}}]$ is identified by (5.8) with the operator $\bigoplus_{r} \mathrm{Ind}_{r,2m}^{r+2m}(\bullet \otimes \mathscr{F}^{2}(\phi_{\lambda}))$ on $\mathbf{\Lambda}$. Since $\phi_{\lambda} = \sum_{\nu} \chi_{\nu}^{\lambda}\, c_{\nu}$, the formula (5.1) yields

$$\bigoplus_{r} \mathrm{Ind}_{r,2m}^{r+2m}(\bullet \otimes \mathscr{F}^{2}(\phi_{\lambda})) = \sum_{\nu} \chi_{\nu}^{\lambda}\, z_{\nu}^{-1}\, b_{2\nu}.$$

We deduce that we have $a_{2\lambda} = [A_{\lambda}]$ and $a_{2\lambda}^{*} = [A_{\lambda}^{*}]$ in $\mathrm{End}([\mathscr{U}_{\mathrm{k}}])$. Now, to prove the lemma it is enough to prove that $a_{2\lambda}^{*}([D]) = 0$ for all partition λ and all cuspidal module D in $\mathrm{Irr}(\mathscr{U}_{\mathrm{k}})$. This follows from Proposition 5.4. □

5.5.1. *The parameters of the ramified Hecke algebras.* It is not obvious that the vector space $[\mathscr{U}_{\mathrm{k}}]^{\mathrm{hw}}$ is spanned by classes of irreducible unipotent modules. Our goal is to determine precisely the cuspidal modules. Let us first recall the following basic facts.

LEMMA 5.9.

(a) *Any unipotent cuspidal pair of* $\mathrm{k}G_n$ *is* G_n*-conjugate to a pair* $(G_r \times GL_{\lambda}\,, D_{\nu} \otimes St_{\lambda})$ *where* $n = r + 2m$, λ *is a cuspidal partition of* m *and* D_{ν} *is a unipotent cuspidal* $\mathrm{k}G_r$*-module.*

(b) *Given a cuspidal pair* $(G_r \times GL_{\lambda}\,, D_{\nu} \otimes St_{\lambda})$ *with* λ *of type* (m_j)*, the ramified Hecke algebra* $\mathscr{H}(\mathrm{k}G_n, D_{\nu} \otimes St_{\lambda})$ *is isomorphic to the* k*-algebra* $\mathbf{H}_{\mathrm{k},m_{-1}}^{Q_t\,;\,q^2} \otimes \bigotimes_{j\geqslant 0} \mathbf{H}_{\mathrm{k},m_j}^{1,1\,;\,1}$*, where the integer* $t \in \mathbb{N}$ *is such that the* e*-core of* ν *is* Δ_t*.*

PROOF. Part (a) follows from Proposition 5.3. To prove (b), let first observe that by [**32**, prop. 4.4] we have an algebra isomorphism

$$\mathscr{H}(\mathrm{k}G_n, D_{\nu} \otimes St_{\lambda}) \simeq \mathbf{H}_{\mathrm{k},m_{-1}}^{P_{-1}\,;\,q^2} \otimes \bigotimes_{j\geqslant 0} \mathbf{H}_{\mathrm{k},m_j}^{P_j\,;\,1},$$

where $P_j = (a_j, b_j)$ is a parameter in $\mathrm{k}^{\times} \times \mathrm{k}^{\times}$ for each $j \geqslant -1$. Using the transitivity and faithfulness of induction, see [**8**, prop. 1.23] for details, we deduce from Corollary 4.22 that $P_{-1} = Q_t$ for some $t \geqslant 0$. By [**38**, prop. 2.3.5][1] the parameter P_j is of the form (a_j, a_j) for some element $a_j \in \mathrm{k}^{\times}$ for each $j \geqslant 0$. □

5.5.2. *The classification of cuspidal unipotent modules.* We can now formulate the theorem comparing cuspidal modules and highest weight vectors.

THEOREM 5.10. *The set* $\{[D] \,|\, D \in \mathrm{Irr}(\mathscr{U}_{\mathrm{k}})$ *is cuspidal*$\}$ *is a basis of* $[\mathscr{U}_{\mathrm{k}}]^{\mathrm{hw}}$.

[1]The reference [**38**, prop. 2.3.5] has been indicated to us by G. Hiss.

PROOF. First, let us fix some notation that we will use in the whole proof. For every sequence $(m_{-1}, m_0, m_1, \dots)$ of integers $\geqslant 0$ we set $m_+ = \sum_{j\geqslant 0} m_j \ell^j$, $m = m_{-1} + em_+$ and $n = r + 2m$. Let λ be the cuspidal partition of m of type $(m_{-1}, m_0, \dots)$ and λ_+ be the cuspidal partition of type $(m_0, m_1, \dots)$. Let $\mathscr{C} \subseteq \mathscr{P}$ be the set of all partitions ν such that the module D_ν is cuspidal. For each $\nu \in \mathscr{C}$ the e-core of ν is a 2-core $\Delta_{t(\nu)}$. We will abbreviate $\Delta_\nu = \Delta_{t(\nu)}$ and $Q_\nu = Q_{t(\nu)}$.

Now, we consider the Harish-Chandra series of $\mathrm{k}G_n$. The crystal basis $B(\Lambda)$ of the $\mathfrak{g}_e$-module $\mathbf{L}(\Lambda)$ is a disjoint union of bases $B(\Lambda)_\alpha$ of the weight subspaces $\mathbf{L}(\Lambda)_{\Lambda-\alpha}$ where α runs over Q_e^+. For each integer $d \geqslant 0$ we set

$$\mathbf{L}(\Lambda)_d = \bigoplus_{\mathrm{ht}(\alpha)=d} \mathbf{L}(\Lambda)_{\Lambda-\alpha}, \quad B(\Lambda)_d = \bigsqcup_{\mathrm{ht}(\alpha)=d} B(\Lambda)_{\Lambda-\alpha}.$$

We first prove the following.

CLAIM 5.11. *Given ν a partition, and $m_{-1}, m_+ \geq 0$, we have a bijection*

$$\bigsqcup_\lambda \mathrm{Irr}(\mathrm{k}G_n, D_\nu \otimes St_\lambda) \stackrel{1:1}{\longleftrightarrow} B(\Lambda_{Q_\nu})_{m_{-1}} \times \mathscr{P}_{m_+}$$

where λ runs over all cuspidal partitions with fixed m_{-1} and m_+.

From Lemma 5.9 and the generalities in Section 3.4, we have a canonical bijection

$$\mathrm{Irr}(\mathrm{k}G_n, D_\nu \otimes St_\lambda) \stackrel{1:1}{\longleftrightarrow} \mathrm{Irr}(\mathbf{H}^{Q_\nu;q^2}_{\mathrm{k},m_{-1}}) \times \prod_{j\geqslant 0} \mathrm{Irr}(\mathbf{H}^{1,1;1}_{\mathrm{k},m_j}). \tag{5.9}$$

By §1.4, there is a canonical bijection

$$B(\Lambda_{Q_\nu})_{m_{-1}} \stackrel{1:1}{\longleftrightarrow} \mathrm{Irr}(\mathbf{H}^{Q_\nu;q^2}_{\mathrm{k},m_{-1}}). \tag{5.10}$$

Further, the elements of $\mathrm{Irr}(\mathbf{H}^{1,1;1}_{\mathrm{k},m_j})$ are labelled by the ℓ-restricted partitions of m_j, see, e.g., [**33**, prop. 4.6.6]. Hence, the unicity of the ℓ-adic expansion of partitions yields the following bijection

$$\mathscr{P}_{m_+} \stackrel{1:1}{\longleftrightarrow} \bigsqcup_{\lambda_+} \prod_{j\geqslant 0} \mathrm{Irr}(\mathbf{H}^{1,1;1}_{\mathrm{k},m_j}), \tag{5.11}$$

where $\lambda_+ = (e^{m_0}, (e\ell)^{m_1}, \dots)$ runs over all cuspidal partitions of m_+ with parts divisible by e (see, e.g., [**8**, lem. 19.26]). The Claim 5.11 follows from (5.9), (5.10) and (5.11).

Now, we consider the complexified Grothendieck group $[\mathscr{U}_\mathrm{k}]$. The modular Harish-Chandra theory yields bijections

$$\mathrm{Irr}(\mathscr{U}_\mathrm{k}) \stackrel{1:1}{\longleftrightarrow} \bigsqcup_{n,\nu,\lambda} \mathrm{Irr}(\mathrm{k}G_n, D_\nu \otimes St_\lambda). \tag{5.12}$$

Next, we identify $[\mathscr{U}_\mathrm{k}]$ with the $\mathfrak{g}_e$-module $\bigoplus_t \mathbf{F}(Q_t)$. Fix a basis B of $[\mathscr{U}_\mathrm{k}]^{\mathrm{hw}}$ containing the family $\{[D_\nu]\,|\,\nu \in \mathscr{C}\}$. By Corollary 4.22 we can identify the $\mathfrak{g}'_e$-submodule of $[\mathscr{U}_\mathrm{k}]$ generated by $[D_\nu]$ with $\mathbf{L}(\Lambda_{Q_\nu})$. We get a canonical isomorphism of $\mathfrak{H}_{2e} \oplus \mathfrak{g}'_e$-modules

$$[\mathscr{U}_\mathrm{k}] = \bigoplus_{\nu\in\mathscr{C}} U(\mathfrak{H}_{2e})(\mathbf{L}(\Lambda_{Q_\nu})) \oplus \bigoplus_{b\in B\setminus\mathscr{C}} U(\mathfrak{H}_{2e})U(\mathfrak{g}'_e)(b). \tag{5.13}$$

For each n let $[\mathscr{U}_{\mathrm{k}}]_n$ be the subspace of $[\mathscr{U}_{\mathrm{k}}]$ given by

$$[\mathscr{U}_{\mathrm{k}}]_n = \bigoplus_{\nu, m_{-1}, m_+, \gamma} b_{2e\gamma}(\mathbf{L}(\Lambda_{Q_\nu})_{m_{-1}}),$$

where $\gamma \in \mathscr{P}_{m_+}$ and $r, m_{-1}, m_+ \in \mathbb{N}$ are such that $n = r + 2m_{-1} + 2em_+$. Comparing (5.12) and (5.13), we deduce that

$$\sum_{\nu,\, \lambda} |\mathrm{Irr}(\mathrm{k}G_n, D_\nu \otimes St_\lambda)| \geqslant \dim([\mathscr{U}_{\mathrm{k}}]_n),$$

and that we have equality for all n if and only if $B = \{[D_\nu] \,|\, \nu \in \mathscr{C}\}$.

Now, any non-zero highest-weight $\mathfrak{H}_{2e}$-module is free as a module over the subalgebra generated by $\{b_{2en} \,|\, n \geqslant 1\}$. Hence, for each n, the Claim 5.11 gives

$$\sum_{\nu,\, \lambda} |\mathrm{Irr}(\mathrm{k}G_n, D_\nu \otimes St_\lambda)| = \dim([\mathscr{U}_{\mathrm{k}}]_n),$$

proving the theorem. □

REMARK 5.12. The theorem implies that

$$\{\, [D] \,|\, D \in \mathrm{Irr}(\mathscr{U}_{\mathrm{k}}) \text{ is cuspidal}\} = \mathrm{Irr}(\mathscr{U}_{\mathrm{k}}) \cap [\mathscr{U}_{\mathrm{k}}]^{\mathrm{hw}}.$$

5.5.3. *Cuspidal modules and FLOTW e-partitions.* Recall that in Theorem 4.20 we have given an explicit isomorphism between the crystal basis $B(\mathscr{U}_{\mathrm{k}})$, whose underlying set is $\mathrm{Irr}(\mathscr{U}_{\mathrm{k}})$, and the abstract crystal B_e of the $\mathfrak{g}_e \times \mathfrak{H}_{2e}$-module $\bigoplus_t \mathbf{F}(Q_t)_e$. Using Theorem 5.10, we deduce that the cuspidal modules in $\mathrm{Irr}(\mathscr{U}_{\mathrm{k}})$ correspond precisely to the elements of B_e whose associated elements in the upper global basis $\mathbf{B}_e^\vee := \bigsqcup_t \mathbf{B}^\vee(s_t)$ of $\bigoplus_t \mathbf{F}(Q_t)_e$, see §2.3.2, are highest weight vectors for the action of both $\mathfrak{g}_e$ and $\mathfrak{H}_{2e}$.

These elements of B_e have been computed recently in [**35**]. To do so, one defines a class of e-partitions called FLOTW, see [**35**, def. 6.22]. Then, the cuspidal modules are related to FLOTW e-partitions as in [**35**, thm. 7.7].

6. Types B and C

This section is devoted to the construction of categorical actions on the category of unipotent representations of the finite groups $\mathrm{SO}_{2n+1}(q)$ and $\mathrm{Sp}_{2n}(q)$. The arguments are similar to those in Section 4 and will yield:

(1) the Hecke algebras associated to weakly cuspidal representations,
(2) the branching graph for the parabolic induction and restriction,
(3) derived equivalences between blocks.

The main difference is the lack of a description in terms of level 1 Fock spaces (given by Ennola duality, see §4). In particular, the grading by the imaginary roots, which in Section 4 was deduced from the grading in the GL_n case, is constructed in types B, C via some explicit combinatorics in terms of Lusztig's symbols.

6.1. Definitions. Throughout this section and the following, q is any prime power. Note however that we will assume that q is odd starting from Section 6.3.

6.1.1. *Odd-dimensional orthogonal groups.* Let J_n be as in §4.1. It is the matrix of a non-degenerate quadratic form. The odd-dimensional orthogonal group $\mathbf{SO}_{2n+1} = \mathrm{SO}_{2n+1}(\overline{\mathbb{F}}_q)$ is

$$\mathrm{SO}_{2n+1}(\overline{\mathbb{F}}_q) = \{g \in \mathbf{SL}_{2n+1} \mid {}^t g J_{2n+1} g = J_{2n+1}\}.$$

It is a connected reductive group of type B_n. The standard Frobenius map $F = F_q$ on $\mathbf{GL}_{2n+1}$ induces a Frobenius endomorphism on $\mathbf{SO}_{2n+1}$. The *finite orthogonal group* $\mathrm{SO}_{2n+1}(q)$ is given by

$$\mathrm{SO}_{2n+1}(q) = (\mathbf{SO}_{2n+1})^F.$$

The subgroup of diagonal matrices $\mathbf{T}$ (resp. upper-triangular matrices $\mathbf{B}$) in $\mathbf{SO}_{2n+1}$ is a split maximal torus (resp. an F-stable Borel subgroup). Note that $\mathbf{T} = \{\mathrm{diag}(t_n, \ldots, t_1, 1, t_1^{-1}, \ldots, t_n^{-1})\}$. The Weyl group W of $(\mathbf{SO}_{2n+1}, \mathbf{T})$ is a Weyl group of type B_n, and F acts trivially on it. For numbering the simple reflections of W we will take the following convention:

(6.1)
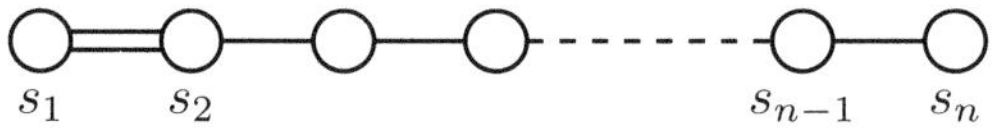

For $i \neq 1$, the action of s_i on $\mathrm{diag}(t_n, \ldots, t_1, 1, t_1^{-1}, \ldots, t_n^{-1})$ swaps t_{i-1} and t_i, whereas s_1 swaps t_1 and t_1^{-1}. For $i \neq 1$, the simple reflection s_i can be lifted to $N_{\mathbf{SO}_{2n+1}}(\mathbf{T})$ as the permutation matrix $(n-i+1, n-i+2)(n+i, n+i+1)$ and s_1 as the signed permutation matrix

$$s_1 = -\left(\begin{array}{c|ccc|c} \mathrm{Id}_{n-1} & & & & \\ \hline & 0 & 0 & 1 & \\ & 0 & 1 & 0 & \\ & 1 & 0 & 0 & \\ \hline & & & & \mathrm{Id}_{n-1} \end{array}\right).$$

6.1.2. *Symplectic groups.* Let $\widetilde{J}_{2n}$ be the $2n \times 2n$ square matrix

$$\widetilde{J}_{2n} = \begin{pmatrix} 0 & J_n \\ -J_n & 0 \end{pmatrix}.$$

It is the matrix of a symplectic form. The symplectic group $\mathbf{Sp}_{2n} = \mathrm{Sp}_{2n}(\overline{\mathbb{F}}_q)$ is

$$\mathbf{Sp}_{2n} = \{g \in \mathbf{GL}_{2n} \mid {}^t g \widetilde{J}_{2n} g = \widetilde{J}_{2n}\}.$$

It is a connected reductive groups of type C_n. The standard Frobenius map $F = F_q$ on $\mathbf{GL}_{2n}$ induces a Frobenius endomorphism on $\mathbf{Sp}_{2n}$. The *finite symplectic group* $\mathrm{Sp}_{2n}(q)$ is given by

$$\mathrm{Sp}_{2n}(q) = (\mathbf{Sp}_{2n})^F.$$

The subgroup of diagonal matrices $\mathbf{T}$ (resp. upper-triangular matrices $\mathbf{B}$) in $\mathbf{Sp}_{2n}$ is a split maximal torus (resp. an F-stable Borel subgroup). Note that $\mathbf{T} = \{\mathrm{diag}(t_n, \ldots, t_1, t_1^{-1}, \ldots, t_n^{-1})\}$. The Weyl group W of $(\mathbf{Sp}_{2n}, \mathbf{T})$ is a Weyl group of type B_n, and F acts trivially on it. For numbering the simple reflections of W we will take the same convention as in (6.1).

As in the case of odd-dimensional orthogonal groups, the action of s_i on the element $\mathrm{diag}(t_n, \ldots, t_1, t_1^{-1}, \ldots, t_n^{-1})$ swaps t_{i-1} and t_i when $i \neq 1$, whereas s_1

swaps t_1 and t_1^{-1}. For $i \neq 1$, we will lift s_i to $N_{\mathbf{Sp}_{2n}}(\mathbf{T})$ as the permutation matrix $(n-i+1, n-i+2)(n+i-1, n+i)$ and s_1 as the signed permutation matrix

$$s_1 = \left(\begin{array}{c|cc|c} \mathrm{Id}_{n-1} & & & \\ \hline & 0 & -1 & \\ & 1 & 0 & \\ \hline & & & \mathrm{Id}_{n-1} \end{array}\right).$$

6.2. The representation datum on RG-mod.

Throughout this section we assume that R is any commutative domain in which $q(q-1)$ is invertible.

Given a positive integer n, we will denote by $\mathbf{G}_n$ and G_n either the odd-dimensional orthogonal groups $\mathbf{SO}_{2n+1}$ and $\mathrm{SO}_{2n+1}(q)$ or the symplectic groups $\mathbf{Sp}_{2n}$ and $\mathrm{Sp}_{2n}(q)$. In addition, we set $G_0 = \{1\}$ by convention.

Given $r, m \in \mathbb{N}$ such that $n = r + m$, let $\mathbf{L}_{r,m}$ and $\mathbf{L}_{r,1^m}$ be the standard Levi subgroups of $\mathbf{G}_n$ corresponding to the sets of simple reflections $\{s_k \mid k \neq r+1\}$, $\{s_k \mid k \leq r\}$ respectively. The corresponding finite groups are $L_{r,m} \simeq G_r \times \mathrm{GL}_m(q)$ and $L_{r,1^m} \simeq G_r \times \mathrm{GL}_1(q)^m$. We will abbreviate $\mathbf{L}_r = \mathbf{L}_{r,1}$. It is isomorphic to $\mathbf{G}_r \times \mathbf{GL}_1$ via the map

$$(g, \lambda) \longmapsto \begin{pmatrix} \lambda & & \\ & g & \\ & & \lambda^{-1} \end{pmatrix}.$$

Let $\mathbf{P}_r \subset \mathbf{G}_{r+1}$ be the corresponding parabolic subgroup and $P_r = \mathbf{P}_r^F$. Let $\mathbf{V}_r$ be the unipotent radical of $\mathbf{P}_r$ and $V_r = \mathbf{V}_r^F$. Let us consider the subgroup $U_r \subset G_{r+1}$ given by

$$U_r = V_r \rtimes \mathbb{F}_q^\times.$$

It is represented by the same figure as in (4.1). For each $r < n$ we set $U_{n,r} = U_{n-1} \rtimes \cdots \rtimes U_r$ and $e_{n,r} = e_{U_{n,r}}$. The embedding of G_r into L_r yields an embedding of G_r into G_{r+1}, and by induction, of G_r into G_n. We obtain functors

$$F_{n,r} = RG_n \cdot e_{n,r} \otimes_{RG_r} \bullet \;:\; RG_r\text{-mod} \to RG_n\text{-mod},$$
$$E_{r,n} = e_{n,r} \cdot RG_n \otimes_{RG_n} \bullet \;:\; RG_n\text{-mod} \to RG_r\text{-mod}.$$

An endomorphism of the functor $F_{n,r}$ can be represented by an (RG_n, RG_r)-bimodule endomorphism of $RG_n \cdot e_{n,r}$, or, equivalently, by an element of $e_{n,r} \cdot RG_n \cdot e_{n,r}$ centralizing RG_r. Thus, the elements

(6.2)
$$X_{r+1,r} = q^r e_{r+1,r}(s_{r+1}s_r \cdots s_1 \cdots s_r s_{r+1})\, e_{r+1,r}, \qquad T_{r+2,r} = q e_{r+2,r} s_{r+2}\, e_{r+2,r}$$

define respectively natural transformations of the functors $F_{r+1,r}$ and $F_{r+2,r}$. Indeed, with our convention $s_{r+1}s_r \cdots s_1 \cdots s_r s_{r+1}$ is one of the matrices

$$-\begin{pmatrix} & & 1 \\ & \mathrm{Id}_{G_r} & \\ 1 & & \end{pmatrix} \quad \text{or} \quad \begin{pmatrix} & & -1 \\ & \mathrm{Id}_{G_r} & \\ 1 & & \end{pmatrix}$$

which centralize G_r. We set

$$F = \bigoplus_{r \geqslant 0} F_{r+1,r}, \qquad X = \bigoplus_{r \geqslant 0} X_{r+1,r}, \qquad T = \bigoplus_{r \geqslant 0} T_{r+2,r}.$$

PROPOSITION 6.1. *The endomorphisms $X \in \mathrm{End}(F)$ and $T \in \mathrm{End}(F^2)$ satisfy the following relations:*

(a) $1_F T \circ T 1_F \circ 1_F T = T 1_F \circ 1_F T \circ T 1_F$,
(b) $(T + 1_{F^2}) \circ (T - q 1_{F^2}) = 0$,
(c) $T \circ (1_F X) \circ T = q X 1_F$.

PROOF. Similar to the proof of [**18**, prop. 4.1]. □

6.3. The categories of unipotent modules $\mathscr{U}_K$ and $\mathscr{U}_{\mathrm{k}}$.

From now on, we fix an ℓ-modular system $(K, \mathscr{O}, \mathrm{k})$ which we assume to be big enough so that every indecomposable unipotent representation of G_n (over K or k) is absolutely indecomposable. Such a modular system exists since unipotent representations of G_n are defined over $\mathbb{Q}_\ell$, see [**55**, cor. 1.12]. In addition, since we will be dealing with representations in non-defining characteristic, we will always assume that $\ell \neq p$. We denote by f (resp. d) the order of q (resp. q^2) in $\mathrm{k}^\times$. If f is odd, we say that the prime number ℓ is *linear*, otherwise we say that ℓ is *unitary*. In the first case we have $f = d$, whereas in the second one $f = 2d$.

From now on, we will always assume that both p and ℓ are odd, and that $f > 1$. In particular $q(q-1) \in \mathscr{O}^\times$ and we can apply the previous constructions with R being any ring among $(K, \mathscr{O}, \mathrm{k})$.

6.3.1. *Parametrization by symbols.* By [**53**], the unipotent characters of G_n are parametrized by symbols. For our purpose it will be more convenient to work with a slightly different notion which we refer to as charged symbols.

A *charged symbol* Θ with charge $s = (s_1, s_2)$ is a pair of charged β-sets $\Theta = (\beta_{s_1}(\mu^1), \beta_{s_2}(\mu^2))$ for some bipartition $\mu = (\mu^1, \mu^2)$. We abbreviate $\Theta = \beta_s(\mu)$. If $\beta_{s_1}(\mu^1) =: X = \{x_1 > x_2 > \cdots\}$ and $\beta_{s_2}(\mu^2) =: Y = \{y_1 > y_2 > \cdots\}$ we write

$$\Theta = (X, Y) = \begin{pmatrix} x_1 & x_2 & \cdots \\ y_1 & y_2 & \cdots \end{pmatrix}.$$

The components X and Y are called the *first and second row* of the symbol. The *defect* of Θ is $D = s_1 - s_2$ and its *rank* is $|\mu| + \lfloor D^2/4 \rfloor$.

A *d-hook* of Θ is a pair of integers $(x, x+d)$ which is either a d-hook of X or a d-hook of Y. The charged symbol obtained by deleting $x+d$ from X (resp. Y) and replacing it by x is said to be gotten from Θ by *removing the d-hook* $(x, x+d)$. A *d-cohook* is a pair of integers $(x, x+d)$ such that $x+d \in X$ and $x \notin Y$, or $x+d \in Y$ and $x \notin X$. The charged symbol obtained by deleting $x+d$ from X (resp. Y) and adding x to Y (resp. X) is said to be gotten from Θ by *removing the d-cohook* $(x, x+d)$. The *d-core* of Θ is obtained by removing recursively all d-hooks from Θ. A similar definition using d-cohooks gives the *d-cocore* of Θ.

We will denote by $\Theta^\dagger = (Y, X)$ the charged symbol of charge (s_2, s_1) obtained by swapping the two β-sets. The defect of $\Theta^\dagger$ is $-D$ but the rank is the same. If one shifts simultaneously the charged β-sets X and Y by an integer m, one obtains a symbol $\Theta[m]$ of charge (s_1+m, s_2+m). This operation does not change the defect nor the rank.

Symbols are orbits of charged symbols under the shift operator and the transformation $\Theta \mapsto \Theta^\dagger$. We write

$$\{X, Y\} = \left\{ \begin{matrix} x_1 & x_2 & \cdots \\ y_1 & y_2 & \cdots \end{matrix} \right\}$$

for the symbol associated with (X, Y). The rank of the symbol is the rank of any charged symbol in its class whereas its defect is $|D|$ where D is the defect of any representative. Removing and adding d-hooks or d-cohooks are well-defined operations on symbols.

We denote by $\mathscr{S}$ the set of symbols and by $\mathscr{S}_{\mathrm{odd}}$ the set of symbols of odd defects.

6.3.2. *The unipotent modules over K.* Fix a positive integer n. The unipotent KG_n-modules were classified in [**53**, thm. 8.2]. They are parametrized in terms of symbols of odd defect and rank n. This parametrization follows from the determination of cuspidal unipotent KG_n-modules and their ramified Hecke algebras.

PROPOSITION 6.2 (Lusztig [**53**]). *Up to isomorphism, there is at most one cuspidal unipotent module in KG_r-mod, and there exists one if and only if $r = t(t+1)$ for some $t \geqslant 0$. It is denoted by E_t.* □

Since there do not exist any cuspidal unipotent character of $\mathrm{GL}_n(q)$ unless $n=1$, we deduce that any cuspidal pair of G_n is conjugate to a pair of the form $(L_{r,1^m}, E_t)$ with $n = r+m$ and $r = t(t+1)$ with $t \geqslant 0$. The general theory recalled in §3.4 implies that the irreducible characters lying in the Harish-Chandra series above E_t are in bijection with the irreducible representations of the ramified Hecke algebra $\mathscr{H}(KG_n, E_t)$. To describe explicitly the latter, we introduce, as in (4.3), the parameters

$$Q_t = \big((-q)^t, (-q)^{-1-t}\big). \tag{6.3}$$

Then $\mathscr{H}(KG_n, E_t)$ is isomorphic to $\mathbf{H}_{K,m}^{Q_t\,;\,q}$ by [**53**, §5,8]. There is a canonical choice for this isomorphism given by Theorem 6.5 below. Consequently, the Harish-Chandra theory yields a canonical bijection

$$\mathrm{Irr}(KG_n, E_t) \overset{1:1}{\longleftrightarrow} \mathrm{Irr}(\mathbf{H}_{K,m}^{Q_t\,;\,q}) \tag{6.4}$$

and hence a parametrization of $\mathrm{Irr}(KG_n, E_t)$. More precisely, given $\mu = (\mu^1, \mu^2)$ a bipartition of m and $t \geqslant 0$ we can associate the symbol

$$\Theta_t(\mu) = \big\{\beta_t(\mu^1), \beta_{-t-1}(\mu^2)\big\}.$$

So, the defect and the rank of $\Theta_t(\mu)$ are

$$D(\Theta_t(\mu)) = 2t+1, \quad \mathrm{rk}(\Theta_t(\mu)) = m + t(t+1).$$

Then we define $E_{\Theta_t(\mu)}$ to be the unipotent KG_r-module corresponding to $S(\mu)_K^{Q_t\,;\,q}$ via the bijection (6.4). This yields a parametrization of the irreducible unipotent KG_n-modules as

$$\{E_{\Theta_t(\mu)} \,|\, t \in \mathbb{N}\,,\, \mu \in \mathscr{P}_m^2\,,\, m + t(t+1) = n\}.$$

We will abbreviate $\Theta_t = \Theta_t(\emptyset)$. Note that $E_t = E_{\Theta_t}$.

PROPOSITION 6.3 (Lusztig [**53**]). *The irreducible unipotent characters are parametrized by symbols of odd defect. If $n = r + m$ and $r = t(t+1)$ with $t \geqslant 0$, then the unipotent module E_Θ lies in the Harish-Chandra series* $\mathrm{Irr}(KG_n, E_t)$ *if and only if Θ has defect $2t+1$ and rank n.* □

It is important to observe that this parametrization is exactly the one described in [**53**, thm. 8.2], which was subsequently used for the determination of blocks in [**23**]. Indeed, the labelling of unipotent characters by bipartitions or symbols in [**53**] (see also [**9**, §13.8]) corresponds to the presentation of the Hecke algebra $\mathscr{H}(KG_n, E_t)$ in terms of the set of generators $(T_0, T_1, \ldots, T_{m-1})$ instead of $(X_1, T_1, \ldots, T_{m-1})$ with $T_0 = (-1)^t q^{t+1} X_1$ so that $(T_0 - q^{2t+1})(T_0 + 1) = 0$. Now the isomorphism $\mathbf{H}_{K,m}^{Q_t\,;\,q} \simeq \mathbf{H}_{K,m}^{(q^{2t+1},-1)\,;\,q}$ induces a bijection on irreducible objects which is just $S(\mu)^{Q_t} \longmapsto S(\mu)^{(q^{2t+1},-1)}$.

REMARK 6.4. Note that $Q_t^\dagger = Q_{-1-t}$, $\Theta_t(\mu) = \Theta_{-1-t}(\mu^\dagger)$ and $t(t+1)$ is invariant under the map $t \longmapsto -t-1$. We will work with symbols $\Theta_t(\mu)$ such that $t \geqslant 0$. We might also have worked with $t < 0$ using the symmetries above.

Recall that $G_0 = \{1\}$. We call the category of unipotent KG-modules the category

$$\mathscr{U}_K = \bigoplus_{n\in\mathbb{N}} KG_n\text{-umod}\,.$$

This category is abelian semisimple. From the previous discussion we have

$$\mathrm{Irr}(\mathscr{U}_K) = \{E_\Theta \,|\, \Theta \in \mathscr{S}_{\mathrm{odd}}\},$$

where by convention $\mathrm{Irr}(KG_0) = \{E_0\}$.

6.3.3. *The unipotent modules over* k. Using the ℓ-modular system we have decomposition maps $d_{\mathscr{O}G_n}$ which by Theorem 3.4 and since ℓ is odd, restrict to linear isomorphisms

$$d_{\mathscr{O}G_n} : [KG_n\text{-umod}] \xrightarrow{\sim} [\mathrm{k}G_n\text{-umod}].$$

We call the category of unipotent kG-modules the category

$$\mathscr{U}_{\mathrm{k}} = \bigoplus_{n\in\mathbb{N}} \mathrm{k}G_n\text{-umod}\,.$$

This is an abelian category which is not semisimple. As above, the decomposition map yields a $\mathbb{Z}$-linear isomorphism $d_{\mathscr{U}} : [\mathscr{U}_K] \xrightarrow{\sim} [\mathscr{U}_{\mathrm{k}}]$. It is conjectured that this map is unitriangular on the basis of irreducible modules, yielding a parametrization of unipotent simple kG-modules (see Conjecture 6.21 below).

6.3.4. *The unipotent blocks.* Recall that d is the order of q^2 modulo ℓ. The partition of the unipotent KG_n-modules into ℓ-blocks was determined in [**23**, (10B), (11E)]. If ℓ is a linear prime (resp. a unitary prime), i.e., if f is odd (resp. even), then two unipotent characters of KG_n belong to the same ℓ-block if and only if their symbols have the same d-core (resp. d-cocore).

In addition, Fong-Srinivasan described the structure of the unipotent blocks of G_n with cyclic defect groups. First, the unipotent ℓ-block containing E_Θ has a cyclic defect if and only if the symbol Θ has a unique d-hook if ℓ is linear or a unique d-cohook if ℓ is unitary. In this case, let $\{X, Y\}$ be the d-core of Θ if ℓ is linear, and the d-cocore of Θ if ℓ is unitary. By [**24**, (5A), (6A)], the Brauer tree of the ℓ-block containing E_Θ is

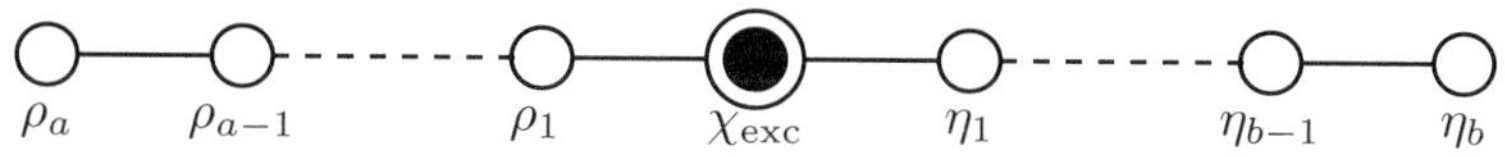

where

(a) if ℓ is linear, then $f = d = a = b$ and ρ_k (resp. η_k) is obtained by adding a d-hook to X (resp. Y) for each $k = 1, \ldots, d$.

(b) if ℓ is unitary, then $f = 2d$, $\rho_1, \ldots, \rho_a$ are the unipotent characters corresponding to the symbols obtained by adding a d-cohook which increases $|X|$ and $\eta_1, \ldots, \eta_b$ are the unipotent characters corresponding to the symbols obtained by adding a d-cohook which increases $|Y|$. In addition, $a = d + D$ and $b = d - D$ where D is the defect of the symbol.

6.4. The $\mathfrak{g}_\infty$-representation on $\mathscr{U}_K$.

We show in this section that the representation datum on KG-mod yields a categorical action of $\mathfrak{g}_\infty$ on $\mathscr{U}_K$.

6.4.1. *The ramified Hecke algebra.* Let $r = t(t+1)$ and $n = r+m$ with $t, m \geqslant 0$. Recall that the inflation from G_r to $L_{r,1^m}$ yields an equivalence between KG_r-umod and $KL_{r,1^m}$-umod which intertwines the functor $F_{n,r}$ with the parabolic induction $R_{L_{r,1^m}}^{G_n}$. In particular, we have a canonical isomorphism

$$\mathscr{H}(KG_n, E_t) := \mathrm{End}_{KG_n}(F^m(E_t))^{\mathrm{op}} \xrightarrow{\sim} \mathrm{End}_{KG_n}(R_{L_{r,1^m}}^{G_n}(E_t))^{\mathrm{op}}.$$

This algebra is isomorphic to the Hecke algebra of type B_m with parameters q and Q_t, giving a parametrisation of the constituents of (the head of) $F^m(E_t)$ in terms of bipartitions. More precisely, recall from §1.3 that to the categorical datum (E, F, X, T) is attached a map $\phi_{F^m} : \mathbf{H}_{K,m}^q \to \mathrm{End}(F^m)$. The evaluation of this map at the module E_t yields a K-algebra homomorphism

$$\phi_{K,m} : \mathbf{H}_{K,m}^q \to \mathscr{H}(KG_n, E_t), \quad X_k \mapsto X_k(E_t), \quad T_l \mapsto T_l(E_t).$$

We show that it induces the aforementioned isomorphism.

THEOREM 6.5. *Let $t, m \geqslant 0$ and $n = t(t+1) + m$. Then the map $\phi_{K,m}$ factors through a K-algebra isomorphism $\mathbf{H}_{K,m}^{Q_t\,;\,q} \xrightarrow{\sim} \mathscr{H}(KG_n, E_t)$.*

PROOF. The proof is similar to the proof of Theorem 4.7, see [**18**, thm. 4.12]. Write $Q_t = (Q_1, Q_2)$ and $X = X(E_t)$. We must check that the operator X on $F(E_t)$ satisfies the relation

$$(X - (-q)^{-1-t})(X - (-q)^t) = 0.$$

Then the invertibility of the morphism $\mathbf{H}_{K,m}^{Q_t\,;\,q} \to \mathscr{H}(KG_n, E_t)$ follows from [**42**]. In fact, it is shown there that X satisfies the relation

$$(X - \epsilon_t\,(-q)^{-1-t})(X - \epsilon_t\,(-q)^t) = 0 \tag{6.5}$$

for some $\epsilon_t = \pm 1$. Therefore we must show that $\epsilon_t = 1$ for all $t \geq 0$, which we will do by induction on t. First observe that the eigenvalues of $X_{1,0}$ on $R_{G_0}^{G_1}(K)$ are 1, $(-q)^{-1}$ and thus are powers of $-q$. Now fix $t \geq 1$ and assume that for all $t > s \geqslant 0$ the eigenvalues of $X(E_s)$ on $F(E_s)$ are powers of $-q$. We will show that the eigenvalues of $X(E_t)$ are also powers of $-q$ using the modular representation theory of G_n.

Recall that K is chosen with respect to an ℓ-modular system $(K, \mathscr{O}, \mathrm{k})$. Since the parametrization of unipotent characters does not depend on ℓ and (E, F, X, T) are defined over $\mathbb{Z}[1/q(q-1)]$, we can first choose a specific prime number ℓ and prove that the eigenvalues of $X(E_t)$ are powers of $-q$ modulo ℓ. We choose ℓ to be odd and such that the order of q in $\mathrm{k}^\times$ is $f := 4t$. Thus the order of q^2 is $d = 2t$. Note that ℓ is a unitary prime.

Set $r = t(t+1)$ and $n = r+1$. The cuspidal representation E_t is attached to the symbol

$$\Theta_t = \begin{Bmatrix} t & t-1 & t-2 & \cdots & -t & -t-1 & \cdots \\ & & & & & -t-1 & \cdots \end{Bmatrix}$$

Since only one d-cohook can be removed from Θ_t, the ℓ-block of E_t has cyclic defect groups. Moreover, the d-cocore of Θ_t equals Θ_{t-1}. Consequently, the Brauer tree of the ℓ-block of E_t is

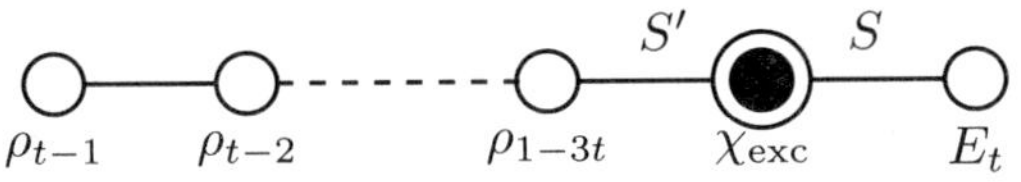

where a symbol Ξ_k of ρ_k is obtained by adding the d-cohook $(k, k+d)$ to Θ_{t-1} for $k \in \{1-3t, \ldots, t-1\}$. Explicitly, we have

$$\Xi_k = \begin{Bmatrix} t-1 & t-2 & \cdots & \widehat{k} & \cdots & \cdots \\ & & k+2t & -t & -t-1 & \cdots \end{Bmatrix}.$$

Here the notation $\widehat{k}$ means that the integer k has been removed. This symbol has defect $|2t-3| = |2(t-2)+1|$. Therefore by Proposition 6.3, the unipotent characters ρ_k all lie in the Harish-Chandra series above E_{t-2} (recall the convention $E_{-1} := E_0$). Furthermore, the bipartition μ_k such that $\Theta_{t-2}(\mu_k) = \Xi_k$ is $((1^{t-1-k}), (k+3t-1))$, except when $t = 1$ in which case $\mu_k = ((k+2), (1^{-k}))$.

The KG_r-module E_t is cuspidal, hence weakly cuspidal. It follows that all the composition factors of any ℓ-reduction of E_t are weakly cuspidal as well. In particular

$$E(S) = 0.$$

For each k we can compute the character $E(\rho_k)$ obtained by removing a 1-hook from Ξ_k. Two cases arise: if $k = 1-3t$ or $k = t-1$, then $E(\rho_k)$ is irreducible, and the corresponding symbols are equal respectively to

$$\begin{Bmatrix} t-1 & t-2 & \cdots & \widehat{-3t+2} & \cdots & \cdots \\ & & -t+1 & -t & -t-1 & \cdots \end{Bmatrix},$$

and

$$\begin{Bmatrix} t-2 & \cdots & \cdots & \cdots & \cdots \\ & 3t-2 & -t & -t-1 & \cdots \end{Bmatrix};$$

otherwise $E(\rho_k)$ has two constituents whose symbols are

$$\begin{Bmatrix} t-1 & t-2 & \cdots & \widehat{k+1} & \cdots \\ & k+2t & -t & -t-1 & \cdots \end{Bmatrix},$$

and

$$\begin{Bmatrix} t-1 & t-2 & \cdots & \widehat{k} & \cdots \\ & k+2t-1 & -t & -t-1 & \cdots \end{Bmatrix}.$$

We deduce that $\sum(-1)^k[E(\rho_k)] = 0$ in $[KG_{r-1}\text{-mod}]$. Since $[S'] = \sum(-1)^k d_{\mathscr{U}}([\rho_k])$ in $[\mathrm{k}G_{r-1}\text{-mod}]$, this implies that

$$E(S') = 0.$$

Therefore, the two composition factors S, S' of the ℓ-reduction of the exceptional characters are weakly cuspidal, which forces the exceptional characters to be weakly cuspidal as well.

Given a symbol Θ, the module $F(E_\Theta)$ is the sum of the unipotent characters associated to the symbols obtained from Θ by adding a 1-hook. Thus, we have $F(E_t) = E_\Xi \oplus E_{\Xi'}$, where

$$\Xi = \left\{ \begin{matrix} t & t-1 & \cdots & -t & -t-1 & -t-2 & \cdots \\ & & & & -t & -t-2 & \cdots \end{matrix} \right\},$$

$$\Xi' = \left\{ \begin{matrix} t+1 & t-1 & t-2 & \cdots & -t-1 & \cdots \\ & & & & -t-1 & \cdots \end{matrix} \right\}.$$

Let B, B' be the ℓ-blocks of E_Ξ and $E_{\Xi'}$ respectively and b, b' be the corresponding idempotents in KG_n. Since the d-cocores of Ξ and Ξ' are different, the idempotents b and b' are orthogonal. Moreover $X(E_t)$ has the eigenvalues $\epsilon_t(-q)^{-1-t}$ and $\epsilon_t(-q)^t$, on $bF(E_t) = E_\Xi$ and $b'F(E_t) = E_{\Xi'}$ respectively. Note that $\epsilon_t(-q)^{-1-t}$ and $\epsilon_t(-q)^t$ are not congruent modulo ℓ since $q^{2t+1} \equiv -q$.

Let χ be an exceptional character. Recall that S is isomorphic to any ℓ-reduction of E_t. Since it is a composition factor of the ℓ-reduction of χ, we deduce that both $bF(\chi)$ and $b'F(\chi)$ are non-zero. Since χ is weakly cuspidal, by the Mackey formula $F(\chi)$ has at most two irreducible constituents and $X(\chi)$ at most two eigenvalues whose product equals $-q^{-1}$. We deduce that $F(\chi) = bF(\chi) \oplus b'F(\chi)$. In addition, the eigenvalues of $X(\chi)$ must be congruent to the eigenvalues of $X(S)$ on $F(S)$, which are equal to $\epsilon_t(-q)^{-1-t}$ and $\epsilon_t(-q)^t$. Now, S' is a composition factor of the ℓ-reduction of χ so one of $bF(S')$ or $b'F(S')$ must be non-zero, and therefore $X(S')$ must have an eigenvalue congruent to $\epsilon_t(-q)^{-1-t}$ or $\epsilon_t(-q)^t$. To conclude that $\epsilon_t = 1$, we must compute the possible eigenvalues of $X(\rho_{1-3t})$ on $bF(\rho_{1-3t})$ and $b'F(\rho_{1-3t})$ and use that S' is a composition factor of the ℓ-reduction of ρ_{1-3t}.

We abbreviate $\rho = \rho_{1-3t}$. The Harish-Chandra induction of ρ is $F(\rho) = E_\Theta \oplus E_{\Theta'} \oplus E_{\Theta''}$ where

$$\Theta = \left\{ \begin{matrix} t & t-2 & t-3 & \cdots & \widehat{1-3t} & \cdots \\ & & & -t+1 & -t & \cdots \end{matrix} \right\},$$

$$\Theta' = \left\{ \begin{matrix} t-1 & t-2 & \cdots & \widehat{-3t} & \cdots & \cdots \\ & & -t+1 & -t & -t-1 & \cdots \end{matrix} \right\},$$

$$\Theta'' = \left\{ \begin{matrix} t-1 & t-2 & \cdots & \widehat{1-3t} & \cdots & \cdots \\ & & -t+2 & -t & -t-1 & \cdots \end{matrix} \right\}.$$

We first observe that the modules E_Θ, $E_{\Theta'}$ belong to the ℓ-blocks B, B' and that $E_{\Theta''}$ is projective. Indeed, the d-cocores of the symbols Θ, Θ' and Θ'' are different and equal respectively to

$$\left\{ \begin{matrix} t & t-2 & t-3 & \cdots & \cdots \\ & & & -t & \cdots \end{matrix} \right\}, \quad \left\{ \begin{matrix} t-1 & t-2 & \cdots & \cdots & \cdots \\ & & -t+1 & -t-1 & \cdots \end{matrix} \right\}, \quad \Theta''.$$

Since ρ belong to the Harish-Chandra series above E_{t-2}, we can compute the eigenvalues of $X(\rho)$ using the eigenvalues of $X(E_{t-2})$. Let $n' = r + m - 4t + 2$ for some integer $m \geqslant 0$. If $t \geq 2$ we can use the induction hypothesis and the map $\phi_{K,m}$ yields a K-algebra isomorphism

$$\mathbf{H}_{K,m}^{Q_{t-2},q} \xrightarrow{\sim} \mathscr{H}(KG_{n'}, E_{t-2}).$$

By the discussion after (6.4), the corresponding bijection

$$\operatorname{Irr}(KG_{n'}, E_{t-2}) \xleftrightarrow{1:1} \operatorname{Irr}(\mathbf{H}_{K,m}^{Q_{t-2},q})$$

takes the module $E_{\Theta_{t-2}(\mu)}$ to $S(\mu)_K^{Q_{t-2},q}$ for each 2-partition μ of m. Under this parametrization, the character ρ of KG_r and the characters E_Θ, $E_{\Theta'}$ and $E_{\Theta''}$ of KG_n are mapped to the modules

$$S(\lambda)_K^{Q_{t-2},q}, \quad S(\mu)_K^{Q_{t-2},q}, \quad S(\mu')_K^{Q_{t-2},q}, \quad S(\mu'')_K^{Q_{t-2},q}$$

labelled by the following 2-partitions

$$\lambda = ((1^{4t-2}), \emptyset), \quad \mu = ((21^{4t-3}), \emptyset), \quad \mu' = ((1^{4t-1}), \emptyset), \quad \mu'' = ((1^{4t-2}), (1)).$$

The (Q_{t-2}, q)-shifted residue of the boxes $Y(\mu)\backslash Y(\lambda)$ and $Y(\mu')\backslash Y(\lambda)$ are $(-1)^t q^{t-1}$ and $(-1)^t q^{-3t}$ respectively. They are congruent to $(-q)^{-1-t}$ and $(-q)^t$ modulo ℓ, because q^{2t} is congruent to -1 modulo ℓ. We deduce that the eigenvalues of the operator $X(\rho_{1-3t})$ on E_Θ, $E_{\Theta'}$ are congruent to $(-q)^{-1-t}$ and $(-q)^t$ modulo ℓ. By the above argument, at least one of these must be congruent to $\epsilon_t(-q)^{-1-t}$ or $\epsilon_t(-q)^t$, which forces $\epsilon_t = 1$.

Finally, if $t < 2$ (in which case $t = 1$) we use the fact that $E_{t-2} = E_{1-t}$ and $Q_{t-2} = Q_{1-t}^\dagger$. In particular, the bijection $\operatorname{Irr}(KG_{n'}, E_{t-2}) \to \operatorname{Irr}(\mathbf{H}_{K,m}^{Q_{1-t},q})$ takes the module $E_{\Theta_{t-2}(\mu)}$ to $S(\mu^\dagger)_K^{(Q_{1-t},q)}$ for any bipartition μ of m. With $t = 1$, we deduce that the unipotent characters ρ, E_Θ, $E_{\Theta'}$ and $E_{\Theta''}$ correspond to Specht modules of $\mathbf{H}_{K,m}^{Q_0,q}$ labelled by the 2-partitions $\lambda = (\emptyset, (1^2))$, $\mu = (\emptyset, (21))$, $\mu' = (\emptyset, (1^3))$ and $\mu'' = ((1), (1^2))$. With $Q_0 = (1, -q^{-1})$, the (Q_0, q)-shifted residues of the boxes $Y(\mu) \setminus Y(\lambda)$ and $Y(\mu') \setminus Y(\lambda)$ are -1 and $-q^{-3}$ respectively, which are congruent to q^{-2} and $-q$ since q^2 is congruent to -1 modulo ℓ. Therefore $\epsilon_1 = 1$. □

6.4.2. *The $\mathfrak{g}_\infty$-representation on $\mathscr{U}_K$.* By Proposition 6.1, the Harish-Chandra induction and restriction yield a representation datum (E, F, X, T) on the category

$$\mathscr{U}_K = \bigoplus_{n \in \mathbb{N}} KG_n\text{-umod}.$$

For any $t, m, n \in \mathbb{N}$, let (KG_n, E_t)-mod be the Serre subcategory of $\mathscr{U}_K$ generated by the modules $F^m(E_t)$ with $n = r + m$ and $r = t(t+1)$. Write

$$\mathscr{U}_{K,t} := \bigoplus_{n \geqslant 0} (KG_n, E_t)\text{-mod}.$$

We have $\operatorname{Irr}((KG_n, E_t)\text{-mod}) = \operatorname{Irr}(KG_n, E_t)$ and $\mathscr{U}_K = \bigoplus_{t \geqslant 0} \mathscr{U}_{K,t}$.

DEFINITION 6.6. Let I_∞ denote the subset $q^{\mathbb{Z}} \sqcup (-q^{\mathbb{Z}})$ of $K^\times$. We define $\mathfrak{g}_\infty$ to be the (derived) Kac-Moody algebra associated to the quiver $I_\infty(q)$.

To avoid cumbersome notation, we will write $I_\infty = I_\infty(q)$. We denote by $\{\Lambda_i\}$, $\{\alpha_i\}$ and $\{\alpha_i^\vee\}$ the fundamental weights, simple roots and simple coroots of $\mathfrak{g}_\infty$. Then $\mathrm{X}_\infty = \mathrm{P}_\infty = \bigoplus \mathbb{Z}\Lambda_i$. There is a Lie algebra isomorphism $(\mathfrak{sl}_{\mathbb{Z}})^{\oplus 2} \xrightarrow{\sim} \mathfrak{g}_\infty$ such that $(\alpha_k^\vee, 0) \mapsto \alpha_{q^k}^\vee$ and $(0, \alpha_k^\vee) \mapsto \alpha_{-q^k}^\vee$. Since the pair Q_t belongs to $(I_\infty)^2$, the $\mathfrak{g}_\infty$-module $\mathbf{F}(Q_t)_\infty$ is well-defined.

THEOREM 6.7. *Let $t \geqslant 0$ and Q_t be as in* (6.3).

(a) *The Harish-Chandra induction and restriction functors yield a representation of $\mathfrak{g}_\infty$ on $\mathscr{U}_{K,t}$ which is isomorphic to $\mathscr{L}(\Lambda_{Q_t})_\infty$.*

(b) *The map $|\mu, Q_t\rangle_\infty \mapsto [E_{\Theta_t(\mu)}]$ gives a $\mathfrak{g}_\infty$-module isomorphism $\mathbf{F}(Q_t)_\infty \xrightarrow{\sim} [\mathscr{U}_{K,t}]$.*

PROOF. The proof is similar to the proof of Theorem 4.10 in [**18**, thm. 4.15]. It is a consequence of Theorem 6.5 and the discussion after (6.4). □

6.5. The $\mathfrak{g}_{2d}$-representation on $\mathscr{U}_{\mathrm{k}}$.

We now turn to the case of unipotent representations in positive characteristic. We first deduce from the previous construction a categorical action of the derived Kac-Moody algebra $\mathfrak{g}'_{2d}$ on $\mathscr{U}_{\mathrm{k}}$. We then show how to extend this action depending on the parity of f. This reflects the difference between the case of linear primes (f odd) and unitary primes (f even).

6.5.1. *The $\mathfrak{g}'_{2d}$-representation on $\mathscr{U}_{\mathrm{k}}$.* Recall that $(K, \mathscr{O}, \mathrm{k})$ is an ℓ-modular system with $K \supset \mathbb{Q}_\ell$ and $\mathrm{k} \supset \overline{\mathbb{F}}_\ell$. We still work under the assumption that $\ell \nmid q$ and ℓ and q are odd. Recall that d (resp. f) is the order of q^2 (resp. q) modulo ℓ.

DEFINITION 6.8. Let I_{2d} be the quiver obtained from I_∞ by specialization $\mathscr{O} \to \mathrm{k}$. We define $\mathfrak{g}'_{2d}$ to be the derived Kac-Moody algebra associated to the quiver I_{2d}.

If f is odd then $f = d$ and -1 cannot be expressed as a power of q in k. In that case the quiver I_{2d} decomposes as $I_{2d} = I_{d,1} \sqcup I_{d,2}$ with $I_{d,1} = q^{\mathbb{Z}}$ and $I_{d,2} = -q^{\mathbb{Z}}$ being both cyclic quivers of size d. This yields a Lie algebra isomorphism

$$(\widetilde{\mathfrak{sl}}_d)^{\oplus 2} \simeq \mathfrak{g}'_{d,1} \oplus \mathfrak{g}'_{d,2} = \mathfrak{g}'_{2d}$$

such that $(\alpha_k^\vee, 0) \mapsto \alpha_{q^k}^\vee$ and $(0, \alpha_k^\vee) \mapsto \alpha_{-q^k}^\vee$.

If f is even, then $f = 2d$ and $q^d = -1$. Hence $I_{2d} = q^{\mathbb{Z}}$ is a cyclic quiver of size $2d$ and we have an isomorphism $\widetilde{\mathfrak{sl}}_f \simeq \mathfrak{g}'_{2d}$ such that $\alpha_k^\vee \mapsto \alpha_{q^k}^\vee$.

The specialization from $\mathscr{O} \subset K$ to k yields a morphism of quivers $\mathrm{sp} : I_\infty \to I_{2d}$ and a morphism of abelian groups $\mathrm{P}_\infty \to \mathrm{P}_{2d}$ such that $\Lambda_i \mapsto \Lambda_{\mathrm{sp}(i)}$. The infinite sums $E_i = \bigoplus_{\mathrm{sp}(j)=i} E_j$ and $F_i = \bigoplus_{\mathrm{sp}(j)=i} F_j$ give well-defined operators on $\mathbf{F}(Q_t)_\infty$. This yields a representation of $\mathfrak{g}'_{2d}$ on $\mathbf{F}(Q_t)_\infty$ such that the linear map

$$\mathrm{sp} \,:\, \mathrm{Res}_{\mathfrak{g}'_{2d}}^{\mathfrak{g}_\infty} \mathbf{F}(Q_t)_\infty \to \mathbf{F}(Q_t)_{2d}$$

given by $|\mu, Q_t\rangle_\infty \mapsto |\mu, Q_t\rangle_{2d}$ is a $\mathfrak{g}'_{2d}$-equivariant isomorphism.

Under the map $d_{\mathscr{U}} : [\mathscr{U}_K] \to [\mathscr{U}_{\mathrm{k}}]$ and the isomorphism $\bigoplus_{t\in\mathbb{N}} \mathbf{F}(Q_t)_\infty \xrightarrow{\sim} [\mathscr{U}_K]$ in Theorem 6.7, the map sp endows $[\mathscr{U}_{\mathrm{k}}]$ with a representation of $\mathfrak{g}'_{2d}$ which is compatible with the representation associated with the representation datum. More precisely, the following hold.

PROPOSITION 6.9. *For each $i \in I_{2d}$, let $\mathrm{k}E_i$ and $\mathrm{k}F_i$ be the generalized i-eigenspace of X on $\mathrm{k}E$ and $\mathrm{k}F$. Then*

(a) $[\mathrm{k}E_i]$, $[\mathrm{k}F_i]$ *endow* $[\mathscr{U}_{\mathrm{k}}]$ *with a structure of* $\mathfrak{g}'_{2d}$*-module,*
(b) $d_{\mathscr{U}}$ *yields a* $\mathfrak{g}'_{2d}$*-module isomorphism* $\mathrm{Res}^{\mathfrak{g}_\infty}_{\mathfrak{g}'_{2d}}[\mathscr{U}_K] \xrightarrow{\sim} [\mathscr{U}_{\mathrm{k}}]$,
(c) *the map* $|\mu, Q_t\rangle_{2d} \mapsto d_{\mathscr{U}}([E_{\Theta_t(\mu)}])$ *yields a* $\mathfrak{g}'_{2d}$*-module isomorphism*

$$\bigoplus_{t\in\mathbb{N}} \mathbf{F}(Q_t)_{2d} \xrightarrow{\sim} [\mathscr{U}_{\mathrm{k}}].$$

□

PROOF. The proposition is a direct application of Theorem 6.7, once we have proved that the decomposition map $d_{\mathscr{U}}$ is a vector space isomorphism. This is known by Theorem 3.4 since ℓ is odd. □

In order to prove that this, and the representation datum (E, F, X, T) on $\mathscr{U}_{\mathrm{k}}$ introduced in §6.2, endow $\mathscr{U}_{\mathrm{k}}$ with a $\mathfrak{g}'_{2d}$-representation, it remains to see that weight spaces are sums of blocks. This follows from the following lemma.

LEMMA 6.10. *Let* $t, s \geqslant 0$ *and* μ, λ *be bipartitions. If* $E_{\Theta_t(\mu)}$ *and* $E_{\Theta_s(\lambda)}$ *are in the same* ℓ*-block then* $|\mu, Q_t\rangle_{2d}$ *and* $|\lambda, Q_s\rangle_{2d}$ *have the same weight for the action of* $\mathfrak{g}'_{2d}$.

PROOF. Recall that if f is even (resp. odd), two unipotent characters of G_n lie in the same ℓ-block if and only if the corresponding symbols have the same d-cocore (resp. d-core), see §6.3.4. Therefore it is enough to show that adding/removing d-cohooks (resp. d-hooks) does not affect the weight of $|\mu, Q_t\rangle_{2d}$ for the action of $\mathfrak{g}'_{2d}$.

Write $Q_t = (Q_1, Q_2)$ and $\mu = (\mu^1, \mu^2)$. Using Proposition 2.2 and (2.4) one can compute the weight of $|\mu, Q_t\rangle_{2d}$ from the β-set $\beta_0(\mu)$. For $i \in I_{2d}$ and N large enough, we have

$$\begin{aligned}\langle \mathrm{wt}(|\mu, Q_t\rangle_{2d}), \alpha_i^\vee\rangle = &\, |\{\beta \in \beta_0(\mu^1),\, \beta \geqslant -N \,|\, q^\beta Q_1 \equiv i \bmod \ell\}| \\ &- |\{\beta \in \beta_0(\mu^1),\, \beta \geqslant -N+1 \,|\, q^\beta Q_1 \equiv qi \bmod \ell\}| \\ &+ |\{\beta \in \beta_0(\mu^2),\, \beta \geqslant -N \,|\, q^\beta Q_2 \equiv i \bmod \ell\}| \\ &- |\{\beta \in \beta_0(\mu^2),\, \beta \geqslant -N+1 \,|\, q^\beta Q_2 \equiv qi \bmod \ell\}|.\end{aligned}$$

This expression becomes simpler when working with the charged β-sets used in the definition of $\Theta_t(\mu)$ in §6.3.2. Write $\Theta_t(\mu) = \{X, Y\}$ where $X = \beta_t(\mu^1)$ and $Y = \beta_{-1-t}(\mu^2)$. Then by definition of Q_t we have, for N large enough

$$\begin{aligned}\langle \mathrm{wt}(|\mu, Q_t\rangle_{2d}), \alpha_i^\vee\rangle = &\, |\{x \in X,\, x \geqslant -N \,|\, q^x \equiv (-1)^t i \bmod \ell\}| \\ &- |\{x \in X,\, x \geqslant -N+1 \,|\, q^x \equiv (-1)^t qi \bmod \ell\}| \\ &+ |\{y \in Y,\, y \geqslant -N \,|\, q^y \equiv (-1)^{t+1} i \bmod \ell\}| \\ &- |\{y \in Y,\, y \geqslant -N+1 \,|\, q^y \equiv (-1)^{t+1} qi \bmod \ell\}|.\end{aligned}$$

If f is odd, then $q^d = q^f = 1$ in k and adding or removing a d-hook has the effect of removing x from X (resp. Y) and adding $x \pm d$ to X (resp. Y), which does not change the congruence of q^x. In particular, this operation does not affect $\mathrm{wt}(|\mu, Q_t\rangle_{2d})$.

If f is even, then $q^d = -1$, $q^f = 1$ and adding or removing a d-cohook has the effect of removing x from X (resp. Y) and adding $x \pm d$ to Y (resp. x). This changes q^x to $q^{x\pm d} \equiv -q^x$. Therefore if $t' = t\pm 1$ and $\Theta_{t'}(\mu')$ is the symbol obtained from $\Theta_t(\mu)$ by adding or removing a d-cohook we have

$$\mathrm{wt}\big(|\mu, Q_t\rangle_{2d}\big), \alpha_i^\vee\rangle = \mathrm{wt}\big(|\mu', Q_{t'}\rangle_{2d}\big), \alpha_{-i}^\vee\rangle.$$

In particular, symbols corresponding to unipotent characters in the same ℓ-block have the same weight. □

Notice that the lemma implies that the simple unipotent modules are weight vectors of the $\mathfrak{g}'_{2d}$-action. From this and the results in the previous section we deduce the expected result.

THEOREM 6.11. *Recall that ℓ and q are odd, and $\ell \nmid q(q^2-1)$. The representation datum on $\mathscr{U}_\mathrm{k}$ yields a $\mathfrak{g}'_{2d}$-representation such that the map $d_{\mathscr{U}} : [\mathscr{U}_K] \to [\mathscr{U}_\mathrm{k}]$ intertwines the representations of $\mathfrak{g}_\infty$ and $\mathfrak{g}'_{2d}$. There is a $\mathfrak{g}'_{2d}$-module isomorphism $\bigoplus_{t\in\mathbb{N}} \mathbf{F}(Q_t)_{2d} \xrightarrow{\sim} [\mathscr{U}_\mathrm{k}]$ sending $|\mu, Q_t\rangle_{2d}$ to $d_{\mathscr{U}}([E_{\Theta_t(\mu)}])$. Further, the classes in $[\mathscr{U}_\mathrm{k}]$ of the simple unipotent modules are weight vectors for the $\mathfrak{g}'_{2d}$-action.* □

6.5.2. *The $\mathfrak{g}_{2d}$-representation on $\mathscr{U}_\mathrm{k}$ in the linear prime case.* In this section we assume that f is odd, so $f = d$. In that case the Kac-Moody algebra $\widetilde{\mathfrak{g}}_{2d}$ associated with the quiver I_{2d} is isomorphic to $(\widehat{\mathfrak{sl}}_d)^{\oplus 2}$. As in the case of unitary groups, see §4.5.1, the action of $\mathfrak{g}'_{2d}$ on $\mathscr{U}_\mathrm{k}$ can be naturally extended to an action of an algebra $\mathfrak{g}_{2d}$ lying between $\mathfrak{g}'_{2d}$ and $\widetilde{\mathfrak{g}}_{2d}$.

Let $\widetilde{\mathrm{X}}_{2d}$ and $\widetilde{\mathrm{X}}_{2d}^\vee$ be the lattices corresponding to $\widetilde{\mathfrak{g}}_{2d}$. Since f is odd, I_{2d} is the disjoint union of two cyclic quivers. We choose α_1 and α_{-1} to be the affine roots attached to each of these quivers. Then we have

$$\widetilde{\mathrm{X}}_{2d} = \mathrm{P}_{2d} \oplus \mathbb{Z}\delta_+ \oplus \mathbb{Z}\delta_-, \quad \widetilde{\mathrm{X}}_{2d}^\vee = \mathrm{Q}_{2d}^\vee \oplus \mathbb{Z}\partial_+ \oplus \mathbb{Z}\partial_-,$$

with $\delta_+ = \sum \alpha_{q^j}$, $\delta_- = \sum \alpha_{-q^j}$, $\partial_+ = \Lambda_1^\vee$ and $\partial_- = \Lambda_{-1}^\vee$. We set $\partial = \partial_+ + \partial_-$ and $\delta = (\delta_+ + \delta_-)/2$. We define $\mathfrak{g}_{2d} := \mathfrak{g}'_{2d} \oplus \mathbb{C}\partial$ and we view it as the Kac-Moody algebra associated with the lattices

$$\mathrm{X}_{2d} := \mathrm{P}_{2d} \oplus \mathbb{Z}\delta \simeq \widetilde{\mathrm{X}}_{2d}/(\delta_+ - \delta_-), \quad \mathrm{X}_{2d}^\vee := \mathrm{Q}_{2d}^\vee \oplus \mathbb{Z}\partial.$$

The pairing $\widetilde{\mathrm{X}}_{2d}^\vee \times \widetilde{\mathrm{X}}_{2d} \longrightarrow \mathbb{Z}$ induces in the obvious way a perfect pairing $\mathrm{X}_{2d}^\vee \times \mathrm{X}_{2d} \longrightarrow \mathbb{Z}$.

For $t \in \mathbb{N}$, the Fock space $\mathbf{F}(Q_t)_{2d}$ has a tensor product decomposition into level 1 Fock spaces

$$\mathbf{F}(Q_t)_{2d} \simeq \mathbf{F}((-q)^t)_d \otimes \mathbf{F}((-q)^{-1-t})_d.$$

Out of the charged Fock spaces $(\mathbf{F}((-q)^t)_d, t)$ and $(\mathbf{F}((-q)^{-1-t})_d, -1-t)$ and the isomorphism $\widetilde{\mathfrak{g}}_{2d} \simeq (\widehat{\mathfrak{sl}}_d)^{\oplus 2}$ (which depends on the parity of t) we can therefore equip $\mathbf{F}(Q_t)_{2d}$ with an action of $\widetilde{\mathfrak{g}}_{2d}$ which in turn restricts to an action of $\mathfrak{g}_{2d}$.

Recall that two unipotent characters are in the same ℓ-block if and only if the corresponding symbols have the same d-core. In particular, the unipotent characters of a given unipotent block all lie in the same Harish-Chandra series. In addition, two unipotent characters lying in different Harish-Chandra series lie in different blocks. Consequently, for each $t \in \mathbb{N}$ we can form the category $\mathscr{U}_{\mathrm{k},t}$ associated with the Harish-Chandra series labelled by t, yielding

$$\mathscr{U}_\mathrm{k} = \bigoplus_{t\in\mathbb{N}} \mathscr{U}_{\mathrm{k},t} \quad \text{with} \quad [\mathscr{U}_\mathrm{k}] \simeq \mathbf{F}(Q_t)_{2d}.$$

Using the action of $\mathfrak{g}_{2d}$ on $\mathbf{F}(Q_t)_{2d}$ defined above we equip each $[\mathscr{U}_{\mathrm{k},t}]$ with a structure of $\mathfrak{g}_{2d}$-module which extends the structure of $\mathfrak{g}'_{2d}$-module defined in §6.5.1. The situation is completely similar to §4.6.1 and we get the following theorem.

THEOREM 6.12. *Recall that ℓ and q are odd, and $\ell \nmid q(q^2-1)$. Assume that f is odd. For each $t \in \mathbb{N}$, the Harish-Chandra induction and restriction functors yield a representation of $\mathfrak{g}_{2d}$ on $\mathscr{U}_{\mathrm{k},t}$ which categorifies $\mathbf{F}(Q_t)_{2d}$.* □

6.5.3. *Combinatorics of d-cohooks and d-cocores.* We now concentrate on the unitary prime case, i.e., the case where f is even, which is the most delicate. We wish to define an action of a (non derived) Kac-Moody algebra $\mathfrak{g}_{2d}$ on $[\mathscr{U}_{\mathrm{k}}]$ which extends the action of $\mathfrak{g}'_{2d}$. Then, we will extend the grading of $[\mathscr{U}_{\mathrm{k}}]$ from P_{2d} to $\mathrm{X}_{2d} = \mathrm{P}_{2d} \oplus \mathbb{Z}\delta/2$. Before this, we must introduce some combinatorial tools related to symbols.

Let $\mu = (\mu^1, \mu^2)$ be a bipartition and $t \in \mathbb{Z}$. Consider the symbol Θ defined by

$$\Theta = \{X\,,\, Y - d\} = \Theta_t(\mu)$$

where $X = \beta_t(\mu^1)$ and $Y = \beta_{d-1-t}(\mu^2)$. With this notation, removing a d-cohook on Θ changes the pair (X, Y) to

$$(X \smallsetminus \{x\}, Y \sqcup \{x\}) \quad \text{or} \quad (X \sqcup \{y-2d\}, Y \smallsetminus \{y\}).$$

We denote by $s = (s_1, \ldots, s_{2d})$ and $r = (r_1, \ldots, r_{2d})$ the $2d$-core of the charged partitions (μ^1, t) and $(\mu^2, d-1-t)$ respectively, and by $(\mu^{1,1}, \ldots, \mu^{1,2d})$ and $(\mu^{2,1}, \ldots, \mu^{2,2d})$ their $2d$-quotients. Hence, we have

$$\tau_{2d}(\mu^1, t) = (\mu^{1,1}, \ldots, \mu^{1,2d}, s_1, \ldots, s_{2d}),$$
$$\tau_{2d}(\mu^2, d-1-t) = (\mu^{2,1}, \ldots, \mu^{2,2d}, r_1, \ldots, r_{2d}).$$

LEMMA 6.13. *The symbol Θ is a d-cocore if and only if*

(a) *μ^1 and μ^2 are $2d$-cores,*
(b) *$r_p - s_p \in \{0,1\}$ for all $p \in \{1, \ldots, 2d\}$.*

PROOF. Let us write $X = \beta_t(\mu^1)$ and $Y = \beta_{d-1-t}(\mu^2)$. Then Θ is a d-cocore if and only if $X \subset Y \subset X + 2d$. In particular, we must have $X \subset X + 2d$ and $Y \subset Y + 2d$, which is equivalent to μ^1 and μ^2 being $2d$-cores. In that case we have

$$X = \bigsqcup_{p=1}^{2d} (p - 2d + 2d\beta_{s_p}(\emptyset)) \quad \text{and} \quad Y = \bigsqcup_{p=1}^{2d} (p - 2d + 2d\beta_{r_p}(\emptyset)).$$

Using the definition of the β-set of an empty partition we deduce that $X \subset Y$ if and only if $s_p \leqslant r_p$ for all p. For the same reason, we have $Y \subset X + 2d$ if and only if $r_p \leqslant s_p + 1$ for all p. The combination of the two conditions gives

$$r_p - s_p \in \{0,1\}, \quad \text{for all } p.$$

□

LEMMA 6.14. *Let Θ, Θ' be two symbols. The following assertions are equivalent:*

(i) *there exists a sequence of symbols $\Theta = \Theta_0, \Theta_1, \ldots, \Theta_m = \Theta'$ where Θ_{i+1} is obtained from Θ_i by adding or removing a d-cohook,*
(ii) *$s + r = s' + r'$.*

PROOF. Let μ be the bipartition such that $\Theta = \Theta_t(\mu)$. We write

$$\beta_t(\mu^1) = \bigsqcup_{p=1}^{2d}(p-2d+2d\beta_{s_p}(\mu^{1,p})) \quad \text{and} \quad \beta_{d-1-t}(\mu^2) = \bigsqcup_{p=1}^{2d}(p-2d+2d\beta_{r_p}(\mu^{2,p})).$$

Therefore adding or removing a d-cohook changes (s_p, r_p) to $(s_p, r_p) \pm (1,-1)$ for some $p \in \{1, \ldots, 2d\}$. This does not change $s+r$.

Let $\Theta_\circ$, $\Theta'_\circ$ be the d-cocore of Θ, Θ', and $(s_\circ, r_\circ)$, $(s'_\circ, r'_\circ)$ be the corresponding $2d$-cores. By the previous argument, we have

$$s+r = s_\circ + r_\circ, \quad s'+r' = s'_\circ + r'_\circ.$$

Now, assertion (i) is equivalent to $\Theta_\circ = \Theta'_\circ$, which yields in particular

$$s+r = s_\circ + r_\circ = s'+r'.$$

Conversely, if we assume (ii), then $s_\circ + r_\circ = s'_\circ + r'_\circ$ which we can write

$$r_\circ - s_\circ = r'_\circ - s'_\circ + 2(r_\circ - r'_\circ).$$

But by Lemma 6.13, both $r_\circ - s_\circ$ and $r'_\circ - s'_\circ$ are $2d$-tuples of integers equal to 0 or 1. This forces $r_\circ = r'_\circ$ and therefore $s_\circ = s'_\circ$. Since $t_\circ = \sum_p s_{\circ,p}$, this determines $\Theta_\circ$ uniquely and therefore we have $\Theta_\circ = \Theta'_\circ$. □

6.5.4. *The weight of a symbol.* We are still working under the assumption that $f = 2d$ is even. In particular $q^d = -1$ in k. Fix an integer $t \in \mathbb{Z}$ and set

$$P_t = q^{c_t} = (q^t, -q^{-1-t}), \quad c_t = (t, d-1-t).$$

In particular $Q_t = (-1)^t P_t$. Let μ be a bipartition and let s and r be the $2d$-cores of the charged partitions (μ^1, t) and $(\mu^2, d-1-t)$. We consider the weight in $\mathrm{X} = \mathrm{P} \oplus \mathbb{Z}\,\delta/2$ given by

$$\begin{aligned} \mathrm{wt}(\mu,t) &= \Lambda_{P_t} - \sum_i n_i(\mu, P_t)\,\alpha_i - \nabla(s,r)\,\delta, \\ \nabla(s,r) &= \Delta(t,2d) + \Delta(d-1-t,2d) + t/2, \end{aligned} \tag{6.6}$$

where $\Delta(t,2d)$ is as in (2.3). For any $2d$-tuple of integers $x = (x_1, \ldots, x_{2d})$, we define as in §2.3.1

$$\pi_x = \sum_{p=1}^{2d}(x_p - x_{p+1})\,\Lambda_{q^p}.$$

Then, we have

$$\begin{aligned} \mathrm{wt}(\mu,t) &\equiv \Lambda_{P_t} - \sum_i n_i(\mu, P_t)\,\alpha_i^{\mathrm{cl}} \bmod \mathbb{Z}\,\delta, \\ &\equiv \pi_{s+r} + 2\Lambda_1 \bmod \mathbb{Z}\,\delta. \end{aligned} \tag{6.7}$$

The effect on the weight of removing a d-cohook is studied in the following lemma.

LEMMA 6.15. *Let t, t' be integers and μ, μ' be bipartitions. Set $\Theta = \Theta_t(\mu)$ and $\Theta' = \Theta_{t'}(\mu')$. If Θ' is obtained from Θ by removing a d-cohook, then* $\mathrm{wt}(\mu,t) = \mathrm{wt}(\mu',t') - \delta/2$.

PROOF. By (2.4), we have

$$\begin{aligned}\mathrm{wt}(\mu,t) &= \mathrm{wt}(|\mu^1,t\rangle_{2d}) + \mathrm{wt}(|\mu^2,d-1-t\rangle_{2d}) + \big(\Delta(t,2d)+\Delta(d-1-t,2d)-\nabla(s,r)\big)\,\delta\\ &= \mathrm{wt}(|\mu^1,t\rangle_{2d}) + \mathrm{wt}(|\mu^2,d-1-t\rangle_{2d}) - t\,\delta/2\\ &= \pi_{s+r} + 2\Lambda_1 - \big(w_{2d}(\mu^1)+w_{2d}(\mu^2)+\Delta(s,1)+\Delta(r,1)+t/2\big)\delta.\end{aligned}$$

where we used the formulas in §2.3.1. Therefore we must study the effect of removing a d-cohook on the integer

$$w_{2d}(\mu^1)+w_{2d}(\mu^2)+\Delta(s,1)+\Delta(r,1)+t/2.$$

Note that by Lemma 6.14, removing a d-cohook has no effect on $s+r$.

First, assume that there is an integer $x\in\beta_t(\mu^1)$ such that $x\notin\beta_{d-1-t}(\mu^2)$ and write

$$\beta_{t'}(\mu'^1)=\beta_t(\mu^1)\setminus\{x\},\quad \beta_{d-1-t'}(\mu'^2)=\beta_{d-1-t}(\mu^2)\sqcup\{x\}.$$

Fix $p\in\{1,\dots,2d\}$ and $z\in\beta_{s_p}(\mu^{1,p})$ such that

$$x=p-2d+2dz.$$

Fix u_0 such that $z=z_{u_0}$ where $\beta_{s_p}(\mu^{1,p})=\{z_1>z_2>\cdots\}$. We have

$$|\mu^{1,p}|=\sum_{u\geqslant 1}(z_u+u-1-s_p).$$

We also have

$$\begin{aligned}|\mu'^{1,p}|-|\mu^{1,p}| &= \sum_{u\geqslant 1}(z'_u+u-1-s_p+1)-\sum_{u\geqslant 1}(z_u+u-1-s_p)\\ &= \sum_{1\leq u<u_0}(z_u+u-s_p)+\sum_{u\geqslant u_0}(z_{u+1}+u-s_p)-\sum_{u\geqslant 1}(z_u+u-1-s_p)\\ &= s_p-z.\end{aligned}$$

We deduce that we have

$$w_{2d}(\mu'^1)=w_{2d}(\mu^1)+s_p-z.$$

A similar computation yields

$$w_{2d}(\mu^2)=w_{2d}(\mu'^2)+r'_p-z=w_{2d}(\mu'^2)+r_p+1-z.$$

We deduce that

$$w_{2d}(\mu'^1)+w_{2d}(\mu'^2)=w_{2d}(\mu^1)+w_{2d}(\mu^2)+s_p-r_p-1.$$

Since we have removed the element x from $\beta_t(\mu^1)$, we have $t'=t-1$.

Next, for any integers v, e with $e\geqslant 1$, we have

$$\Delta(v,e)-\Delta(v-1,e)=\left\lfloor\frac{v-1}{e}\right\rfloor.$$

We deduce that

$$\Delta(s',1)-\Delta(s,1)=1-s_p,\quad \Delta(r',1)-\Delta(r,1)=r_p.$$

In particular, the value of

$$w_{2d}(\mu^1)+w_{2d}(\mu^2)+\Delta(s,1)+\Delta(r,1)+t/2$$

decreases by $1/2$.

Now assume that there is an integer $y \in \beta_{d-1-t}(\mu^2)$ such that $y - 2d \notin \beta_t(\mu^1)$ and let us write

$$\beta_{t'}(\mu'^1) = \beta_t(\mu^1) \sqcup \{y - 2d\}, \quad \beta_{d-1-t'}(\mu'^2) = \beta_{d-1-t}(\mu^2) \setminus \{y\}.$$

Fix $p \in \{1, \ldots, 2d\}$ and $z \in \beta_{r_p}(\mu^{2,p})$ such that

$$y = p - 2d + 2dz.$$

The same computation as above gives

$$w_{2d}(\mu'^2) = w_{2d}(\mu^2) + r_p - z, \quad w_{2d}(\mu^1) = w_{2d}(\mu'^1) + (s_p + 1) - (z - 1),$$

from which we deduce that

$$w_{2d}(\mu'^1) + w_{2d}(\mu'^2) = w_{2d}(\mu^1) + w_{2d}(\mu^2) + r_p - s_p - 2.$$

Since $t' = t + 1$, the value of

$$w_{2d}(\mu^1) + w_{2d}(\mu^2) + \Delta(s, 1) + \Delta(r, 1) + t/2$$

decreases by 1/2. □

6.5.5. *The $\mathfrak{g}_{2d}$-representation on $\mathscr{U}_\mathrm{k}$ in the unitary case.* We now have all the tools to extend the action of $\mathfrak{g}'_{2d}$ to $\mathfrak{g}_{2d}$ on $\mathscr{U}_\mathrm{k}$. Recall that we assume that f is even, so $f = 2d$. Hence I_{2d} is a cyclic quiver of size $2d$ and we have an isomorphism $\widetilde{\mathfrak{sl}}_f \xrightarrow{\sim} \mathfrak{g}'_{2d}$ sending $\alpha_k^\vee$ to $\alpha_{q^k}^\vee$. Let $\mathfrak{g}_{2d}$ be the Kac-Moody algebra associated with the quiver I_{2d}, i.e., we have $\widehat{\mathfrak{sl}}_f \simeq \mathfrak{g}_{2d}$, and set $\mathrm{X}_{2d} = \mathrm{P}_{2d} \oplus \mathbb{Z}\delta/2$.

Let σ be the automorphism of the Lie algebra $\mathfrak{g}_{2d}$ given by the quiver automorphism of I_{2d} such that $i \mapsto -i$. In particular, we have $\sigma(\alpha_i^\vee) = \alpha_{-i}^\vee$.

DEFINITION 6.16. We define the weight of a symbol by $\mathrm{wt}(\Theta_t(\mu)) = \sigma_*^t(\mathrm{wt}(\mu, t))$, where $\mathrm{wt}(\mu, t)$ is as in (6.6).

We can now prove the following.

PROPOSITION 6.17. *Let μ be a bipartition and $t \in \mathbb{Z}$. Let Θ, Θ' be symbols.*
(a) $\mathrm{wt}(\Theta_t(\mu)) \equiv \mathrm{wt}(|\mu, Q_t\rangle_{2d})$ *mod* $\mathbb{Z}\,\delta$,
(b) *if Θ' is obtained from Θ by removing a d-hook, then* $\mathrm{wt}(\Theta) = \sigma_*\mathrm{wt}(\Theta') - \delta/2$,
(c) *E_Θ, $E_{\Theta'}$ belong to the same ℓ-block if and only if we have* $\mathrm{wt}(\Theta) = \mathrm{wt}(\Theta')$.

PROOF. We have $Q_t = (-1)^t P_t$. Further,

$$\mathrm{wt}(\mu, t) \equiv \Lambda_{P_t} - \sum_i n_i(\mu, P_t)\, \alpha_i^{\mathrm{cl}} \bmod \mathbb{Z}\,\delta,$$

$$\mathrm{wt}(|\mu, Q_t\rangle_{2d}) \equiv \Lambda_{Q_t} - \sum_i n_i(\mu, Q_t)\, \alpha_i^{\mathrm{cl}} \bmod \mathbb{Z}\,\delta.$$

This implies part (a) because

$$n_{(-1)^t i}(\mu, P_t) = n_i(\mu, (-1)^t P_t) = n_i(\mu, Q_t).$$

Part (b) follows from Lemma 6.15.

Now, we prove part (c). Recall that the representations E_Θ and $E_{\Theta'}$ are in the same ℓ-block if and only if Θ and Θ' have the same rank and the same d-cocore $\Theta_\circ$, see §6.3.4.

Assume that this is the case. Since the number of d-cohooks one needs to add to $\Theta_\circ$ to obtain Θ and Θ' depends only on the rank of the symbols, we deduce from Lemma 6.15 that Θ and Θ' have the same weight.

Conversely, assume now that $\mathrm{wt}(\Theta) = \mathrm{wt}(\Theta')$. Then (6.7) implies that

$$\pi_{s+r} = \pi_{s'+r'}. \tag{6.8}$$

By definition of s, r, s' and r' we have

$$\sum_p s_p = t, \quad \sum_p r_p = d-1-t, \quad \sum_p s'_p = t', \quad \sum_p r'_p = d-1-t',$$

hence

$$\sum_p (s_p + r_p) = \sum_p (s'_p + r'_p) = d-1. \tag{6.9}$$

From (6.8) and (6.9), we deduce that the $2d$-tuples $s+r$ and $s'+r'$ are equal. Hence, by Lemma 6.14, the symbols Θ and Θ' have the same d-cocore $\Theta_\circ$. Let w and w' be the number of d-cohooks we need to add to $\Theta_\circ$ to obtain Θ and Θ' respectively. Since $\sigma_*(\delta) = \delta$, by part (b) we have

$$\mathrm{wt}(\Theta) = (\sigma_*)^w \, \mathrm{wt}(\Theta_\circ) - w\,\delta/2,$$
$$\mathrm{wt}(\Theta') = (\sigma_*)^{w'} \, \mathrm{wt}(\Theta_\circ) - w'\,\delta/2.$$

Since $\sigma_*(\delta) = \delta$ and $\mathrm{wt}(\Theta) = \mathrm{wt}(\Theta')$, we deduce that $w = w'$. In particular, the symbols Θ and Θ' have the same rank, which shows that the corresponding representations lie in the same ℓ-block. □

We have proved the following refinement of Theorem 6.11.

THEOREM 6.18. *Recall that ℓ and q are odd, and $\ell \nmid q(q^2-1)$. Assume that f is even. The representation datum on $\mathscr{U}_\mathrm{k}$ gives a $\mathfrak{g}_{2d}$-representation yielding a $\mathfrak{g}'_{2d}$-module isomorphism $\bigoplus_{t\in\mathbb{N}} \mathbf{F}(Q_t)_{2d} \xrightarrow{\sim} [\mathscr{U}_\mathrm{k}]$ such that $|\mu, Q_t\rangle_{2d} \mapsto d_{\mathscr{U}}([E_{\Theta_t(\mu)}])$. The weights of the $\mathfrak{g}_{2d}$-module $[\mathscr{U}_\mathrm{k}]$ belong to the lattice X_{2d}. The weight of $d_{\mathscr{U}}([E_{\Theta_t(\mu)}])$ is equal to $\mathrm{wt}(\Theta_t(\mu))$. The classes in $[\mathscr{U}_\mathrm{k}]$ of the simple unipotent modules are weight vectors. Two simple unipotent modules belong to the same block if and only if their classes have the same weight in X_{2d}.* □

6.5.6. *Determination of the ramified Hecke algebras.* As in the case of finite unitary groups, see §4.7.2, we can invoke Proposition 1.5 to show that every ramified Hecke algebra associated to a weakly cuspidal module is indeed a Hecke algebra, whose parameters can be read off from the weight of the cuspidal module.

COROLLARY 6.19. *Let D be a weakly cuspidal $\mathrm{k}G_r$-module. Then*

(a) *the weight of the class $[D]$ with respect to the $\mathfrak{g}'_{2d}$-action on $[\mathscr{U}_\mathrm{k}]$ is of the form Λ_Q for some element $Q = (Q_1, Q_2)$ in I^2_{2d},*
(b) *for each $m \geqslant 0$ and $n = r + 2m$, the map $\phi_{\mathrm{k},m} : \mathbf{H}^{q^2}_{\mathrm{k},m} \longrightarrow \mathrm{End}_{\mathrm{k}G_n}(F^m)^{\mathrm{op}}$ factors to an algebra isomorphism $\mathbf{H}^{Q;\,q^2}_{\mathrm{k},m} \xrightarrow{\sim} \mathscr{H}(\mathrm{k}G_n, D)$.* □

In order to apply this corollary efficiently, we must determine explicitly the weakly cuspidal modules and their weights. In the linear prime case, this is known by [**39**]. In the unitary prime case, by Theorem 6.18, the weakly cuspidal modules form a basis of the subspace of highest weight vectors of the representation of $\mathfrak{g}'_{2d}$ on $[\mathscr{U}_\mathrm{k}]$, hence all ramified Hecke algebras can be explicitly computed. If, moreover, we assume that the decomposition matrix is unitriangular with respect to the a-function, see Conjecture 6.21 below, then the weakly cuspidal modules and their weights can also be read off from the crystal isomorphism in Theorem 6.22 below.

More precisely, any weakly cuspidal module is of the form $D = D_\Theta$ for some symbol $\Theta = \Theta_t(\mu)$. The parameter Q can be explicitly computed from μ and t according to the rule

$$\begin{aligned}\mathrm{wt}(D_\Theta) &\equiv \Lambda_{Q_t} - \sum_i n_i(\mu, Q_t)\,\alpha_i^{\mathrm{cl}} \bmod \mathbb{Z}\,\delta,\\ &\equiv \Lambda_{Q_1} + \Lambda_{Q_2} \bmod \mathbb{Z}\,\delta.\end{aligned}$$

Since the isomorphism in Proposition 6.9 identifies the set

$$\{[D_{\Theta_t(\mu)}]\,;\, \mu \in \mathscr{P}^2,\ D_{\Theta_t(\mu)} \text{ is weakly cuspidal}\}$$

with a basis of the space of highest weight vectors

$$\{x \in \mathbf{F}(Q_t)_{2d}\,;\, E_i(x) = 0,\ \forall i = 1, 2, \ldots, 2d\},$$

the weakly cuspidal modules correspond precisely to the elements in the crystal $B(s_t)_{2d}$ which are killed by the operators $\widetilde{E}_1, \ldots, \widetilde{E}_{2d}$, with the notation in §6.7.3 below. These elements have been computed recently in [**43**, thm. 5.9] in terms of bipartitions μ whose abacus is totally $2d$-periodic. N

6.6. Derived equivalences.

Blocks of $\mathscr{U}_{\mathrm{k}}$ are parametrized by pairs (Θ, w) where Θ is a d-core (resp. a d-cocore) when f is odd (resp. f is even) and w is a non-negative integer called the *degree* of the block. The unipotent characters lying in $B_{\Theta,w}$ are parametrized by the symbols which are obtained from Θ by adding w successive d-hooks (resp. d-cohooks) if f is odd (resp. f is even).

Our categorification theorems (see Theorem 6.18 and 6.12) show that blocks of $\mathscr{U}_{\mathrm{k}}$ and $\mathscr{U}_{\mathrm{k},t}$ correspond to weight spaces for the action of the Kac-Moody algebra $\mathfrak{g}_{2d}$. Using the work of Chuang-Rouquier [**10**] (see also Theorem 1.11) we obtain derived equivalences between blocks of $\mathscr{U}_{\mathrm{k}}$ which are in the same orbit under the action of the affine Weyl group W_{2d}.

If ℓ is a linear prime (i.e. f is odd), the situation is similar to the case of finite unitary groups. The d-cores, which are formed by pairs of partitions which are both d-cores, form a single orbit under the action of W_{2d}. In particular, orbits of blocks of $\mathscr{U}_{\mathrm{k},t}$ under W_{2d} are parametrized by their degree. Let $w \in \mathbb{N}$. Livesey showed in [**51**, thm. 7.1] how to construct a *good* block of $\mathscr{U}_{\mathrm{k},t}$ of degree w which is derived equivalent to its Brauer correspondent. We deduce that Broué's abelian defect group conjecture holds for groups of type BC at odd linear primes.

THEOREM 6.20. *Recall that ℓ and q are odd. Assume that the order f of q modulo ℓ is even. Let B be a unipotent block of G_n over* k *or $\mathscr{O}$, and D be a defect group of B. If D is abelian, then B is derived equivalent to its Brauer correspondent in $N_{G_n}(D)$.* □

6.7. The crystal of $\mathscr{U}_{\mathrm{k}}$.

As in the case of finite unitary groups, see §4.7, our goal is to compare the crystal of the categorical representation on $\mathscr{U}_{\mathrm{k}}$ with the crystal of the $\mathfrak{g}_{2d}$-module $\bigoplus_t \mathbf{F}(Q_t)_{2d}$. In order to get an explicit crystal isomorphism, we will assume that the ℓ-decomposition matrices of unipotent blocks are unitriangular with respect to the a-function. This gives a parametrization of the simple objects of $\mathscr{U}_{\mathrm{k}}$ and a combinatorial way (yet conjectural) to compute the (weak) Harish-Chandra branching graph.

Throughout this section we will assume that ℓ is a *unitary prime*, which means that the order f of q modulo ℓ is even. We will write $f = 2d$.

6.7.1. *Ordering symbols.* For $t \geqslant 0$, we define the charge

$$s_t = \begin{cases} (t, d-1-t) & \text{if } t \text{ is even,} \\ (t-d, -1-t) & \text{if } t \text{ is odd.} \end{cases} \tag{6.10}$$

The symbol associated with a bipartition μ is $\Theta = \Theta_t(\mu) = \{\beta_t(\mu^1), \beta_{-1-t}(\mu^2)\}$. As in [**33**, §5.5.10], we can consider the union of the two β-sets $\beta_t(\mu^1)$ and $\beta_{-1-t}(\mu^2)$ (allowing multiplicities) and write their elements in non-increasing order

$$\kappa_{m_t}(\mu) := (\kappa_1 \geqslant \kappa_2 \geqslant \kappa_3 \geqslant \cdots),$$

where the parameter m_t is given by

$$m_t = (t, -1-t).$$

We can check that $\kappa_i = \lfloor i/2 \rfloor$ for i large enough. Then, we define the a-value of the symbol Θ by

$$a(\Theta_t(\mu)) = a(\Theta) := \sum_{i=1}^{\infty} (i-1)\left(\kappa_i - \left\lfloor \frac{i}{2} \right\rfloor\right).$$

This differs slightly from the a-value $a^{(m_t,1)}(\mu)$ defined in [**33**] : from [**33**, prop. 5.5.11] we have $a(\Theta_t(\mu)) = a^{(m_t,1)}(\mu) + a(\Theta_t(\emptyset))$.

On the set of symbols in the same Harish-Chandra series and of the same rank we define as in [**33**, §5.7.5] a partial order by

$$\Theta_t(\mu) \preccurlyeq \Theta_t(\lambda) \iff \mu \ll_{m_t} \lambda \iff \lambda = \mu \text{ or } \kappa_{m_t}(\lambda) < \kappa_{m_t}(\mu).$$

We deduce from [**33**, prop. 5.7.7] that

$$\Theta_t(\mu) \prec \Theta_t(\lambda) \implies a(\Theta_t(\mu)) < a(\Theta_t(\lambda)). \tag{6.11}$$

6.7.2. *Parametrization of unipotent modules.* By Theorem 3.4 we know that when ℓ is odd, there are as many unipotent $\mathrm{k}G_n$-modules as symbols of odd defect and rank n. However there is no known natural parametrization of these modules in terms of symbols unless the decomposition matrix has unitriangular shape. This property has been proved when f is odd [**39**], i.e., when ℓ is a linear prime, or when f and n are small. Geck conjectured in [**30**, conj. 3.4] that it should always hold.

CONJECTURE 6.21 (Geck). *Assume ℓ is odd. There exists a parametrization*

$$\mathrm{Irr}(\mathscr{U}_{\mathrm{k}}) = \{[D_\Theta] \,|\, \lambda \in \mathscr{S}_{odd}\}$$

of the simple unipotent modules and an ordering $\leqslant$ on symbols such that the decomposition map satisfies the following unitriangularity property

$$d_{\mathscr{U}}([E_\Theta]) \in [D_\Theta] + \sum_{\Theta' < \Theta} \mathbb{Z}[D_{\Theta'}].$$

Moreover, the ordering $\leqslant$ can be chosen to be compatible with the a-function so that $\Theta' \leqslant \Theta$ implies $a(\Theta') \leqslant a(\Theta)$.

This conjecture is expected to hold for any finite reductive group as long as ℓ is a very good prime. In that case the ordering $\leqslant$ should be induced from the partial order on families of unipotent characters defined by Lusztig.

6.7.3. *Comparison of the crystals.* Recall that to any categorical representation one can associate a perfect basis, and hence an abstract crystal, see Proposition 1.9. In §6.5.1 we constructed a categorical representation of $\mathfrak{g}'_{2d}$ on the category $\mathscr{U}_{\mathrm{k}}$ of unipotent representations over k. From this categorical representation we get an abstract crystal $B(\mathscr{U}_{\mathrm{k}}) = \big(\mathrm{Irr}(\mathscr{U}_{\mathrm{k}}), \widetilde{E}_i, \widetilde{F}_i\big)$. As for finite unitary groups, this crystal is related to the (weak) Harish-Chandra series. More precisely, the (uncolored) crystal graph associated with $B(\mathscr{U}_{\mathrm{k}})$ coincides with the weak Harish-Chandra branching graph, and its connected components with the weak Harish-Chandra series. See §4.7.1 for more details.

Recall from Theorem 6.11 that $[\mathscr{U}_{\mathrm{k}}]$ is isomorphic to a direct sum of Fock spaces $\mathbf{F}(Q_t)_{2d}$ where t runs over $\mathbb{N}$ and $Q_t = (Q_1, Q_2) = ((-q)^t, (-q)^{-1-t})$. Recall from (6.10) that we have defined the charge $s_t = (s_1, s_2)$ by

$$s_t = \begin{cases} (t, d-1-t) & \text{if } t \text{ is even,} \\ (t-d, -1-t) & \text{if } t \text{ is odd} \end{cases}$$

and $m_t = (t, -1-t)$. Since $q^d = -1$ in k, we have $Q_p = q^{2s_p}$ for each $p = 1, 2$. In other words s_t is a charge for $\mathbf{F}(Q_t)_{2d}$ with respect to q. Let $B(s_t)_{2d}$ be the corresponding abstract crystal of $\mathbf{F}(Q_t)_{2d}$, with the canonical labeling $B(s_t)_{2d} = \{b(\mu, s_t) \,|\, \mu \in \mathscr{P}^2,\ t \in \mathbb{N}\}$. Finally, we set $B_{2d} = \bigsqcup_{t\in\mathbb{N}} B(s_t)_{2d}$.

Under the assumption that the decomposition matrix is unitriangular with respect to the a-function, we can compare the crystal B_{2d} with the abstract crystal $B(\mathscr{U}_{\mathrm{k}})$ coming from the categorical action of $\widehat{\mathfrak{sl}}_{2d}$ on $\mathscr{U}_{\mathrm{k}}$.

THEOREM 6.22. *Recall that ℓ, q are odd and f is even. If Conjecture 6.21 holds, then the map $b(\mu, s_t) \mapsto [D_{\Theta_t(\mu)}]$ is a crystal isomorphism $B_{2d} \overset{\sim}{\to} B(\mathscr{U}_{\mathrm{k}})$.*

PROOF. Let $t \in \mathbb{Z}$ and μ be a bipartition. By [**33**, thm. 6.6.12] the Uglov basis satisfies

$$b(\mu, s_t) \in |\mu, s_t\rangle + \sum_{\substack{\lambda\neq\mu \\ \lambda \ll_{m_t} \mu}} \mathbb{Z}|\lambda, s_t\rangle.$$

Let $\phi : |\mu, s_t\rangle \mapsto d_{\mathscr{U}}([E_{\Theta_t(\mu)}])$ be the morphism of $\mathfrak{g}'_{2d}$-modules studied in §6.5.1. By the previous formula and the definition of $\preccurlyeq$ we have

$$\phi(b(\mu, s_t)) \in d_{\mathscr{U}}([E_{\Theta_t(\mu)}]) + \sum_{\substack{\Theta'\neq\Theta_t(\mu) \\ \Theta' \preccurlyeq \Theta_t(\mu)}} \mathbb{Z} d_{\mathscr{U}}([E_{\Theta'}]).$$

Let n be the rank of the symbol $\Theta_t(\mu)$. If Conjecture 6.21 holds, then we can choose a total order $\leqslant$ on symbols of rank n and odd defect, compatible with the a-function and for which the decomposition map $d_{\mathscr{U}}$ is unitriangular. In particular, using (6.11) we have

$$\Theta' \neq \Theta_t(\mu) \text{ and } \Theta' \preccurlyeq \Theta_t(\mu) \implies a(\Theta') < a(\Theta_t(\mu)) \implies \Theta' \leqslant \Theta_t(\mu).$$

This, together with the previous formula shows that

$$\phi(b(\mu, s_t)) \in [D_{\Theta_t(\mu)}] + \sum_{\substack{\Theta'\neq\Theta_t(\mu) \\ \Theta' \leqslant \Theta_t(\mu)}} \mathbb{Z}[D_{\Theta'}]$$

and we can invoke Proposition 1.10 to conclude. □

REMARK 6.23. Note that the definition of s_t makes sense for negative integers t as well. Then $b(\mu, s_t) \longmapsto b(\mu^\dagger, s_{-1-t})$ induces an isomorphism of crystals. This is a particular case of changing the charge (a, b) to $(b-d, a+d)$.

References

[1] A. Berenstein and D. Kazhdan, *Geometric and unipotent crystals. II. From unipotent bicrystals to crystal bases*, Quantum groups, Contemp. Math., vol. 433, Amer. Math. Soc., Providence, RI, 2007, pp. 13–88, DOI 10.1090/conm/433/08321. MR2349617

[2] C. Bonnafé and J. Michel, *Computational proof of the Mackey formula for $q > 2$*, J. Algebra **327** (2011), 506–526, DOI 10.1016/j.jalgebra.2010.10.030. MR2746047

[3] C. Bonnafé and R. Rouquier, *Catégories dérivées et variétés de Deligne-Lusztig* (French), Publ. Math. Inst. Hautes Études Sci. **97** (2003), 1–59, DOI 10.1007/s10240-003-0013-3. MR2010739

[4] M. Broué, G. Malle, and J. Michel, *Generic blocks of finite reductive groups*, Astérisque **212** (1993), 7–92. Représentations unipotentes génériques et blocs des groupes réductifs finis. MR1235832

[5] M. Broué and J. Michel, *Blocs et séries de Lusztig dans un groupe réductif fini* (French), J. Reine Angew. Math. **395** (1989), 56–67, DOI 10.1515/crll.1989.395.56. MR983059

[6] J. Brundan, R. Dipper, and A. Kleshchev, *Quantum linear groups and representations of* $\mathrm{GL}_n(\mathbf{F}_q)$, Mem. Amer. Math. Soc. **149** (2001), no. 706, viii+112, DOI 10.1090/memo/0706. MR1804485

[7] M. Cabanes and M. Enguehard, *On unipotent blocks and their ordinary characters*, Invent. Math. **117** (1994), no. 1, 149–164, DOI 10.1007/BF01232237. MR1269428

[8] M. Cabanes and M. Enguehard, *Representation theory of finite reductive groups*, New Mathematical Monographs, vol. 1, Cambridge University Press, Cambridge, 2004. MR2057756

[9] R. W. Carter, *Finite groups of Lie type*, Pure and Applied Mathematics (New York), John Wiley & Sons, Inc., New York, 1985. Conjugacy classes and complex characters; A Wiley-Interscience Publication. MR794307

[10] J. Chuang and R. Rouquier, *Derived equivalences for symmetric groups and* $\mathfrak{sl}_2$*-categorification*, Ann. of Math. (2) **167** (2008), no. 1, 245–298, DOI 10.4007/annals.2008.167.245. MR2373155

[11] C. W. Curtis and I. Reiner, *Methods of representation theory. Vol. I*, John Wiley & Sons, Inc., New York, 1981. With applications to finite groups and orders; Pure and Applied Mathematics; A Wiley-Interscience Publication. MR632548

[12] F. Digne and J. Michel, *Representations of finite groups of Lie type*, London Mathematical Society Student Texts, vol. 21, Cambridge University Press, Cambridge, 1991. MR1118841

[13] R. Dipper, *On the decomposition numbers of the finite general linear groups*, Trans. Amer. Math. Soc. **290** (1985), no. 1, 315–344, DOI 10.2307/1999797. MR787968

[14] R. Dipper, *On the decomposition numbers of the finite general linear groups. II*, Trans. Amer. Math. Soc. **292** (1985), no. 1, 123–133, DOI 10.2307/2000173. MR805956

[15] R. Dipper and G. James, *The q-Schur algebra*, Proc. London Math. Soc. (3) **59** (1989), no. 1, 23–50, DOI 10.1112/plms/s3-59.1.23. MR997250

[16] R. Dipper and J. Du, *Harish-Chandra vertices and Steinberg's tensor product theorems for finite general linear groups*, Proc. London Math. Soc. (3) **75** (1997), no. 3, 559–599, DOI 10.1112/S0024611597000439. MR1466661

[17] O. Dudas and G. Malle, *Decomposition matrices for low-rank unitary groups*, Proc. Lond. Math. Soc. (3) **110** (2015), no. 6, 1517–1557, DOI 10.1112/plms/pdv008. MR3356813

[18] O. Dudas, M. Varagnolo, and E. Vasserot, *Categorical actions on unipotent representations I. Finite unitary groups,* arXiv:1509.03269.

[19] O. Dudas, M. Varagnolo, and E. Vasserot, *Categorical actions from Lusztig induction and restriction on finite linear groups,* in Representation theory - current trends and perspectives, EMS Ser. Congr. Rep., Eur. Math. Soc., Zürich (2017), pp. 59–74.

[20] C. Dunkl and S. Griffeth, *Generalized Jack polynomials and the representation theory of rational Cherednik algebras*, Selecta Math. (N.S.) **16** (2010), no. 4, 791–818, DOI 10.1007/s00029-010-0032-0. MR2734331

[21] O. Foda, B. Leclerc, M. Okado, J.-Y. Thibon, and T. A. Welsh, *Branching functions of $A_{n-1}^{(1)}$ and Jantzen-Seitz problem for Ariki-Koike algebras*, Adv. Math. **141** (1999), no. 2, 322–365, DOI 10.1006/aima.1998.1783. MR1671762

[22] P. Fong and B. Srinivasan, *The blocks of finite general linear and unitary groups*, Invent. Math. **69** (1982), no. 1, 109–153, DOI 10.1007/BF01389188. MR671655

[23] P. Fong and B. Srinivasan, *The blocks of finite classical groups*, J. Reine Angew. Math. **396** (1989), 122–191. MR988550

[24] P. Fong and B. Srinivasan, *Brauer trees in classical groups*, J. Algebra **131** (1990), no. 1, 179–225, DOI 10.1016/0021-8693(90)90172-K. MR1055005

[25] M. Geck, *An introduction to algebraic geometry and algebraic groups*, Oxford Graduate Texts in Mathematics, vol. 10, Oxford University Press, Oxford, 2003. MR2032320

[26] M. Geck, *On the decomposition numbers of the finite unitary groups in nondefining characteristic*, Math. Z. **207** (1991), no. 1, 83–89, DOI 10.1007/BF02571376. MR1106814

[27] M. Geck, *Basic sets of Brauer characters of finite groups of Lie type. II*, J. London Math. Soc. (2) **47** (1993), no. 2, 255–268, DOI 10.1112/jlms/s2-47.2.255. MR1207947

[28] M. Geck, *Modular Harish-Chandra series, Hecke algebras and (generalized) q-Schur algebras*, Modular representation theory of finite groups (Charlottesville, VA, 1998), de Gruyter, Berlin, 2001, pp. 1–66. MR1889339

[29] M. Geck and G. Hiss, *Basic sets of Brauer characters of finite groups of Lie type*, J. Reine Angew. Math. **418** (1991), 173–188. MR1111205

[30] M. Geck and G. Hiss, *Modular representations of finite groups of Lie type in non-defining characteristic*, Finite reductive groups (Luminy, 1994), Progr. Math., vol. 141, Birkhäuser Boston, Boston, MA, 1997, pp. 195–249. MR1429874

[31] M. Geck, G. Hiss, and G. Malle, *Cuspidal unipotent Brauer characters*, J. Algebra **168** (1994), no. 1, 182–220, DOI 10.1006/jabr.1994.1226. MR1289097

[32] M. Geck, G. Hiss, and G. Malle, *Towards a classification of the irreducible representations in non-describing characteristic of a finite group of Lie type*, Math. Z. **221** (1996), no. 3, 353–386, DOI 10.1007/PL00004253. MR1381586

[33] M. Geck and N. Jacon, *Representations of Hecke algebras at roots of unity*, Algebra and Applications, vol. 15, Springer-Verlag London, Ltd., London, 2011. MR2799052

[34] M. Geck and G. Pfeiffer, *Characters of finite Coxeter groups and Iwahori-Hecke algebras*, London Mathematical Society Monographs. New Series, vol. 21, The Clarendon Press, Oxford University Press, New York, 2000. MR1778802

[35] T. Gerber, *Triple crystal action in Fock spaces*, arXiv:1601.00581.

[36] T. Gerber, G. Hiss, and N. Jacon, *Harish-Chandra series in finite unitary groups and crystal graphs*, Int. Math. Res. Not. IMRN **22** (2015), 12206–12250. MR3456719

[37] T. Gerber and G. Hiss, *Branching graphs for finite unitary groups in non-defining characteristic*, Comm. Algebra **45** (2015), pp. 561–574.

[38] J. Gruber, *Green vertex theory, Green correspondence, and Harish-Chandra induction*, J. Algebra **186** (1996), no. 2, 476–521, DOI 10.1006/jabr.1996.0384. MR1423273

[39] J. Gruber and G. Hiss, *Decomposition numbers of finite classical groups for linear primes*, J. Reine Angew. Math. **485** (1997), 55–91, DOI 10.1515/crll.1997.485.55. MR1442189

[40] G. Hiss, *Regular and semisimple blocks of finite reductive groups*, J. London Math. Soc. (2) **41** (1990), no. 1, 63–68, DOI 10.1112/jlms/s2-41.1.63. MR1063543

[41] G. Hiss, *Harish-Chandra series of Brauer characters in a finite group with a split BN-pair*, J. London Math. Soc. (2) **48** (1993), no. 2, 219–228, DOI 10.1112/jlms/s2-48.2.219. MR1231711

[42] R. B. Howlett and G. I. Lehrer, *Induced cuspidal representations and generalised Hecke rings*, Invent. Math. **58** (1980), no. 1, 37–64, DOI 10.1007/BF01402273. MR570873

[43] N. Jacon and C. Lecouvey, *A combinatorial decomposition of higher level Fock spaces*, Osaka J. Math. **50** (2013), no. 4, 897–920. MR3161420

[44] M. Jimbo, K. C. Misra, T. Miwa, and M. Okado, *Combinatorics of representations of $U_q(\widehat{\mathfrak{sl}}(n))$ at $q = 0$*, Comm. Math. Phys. **136** (1991), no. 3, 543–566. MR1099695

[45] V. G. Kac, *Infinite-dimensional Lie algebras*, 3rd ed., Cambridge University Press, Cambridge, 1990. MR1104219

[46] S.-J. Kang and M. Kashiwara, *Categorification of highest weight modules via Khovanov-Lauda-Rouquier algebras*, Invent. Math. **190** (2012), no. 3, 699–742, DOI 10.1007/s00222-012-0388-1. MR2995184

[47] S.-J. Kang, S.-j. Oh, and E. Park, *Perfect bases for integrable modules over generalized Kac-Moody algebras*, Algebr. Represent. Theory **14** (2011), no. 3, 571–587, DOI 10.1007/s10468-009-9204-z. MR2785923

[48] M. Kashiwara, *On crystal bases*, Representations of groups (Banff, AB, 1994), CMS Conf. Proc., vol. 16, Amer. Math. Soc., Providence, RI, 1995, pp. 155–197. MR1357199

[49] M. Kashiwara, *Biadjointness in cyclotomic Khovanov-Lauda-Rouquier algebras*, Publ. Res. Inst. Math. Sci. **48** (2012), no. 3, 501–524, DOI 10.2977/PRIMS/78. MR2973390

[50] A. Lascoux, B. Leclerc, and J.-Y. Thibon, *Hecke algebras at roots of unity and crystal bases of quantum affine algebras*, Comm. Math. Phys. **181** (1996), no. 1, 205–263. MR1410572

[51] M. Livesey, *On Rouquier blocks for finite classical groups at linear primes*, J. Algebra **432** (2015), 91–128, DOI 10.1016/j.jalgebra.2014.10.041. MR3334143

[52] M. Livesey, *A note on perfect isometries between finite general linear and unitary groups at unitary primes*, arXiv:1411.6862.

[53] G. Lusztig, *Irreducible representations of finite classical groups*, Invent. Math. **43** (1977), no. 2, 125–175. MR0463275

[54] G. Lusztig, *Characters of reductive groups over a finite field*, Annals of Mathematics Studies, vol. 107, Princeton University Press, Princeton, NJ, 1984. MR742472

[55] G. Lusztig, *Rationality properties of unipotent representations*, J. Algebra **258** (2002), no. 1, 1–22, DOI 10.1016/S0021-8693(02)00514-8. Special issue in celebration of Claudio Procesi's 60th birthday. MR1958895

[56] G. Lusztig and B. Srinivasan, *The characters of the finite unitary groups*, J. Algebra **49** (1977), no. 1, 167–171. MR0453886

[57] I. G. Macdonald, *Symmetric functions and Hall polynomials*, 2nd ed., Oxford Mathematical Monographs, The Clarendon Press, Oxford University Press, New York, 1995. With contributions by A. Zelevinsky; Oxford Science Publications. MR1354144

[58] A. Mathas, *The representation theory of the Ariki-Koike and cyclotomic q-Schur algebras*, Representation theory of algebraic groups and quantum groups, Adv. Stud. Pure Math., vol. 40, Math. Soc. Japan, Tokyo, 2004, pp. 261–320. MR2074597

[59] R. Rouquier, *2-Kac-Moody algebras*, arXiv:0812.5023v1.

[60] R. Rouquier, P. Shan, M. Varagnolo, and E. Vasserot, *Categorifications and cyclotomic rational double affine Hecke algebras*, Invent. Math. **204** (2016), no. 3, 671–786, DOI 10.1007/s00222-015-0623-7. MR3502064

[61] P. Shan, *Crystals of Fock spaces and cyclotomic rational double affine Hecke algebras* (English, with English and French summaries), Ann. Sci. Éc. Norm. Supér. (4) **44** (2011), no. 1, 147–182. MR2760196

[62] P. Shan and E. Vasserot, *Heisenberg algebras and rational double affine Hecke algebras*, J. Amer. Math. Soc. **25** (2012), no. 4, 959–1031, DOI 10.1090/S0894-0347-2012-00738-3. MR2947944

[63] B. Srinivasan, *On CRDAHA and finite general linear and unitary groups*, Pacific J. Math. **279** (2015), no. 1-2, 465–480, DOI 10.2140/pjm.2015.279.465. MR3437787

[64] M. Takeuchi, *The group ring of* $\mathrm{GL}_n(q)$ *and the q-Schur algebra*, J. Math. Soc. Japan **48** (1996), no. 2, 259–274, DOI 10.2969/jmsj/04820259. MR1376081

[65] D. Uglov, *Canonical bases of higher-level q-deformed Fock spaces and Kazhdan-Lusztig polynomials*, Physical combinatorics (Kyoto, 1999), Progr. Math., vol. 191, Birkhäuser Boston, Boston, MA, 2000, pp. 249–299. MR1768086

[66] X. Yvonne, Bases canoniques d'espaces de Fock de niveau supérieur, PhD thesis, University of Caen (2005).

University Paris Diderot, UMR CNRS 7586, F-75205 Paris cedex 13
E-mail address: `olivier.dudas@imj-prg.fr`

University of Cergy-Pontoise, UMR CNRS 8088, F-95000 Cergy-Pontoise
E-mail address: `michela.varagnolo@u-cergy.fr`

University Paris Diderot, UMR CNRS 7586, F-75205 Paris cedex 13
E-mail address: `eric.vasserot@imj-prg.fr`

Contemporary Mathematics
Volume **683**, 2017
http://dx.doi.org/10.1090/conm/683/13724

Categorical actions and crystals

Jonathan Brundan and Nicholas Davidson

ABSTRACT. This is an expository article developing some aspects of the theory of categorical actions of Kac-Moody algebras in the spirit of works of Chuang–Rouquier, Khovanov–Lauda, Webster, and many others.

1. Introduction

This work is a contribution to the study of categorifications of Kac-Moody algebras and their integrable modules. The subject has its roots in Lusztig's construction of canonical bases of quantum groups using geometry of quiver varieties [**Lu1**] (which happened around 1990). It is intimately connected to the rich combinatorial theory of crystal bases initiated at the same time by Kashiwara [**K1**]. In the decade after that, several other examples were studied related to the representation theory of the symmetric group and associated Hecke algebras [**LLT**, **A**, **G**] (building in particular on ideas of Bernstein and Zelevinsky [**BZ**]), rational representations of the general linear group [**BK**], and the Bernstein-Gelfand-Gelfand category $\mathcal{O}$ associated to the general linear Lie (super)algebra [**BFK**, **B1**]. The first serious attempt to put these examples into a unified axiomatic framework was undertaken by Chuang and Rouquier [**CR**]. They built a powerful structure theory for studying categorical actions of $\mathfrak{sl}_2$, which they applied notably to prove Broué's Abelian Defect Conjecture for the symmetric groups.

Another major breakthrough came in 2008, when Khovanov and Lauda [**KL1**, **KL2**, **KL3**] and Rouquier [**R1**] independently introduced some new algebras called *quiver Hecke algebras*, and used them to construct *Kac-Moody 2-categories* associated to arbitrary Kac-Moody algebras. The definitions of Kac-Moody 2-categories given by Khovanov and Lauda and by Rouquier look quite different, so that for a while subsequent works split into two different schools according to which definition they were following. In fact, Rouquier's and Khovanov and Lauda's definitions are equivalent, as was established by the first author [**B3**].

In this (mostly expository) article, we will revisit some of Rouquier's foundational definitions in the light of [**B3**]. We do this using the diagrammatic formalism of Khovanov and Lauda wherever possible. From the outset, we have systematically incorporated the better choice of normalization for the second adjunction of

2010 *Mathematics Subject Classification.* Primary 17B10, 18D10.
Research supported in part by NSF grant DMS-1161094.

the Kac-Moody 2-category suggested by [**BHLW**]. For a survey with greater emphasis on the connections to geometry, we refer the reader to Kamnitzer's text [**Kam**].

Another of our goals is to extend several of the existing results so that they may be applied in some more general situations. To explain the novelty, we need some definitions. Let $\Bbbk$ be an algebraically closed field and $\mathcal{V}ec$ be the category of (small) vector spaces. A *finite-dimensional category* is a small $\Bbbk$-linear category $\mathcal{A}$ all of whose morphism spaces are finite-dimensional. Let Mod-$\mathcal{A}$ denote the functor category $\mathcal{H}om(\mathcal{A}^{\mathrm{op}}, \mathcal{V}ec)$ of right modules over $\mathcal{A}$. We say that $\mathcal{A}$ is *Artinian* if all of the finitely generated objects and the finitely cogenerated objects in Mod-$\mathcal{A}$ have finite length (see also Remark 2.1). A *locally Schurian category* is an Abelian category that is equivalent to Mod-$\mathcal{A}$ for some finite-dimensional category $\mathcal{A}$. If in addition $\mathcal{A}$ is Artinian, then the full subcategory of Mod-$\mathcal{A}$ consisting of all objects of finite length is a *Schurian category* in the sense of [**BLW**, §2.1].

In the Abelian setting, the general structural results about 2-representations of Kac-Moody 2-categories obtained in [**CR**, **R1**, **R2**] typically only apply to categories in which all objects have finite length and whose irreducible objects satisfy Schur's Lemma. If one wants there to be enough projectives and injectives too, this means that one is working in a Schurian category in the sense just defined. The main new contribution of this paper is to extend some of these structural results to locally Schurian categories.

The motivation for doing this from a Lie theoretic perspective is as follows. Let $\mathfrak{g}$ be a symmetrizable Kac-Moody algebra with Chevalley generators $\{e_i, f_i \,|\, i \in I\}$, weight lattice P, etc... Recall that a $\mathfrak{g}$-module V is *integrable* if it decomposes into weight spaces as $V = \bigoplus_{\lambda \in P} V_\lambda$, and each e_i and f_i acts locally nilpotently. In order to categorify an integrable module with finite-dimensional weight spaces, it is reasonable to hope that one can use a finite-dimensional category whose blocks are finite-dimensional algebras, in which case all subsequent constructions can be performed in the Schurian category consisting of finite-dimensional modules over these algebras. Examples include the *minimal categorification* $\mathcal{L}_{\min}(\kappa)$ of the integrable highest weight module $L(\kappa)$ of (dominant) highest weight κ defined already by Khovanov, Lauda and Rouquier via cyclotomic quiver Hecke algebras, and the minimal categorifications $\mathcal{L}_{\min}(\kappa_1, \dots, \kappa_n)$ of tensor products $L(\kappa_1) \otimes \cdots \otimes L(\kappa_n)$ of integrable highest weight modules introduced by Webser in [**W2**].

In [**W1**], Webster also investigated categorifications of more general tensor products involving both integrable lowest weight and highest weight modules; see also [**BW**] for the construction of canonical bases in such mixed tensor products. Away from finite type, these modules have infinite-dimensional weight spaces. The candidates for their minimal categorifications suggested by Webster are finite-dimensional categories which are not Artinian in general, so that the locally Schurian setting becomes essential. In type A, there are some closely related examples arising from the cyclotomic oriented Brauer categories of [**BCNR**], which in level one are Deligne's categories $\underline{\mathrm{Rep}}(GL_t)$ (e.g. see [**EHS**]). These also fit into the framework of this article.

Here is a guide to the organization of the remainder of the article.

In Section 2, we set up the basic algebraic foundations of locally Schurian categories. Everything here is either well known (e.g. see [**M**]), or it is an obvious extension of classical results. However our language is new.

Section 3 is an exposition of the definition of Kac-Moody 2-category, based mainly on [**B3**]. We also discuss briefly the graded version of the Kac-Moody 2-category. This is important as it makes the connection to quantum groups, although we will not emphasize it elsewhere in the article.

Section 4 begins with a review of Rouquier's theory of 2-representations of Kac-Moody 2-categories. We recall his definition of the universal categorification $\mathcal{L}(\kappa)$ of $L(\kappa)$ from [**R2**, §4.3.3]. The minimal categorification $\mathcal{L}_{\min}(\kappa)$ is a certain finite-dimensional specialization of $\mathcal{L}(\kappa)$; it can be realized equivalently in terms of cyclotomic quiver Hecke algebras. We also introduce a 2-representation $\mathcal{L}(\kappa'|\kappa)$, which is expected to play the role of universal categorification for the tensor product $L(\kappa'|\kappa) := L'(\kappa') \otimes L(\kappa)$ of the integrable lowest weight module $L'(\kappa')$ of (anti-dominant) lowest weight κ' with the integrable highest weight module $L(\kappa)$ (see Construction 4.13). The minimal categorification $\mathcal{L}_{\min}(\kappa'|\kappa)$ from [**W1**, Proposition 5.6] is a certain finite-dimensional specialization of $\mathcal{L}(\kappa'|\kappa)$. After that, we focus on nilpotent categorical actions on locally Schurian categories. Any such structure has an *associated crystal* in the sense of Kashiwara; for example, the crystal associated to $\mathcal{L}_{\min}(\kappa)$ is the highest weight crystal $\mathbf{B}(\kappa)$. This has already found many striking applications in classical representation theory; e.g. see [**FK**] (the oldest) and [**DVV**] (the most recent at the time of writing).

Acknowledgements. We thank Ben Webster for sharing his ideas in [**W3**], and for suggesting the reduction argument used in the proof of Theorem 4.27. Also we thank Aaron Lauda for giving us the opportunity to write this survey, and the referee for many helpful suggestions.

Notation. Throughout, we work over an algebraically closed field $\Bbbk$. This means that all (2-)categories and (2-)functors will be assumed to be $\Bbbk$-linear by default.

2. Locally Schurian categories

In this section, we introduce our language of locally Schurian categories.

2.1. Locally unital algebras. A *locally unital algebra* is an associative (not necessarily unital) algebra A equipped with a small family $(1_x)_{x\in X}$ of mutually orthogonal idempotents such that

$$A = \bigoplus_{x,y\in X} 1_y A 1_x.$$

A *locally unital homomorphism* (resp. *isomorphism*) between two locally unital algebras is an algebra homomorphism (resp. isomorphism) which takes distinguished idempotents to distinguished idempotents. Also, we say that A is a *contraction* of B if there is an algebra isomorphism $A \xrightarrow{\sim} B$ sending each distinguished idempotent in A to a sum of distinguished idempotents in B.

We say that A is *locally Noetherian* (resp. *locally Artinian*) if all of the left ideals $A1_x$ and all of the right ideals 1_yA satisfy the Ascending Chain Conditon (resp. the Descending Chain Condition). One can also define analogs of *left* (resp. *right*) Noetherian or Artinian for locally unital algebras, requiring just that all the left ideals $A1_x$ (resp. the right ideals 1_yA) satisfy the appropriate chain condition. Unlike in the unital setting, locally left/right Artinian does not imply locally left/right Noetherian (but see Lemma 2.8 below). The following example of a locally unital algebra that is locally left Artinian but not locally left Noetherian is taken from the end of [**M**, §3]: consider the locally unital algebra of upper triangular matrices over $\Bbbk$ with rows and columns indexed by the totally ordered set $\mathbb{N} \cup \{\infty\}$, all but finitely many of whose entries are zero.

All *modules* over a locally unital algebra will be assumed to be locally unital without further mention; for a right module V this means that $V = \bigoplus_{x \in X} V1_x$ as a direct sum of subspaces. If V is any A-module satisfying ACC, it is clearly finitely generated. Conversely, assuming that A is locally Noetherian (resp. locally Artinian), finitely generated modules satisfy ACC (resp. DCC). We deduce in the locally Noetherian case that submodules of finitely generated modules are finitely generated.

Let Mod-A be the category of all right A-modules. We'll also need the following full subcategories of Mod-A:

- lfdMod-A consisting of all *locally finite-dimensional modules*, i.e. right modules V with $\dim V1_x < \infty$ for all $x \in X$;
- fgMod-A consisting of all finitely generated modules;
- pMod-A consisting of all finitely generated projective modules.

Replacing "right" with "left" everywhere here, we obtain analogous categories A-Mod, A-lfdMod, A-fgMod and A-pMod of left modules. There are contravariant equivalences

$$\circledast : \text{lfdMod-}A \to A\text{-lfdMod}, \qquad \# : \text{pMod-}A \to A\text{-pMod}$$

defined as follows: the *dual* $V^{\circledast}$ of a locally finite-dimensional right module V is the left module $\bigoplus_{x \in X} \mathrm{Hom}_{\Bbbk}(V1_x, \Bbbk)$; the *dual* $P^{\#}$ of a finitely generated projective right module P is the left module $\mathrm{Hom}_A(P, A)$. If V and P are left modules instead, their duals ${}^{\circledast}V$ and ${}^{\#}P$ are the right modules defined analogously.

REMARK 2.1. The data of a locally unital algebra A is the same as the data of a small category $\mathcal{A}$ with object set X and morphisms $\mathrm{Hom}_{\mathcal{A}}(x, y) := 1_y A 1_x$. In this incarnation, locally unital algebra homomorphisms correspond to functors. A right A-module becomes a functor $\mathcal{A}^{\mathrm{op}} \to \mathcal{V}ec$, and then a module homomorphism is a natural transformation of functors. We say $\mathcal{A}$ is *Noetherian* (resp. *Artinian*) if A is locally Noetherian (resp. locally Artinian) in the sense already defined. All of the other notions introduced in this subsection can be recast in this more categorical language too, as was done in [**M**]. For example, the projective module $1_x A$ corresponds to the functor $\mathrm{Hom}_{\mathcal{A}}(-, x) : \mathcal{A}^{\mathrm{op}} \to \mathcal{V}ec$. Then the Yoneda Lemma asserts that there is a fully faithful functor $\mathcal{A} \to$ pMod-A sending $x \in \mathrm{ob}\,\mathcal{A}$ to $1_x A$, and $a \in \mathrm{Hom}_{\mathcal{A}}(x, y)$ to the homomorphism $1_x A \to 1_y A$ defined by left multiplication by $a \in 1_y A 1_x$. This extends canonically to an equivalence of categories

$$\dot{\mathcal{A}} \to \text{pMod-}A, \tag{2.1}$$

where $\dot{\mathcal{A}}$ denotes the *additive Karoubi envelope* of $\mathcal{A}$, that is, the idempotent completion of the additive envelope of $\mathcal{A}$.

For locally unital algebras A and B with distinguished idempotents $(1_x)_{x \in X}$ and $(1_y)_{y \in Y}$, respectively, an (A, B)-bimodule $M = \bigoplus_{x \in X, y \in Y} 1_x M 1_y$ determines an adjoint pair (T_M, H_M) of functors

$$T_M := - \otimes_A M : \text{Mod-}A \to \text{Mod-}B,$$

$$H_M := \bigoplus_{x \in X} \operatorname{Hom}_B(1_x M, -) : \text{Mod-}B \to \text{Mod-}A.$$

Here are a couple of useful facts about tensoring with bimodules. First, there is a natural isomorphism

$$V \otimes_A \operatorname{Hom}_B(Q, M) \xrightarrow{\sim} \operatorname{Hom}_B(Q, V \otimes_A M), \qquad v \otimes f \mapsto (q \mapsto v \otimes f(q)) \tag{2.2}$$

for all right A-modules V and finitely generated projective right B-modules Q; cf. [**AF**, 20.10]. Also, given another locally unital algebra C and a right exact functor $E : \text{Mod-}B \to \text{Mod-}C$ commuting with direct sums, EM is an (A, C)-bimodule, and there is a natural isomorphism of right C-modules

$$V \otimes_A EM \xrightarrow{\sim} E(V \otimes_A M), \qquad v \otimes n \mapsto E(f_v)(n) \tag{2.3}$$

for all right A-modules V, where $f_v : M \to V \otimes_A M$ is the right C-module homomorphism $m \mapsto v \otimes m$. The proof of this involves a reduction to the case $V = A$ using the Five Lemma.

DEFINITION 2.2. We say that an (A, B)-bimodule M is *left rigid* if it has a left dual in the 2-category of bimodules; see [**EGNO**, §2.10] for our conventions here. It means that there exists a (B, A)-bimodule $M^\#$, an (A, A)-bimodule homomorphism $\operatorname{coev} : A \to M \otimes_B M^\#$, and a (B, B)-bimodule homomorphism $\operatorname{ev} : M^\# \otimes_A M \to B$, such that the compositions

$$M \xrightarrow{\text{can}} A \otimes_A M \xrightarrow{\operatorname{coev} \otimes 1} M \otimes_B M^\# \otimes_A M \xrightarrow{1 \otimes \operatorname{ev}} M \otimes_B B \xrightarrow{\text{can}} M,$$

$$M^\# \xrightarrow{\text{can}} M^\# \otimes_A A \xrightarrow{1 \otimes \operatorname{coev}} M^\# \otimes_A M \otimes_B M^\# \xrightarrow{\operatorname{ev} \otimes 1} B \otimes_B M^\# \xrightarrow{\text{can}} M^\#$$

are the identities. In other words, $T_{M^\#}$ is right adjoint to T_M. We say that M is *right rigid* if it has a right dual, i.e. there exists a (B, A)-bimodule ${}^\# M$ such that $T_{{}^\# M}$ is left adjoint to T_M. Finally, we say M is *rigid* if it is both left and right rigid, and *sweet*[1] if in addition $M^\# \cong {}^\# M$ as (B, A)-bimodules.

The following is essentially [**EGNO**, Ex. 2.10.16].

LEMMA 2.3. *Let A and B be locally unital algebras with distinguished idempotents $(1_x)_{x \in X}$ and $(1_y)_{y \in Y}$, respectively. Let M be an (A, B)-bimodule.*

(1) The bimodule M is left rigid if and only if $1_x M$ is finitely generated and projective as a right B-module for each $x \in X$. In that case, $M^\# \cong \bigoplus_{x \in X} (1_x M)^\#$.

(2) It is right rigid if and only if $M 1_y$ is finitely generated and projective as a left A-module for each $y \in Y$. In that case, ${}^\# M \cong \bigoplus_{y \in Y} {}^\# (M 1_y)$.

PROOF. (1) Suppose that M possesses a left dual $M^\#$. Then T_M has a right exact right adjoint $T_{M^\#}$, so T_M sends projectives to projectives. Hence, $1_x M \cong T_M(1_x A)$ is projective for each $x \in X$. Let $\operatorname{coev}(1_x) = \sum_{i=1}^{n_x} v_{x,i} \otimes f_{x,i}$ for $v_{x,i} \in 1_x M$ and $f_{x,i} \in (M^\#) 1_x$. Then any $v \in 1_x M$ is equal to $(1 \otimes \operatorname{ev})(v_{x,i} \otimes f_{x,i} \otimes v) \in \sum_{i=1}^{n_i} v_{x,i} A$. This shows that $1_x M$ is finitely generated.

Conversely, suppose that each $1_x M$ is finitely generated and projective as a right module. Then (2.2) implies that $H_M \cong T_{M^\#}$ where $M^\# := \bigoplus_{x \in X} (1_x M)^\#$.

[1]The language "sweet bimodule" appears in [**K**, §2.6], but (in view of Lemma 2.3) the ones defined there are just what we call rigid bimodules, since it is not assumed that ${}^\# M \cong M^\#$. Note though that the important examples constructed in [**K**] do satisfy this extra hypothesis, so they are sweet in our sense too.

Hence, we have constructed a bimodule $M^\#$ such that $T_{M^\#}$ is right adjoint to T_M, proving that M is left rigid.

(2) Similar (working with left modules instead of right ones). □

By a *projective generating family* for an Abelian category $\mathcal{C}$, we mean a small family $(P(x))_{x\in X}$ of compact[2] projective objects such that for each $V \in \operatorname{ob}\mathcal{C}$ there is some $x \in X$ with $\operatorname{Hom}_{\mathcal{C}}(P(x), V) \neq 0$. Just like in [**F**, Exercise 5.F], one can show that an Abelian category $\mathcal{C}$ is equivalent to Mod-A for some locally unital algebra A if and only if $\mathcal{C}$ possesses arbitrary direct sums and has a projective generating family; see also [**M**, Theorem 3.1]. We just need this in the following special case, which is the locally unital analog of the classical Morita Theorem:

THEOREM 2.4. *Let B be a locally unital algebra. Suppose that $(P(x))_{x\in X}$ is a projective generating family for* Mod-B. *Let*

$$A := \bigoplus_{x,y\in X} \operatorname{Hom}_B(P(x), P(y)),$$

viewed as a locally unital algebra with distinguished idempotents $(1_x := 1_{P(x)})_{x\in X}$. Let $P := \bigoplus_{x\in X} P(x)$, which is an (A,B)-bimodule.

1. *The functors $T_P = - \otimes_A P$ and $H_P = \bigoplus_{x\in X} \operatorname{Hom}_B(1_x P, -)$ are quasi-inverse equivalences between the categories* Mod-A *and* Mod-B.
2. *We have that $H_P \cong T_Q$ where $Q := P^\#$.*

Thus, we have constructed a sweet (A,B)-bimodule P and a sweet (B,A)-bimodule Q such that $P \otimes_B Q \cong A$ and $Q \otimes_A P \cong B$ as bimodules.

PROOF. The fact that $H_P \circ T_P \cong 1_{\text{Mod-}A}$ follows from (2.2). Then one deduces that $T_P \circ H_P \cong 1_{\text{Mod-}B}$ too by a standard argument; cf. [**AF**, 22.2]. Finally Lemma 2.3(1) implies that $H_P \cong T_Q$. □

COROLLARY 2.5. *For locally unital algebras A and B, the following are equivalent:*

1. *the categories* Mod-A *and* Mod-B *are equivalent;*
2. *the categories* pMod-A *and* pMod-B *are equivalent;*
3. *the categories* A-pMod *and* B-pMod *are equivalent;*
4. *the categories* A-Mod *and* B-Mod *are equivalent.*

PROOF. (1) $\Rightarrow$ (2). The restriction of an equivalence Mod-$A \to$ Mod-B gives an equivalence pMod-$A \to$ pMod-B.

(2) $\Rightarrow$ (1). Let F : pMod-$A \to$ pMod-B be an equivalence of categories. Let $(1_x)_{x\in X}$ be the distinguished idempotents in A. Then $(P(x) := F(1_x A))_{x\in X}$ is a projective generating family for Mod-B such that $A \cong \bigoplus_{x,y\in X} \operatorname{Hom}_B(P(x), P(y))$. Now apply Theorem 2.4.

(3) $\Leftrightarrow$ (4). This is the same as (1) $\Leftrightarrow$ (2) with A and B replaced by the opposite algebras.

(2) $\Leftrightarrow$ (3). This follows as pMod-A (resp. pMod-B) is contravariantly equivalent to A-pMod (resp. B-pMod). □

[2]For categories of the form Mod-A for some locally unital algebra A, a projective is compact if and only if it is finitely generated.

Two locally unital algebras A and B are said to be *Morita equivalent* if the conditions of Corollary 2.5 are satisfied. For example, if A is a contraction of B, then the categories Mod-A and Mod-B are obviously isomorphic. Hence, A and B are Morita equivalent. For another simple example, let N be any (possibly infinite but small) set and $M_N(\Bbbk)$ be the algebra of $N \times N$ matrices with entries in $\Bbbk$, all but finitely many of which are zero. This is a locally unital algebra with distinguished idempotents given by the diagonal matrix units $\{e_{i,i} | i \in N\}$. Applying Theorem 2.4 with $B := \Bbbk, X := N$ and taking each $P(x)$ to be a copy of $\Bbbk$, we see that $M_N(\Bbbk)$ is Morita equivalent to the ground field $\Bbbk$.

REMARK 2.6. Suppose that $\mathcal{A}$ and $\mathcal{B}$ are the categories associated to locally unital algebras A and B as in Remark 2.1. We say that $\mathcal{A}$ and $\mathcal{B}$ are *Morita equivalent* if their additive Karoubi envelopes $\dot{\mathcal{A}}$ and $\dot{\mathcal{B}}$ are equivalent. In view of Corollary 2.5 and (2.1), this is equivalent to the algebras A and B being Morita equivalent as above.

The final theorem in this subsection is concerned with adjoint functors. Again this is classical in the unital setting.

THEOREM 2.7. *Let B, $P = \bigoplus_{x \in X} P(x)$ and A be as in Theorem 2.4, so that H_P : Mod-$B \to$ Mod-A is an equivalence of categories. Suppose we are given a functor E : Mod-$B \to$ Mod-B. Then E possesses a right adjoint if and only if it is right exact and commutes with direct sums. In that case, let*

$$M := \bigoplus_{x,y \in X} \mathrm{Hom}_B(P(x), EP(y))$$

viewed as an (A, A)-bimodule in the natural way. Then the diagram

$$\begin{array}{ccc} \text{Mod-}B & \xrightarrow{H_P} & \text{Mod-}A \\ {\scriptstyle E}\big\downarrow & & \big\downarrow{\scriptstyle T_M} \\ \text{Mod-}B & \xrightarrow[H_P]{} & \text{Mod-}A \end{array}$$

commutes up to a canonical isomorphism.

PROOF. It is standard that functors possessing a right adjoint are right exact and commute with direct sums. Conversely, suppose that E is right exact and commutes with direct sums. Using (2.3) then (2.2), we get that

$$H_P \circ E \circ T_P \cong H_P \circ T_{EP} \cong T_{H_P(EP)},$$

which is isomorphic to T_M as $H_P(EP) \cong M$. Thus $H_P \circ E \circ T_P \cong T_M$. Composing on the right with the quasi-inverse H_P of T_P, we deduce that $H_P \circ E \cong T_M \circ H_P$. This proves the final part of the theorem. Hence E has a right adjoint as T_M has the right adjoint H_M. □

2.2. Finite-dimensional categories. Let A be a locally unital algebra with distinguished idempotents $(1_x)_{x \in X}$. We assume in this subsection that A is *locally finite-dimensional*, i.e. each subspace $1_y A 1_x$ is finite-dimensional. Equivalently, the associated category $\mathcal{A}$ from Remark 2.1 is a *finite-dimensional category*, i.e. it is a small $\Bbbk$-linear category all of whose morphism spaces are finite-dimensional. All of the right ideals $1_x A$ and the left ideals $A 1_x$ are locally finite-dimensional. Hence, so are their duals $(A1_x)^{\circledast}$ and $(1_xA)^{\circledast}$. Consequently, all finitely generated modules are locally finite-dimensional, as are all finitely cogenerated modules.

LEMMA 2.8. *For a locally finite-dimensional locally unital algebra A, locally Artinian implies locally Noetherian.*

PROOF. As $A1_x$ satisfies DCC, its dual $(A1_x)^{\circledast}$ satisfies ACC. Hence, $(A1_x)^{\circledast}$ is finitely generated, so satisfies DCC. Hence, $A1_x$ satisfies ACC. Similarly each 1_xA has ACC. □

Here are various other basic facts about modules over a locally finite-dimensional locally unital algebra A. For the most part, these are proved by mimicking the usual proofs in the setting of finite-dimensional algebras, so we will be quite brief. Fix representatives $\{L(b) \mid b \in \mathbf{B}\}$ for the isomorphism classes of irreducible right A-modules[3].

(L1) If V is finitely generated (resp. locally finite-dimensional) and W is locally finite-dimensional (resp. finitely cogenerated) then $\mathrm{Hom}_A(V, W)$ is finite-dimensional.

(L2) *Schur's Lemma* holds: $\mathrm{End}_A(L(b)) \cong \Bbbk$ for each $b \in \mathbf{B}$.

(L3) Any finitely generated (resp. finitely cogenerated) module satisfies the *Krull-Schmidt Theorem.*

(L4) The category Mod-A is a Grothendieck category, i.e. it is Abelian, it possesses arbitrary direct sums, direct limits of short exact sequences are exact, and there is a generator (namely, the regular module A itself). Hence, by the general theory of Grothendieck categories, every A-module has an injective hull; moreover, a module V is finitely cogenerated if and only if its *socle* $\mathrm{soc}(V)$, i.e. the sum of the irreducible submodules of V, is an essential submodule of V of finite length.

(L5) For $b \in \mathbf{B}$, let $A_b := A/\mathrm{Ann}_A(L(b))$, which is a locally unital algebra with distinguished idempotents $(1_x)_{x\in X}$ that are the images of the ones in A. Also let $M_b := \bigoplus_{x,y\in X} \mathrm{Hom}_{\Bbbk}(L(b)1_x, L(b)1_y)$, viewed as a locally unital algebra with multiplication that is the opposite of composition. Note that M_b is a contraction of $M_N(\Bbbk)$ where $N := \{(x, i) \mid x \in X, 1 \leq i \leq \dim L(b)1_x\}$. Then the natural right action of A on $L(b)$ induces a locally unital isomorphism $A_b \xrightarrow{\sim} M_b$. Hence, A_b is a contraction of a locally unital matrix algebra. Next, let $J := \bigcap_{b\in\mathbf{B}} \mathrm{Ann}_A(L(b))$ be the *Jacobson radical* of A. The map

$$A/J \to \bigoplus_{b\in\mathbf{B}} A_b, \qquad a + J \mapsto (a + \mathrm{Ann}_A(L(b)))_{b\in\mathbf{B}}$$

is a well-defined algebra isomorphism. Hence, A/J is a contraction of a (possibly infinite) direct sum of locally unital matrix algebras. It follows that A/J is *semisimple*, i.e. every A/J-module is completely reducible. Moreover, J is the smallest two-sided ideal of A with this property.

(L6) For a right A-module V, its *radical* $\mathrm{rad}(V) := VJ$ is the intersection of all of its proper maximal submodules; its *head* $\mathrm{hd}(V) := V/\mathrm{rad}(V)$ is its largest completely reducible quotient. Applying $\circledast$ to the statements made in (L4), we deduce that every finitely generated A-module has a

[3]We use the notation $\mathbf{B}$ here as ultimately this set will carry a crystal structure; cf. Definition 4.30.

projective cover; moreover, V is finitely generated if and only if $\mathrm{rad}(V)$ is a superfluous submodule and $\mathrm{hd}(V)$ is of finite length[4].

(L7) Let $P(b)$ be a projective cover of $L(b)$. For any right A-module V, the *composition multiplicity* $[V : L(b)]$ is defined as usual to be the supremum of the multiplicities $\#\{i = 1, \dots, n \mid V_i/V_{i-1} \cong L(b)\}$ taken over all filtrations $0 = V_0 < \cdots < V_n = V$ and all $n \in \mathbb{N}$. By Schur's Lemma, we have that $[V : L(b)] = \dim \mathrm{Hom}_A(P(b), V) \in \mathbb{N} \cup \{\infty\}$. Noting that $\mathrm{Hom}_A(1_x A, L(b)) \cong L(b)1_x$, the projective module $1_x A$ decomposes as

$$1_x A \cong \bigoplus_{b \in \mathbf{B}} P(b)^{\oplus \dim L(b)1_x}.$$

All but finitely many summands on the right hand side are zero, so there are only finitely many $b \in \mathbf{B}$ such that $L(b)1_x \neq 0$. Hence, for any V, we have that

$$\dim V1_x = \sum_{b \in \mathbf{B}} [V : L(b)] \dim L(b)1_x.$$

In particular, we get from this that V is locally finite-dimensional if and only if $[V : L(b)] < \infty$ for all $b \in \mathbf{B}$.

(L8) Given $b \in \mathbf{B}$, we choose $x \in X$ so that $L(b)1_x \neq 0$. The decomposition of $1_x A$ derived in (L7) implies that there exists a primitive idempotent $1_b \in 1_x A 1_x$ such that $P(b) \cong 1_b A$. Then $I(b) := (A1_b)^{\circledast}$ is an injective hull of $L(b)$. For any V, we have that $[V : L(b)] = \dim \mathrm{Hom}_A(V, I(b))$.

(L9) Suppose we are given a family $(A_i)_{i \in I}$ of locally finite-dimensional locally unital algebras, with the distinguished idempotents in A_i indexed by X_i. Then $A := \bigoplus_{i \in I} A_i$ is a locally finite-dimensonal locally unital algebra with distinguished idempotents indexed by $X := \bigsqcup_{i \in I} X_i$. Moreover there is an equivalence of categories $\prod_{i \in I} \text{Mod-}A_i \to \text{Mod-}A$ which sends an object $(V_i)_{i \in I}$ of $\prod_{i \in I} \text{Mod-}A_i$ to the A-module $\bigoplus_{i \in I} V_i$.

(L10) Suppose $\mathbf{B} = \bigsqcup_{i \in I} \mathbf{B}_i$ is a partition such that that $\mathrm{Hom}_A(P(b), P(c)) = 0$ for all $b \in \mathbf{B}_i, c \in \mathbf{B}_j$ and $i \neq j$. For $x \in X$, we can write 1_x uniquely as a sum of mutually orthogonal idempotents $1_x = \sum_{i \in I} 1_{(x,i)}$ so that $1_{(x,i)} A \cong \bigoplus_{b \in \mathbf{B}_i} P(b)^{\oplus \dim L(b)1_x}$. Let $A_i := \bigoplus_{x,y \in X} 1_{(y,i)} A 1_{(x,i)}$, which is itself a locally unital algebra with idempotents $(1_{(x,i)})_{x \in X}$ and irreducibles represented by $\{L(b) \mid b \in \mathbf{B}_i\}$. Then we have that $A = \bigoplus_{i \in I} A_i$. Hence, A is a contraction of $\bigoplus_{i \in I} A_i$. If none of the $\mathbf{B}_i$ can be partitioned any further in this way, we call this the *block decomposition* of A, and refer to indecomposable subalgebras A_i as *blocks*.

2.3. Locally Schurian categories. Later in the article, we will be interested in categorical actions on categories of the following form:

DEFINITION 2.9. We say that a category $\mathcal{C}$ is *locally Schurian* if it is equivalent to Mod-A for some locally finite-dimensional locally unital algebra A.

Given a locally Schurian category $\mathcal{C}$, Theorem 2.4 gives a recipe for constructing a locally finite-dimensional locally unital algebra A such that $\mathcal{C}$ is equivalent to

[4]One can give a direct proof of this using the fact that J is *locally nilpotent* in the sense that eJe is a nilpotent ideal of the finite-dimensional algebra eAe for any idempotent $e \in A$.

Mod-A: choose a projective generating family $(P(x))_{x\in X}$ for $\mathcal{C}$; set

$$A := \bigoplus_{x,y\in X} \mathrm{Hom}_{\mathcal{C}}(P(x), P(y)) \tag{2.4}$$

viewed as a locally unital algebra with distinguished idempotents $(1_x := 1_{P(x)})_{x\in X}$; then the functor

$$H := \bigoplus_{x\in X} \mathrm{Hom}_{\mathcal{C}}(P(x), -) : \mathcal{C} \to \text{Mod-}A \tag{2.5}$$

is an equivalence of categories. Often it is convenient to proceed by choosing representatives $\{L(b) \mid b \in \mathbf{B}\}$ for the isomorphism classes of irreducible object in $\mathcal{C}$ and letting $P(b)$ (resp. $I(b)$) be a projective cover (resp. an injective hull) of $L(b)$. Then we call $(P(b))_{b\in\mathbf{B}}$ a *minimal* projective generating family for $\mathcal{C}$, and $\mathcal{C}$ is equivalent to Mod-B where

$$B := \bigoplus_{b,c\in\mathbf{B}} \mathrm{Hom}_{\mathcal{C}}(P(b), P(c)). \tag{2.6}$$

This a *basic* locally unital algebra: the irreducible B-modules are all one dimensional.

An object V in $\mathcal{C}$ is *locally finite-dimensional* if and only if all its composition multiplicities are finite. We let $\mathrm{p}\mathcal{C} \subseteq \mathrm{fg}\mathcal{C} \subseteq \mathrm{lfd}\mathcal{C}$ be the full subcategories of $\mathcal{C}$ consisting of finitely generated projective, finitely generated, and locally finite-dimensional objects; for A as in (2.4), these are equivalent to the subcategories pMod-A, fgMod-A and lfdMod-A of Mod-A.

We say that $\mathcal{C}$ is *Noetherian* if all finitely generated (resp. cogenerated) objects satisfy ACC (resp. DCC); equivalently, the algebra A from (2.4) is locally Noetherian. We say that $\mathcal{C}$ is *Artinian* if all finitely generated (resp. cogenerated) objects satisfy DCC (resp. ACC); equivalently, the algebra A is locally Artinian. We say that $\mathcal{C}$ is *finite* if there are only finitely many isomorphism classes of irreducible object; equivalently, the basic algebra B from (2.6) is finite-dimensional. By Lemma 2.8, Artinian implies Noetherian.

If $\mathcal{C}$ is Artinian then $\mathrm{fg}\mathcal{C}$ is a *Schurian category* in the sense of [**BLW**, §2.1]: it is Abelian, all objects are of finite length, there are enough projectives and injectives, and the endomorphism algebras of the irreducible objects are one dimensional. Moreover, for $V \in \mathrm{ob}\,\mathcal{C}$, the following are equivalent:

- V is finitely generated;
- V is finitely cogenerated;
- V has finite length.

2.4. Sweet endofunctors. We will consider categorical actions on locally Schurian categories involving functors of the following form:

DEFINITION 2.10. Let $\mathcal{C}$ be a locally Schurian category. We say that an endofunctor E of $\mathcal{C}$ is *sweet* if there is an endofunctor F which is biadjoint to E.

Recalling the definition of sweet bimodule from Definition 2.2, the following theorem gives an algebraic characterization of sweet endofunctors.

THEOREM 2.11. *Let E be an endofunctor of a locally Schurian category $\mathcal{C}$. Fix an equivalence $H : \mathcal{C} \to$ Mod-A as in (2.5). Then E is sweet if and only if there is* *a sweet bimodule M such that $H \circ E \cong T_M \circ H$. In that case, we have that*

$$M \cong \bigoplus_{x,y \in X} \mathrm{Hom}_{\mathcal{C}}(P(x), EP(y)). \tag{2.7}$$

Moreover, E is exact, continuous and cocontinuous, and it preserves the sets of locally finite-dimensional, finitely generated, finitely cogenerated, projective and injective objects.

PROOF. If M is a sweet bimodule such that $H \circ E \cong T_M \circ H$, then E is sweet since T_M is a sweet endofunctor of Mod-A and H is an equivalence. Conversely, suppose that E possesses a biadjoint F. Theorem 2.7 shows that $H \circ E \cong T_M \circ H$ for M as in (2.7); similarly $H \circ F \cong T_N \circ H$ for some bimodule N. Then T_M and T_N are biadjoint. Hence, N is both right and left dual to M, i.e. M is a sweet bimodule. It follows at once that E and F both send finitely generated objects to finitely generated objects, as T_M and T_N clearly do.

Since E has a biadjoint, it is exact, continuous and cocontinous. Also F is exact, so E preserves projectives and injectives. To see that E preserves locally finite-dimensional objects, we observe for locally finite-dimensional V that

$$\dim \mathrm{Hom}_{\mathcal{C}}(P(b), EV) = \dim \mathrm{Hom}_{\mathcal{C}}(FP(b), V) < \infty$$

for all $b \in \mathbf{B}$. Similarly F preserves locally finite-dimensional objects. Finally, since E preserves finitely generated objects, we have that

$$\dim \mathrm{Hom}_{\mathcal{C}}(EP(b), L(c)) = \dim \mathrm{Hom}_{\mathcal{C}}(P(b), FL(c)) = [FL(c) : L(b)]$$

is zero for all but finitely many c. Hence,

$$\dim \mathrm{Hom}_{\mathcal{C}}(L(c), EI(b)) = \dim \mathrm{Hom}_{\mathcal{C}}(FL(c), I(b)) = [FL(c) : L(b)]$$

is zero for all but finitely many c. This implies that $EI(b)$ is a finite direct sum of $I(c)$'s. Hence, E preserves finitely cogenerated objects, and similarly for F. □

LEMMA 2.12. *Suppose that F and G are sweet endofunctors of a locally Schurian category $\mathcal{C}$. Let $\eta : F \Rightarrow G$ be a natural transformation. If $\eta_L : FL \to GL$ is an isomorphism for each irreducible object $L \in \mathrm{ob}\,\mathcal{C}$, then η is an isomorphism.*

PROOF. We may assume that $\mathcal{C} = $ Mod-B for a basic locally unital algebra B. This means that the irreducible B-modules are parametrized by the same set $\mathbf{B}$ as indexes its distinguished idempotents, and $[V : L(b)] = \dim \mathrm{Hom}_B(P(b), V) = \dim V 1_b$ for each $b \in \mathbf{B}$. The main step is to prove that $\eta_V : FV \to GV$ is an isomorphism for each locally finite-dimensional B-module V. Assuming this, the lemma may be deduced as follows: given any B-module V, consider a two-step projective resolution $Q \to P \to V \to 0$; since P and Q are direct sums of finitely generated projectives, and F and G commute with arbitrary direct sums, the locally finite-dimensional result shows that η_P and η_Q are isomorphisms. Hence, η_V is an isomorphism too by the Five Lemma.

So now suppose that V is locally finite-dimensional. It suffices to show for each fixed $a \in \mathbf{B}$ that the restriction of η_V defines a linear isomorphism between $(FV)1_a$ and $(GV)1_a$. Let

$$X := \{b \in \mathbf{B} \mid (FL(b))1_a \neq 0\} = \{b \in \mathbf{B} \mid \mathrm{Hom}_B(P(a), FL(b)) \neq 0\}.$$

Fixing a left adjoint E to F, we have that $X = \{b \in \mathbf{B} \mid \mathrm{Hom}_B(EP(a), L(b)) \neq 0\}$. Since $EP(a)$ is a finitely generated projective, we deduce from this that X is finite;

moreover, $EP(a)$ a direct sum of indecomposable projectives of the form $P(b)$ for $b \in X$. Now we proceed by induction on $n := \sum_{b \in X} \dim V1_b \in \mathbb{N}$. In case $n = 0$, we have that $\operatorname{Hom}_B(P(a), FV) \cong \operatorname{Hom}_B(EP(a), V) = 0$. Hence, $(FV)1_a = 0$; similarly, $(GV)1_a = 0$. So the desired conclusion that $(FV)1_a \cong (GV)1_a$ is trivial. For the induction step, we take a vector $0 \neq v \in V1_b$ for some $b \in X$. Let $W := vB$ and $W' := \operatorname{rad}(W)$, so that we have a filtration $0 \leq W' < W \leq V$ with $W/W' \cong L(b)$. By induction, $\eta_{W'}$ and $\eta_{V/W}$ restrict to isomorphisms $(FW')1_a \xrightarrow{\sim} (GW')1_a$ and $(FV/FW)1_a \xrightarrow{\sim} (GV/GW)1_a$. Also $\eta_{W/W'}$ is an isomorphism as W/W' is irreducible. Hence, η_V defines an isomorphism $(FV)1_a \xrightarrow{\sim} (GV)1_a$ as required. □

2.5. Serre quotients. Finally in this section, we review briefly the standard notions of *Serre subcategory* and *Serre quotient category* in the setting of locally Schurian categories. Let $\mathcal{C}$ be a locally Schurian category with irreducible objects represented by $\{L(b) \mid b \in \mathbf{B}\}$ as above. Let $\mathbf{B}'$ be any subset of $\mathbf{B}$ and $\mathcal{C}'$ be the full subcategory of $\mathcal{C}$ consisting of all the objects whose irreducible subquotients are isomorphic to $L(b)$ for $b \in \mathbf{B}'$. It is a Serre subcategory of $\mathcal{C}$, i.e. it is closed under taking subobjects, quotients and extensions. Moreover it is itself a locally Schurian category with irreducible objects represented by $\{L(b) \mid b \in \mathbf{B}'\}$. To see this, define B according to (2.6) so that $\mathcal{C}$ is equivalent to Mod-B. Then $\mathcal{C}'$ is equivalent to Mod-B' where B' is the quotient of B by the two-sided ideal generated by the idempotents $\{1_b \mid b \in \mathbf{B} \setminus \mathbf{B}'\}$. The exact inclusion functor $\iota : \mathcal{C}' \to \mathcal{C}$ corresponds to the natural inflation functor from Mod-B' to Mod-B, and it has a left adjoint $\iota^!$ (resp. a right adjoint ι^*) which sends an object to its largest quotient (resp. subobject) belonging to $\mathcal{C}'$. We have that $\iota^! \circ \iota \cong 1_{\mathcal{C}'} \cong \iota^* \circ \iota$.

The *Serre quotient* $\mathcal{C}/\mathcal{C}'$ is an Abelian category equipped with an exact *quotient functor* $\pi : \mathcal{C} \to \mathcal{C}/\mathcal{C}'$ satisfying the following universal property: if $F : \mathcal{C} \to \mathcal{D}$ is any exact functor to an Abelian category $\mathcal{D}$ then there is a unique functor $\bar{F} : \mathcal{C}/\mathcal{C}' \to D$ such that $F = \bar{F} \circ \pi$. In fact $\mathcal{C}/\mathcal{C}'$ is another locally Schurian category with irreducibles represented by $\{\pi L(b) \mid b \in \mathbf{B} \setminus \mathbf{B}'\}$. Again this is easy to see in terms of the algebra B: the category $\mathcal{C}/\mathcal{C}'$ is equivalent to modules over the algebra $eBe := \bigoplus_{b,c \in \mathbf{B} \setminus \mathbf{B}'} 1_b B 1_c$. The quotient functor π corresponds to the obvious truncation functor $e :$ Mod-$B \to$ Mod-eBe sending a B-module V to $Ve := \bigoplus_{b \in \mathbf{B} \setminus \mathbf{B}'} V1_b$. Consequently, π has a left adjoint $\pi^! : \mathcal{C}/\mathcal{C}' \to \mathcal{C}$ and a right adjoint $\pi^* : \mathcal{C}/\mathcal{C}'$ corresponding to the functors $- \otimes_{eBe} eB$ and $\bigoplus_{b \in \mathbf{B} \setminus \mathbf{B}'} \operatorname{Hom}_{eBe}(B1_b, -)$, respectively. We have that $\pi \circ \pi^! \cong 1_{\mathcal{C}/\mathcal{C}'} \cong \pi \circ \pi^*$.

LEMMA 2.13. *In the above setup, assume that we are given $V, W \in \operatorname{ob} \mathcal{C}$ such that V is finitely generated, W is finitely cogenerated, and all constituents of* $\operatorname{hd}(V)$ *and* $\operatorname{soc}(W)$ *are of the form $L(b)$ for $b \in \mathbf{B} \setminus \mathbf{B}'$. Then the functor π induces an isomorphism*

$$\operatorname{Hom}_{\mathcal{C}}(V, W) \xrightarrow{\sim} \operatorname{Hom}_{\mathcal{C}/\mathcal{C}'}(\pi V, \pi W).$$

PROOF. The counit of adjunction defines a morphism $f : \pi^! \pi V \to V$. By the assumptions on V, f is an epimorphism. Moreover f becomes an isomorphism on applying π, hence $\ker f$ belongs to $\mathcal{C}'$. Using also the assumptions on W, we deduce that $\operatorname{Hom}_{\mathcal{C}/\mathcal{C}'}(\pi V, \pi W) \cong \operatorname{Hom}_{\mathcal{C}}(\pi^! \pi V, W) \cong \operatorname{Hom}_{\mathcal{C}}(V, W)$. □

3. Kac-Moody 2-categories

In this section, we review Rouquier's definition of Kac-Moody 2-category from [**R1**]. Then, following [**B3**], we explain the relationship between this and the 2-category introduced by Khovanov and Lauda in [**KL3**], and discuss the graded version. Note our exposition uses a slightly different normalization for the second adjunction compared to [**B3**] based on the idea of [**BHLW**].

3.1. Kac-Moody data. Let I be a finite index set[5] and $A = (-d_{ij})_{i,j\in I}$ be a generalized Cartan matrix, so $d_{ii} = -2$, $d_{ij} \geq 0$ for $i \neq j$, and $d_{ij} = 0 \Leftrightarrow d_{ji} = 0$. We assume that A is *symmetrizable*, so that there exist positive integers $(d_i)_{i\in I}$ such that $d_i d_{ij} = d_j d_{ji}$ for all $i, j \in I$. Pick a finite-dimensional complex vector space $\mathfrak{h}$ and linearly independent subsets $\{\alpha_i \mid i \in I\}$ and $\{h_i \mid i \in I\}$ of $\mathfrak{h}^*$ and $\mathfrak{h}$, respectively, such that $\langle h_i, \alpha_j\rangle = -d_{ij}$ for all $i, j \in I$. Let

$$P := \{\lambda \in \mathfrak{h}^* \mid \langle h_i, \lambda\rangle \in \mathbb{Z} \text{ for all } i \in I\}, \qquad Q := \bigoplus_{i\in I} \mathbb{Z}\alpha_i,$$

$$P^+ := \{\lambda \in \mathfrak{h}^* \mid \langle h_i, \lambda\rangle \in \mathbb{N} \text{ for all } i \in I\}, \qquad Q^+ := \bigoplus_{i\in I} \mathbb{N}\alpha_i.$$

We refer to P and Q as the *weight lattice* and the *root lattice*, respectively. We view P as an interval-finite poset via the usual *dominance ordering*: $\lambda \leq \mu \Leftrightarrow \mu - \lambda \in Q^+$.

Let $\mathfrak{g}$ be the associated *Kac-Moody algebra* with Cartan subalgebra $\mathfrak{h}$. Thus, $\mathfrak{g}$ is the Lie algebra generated by $\mathfrak{h}$ and elements e_i, f_i $(i \in I)$ subject to the usual Serre relations: for $h, h' \in \mathfrak{h}$ and $i, j \in I$ we have that

$$[h, h'] = 0, \qquad [e_i, f_j] = \delta_{i,j} h_i, \tag{3.1}$$

$$[h, e_i] = \langle h, \alpha_i\rangle e_i, \qquad [h, f_i] = -\langle h, \alpha_i\rangle f_i, \tag{3.2}$$

$$(\text{ad } e_i)^{d_{ij}+1}(e_j) = 0, \qquad (\text{ad } f_i)^{d_{ij}+1}(f_j) = 0. \tag{3.3}$$

Let $U(\mathfrak{g})$ be its *universal enveloping algebra.* Actually it is often more convenient to work with the idempotented version $\dot{U}(\mathfrak{g})$ of $U(\mathfrak{g})$, which is a certain locally unital algebra with distinguished idempotents $(1_\lambda)_{\lambda\in P}$. It is defined by analogy with [**Lu2**, §23.1] (which treats the quantum case). As well as being an algebra, $\dot{U}(\mathfrak{g})$ is a $(U(\mathfrak{g}), U(\mathfrak{g}))$-bimodule with $h1_\lambda = \langle h, \lambda\rangle 1_\lambda = 1_\lambda h$, $e_i 1_\lambda = 1_{\lambda+\alpha_i} e_i$ and $f_i 1_\lambda = 1_{\lambda-\alpha_i} f_i$. The elements $\{1_\lambda, e_i 1_\lambda, f_i 1_\lambda \mid i \in I, \lambda \in P\}$ generate $\dot{U}(\mathfrak{g})$ subject to the relations derived from (3.1)–(3.3). Weight modules for $\mathfrak{g}$, i.e. $\mathfrak{g}$-modules V such that $V = \bigoplus_{\lambda\in P} V_\lambda$, are just the same as (locally unital) $\dot{U}(\mathfrak{g})$-modules.

By an *upper* (resp. *lower*) *integrable module*, we mean a weight module with finite-dimensional weight spaces all of whose weights lie in a finite union of sets of the form $\lambda - Q^+$ (resp. $\lambda + Q^+$). By the classical theory from [**Kac**, Chapters 9–10], a $\mathfrak{g}$-module is upper integrable if and only if it is isomorphic to a finite direct sum of the irreducible modules[6]

$$L(\kappa) := \dot{U}(\mathfrak{g})1_\kappa / \langle e_i 1_\kappa, f_i^{1+\langle h_i,\kappa\rangle} 1_\kappa \mid i \in I\rangle \tag{3.4}$$

for $\kappa \in P^+$; these are the *integrable highest weight modules.* Similarly, a $\mathfrak{g}$-module is lower integrable if and only if it is a finite direct sum of the (irreducible) *integrable*

[5] With minor adjustments to the basic definitions, everything here can be extended to infinite I too; type A_∞ is particularly important in applications.

[6] All of these assertions depend on the assumption that A is symmetrizable.

lowest weight modules

$$L'(\kappa') := \dot{U}(\mathfrak{g})1_{\kappa'}/\langle e_i^{1-\langle h_i,\kappa'\rangle}1_{\kappa'}, f_i 1_{\kappa'} \mid i \in I\rangle. \tag{3.5}$$

for $\kappa' \in -P^+$. Generalizing (3.4)–(3.5), we let

$$L(\kappa'|\kappa) := \dot{U}(\mathfrak{g})1_{\kappa+\kappa'}/\langle e_i^{1-\langle h_i,\kappa'\rangle}1_{\kappa+\kappa'}, f_i^{1+\langle h_i,\kappa\rangle}1_{\kappa+\kappa'} \mid i \in I\rangle \tag{3.6}$$

for $\kappa \in P^+$ and $\kappa' \in -P^+$. These modules are not so well studied, but they play an important role in Lusztig's construction of canonical bases for $\dot{U}(\mathfrak{g})$ from [**Lu2**, Part IV]. They are integrable modules but they may have infinite-dimensional weight spaces outside of finite type, so that they are neither upper nor lower integrable. The next lemma is the classical counterpart of [**Lu2**, Proposition 23.3.6].

LEMMA 3.1. *There is a $\mathfrak{g}$-module isomorphism $L(\kappa'|\kappa) \xrightarrow{\sim} L(\kappa') \otimes L(\kappa)$ such that $u\bar{1}_{\kappa+\kappa'} \mapsto u(\bar{1}_{\kappa'} \otimes \bar{1}_\kappa)$ for $u \in \dot{U}(\mathfrak{g})$.*

The following lemma about annihilators will be useful later on.

LEMMA 3.2. *We have that*

$$\bigcap_{\substack{\kappa\in P^+\\ \kappa'\in -P^+}} \mathrm{Ann}_{\dot{U}(\mathfrak{g})}(L(\kappa'|\kappa)) = \{0\}.$$

Moreover if $\mathfrak{g}$ is of finite type then

$$\bigcap_{\lambda\in P^+} \mathrm{Ann}_{\dot{U}(\mathfrak{g})}(L(\lambda)) = \{0\}.$$

PROOF. The proof of the first statement reduces using Lemma 3.1 to checking that the maps

$$U^-(\mathfrak{g}) \to \bigoplus_{\kappa\in P^+} L(\kappa), \qquad u \mapsto (u\bar{1}_\kappa)_{\kappa\in P^+},$$
$$U^+(\mathfrak{g}) \to \bigoplus_{\kappa'\in -P^+} L'(\kappa'), \qquad u \mapsto (u\bar{1}_{\kappa'})_{\kappa'\in -P^+}$$

are injective, where $U^\pm(\mathfrak{g})$ are the positive and negative parts of $U(\mathfrak{g})$ generated by the e_i and f_i, respectively. These are well-known facts; e.g. they can be deduced in a non-classical way from [**Lu2**, Proposition 19.3.7]. The second statement (which is even better known) follows from the first since each $L(\kappa'|\kappa)$ is a finite direct sum of $L(\lambda)$'s in view of Lemma 3.1 and complete reducibility. □

3.2. Strict 2-categories. Let $\mathcal{C}at$ be the category of (small) $\Bbbk$-linear categories and $\Bbbk$-linear functors. It is a monoidal category with tensor functor $\boxtimes$: $\mathcal{C}at \times \mathcal{C}at \to \mathcal{C}at$ defined on objects (= categories) by letting $\mathcal{C} \boxtimes \mathcal{C}'$ be the category with objects that are pairs $(\lambda, \lambda') \in \mathrm{ob}\,\mathcal{C} \times \mathrm{ob}\,\mathcal{C}'$, morphisms

$$\mathrm{Hom}_{\mathcal{C}\boxtimes\mathcal{C}'}((\lambda,\lambda'),(\mu,\mu')) := \mathrm{Hom}_{\mathcal{C}}(\lambda,\mu) \otimes_\Bbbk \mathrm{Hom}_{\mathcal{C}}(\lambda',\mu'),$$

and composition law defined by $(g \otimes g') \circ (f \otimes f') := (g \circ f) \otimes (g' \circ f')$. The definition of $\boxtimes$ on morphisms (= functors) is obvious: $F \boxtimes F'$ is the functor that sends $(\lambda, \lambda') \mapsto (F\lambda, F'\lambda')$ and $g \otimes g' \mapsto Fg \otimes F'g'$.

DEFINITION 3.3. A *strict 2-category* is a category enriched in $\mathcal{C}at$. Thus, for objects λ, μ in a strict 2-category $\mathfrak{C}$, there is given a category $\mathcal{H}om_{\mathfrak{C}}(\lambda, \mu)$ of morphisms from λ to μ, whose objects F, G are the *1-morphisms* of $\mathfrak{C}$, and whose morphisms $x : F \Rightarrow G$ are the *2-morphisms* of $\mathfrak{C}$.

For example, $\mathcal{C}at$ can be viewed as a strict 2-category $\mathfrak{Cat}$ with 2-morphisms being natural transformations.

Given a strict 2-category $\mathfrak{C}$, we use the shorthand $\mathrm{Hom}_{\mathfrak{C}}(F, G)$ for the vector space $\mathrm{Hom}_{\mathcal{H}om_{\mathfrak{C}}(\lambda,\mu)}(F, G)$ of all 2-morphisms $x : F \Rightarrow G$. Let us also briefly recall the "string calculus" for 2-morphisms in $\mathfrak{C}$; e.g. see [**L2**, §2]. We represent a 2-morphism $x \in \mathrm{Hom}_{\mathfrak{C}}(F, G)$ by the picture

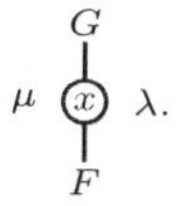

The vertical composition $y \circ x$ of x with another 2-morphism $y \in \mathrm{Hom}_{\mathfrak{C}}(G, H)$ is obtained by vertically stacking pictures:

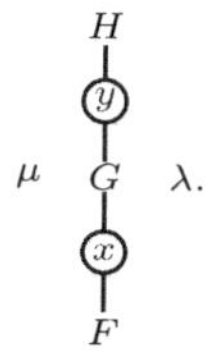

Given 2-morphisms $x : F \Rightarrow H, y : G \Rightarrow K$ between 1-morphisms $F, H : \lambda \to \mu, G, K : \mu \to \nu$, we denote their horizontal composition by $yx : GF \Rightarrow KH$, and represent it by horizontally stacking pictures:

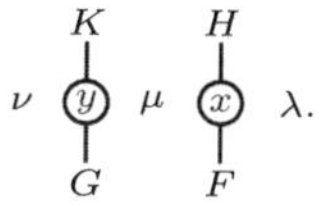

When confusion seems unlikely, we will use the same notation for a 1-morphism F as for its identity 2-morphism. With this convention, we have that $yH \circ Gx = yx = Kx \circ yF$, or in pictures:

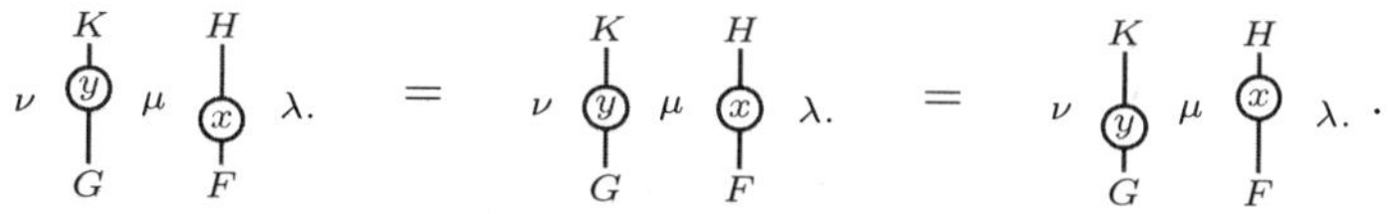

This is the *interchange law*; it means that diagrams for 2-morphisms are invariant under rectilinear isotopy.

We note that any strict 2-category $\mathfrak{C}$ has an *additive envelope* constructed by taking the additive envelope of each of the morphism categories in $\mathfrak{C}$. The *additive Karoubi envelope* $\dot{\mathfrak{C}}$ is the strict 2-category obtained by taking idempotent completions after that. Finally, we define the *Grothendieck ring*

$$K_0(\dot{\mathfrak{C}}) := \bigoplus_{\lambda,\mu \in \mathrm{ob}\,\mathfrak{C}} K_0(\mathcal{H}om_{\dot{\mathfrak{C}}}(\lambda, \mu)), \tag{3.7}$$

where the latter K_0 denotes the usual split Grothendieck group of an additive category. Horizontal composition induces a multiplication making $K_0(\dot{\mathfrak{C}})$ into a locally unital ring with distinguished idempotents $(1_\lambda)_{\lambda \in \mathrm{ob}\,\mathfrak{C}}$.

3.3. Quiver Hecke categories and the nil Hecke algebra. The data of a strict monoidal category $\mathcal{C}$ is equivalent to that of a strict 2-category $\mathfrak{C}$ with one object; the objects and morphisms in $\mathcal{C}$ correspond to the 1-morphisms and 2-morphisms in $\mathfrak{C}$. For strict monoidal categories, we will use the same diagrammatic formalism as explained in the previous subsection; the only difference is that there no need to label the regions of the diagrams by objects since there is only one.

In the next definition, we introduce the *quiver Hecke category*, which is "half" of the Kac-Moody 2-category $\mathfrak{U}(\mathfrak{g})$ to be defined in the next subsection. Everything from this point on depends on the Kac-Moody data from §3.1 plus some additional parameters: we fix

- units $t_{ij} \in \Bbbk^\times$ such that $t_{ii} = 1$ and $d_{ij} = 0 \Rightarrow t_{ij} = t_{ji}$;
- scalars[7] $s_{ij}^{pq} \in \Bbbk$ for $0 < p < d_{ij}$, $0 < q < d_{ji}$ such that $s_{ij}^{pq} = s_{ji}^{qp}$.

DEFINITION 3.4. The (positive) *quiver Hecke category* $\mathcal{H}$ is the strict monoidal category generated by objects I and morphisms  $: i \to i$ and  $: i \otimes j \to j \otimes i$ subject to the following relations:

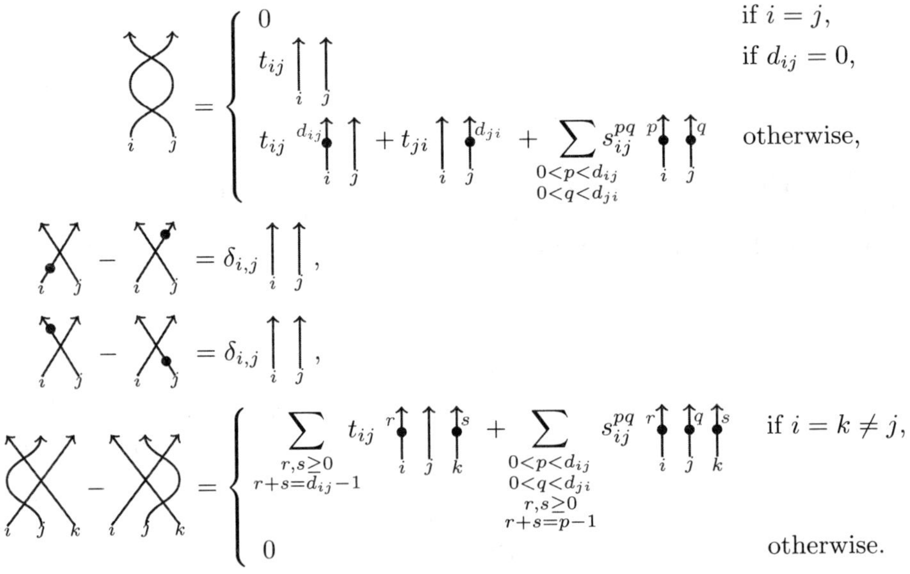

(We depict the nth power of  under vertical composition by labelling the dot with n.) There is also the (negative) *quiver Hecke category* $\mathcal{H}'$ generated by objects I and morphisms  $: i \to i$ and 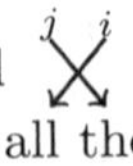 $: i \otimes j \to j \otimes i$ subject to the relations obtained by reversing the directions of all the arrows in the above, then switching the order of the terms on the left hand sides of the second, third and fourth relations. In fact, $\mathcal{H}'$ is isomorphic to $\mathcal{H}$, but the different normalization of generators is sometimes more convenient.

[7]In [**B3**] and elsewhere, scalars s_{ij}^{pq} are incorporated into the relations also for $p = 0$ or $q = 0$; we don't allow so much freedom here because it makes it impossible to prove that dots are nilpotent in the cyclotomic quotients discussed below (see Lemma 4.16).

For objects $\boldsymbol{i} = i_n \otimes \cdots \otimes i_1 \in I^{\otimes n}$ and $\boldsymbol{j} = j_m \otimes \cdots \otimes j_1 \in I^{\otimes m}$, there are no morphisms $\boldsymbol{i} \to \boldsymbol{j}$ in $\mathcal{H}$ unless $m = n$. The endomorphism algebra

$$H_n := \bigoplus_{\boldsymbol{i},\boldsymbol{j} \in I^{\otimes n}} \operatorname{Hom}_{\mathcal{H}}(\boldsymbol{i},\boldsymbol{j}) \tag{3.8}$$

is the (positive) *quiver Hecke algebra* which was introduced independently in [**R1**] and [**KL1**]. There is also the negative version $H'_n := \bigoplus_{\boldsymbol{i},\boldsymbol{j} \in I^{\otimes n}} \operatorname{Hom}_{\mathcal{H}'}(\boldsymbol{i},\boldsymbol{j})$, which is isomorphic to H_n with a different normalization of generators.

In the special case that I is a singleton, the quiver Hecke algebra H_n is the *nil Hecke algebra* NH_n, which plays a crucial role in the general theory. Numbering strands by $1, \dots, n$ from right to left, let us write X_i for the element of NH_n corresponding to a dot on the ith strand, and T_i for the element corresponding to the crossing of the ith and $(i+1)$th strands. Let S_n be the symmetric group with its usual simple reflections $s_1, \dots, s_{n-1}$, length function ℓ and longest element w_n. It acts naturally on the polynomial algebra $\operatorname{Pol}_n := \Bbbk[X_1, \dots, X_n]$; we write Sym_n for the subalgebra of invariants. The following are well known; e.g. see [**KL1**, §2], [**R2**, §2] or [**B2**, §2].

(N1) There is a faithful representation of NH_n on Pol_n defined by $X_i \cdot f := X_i f$ and $T_i \cdot f := \frac{s_i(f) - f}{X_i - X_{i+1}}$ (the ith *Demazure operator*).
(N2) For any $w \in S_n$, let T_w be the corresponding element of NH_n defined via a reduced expression for w. Then NH_n is a free left Pol_n-module on basis $\{T_w \mid w \in S_n\}$. In particular, $\operatorname{Pol}_n \hookrightarrow NH_n$.
(N3) The algebra Pol_n is a free Sym_n-module on basis $\{b_w \mid w \in S_n\}$ where $b_w := (-1)^{\ell(w)} T_w \cdot X_1^{n-1} X_2^{n-2} \cdots X_{n-1}$. Moreover $b_{w_n} = 1$.
(N4) There is an algebra isomorphism $NH_n \xrightarrow{\sim} \operatorname{End}_{\operatorname{Sym}_n}(\operatorname{Pol}_n)$ induced by the action of NH_n on Pol_n. Hence, $NH_n \cong M_{n!}(\operatorname{Sym}_n)$ and $Z(NH_n) = \operatorname{Sym}_n$.
(N5) The element $\pi_n := (-1)^{\ell(w_n)} X_1^{n-1} X_2^{n-2} \cdots X_{n-1} T_{w_n}$ acts on the basis for Pol_n from (N3) by $\pi_n \cdot b_w = \delta_{w,1} b_w$. Hence, it is a primitive idempotent in NH_n, and $NH_n \cong (NH_n \pi_n)^{\oplus n!}$ as a left NH_n-module.

In the remainder of the subsection, we wish to give a first indication of the power of the quiver Hecke relations. Let $\mathcal{C}$ be some category which is additive and idempotent-complete. Suppose that we are given a categorical action of the quiver Hecke category $\mathcal{H}$ on $\mathcal{C}$, i.e. there is a strict monoidal functor

$$\Phi : \mathcal{H} \to \mathcal{E}nd(\mathcal{C}) \tag{3.9}$$

where $\mathcal{E}nd(\mathcal{C})$ denotes the strict monoidal category of all endofunctors of $\mathcal{C}$. Let $E_i := \Phi(i)$ and $x_i := \Phi($  $) \in \operatorname{End}(E_i)$. Since E_i is $\Bbbk$-linear (as always), it is additive, hence it induces an endomorphism $e_i := [E_i]$ of the split Grothendieck group $K_0(\mathcal{C})$. More generally, for $i \in I$ and $n \geq 1$, we can obviously identify the nil Hecke algebra NH_n with $\operatorname{End}_{\mathcal{H}}(i^{\otimes n})$; then the image under Φ of the idempotent π_n from (N5) gives us an idempotent $\pi_{i,n} \in \operatorname{End}(E_i^n)$. Let $E_i^{(n)} := \pi_{i,n} E_i^n$, i.e. it is the endofunctor of $\mathcal{C}$ that sends an object V to the image of $(\pi_{i,n})_V \in \operatorname{End}_{\mathcal{C}}(E_i^n V)$, and a morphism $f : V \to W$ to the restriction of $E_i^n f$. The following lemma shows that this categorifies the divided power $e_i^{(n)} := e_i^n / n!$.

LEMMA 3.5. *We have that* $E_i^n \cong \left(E_i^{(n)}\right)^{\oplus n!}$.

PROOF. By (N5), the identity element of NH_n is a sum of $n!$ primitive idempotents, each of which is conjugate to π_n. □

LEMMA 3.6. *Suppose for some $V \in \mathrm{ob}\,\mathcal{C}$ that there is a monic polynomial $f(t)$ of degree n such that $f((x_i)_V) = 0$. Then $E_i^{(n+1)}V = 0$. In particular, if $\mathcal{C}$ is finite-dimensional, then all e_i act locally nilpotently on $K_0(\mathcal{C})$.*

PROOF. The second statement follows from the first: if $\mathcal{C}$ is finite-dimensional then, for any $V \in \mathrm{ob}\,\mathcal{C}$, we have that $(x_i)_V$ is an element of the finite-dimensional algebra $\mathrm{End}_{\mathcal{C}}(E_iV)$. Hence, it certainly satisfies some polynomial relation. To prove the first statement, we first note the following identity in NH_{n+1}:

$$(-1)^{\ell(w_{n+1})}\pi_{n+1}f(X_1)X_2^{n-1}\cdots X_nT_{w_{n+1}} = \pi_{n+1}.$$

This holds because $\pi_{n+1}^2 = \pi_{n+1}$ and moreover $T_{w_{n+1}}X_1^mX_2^{n-1}\cdots X_nT_{w_{n+1}} = 0$ for $m < n$ by degree considerations. Now as above we identify NH_{n+1} with $\mathrm{End}_{\mathcal{H}}(i^{\otimes(n+1)})$, apply Φ to our identity, then evaluate the resulting natural transformations at V. By assumption, $\Phi(f(X_1))_V = E_i^n f((x_i)_V) = 0$. Hence, the left hand side vanishes, and we deduce that $\Phi(\pi_{n+1})_V = 0$. This is the identity endomorphism of $E_i^{(n+1)}V$, so the latter object is isomorphic to zero. □

Perhaps most striking of all, we have the following, which is an immediate consequence of the even stronger categorical Serre relations from [**R1**, Proposition 4.2]:

LEMMA 3.7. *The endomorphisms e_i of $K_0(\mathcal{C})$ satisfy the Serre relations from (3.3).*

3.4. Kac-Moody 2-categories. We are ready to formulate Rouquier's definition of the Kac-Moody 2-category $\mathfrak{U}(\mathfrak{g})$; cf. [**R1**, §4.1.3].

DEFINITION 3.8. The *Kac-Moody 2-category* $\mathfrak{U}(\mathfrak{g})$ is the strict 2-category[8] with objects P; generating 1-morphisms $E_i1_\lambda : \lambda \to \lambda{+}\alpha_i$ and $F_i1_\lambda : \lambda \to \lambda{-}\alpha_i$ for each $i \in I$ and $\lambda \in P$, whose identity 2-morphisms will be represented diagrammatically by $\lambda{+}\alpha_i \underset{i}{\uparrow} \lambda$ and $\lambda{-}\alpha_i \overset{i}{\downarrow} \lambda$, respectively; and generating 2-morphisms $\underset{i}{\blacklozenge\!\uparrow} \lambda : E_i1_\lambda \to E_i1_\lambda$, $\underset{i\ j}{\times} \lambda : E_iE_j1_\lambda \to E_jE_i1_\lambda$, $\overset{i}{\cup}\lambda : 1_\lambda \to F_iE_i1_\lambda$ and $\underset{i}{\cap}\lambda : E_iF_i1_\lambda \to 1_\lambda$.

The generating 2-morphisms are subject to the following relations. First, we have the *positive quiver Hecke relations* (cf. Definition 3.4):

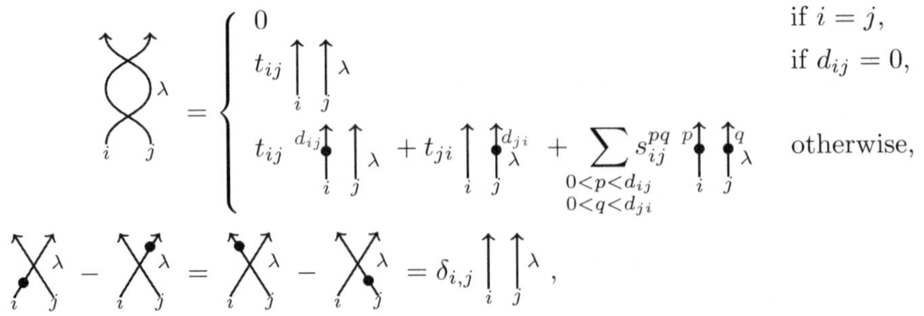

[8]Some authors require it is additive from the outset but we don't assume this.

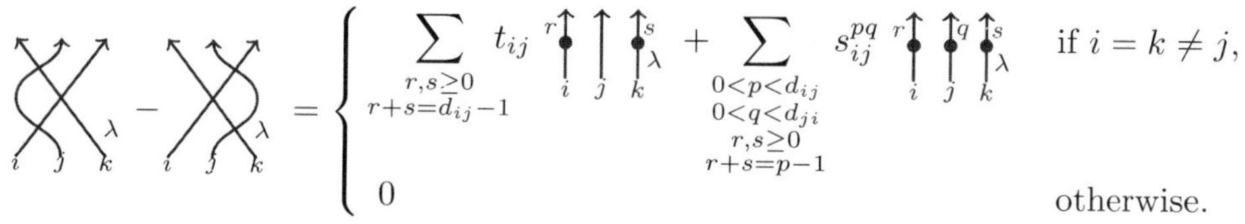

Next we have the *right adjunction relations*

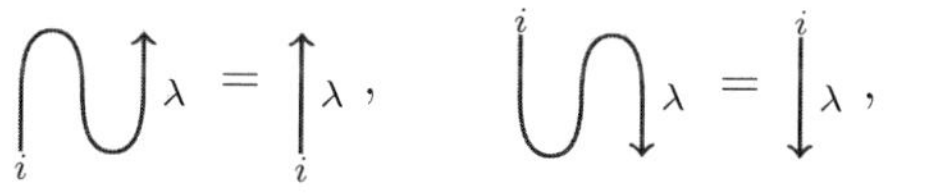

which imply that $F_i 1_{\lambda+\alpha_i}$ is a right dual of $E_i 1_\lambda$. Finally there are some *inversion relations*. To formulate these, define a new 2-morphism  $\lambda : E_j F_i 1_\lambda \to F_i E_j 1_\lambda$ by setting

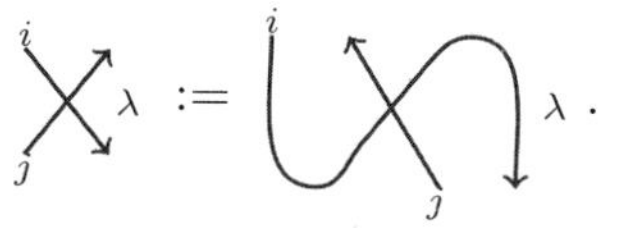

Then the inversion relations assert that the following are isomorphisms:

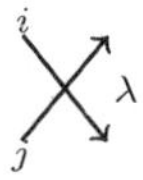 $: E_j F_i 1_\lambda \xrightarrow{\sim} F_i E_j 1_\lambda$ if $i \neq j$,

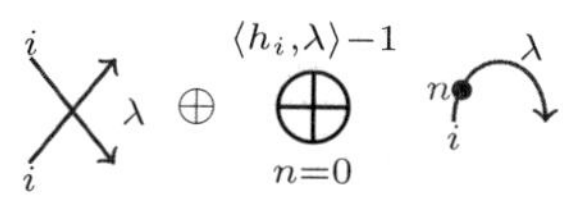 $: E_i F_i 1_\lambda \xrightarrow{\sim} F_i E_i 1_\lambda \oplus 1_\lambda^{\oplus \langle h_i, \lambda \rangle}$ if $\langle h_i, \lambda \rangle \geq 0$,

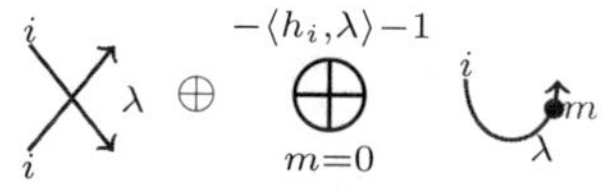 $: E_i F_i 1_\lambda \oplus 1_\lambda^{\oplus -\langle h_i, \lambda \rangle} \xrightarrow{\sim} F_i E_i 1_\lambda$ if $\langle h_i, \lambda \rangle \leq 0$,

the last two being 2-morphisms in the additive envelope of $\mathfrak{U}(\mathfrak{g})$.

REMARK 3.9. More formally, the inversion relations mean that there are some additional generating 2-morphisms 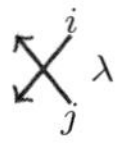 $\lambda : F_i E_j 1_\lambda \to E_j F_i 1_\lambda$,  $\lambda : 1_\lambda \to E_i F_i 1_\lambda$ and 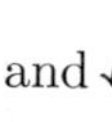 $\lambda : F_i E_i 1_\lambda \to 1_\lambda$ for $0 \leq n < \langle h_i, \lambda \rangle$ and $0 \leq m < -\langle h_i, \lambda \rangle$ such that the following hold:

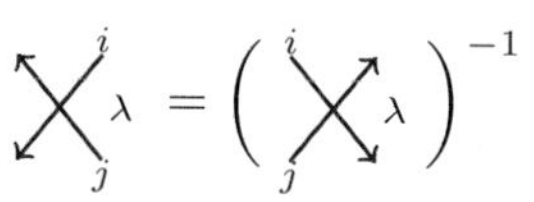 if $i \neq j$,

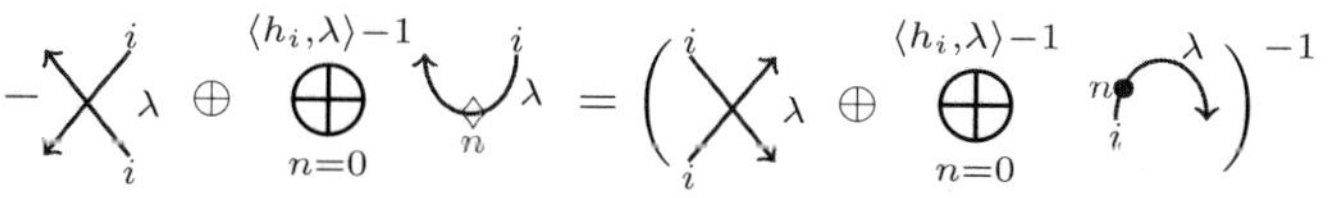 if $\langle h_i, \lambda \rangle \geq 0$,

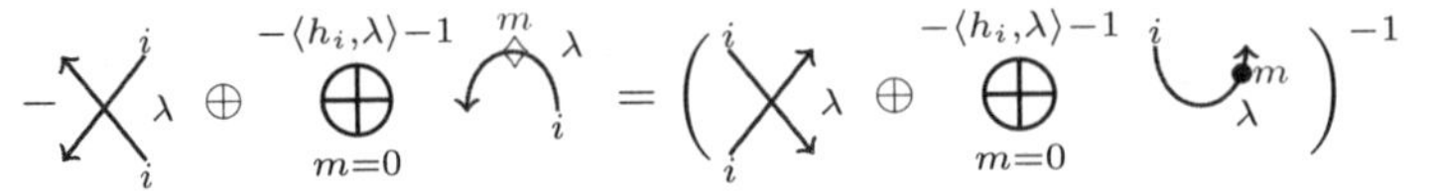

if $\langle h_i, \lambda\rangle \leq 0$.

As usual with objects defined by generators and relations, one then needs to play the game of deriving consequences from the defining relations. Here we record some which were established in [**B3**]; we also cite below the more recent exposition in [**BE2**] since that uses exactly the same normalization as here.

(K1) *Negative quiver Hecke relations.* Define 2-morphisms  $\lambda : F_i 1_\lambda \to F_i 1_\lambda$ and  $\lambda : F_j F_i 1_\lambda \to F_i F_j 1_\lambda$ by setting

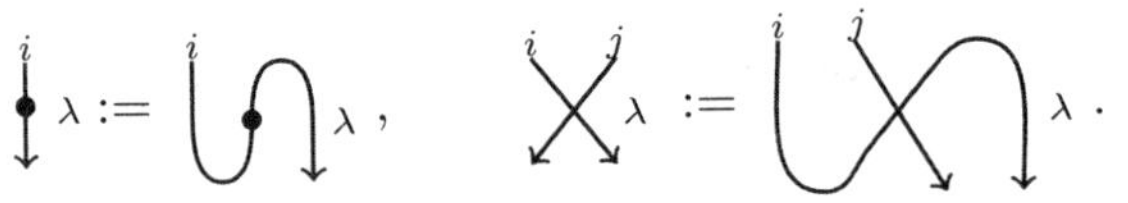

On rotating the positive quiver Hecke relations clockwise through 180°, one sees that these satisfy the negative quiver Hecke relations:

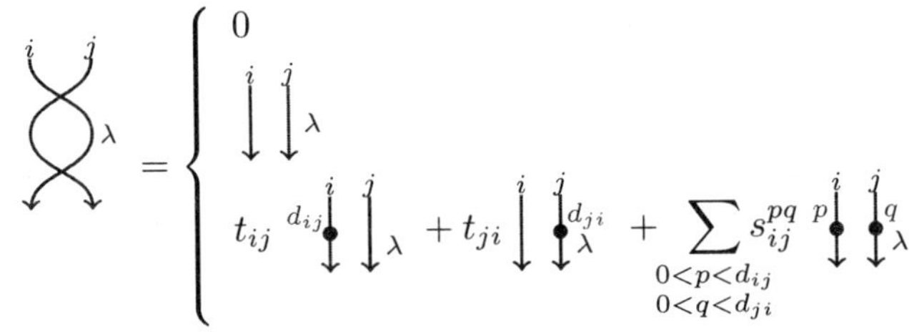

if $i = j$,

if $d_{ij} = 0$,

otherwise,

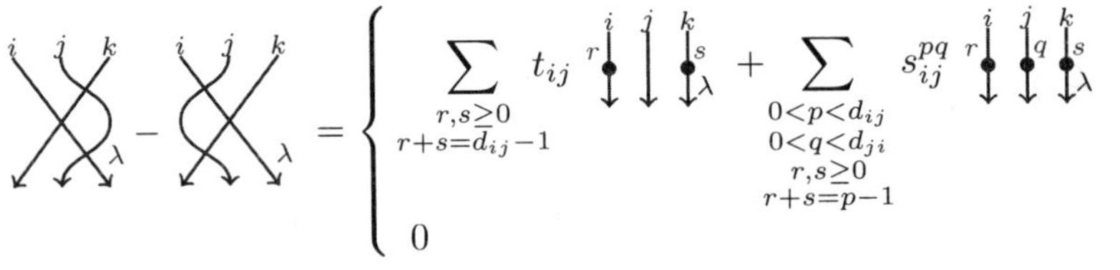

if $i = k \neq j$,

otherwise.

(K2) *Second adjunction.* We next introduce 2-morphisms 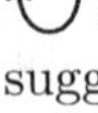 $\lambda : 1_\lambda \to E_i F_i 1_\lambda$ and 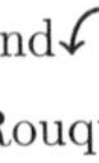 $\lambda : F_i E_i 1_\lambda \to 1_\lambda$. The definition of these was suggested already by Rouquier in [**R1**, §4.1.4]. Following the idea of [**BHLW**] we will normalize them in a different way which depends on an additional choice of units $c_{\lambda;i} \in \Bbbk^\times$ for each $i \in I$ and $\lambda \in P$ such that $c_{\lambda+\alpha_j;i} = t_{ij} c_{\lambda;i}$. Fixing such a choice from now on, we set

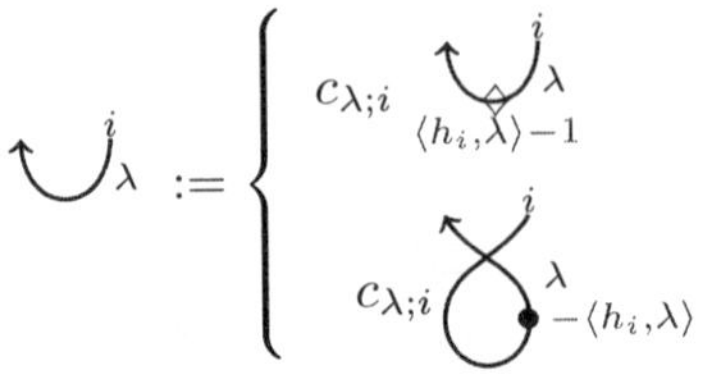

if $\langle h_i, \lambda\rangle > 0$,

if $\langle h_i, \lambda\rangle \leq 0$,

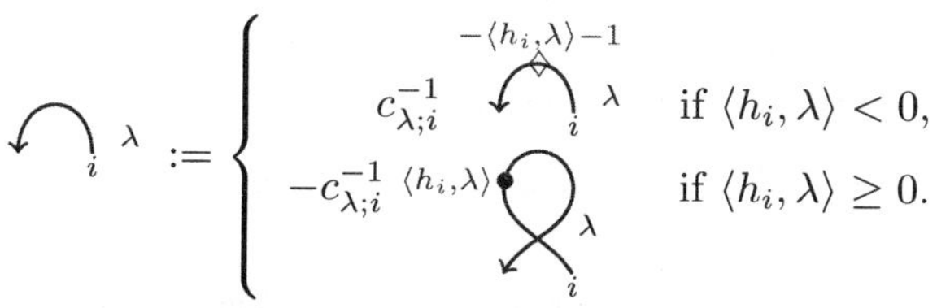

Then by [**B3**, Theorem 4.3] (or [**BE2**, Proposition 6.2] with the present normalization) we have the *left adjunction relations*:

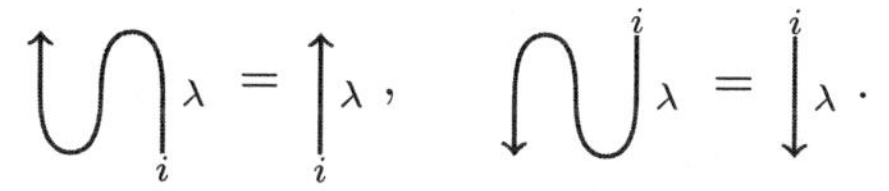

This means that $F_i 1_{\lambda+\alpha_i}$ is also a left dual of $E_i 1_\lambda$.

(K3) *Pitchfork relations and cyclicity.* The following relations follow from the definitions and [**B3**, Theorem 5.3] (or [**BE2**, Proposition 4.1] and [**BE2**, Proposition 7.2]):

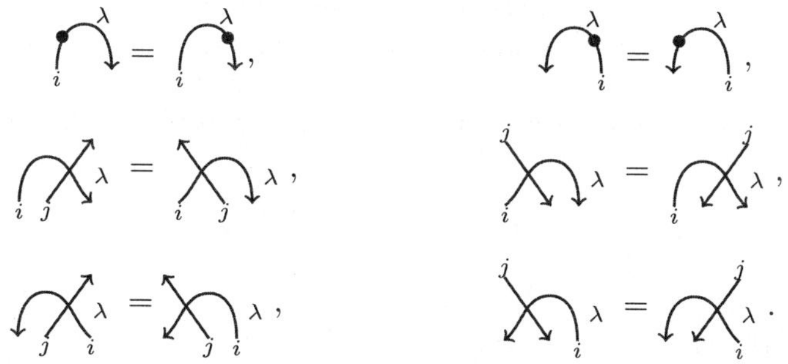

It follows that

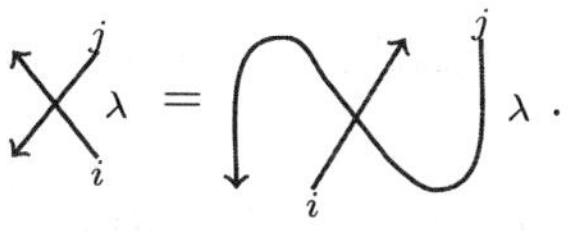

Moreover, the 2-morphisms 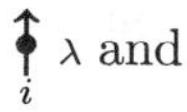 and 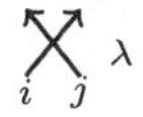 are *cyclic* i.e. their right mates are equal to their left mates:

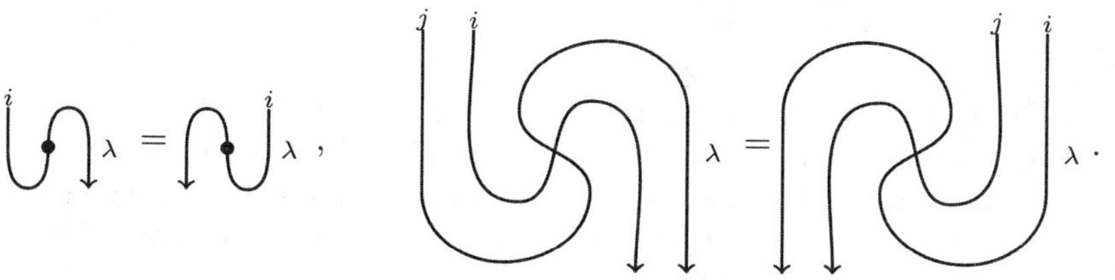

This is the main advantage of the normalization of the second adjunction from [**BHLW**] as chosen in (K2).

(K4) *Infinite Grassmannian relations.* Let Sym be the algebra of symmetric functions over $\Bbbk$. Recall Sym is generated both by the elementary symmetric functions e_n ($n \geq 1$) and by the complete symmetric functions

h_n ($n \geq 1$). Adopting the convention that $\mathrm{e}_0 = \mathrm{h}_0 = 1$ and $\mathrm{e}_n = \mathrm{h}_n = 0$ for $n < 0$, these two families of generators are related by the equation

$$\sum_{r+s=n} (-1)^r \mathrm{e}_r \mathrm{h}_s = 0 \text{ for all } n > 0. \tag{3.10}$$

Take $i \in I$, $\lambda \in P$ and set $h := \langle h_i, \lambda \rangle$. Then the infinite Grassmannian relations assert that there is a well-defined homomorphism

$$\beta_{\lambda;i} : \mathrm{Sym} \to \mathrm{End}_{\mathfrak{U}(\mathfrak{g})}(1_\lambda)$$

such that

(3.11) $\beta_{\lambda;i}(\mathrm{h}_n) = c_{\lambda;i}^{-1}$ 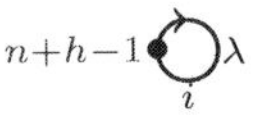 if $n > -h$,

(3.12) $\beta_{\lambda;i}(\mathrm{e}_n) = (-1)^n c_{\lambda;i}$  if $n > h$.

This was proved originally by Lauda[9] in [**L1**, Proposition 8.2]; see [**BE2**, Proposition 5.1] where it is established using our normalization. It motivated Lauda's introduction also of certain *negatively dotted bubbles*, which are 2-morphisms in $\mathrm{End}_{\mathfrak{U}(\mathfrak{g})}(1_\lambda)$ defined so that (3.11)–(3.12) hold for all $n \in \mathbb{Z}$.

(K5) *Dual inversion relations.* The following 2-morphisms are invertible:

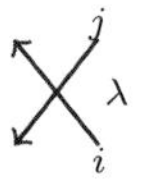 $: F_j E_i 1_\lambda \xrightarrow{\sim} E_i F_j 1_\lambda$ if $i \neq j$,

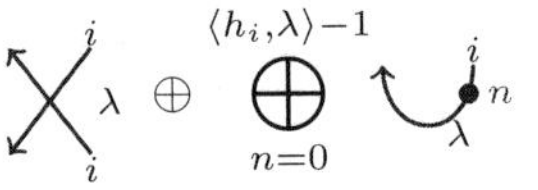 $: F_i E_i 1_\lambda \oplus 1_\lambda^{\oplus \langle h_i, \lambda \rangle} \xrightarrow{\sim} E_i F_i 1_\lambda$ if $\langle h_i, \lambda \rangle \geq 0$,

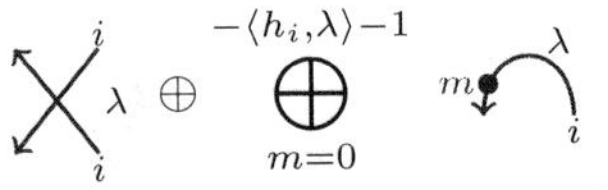 $: F_i E_i 1_\lambda \xrightarrow{\sim} E_i F_i 1_\lambda \oplus 1_\lambda^{\oplus -\langle h_i, \lambda \rangle}$ if $\langle h_i, \lambda \rangle \leq 0$.

This may be deduced from the definitions above using also the following relations proved in [**B3**, Corollary 3.3] (or [**BE2**, Corollary 5.2]):

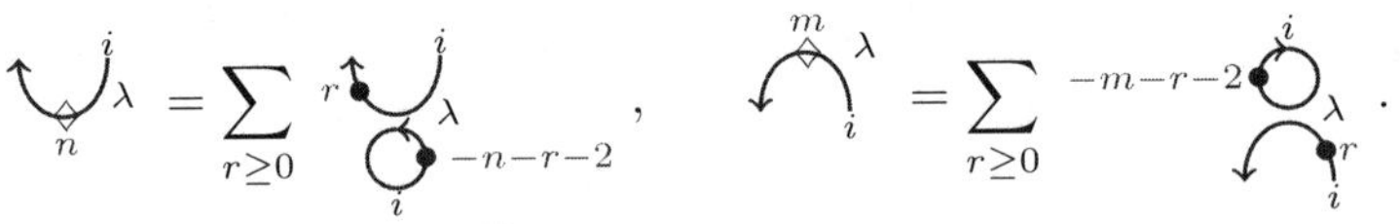

Note here we are using the negatively dotted bubbles. Another consequence of the last relations displayed, plus the description of the leftward crossing given in (K3), is that all 2-morphisms in $\mathfrak{U}(\mathfrak{g})$ are generated (under both vertical and horizontal composition) by upward dots, upward crossings, and leftward and rightward cups and caps.

[9]In fact, Lauda showed for $\mathfrak{g} = \mathfrak{sl}_2$ that this homomorphism is an isomorphism. In general, the product of the homomorphisms $\beta_{\lambda;i}$ over all $i \in I$ should given an isomorphism $\bigotimes_{i \in I} \mathrm{Sym} \xrightarrow{\sim} \mathrm{End}_{\mathfrak{U}(\mathfrak{g})}(1_\lambda)$, but the proof of this assertion depends on the Nondegeneracy Condition discussed later in the subsection.

(K6) *Curl relations.* For $n \geq 0$ we have:

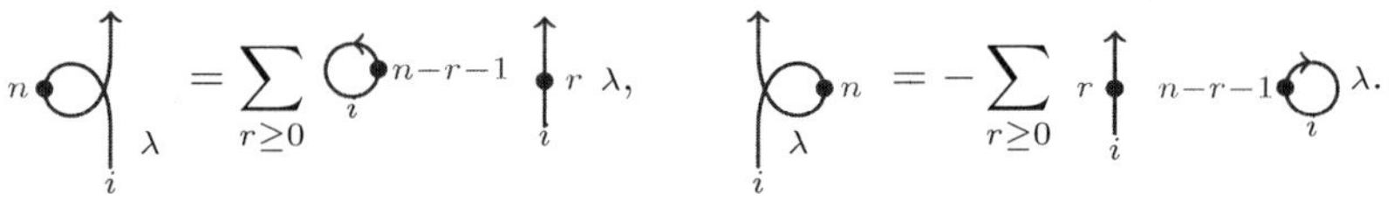

These are proved e.g. in [**BHLW**, Lemma 3.2] or [**BE2**, Corollary 5.4].

(K7) *Bubble slides.* For the next relations, we adopt the following convenient shorthand:

(3.13)

Then following hold for all $n \in \mathbb{Z}$. First:

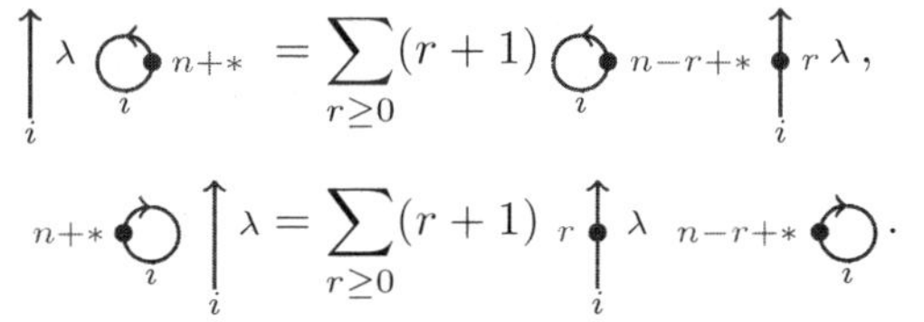

Then for $i \neq j$ with $d_{ij} = 0$ we have:

Finally for $i \neq j$ with $d_{ij} > 0$ we have:

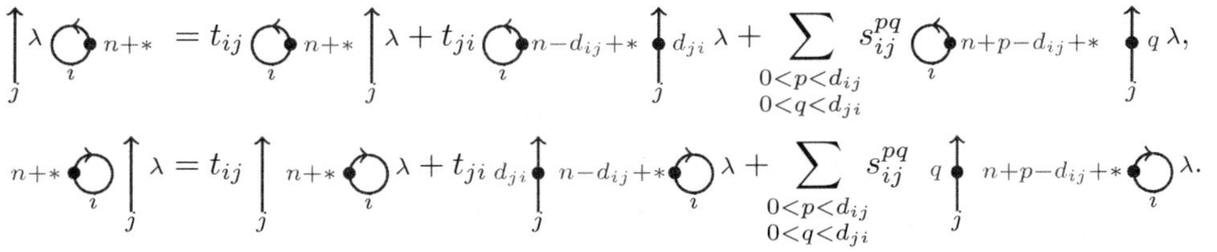

In the simply-laced case, these were proved in [**KL3**, Propositions 3.3–3.5]. In general, they are recorded in various places in the literature; e.g. see [**W2**, Proposition 2.8] or [**BHLW**, §3.2]. When $d_{ij} > 1$, the proof of the final two bubble slides above is not as straightforward as those references may suggest as it requires also an application of the deformed braid relation; we refer to [**BE2**, Proposition 7.3] for the detailed argument.

(K8) *Alternating crossings.* Finally we record the following relations from [**BE2**, Corollary 5.3] and [**BE2**, Proposition 7.6]; the first two are immediate from the definitions and the "diamond relations" recorded in (K5); the last one was derived already in [**KL3**].

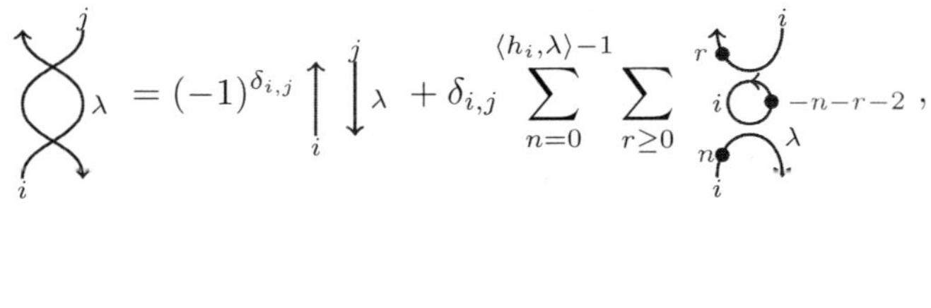

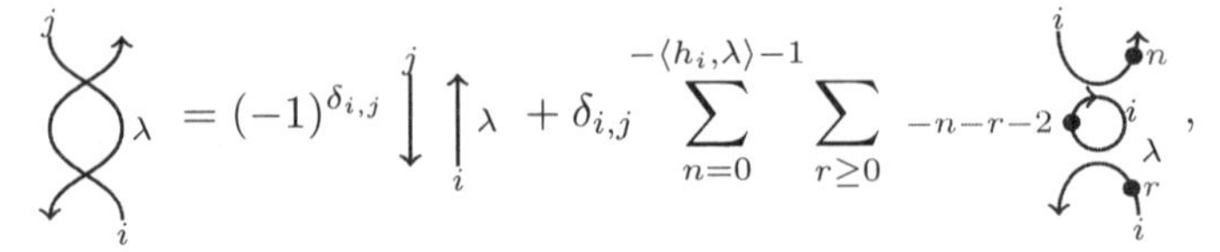

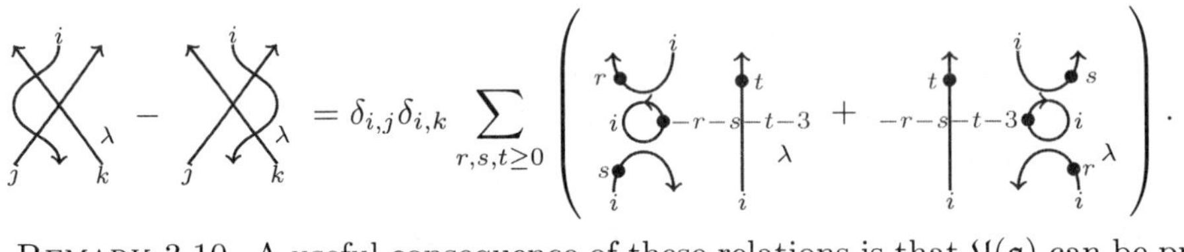

REMARK 3.10. A useful consequence of these relations is that $\mathfrak{U}(\mathfrak{g})$ can be presented equivalently with generating 2-morphisms 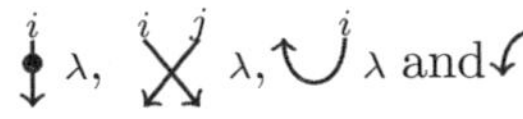 subject to the negative quiver Hecke relations from (K1), the left adjunction relations from (K2), and replacing the original inversion relations by the dual inversion relations from (K5) where

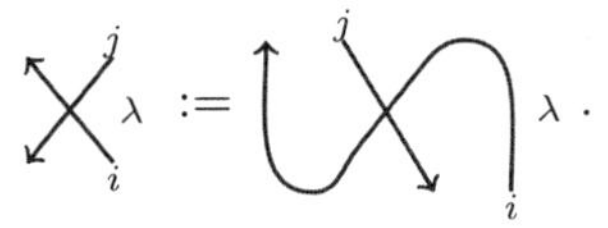

In pictures, this new presentation is the original presentation from Definition 3.8 rotated through 180°.

It still seems somewhat remarkable that all of the above relations (K1)–(K8) can be derived from Rouquier's minimalistic definition. Independently, Khovanov and Lauda [**KL3**] introduced a strict 2-category incorporating extra generators and relations essentially matching the ones above; see also [**CL**] which extended the definition in [**KL3**] to more general parameters. The following is a consequence of (K1)–(K8).

THEOREM 3.11 ([**B3**, Main Theorem]). *Rouquier's Kac-Moody 2-category* $\mathfrak{U}(\mathfrak{g})$ *from Definition* 3.8 *is isomorphic to the 2-category introduced in* [**KL3, CL**].

Khovanov and Lauda exploited their extra generators and relations to construct some explicit sets of 2-morphisms[10] which they showed span the 2-morphism spaces in $\mathfrak{U}(\mathfrak{g})$; see [**KL3**, Proposition 3.11]. They then conjectured that these spanning sets actually give bases for the 2-morphism spaces in $\mathfrak{U}(\mathfrak{g})$. This is the *Nondegeneracy Condition* from [**KL3**, §3.2.3]. For $\mathfrak{g} = \mathfrak{sl}_n$, it is known that the Nondegeneracy Condition holds, thanks to [**KL3**, §6.4].

To explain the significance of the Nondegeneracy Condition, let $\dot{\mathfrak{U}}(\mathfrak{g})$ be the additive Karoubi envelope of $\mathfrak{U}(\mathfrak{g})$ and $K_0(\dot{\mathfrak{U}}(\mathfrak{g})) = \bigoplus_{\lambda,\mu \in P} 1_\mu K_0(\dot{\mathfrak{U}}(\mathfrak{g}))1_\lambda$ be its Grothendieck ring as in (3.7). Also let $\dot{U}(\mathfrak{g})_\mathbb{Z}$ be the Kostant $\mathbb{Z}$-form for $\dot{U}(\mathfrak{g})$, i.e. the subring generated by the idempotents 1_λ and the divided powers $e_i^{(r)}1_\lambda, f_i^{(r)}1_\lambda$ for $\lambda \in P, i \in I$ and $r \geq 1$. Then the arguments used to prove [**KL3**, Theorem 1.1] (which are based ultimately on Lemmas 3.5 and 3.7) show that there is a unique surjective locally unital homomorphism

$$\gamma : \dot{U}(\mathfrak{g})_\mathbb{Z} \twoheadrightarrow K_0(\dot{\mathfrak{U}}(\mathfrak{g})) \tag{3.14}$$

[10] They worked with a restricted choice of parameters compared to here, but it is clear how to extend their constructions to the general case using (K1)–(K8).

which sends $e_i 1_\lambda$ and $f_i 1_\lambda$ to $[E_i 1_\lambda]$ and $[F_i 1_\lambda]$, respectively. Moreover, *assuming the parameters satisfy the homogeneity property*

$$s_{ij}^{pq} \neq 0 \Rightarrow p d_{ji} + q d_{ij} = d_{ij} d_{ji} \tag{3.15}$$

and *providing that the Nondegeneracy Condition holds*, [**KL3**, Theorem 1.2] implies that γ is an isomorphism. This makes precise the sense in which $\mathfrak{U}(\mathfrak{g})$ should categorify the universal enveloping algebra of $\mathfrak{g}$.

REMARK 3.12. In finite type, it is known that γ is an isomorphism (regardless of whether (3.15) holds); see Corollary 4.21 below.

REMARK 3.13. As we were finalizing this article, Webster released a preliminary version of [**W3**]. In this work, he appears to have found a general proof of the Nondegeneracy Condition valid for all types and all choices of parameters satisfying (3.15).

3.5. Gradings. In this subsection, we discuss the graded version $\mathfrak{U}_q(\mathfrak{g})$ of $\mathfrak{U}(\mathfrak{g})$ and its connection to quantum groups. Our language is based on [**BE1**, §6], and is slightly different to that of [**KL3**, **R1**].

Let $\underline{\mathcal{GV}ec}$ be the symmetric monoidal category of (small) $\mathbb{Z}$-graded vector spaces and degree-preserving linear maps. The grading shift functor gives an automorphism

$$Q : \underline{\mathcal{GV}ec} \to \underline{\mathcal{GV}ec},$$

our convention for this being that $(QV)_n = V_{n-1}$. By a *graded category*, we mean a category enriched in $\underline{\mathcal{GV}ec}$. Thus, the morphism spaces in a graded category are equipped with a $\mathbb{Z}$-grading in a way that is compatible with composition. If $\mathcal{C}$ is any graded category, the *underlying category* $\underline{\mathcal{C}}$ is the category with the same objects as $\mathcal{C}$, but only the homogeneous morphisms of degree zero. For example, $\underline{\mathcal{GV}ec}$ is the underlying category of the graded category $\mathcal{GV}ec$ whose objects are (small) $\mathbb{Z}$-graded vector spaces and whose morphisms are sums of homogeneous linear maps of various degrees.

Let $\mathcal{GC}at$ be the category of all (small) graded categories. It is monoidal with product $\boxtimes$ defined just like in §3.2. A *strict graded 2-category* is a category enriched in $\mathcal{GC}at$; cf. Definition 3.3. If $\mathfrak{C}$ is any graded 2-category, the *underlying 2-category* $\underline{\mathfrak{C}}$ is the 2-category with the same objects and 1-morphisms as $\mathfrak{C}$, but only the homogeneous 2-morphisms of degree zero. Also let $\dot{\underline{\mathfrak{C}}}$ be the additive Karoubi envelope of $\underline{\mathfrak{C}}$, and $K_0(\dot{\underline{\mathfrak{C}}})$ be its Grothendieck ring defined as in (3.7).

For any graded 2-category $\mathfrak{C}$, there is a universal construction of another graded 2-category $\mathfrak{C}_q$, which we call the *Q-envelope* of $\mathfrak{C}$. It has the same object set as $\mathfrak{C}$. Given objects λ, μ, the 1-morphisms $\lambda \to \mu$ in $\mathfrak{C}_q$ are defined formally to be symbols $Q^n F$ for all 1-morphisms $F : \lambda \to \mu$ in $\mathfrak{C}$ and all $n \in \mathbb{Z}$. Then the 2-morphisms in $\mathfrak{C}_q$ are defined from

$$\mathrm{Hom}_{\mathfrak{C}_q}(Q^n F, Q^m G) := Q^{m-n}\,\mathrm{Hom}_{\mathfrak{C}}(F, G),$$

where the Q^{m-n} on the right hand side is the grading shift in $\underline{\mathcal{GV}ec}$. Horizontal composition of 1-morphisms in $\mathfrak{C}_q$ is induced by the horizontal composition in $\mathfrak{C}$ so that $(Q^n F)(Q^m G) := Q^{n+m} FG$. Similarly, the horizontal and vertical compositions of 2-morphisms in $\mathfrak{C}_q$ are induced in an obvious way by the horizontal and vertical compositions in $\mathfrak{C}$. Note also for each 1-morphism F of $\mathfrak{C}$ and $n \in \mathbb{Z}$ that 1_F

defines a canonical 2-isomorphism $Q^n F \xrightarrow{\sim} F$ in $\mathfrak{C}_q$ that is homogeneous of degree $-n$.

The point of the construction of Q-envelope is that $\mathfrak{C}_q$ (hence, the underlying 2-category $\underline{\mathfrak{C}}_q$) is equipped with distinguished 1-morphisms $q_\lambda := Q1_\lambda : \lambda \to \lambda$ for each object λ, such that $q_\mu F = F q_\lambda$ for each 1-morphism $F : \lambda \to \mu$. In particular, the Grothendieck ring $K_0(\dot{\underline{\mathfrak{C}}}_q)$ is actually a $\mathbb{Z}[q, q^{-1}]$-algebra, with q acting on $1_\mu K_0(\dot{\underline{\mathfrak{C}}}_q)1_\lambda$ by left multiplication by $[q_\mu]$ (= right multiplication by $[q_\lambda]$).

DEFINITION 3.14. Assume that the parameters fixed in §3.3 satisfy (3.15). Then the relations defining the Kac-Moody 2-category $\mathfrak{U}(\mathfrak{g})$ are all homogeneous, so we can make $\mathfrak{U}(\mathfrak{g})$ into a graded 2-category by declaring that the generating 2-morphisms  λ,  λ,  λ and  λ are of degrees $2d_i$, $d_i d_{ij}$, $d_i(1 + \langle h_i, \lambda\rangle)$ and $d_i(1 - \langle h_i, \lambda\rangle)$, respectively. Then we define the *graded Kac-Moody 2-category* $\mathfrak{U}_q(\mathfrak{g})$ to be the Q-envelope of this graded 2-category.

Using Theorem 3.11, it is easy to see that the underlying 2-category $\underline{\mathfrak{U}}_q(\mathfrak{g})$ is isomorphic to the graded version of the Kac-Moody 2-category as defined in [**R1**, **KL3**]. Then [**KL3**, Theorem 1.1] shows that there is a surjective $\mathbb{Z}[q, q^{-1}]$-algebra homomorphism

$$\gamma_q : \dot{U}_q(\mathfrak{g})_{\mathbb{Z}} \twoheadrightarrow K_0(\dot{\underline{\mathfrak{U}}}_q(\mathfrak{g})), \tag{3.16}$$

where $\dot{U}_q(\mathfrak{g})_{\mathbb{Z}}$ denotes the Kostant-Lusztig $\mathbb{Z}$-form of the idempotented version of the quantized enveloping algebra $U_q(\mathfrak{g})$. Moreover, *providing that the Nondegeneracy Condition holds*, [**KL3**, Theorem 1.2] shows that γ_q is an isomorphism. The injectivity of γ claimed earlier in the ungraded setting follows from this on specializing q to 1.

4. Categorical actions and crystals

In this final section, we focus on 2-representations of Kac-Moody 2-categories. We will review the existing results mostly following Rouquier [**R2**]. After that, we focus on the locally Schurian case, explaining various results in that setting which generalize aspects of [**CR**, **R1**].

4.1. 2-representations. We keep the choices of Kac-Moody data and parameters as in the previous section. The following is [**R1**, Definition 5.1.1].

DEFINITION 4.1. A *2-representation* of $\mathfrak{U}(\mathfrak{g})$ is the following data:

(M1) a category $\mathcal{R}$ with a given decomposition $\mathcal{R} = \coprod_{\lambda \in P} \mathcal{R}_\lambda$ into *weight subcategories*;

(M2) endofunctors E_i and F_i of $\mathcal{R}$ for each $i \in I$, such that $E_i|_{\mathcal{R}_\lambda} : \mathcal{R}_\lambda \to \mathcal{R}_{\lambda+\alpha_i}$ and $F_i|_{\mathcal{R}_\lambda} : \mathcal{R}_\lambda \to \mathcal{R}_{\lambda-\alpha_i}$;

(M3) natural transformations $x_i : E_i \to E_i$ and $\tau_{ij} : E_i E_j \to E_j E_i$ satisfying the positive quiver Hecke relations from Definition 3.4, i.e. so that there is a strict monoidal functor $\Phi : \mathcal{H} \to \mathcal{E}nd(\mathcal{R})$ with $\Phi(i) = E_i$, $\Phi($ $) = x_i$ and $\Phi($  $) = \tau_{ij}$ for all $i, j \in I$;

(M4) the unit $\eta_i : 1_{\mathcal{R}} \to F_i E_i$ and counit $\varepsilon_i : E_i F_i \to 1_{\mathcal{R}}$ of an adjunction making F_i into a right adjoint of E_i.

Then we require that the following axiom holds:

(M5) all of the natural transformations σ_{ij} $(i \neq j)$ and $\rho_{i,\lambda}$ are invertible, where

$$\sigma_{ij} := F_i E_j \varepsilon_i \circ F_i \tau_{ij} F_i \circ \eta_i E_j F_i : E_j F_i \to F_i E_j,$$

$$\rho_{i,\lambda} := \begin{cases} \sigma_{ii} \oplus \bigoplus_{n=0}^{h-1} \varepsilon_i \circ (x_i F_i)^{\circ n} : E_i F_i|_{\mathcal{R}_\lambda} \to F_i E_i|_{\mathcal{R}_\lambda} \oplus 1_{\mathcal{R}_\lambda}^{\oplus h} & \text{if } h \geq 0, \\ \sigma_{ii} \oplus \bigoplus_{m=0}^{-h-1} (F_i x_i)^{\circ m} \circ \eta_i : E_i F_i|_{\mathcal{R}_\lambda} \oplus 1_{\mathcal{R}_\lambda}^{\oplus -h} \to F_i E_i|_{\mathcal{R}_\lambda} & \text{if } h \leq 0, \end{cases}$$

for $h := \langle h_i, \lambda \rangle$.

We say that a 2-representation $\mathcal{R}$ is *small*, *finite-dimensional*, *additive*, *Abelian* etc... if all of the categories $\mathcal{R}_\lambda$ $(\lambda \in P)$ are small, finite-dimensional, additive, Abelian etc... In the additive case, the functors E_i and F_i extend to $\bigoplus_{\lambda \in P} \mathcal{R}_\lambda$, and it is more convenient to denote this by $\mathcal{R}$ in place of $\coprod_{\lambda \in P} \mathcal{R}_\lambda$. If $\mathcal{R}$ is not additive, one can always replace it by its additive envelope, or indeed its additive Karoubi envelope $\dot{\mathcal{R}}$; the endofunctors E_i and F_i extend canonically to make these into 2-representations too.

The point of Definition 4.1 is that a small 2-representation $\mathcal{R}$ is exactly the same data as a strict 2-functor $\mathbb{R} : \mathfrak{U}(\mathfrak{g}) \to \mathfrak{Cat}$. The dictionary for going between the two notions is given by $\mathcal{R}_\lambda = \mathbb{R}(\lambda)$, $E_i|_{\mathcal{R}_\lambda} = \mathbb{R}(E_i 1_\lambda)$, $F_i|_{\mathcal{R}_\lambda} = \mathbb{R}(F_i 1_\lambda)$, $x_i|_{E_i|_{\mathcal{R}_\lambda}} =$ $\mathbb{R}($  $\lambda)$, $\tau_{ij}|_{E_i E_j|_{\mathcal{R}_\lambda}} = \mathbb{R}($ 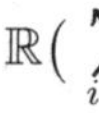 $\lambda)$, $\eta_i|_{1_{\mathcal{R}_\lambda}} = \mathbb{R}($ 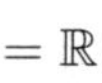 $\lambda)$ and $\varepsilon_i|_{F_i E_i|_{\mathcal{R}_\lambda}} = \mathbb{R}($  $\lambda)$.

The following strengthens [**R1**, Theorem 5.16].

LEMMA 4.2. *Suppose that $\mathcal{R}$ is a 2-representation in the sense of Definition 4.1. Then there is a canonical choice for the unit $\eta'_i : 1_\mathcal{R} \to E_i F_i$ and counit $\varepsilon'_i : F_i E_i \to 1_\mathcal{R}$ of an adjunction making F_i into a left adjoint of E_i.*

PROOF. We may assume that $\mathcal{R}$ is small, so that there is a corresponding strict 2-functor $\mathbb{R} : \mathfrak{U}(\mathfrak{g}) \to \mathfrak{Cat}$. Then let $\eta'_i|_{1_{\mathcal{R}_\lambda}} := \mathbb{R}($ 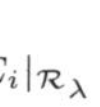 $\lambda)$ and $\varepsilon'_i|_{F_i E_i|_{\mathcal{R}_\lambda}} := \mathbb{R}($ $\lambda)$, where notation is as in (K2) from §3.4. □

The other generators and relations from (K1)–(K8) can be transported to any 2-representation $\mathcal{R}$ in a similar way. For example, from the images of the downward dots and crossings from (K1), one obtains canonical natural transformations $x'_i : F_i \to F_i$ and $\tau'_{ij} : F_i F_j \to F_j F_i$ which satisfy the negative quiver Hecke relations.

REMARK 4.3. Definition 4.1 can be formulated equivalently by replacing (M3)–(M5) by

(M3′) natural transformations $x'_i : F_i \to F_i$ and $\tau'_{ij} : F_i F_j \to F_j F_i$ satisfying the negative quiver Hecke relations, i.e. so that there is a strict monoidal functor $\Psi : \mathcal{H}' \to \mathcal{E}nd(\mathcal{R})$ with $\Psi(i) = F_i$, $\Psi($  $) = x'_i$ and $\Phi($  $) = \tau'_{ij}$ for all $i, j \in I$;

(M4′) the unit $\eta'_i : 1_\mathcal{R} \to E_i F_i$ and counit $\varepsilon'_i : F_i E_i \to 1_\mathcal{R}$ of an adjunction making F_i into a left adjoint of E_i;

(M5′) all of the natural transformations σ'_{ij} $(i \neq j)$ and $\rho'_{i,\lambda}$ are invertible, where

$$\sigma'_{ij} := E_i F_j \varepsilon'_i \circ E_i \tau'_{ij} E_i \circ \eta'_i F_j E_i : F_j E_i \to E_i F_j,$$

$$\rho'_{i,\lambda} := \begin{cases} \sigma'_{ii} \oplus \bigoplus_{n=0}^{h-1} (E_i x'_i)^{\circ n} \circ \eta'_i : F_i E_i|_{\mathcal{R}_\lambda} \oplus 1_{\mathcal{R}_\lambda}^{\oplus h} \to E_i F_i|_{\mathcal{R}_\lambda} & \text{if } h \geq 0, \\ \sigma'_{ii} \oplus \bigoplus_{m=0}^{-h-1} \varepsilon'_i \circ (x'_i E_i)^{\circ m} : F_i E_i|_{\mathcal{R}_\lambda} \to E_i F_i|_{\mathcal{R}_\lambda} \oplus 1_{\mathcal{R}_\lambda}^{\oplus -h} & \text{if } h \leq 0, \end{cases}$$

for $h := \langle h_i, \lambda \rangle$.

This follows because, in view of the alternative presentation of $\mathfrak{U}(\mathfrak{g})$ from Remark 3.10, the new formulation is also the data of a strict 2-functor.

If $\mathcal{R}$ is a 2-representation, the endofunctors E_i and F_i induce endomorphisms $[E_i]$ and $[F_i]$ of the split Grothendieck group

$$K_0(\dot{\mathcal{R}}) = \bigoplus_{\lambda \in P} K_0(\dot{\mathcal{R}}_\lambda) \tag{4.1}$$

of its additive Karoubi envelope.

LEMMA 4.4. *Given a 2-representation $\mathcal{R}$, there is a unique way to make $K_0(\dot{\mathcal{R}})$ into a module over the Kostant $\mathbb{Z}$-form $U(\mathfrak{g})_{\mathbb{Z}}$ of the universal enveloping algebra of $\mathfrak{g}$ so that the Chevalley generators e_i, f_i act as $[E_i], [F_i]$, respectively, and (4.1) is its decomposition into weight spaces.*

PROOF. We may assume that $\mathcal{R}$ (hence, $\dot{\mathcal{R}}$) is small, so that there is a corresponding strict 2-functor $\dot{\mathbb{R}} : \dot{\mathfrak{U}}(\mathfrak{g}) \to \mathfrak{Cat}$ with $\dot{\mathbb{R}}(\lambda) = \dot{\mathcal{R}}_\lambda$, etc... The definition of strict 2-functor then ensures that $K_0(\dot{\mathcal{R}})$ is a module over $K_0(\dot{\mathfrak{U}}(\mathfrak{g}))$: for $F \in \operatorname{ob} \mathcal{H}om_{\dot{\mathfrak{U}}(\mathfrak{g})}(\lambda, \mu)$ defining $[F] \in K_0(\dot{\mathfrak{U}}(\mathfrak{g}))$, and $P \in \operatorname{ob} \dot{\mathcal{R}}_\lambda$ defining $[P] \in K_0(\dot{\mathcal{R}}_\lambda)$, we set $[F][P] := [\dot{\mathbb{R}}(F)(P)] \in K_0(\dot{\mathcal{R}}_\mu)$. It remains to lift the action of $K_0(\dot{\mathfrak{U}}(\mathfrak{g}))$ to $\dot{U}(\mathfrak{g})_{\mathbb{Z}}$ using the homomorphism γ from (3.14); this does not depend on the injectivity of γ. □

Lemma 4.4 shows for any 2-representation $\mathcal{R}$ that $\mathbb{C} \otimes_{\mathbb{Z}} K_0(\dot{\mathcal{R}})$ is a $\mathfrak{g}$-module in a canonical way. Typically, it is an *integrable* $\mathfrak{g}$-module in view of the next lemma.

LEMMA 4.5. *If $\mathcal{R}$ is a finite-dimensional 2-representation, then $\mathbb{C} \otimes_{\mathbb{Z}} K_0(\dot{\mathcal{R}})$ is an integrable $\mathfrak{g}$-module.*

PROOF. The local nilpotency of e_i follows from Lemma 3.6. Similar arguments with the negative nil Hecke category show that f_i is locally nilpotent too. □

DEFINITION 4.6. Let $\mathcal{R}$ and $\mathcal{S}$ be 2-representations. A *strongly equivariant functor* $G : \mathcal{R} \to \mathcal{S}$ is a functor such that $G|_{\mathcal{R}_\lambda} : \mathcal{R}_\lambda \to \mathcal{S}_\lambda$ for each $\lambda \in P$, plus natural isomorphisms $\zeta_i = \zeta_{G,i} : E_i G \Rightarrow G E_i$, such that the following hold for each $i \in I$:

(E1) the natural transformation $F_i G \varepsilon_i \circ F_i \zeta_i F_i \circ \eta_i G F_i : G F_i \Rightarrow F_i G$ is invertible, with inverse denoted $\zeta'_i : F_i G \Rightarrow G F_i$;
(E2) we have that $G x_i \circ \zeta_i = \zeta_i \circ x_i G$;
(E3) we have that $G \tau_{ij} \circ \zeta_i E_j \circ E_i \zeta_j = \zeta_j E_i \circ E_j \zeta_i \circ \tau_{ij} G$.

There is an obvious way to make the composition GH of two strongly equivariant functors into a strongly equivariant functor in its own right: set $\zeta_{GH,i} := \zeta_{G,i} H \circ G \zeta_{H,i}$. Also the identity functor 1 is strongly equivariant with $\zeta_{1,i} := 1_{E_i}$. Let $\mathcal{R}ep(\mathfrak{U}(\mathfrak{g}))$ be the resulting category of (small) 2-representations and strongly equivariant functors.

DEFINITION 4.7. A *strongly equivariant equivalence* is an equivalence of categories $G : \mathcal{R} \to \mathcal{S}$ such that $G|_{\mathcal{R}_\lambda} : \mathcal{R}_\lambda \to \mathcal{S}_\lambda$ for each λ, plus isomorphisms $\zeta_i : E_i G \Rightarrow G E_i$ satisfying (E2)–(E3) for each $i \in I$.

REMARK 4.8. In the situation of Definition 4.7, the axiom (E1) holds automatically, i.e. strongly equivariant equivalences are strongly equivariant functors. To see this, fix a right adjoint H to G. The given isomorphism $\zeta_i : E_i G \Rightarrow G E_i$ induces a canonical isomorphism $F_i H \Rightarrow H F_i$ between the right adjoints of GE_i and $E_i G$. Horizontally composing on the left and right with G and using the isomorphisms $1 \Rightarrow HG$ and $GH \Rightarrow 1$ defined by the adjunction, we get an isomorphism $GF_i \Rightarrow F_i G$. This is precisely the natural transformation written down in (E1).

It is helpful to interpret Definition 4.6 in terms of the string calculus for 2-morphisms in $\mathfrak{Cat}$. We represent ζ_i by  (so the dotted line is the identity morphism 1_G, to the right of which is the category $\mathcal{R}$ and to the left is $\mathcal{S}$). Then it is natural to denote its inverse by . The 2-morphism in (E1) is 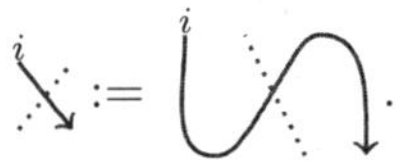.
We denote its inverse ζ_i' by . The axioms (E2) and (E3) are the identities

(4.2) 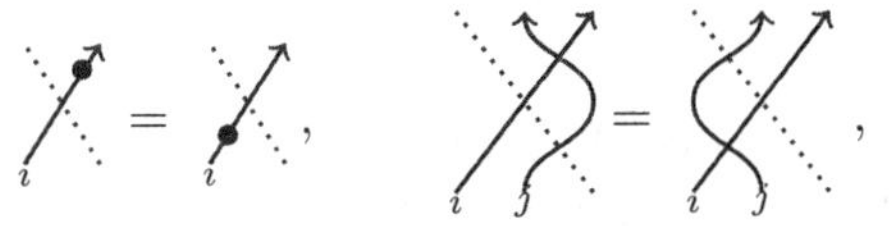

where the dots and solid crossings represent x_i and τ_{ij}, respectively. Representing $\eta_i, \eta_i', \varepsilon_i$ and ε_i' by oriented cups and caps as usual, one can check further that

(4.3) 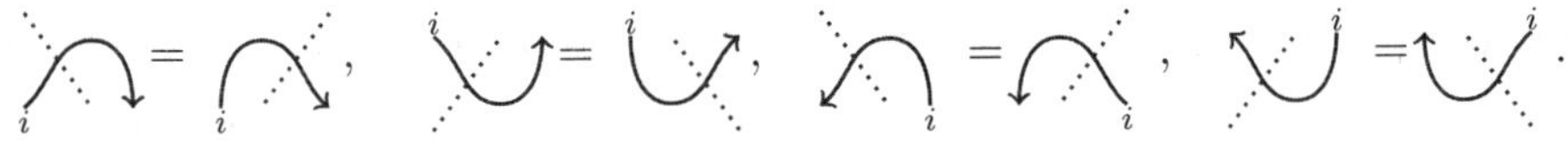

The first two of these are almost immediate; the second two follow using the inversion relations and the definitions of the leftward cups and caps in $\mathfrak{U}(\mathfrak{g})$.

REMARK 4.9. If $\mathcal{R}, \mathcal{S}$ are small 2-representations with associated 2-functors $\mathbb{R}$, $\mathbb{S}$, the data of a strongly equivariant functor $G : \mathcal{R} \to \mathcal{S}$ is the same as the data of a morphism of 2-functors $G : \mathbb{R} \to \mathbb{S}$ as in [**R1**, Definition 2.3] (with $G(\lambda) = G|_{\mathcal{R}_\lambda}$). Indeed, given any 1-morphism $u : \lambda \to \mu$ in $\mathfrak{U}(\mathfrak{g})$, i.e. an appropriate composition of several E_i and F_j applied to 1_λ, the corresponding horizontal composition of ζ_i and ζ_j' defines a natural isomorphism $\zeta(u) : \mathbb{S}(u)G(\lambda) \Rightarrow G(\mu)\mathbb{R}(u)$ satisfying the axioms of a morphism of 2-functors. The non-trivial part about this assertion is the naturality of $\zeta(u)$, i.e. the statement that $G(\mu)\mathbb{R}(\xi) \circ \zeta(u) = \zeta(v) \circ \mathbb{S}(\xi)G(\lambda)$ for all 2-morphisms $\xi : u \Rightarrow v$. In pictures:

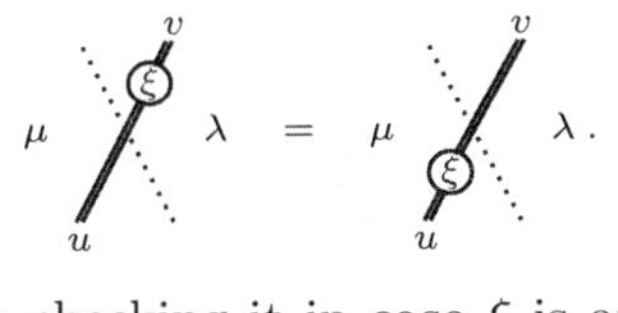

The proof of this reduces to checking it in case ξ is an upward dot or crossing or any cup or cap, since these generate all 2-morphisms in $\mathfrak{U}(\mathfrak{g})$ thanks to (K5). These cases are covered by (4.2)–(4.3).

REMARK 4.10. Using (4.3), we see in particular that 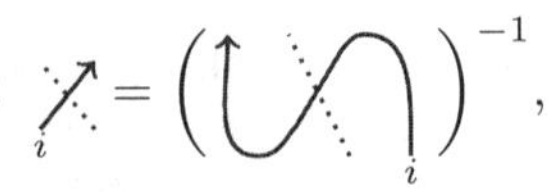, i.e. the natural transformation ζ_i' determines ζ_i. Indeed, using also Remark 4.9, one can reformulate Definition 4.6 equivalently in terms of isomorphisms $\zeta_i' : F_iG \Rightarrow GF_i$. The axioms (E1)–(E3) become:

(E1′) the natural transformation $E_iG\varepsilon_i' \circ E_i\zeta_i'E_i \circ \eta_i'GE_i : GE_i \Rightarrow E_iG$ is invertible, with inverse denoted $\zeta_i : E_iG \Rightarrow GE_i$;

(E2′) we have that $Gx_i' \circ \zeta_i' = \zeta_i' \circ x_i'G$;

(E3′) we have that $G\tau_{ij}' \circ \zeta_i'F_j \circ F_i\zeta_j' = \zeta_j'F_i \circ F_j\zeta_i' \circ \tau_{ij}'G$.

This version is compatible with the definition of 2-representation from Remark 4.3.

DEFINITION 4.11. A *strongly equivariant natural transformation* between strongly equivariant functors $G, H : \mathcal{R} \to \mathcal{S}$ is a natural transformation $\pi : G \Rightarrow H$ such that $\pi E_i \circ \zeta_{G,i} = \zeta_{H,i} \circ E_i\pi : E_iG \Rightarrow HE_i$. Let $\mathfrak{Rep}(\mathfrak{U}(\mathfrak{g}))$ be the strict 2-category of (small) 2-representations, strongly equivariant functors and strongly equivariant natural transformations. We denote the morphism categories in this 2-category by $\mathcal{H}om_{\mathfrak{U}(\mathfrak{g})}(\mathcal{R}, \mathcal{S})$.

REMARK 4.12. In the setup of Remark 4.9, a strongly equivariant natural transformation is the same as a modification between morphisms of 2-functors in the sense of [**R1**, Definition 2.4].

4.2. Generalized cyclotomic quotients. In this subsection, we define some important examples of 2-representations. We need a couple more basic notions to prepare for this.

Fix a weight $\kappa \in P$. Then there is a 2-representation[11] $\mathcal{R}(\kappa)$ of $\mathfrak{U}(\mathfrak{g})$ defined as follows: $\mathcal{R}(\kappa)_\lambda := \mathcal{H}om_{\mathfrak{U}(\mathfrak{g})}(\kappa, \lambda)$; E_i (resp. F_i) is the functor defined by horizontally composing 1-morphisms on the left by E_i (resp. F_i) and 2-morphisms on the left by $\underset{i}{\uparrow}\lambda$ (resp. $\overset{i}{\downarrow}\lambda$); x_i, τ_{ij}, η_i and ε_i are the natural transformations defined by horizontally composing on the left by , ,  and 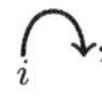, respectively.

An *invariant ideal*[12] $\mathcal{I}$ of a 2-representation $\mathcal{R}$ is a family of subspaces $\mathcal{I}(b, c) \leq \mathrm{Hom}_{\mathcal{R}}(b, c)$ for each $b, c \in \mathrm{ob}\,\mathcal{R}$ such that

- $f \in \mathrm{Hom}_{\mathcal{R}}(a, b)$ and $g \in \mathcal{I}(b, c) \Rightarrow g \circ f \in \mathcal{I}(a, c)$;
- $h \in \mathrm{Hom}_{\mathcal{R}}(c, d)$ and $g \in \mathcal{I}(b, c) \Rightarrow h \circ g \in \mathcal{I}(b, d)$;
- $g \in \mathcal{I}(b, c)$ and $i \in I \Rightarrow E_i(g) \in \mathcal{I}(E_i(b), E_i(c))$ and $F_i(g) \in \mathcal{I}(F_i(b), F_i(c))$.

Given an invariant ideal $\mathcal{I}$, the quotient category $\mathcal{R}/\mathcal{I}$ is the category with the same objects as $\mathcal{R}$ and morphisms $\mathrm{Hom}_{\mathcal{R}/\mathcal{I}}(b, c) := \mathrm{Hom}_{\mathcal{R}}(b, c)/\mathcal{I}(b, c)$. It has a naturally induced structure of 2-representation in its own right.

Now we are ready for the main construction.

CONSTRUCTION 4.13. Fix weights $\kappa \in P^+$, $\kappa' \in -P^+$. Let $k_i := \langle h_i, \kappa \rangle$, $k_i' := -\langle h_i, \kappa' \rangle$, and take a family of indeterminates $\{z_{i,r}, z_{j,s}' \mid i, j \in I, 1 \leq r \leq k_i, 1 \leq s \leq k_j'\}$. Let $\Bbbk[z] := \Bbbk[z_{i,r}, z_{j,s}' \mid i, j \in I, 1 \leq r \leq k_i, 1 \leq k_j']$ be the

[11] In [**R2**, §4.3.3], Rouquier denotes this by $\mathcal{M}(\kappa)$, but that seems confusing notation since it is a categorification of the left ideal $\dot{U}(\mathfrak{g})1_\kappa$ rather than the Verma module $M(\kappa)$.

[12] In Rouquier's language, an invariant ideal is the data of a full sub-2-representation.

corresponding polynomial algebra. Adopting the convention that $z_{i,0} = z'_{i,0} := 1$, we define new variables $\delta_{i,s}, \delta'_{i,s} \in \Bbbk[z]$ for $s \geq 0$ from the generating functions

$$\sum_{s\geq 0} \delta_{i,s} t^s := c_{\kappa+\kappa';i} \frac{\sum_{r=0}^{k'_i} z'_{i,r} t^r}{\sum_{r=0}^{k_i} z_{i,r} t^r}, \qquad \sum_{s\geq 0} \delta'_{i,s} t^s := c^{-1}_{\kappa+\kappa';i} \frac{\sum_{r=0}^{k_i} z_{i,r} t^r}{\sum_{r=0}^{k'_i} z'_{i,r} t^r}. \tag{4.4}$$

Here, we are working in $\Bbbk[z][[t]]$ where t is a formal parameter. Let $\mathcal{R} := \mathcal{R}(\kappa + \kappa') \otimes_\Bbbk \Bbbk[z]$ be the $\Bbbk[z]$-linear 2-representation obtained from $\mathcal{R}(\kappa+\kappa')$ by extending scalars in the obvious way. Let $\mathcal{I}$ be the $\Bbbk[z]$-linear invariant ideal of $\mathcal{R}$ generated by the morphisms

$$\sum_{r=0}^{k_i} \left({}_{k_i - r}\overset{i}{\underset{\bullet}{\downarrow}}\; {}_{\kappa+\kappa'} \right) z_{i,r}, \tag{4.5}$$

$$\kappa+\kappa' \; \underset{i}{\circlearrowleft}\!\bullet^{\,s+*} - 1_{\kappa+\kappa'} \delta_{i,s} \tag{4.6}$$

for all $i \in I$ and $s = 1, \dots, k'_i$, using the shorthand (3.13). Taking the quotient, we obtain the $\Bbbk[z]$-linear 2-representation

$$\mathcal{L}(\kappa'|\kappa) := \mathcal{R}/\mathcal{I}. \tag{4.7}$$

Finally, we viewing $\Bbbk$ as a $\Bbbk[z]$-algebra so that each $z_{i,r}, z'_{i,r}$ act as zero, we have the *minimal specialization*

$$\mathcal{L}_{\min}(\kappa'|\kappa) := \mathcal{L}(\kappa'|\kappa) \otimes_{\Bbbk[z]} \Bbbk. \tag{4.8}$$

LEMMA 4.14. *The ideal $\mathcal{I}$ in Construction* 4.13 *is generated also by the morphisms*

$$\sum_{r=0}^{k'_i} \left({}_{k'_i - r}\underset{i}{\overset{\bullet}{\uparrow}}\; {}_{\kappa+\kappa'} \right) z'_{i,r}, \tag{4.9}$$

$$\kappa+\kappa' \; \underset{i}{\circlearrowright}\!\bullet^{\,s+*} - 1_{\kappa+\kappa'} \delta'_{i,s}, \tag{4.10}$$

*for all $i \in I$ and $s = 1, \dots, k_i$. Moreover, it contains (*4.6*) and (*4.10*) for all $s \geq 0$.*

PROOF. We first show that the images of the elements (4.6) are zero in $\mathcal{L}(\kappa'|\kappa)$ for all $s \geq 0$. For the induction step, we may assume that $s > k'_i$. Note by the definition from (K7) that

$$\kappa+\kappa' \; \underset{i}{\circlearrowleft}\!\bullet^{\,s+*} = \kappa+\kappa' \; \underset{i}{\circlearrowleft}\!\bullet^{\,s+k_i-k'_i-1},$$

so for $s > k'_i$ there are $\geq k_i$ dots on the right hand side here. Therefore using the relation (4.5), we get that

$$\sum_{r=0}^{k_i} \kappa+\kappa' \; \underset{i}{\circlearrowleft}\!\bullet^{\,s-r+*} z_{i,r} = 0,$$

working in the quotient $\mathcal{L}(\kappa'|\kappa)$. Applying the inductive hypothesis, we deduce that

$$\kappa+\kappa' \; \underset{i}{\circlearrowleft}\!\bullet^{\,s+*} + \sum_{r=1}^{k_i} 1_{\kappa+\kappa'} \delta_{i,s-r} z_{i,r} = 0.$$

It remains to observe that $\sum_{r=0}^{k_i} \delta_{i,s-r} z_{i,r} = 0$ already in $\Bbbk[z]$ when $s > k_i'$. This follows because $\left(\sum_{s\geq 0} \delta_{i,s} t^s\right)\left(\sum_{r=0}^{k_i} z_{i,r} t^r\right)$ is a polynomial of degree k_i' by the definition (4.4).

Now let $\mathrm{e}(t) := \sum_{r\geq 0} \mathrm{e}_r t^r$ and $\mathrm{h}(t) := \sum_{r \geq 0} \mathrm{h}_r t^r$, so that $\mathrm{e}(-t)\mathrm{h}(t) = 1$ by (3.10); remember also the definitions (3.11)–(3.12). Also set $\delta_i(t) := \sum_{s\geq 0} \delta_{i,s} t^s$ and $\delta_i'(t) := \sum_{s \geq 0} \delta_{i,s}' t^s$, so that $\delta_i'(t)\delta_i(t) = 1$ by (4.4). In the previous paragraph, we have shown that the image of $\beta_{\kappa+\kappa';i}(\mathrm{h}(t))$ is $c_{\kappa+\kappa';i}^{-1} 1_{\kappa+\kappa'}\delta_i(t)$. Hence, the image of $\beta_{\kappa+\kappa';i}(\mathrm{e}(-t))$ is $c_{\kappa+\kappa';i} 1_{\kappa+\kappa'}\delta_i'(t)$. This shows that (4.10) belongs to $\mathcal{I}$ for all $s \geq 0$.

In this paragraph we show that (4.9) belongs to $\mathcal{I}$ too. Working in $\mathcal{L}(\kappa'|\kappa)$ once again, we have by (4.5) and (K6) that

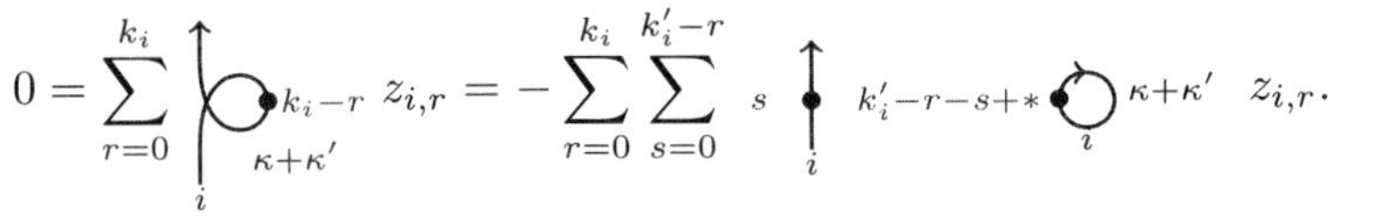

Changing the summation using also (4.6), we have shown that

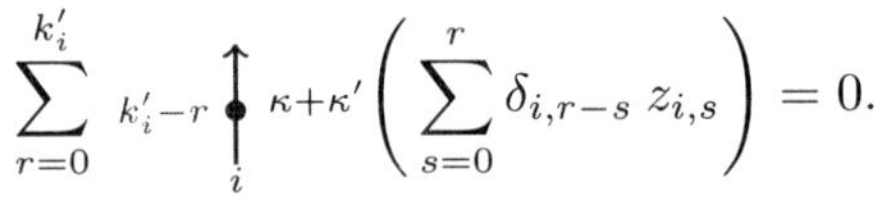

It remains to apply (4.4) to simplify this to (4.9).

Conversely, one checks by similar arguments that the $\Bbbk[z]$-linear invariant ideal $\mathcal{I}'$ generated by the elements (4.9)–(4.10) contains (4.5)–(4.6). $\square$

LEMMA 4.15. *There is a unique $\mathfrak{g}$-module homomorphism*

$$L(\kappa'|\kappa) \to \mathbb{C} \otimes_{\mathbb{Z}} K_0(\dot{\mathcal{L}}_{\min}(\kappa'|\kappa)), \qquad \bar{1}_{\kappa+\kappa'} \mapsto [1_{\kappa+\kappa'}].$$

PROOF. We need to show that the homomorphism

$$\dot{U}(\mathfrak{g})1_{\kappa+\kappa'} \to \mathbb{C} \otimes_{\mathbb{Z}} K_0(\dot{\mathcal{L}}_{\min}(\kappa'|\kappa))$$

sending $1_{\kappa+\kappa'} \mapsto [1_{\kappa+\kappa'}]$ factors through the quotient $L(\kappa'|\kappa)$ from (3.6). This follows because $E_i^{(1+k_i')}1_{\kappa+\kappa'} = 0 = F_i^{(1+k_i)}1_{\kappa+\kappa'}$ in $\dot{\mathcal{L}}_{\min}(\kappa'|\kappa)$. The first equality here follows from Lemma 3.6 and the defining relation (4.9); the second one follows similarly using (4.5) and a rotated version of Lemma 3.6. $\square$

LEMMA 4.16. *The 2-representation $\mathcal{L}_{\min}(\kappa'|\kappa)$ is* nilpotent *in the sense that $(x_i)_u \in \mathrm{End}_{\mathcal{L}_{\min}(\kappa'|\kappa)}(E_i u)$ and $(x_i')_u \in \mathrm{End}_{\mathcal{L}_{\min}(\kappa'|\kappa)}(F_i u)$ are nilpotent for all $i \in I$ and $u \in \mathrm{ob}\,\mathcal{L}_{\min}(\kappa'|\kappa)$.*

PROOF. We show by induction on r that $(x_i)_u$ is nilpotent for any object u that is a monomial obtained by applying r of the generating E's and F's to $1_{\kappa+\kappa'}$; similar arguments give the nilpotency of $(x_i')_u$ too. The base case $r = 0$ follows from (4.5) and (4.9), since they show that 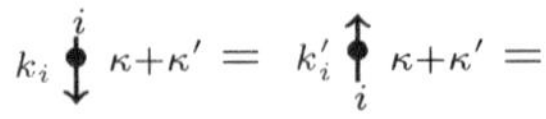 in $\mathcal{L}_{\min}(\kappa'|\kappa)$.

For the induction step, we consider $(x_i)_u$ for a monomial u of length $(r+1)$. There are three cases:

Case one: $u = E_i 1_\lambda v$. Introduce the *intertwiner*  $\lambda :=$  $\lambda -$  λ. Using the relations, one checks that 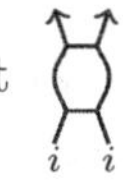 $\lambda =$  λ and 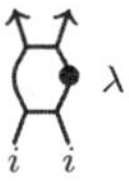 $\lambda =$  λ. Hence, 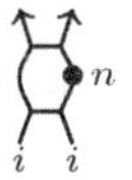 $\lambda =$  λ. As $(x_i^n)_v \in \mathrm{End}_{\mathcal{L}_{\min}(\kappa'|\kappa)}(E_i v)$ is zero for some n by induction, we deduce that $(x_i^n)_u \in \mathrm{End}_{\mathcal{L}_{\min}(\kappa'|\kappa)}(E_i u)$ is zero too.

Case two: $u = E_j 1_\lambda v$ *for* $j \neq i$. By induction, we have that $(x_i^n)_v = (x_j^n)_v = 0$ for some n. Let $m := (d_{ij} + 1)n$. Then we can use the defining relations to rewrite 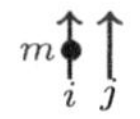 λ as a linear combination of 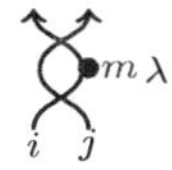 λ and terms of the form 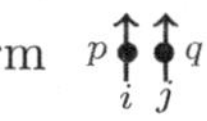 λ with $p \geq nd_{ij}$ and $q \geq 1$. Repeating the calculation $(n-1)$ more times, we get that 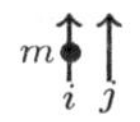 λ is a linear combination of 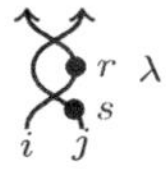 λ and terms of the form 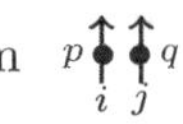 λ with $r \geq n, s \geq 0, p \geq d_{ij}$ and $q \geq n$. It remains to act on v. All the terms vanish and we have shown that $(x_i^m)_u = 0$.

Case three: $u = F_j 1_\lambda v$. Use the first alternating crossing relation from (K8) to move dots to the right in a similar way. □

COROLLARY 4.17. *$\mathcal{L}_{\min}(\kappa'|\kappa)$ is a finite-dimensional category.*

PROOF. We take $u, v \in \mathrm{ob}\,\mathcal{L}_{\min}(\kappa'|\kappa)_\lambda$, i.e. 1-morphisms $\kappa + \kappa' \to \lambda$ in $\mathfrak{U}(\mathfrak{g})$ for some $\lambda \in P$, and consider the explicit spanning set for $\mathrm{Hom}_{\mathfrak{U}(\mathfrak{g})}(u, v)$ constructed as in [**KL3**, Proposition 3.11]; we arrange this so that all the dotted bubbles appear at the right hand side of the diagrams. We need to show that the image of this set spans a finite-dimensional vector space when we pass to the quotient $\mathcal{L}_{\min}(\kappa'|\kappa)$. This is clear because any diagram with a bubble vanishes in the quotient by the relations (4.6) and (4.10), and any diagram with too many dots on any given strand vanishes by the nilpotency established in Lemma 4.16. □

REMARK 4.18. The category $\mathcal{L}(\kappa'|\kappa)$ is not finite-dimensional. However, each of its morphism spaces is *finitely generated* as a $\Bbbk[z]$-module. This follows by a similar but more delicate inductive argument compared to the proof of Corollary 4.17.

Taking $\kappa' = 0$ in Construction 4.13, we obtain the $\Bbbk[z]$-linear 2-representation

$$\mathcal{L}(\kappa) := \mathcal{L}(0|\kappa). \tag{4.11}$$

Note in this situation that $\Bbbk[z] = \Bbbk[z_{i,r} \mid i \in I, r = 1, \dots, k_i]$ and $z_{i,r} = c_{\kappa;i}\delta'_{i,r}$ for $r = 1, \dots, k_i$. Since each $k'_i = 0$, Lemma 4.14 shows that $\mathcal{L}(\kappa)$ is the quotient of $\mathcal{R} := \mathcal{R}(\kappa) \otimes_\Bbbk \Bbbk[z]$ by the $\Bbbk[z]$-linear invariant ideal generated by the morphisms

(4.12) 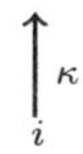 κ,

(4.13) $1_\kappa z_{i,r} - c_{\kappa;i}$ 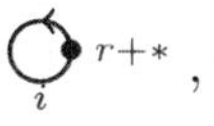 ,

for $i \in I$ and $r = 1, \dots, k_i$. In view of (4.13), there is no need to extend scalars to $\Bbbk[z]$ after all: we could equivalently define $\mathcal{L}(\kappa)$ to be the quotient of $\mathcal{R}(\kappa)$ by the invariant ideal generated by the morphisms (4.12) for all $i \in I$, viewing it as a

$\Bbbk[z]$-linear 2-representation so that each $z_{i,r}$ acts by horizontally composing on the right with $c_{\kappa;i}$ 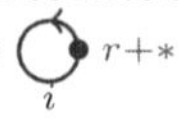 .

The discussion in the previous paragraph shows that $\mathcal{L}(\kappa)$ is Rouquier's *universal categorification* of $L(\kappa)$ from [**R2**, §4.3.3]. These 2-representations play a fundamental role in his general structure theory for upper integrable 2-representations. To start with, they formally satisfy the following universal property: for any 2-representation $\mathcal{V}$, let

$$\mathcal{V}_\kappa^{\mathrm{hw}} := \{V \in \mathrm{ob}\,\mathcal{V}_\kappa \mid E_i V = 0 \text{ for all } i \in I\}$$

which is a full subcategory of $\mathcal{V}_\kappa$; then there is an equivalence of categories

$$\mathrm{ev}_{1_\kappa} : \mathcal{H}om_{\mathfrak{U}(\mathfrak{g})}(\mathcal{L}(\kappa), \mathcal{V}) \to \mathcal{V}_\kappa^{\mathrm{hw}}, \qquad G \mapsto G1_\kappa, \quad \pi \mapsto \pi 1_\kappa. \tag{4.14}$$

This is a key ingredient in [**R2**, Theorem 4.22], which shows that any upper integrable, additive, idempotent-complete 2-representation has a finite filtration whose sections are specializations of the $\dot{\mathcal{L}}(\kappa)$'s ordered in some way refining the dominance order (most dominant at the bottom). This result is a categorical analog of the filtration of a based module constructed by Lusztig in [**Lu2**, Ch. 27].

Using a fundamental theorem of Kang and Kashiwara [**KK**, Theorem 5.2], Rouquier has also given an equivalent realization of $\dot{\mathcal{L}}(\kappa)$ as follows. Recalling Definition 3.4, introduce the (negative) *cyclotomic quiver Hecke category* $\mathcal{H}'(\kappa)$, namely, the quotient of the $\Bbbk[z]$-linear monoidal category $\mathcal{H}' \otimes_\Bbbk \Bbbk[z]$ by the $\Bbbk[z]$-linear left tensor ideal generated by the morphisms

$$\sum_{r=0}^{k_i} \left({}_{k_i-r}\bullet\!\!\downarrow^{i} \right) z_{i,r} \qquad (i \in I). \tag{4.15}$$

The endomorphism algebras

$$\widehat{H}'_n(\kappa) := \bigoplus_{\boldsymbol{i},\boldsymbol{j} \in I^{\otimes n}} \mathrm{Hom}_{\mathcal{H}'(\kappa)}(\boldsymbol{i}, \boldsymbol{j}) \tag{4.16}$$

are Rouquier's *deformed cyclotomic quiver Hecke algebras*; in particular, $\widehat{H}'_0(\kappa)$, the endomorphism algebra of the unit object $\varnothing$ of $\mathcal{H}'(\kappa)$, is $\Bbbk[z]$. The theorem of Kang and Kashiwara shows that the additive Karoubi envelope $\dot{\mathcal{H}}'(\kappa)$ can be endowed with the structure of a $\Bbbk[z]$-linear 2-representation with E_i and F_i arising from certain restriction and induction functors[13]. Applying the universal property from (4.14), we get a canonical strongly equivariant functor $G : \mathcal{L}(\kappa) \to \dot{\mathcal{H}}'(\kappa)$ such that $\mathrm{ev}_{1_\kappa}(G) = \varnothing$.

THEOREM 4.19 ([**R2**, Theorem 4.25]). *The functor G is $\Bbbk[z]$-linear and it induces a strongly equivariant equivalence $G : \dot{\mathcal{L}}(\kappa) \to \dot{\mathcal{H}}'(\kappa)$. Hence,* $\mathrm{End}_{\mathcal{L}(\kappa)}(1_\kappa) \cong \Bbbk[z]$.

It follows immediately that the *minimal categorification*

$$\mathcal{L}_{\min}(\kappa) := \mathcal{L}_{\min}(0|\kappa) \tag{4.17}$$

[13]The equivalent formulation of the definition of 2-representation from Remark 4.3 is convenient here since induction is obviously left adjoint to restriction; Lemma 4.2 implies that it is right adjoint too, but this is far from clear from the outset.

is Morita equivalent to the quotient $\mathcal{H}'_{\min}(\kappa)$ of $\mathcal{H}'$ by the left tensor ideal generated by $k_i \overset{i}{\downarrow}$ $(i \in I)$. The endomorphism algebras

$$H'_n(\kappa) := \bigoplus_{\boldsymbol{i},\boldsymbol{j} \in I^{\otimes n}} \mathrm{Hom}_{\mathcal{H}'_{\min}(\kappa)}(\boldsymbol{i},\boldsymbol{j}) \tag{4.18}$$

are the *cyclotomic quiver Hecke algebras* introduced by Khovanov and Lauda in [**KL1**]. They are finite-dimensional algebras, so the blocks of $\mathcal{L}_{\min}(\kappa)$ are finite-dimensional algebras too; in particular, $\mathcal{L}_{\min}(\kappa)$ is Artinian. Moreover,

$$\mathrm{End}_{\mathcal{L}_{\min}(\kappa)}(1_\kappa) \cong H'_0(\kappa) \cong \Bbbk.$$

This shows that 1_κ is non-zero in $\mathcal{L}_{\min}(\kappa)$, which is the crucial point needed in order to deduce the following theorem, which was established already in [**KK**, Theorem 6.2]. Note also that Webster has given a different proof of all of these results in [**W2**, §3].

THEOREM 4.20 ([**KK**, **W2**]). *For any $\kappa \in P^+$, the homomorphism*

$$L(\kappa) \to \mathbb{C} \otimes_{\mathbb{Z}} K_0(\dot{\mathcal{L}}_{\min}(\kappa))$$

from Lemma 4.15 is an isomorphism.

PROOF. We just saw that the homomorphism is non-zero. Hence, it is injective. It is surjective too by a standard argument recalled in Corollary 4.34 below. □

COROLLARY 4.21. *If $\mathfrak{g}$ is of finite type, then the homomorphism $\gamma : \dot{U}(\mathfrak{g})_{\mathbb{Z}} \to K_0(\dot{\mathfrak{U}}(\mathfrak{g}))$ from (3.14) is an isomorphism; similarly, so is γ_q from (3.16).*

PROOF. It remains to show that γ is injective. Take $u \in \dot{U}(\mathfrak{g})_{\mathbb{Z}}$ with $\gamma(u) = 0$. By its definition, the isomorphism $L(\kappa) \overset{\sim}{\to} \mathbb{C} \otimes_{\mathbb{Z}} K_0(\dot{\mathcal{L}}_{\min}(\kappa))$ from Theorem 4.20 intertwines the action of u on the left hand space with the action of $\gamma(u)$ on the right hand space. Hence, u annihilates $L(\kappa)$. This is true for each $\kappa \in P^+$, so we get that $u = 0$ by the second statement of Lemma 3.2. □

REMARK 4.22. The results just explained obviously have lowest weight analogs too. Taking $\kappa = 0$ in Construction 4.13, the universal and minimal categorifications of $L'(\kappa')$ are $\mathcal{L}'(\kappa') := \mathcal{L}(\kappa'|0)$ and $\mathcal{L}'_{\min}(\kappa') := \mathcal{L}_{\min}(\kappa'|0)$, respectively. They are Morita equivalent to analogously defined cyclotomic quotients $\mathcal{H}(\kappa')$ and $\mathcal{H}_{\min}(\kappa')$ of the quiver Hecke category $\mathcal{H}$. Moreover $\mathrm{End}_{\mathcal{L}'(\kappa')}(1_{\kappa'}) \cong \Bbbk[z]$, the finite-dimensional category $\mathcal{L}'_{\min}(\kappa')$ is Artinian, and $\mathbb{C} \otimes_{\mathbb{Z}} K_0(\dot{\mathcal{L}}'_{\min}(\kappa')) \cong L(\kappa')$.

REMARK 4.23. The general definition of the 2-representations $\mathcal{L}(\kappa'|\kappa)$ for $\kappa, \kappa' \neq 0$ given in Construction 4.13 is new, but their minimal specializations $\mathcal{L}_{\min}(\kappa'|\kappa)$ appear already[14] in [**W1**, Proposition 5.6]. In [**W1**, Theorem 5.13], Webster asserts that the homomorphism in Lemma 4.15 is an isomorphism for all κ, κ', so that $\mathcal{L}_{\min}(\kappa'|\kappa)$ is a categorification of $L(\kappa'|\kappa) \cong L'(\kappa') \otimes L(\kappa)$ in general. Like in the proof of Corollary 4.21 (using the first statement of Lemma 3.2 instead of the second), such a result implies that $\dot{U}(\mathfrak{g})_{\mathbb{Z}} \cong K_0(\dot{\mathfrak{U}}(\mathfrak{g}))$ for arbitrary $\mathfrak{g}$. However, Webster's proof of this is intertwined with his new approach in [**W3**] to verifying the Nondegeneracy Condition; cf. Remark 3.13. Actually, [**W3**] is based on some even more general deformations, which should be closely related to our $\mathcal{L}(\kappa'|\kappa)$ when there is just one lowest and one highest weight tensor factor.

[14] Note that Webster's diagrams are the mirror images of ours in a vertical axis.

REMARK 4.24. The finite-dimensional category $\mathcal{L}_{\min}(\kappa'|\kappa)$ is not Artinian in general (outside of finite type). We conjecture that it is always Noetherian.

4.3. Categorical actions. Henceforth, $\mathcal{C}$ denotes a locally Schurian category in the sense of Definition 2.9. We fix a set of representatives $\{L(b) \mid b \in \mathbf{B}\}$ for the isomorphism classes of irreducible objects, and let $P(b)$ be a projective cover of $L(b)$. Let $K_0(\mathrm{p}\mathcal{C})$ denote the split Grothendieck group of the additive category $\mathrm{p}\mathcal{C}$ (= finitely generated projectives in $\mathcal{C}$). The classes $\{[P(b)] \mid b \in \mathbf{B}\}$ give a distinguished basis for $\mathbb{C} \otimes_{\mathbb{Z}} K_0(\mathrm{p}\mathcal{C})$.

DEFINITION 4.25. A *categorical action* of $\mathfrak{g}$ on $\mathcal{C}$ is the following additional data:

(A1) a partition $\mathbf{B} = \bigsqcup_{\lambda \in P} \mathbf{B}_\lambda$ inducing a decomposition $\mathcal{C} = \prod_{\lambda \in P} \mathcal{C}_\lambda$ as in (L10) from §2.2;

(A2) sweet endofunctors E_i of $\mathcal{C}$ for each $i \in I$ such that $E_i|_{\mathcal{C}_\lambda} : \mathcal{C}_\lambda \rightarrow \mathcal{C}_{\lambda+\alpha_i}$ (recall Definition 2.10);

(A3) a strict monoidal functor $\Phi : \mathcal{H} \rightarrow \mathcal{E}nd(\mathcal{C})$ with $\Phi(i) = E_i$ for each i, where $\mathcal{H}$ is the quiver Hecke category from Definition 3.4.

Let $x_i := \Phi(\,\underset{i}{\uparrow\!\!\bullet}\,)$ and $\tau_{ij} := \Phi(\,\underset{i\quad j}{\times}\,)$. We also fix the choice of a right adjoint F_i to E_i for each $i \in I$, and set $e_i := [E_i]$ and $f_i := [F_i]$, which are endomorphisms of $K_0(\mathrm{p}\mathcal{C})$. Then we impose the following axiom:

(A4) for each $i, j \in I$ and $\lambda \in P$, the commutator $[e_i, f_j]$ acts on $K_0(\mathrm{p}\mathcal{C}_\lambda)$ as multiplication by the scalar $\delta_{i,j}\langle h_i, \lambda\rangle$.

We say that the categorical action is *nilpotent*[15] if $(x_i)_V$ is a nilpotent element of the finite-dimensional algebra $\mathrm{End}_{\mathcal{C}}(E_i V)$ for all $i \in I$ and $V \in \mathrm{ob}\,\mathrm{fg}\mathcal{C}$.

The following gives a general recipe producing a categorical action on the locally Schurian category Mod-$\mathcal{A}$ for any finite-dimensional 2-representation $\mathcal{A}$. In particular, applying it to the finite-dimensional 2-representation $\mathcal{L}_{\min}(\kappa'|\kappa)$ from Construction 4.13, this shows that Mod-$\mathcal{L}_{\min}(\kappa'|\kappa)$ admits a categorical action; this example is also nilpotent thanks to Lemma 4.16.

CONSTRUCTION 4.26. Let $\mathcal{A} = \coprod_{\lambda \in P} \mathcal{A}_\lambda$ be a finite-dimensional 2-representation. Let $A = \bigoplus_{\lambda \in P} A_\lambda$ be the associated locally unital algebra as in Remark 2.1; the distinguished idempotents in A are indexed by $X = \bigsqcup_{\lambda \in P} X_\lambda$ where $X_\lambda := \mathrm{ob}\,\mathcal{A}_\lambda$. We define a categorical action on $\mathcal{C} :=$ Mod-A as follows.

- Fix representatives $\{L(b) \mid b \in \mathbf{B}_\lambda\}$ for the isomorphism classes of irreducible A_λ-modules, and let $\mathbf{B} := \bigsqcup_{\lambda \in P} \mathbf{B}_\lambda$. This partition induces the decomposition $\mathcal{C} = \prod_{\lambda \in P} \mathcal{C}_\lambda$ required for (A1); of course, we have that $\mathcal{C}_\lambda =$ Mod-A_λ.
- The functor E_i defines locally unital homomorphisms $e_i : A_\lambda \rightarrow A_{\lambda+\alpha_i}$ for each $\lambda \in P$; moreover, $e_i(1_u) = 1_{E_i u}$ for each $u \in X_\lambda$. Let $e_i^* A_{\lambda+\alpha_i}$ be the $(A_\lambda, A_{\lambda+\alpha_i})$-bimodule obtained from $A_{\lambda+\alpha_i}$ by restricting the natural left action through this homomorphism. Tensoring on the right with this bimodule defines a functor $\hat{E}_i :$ Mod-$A_\lambda \rightarrow$ Mod-$A_{\lambda+\alpha_i}$ for each $\lambda \in P$. This is the data required for (A2). The endofunctor $\hat{E}_i$ is sweet because the functor F_i extends similarly to a functor $\hat{F}_i :$ Mod-$A_{\lambda+\alpha_i} \rightarrow$ Mod-A_λ which is biadjoint to $\hat{E}_i$ thanks to Lemma 4.2.

[15] Just as good would be to assume there is $c \in \Bbbk$ that $(x_i - c1)_V$ is nilpotent for all i and V.

- The natural transformation $x_i : E_i \to E_i$ on each $u \in X_\lambda$ produces a family of elements $x_{i;u} \in 1_{e_i u} A_{\lambda+\alpha_i} 1_{e_i u}$ $(u \in X_\lambda)$ such that $e_i(f) x_{i;u} = x_{i;v} e_i(f)$ for all $f \in 1_v A_\lambda 1_u$. Hence, there is a bimodule homomorphism $e_i^* A_{\lambda+\alpha_i} \to e_i^* A_{\lambda+\alpha_i}$ defined on $1_{e_i u} A_{\lambda+\alpha_i}$ by left multiplication by $x_{i;u}$, from which we get $\hat{x}_i : \hat{E}_i \to \hat{E}_i$. Similarly, $\tau_{ij} : E_i E_j \to E_j E_i$ translates to $\tau_{ij;u} \in 1_{e_j e_i u} A_{\lambda+\alpha_i+\alpha_j} 1_{e_i e_j u}$ such that $(e_j e_i(f))\tau_{ij;u} = \tau_{ji;v}(e_i e_j(f))$ for all $f \in 1_v A_\lambda 1_u$ and $u, v \in X_\lambda$. Left multiplication by these elements defines a bimodule homomorphism $(e_i e_j)^* A_{\lambda+\alpha_i+\alpha_j} \to (e_j e_i)^* A_{\lambda+\alpha_i+\alpha_j}$. The composite functor $\hat{E}_i \hat{E}_j$ is defined by tensoring with $(e_i e_j)^* A_{\lambda+\alpha_i+\alpha_j} \cong e_j^* A_{\lambda+\alpha_j} \otimes_{A_{\lambda+\alpha_j}} e_i^* A_{\lambda+\alpha_i+\alpha_j}$, so this is what we need to get $\hat{\tau}_{ij} : \hat{E}_i \hat{E}_j \to \hat{E}_j \hat{E}_i$. Thus we have the data for (A3).
- Finally the axiom (A4) follows from Lemma 4.4 and (2.1).

In fact, assuming nilpotency, the notion of a finite-dimensional 2-representation is equivalent to the notion of categorical action on a locally Schurian category. This depends on the following theorem, which is a variation on [**R1**, Theorem 5.27]. It is a remarkable example of relations on the Grothendieck group (specifically, axiom (A4)) implying relations between 2-morphisms (specifically, axiom (M5)); Rouquier refers to this as "control by K_0." It is very useful since (M5) can be very difficult to check directly.

THEOREM 4.27. *Suppose that we are given a nilpotent categorical action on some locally Schurian category $\mathcal{C}$. Then the natural transformations σ_{ij} $(i \neq j)$ and $\rho_{i,\lambda}$ from Definition* 4.1 *are invertible. Hence, $\mathcal{C}$ is a locally Schurian 2-representation.*

PROOF. Note to start with that $\mathbb{C} \otimes_{\mathbb{Z}} K_0(\mathrm{p}\mathcal{C})$ is an integrable $\mathfrak{g}$-module, by the same argument as in the proof of Lemma 4.5.

In this paragraph, we explain how to see that $\rho_{i,\lambda}$ is invertible (for fixed i and λ). By Lemma 2.12, it suffices to show that $\rho_{i,\lambda}$ is invertible on $L(b)$ for $b \in \mathbf{B}_\lambda$. In the Artinian case, this follows immediately from [**R1**, Theorem 5.22]. The proof in general reduces to the Artinian case as follows. By (A4) and integrability, the set

$$\mathbf{B}'' := \{a \in \mathbf{B} \,|\, P(a) \text{ is a summand of some sequence of } E_i \text{ and } F_i \text{ applied to } P(b)\}$$

is finite. Of course, $b \in \mathbf{B}''$. Let $\mathbf{B}' := \mathbf{B} \setminus \mathbf{B}''$ and $\pi : \mathcal{C} \to \mathcal{C}/\mathcal{C}'$ be the corresponding Serre quotient as in §2.5. The isomorphism classes of irreducible objects of $\mathcal{C}/\mathcal{C}'$ are represented by $\{\pi L(a) \mid a \in \mathbf{B}''\}$; in particular, $\mathcal{C}/\mathcal{C}'$ is finite. Moreover, using the left adjoint functor $\pi^!$, we may identify its complexified K_0 with the subspace of $\mathbb{C} \otimes_{\mathbb{Z}} K_0(\mathrm{p}\mathcal{C})$ spanned by $\{[P(a)] \mid a \in \mathbf{B}''\}$. The definition of $\mathbf{B}''$ ensures that this subspace is stable under the action of e_i and f_i. Consequently, the functors E_i and F_i preserve the subcategory $\mathcal{C}'$; for example, for E_i, this follows because $\mathrm{Hom}_{\mathcal{C}}(P(c), E_i L(d)) \cong \mathrm{Hom}_{\mathcal{C}}(F_i P(c), L(d)) = 0$ for all $c \in \mathbf{B}'', d \in \mathbf{B}'$. Hence, E_i and F_i induce endofunctors of $\mathcal{C}/\mathcal{C}'$, showing that $\mathcal{C}/\mathcal{C}'$ admits a categorical $\mathfrak{sl}_2$-action. By [**R1**, Theorem 5.22], $\rho_{i,\lambda}$ is invertible on $\pi L(b) \in \mathrm{ob}\,\mathcal{C}/\mathcal{C}'$. It just remains to invoke Lemma 2.13 to deduce that $\rho_{i,\lambda}$ is invertible on $L(b) \in \mathrm{ob}\,\mathcal{C}$ too. In order to check the hypotheses of Lemma 2.13 here, we should note that if X is any functor obtained by taking a finite composition and/or direct sum of the categorification functors E_i and F_i, then $XL(b)$ is finitely generated and cogenerated by Theorem 2.11. Moreover all constituents of $\mathrm{soc}(XL(b))$ and $\mathrm{hd}(XL(b))$ are $\cong L(c)$ for

$c \in \mathbf{B}''$; for example, to see this for the head, the right adjoint Y to X preserves $\mathcal{C}'$ as before, so for $c \in \mathbf{B}'$ we get that $\mathrm{Hom}_{\mathcal{C}}(XL(b), L(c)) \cong \mathrm{Hom}_{\mathcal{C}}(L(b), YL(c)) = 0$.

It remains to show that $\sigma_{ij} (i \neq j)$ is invertible. For this, we appeal to the proof of [**R1**, Theorem 5.25] to get that σ_{ij} is invertible on $E_i^r K$ for all $r \geq 0$ and any irreducible K with $F_i K = 0$; this is a very general result which requires no finiteness assumptions about $\mathcal{C}$ other than integrability. To deduce the invertibility of σ_{ij} on arbitrary objects, we claim that every irreducible object $L \in \mathrm{ob}\,\mathcal{C}$ can be realized as a quotient (resp. subobject) of some such object $E_i^r K$. Using the claim and naturality, the invertibility of σ_{ij} on $E_i^r K$ implies the surjectivity (resp. injectivity) of σ_{ij} on L too. Then we apply Lemma 2.12 to get that σ_{ij} is invertible on arbitary objects. Finally, we must prove the claim. By integrability, there is a unique $r \geq 0$ such that $F_i^r L \neq 0$ but $F_i^{r+1} L = 0$. Then we let K be any irreducible constituent of the socle (resp. head) of $F_i^r L$, so that $K \hookrightarrow F_i^r L$ (resp. $F_i^r L \twoheadrightarrow K$); this relies on the fact that $F_i^r L$ is finitely cogenerated (resp. generated) according to Theorem 2.11. Applying adjointness, we get that there is a non-zero homomorphism $E_i^r K \twoheadrightarrow L$ (resp. $L \hookrightarrow E_i^r K$), as required. □

REMARK 4.28. The proof of Theorem 4.27 relies ultimately on [**CR**, Theorem 5.27], in which nilpotency is certainly assumed. However, we expect that this result can be generalized, so that the nilpotency assumption in the statement of Theorem 4.27 (and in the remainder of this subsection) should actually be unnecessary.

CONSTRUCTION 4.29. Let $\mathcal{C}$ be a locally Schurian category admitting a nilpotent categorical action. Fix a set X_0 indexing finitely generated projective objects $(P_x)_{x \in X_0}$ such that each P_x belongs to some weight subcategory of $\mathcal{C}$. For $n \geq 1$, define X_n and $(P_x)_{x \in X_n}$ recursively by letting X_n consist of the symbols $e_i x, f_i x$ for all $x \in X_{n-1}$ and $i \in I$, and setting $P_{e_i x} := E_i P_x, P_{f_i x} := F_i P_x$. Let $X := \bigsqcup_{n \geq 0} X_n$. We assume further that $(P_x)_{x \in X}$ is a projective generating family for $\mathcal{C}$. Having made this choice, we can define a finite-dimensional 2-representation $\mathcal{A}$ as follows.

- Let $\mathcal{A}$ be the finite-dimensional category with object set X, $\mathrm{Hom}_{\mathcal{A}}(x, y) := \mathrm{Hom}_{\mathcal{C}}(P_x, P_y)$, and composition induced by composition in $\mathcal{C}$. Note that $\mathcal{A} = \coprod_{\lambda \in P} \mathcal{A}_\lambda$ where $\mathcal{A}_\lambda$ is the full subcategory generated by $X_\lambda := \{x \in X \mid P_x \in \mathrm{ob}\,\mathcal{C}_\lambda\}$.
- Let $\bar{E}_i, \bar{F}_i : \mathcal{A} \to \mathcal{A}$ be the endofunctors defined on objects by $\bar{E}_i x := e_i x, \bar{F}_i x := f_i x$. On morphisms, $\bar{E}_i$ and $\bar{F}_i$ are defined by applying the given categorification functors E_i and F_i in $\mathcal{C}$.
- Let $\bar{x}_i : \bar{E}_i \Rightarrow \bar{E}_i, \bar{\tau}_{ij} : \bar{E}_i \bar{E}_j \Rightarrow \bar{E}_j \bar{E}_i, \bar{\eta}_i : 1_{\mathcal{A}} \Rightarrow \bar{F}_i \bar{E}_i$ and $\bar{\varepsilon}_i : \bar{E}_i \bar{F}_i \Rightarrow 1_{\mathcal{A}}$ be the natural transformations obtained by restricting x_i, τ_{ij}, η_i and ε_i in the obvious way.

This produces all of the data required by Definition 4.1(M1)–(M4). The final axiom (M5) is satisfied thanks to Theorem 4.27.

We leave it as an exercise for the reader to show that Constructions 4.26 and 4.29 are quasi-inverses in the appropriate sense. In particular, if one starts with $\mathcal{C}$ equipped with a nilpotent categorical action, applies Construction 4.29 to obtain $\mathcal{A}$, then applies Construction 4.26 to define a categorical action on Mod-$\mathcal{A}$, then $\mathcal{C}$ and Mod-$\mathcal{A}$ are strongly equivariantly equivalent.

4.4. Associated crystals. Finally we recall a definition of Kashiwara; e.g. see [**K2**].

DEFINITION 4.30. A *normal crystal* is a set $\mathbf{B}$ with a decomposition $\mathbf{B} = \bigsqcup_{\lambda \in P} \mathbf{B}_\lambda$, plus *crystal operators* $\tilde{e}_i, \tilde{f}_i : \mathbf{B} \to \mathbf{B} \sqcup \{0\}$ for each $i \in I$ satisfying the following axioms:

(C1) for every $\lambda \in P$, the crystal operator $\tilde{e}_i$ restricts to a map $\mathbf{B}_\lambda \to \mathbf{B}_{\lambda+\alpha_i} \sqcup \{0\}$;
(C2) for $b \in \mathbf{B}$, we have that $\tilde{e}_i(b) = b' \neq 0$ if and only if $\tilde{f}_i(b') = b \neq 0$;
(C3) for every $b \in \mathbf{B}$, there is an $r \in \mathbb{N}$ such that $\tilde{e}_i^r(b) = \tilde{f}_i^r(b) = 0$.

For each i, define functions $\varepsilon_i, \varphi_i : \mathbf{B} \to \mathbb{N}$ by

$$\varepsilon_i(b) = \max\{r \in \mathbb{N} \mid \tilde{e}_i^r(b) \neq 0\}, \qquad \varphi_i(b) = \max\{r \in \mathbb{N} \mid \tilde{f}_i^r(b) \neq 0\}.$$

Then we also require that

(C4) $\varphi_i(b) - \varepsilon_i(b) = \langle h_i, \lambda \rangle$ for each $b \in \mathbf{B}_\lambda$ and $i \in I$.

The following theorem is essentially due to Chuang and Rouquier [**CR**], but it has its origins in [**GV, G**]. It shows that every nilpotent categorical action has a canonical *associated crystal.*

THEOREM 4.31. *Suppose that we are given a nilpotent categorical action on a locally Schurian category $\mathcal{C}$ with irreducible objects $\{L(b) \mid b \in \mathbf{B}\}$ as in Definition 4.25. There is a unique structure of normal crystal on $\mathbf{B} = \bigsqcup_{\lambda \in P} \mathbf{B}_\lambda$ such that*

(1) $\tilde{e}_i b \neq 0 \Leftrightarrow E_i L(b) \neq 0 \Rightarrow \operatorname{soc}(E_i L(b)) \cong \operatorname{hd}(E_i L(b)) \cong L(\tilde{e}_i b)$;
(2) $\tilde{f}_i b \neq 0 \Leftrightarrow F_i L(b) \neq 0 \Rightarrow \operatorname{soc}(F_i L(b)) \cong \operatorname{hd}(F_i L(b)) \cong L(\tilde{f}_i b)$.

Moreover, the following hold for any $b \in \mathbf{B}$, $i \in I$ and $0 \leq n \leq m := \varepsilon_i(b)$:

(3) $E_i^{(n)} L(b)$ *has irreducible socle and head both isomorphic to* $L(\tilde{e}_i^n b)$;
(4) $[E_i^{(n)} L(b) : L(\tilde{e}_i^n b)] = \binom{m}{n}$, *and all irreducible subquotients of* $E_i^{(n)} L(b)$ *other than* $L(\tilde{e}_i^n b)$ *are of the form* $L(c)$ *for* c *with* $\varepsilon_i(c) < m - n$;
(5) the natural action of $Z(NH_n)$ *on* $E_i^{(n)} L(b)$ *induces an isomorphism*

$$\operatorname{Sym}_n / \langle \mathrm{h}_{m-n+1}, \mathrm{h}_{m-n+2}, \ldots, \mathrm{h}_m \rangle \xrightarrow{\sim} \operatorname{End}_{\mathcal{C}}(E_i^{(n)} L(b)),$$

where h_r *denotes the* r*th elementary symmetric polynomial in* $X_1, \ldots, X_n$.

Analogous statements to (3)–(5) with E *and* ε *replaced by* F *and* φ *also hold.*

PROOF. Note this only involves some fixed i, so we are reduced immediately to the case that $\mathfrak{g}$ is of rank one. In the Artinian case, it suffices to work in the Schurian category fg$\mathcal{C}$, and then everything that we need is a consequence of [**CR**, Proposition 5.20] and the construction of [**R1**, Theorem 5.22]. To give a little more detail, *loc. cit.* shows that $E_i L(b)$ is either zero, or it has irreducible head and socle which are isomorphic, and similarly for $F_i L(b)$. Hence we can use (1)–(2) to define $\tilde{e}_i, \tilde{f}_i : \mathbf{B} \to \mathbf{B} \sqcup \{0\}$. The axiom (C1) is clear, while (C2) follows by an adjunction argument. Temporarily redefining $\varepsilon_i(b)$ from $\varepsilon_i(b) := \max\{r \in \mathbb{N} \mid E_i^{(r)} L(b) \neq 0\}$, we get that properties (3) and (4) hold by [**CR**, Proposition 3.20] again. Using them, an easy induction on $\varepsilon_i(b)$ shows that (C3) holds and that $\varepsilon_i(b)$ agrees with the function from Definition 4.30. Similarly,

$$\phi_i(b) := \max\{n \in \mathbb{N} \mid F_i^{(n)} L(b) \neq 0\} = \max\{n \in \mathbb{N} \mid \tilde{f}_i^n b \neq 0\}.$$

Now we can establish the final axiom (C4). Suppose that $b \in \mathbf{B}_\lambda$ and set $c := \tilde{e}_i^m b$. We have that $\varepsilon_i(c) = 0$, hence $E_i L(c) = 0$. Thus, in the Grothendieck group, the class of $L(c)$ is an $\mathfrak{sl}_2$-highest weight vector. By $\mathfrak{sl}_2$-theory, we deduce that $\phi_i(c) = \langle h_i, \lambda + m\alpha_i\rangle$. Hence, $\phi_i(b) - \varepsilon_i(b) = \phi_i(c) - 2m = \langle h_i, \lambda\rangle$ as required. Finally, for (5), the proof of [**CR**, Proposition 3.20] shows that the natural action of NH_n on $E_i^n L(b)$ induces an isomorphism $NH_n/\langle X_1^m\rangle \xrightarrow{\sim} \mathrm{End}_{\mathcal{C}}(E_i^n L(b))$. By an elementary relation chase (omitted), the two-sided ideal of NH_n generated by X_1^m is also generated by $\mathrm{h}_{m-n+1}, \dots, \mathrm{h}_m$. Recalling that NH_n is a matrix algebra over its center Sym_n, we deduce on truncating with the idempotent $\pi_{i,n}$ that $\mathrm{End}_{\mathcal{C}}(E_i^{(n)} L(b)) \cong \mathrm{Sym}_n / \langle \mathrm{h}_{m-n+1}, \dots, \mathrm{h}_m\rangle$.

To extend the result to the general locally Schurian case, we make a reduction similar to the one made in the second paragraph of the proof of Theorem 4.27. Fix $b \in \mathbf{B}$ and define $\mathbf{B}'', \mathbf{B}', \mathcal{C}'$ and the quotient functor $\pi : \mathcal{C} \to \mathcal{C}/\mathcal{C}'$ exactly as there. As we explained already, E_i and F_i induce endofunctors of $\mathcal{C}/\mathcal{C}'$, hence giving a categorical $\mathfrak{sl}_2$-action on $\mathcal{C}/\mathcal{C}'$, which is finite. Moreover all $E_i^{(n)} L(b)$ and $F_i^{(n)} L(b)$ are finitely generated and cogenerated, and their socles and heads have constituents only of the form $L(c)$ for $c \in \mathbf{B}''$. So we can use Lemma 2.13 to transport the results from $\mathcal{C}/\mathcal{C}'$ established in the previous paragraph to $\mathcal{C}$, and the general result follows. Perhaps the only statement that requires additional comment is the second assertion of (4). For this, the other properties imply that $E_i^{(m-n)}$ annihilates all composition factors $L(c)$ of $E_i^{(n)} L(b)$ different from $L(\tilde{e}_i^n b)$, hence we get that $\varepsilon_i(c) < m - n$. □

REMARK 4.32. By a classical result, the algebra $\mathrm{Sym}_n / \langle \mathrm{h}_{m-n+1}, \mathrm{h}_{m-n+2}, \dots, \mathrm{h}_m\rangle$ in Theorem 4.31(5) is isomorphic to the cohomology of the Grassmannian $\mathrm{Gr}_{n,m}$. This is explained in [**CR**, §3.3.2].

REMARK 4.33. It is interesting to consider what happens in Theorem 4.31 if the nilpotency assumption is dropped. In general, one still obtains a crystal structure on $\mathbf{B}$, but for a certain unfurling $\widetilde{\mathfrak{g}}$ of $\mathfrak{g}$ in the sense of [**W3**]. This is a consequence of the isomorphism theorem established in [**W3**, §3].

The following well-known corollary is a first application; again this argument appeared already in a special case in [**G**].

COROLLARY 4.34. *For $\kappa \in P^+$, the Grothedieck group $K_0(\dot{\mathcal{L}}_{\min}(\kappa))$ is the $\mathbb{Z}$-span of the vectors $f_{i_n}^{(r_n)} \cdots f_{i_1}^{(r_1)}[1_\kappa]$ for $i_1, \dots, i_n \in I$ and $r_1, \dots, r_n \geq 1$.*

PROOF. We apply Theorem 4.31 with $\mathcal{C} := \text{Mod-}\mathcal{L}_{\min}(\kappa)$; cf. Construction 4.26. Let M be the span of all the vectors $f_{i_n}^{(r_n)} \cdots f_{i_1}^{(r_1)}[1_\kappa]$. Proceeding by downward induction on weight, consider some $\lambda < \kappa$. We need to show for each $b \in \mathbf{B}_\lambda$ that $[P(b)] \in M$. Pick i so that $m := \varepsilon_i(b) \neq 0$. Using Theorem 4.31 and an argument with adjunctions, one shows that $F_i^{(m)} P(\tilde{e}_i^m(b)) \cong P(b) \oplus (*)$ where $(*)$ is a direct sum of projectives of the form $P(b')$ for $b' \in \mathbf{B}_\lambda$ with $\varepsilon_i(b') > \varepsilon_i(b)$. By downward induction on $\varepsilon_i(b)$, we may assume that all of these $[P(b')]$ lie in M. Hence, we get that $[P(b)] \in M$ too. □

The proof of Corollary 4.34 implicitly uses the defining property of a *dual perfect basis* from [**KKKS**, Definition 4.2]. In fact, Theorem 4.31 easily implies for any nilpotent locally Schurian categorical action that $\{[P(b)] \mid b \in \mathbf{B}\}$ is a dual perfect basis for $\mathbb{C} \otimes_{\mathbb{Z}} K_0(\mathrm{p}\mathcal{C})$. In particular, we recover the following well-known result on appealing also to [**KKKS**, Theorem 6.1]; this was originally proved in [**LV**] by a different method.

COROLLARY 4.35. *For $\kappa \in P^+$, the crystal associated to the minimal categorification $\mathcal{L}_{\min}(\kappa)$ is a copy of Kashiwara's highest weight crystal $\mathbf{B}(\kappa)$.*

REMARK 4.36. If $\mathcal{C}$ is Artinian, one can also show that $\{[L(b)] \mid b \in \mathbf{B}\}$ is a *perfect basis* for the complexified Grothendieck group of the Schurian category fg$\mathcal{C}$ in the (older) sense of [**BeK**, Definition 5.49]. This was observed originally by Shan [**S**, Proposition 6.2]. Combined with [**BeK**, Theorem 5.37] (in place of [**KKKS**, Theorem 6.1]), Corollary 4.35 may also be deduced from this; cf. [**K**, Remark 10.3.6]. However, perfect bases are not a natural thing to consider in the locally Schurian setup: in general it is not even clear that E_i and F_i send irreducible objects to objects of finite length.

REMARK 4.37. It is natural to expect that the crystal associated to $\mathcal{L}_{\min}(\kappa'|\kappa)$ is Kashiwara's tensor product $\mathbf{B}'(\kappa') \otimes \mathbf{B}(\kappa)$ of the lowest weight crystal $\mathbf{B}'(\kappa')$ with the highest weight crystal $\mathbf{B}(\kappa)$. We hope to prove this in subsequent work using some of Losev's techniques from [**L**].

References

[AF] F. W. Anderson and K. R. Fuller, *Rings and categories of modules*, 2nd ed., Graduate Texts in Mathematics, vol. 13, Springer-Verlag, New York, 1992. MR1245487

[A] S. Ariki, *On the decomposition numbers of the Hecke algebra of $G(m, 1, n)$*, J. Math. Kyoto Univ. **36** (1996), no. 4, 789–808. MR1443748

[BW] H. Bao and W. Wang, Canonical bases in tensor products revisited, to appear in *Amer. J. Math.*.

[BHLW] A. Beliakova, K. Habiro, A. D. Lauda, and B. Webster, *Cyclicity for categorified quantum groups*, J. Algebra **452** (2016), 118–132, DOI 10.1016/j.jalgebra.2015.11.041. MR3461059

[BeK] A. Berenstein and D. Kazhdan, *Geometric and unipotent crystals. II. From unipotent bicrystals to crystal bases*, Quantum groups, Contemp. Math., vol. 433, Amer. Math. Soc., Providence, RI, 2007, pp. 13–88, DOI 10.1090/conm/433/08321. MR2349617

[BFK] J. Bernstein, I. Frenkel, and M. Khovanov, *A categorification of the Temperley-Lieb algebra and Schur quotients of $U(\mathfrak{sl}_2)$ via projective and Zuckerman functors*, Selecta Math. (N.S.) **5** (1999), no. 2, 199–241, DOI 10.1007/s000290050047. MR1714141

[BZ] I. N. Bernstein and A. V. Zelevinsky, *Induced representations of reductive $\mathfrak{p}$-adic groups. I*, Ann. Sci. École Norm. Sup. (4) **10** (1977), no. 4, 441–472. MR0579172

[B1] J. Brundan, *Kazhdan-Lusztig polynomials and character formulae for the Lie superalgebra $\mathfrak{gl}(m|n)$*, J. Amer. Math. Soc. **16** (2003), no. 1, 185–231, DOI 10.1090/S0894-0347-02-00408-3. MR1937204

[B2] J. Brundan, *Quiver Hecke algebras and categorification*, Advances in representation theory of algebras, EMS Ser. Congr. Rep., Eur. Math. Soc., Zürich, 2013, pp. 103–133. MR3220535

[B3] J. Brundan, *On the definition of Kac-Moody 2-category*, Math. Ann. **364** (2016), no. 1-2, 353–372, DOI 10.1007/s00208-015-1207-y. MR3451390

[BCNR] J. Brundan, J. Comes, D. Nash and A. Reynolds, A basis theorem for the oriented Brauer category and its cyclotomic quotients, to appear in *Quantum Top.*.

[BE1] J. Brundan and A. Ellis, Monoidal supercategories; `arXiv:1603.05928`.

[BE2] J. Brundan and A. Ellis, Super Kac-Moody 2-categories; in preparation.

[BK] J. Brundan and A. Kleshchev, *On translation functors for general linear and symmetric groups*, Proc. London Math. Soc. (3) **80** (2000), no. 1, 75–106, DOI 10.1112/S0024611500012132. MR1719176

[BLW] J. Brundan, I. Losev and B. Webster, Tensor product categorifications and the super Kazhdan-Lusztig conjecture, to appear in *Int. Math. Res. Notices.*

[CL] S. Cautis and A. D. Lauda, *Implicit structure in 2-representations of quantum groups*, Selecta Math. (N.S.) **21** (2015), no. 1, 201–244, DOI 10.1007/s00029-014-0162-x. MR3300416

[CR] J. Chuang and R. Rouquier, *Derived equivalences for symmetric groups and $\mathfrak{sl}_2$-categorification*, Ann. of Math. (2) **167** (2008), no. 1, 245–298, DOI 10.4007/annals.2008.167.245. MR2373155

[DVV] O. Dudas, M. Varagnolo and E. Vasserot, Categorical actions on unipotent representations of finite classical groups; `arXiv:1603.00742`.

[EHS] I. Entova-Aizenbud, V. Hinich and V. Serganova, Deligne categories and the limit of categories $\mathrm{Rep}(GL(m|n))$; `arXiv:1511.07699`.

[EGNO] P. Etingof, S. Gelaki, D. Nikshych, and V. Ostrik, *Tensor categories*, Mathematical Surveys and Monographs, vol. 205, American Mathematical Society, Providence, RI, 2015. MR3242743

[FK] B. Ford and A. S. Kleshchev, *A proof of the Mullineux conjecture*, Math. Z. **226** (1997), no. 2, 267–308, DOI 10.1007/PL00004340. MR1477629

[F] P. Freyd, Abelian categories, *Reprints in Theory and Applications of Categories* **3** (2003), 23–164.

[G] I. Grojnowski, Affine $\mathfrak{sl}_p$ controls the representation theory of the symmetric group and related Hecke algebras; `arXiv:math.RT/9907129`.

[GV] I. Grojnowski and M. Vazirani, *Strong multiplicity one theorems for affine Hecke algebras of type A*, Transform. Groups **6** (2001), no. 2, 143–155, DOI 10.1007/BF01597133. MR1835669

[Kac] V. G. Kac, *Infinite-dimensional Lie algebras*, 3rd ed., Cambridge University Press, Cambridge, 1990. MR1104219

[KKKS] B. H. Kahng, S.-J. Kang, M. Kashiwara, and U. R. Suh, *Dual perfect bases and dual perfect graphs* (English, with English and Russian summaries), Mosc. Math. J. **15** (2015), no. 2, 319–335, 405. MR3427426

[Kam] J. Kamnitzer, *Categorification of Lie algebras (after Rouquier, Khovanov-Lauda, ...)*, Astérisque **361** (2014), Exp. No. 1072, x, 397–419. MR3289289

[KK] S.-J. Kang and M. Kashiwara, *Categorification of highest weight modules via Khovanov-Lauda-Rouquier algebras*, Invent. Math. **190** (2012), no. 3, 699–742, DOI 10.1007/s00222-012-0388-1. MR2995184

[K1] M. Kashiwara, *Crystalizing the q-analogue of universal enveloping algebras*, Comm. Math. Phys. **133** (1990), no. 2, 249–260. MR1090425

[K2] M. Kashiwara, *On crystal bases*, Representations of groups (Banff, AB, 1994), CMS Conf. Proc., vol. 16, Amer. Math. Soc., Providence, RI, 1995, pp. 155–197. MR1357199

[K] M. Khovanov, *A functor-valued invariant of tangles*, Algebr. Geom. Topol. **2** (2002), 665–741, DOI 10.2140/agt.2002.2.665. MR1928174

[KL1] M. Khovanov and A. D. Lauda, *A diagrammatic approach to categorification of quantum groups. I*, Represent. Theory **13** (2009), 309–347, DOI 10.1090/S1088-4165-09-00346-X. MR2525917

[KL2] M. Khovanov and A. D. Lauda, *A diagrammatic approach to categorification of quantum groups II*, Trans. Amer. Math. Soc. **363** (2011), no. 5, 2685–2700, DOI 10.1090/S0002-9947-2010-05210-9. MR2763732

[KL3] M. Khovanov and A. D. Lauda, *A categorification of quantum* $\mathrm{sl}(n)$, Quantum Topol. **1** (2010), no. 1, 1–92, DOI 10.4171/QT/1. MR2628852

[K] A. Kleshchev, *Linear and projective representations of symmetric groups*, Cambridge Tracts in Mathematics, vol. 163, Cambridge University Press, Cambridge, 2005. MR2165457

[LLT] A. Lascoux, B. Leclerc, and J.-Y. Thibon, *Hecke algebras at roots of unity and crystal bases of quantum affine algebras*, Comm. Math. Phys. **181** (1996), no. 1, 205–263. MR1410572

[L1] A. D. Lauda, *A categorification of quantum* $\mathrm{sl}(2)$, Adv. Math. **225** (2010), no. 6, 3327–3424, DOI 10.1016/j.aim.2010.06.003. MR2729010

[L2] A. D. Lauda, *An introduction to diagrammatic algebra and categorified quantum* $\mathfrak{sl}_2$, Bull. Inst. Math. Acad. Sin. (N.S.) **7** (2012), no. 2, 165–270. MR3024893

[LV] A. D. Lauda and M. Vazirani, *Crystals from categorified quantum groups*, Adv. Math. **228** (2011), no. 2, 803–861, DOI 10.1016/j.aim.2011.06.009. MR2822211

[L] I. Losev, *Highest weight $\mathfrak{sl}_2$-categorifications I: crystals*, Math. Z. **274** (2013), no. 3-4, 1231–1247, DOI 10.1007/s00209-012-1114-9. MR3078265

[LW] I. Losev and B. Webster, *On uniqueness of tensor products of irreducible categorifications*, Selecta Math. (N.S.) **21** (2015), no. 2, 345–377, DOI 10.1007/s00029-014-0172-8. MR3338680

[Lu1] G. Lusztig, *Canonical bases arising from quantized enveloping algebras*, J. Amer. Math. Soc. **3** (1990), no. 2, 447–498, DOI 10.2307/1990961. MR1035415

[Lu2] G. Lusztig, *Introduction to quantum groups*, Progress in Mathematics, vol. 110, Birkhäuser Boston, Inc., Boston, MA, 1993. MR1227098

[M] B. Mitchell, *Rings with several objects*, Advances in Math. **8** (1972), 1–161. MR0294454

[R1] R. Rouquier, 2-Kac-Moody algebras; `arXiv:0812.5023`.

[R2] R. Rouquier, *Quiver Hecke algebras and 2-Lie algebras*, Algebra Colloq. **19** (2012), no. 2, 359–410, DOI 10.1142/S1005386712000247. MR2908731

[S] P. Shan, *Crystals of Fock spaces and cyclotomic rational double affine Hecke algebras* (English, with English and French summaries), Ann. Sci. Éc. Norm. Supér. (4) **44** (2011), no. 1, 147–182. MR2760196

[W1] B. Webster, *Canonical bases and higher representation theory*, Compos. Math. **151** (2015), no. 1, 121–166, DOI 10.1112/S0010437X1400760X. MR3305310

[W2] B. Webster, Knot invariants and higher representation theory, to appear in *Mem. Amer. Math. Soc.*.

[W3] B. Webster, Unfurling Khovanov-Lauda-Rouquier algebras; `arXiv:1603.06311`.

Department of Mathematics, University of Oregon, Eugene, Oregon 97403
E-mail address: `brundan@uoregon.edu`

Department of Mathematics, University of Oklahoma, Norman, OK 73019-3103
E-mail address: `ndavidson@math.ou.edu`

Contemporary Mathematics
Volume **683**, 2017
http://dx.doi.org/10.1090/conm/683/13723

On the 2-linearity of the free group

Anthony M. Licata

1. Introduction

Let Γ be a group, and $\mathcal{D}$ a category. A weak action of Γ on $\mathcal{D}$ is a map

$$\psi : \Gamma \longrightarrow \mathrm{Aut}(\mathcal{D}), \quad g \mapsto \psi_g,$$

from Γ to the auto-equivalences of $\mathcal{D}$, such that for all $h, g \in \Gamma$, $\psi_g \psi_h \cong \psi_{gh}$. In a genuine action of Γ, one also specifies isomorphisms $\alpha_{g,h} : \psi_g\psi_h \to \psi_{gh}$ and requires them to satisfy an associativity constraint. Though the entirety of this paper is concerned only with a single weak action, the one we consider can be made genuine, and it is worth noting from the outset the importance of genuine actions. If $\Bbbk$ is a field and the category $\mathcal{D}$ is enriched in finite dimensional $\Bbbk$-vector spaces, we say that ψ is a *2-representation* of Γ.

Let $[\mathrm{Aut}(\mathcal{D})]$ denote the group of isomorphism classes of auto-equivalences of $\mathcal{D}$. Given a weak action of Γ on $\mathcal{D}$, the induced map $[\psi] : \Gamma \longrightarrow [\mathrm{Aut}(\mathcal{D})]$ is a group homomorphism. It often happens in examples that the category $\mathcal{D}$ is also triangulated, with Γ acting by triangulated auto-equivalences. We can then restrict even more and ask that the triangulated category $\mathcal{D}$ be finitely-generated, which means that there are finitely many objects of $\mathcal{D}$ with the property that the smallest full triangulated subcategory containing all of them is $\mathcal{D}$ itself. We then arrive at the following definition.

DEFINITION 1. *A group Γ is* **2-linear** *if Γ admits a 2-representation ψ on a finitely-generated $\Bbbk$-linear triangulated category such that the associated group homomorphism $[\psi]$ is injective.*

This notion of a 2-linear group is analogous to the notion of a linear group, which is one that embeds into the group of invertible endomorphisms of a finite-dimensional $\Bbbk$-vector space. In the last twenty years a number of explicit examples of 2-representations of groups on triangulated categories have appeared in representation theory, symplectic and algebraic geometry, and mathematical physics. Often in these examples, however, it is the triangulated category that is of interest, with the group and its 2-representation appearing mainly as organizing objects. Moreover, despite the existence of interesting examples, the class of 2-linear groups remains somewhat understudied as a chapter of combinatorial and geometric group theory. For example, almost nothing is known at present about the group theoretic

2010 *Mathematics Subject Classification.* Primary 16G10; Secondary 18G50, 20J05, 57M05.

consequences of 2-linearity, and it is not at all clear how different the classes of 2-linear and linear groups are from one another. Nevertheless, it appears that the class of 2-linear groups is very rich. In particular, the following groups are either known or conjectured to be 2-linear.

- *Coxeter groups and their braid groups.* Coxeter groups themselves are 2-linear [1] while their groups are known to be 2-linear in some important special cases [**9**, **15**, **16**, **18**, **20**, **26**]. In [**25**], Rouquier conjectures that the 2-representation of an arbitrary Artin-Tits braid group on the homotopy category of Soergel bimodules is faithful.
- *Mapping class groups of surfaces.* The theory of bordered Floer-homology of Lipshitz-Oszvath-Thurston provides faithful 2-representations of mapping class groups of surfaces with boundary [**22**]. As far as we are aware, it is an open problem to show that the mapping class groups of closed surfaces are 2-linear. The question of group-theoretic consequences of 2-linearity was first posed by Lipshitz-Oszvath-Thurston.
- *Fundamental groups of complexified hyperplane arrangements.* Many of the organizing functors (twisting, shuffling, wall-crossing, cross-walling,...) of geometric representation theory and symplectic-algebraic geometry produce 2-representations of these fundamental groups. Of particular interest here are the 2-representations of fundamental groups on the derived category of category $\mathcal{O}$ for a symplectic resolution [**5**, **7**, **8**, **23**]. It is an open question to determine when these 2-representations are faithful.

The n-strand Artin braid group B_n is perhaps the best studied example from the point of view of 2-linearity, and B_n appears as an entry in each of the three classes described above. The existing literature on 2-representations of B_n also motivates the interest in studying explicit faithful 2-representations in detail. For the first non-trivial example of a representation of B_n on a vector space – the Burau representation – is not faithful except for small n. Moreover, although the group B_n is known to be linear by work of Bigelow and Krammer [**6**, **19**], studying B_n via the Bigelow-Krammer representation is difficult. On the other hand, the first non-trivial example of a 2-representation of B_n – the categorification of the Burau representation – is already faithful, as was shown first by Khovanov-Seidel [**18**] and, for $n = 3$, Rouquier-Zimmermann [**26**]. Moreover, much of the rich mathematics related to the braid group is readily visible from the lens of this first faithful 2-representation. (See, for example, [**3**, **9**, **18**, **26**, **28**], each of which contains a explanation of how some structure of interest in the study of the braid group appears in the categorified Burau representation.)

The goal of the current paper is to define and study what we feel is a basic and prototypical example of a faithful 2-representation of a group. We construct a faithful 2-representation of the free group F_n and explore how some of the structure of F_n appears from the homological algebra of this 2-representation. Our underlying category $\mathcal{D}$ is the homotopy category of projective modules over a finite dimensional

[1]The 2-linearity of a Coxeter group W of simply-laced type follows, for example, from work of Frenkel-Khovanov-Schiffmann [**12**]. The underlying triangulated category in the FKS 2-representation is a category of matrix factorizations over a non-commutative zigzag algebra. The Grothendieck group of this 2-representation is the reflection representation of W, which is known to be faithful. One can give a similar construction for other Coxeter groups using matrix factorizations of Soergel bimodules.

$\Bbbk$-algebra $\widehat{A}$, where $\Bbbk$ is a field of characteristic 0. This algebra, which is an example of a *zigzag algebra*, is the Koszul dual of the preprojective algebra of the doubled complete graph (see Section 4.1.) Our main result, which establishes the 2-linearity of F_n, is the following.

THEOREM 1.1. *The action of F_n on $\mathcal{D}$ is faithful.*

The zigzag algebra $\widehat{A}$ has n indecomposable projective modules $P_1, \ldots, P_n$, and the Seidel-Thomas spherical twists in these objects define the action of the standard generators of $\sigma_1, \ldots, \sigma_n$ of F_n. If spherical objects E_1, E_2 have no morphisms between them, then the associated spherical twists Φ_{E_1} and Φ_{E_2} will commute; when there is a one-dimensional space of morphisms between them, on the other hand, the spherical twists will braid. Thus, if the spherical twists in objects $P_1, \ldots, P_n$ are to generate a free action of F_n, the dimensions of morphism spaces between each pair of the generating spherical objects must be at least 2. Our algebra $\widehat{A}$ is constructed so that $\dim \mathrm{Hom}(P_i, P_j) = 2$ for all $i, j = 1, \ldots, n$. From this point of view, our main result Theorem 1.1 below is somewhat analogous to a theorem of Keating [**17**], who proves that if L and L' are two non-quasi-isomorphic Lagrangian spheres in a symplectic manifold M with $\dim \mathrm{Hom}(L, L') = 2$ in the Fukaya category of M, then the Dehn twists in L and L' generate a free froup of rank 2 inside the symplectomorphism group of M.

Our proof of Theorem 1.1 is by ping pong. After establishing Theorem 1.1, we devote the remaining sections to describing how other structure of interest in the mathematics of the free group appears in the representation theory of the zigzag algebra $\widehat{A}$. For instance, we describe is how two different monoids in F_n arise as stabilizers of "non-negative" subcategories of the category $\mathcal{D}$. These non-negative subcategories, in turn, arise from non-negative gradings on the zigzag algebra. For example, one grading on $\widehat{A}$ – the $\widetilde{o}$-grading – gives rise to a subcategory $\mathcal{D}^{\widetilde{o}}_{\geq 0} \subset \mathcal{D}$ with the property that $g \in F_n$ preserves $\mathcal{D}^{\widetilde{o}}_{\geq 0}$ if and only if g is in the standard positive monoid generated by $\sigma_1, \ldots, \sigma_n$. This particular grading is used crucially in our ping pong proof of Theorem 1.1. Another grading on $\widehat{A}$ – the $\vec{o}$-grading – gives rise to a different subcategory $\mathcal{D}^{\vec{o}}_{\geq 0}$ with the property that $g(\mathcal{D}^{\vec{o}}_{\geq 0}) \subset \mathcal{D}^{\vec{o}}_{\geq 0}$ if and only if g is in the dual positive monoid F_n^+ of F_n defined by Bessis in [**4**]. In fact, the ping pong construction used in the proof Theorem 1.1, which uses the baric structure determined by the $\widetilde{o}$-grading, has a dual analog which uses the $\vec{o}$-grading and the Bessis monoid. The appearance of the Bessis monoid in relation to the $\vec{o}$-grading hints at a close relationship between the $\vec{o}$-graded representation theory of $\widehat{A}$ and the topology of the punctured disk (whose fundamental group is the free group). This relationship is partially explored in Section 6 (see, for example, Proposition 6.4).

The categories $\mathcal{D}_{\geq 0}(\widetilde{o})$ and $\mathcal{D}_{\geq 0}(\vec{o})$ are both the non-negative parts of *baric structures* in the sense of Achar-Treumann [**1**]. At the end of the paper use the compatability of monoids and baric structures to give homological interpretations of the standard and Bessis dual word-length metrics on F_n in Theorems 7.1 and 7.2 by equating the word length of $g \in F_n$ with the "baric spreads" of g in its action on $\mathcal{D}$.

A third grading on $\widehat{A}$, the Koszul path-length grading, gives rise to a t-structure $(\mathcal{D}^{\geq 0}, \mathcal{D}^{\leq 0})$. Just as for the baric subcategory $\mathcal{D}_{\geq 0}(\widetilde{o})$, a group element $g \in F_n$ preserves the subcategory $\mathcal{D}^{\geq 0}$ if and only if g is in the standard positive monoid.

However, the metric on the free group determined by this t structure is different than the metrics defined by baric structures, and the little bit we know about it is mentioned at the end of the paper. It would be interesting to give a combinatorial description, without direct reference to the category $\mathcal{D}$, of this exotic metric.

The construction of the zigzag algebra $\widehat{A}$ and the action of F_n on the homotopy category of projective $\widehat{A}$ modules is very similar to a number of earlier constructions of braid and Artin-Tits group actions on representation categories of quivers [**14, 18, 25, 26**]. Analogs of some aspects of this paper should exist for arbitrary Artin-Tits braid groups, and we plan to write about special cases of spherical and right-angled Artin-Tits groups in [**21**] and [**20**], respectively. In the remainder of the introduction we give a more detailed explanation of the contents of the paper.

Contents. In Section 2 we collect some information about the free group to be used later, including the definition of reflections in F_n, the Bessis monoid F_n^+, and the Hurwitz action of B_n on reduced expressions for the Coxeter element $\gamma = \sigma_1 \dots \sigma_n$.

In Section 3, we recall Klein's ping pong lemma and state a dual ping pong lemma using the Bessis monoid F_n^+. The point of both of these lemmas is to formulate conditions which guarantee that an action of F_n on a set is free. A notable point about the dual ping pong lemma is that it only requires studying the positive Bessis monoid F_n^+ and not the entire free group. Our formulation of dual ping pong using F_n^+ is motivated by a similar formulation for the positive monoid of the braid group B_n used by Krammer in [**19**].

In Section 4 we begin by defining the zigzag algebra $\widehat{A}$, introduce several gradings on it, and discuss some basic properties of the homotopy category of (graded) projective $\widehat{A}$ modules. We then define the relevant baric and t-structures on this triangulated category. Finally, the 2-representation $\Psi : F_n \to \mathrm{Aut}(\mathcal{D})$ is defined in Section 4.8.

The proof of Theorem 1.1 is given in Section 5, and it is by ping pong. Crucial in the construction is the non-negative $\widetilde{o}$-grading of the zigzag algebra arising from a specific orientation $\widetilde{o}$ of the doubled complete graph; this grading induces a baric structure on $\mathcal{D}$, and we use the slices of this baric structure to construct the non-empty disjoint sets required for the application of the ping pong lemma. As a consequence, we obtain a homological realization of the standard word-length metric on F_n, which we formulate at the end of the paper in Theorem 7.1.

In Section 6, we revisit the relationship between the free group and the category $\mathcal{D}$ from the point of view of the $\vec{o}$-grading of the zigzag algebra and the associated baric structure on $\mathcal{D}$. Here the appearance of the free group as the fundamental group of the punctured disk becomes important. The main result here is Proposition 6.6, which plays dual ping pong using complexes of projective modules. (Since they are not needed for the proof of Theorem 1.1, a few of the proofs in Section 6 are sketched rather than written out completely.) A corollary of Proposition 6.6 is that we obtain a homological realization of the word-length metric in the simple Bessis generators of F_n; the precise statement is written in Theorem 7.2. An important part of the setup for dual pingpong is the Hurwitz action of the braid group B_n on reduced reflection expressions for $\gamma = \sigma_1 \dots \sigma_n$ and a parallel action of B_n on spherical collections in $\mathcal{D}$ (see Section 6.1). The action of B_n on spherical collections in $\mathcal{D}$ is similar to (and motivated by) constructions of Bridgeland in [**10**]. The final subsection of the paper constructs the exotic metric on F_n, explains the little bit we know about it, and poses the question of describing it combinatorially.

Acknowledgements. We would like to thank Hoel Queffelec for a number of helpful conversations about this paper and our related ongoing collaboration, and the referee for a number of helpful comments. This paper was written for the proceedings of the conference on Categorification in Algebra, Geometry and Physics, held in Corsica in May, 2015 in honor of the 60th birthday of Christian Blanchet. We thank the organizers of that conference for the opportunity to speak about this work there, and the editors of this conference proceedings for their patience.

2. The Free group

In this section we collect some information about the free group. We recall the definitions of reflections in F_n, the positive monoid $F_n^+ \subset F_n$ defined by Bessis [**4**], and the Hurwitz action of the n-strand Artin braid group B_n on reduced reflection expressions for the Coxeter element $\gamma = \sigma_1 \dots \sigma_n$.

2.1. The positive monoid of Bessis. Our presentation in this section closely follows that of [**4**], and we refer the reader there for further information regarding the positive monoid and quasi-Garside structure of F_n. We fix $n+1$ distinct points $x_0, x_1, \dots, x_n \in \mathbb{C}$, and set

$$F_n := \pi_1(\mathbb{C} - \{x_1, \dots, x_n\}, x_0).$$

It is convenient to choose $x_0 = -1$, and x_i for $i \neq 0$ purely imaginary, with

$$-1 < im(x_1) < \cdots < im(x_n) < 1,$$

so that the points $\{x_1, \dots, x_n\}$ lie on a vertical line segment inside the unit disc. The unit circle itself, traversed counterclockwise starting at $x_0 = -1$, defines an element $\gamma \in F_n$. We refer to γ as a *Coxeter element* of F_n, since it's image in the Coxeter group $W_n = F_n/\langle \sigma_i^2 \rangle$ is a Coxeter element of W_n. In terms of the standard generators $\sigma_1, \dots, \sigma_n$, we have

$$\gamma = \sigma_1 \sigma_2 \dots \sigma_n.$$

A *connecting path* is a continuous map $f : [0,1] \to \mathbb{C}$ such that $f(0) = x_0$, $f(1) = x_i$ for some $i = 1, \dots, n$, and $t \neq 1 \implies f(t) \notin \{x_1, \dots, x_n\}$. Given a connecting path f, we may assign an element $r_f \in F_n$ as follows: begin at $f(0) = x_0$, travel close to $f(1)$ via the connecting path, make a counterclockwise turn around a small circle centred at $f(1)$, and then return to x_0 by traversing f backwards. A *γ-reflection* of F_n is an element $r \in F_n$ such that there is a connecting path f such that $r = r_f$. For example, the standard generators $\sigma_1, \dots, \sigma_n$ of F_n are γ-reflections, and the corresponding connecting paths are drawn in Figure 1. (Note that the inverses $\sigma_1^{-1}, \dots, \sigma_n^{-1}$ are *not* γ-reflections.) Another connecting path f given in Figure 2; it corresponds to the γ-reflection $g^{-1}\sigma_2 g$, with $g = \sigma_3^{-1}\sigma_4\sigma_3\sigma_2\sigma_1$.

We denote by F_n^+ the *Bessis monoid*, which is the submonoid of F_n generated by the γ-reflections. The Bessis monoid F_n^+ has a divisibility partial ordering: for $f, g \in F_n^+$,

$$f \leq g \iff \exists y \in F_n^+, fh = g.$$

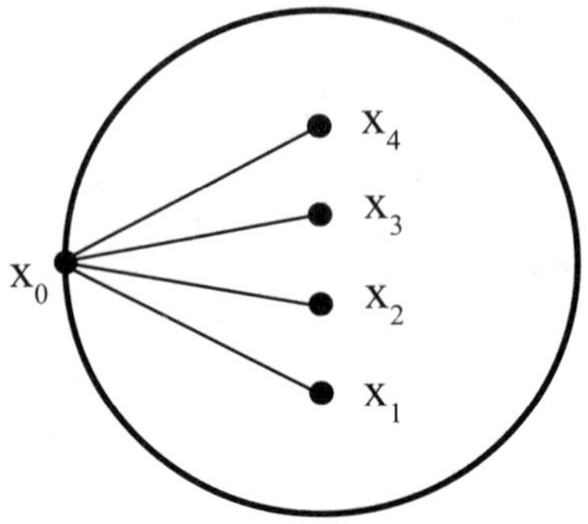

FIGURE 1. The standard coordinate system for F_n.

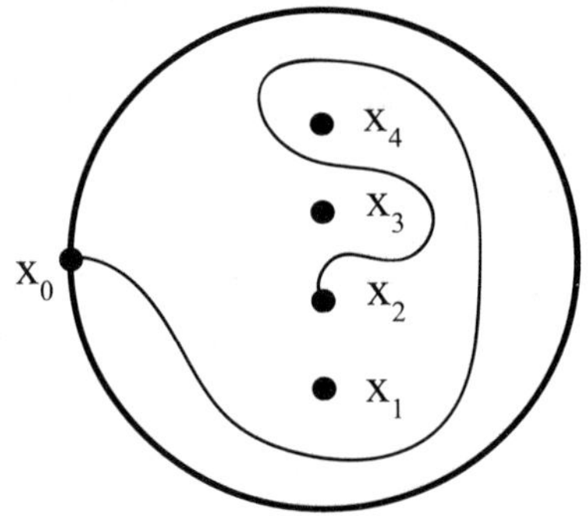

FIGURE 2. The γ-reflection $(\sigma_1^{-1}\sigma_2^{-1}\sigma_3^{-1}\sigma_4^{-1}\sigma_3)\sigma_2(\sigma_3^{-1}\sigma_4\sigma_3\sigma_2\sigma_1)$

In fact, there exists $h \in F_n^+$ with $fh = g$ if and only if there exists $h' \in F_n^+$ with $h'f = g$, so the partial order defined above does not depend on the choice of left or right-divisibility. If $f \leq g$, we say that f *divides* g.

A slightly larger generating set for F_n^+, which includes the γ-reflections and which plays a prominent role later, is the interval in the Bessis monoid between 1 and the Coxeter element. We denote this set by $\mathcal{B}e^+$:

$$\mathcal{B}e^+ := \{g \in F_n^+ : 1 \leq g \leq \gamma\}.$$

We refer to elements of $\mathcal{B}e^+$ as *simple Bessis elements* of F_n^+.

A crucial property of the Bessis monoid F_n^+ is that it is a lattice, that is, has greatest common divisors and least common multiples. In particular, for any element $g \in F_n^+$, there exists a unique left factor $lf(g) \in \mathcal{B}e^+$ such that

- the left factor $lf(g)$ divides g, and
- for all simple Bessis elements $x \in \mathcal{B}e^+$, if x divides g, then x divides $lf(g)$.

The corresponding *left-greedy normal form* of $g \in F_n^+$ is defined to be the unique expression for g as a product

$$g = y_1 y_2 \dots y_k$$

with $y_i \in \mathcal{B}e^+$ and $y_i = lf(y_i y_{i+1} \dots y_k)$ for all $i = 1, \dots, k$.

2.2. The Hurwitz action of B_n on reduced reflection expressions of $\gamma = \sigma_1 \dots \sigma_n$. For any $(g_1, \dots, g_n) \in F_n^{\times n}$, set

$$\tau_i(g_1, \dots, g_n) = (g_1, \dots, g_{i-1}, g_i g_{i+1} g_i^{-1}, g_i, g_{i+2}, \dots, g_n).$$

This defines *the Hurwitz action* of the n-strand Artin braid group B_n on $F_n^{\times n}$. We are interested in the B_n action on n-tuples of γ-reflections whose product is the Coxeter element γ.

DEFINITION 2. *A length k reflection expression for γ is a k-tuple $(r_1, \dots, r_k)$ of γ-reflections whose product $r_1 \dots r_k = \gamma$. Since $k = n$ is the minimal possible length of a reflection expression for γ, we refer to a length n reflection expression for γ as a minimal length reflection expression. We denote by $Red(\gamma)$ the set of minimal length reflection expressions for γ.*

The following theorem is then due to Artin [**2**], and we refer to [**4**] for a proof.

THEOREM 2.1. *The Hurwitz action of the Artin braid group B_n on $Red(\gamma)$ is simple and transitive.*

3. Ping pong and dual ping pong

In this section we recall the version of the ping pong lemma which we will use to prove the faithfulness Theorem 1.1 in Section 5. We also make use of the Bessis monoid F_n^+ to formulate a dual ping pong lemma, which we will use later in Section 6.

3.1. Ping pong. The following lemma, versions of which are ubiquitous in combinatorial and geometric group theory, is usually attributed to Klein.

LEMMA 3.1. *Fix $n \geq 2$. Let Y be a set, and let*

$$g_1, \dots, g_n : Y \to Y$$

be bijections. Let $X_1^+, \dots, X_n^+, X_1^-, \dots X_n^-$ be subsets of Y, and denote by $X \subset Y$ their union. Suppose that

- *the subsets $X_1^+, \dots, X_n^+, X_1^-, \dots X_n^-$ are pairwise disjoint;*
- *there exist points $y_1, \dots, y_n \in Y \setminus X$ such that for $j \neq i$, $g_i^{\pm 1}(y_j) \in X_i^{\pm}$, and*
- *$x \in X \setminus X_i^{\mp} - \implies g_i(x) \in X_i^{\pm}$.*

Then the group homomorphism

$$F_n \to \{\textit{Bijections } g : Y \to Y\}, \quad \sigma_i \mapsto g_i$$

is injective.

PROOF. Let $\sigma = \sigma_{i_k}^{\epsilon_k} \dots \sigma_{i_1}^{\epsilon_1}$ be a reduced word in F_n, so that no σ_i is adjacent to a σ_i^{-1}, and let $g = g_{i_k}^{\epsilon_k} \dots g_{i_1}^{\epsilon_1}$ be the image of σ under the homomorphism $\sigma_i \mapsto g_i$. We must show that g is not the identity map 1_Y.

If we can find a point $y \in Y$ such that $g_{i_1}^{\epsilon_1}(y) \in X_{i_1}^{\epsilon_1}$, then it follows from the assumptions of the lemma that $g(y) \in X_{i_k}^{\epsilon_k}$. So if we can find a y such that $g_{i_1}^{\epsilon_1}(y) \in X_{i_1}^{\epsilon_1}$ but $y \notin X_{i_k}^{\epsilon_k}$, we will have $g(y) \neq y$. Now the assumption on the points $y_j \in Y \setminus X$ guarantees that for $j \neq i_1$,

$$g(y_j) \neq y_j.$$

Thus that $g \neq 1$. □

3.2. Dual Ping pong. The Bessis monoid F_n^+ has the pleasant feature that any element $\beta \in F_n$ can be written

$$\beta = \alpha_1^{-1}\alpha_2,$$

where $\alpha_1, \alpha_2 \in F_n^+$ are in the positive Bessis monoid. This implies the following.

LEMMA 3.2. *Suppose that $\psi : F_n \to \{Bijections\ g : Y \to Y\}$ is an action of F_n on a set Y which distinguishes elements of the Bessis monoid F_n^+. Then ψ is injective.*

PROOF. Suppose that $\psi(\beta) = 1_Y$ for some $\beta \in F_n$. Writing $\beta = \alpha_1^{-1}\alpha_2$ with $\alpha_1, \alpha_2 \in F_n^+$, we see that $\psi(\alpha_2) = \psi(\alpha_1)$. But by assumption, ψ distinguishes elements of F_n^+, so that $\alpha_1 = \alpha_2$. Thus $\beta = 1$. □

Lemma 3.2 shows that one can reduce the question of whether or not an action of the free group is faithful to a question about whether or not the action distinguishes elements of the Bessis monoid. As a result, a "dual" analogue of the ping pong Lemma 3.1 can be formulated using only the simple Bessis elements $\mathcal{B}e^+ = \{w : 1 \leq w \leq \gamma\}$. More precisely, we have the following, which we refer to as a dual ping pong lemma for the free group.

LEMMA 3.3. *Let Y be a set, and let $\psi : F_n \longrightarrow \{Bijections\ g : Y \to Y\}$ be an action. Denote by ψ_β the image of an element $\beta \in F_n$ under this homomorphism. Suppose that there exist non-empty disjoint subsets $\{X_w\}_{w \in \mathcal{B}e^+}$ of Y, such that for all simple Bessis elements $u, w \in \mathcal{B}e^+$,*

$$\psi_u(X_w) \subset X_{lf(uw)}.$$

Then the group homomorphism ψ is injective.

PROOF. Suppose that $\beta \in F_n^+$. Let $\beta = u_1 \ldots u_k$ be the left greedy normal form of β. We will show that the left factor u_1 of β is determined by the bijection ψ_β. Once we show this, it follows by induction that the entire left greedy normal form of β can be read off from the bijection ψ_β; in particular, ψ will distinguish elements of the Bessis monoid F_n^+, and the statement then follows from Lemma 3.2.

To see that u_1 is determined by the bijection ψ_β, choose $x \in X_{u_k}$, where u_k is the rightmost term in the left greedy normal form of β. By the assumptions of the lemma, we have $\psi_\beta(x) \in X_{u_1}$, so that ψ_β determines u_1. □

REMARK 1. *One can also show that in checking the condition*

$$\psi_u(X_w) \subset X_{lf(uw)}$$

for all simple Bessis elements $u, w \in \mathcal{B}e^+$, it suffices to check that $\psi_t(X_w) \subset X_{lf(tw)}$ for all $w \in \mathcal{B}e^+$ and t a γ-reflection.

4. A 2-representation the free group

The goal of this section is to define a 2-representation of F_n. The underlying category of the 2-representation will be the homotopy category of projective modules over a finite dimensional algebra $\widehat{A}$. The algebra $\widehat{A}$ is itself a quotient of the the path algebra of a quiver, so we begin by introducing the required quiver notation.

4.1. Quivers associated to the doubled complete graph. Let Q denote the doubled complete graph on n vertices. We number the vertices of Q by the integers $1, 2, \ldots, n$. In the doubled complete graph, for any pair of distinct vertices i, j there are two (unoriented) edges between i and j. To distinguish these two edges from each other, we will name one of them the x-edge from i to j and the other the y edge from i to j. The graph Q for $n = 4$ is depicted below.

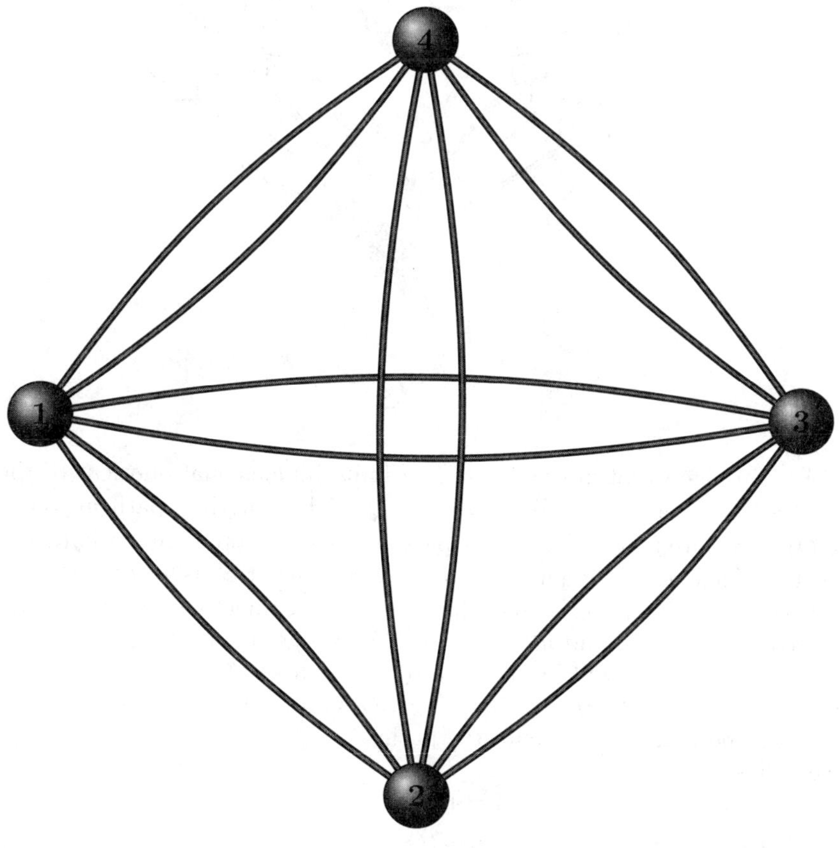

4.2. The zigzag algebra $\widehat{A}$. We now fix a field $\Bbbk$ of characteristic 0. For a quiver X, we denote by $\mathrm{Path}(X)$ the path algebra over $\Bbbk$.

Let $\widehat{Q}$ be the Ginzburg double of Q: in the quiver $\widehat{Q}$, each unoriented edge of Q has been replaced by two oriented edges, one in each direction; thus, for each $i \neq j$ there are now four oriented edges between i and j, two of them pointing $i \to j$ and two pointing $j \to i$. In addition, we put a loop, labeled z, at each vertex. For each pair of vertices i, j with $i < j$, we label the edges from $i \to j$ by x and y, and the edges pointed $j \to i$ by x^* and y^*. The quiver $\widehat{Q}$ for $n = 3$, with its edge labels, is drawn below.

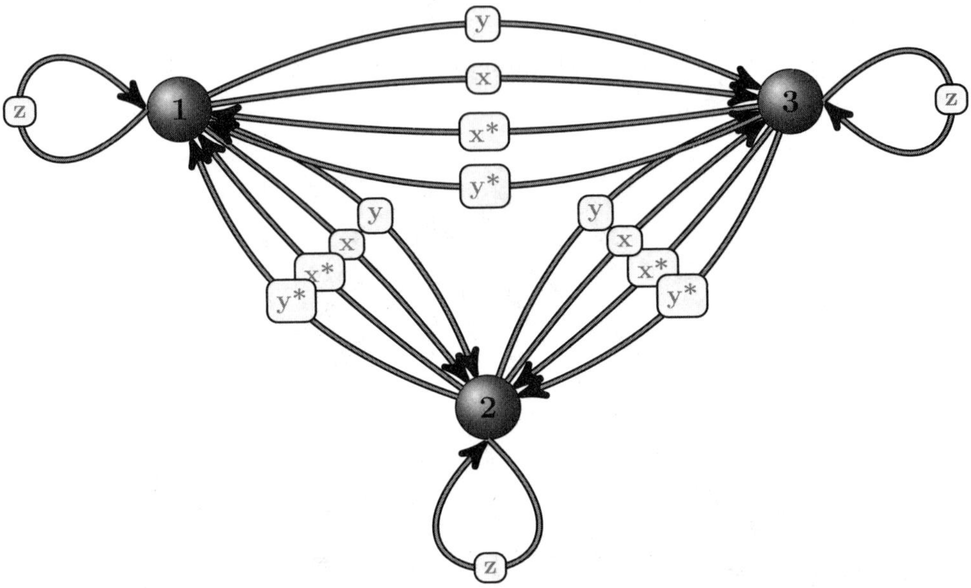

The main algebra of interest here is a finite-dimensional quotient of the path algebra $\mathrm{Path}_{\Bbbk}(\widehat{Q})$ of the Ginzburg quiver $\widehat{Q}$. The length 0 path at vertex i in $\mathrm{Path}_{\Bbbk}(\widehat{Q})$ is denoted by e_i, and this path is an idempotent in $\mathrm{Path}_{\Bbbk}(\widehat{Q})$. When regarded as elements of the path algebra, we somewhat abusively refer to edges of $\widehat{Q}$ by their labels. So, for example, the four edges connecting vertices 1 and 2 will be referred to as x, x^*, y and y^*, as will the four edges connecting vertices 2 and 3, while each of the loops will be denoted by z. As a result, in order for a string of letters in the alphabet $\{x, y, x^*, y^*, z\}$ to give a well-defined path, we should specify the starting and ending vertices of the path.

We define

$$\widehat{A} = \mathrm{Path}_{\Bbbk}(\widehat{Q})/I,$$

where I is the two-sided ideal in $\mathrm{Path}_{\Bbbk}(\widehat{Q})$ generated by

- all length two paths which start and end at distinct vertices;
- the following linear combinations of length two paths which start and end at the same vertex:

$$e_i a^* b e_i - \delta_{a,b} e_i z e_i, \quad a, b \in \{x, y, x^*, y^*\}.$$

In the last relation it is understood that $x^{**} = x$ and $y^{**} = y$.

A few words about these relations: first, note that as a consequence of the second relation we could have presented the algebra $\widehat{A}$ as a quotient of the path algebra of the quiver without using the loops z at all: for example, the loop $e_1 z e_1$ at vertex 1 is equal to the length 2 path $e_1 x x^* e_1$ starting and ending at vertex 1 and passing through vertex 3; second, since all of the length two relations involving a path $i \to j \to k$ do not depend on the middle vertex j, we have also omitted notation for this middle vertex in the relations above. Finally, a special case of the last relation says that a length two path of the form $e_i x^* y e_i$ starting and ending at vertex i – regardless of which other vertex it travels through – is zero. (Similarly, the length two paths $e_i x y^* e_i$, $e_i y x^* e_i$ and $e_i y^* x e_i$ are zero.) The loop $e_i z e_i$ at vertex i is non-zero, but it follows from relations that any path through i which involves both z and another edge in the quiver must be equal to zero in $\widehat{A}$.

The case $n = 1$ is slightly degenerate, since when $n = 1$ there are no edges in Q; our convention when $n = 1$ is to set $\widehat{A} \cong \Bbbk[z]/z^2$. By analogy with the case of simply-laced quivers considered in [**14**], we call the algebra $\widehat{A}$ the *zigzag algebra* of Q.

The following lemma is immediate from the definition of the algebra $\widehat{A}$. (Essentially, the relations in $\widehat{A}$ were chosen so that the lemma below would hold.)

LEMMA 4.1. *For $i < j$:*

$$e_i \widehat{A} e_j = span_{\Bbbk}\{x, y\} \cong \Bbbk^2,$$
$$e_j \widehat{A} e_i = span_{\Bbbk}\{x^*, y^*\} \cong \Bbbk^2,$$

and

$$e_i \widehat{A} e_i = span_{\Bbbk}\{e_i, z\} \cong \Bbbk^2.$$

In particular, $\widehat{A}$ is a finite-dimensional $\Bbbk$-algebra of dimension $2n^2$.

The left $\widehat{A}$ module $P_i = \widehat{A} e_i$ is indecomposable and projective, and the n modules $\{P_i\}_{i=1}^n$ form a complete collection of pairwise distinct isomorphism classes of indecomposable projective left $\widehat{A}$ modules. From Lemma 4.1, we see that the endomorphism algebra of P_i is given by

$$\mathrm{End}_{\widehat{A}}(P_i, P_i) \cong e_i \widehat{A} e_i \cong \Bbbk[z]/z^2,$$

While all the other hom spaces $\mathrm{Hom}_{\widehat{A}}(P_i, P_j) \cong e_i \widehat{A} e_j$ are also two-dimensional vector spaces with a distinguished basis.

4.3. Gradings on $\widehat{A}$. In the sections to come we will consider several different non-negative gradings on $\widehat{A}$. Of primary importance are

- the path length grading, given by setting

$$\deg_{path}(e_i) = 0, \quad \deg_{path}(z) = 2,$$
$$\deg_{path}(x) = \deg_{path}(x^*) = \deg_{path}(y) = \deg_{path}(y^*) = 1;$$

- the orientation grading $\deg_o$ corresponding to an orientation o of the graph Q. A choice of orientation o of Q is a choice of head and tail for each edge of Q, and choosing such an orientation gives us a quiver Q_o. The quiver Q_o gives rise to a grading on $\widehat{A}$ by setting, for $i < j$,

$$\deg_o(x) = \begin{cases} 1, & \text{if the } x \text{ edge of } Q \text{ between } i \text{ and } j \text{ goes from } i \to j \text{ in } Q_o; \\ 0, & \text{otherwise.} \end{cases}$$

$$\deg_o(y) = \begin{cases} 1, & \text{if the } y \text{ edge of } Q \text{ between } i \text{ and } j \text{ goes from } i \to j \text{ in } Q_o; \\ 0, & \text{otherwise.} \end{cases}$$

$$\deg_o(x^*) = 1 - \deg_o(x);$$
$$\deg_o(y^*) = 1 - \deg_o(y).$$
$$\deg_o(z) = 1.$$

These all non-negative integral gradings on the algebra $\widehat{A}$. For an orientation o of Q, let $o^\vee$ be the opposite orientation. The degree 0 subalgebras of $\widehat{A}$ for various gradings can be described as follows.

- When endowed with the path-length grading, the degree 0 subalgebra of $\widehat{A}$ is semi-simple and isomorphic to $\Bbbk^n = \mathrm{span}_\Bbbk\{e_1, \ldots, e_n\}$.
- When endowed with the o-orientation grading, the degree 0 subalgebra of $\widehat{A}$ is isomorphic to the quotient of the path algebra $\mathrm{Path}_\Bbbk(Q_{o^\vee})$ by the two-sided ideal generated by all length two paths in $Q_{o^\vee}$.

We introduce special notation for the two specific orientations that will play an important role later:

- Let $\widetilde{o}$ denote the orientation of Q obtained by orienting one edge (the x edge) between i and j as an arrow $i \to j$ and orienting the other edge (the y edge) of Q as an arrow $j \to i$. We denote by $\widetilde{Q} = Q_{\widetilde{o}}$ the associated quiver, given by endowing Q with the orientation $\widetilde{o}$. The quiver $\widetilde{Q}$ for $n = 4$ is drawn below.

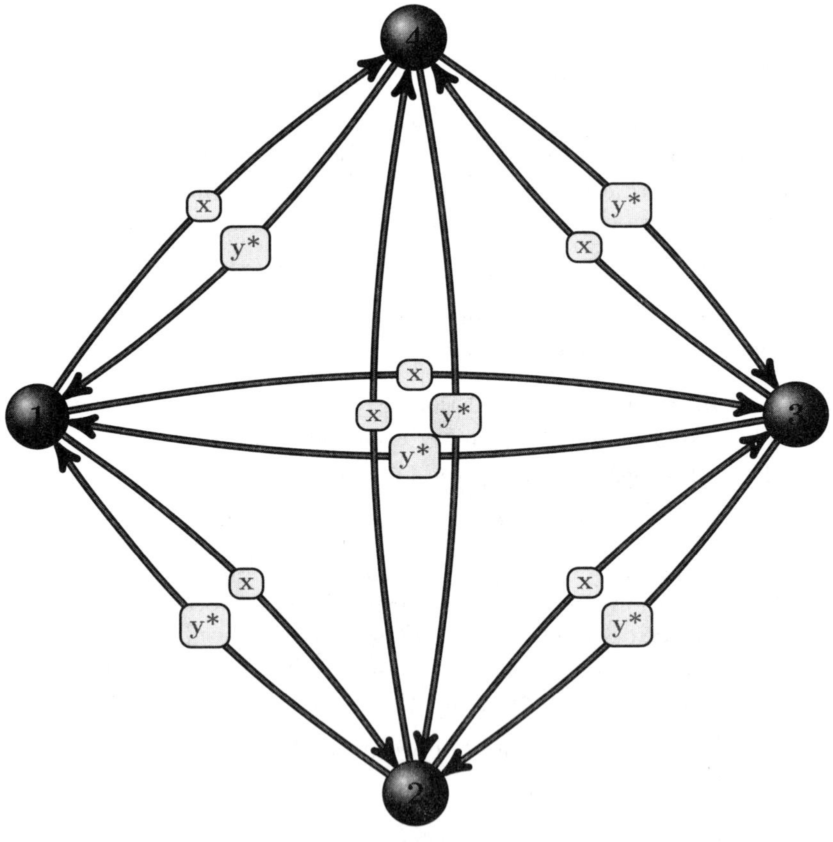

The edge labels on the above quiver specify the natural embedding of $\widetilde{Q}$ as a sub quiver of the Ginzburg double $\widehat{Q}$. In the zigzag algebra $\widehat{A}$, the edges of the Ginzburg double $\widehat{Q}$ which are in the image of this embedding (the x and y^* edges) will have $\widetilde{o}$-degree 1 in the zigzag algebra $\widehat{A}$, whereas the dual edges (x^* and y) will have $\widetilde{o}$-degree 0.

- Let $\vec{o}$ denote the orientation of Q such that, for $i < j$, both edges between i and j are oriented to point $i \to j$ (thus the quiver $\vec{Q} = Q_{\vec{o}}$ has no oriented cycles). The quiver $\vec{Q}$ for $n = 4$ is drawn below.

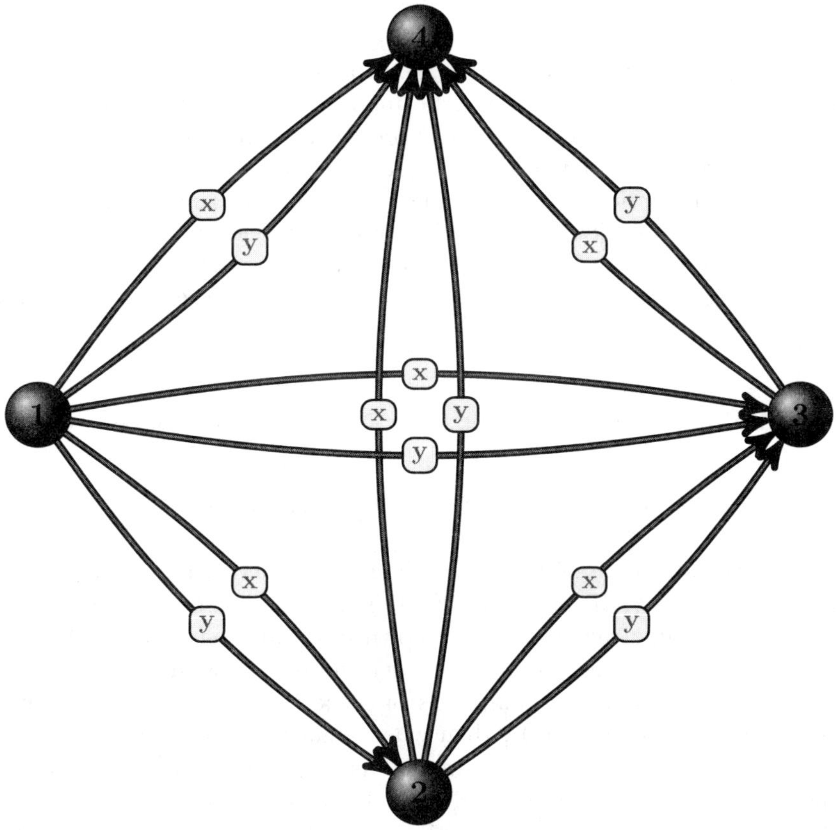

The edge labels on the above quiver specify the embedding of $\vec{Q}$ as a sub quiver of the Ginzburg double $\widehat{Q}$. In the zigzag algebra $\widehat{A}$, the edges of the Ginzburg double $\widehat{Q}$ which are in the image of this embedding (the x and y edges) will have $\vec{o}$-degree 1, whereas the dual edges (x^* and y^*) will have $\vec{o}$-degree 0.

4.4. Categories of complexes. We denote by $\mathcal{D}$ the homotopy category of bounded complexes of finitely-generated projective $\widehat{A}$-modules. An object of $\mathcal{D}$ is a bounded complex of finitely generated projective modules

$$Y = (Y^m, \partial^m), \quad \partial^m : Y^m \to Y^{m+1}, \quad \partial^{m+1} \circ \partial^m = 0.$$

A morphism f from X to Y is a collection of $\widehat{A}$ module maps $f^i : X^i \to Y^i$ intertwining the differentials. Two maps $f, g : X \to Y$ are equal in $\mathcal{D}$ if they are homotopic when regarded as maps of chain complexes. We let $[k]$ denote the auto-equivalence of $\mathcal{D}$ which shifts a complex k degrees to the left:

$$Y[k]^m = Y^{k+m}, \quad \partial_{Y[k]} = (-1)^k \partial_Y.$$

The pair $(\mathcal{D}, [1])$ is a finitely-generated, $\Bbbk$-linear triangulated category.

Given a map of complexes $f : X \to Y$, the cone of f is the complex $X[1] \oplus Y$ with the differential

$$\partial(x, y) = (-\partial_X(x), f(x) + \partial_Y(y)).$$

In the sequel we will often endow $\widehat{A}$ with one of the non-negative gradings defined earlier, in which case we will be interested in the homotopy category of bounded complexes of finitely-generated projective *graded* $\widehat{A}$-modules. Since different gradings will be used in different sections, we will state explicitly at the beginning of each section which grading we are considering. Abusing notation slightly, we will still use $\mathcal{D}$ to denote the homotopy category of bounded complexes of finitely-generated graded modules in each of these sections; $[k]$ always denotes a shift in homological degree; we will always denote by $\langle k \rangle$ the shift in the relevant internal grading.

4.5. Minimal complexes. Let $Y \in \mathcal{D}$ be an indecomposable complex, so that

$$Y \cong Y_1 \oplus Y_2 \in \mathcal{D} \implies Y_1 \cong 0 \text{ or } Y_2 \cong 0.$$

Let $\tilde{Y}$ be a representative of Y in the $\Bbbk$-linear category of bounded complexes of projective $\widehat{A}$ modules. In other words, $\tilde{Y}$ is a bounded complex of $\widehat{A}$ modules which is homotopy equivalent to Y, but not necessarily isomorphic to Y as a chain complex. We say that $\tilde{Y}$ is a *minimal complex* for Y if $\tilde{Y}$ is indecomposable in the additive category of bounded complexes of projective $\widehat{A}$ modules. In particular, if $\tilde{Y}$ is a minimal complex for Y, then when regarded as a complex of projective $\widehat{A}$ modules, $\tilde{Y}$ has no contractible summands. Moreover, the chain groups of a minimal complex are themselves determined up to isomorphism as projective $\widehat{A}$ modules. We thus refer to $\tilde{Y}$ as *the* minimal complex of Y, with the understanding that the chain groups of $\tilde{Y}$ are only determined up to isomorphism. When Y is not necessarily indecomposable, the minimal complex of Y is defined to be the direct sum of the minimal complexes of the indecomposable summands of Y. As every complex Y is homotopy equivalent to a minimal complex, in some arguments it will be convenient to study Y by assuming it is already minimal.

4.6. Slicing complexes. The goal of this section is to briefly describe two ways to use a grading on $\widehat{A}$ to "slice" objects of $\mathcal{D}$ into homogeneous pieces. The first way, which is very well known, is via a bounded t-structure. We assume some familiarity with the notion of a t-structure here, and refer the reader to [**13**] for further information. For now, it suffices to note that a bounded t-structure on $\mathcal{D}$ defines a collection of abelian subcategories $\{\mathcal{D}^k\}_{k\in\mathbb{Z}}$. The subcategory $\mathcal{D}^0$ is called the *heart* of the t-structure, and a bounded t-structure on $\mathcal{D}$ is determined by its heart. Moreover, a collection of full $\Bbbk$-linear subcategories $\{\mathcal{D}^k\}_{k\in\mathbb{Z}}$ determines a t-structure provided the following conditions hold:

(a) for all $k \in \mathbb{Z}$, $\mathcal{D}^k[1] = \mathcal{D}^{k+1}$,
(b) if $X \in \mathcal{D}^k$ and $Y \in \mathcal{D}^l$ with $k > l$, then $\mathrm{Hom}_{\mathcal{D}}(X, Y) = 0$, and

(c) for every nonzero object $E \in \mathcal{D}$ there are integers $m < n$ and a collection of triangles

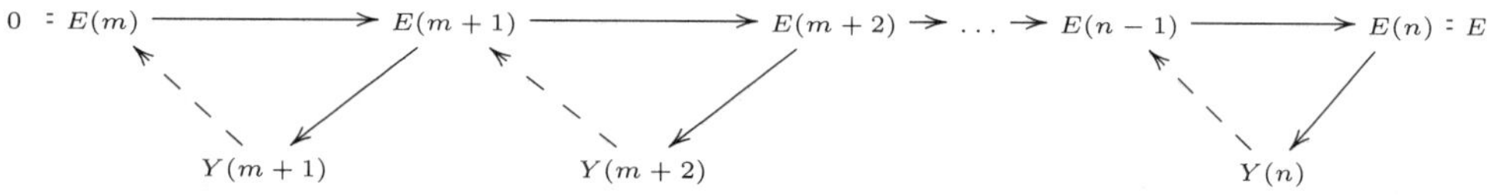

with $Y(k) \in \mathcal{D}^k$ for all k.

We will refer to the terms $Y(k)$ appearing above as the *t-slices* of the object E.

A non-negative grading on the algebra $\widehat{A}$ gives rise to a bounded t-structure on the homotopy category of graded $\widehat{A}$ modules $\mathcal{D}$ whose heart is the category of *linear complexes of projectives.* By definition, a linear complexes of projective modules is a complex $Y \in \mathcal{D}$ such that the term Y^k in homological degree k is a direct sum of projective modules $\oplus P_j\langle k\rangle$, so that the homological degree matches the internal grading shift on the nose. Thus gradings on $\widehat{A}$ give rise to t-structures, and hence to "slicings" of complexes of graded projective $\widehat{A}$ modules. We refer the reader to [**24**] for a detailed discussion of linear complexes of projective modules over finite dimensional non-negatively-graded algebras.

Besides using t-structures, there is another somewhat simpler way to slice an object of $\mathcal{D}$ into homogeneous pieces, which is to ignore the homological degree completely and slice the minimal complex of E using only an internal grading on $\widehat{A}$-modules. The resulting decomposition of can be formalised in the language of *baric structures*, introduced by Achar and Treumann in [**1**].

In more detail, fix an orientation o of Q and let $\mathcal{D}$ denote the homotopy category of complexes of o-graded $\widehat{A}$-modules, where $\widehat{A}$ is endowed with the o-grading defined in Section 4.3. For $k \in \mathbb{Z}$, let $\mathcal{D}_k$ denote the full subcategory of $\mathcal{D}$ consisting of complexes $Y = (Y^m, \partial^m)$ such that, in the minimal complex of Y, the chain groups are direct sums of projective modules whose internal grading shift is exactly k in every homological degree:

$$Y^m \cong \oplus_i P_i^{\oplus n_i}\langle k\rangle \text{ for all } m \in \mathbb{Z}.$$

The subcategories $\{\mathcal{D}_k\}$ define a baric structure in the sense of [**1**]. Since the baric structures we consider are essentially just gradings on projective $\widehat{A}$-modules, rather than give the full abstract definition of a baric structure on a triangulated category here, we prefer to simply note that the subcategories $\{\mathcal{D}_k\}$ defined above satisfy the following properties (all of which follow directly from properties of the non-negative grading on $\widehat{A}$):

(a) for all $k \in \mathbb{Z}$, $\mathcal{D}_k[1] = \mathcal{D}_k$,
(b) if $X \in \mathcal{D}_k$ and $Y \in D_l$ with $k > l$, then $\mathrm{Hom}_{\mathcal{D}}(X, Y) = 0$, and
(c) for every non-zero object $E \in \mathcal{D}$ there are integers $m < n$ and a collection of triangles

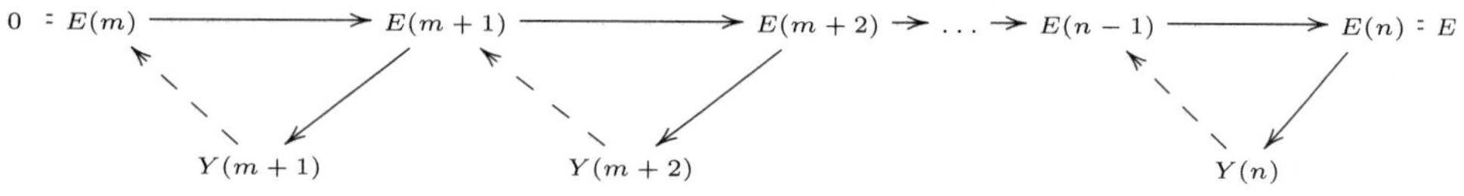

with $Y(k) \in \mathcal{D}_k$ for all k.

We refer to the subcategories $\{\mathcal{D}_k\}$ induced from the o-grading on $\widehat{A}$ as the *o-baric structure* on $\mathcal{D}$, and to the objects $Y(k) \in \mathcal{D}_k$ appearing above as the *o-baric slices* of the object E.

Given an interval $[k,l] \subset \mathbb{Z}$, we set $\mathcal{D}_{[k,l]}$ to be the full subcategory of $\mathcal{D}$ consisting of those complexes Y whose non-zero baric slices live in $\mathcal{D}_m$ for $k \leq m \leq l$. Similarly, $\mathcal{D}^{[k,l]}$ is the full subcategory of $\mathcal{D}$ whose non-zero t-slices live in $\mathcal{D}^m$ for $k \leq m \leq l$. The subcategories $\mathcal{D}_{\leq k}$, $\mathcal{D}_{\geq k}$, $\mathcal{D}^{\leq k}$, $\mathcal{D}^{\geq k}$ are defined similarly.

We have chosen to use the same notation $Y(k)$ for both baric slices and for t-slices; with the exception of a brief point in the last section of paper we will on make use of baric slices, and in any case it will always be made clear in each section which structures – baric or t – we are considering. Despite some similarity in the properties shared by baric and t-structures, there are important differences between the two notions. For instance, in a t-structure, the heart $\mathcal{D}^0$ is an abelian subcategory of the triangulated category $\mathcal{D}$, and this subcategory is not preserved by the homological shift [1]; by contrast, the baric heart $\mathcal{D}_0$ will not in general be an abelian subcategory of $\mathcal{D}$, though in the cases we consider it will be closed under homological shifts.

4.7. Top and bottom slices. Fix an orientation o of $\mathcal{D}$, and let $Y \in \mathcal{D}$. Using the o-baric slices $\{Y(k)\}_{k\in\mathbb{Z}}$, we may define integers $\phi_+(Y)$ by

$$\phi_+(Y) = sup\{k \in \mathbb{Z} : Y(k) \neq 0\}$$

and $\phi_-(Y)$ by

$$\phi_-(Y) = inf\{k \in \mathbb{Z} : Y(k) \neq 0\}.$$

As the minimal complex of Y is a bounded complex of finitely-generated graded $\widehat{A}$-modules, the chain groups of the minimal complex are all supported in finitely many o-degrees, so that slices $Y(k)$ will be non-zero for only finitely many k. We may also define complexes $Y(\phi_+)$ and $Y(\phi_-)$ – the *top and bottom slices of* Y – by

$$Y(\phi_+) := Y(\phi_+(Y)), \quad Y(\phi_-) := Y(\phi_-(Y)).$$

These top and bottom slices will play an essential role in the ping pong and dual ping pong constructions in latter sections.

4.8. The functors Σ_i. In this section we define our main object of interest, the 2-representation

$$\Psi : F_n \longrightarrow \mathrm{Aut}(\mathcal{D}).$$

Define homomorphisms f_i, g_i, $1 \leq i \leq n$ of $(\widehat{A}, \widehat{A})$ bimodules

$$f_i : \widehat{A}e_i \otimes_{\Bbbk} e_i\widehat{A} \longrightarrow \widehat{A},$$

$$g_i : \widehat{A} \longrightarrow \widehat{A}e_i \otimes_{\Bbbk} e_i\widehat{A}$$

by

$$f_i : e_i \otimes e_i \mapsto e_i,$$

$$g_i : 1 \mapsto z \otimes e_i + e_i \otimes z + \sum x \otimes x^* + x^* \otimes x + y \otimes y^* + y^* \otimes y,$$

where z denotes the loop at i and the sum is over all x, x^*, y, y^* edges out of vertex i.

Let Σ_i denote the two-term complex of $(\widehat{A}, \widehat{A})$ bimodules

$$\Sigma_i := \widehat{A} \xrightarrow{g_i} \widehat{A}e_i \otimes_{\Bbbk} e_i\widehat{A},$$

with the term $\widehat{A}$ sitting in homological degree 0. Let Σ_i^{-1} denote the two-term complex of $(\widehat{A}, A)$ bimodules

$$\Sigma_i^{-1} := \widehat{A}e_i \otimes_{\Bbbk} e_i\widehat{A} \xrightarrow{f_i} \widehat{A},$$

with the term $\widehat{A}$ sitting in homological degree 0. The functors

$$Y \mapsto \Sigma_i \otimes_{\widehat{A}} Y, \quad Y \mapsto \Sigma_i^{-1} \otimes_{\widehat{A}} Y,$$

are endofunctors of $\mathcal{D}$; we somewhat abusively denote these endofunctors by Σ_i, Σ_i^{-1}, too.

The proof of the following proposition follows exactly as in Proposition 2.4 of [**18**] (see also [**14**]).

PROPOSITION 4.2. *For each $i = 1, \ldots, n$, the functors Σ_i and Σ_i^{-1} are mutually inverse equivalences of $\mathcal{D}$, i.e., there are functor isomorphisms*

$$\Sigma_i\Sigma_i^{-1} \cong 1_{\mathcal{D}} \cong \Sigma_i^{-1}\Sigma_i.$$

Thus the assignment

$$\Psi : F_n \longrightarrow Aut(\mathcal{D}), \quad \sigma_i \mapsto \Sigma_i, \quad \sigma_i^{-1} \mapsto \Sigma_i^{-1},$$

is a (weak) 2-representation of the free group F_n.

An important point which emerges in the proof of the above proposition is that the functor isomorphisms

$$\Sigma_i\Sigma_i^{-1} \cong 1_{\mathcal{D}} \cong \Sigma_i^{-1}\Sigma_i$$

will not hold if $\mathcal{D}$ is replaced by the additive category of complexes of finitely-generated projective $\widehat{A}$-modules; these isomorphisms only hold after passing to the homotopy category.

When we regard $\widehat{A}$ as a graded algebra, the bimodule map g_i used in the definition of the auto-equivalence Σ_i is not a map of degree 0, but rather a map of degree 1 for the o-grading for an orientation o and a map of degree 2 for the path-length grading. Thus, in order to define auto-equivalences of the homotopy category of graded modules, we need to add a grading shift to one of the bimodules used in the definition of Σ_i. When we consider $\widehat{A}$ as a graded algebra with the o-grading coming from an orientation, the auto-equivalence Σ_i will be defined as tensoring with the complex

$$\widehat{A} \xrightarrow{g_i} (\widehat{A}e_i \otimes_{\Bbbk} e_i\widehat{A})\langle 1 \rangle,$$

while when we consider $\widehat{A}$ as a graded algebra with the path-length grading, Σ_i will be given by tensoring with

$$\widehat{A} \xrightarrow{g_i} (\widehat{A}e_i \otimes_{\Bbbk} e_i\widehat{A})\langle 2 \rangle.$$

With these additional grading shifts, the statement of Proposition 4.2 also holds when Σ_i and Σ_i^{-1} are regarded as endofunctors of the homotopy category of graded projective $\widehat{A}$ modules.

We record here basic example calculations of the complexes $\Sigma_i(P_j)$, taking into account the different possible gradings.

- If we endow $\widehat{A}$ with the symmetric $\widetilde{o}$ grading and let P_i denote the $\widetilde{o}$-graded projective $\widehat{A}$ module with e_i sitting in $\widetilde{o}$-degree 0, then for $i \neq j$,

$$\Sigma_i(P_j) \cong \big(P_j \to P_i\langle 1 \rangle \oplus P_i\big) \text{ and } \Sigma_i(P_i) \cong P_i\langle 1 \rangle[-1].$$

- If we endow $\widehat{A}$ with the ordered $\vec{o}$-grading and now let P_i denote the $\vec{o}$-graded projective $\widehat{A}$ module with e_i sitting in degree 0, then
$$\Sigma_i(P_i) \cong P_i\langle 1\rangle[-1];$$
for $i > j$,
$$\Sigma_i(P_j) \cong \big(P_j \to P_i\langle 1\rangle \oplus P_i\langle 1\rangle\big);$$
for $i < j$,
$$\Sigma_i(P_j) \cong \big(P_j \to P_i \oplus P_i\big).$$
- If we endow $\widehat{A}$ with the path-length grading and let P_i denote the path-graded projective $\widehat{A}$ module with e_i sitting in degree 0, then we have
$$\Sigma_i(P_j) \cong \big(P_j \to P_i\langle 1\rangle \oplus P_i\langle 1\rangle\big) \text{ and } \Sigma_i(P_i) \cong P_i\langle 2\rangle[-1]$$
for all $i \neq j$.

5. 2-linearity via ping pong and the $\widetilde{o}$-grading on $\widehat{A}$

In this section, we prove Theorem 1.1, which says that the group homomorphism
$$\psi : F_n \longrightarrow [\mathrm{Aut}(\mathcal{D})]$$
is injective. *In the entirety of this section, we endow all algebras, modules, and complexes with the $\widetilde{o}$-grading determined by the symmetric orientation $\widetilde{o}$ on the doubled complete graph Q.* In particular, in this section $\mathcal{D}$ will denote the homotopy category of bounded complexes of projective $\widetilde{o}$-graded $\widehat{A}$ modules. The main ingredients in the proof of Theorem 1.1 are the symmetric orientation $\widetilde{o}$, the top and bottom $\widetilde{o}$-baric slices $Y(\phi_+), Y(\phi_-)$ of $Y \in \mathcal{D}$ defined in Section 4.6, and the ping pong Lemma 3.1.

5.1. The sets $X_i^{\pm}$. For $i \in I$, we set
$$\mathfrak{P}_i = \bigoplus_{k\in\mathbb{Z}} P_i[k].$$
Thus, for $m \in \mathbb{Z}$, the projective module $\mathfrak{P}_i\langle m\rangle$ is the direct sum of all possible shifts of P_i which lie in the subcategory $\mathcal{D}_m$ for the $\widetilde{o}$ baric structure on $\mathcal{D}$. Technically, the module $\mathfrak{P}_i$ is not an object of $\mathcal{D}$, as it is not homotopy equivalent to a bounded complex of finitely-generated modules. Nevertheless, the summands of $\mathfrak{P}_i$ are all objects of $\mathcal{D}$, and for $Y \in \mathcal{D}$, the hom space
$$\mathrm{Hom}(Y, \mathfrak{P}_i) = \bigoplus_{k\in\mathbb{Z}} \mathrm{Hom}(Y, P_i[k])$$
is a well-defined finite-dimensional vector space.

The following is the key definition in the proof of Theorem 1.1. For $i = 1, \ldots, n$, we define subsets X_i^+ and X_i^- of objects of $\mathcal{D}$ by declaring that

(1) X_i^+ consists of those indecomposable complexes Y such that
- the top slice $Y(\phi_+)$ is isomorphic to a direct sum of shifts of P_i;
- $\mathrm{Hom}(Y, \mathfrak{P}_j\langle\phi_-(Y)\rangle) = 0 \iff j = i$;

(2) X_i^- consists of those indecomposable complexes Y such that
- The bottom slice $Y(\phi_-)$ is isomorphic to a direct sum of shifts of P_i;
- $\mathrm{Hom}(\mathfrak{P}_j\langle\phi_+(Y)\rangle, Y) = 0 \iff j = i$.

We now check that the sets $X_i^{\pm}$ and the points $z_i = P_i$ satisfy the assumptions of the ping pong Lemma 3.1.

PROPOSITION 5.1. *We have the following:*

(1) *for all* $j \neq i$, $\Sigma_i^{\pm 1}(P_j) \in X_i^{\pm}$;
(2) *the sets* $X_1^-, \dots, X_n^-, X_1^+, \dots, X_n^+$ *are pairwise disjoint;*
(3) *for all* $i \neq j$, $Y \in X_j^- \implies \Sigma_i(Y) \in X_i^+$;
(4) *for all* $i \neq j$, $Y \in X_j^+ \implies \Sigma_i^{-1}(Y) \in X_i^-$;
(5) *for all* i, j, $Y \in X_j^+ \implies \Sigma_i(Y) \in X_i^+$;
(6) *for all* i, j, $Y \in X_j^- \implies \Sigma_i^{-1}(Y) \in X_i^-$

PROOF. The fact that $\Sigma_i(P_j) \in X_i^+$ when $j \neq i$ follows by direct computation, as

$$\Sigma_i(P_j) \cong \big(P_j \to P_i\langle 1\rangle \oplus P_i\big),$$

and the properties required for membership in X_i^+ are easily checked for this complex. Similarly, when $j \neq i$, it is straightforward to see that $\Sigma_i^{-1}(P_j) \in X_i^-$. It is also clear from the definition of the sets X_i^+ that $X_i^+ \cap X_j^+ = \emptyset$ when $i \neq j$, since the top slice of a complex in X_i^+ is a sum of shifts of P_i and the top slice of a complex in X_j^+ is a sum of shifts of P_j. Similarly, $X_i^- \cap X_j^- = \emptyset$ when $i \neq j$.

We now show that $X_i^+ \cap X_j^- = \emptyset$. To see this, note that by definition complexes in X_k^- have a shift of P_k as a direct summand in the bottom $\widetilde{o}$-baric slice. Thus it suffices to note that if $Y \in X_i^+$, then the bottom slice $Y(\phi_-)$ does not have any direct summands isomorphic to a shift of P_k for any $k = 1, \dots, n$. For suppose that $Y(\phi_-)$ has a direct summand $N \cong P_k\langle\phi_-(Y)\rangle$, which for simplicity we assume is in homological degree 0. Since

$$\mathrm{Hom}(P_k\langle\phi_-(Y)\rangle, \mathfrak{P}_i\langle\phi_-(Y)\rangle) \cong \mathrm{Hom}(P_k, \mathfrak{P}_i) \neq 0,$$

it follows that $\mathrm{Hom}(Y, \mathfrak{P}_i\langle\phi_-(Y)\rangle) \neq 0$, too. To see this, let

$$0 \neq f \in \mathrm{Hom}(Y, \mathfrak{P}_i\langle\phi_-(Y)\rangle)$$

be the map on N is given by a $\widetilde{o}$-degree 0 edge of the quiver, extended by 0 to all of Y. (The restriction of f to N is a scalar multiple of either an x^*, a y, or an e_i, depending on whether $i < k$, $i > k$, or $i = k$.) Let $f^* \in \mathrm{Hom}(P_i\langle\phi_-(Y) - 1\rangle, Y)$ denote the dual morphism (either an x, a y^*, or a z) from $P_i\langle\phi_-(Y)\rangle$ into the summand N. Since

$$f \circ f^* = z \in \mathrm{Hom}(P_i\langle\phi_-(Y) - 1\rangle, \mathfrak{P}_i\langle\phi_-(Y)\rangle)$$

is nonzero, it follows that $f \neq 0$. But this implies that $Y \notin X_i^+$. Thus we have shown that the sets

$$X_1^-, \dots, X_n^-, X_1^+, \dots, X_n^+$$

are pairwise disjoint. This completes the proof of (1) and (2).

We now show statement (5):

$$Y \in X_j^+ \implies \Sigma_i(Y) \in X_i^+.$$

To see that $\Sigma_i(Y)(\phi_+)$ is isomorphic to a sum of shifts of P_i, it suffices to note that

$$\phi_+(\Sigma_i(Y)) > \phi_+(Y),$$

as the only complex that can appear as an $\widetilde{o}$-baric slice of $\Sigma_i(Y)$ in a level greater than $\phi_+(Y)$ is a direct sum of shifts of P_i. To show $\phi_+(\Sigma_i(Y)) > \phi_+(Y)$, note that by assumption $Y(\phi_+)$ is isomorphic to a direct sum of shifts of P_j. Let M_j be such a summand, which for simplicity we assume lives in homological degree 0. Now

$$\mathrm{Hom}(P_i\langle\phi_+(Y)\rangle, P_j\langle\phi_+(Y)\rangle) \cong \mathrm{Hom}(P_i, P_j)$$

is clearly non-zero, and if $0 \neq f \in \mathrm{Hom}(P_i\langle\phi_+(Y)\rangle, P_j\langle\phi_+(Y)\rangle)$, the map f' : $P_i\langle\phi_+(Y)\rangle \to Y$ which maps into M_j via f cannot be homotopic to 0. It follows that

$$\mathrm{Hom}(P_i\langle\phi_+(Y)\rangle, Y) \neq 0,$$

and by adjunction

$$\mathrm{Hom}(P_i\langle\phi_+(Y)+1\rangle[-1], \Sigma_i(Y)) \cong \mathrm{Hom}(P_i\langle\phi_+(Y)\rangle, Y) \neq 0,$$

too. This shows that $\phi_+(\Sigma_i(Y)) > \phi_+(Y)$, so that the top level of $\Sigma_i(Y)$ is isomorphic to a direct sum of shifts of P_i.

We must now show that $\Sigma_i(Y)$ satisfies the second requirement for membership in X_i^+, namely, that

$$\mathrm{Hom}(\Sigma_i(Y), \mathfrak{P}_k\langle\phi_-(\Sigma_i(Y))\rangle = 0 \iff k = i.$$

To this end, note that $\mathrm{Hom}(Y, \mathfrak{P}_j\langle\phi_-(Y)\rangle) = 0$ by assumption. Thus the bottom slice $Y(\phi_-)$ does not have a summand of the form $P_k\langle\phi_-(Y)\rangle$ for any k. In particular, we have $\phi_-(\Sigma_i(Y)) = \phi_-(Y)$, since these two numbers would be different precisely when the bottom slice of Y is a direct sum of shifts of P_i. Since $\phi_-(\Sigma_i(Y)) = \phi_-(Y)$, we must show that

$$\mathrm{Hom}(\Sigma_i(Y), \mathfrak{P}_k\langle\phi_-(Y)\rangle = 0 \iff k = i.$$

For this we check the cases $k = i$ and $k \neq i$ separately.

- For $k = i$, we have

$$\mathrm{Hom}(\Sigma_i(Y), \mathfrak{P}_i\langle\phi_-(Y)\rangle = \mathrm{Hom}(Y, \Sigma_i^{-1}\mathfrak{P}_i\langle\phi_-(Y)\rangle) = \mathrm{Hom}(Y, \mathfrak{P}_i\langle\phi_-(Y)-1\rangle) = 0,$$

  since there are no nonzero maps from Y to any complex in $\mathcal{D}_{<\phi_-(Y)}$.
- Now suppose $k \neq i$. We need to show that $\mathrm{Hom}(\Sigma_i(Y), \mathfrak{P}_k\langle\phi_-(Y)\rangle) \neq 0$. We have

$$\begin{aligned}\mathrm{Hom}(\Sigma_i(Y), \mathfrak{P}_k\langle\phi_-(Y)\rangle &\cong \mathrm{Hom}(Y, \Sigma_i^{-1}(\mathfrak{P}_k)\langle\phi_-(Y)\rangle) \\ &\cong \mathrm{Hom}(Y, (\mathfrak{P}_i\langle-1\rangle \oplus \mathfrak{P}_i \to \mathfrak{P}_k)\langle\phi_-(Y)\rangle).\end{aligned}$$

  There are no non-zero $\widehat{A}$-module maps – much less chain maps – from the minimal complex of Y to the lowest term $\mathfrak{P}_i\langle-1+\phi_-(Y)\rangle$ above; from this we see that

$$\mathrm{Hom}(\Sigma_i(Y), \mathfrak{P}_k\langle\phi_-(Y)\rangle \cong \mathrm{Hom}(Y, (\mathfrak{P}_i \to \mathfrak{P}_k)\langle\phi_-(Y)\rangle),$$

  so we must show that the right hand side is non-zero. We do this by finding an explicit non-zero element in this hom space. We consider the the minimal complex of the bottom slice $Y(\phi_-)$ of Y:

$$Y(\phi_-) = \big(\dots Y_{s-1} \to Y_s \to 0 \to \dots\big),$$

  so that Y_s is the rightmost non-zero term in the minimal complex of the bottom slice.

  Suppose first that the chain group Y_s has a summand $M_l \cong P_l\langle\phi_-(Y)\rangle$ with $l \neq i$, and write $Y_s \cong M_l \oplus Y_s'$. Let f be a degree 0 morphism (either an x^* or y) from M_l to $P_i\langle\phi_-(Y)\rangle$. We extend f by 0 to a map on chain groups:

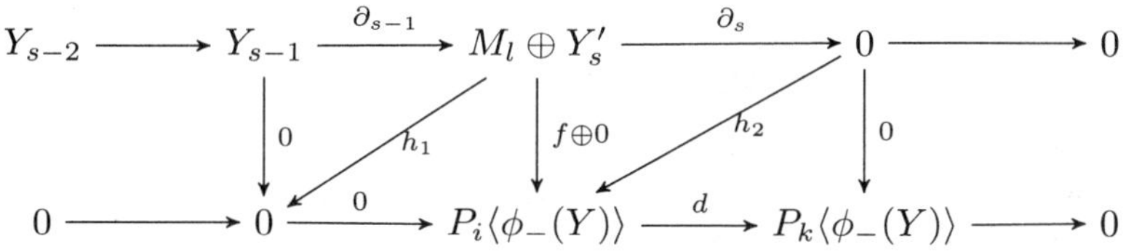

Since the differential d above is of $\widetilde{\sigma}$-baric degree 0 and is an x^* or a y, the relations in the algebra $\widehat{A}$ imply that the composition of $d \circ f$ is zero; by the same reasoning, the composition $f \circ \partial_{s-1} = 0$. Thus the map f may be regarded a well-defined element of $\mathrm{Hom}(Y, (\mathfrak{P}_i \to \mathfrak{P}_k)\langle\phi_-(Y)\rangle)$. But, as Y_s is the rightmost term in the bottom slice of the minimal complex, there is no possible homotopy $h = (h_1, h_2)$ between f and the 0 map. Thus in this case $\mathrm{Hom}(Y, (\mathfrak{P}_i \to \mathfrak{P}_k)\langle\phi_-(Y)\rangle) \neq 0$.

On the other hand, suppose that $Y_s \cong P_i^{\oplus m}\langle\phi_-(Y)\rangle$. We choose a single summand from Y_s and denote it by N_i, so that $Y_s \cong N_i \oplus Y'_s$. Let g be the degree 0 map from N_i to $P_k\langle\phi_-(Y)\rangle$ (so g is either an x^* or a y). This g extends to a chain map from Y to $(\mathfrak{P}_i \to \mathfrak{P}_k)\langle\phi_-(Y)\rangle$, which we claim cannot be homotopic to 0.

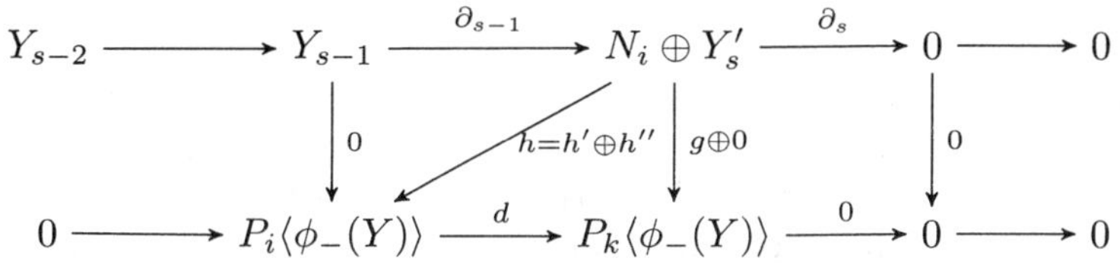

To see this, note first that the only possible components to use for a homotopy $h = (h', h'')$ must be scalar multiples of the $\widetilde{\sigma}$-degree 0 idempotent e_i. Now, Y'_s is isomorphic to a direct sum of $P_i\langle\phi_-(Y)\rangle$'s, h'' is a map of $\widehat{A}$ modules whose matrix entries are scalar multiples of the idempotent e_i, and the differential d satisfies

$$d \circ e_i = d \neq 0.$$

From this we see that $d \circ h'' = 0 \implies h'' = 0$. But once $h'' = 0$, the homotopy relations imply that we must have $h' \circ \partial_{s-1} = 0$, too. Thus the composition of ∂_{s-1} with the projection $Y_s \to N_i$ must be 0, and N_i is a direct summand of the bottom slice $Y(\phi(-))$ of Y in the homotopy category. But, as we have observed earlier, if $Y \in X_j^+$, then the bottom slice of Y cannot have any summands isomorphic to $P_k\langle\phi_-(Y)\rangle$ for any k. Thus we conclude that the homotopy h cannot exist, and therefore that

$$\mathrm{Hom}(Y, (\mathfrak{P}_i \to \mathfrak{P}_k)\langle\phi_-(Y)\rangle) \neq 0.$$

This concludes the check of (5).

We now show (3), namely, that $Y \in X_j^-$ and $j \neq i$ implies $\Sigma_i(Y) \in X_i^+$. As before, in order to show that the top slice of $\Sigma_i(Y)$ is a direct sum of shifts of P_i, it suffices to note that $\phi_+(\Sigma_i(Y)) > \phi_+(Y)$. Since

$$\mathrm{Hom}(\mathfrak{P}_i\langle\phi_+(Y)+1\rangle, \Sigma_i(Y)) \cong \mathrm{Hom}(\mathfrak{P}_i\langle\phi_+(Y)\rangle, Y) \neq 0$$

by assumption, we must have $\phi_+(\Sigma_i(Y)) > \phi_+(Y)$. What remains is to show that

$$\mathrm{Hom}(\Sigma_i(Y), \mathfrak{P}_k\langle\phi_-(\Sigma_i(Y))\rangle) = 0 \iff k = i.$$

But this is clear: by assumption $Y(\phi_-)$ is isomorphic to a direct sum of shifts of P_j, and it follows immediately (e.g. from the example computations in Section 4.8) that $\Sigma_i(Y)(\phi_-)$ is isomorphic to a direct sum of shifts of the complex $(P_j \to P_i)$. Since maps from $\Sigma_i(Y)$ to objects of $D_{\leq \phi_-(\Sigma_i(Y))}$ are determined completely by maps out of the bottom slice $\Sigma_i(Y)(\phi_-)$, it follows that $\mathrm{Hom}(\Sigma_i(Y), \mathfrak{P}_k\langle\phi_-(\Sigma_i(Y))\rangle)$ is isomorphic to a direct sum of hom spaces of the form

$$\mathrm{Hom}((P_j \to P_i)\langle\phi_-(\Sigma_i(Y))\rangle, \mathfrak{P}_k\langle\phi_-(\Sigma_i(Y))\rangle) \cong \mathrm{Hom}((P_j \to P_i), \mathfrak{P}_k)$$

But now one can see directly using the relations in the algebra $\widehat{A}$ that the hom space on the right is zero if and only if $k = i$. The remaining claims (4) and (6) in the proposition now follow by symmetric arguments. □

As a corollary, we obtain Theorem 1.1 from the introduction.

COROLLARY 5.2. *The group homomorphism*

$$\psi : F_n \longrightarrow [Aut(\mathcal{D})]$$

from the free group to the isomorphism classes of auto-equivalences of $\mathcal{D}$, is injective.

We have proven that F_n acts faithfully on the homotopy category of *graded* projective $\widehat{A}$ modules, where the grading is the one induced by the $\widetilde{o}$ orientation of Q. Of course, one may ask about the faithfulness of the F_n action on the homotopy category of graded modules for one of our other gradings, or about the faithfulness of the action on the homotopy category of ungraded projective modules. In fact Corollary 5.2 implies that F_n acts faithfully on all of these different triangulated categories. The main point is that while not all complexes of ungraded $(\widehat{A}, \widehat{A})$ bimodules have graded lifts, the minimal complexes of ungraded $(\widehat{A}, \widehat{A})$ bimodules of the form Ψ_g do all have canonical graded lifts, at least for the gradings we are considering. From this, one can show directly that if Ψ_g is homotopic to $\widehat{A}$ as a complex of ungraded bimodules, then there must actually exist an graded homotopy between Ψ_g and $\widehat{A}$. Thus faithfulness on the homotopy category of $\widetilde{o}$-graded projective modules implies faithfulness on the homotopy category of ungraded projective modules (and therefore on the homotopy category of graded projective modules for any of our other gradings.)

6. The Bessis monoid, the $\vec{o}$-grading on $\widehat{A}$, and dual ping pong

In this section we describe the relationship between the Bessis monoid F_n^+ of F_n and the triangulated category $\mathcal{D}$. The basic ingredients are the 2-representation Ψ and the $\vec{o}$-grading on the zigzag algebra $\widehat{A}$.

To begin, we explain how the Hurwitz action of the Artin braid group B_n appears naturally from the point of view of the 2-representation Ψ. Our description is very similar to the well-known appearance of B_n in the theory of exceptional sequences of representations of quivers [**11**]. Actions of B_n on spherical collections in triangulated categories have also been considered before, see for example [**10**] and references therein. Our main goal here is to directly relate these kinds of actions to the Hurwitz action of B_n on reduced reflection expressions of $\gamma = \sigma_1 \dots \sigma_n$.

In this section the grading shift $\langle \bullet \rangle$ will denote a shift in the $\vec{\sigma}$-grading coming from the orientation $\vec{\sigma}$ of Q, and $\mathcal{D}$ will denote the homotopy category of bounded complexes of finitely-generated $\vec{\sigma}$-graded projective modules. Given $Y \in \mathcal{D}$, $Y(k)$ will denote the $\vec{\sigma}$-baric slice of Y.

6.1. The Hurwitz action of the braid group on spherical collections. One special feature of complexes in $\mathcal{D}$ of the form $\Psi_g(P_i)$ is that their endomorphism algebras are isomorphic to $\Bbbk[z]/z^2$, the homology ring of the sphere. For this reason, we refer to these objects as *spherical.* [2]If $E \cong \Psi_g(P_k)$ is spherical, we denote by $\Phi_E \cong \Psi_g \Sigma_k \Psi_g^{-1}$ the associated *spherical twist,* which is an auto-equivalence of $\mathcal{D}$.

A *spherical collection* in $\mathcal{D}$ is a collection

$$\{E_1, E_2, \dots, E_n\}$$

of finitely-many spherical objects E_i. The auto-equivalence group of $\mathcal{D}$ acts on the set of spherical collections, with an auto-equivalence Φ acting by

$$\Phi \cdot \{E_1, \dots, E_n\} = \{\Phi(E_1), \dots, \Phi(E_n)\}.$$

It is sometimes convenient to fix an order $(E_1, E_2, \dots, E_n)$ on the objects in a spherical collection; after doing so the Artin braid group B_n then acts on the set of spherical collections, where the action of the generator $\tau_i \in B_n$ is defined by

$$\tau_i(E_1, \dots, E_i, E_{i+1}, \dots, E_n) = (E_1, \dots, \Phi_{E_i}(E_{i+1}), E_i \dots, E_n).$$

(See, for example, [**10**]). We denote the set of ordered spherical collections in $\mathcal{D}$ by $\mathcal{O}_{\mathcal{D}}$, and, in anticipation of Lemma 6.2, we refer to this B_n action as the *Hurwitz action of the braid group on* $\mathcal{O}_{\mathcal{D}}$. An interesting question is to describe the orbits of B_n on ordered spherical collections in $\mathcal{D}$; in particular, it is desirable to find orbits where the B_n action is free. It is also desirable to find conditions on the morphism spaces between objects in a collection which are preserved by the Hurwitz action. This motivates the following definition.

DEFINITION 3. *An* $\vec{\sigma}$-**spherical collection** *in* $\mathcal{D}$ *is an* n*-tuple*

$$(E_1, E_2, \dots, E_n)$$

such that

- *each* E_i *is spherical;*
- *each* $E_i \in \mathcal{D}_0$*, where* $\mathcal{D}_0$ *denotes the heart of the* $\vec{\sigma}$*-baric structure on* $\mathcal{D}$ *(see Section* 4.6*);*
- *for* $i < j$ *and* $l \in \mathbb{Z}$*,*
$$\mathrm{Hom}(E_i, E_j\langle k\rangle[l]) \neq 0 \implies k = 1;$$
- *for* $i > j$ *and* $l \in \mathbb{Z}$*,*
$$\mathrm{Hom}(E_i, E_j\langle k\rangle[l]) \neq 0 \implies k = 0;$$

[2]The terms *spherical object* and *spherical twist* are due to Seidel-Thomas [**27**], who define a spherical object to be an object E such that $\oplus_k \mathrm{Hom}(E, E[k])$ is isomorphic to the homology ring of a sphere. The notion of spherical object we use here is quadratic dual to theirs. In particular, if we choose to work instead with the corresponding simple modules over the quadratic dual of $\widehat{A}$, these simple modules will be spherical in the original sense of Seidel-Thomas.

The basic example of an $\vec{o}$-spherical collection is

$$(P_1, P_2, \ldots, P_n).$$

The proof of the following lemma is almost immediate from the definitions.

LEMMA 6.1. *The Hurwitz action of B_n preserves the set of $\vec{o}$-spherical collections.*

Let $\mathcal{O}^{\vec{o}}_{\mathcal{D}}(\Psi_\gamma)$ denote the set of $\vec{o}$-spherical collections $(E_1, \ldots, E_n)$ such that the composition of their spherical twists is given by the action of the Coxeter element

$$\Phi_{E_1} \ldots \Phi_{E_n} \cong \Psi_\gamma \in \mathrm{Aut}(\mathcal{D}).$$

Recall that $Red(\gamma)$ denotes the set of reduced γ-reflection expressions for $\gamma \in F_n$. A comparison of the definition of the Hurwitz action on $\vec{o}$-spherical collections to that of the Hurwitz action on reduced γ-reflection expressions for γ yields the following.

LEMMA 6.2. *The assignment $(\sigma_1, \ldots, \sigma_n) \mapsto (P_1, \ldots, P_n)$ extends uniquely to a map*

$$f : Red(\gamma) \longrightarrow \mathcal{O}^{\vec{o}}_{\mathcal{D}}(\Psi_\gamma)$$

which intertwines the B_n actions. If $f : (t_1, \ldots, t_n) \mapsto (E_1, \ldots, E_n)$, then for all $k = 1, \ldots, n$ the auto-equivalence Ψ_{t_k} is isomorphic to the spherical twist Φ_{E_k}.

PROPOSITION 6.3. *The Hurwitz action of the braid group B_n on the orbit of the spherical collection $(P_1, \ldots, P_n)$ is free. Thus the assignment $(\sigma_1, \ldots, \sigma_n) \mapsto (P_1, \ldots, P_n)$ extends uniquely to an isomorphism of sets with a B_n action.*

PROOF. Let $1 \neq \beta \in B_n$, and suppose that $\beta(P_1, \ldots, P_n) = (E_1, \ldots, E_n)$. We must show that for some i, E_i is not isomorphic to P_i. To see this, consider the Hurwitz action of B_n on the reduced γ-reflection expressions for γ, and let

$$\beta(\sigma_1, \ldots, \sigma_n) = (t_1, \ldots, t_n).$$

In particular, in the 2-representation Ψ, the reflection t_i acts by the spherical twist Φ_{E_i}. By Artin's Theorem 2.1

$$(t_1, \ldots, t_n) = (\sigma_1, \ldots, \sigma_n) \iff \beta = 1.$$

Thus, if $\beta \neq 1$, at least one of the reflections t_i is not equal to σ_i. Now, by Theorem 1.1, it follows that the spherical twists Φ_{E_i} and $\Phi_{P_i} = \Sigma_i$ are not isomorphic, and thus E_i is not isomorphic to P_i. □

6.2. Reflection complexes. Let $t \in F_n^+$ be a γ-reflection. To t we may associate a complex $\mathcal{C}_t$ whose associated twist is t. We call $\mathcal{C}_t$ the *reflection complex* associated to t, and in this section we study the basic properties of reflection complexes. The most important point is that there is a close connection between the morphism spaces between reflection complexes and the topology of the punctured disc. The basic features of this connection will be summarized in Proposition 6.4 below.

Let $\beta \in B_n$ be a braid such that

$$\beta \cdot (\sigma_1, \ldots, \sigma_n) = (t, x_2, \ldots, x_n),$$

where B_n acts on n-tuples of γ-reflections via the Hurwitz action. Similarly, there exists a complex $\mathcal{C}_t$ with

$$\beta \cdot (P_1, \ldots, P_n) = (\mathcal{C}_t, \mathcal{C}_{x_2}, \ldots, \mathcal{C}_{x_n}).$$

The complex $\mathcal{C}_t$ is well defined independent of the choice of $\beta \in B_n$ used to obtain an n-tuple of the form $(t, x_2, \ldots, x_n)$. The complex $\mathcal{C}_t$, which we call the *reflection complex* associated to t, lives in the heart $\mathcal{D}_0$ of the $\vec{\sigma}$-baric structure on $\mathcal{D}$.

As an example, let t be the γ-reflection $\sigma_1\sigma_2\sigma_1^{-1}$. The associated reflection complex is

$$\mathcal{C}_{\sigma_1\sigma_2\sigma_1^{-1}} \cong \Sigma_1(P_2) \cong \big(P_2 \xrightarrow{x^*\oplus y^*} P_1^{\oplus 2}\big).$$

An alternative description of the reflection complex $\mathcal{C}_t$ is as follows: any γ-reflection t may be written as $t = g\sigma_i g^{-1}$ for some $g \in F_n$ and some $i = 1, \ldots, n$, with g determined uniquely up to multiplication on the right by a power of σ_i. The complex $\Psi_g(P_i)$ will agree up to a grading and homological shift with the reflection complex $\mathcal{C}_t$ defined above, where the "up to a shift" is a result of the ambiguity in the choice of g. The advantage of our first definition above is that it uses the Hurwitz action to fix this choice of g uniformly for all γ-reflections. However, for some statements it seems more natural to allow for an "up to shift" ambiguity in the definition of $\mathcal{C}_t$. In particular, for t a γ-reflection, we define a complex $\mathfrak{C}_t$ by

$$\mathfrak{C}_t = \bigoplus_{k\in\mathbb{Z}} \mathcal{C}_t[k].$$

Thus $\mathfrak{C}_t$ is the direct sum of all homological shifts of $\mathcal{C}_t$ which live in the $\vec{\sigma}$-baric heart $\mathcal{D}_0$. Note that changing the definition of $\mathcal{C}_t$ by homological shift does not change the object $\mathfrak{C}_t$. The complexes $\mathfrak{C}_t$ are not objects of $\mathcal{D}$, though the summands of $\mathfrak{C}_t$ are. In particular, for any $Y \in \mathcal{D}$, the space $\operatorname{Hom}(Y, \mathfrak{C}_t)$ is a well-defined finite dimensional $\Bbbk$-vector space.

In the statement of the Proposition 6.4 below, we find it convenient to introduce the notation $\doteq$ by writing, for complexes Y and B,

$$Y \doteq B \iff \text{there exists an integer } m \text{ such that } Y \cong B[m].$$

When we have fixed a direct sum decomposition $B \cong B_1 \oplus \cdots \oplus B_k$, we will also write

$$Y \doteq B_1 \oplus \cdots \oplus B_k \iff \text{there exist integers } m_1, \ldots, m_k \text{ such that } Y \cong B_1[m_1] \oplus \cdots \oplus B_k[m_k].$$

In the special case $\mathcal{C}_{\sigma_i} \cong P_i$, we follow the notation of Section 5 and write $\mathfrak{P}_i := \mathfrak{C}_{\sigma_i}$ (though we remind the reader again that in this section the module P_i is endowed with its $\vec{\sigma}$-grading, while in Section 5 the notations P_i and $\mathfrak{P}_i$ were used to denote $\tilde{\sigma}$-graded objects).

PROPOSITION 6.4. *Let t and u be γ-reflections. The following are equivalent:*

(1) *$tu \in \mathcal{B}e^+$ is a simple Bessis element;*
(2) *tut^{-1} is a γ-reflection;*
(3) *$u^{-1}tu$ is a γ-reflection;*
(4) *The connecting paths f_t and f_u have isotopic representatives in the unit disc which intersect only at the base point x_0, and such that in a sufficiently small neighbourhood of x_0, the curve f_t lies below f_u.*
(5) $\operatorname{Hom}(\mathcal{C}_u, \mathfrak{C}_t\langle k\rangle) = 0$ *for* $k \neq 0$.
(6) $\operatorname{Hom}(\mathcal{C}_t, \mathfrak{C}_u\langle k\rangle) = 0$ *for* $k \neq 1$.
(7) $\Psi_t(\mathcal{C}_u) \cong \mathcal{C}_{tut^{-1}} \in \mathcal{D}_0$;
(8) $\Psi_u^{-1}(\mathcal{C}_t) \cong \mathcal{C}_{u^{-1}tu} \subset \mathcal{D}_0$;

(9) $\Psi_u(\mathcal{C}_t) \in \mathcal{D}_{[0,1]}$ *with $\vec{o}$-baric slices*

$$\Psi_u(\mathcal{C}_t)(\phi_+) \doteq \oplus_{i=1}^r \mathcal{C}_u\langle 1\rangle \text{ and } \Psi_u(\mathcal{C}_t)(\phi_-) \doteq \mathcal{C}_t;$$

(10) $\Psi_t^{-1}(\mathcal{C}_u) \in \mathcal{D}_{[-1,0]}$ *with $\vec{o}$-baric slices*

$$\Psi_t^{-1}(\mathcal{C}_u)(\phi_-) \doteq \oplus_{i=1}^r \mathcal{C}_t\langle -1\rangle \text{ and } \Psi_t^{-1}(\mathcal{C}_u)(\phi_+) \doteq \mathcal{C}_u;$$

PROOF. We only give a sketch of the proof here (see also [**20**]). The equivalence of the first four claims is immediate from Bessis' work [**4**], and does not involve the 2-representation Ψ. The equivalence of the remaining statements with each other and with items (1), (2), (3), (4) can be shown using the Hurwitz action on the set of $\vec{o}$-spherical collections.

For example, we show the equivalence of (1) and (5). Recall that if $(\mathcal{C}_{t_1}, \ldots, \mathcal{C}_{t_n}) \in \mathcal{O}_{\mathcal{D}}^{\vec{o}}(\Psi_\gamma)$ is a $\vec{o}$-spherical collection of reflection complexes corresponding to a minimal length reflection expression

$$\gamma = t_1 t_2 \ldots t_n,$$

then it follows from the definition of a $\vec{o}$-spherical collection that for $i < j$

$$\mathrm{Hom}(\mathcal{C}_{t_i}, \mathfrak{C}_{t_j}\langle k\rangle) \neq 0 \implies k = 1,$$

while for for $i > j$

$$\mathrm{Hom}(\mathcal{C}_{t_i}, \mathfrak{C}_{t_j}\langle k\rangle) \neq 0 \implies k = 0.$$

Now suppose that tu is a simple Bessis element, and $(t, u, x_1 \ldots x_{n-2})$ be a minimal length γ-reflection expression for the Coxeter element γ. Then there is some braid $\beta \in B_n$ such that

$$\beta(P_1, P_2, \ldots, P_n) \cong (\mathcal{C}_t, \mathcal{C}_u, \ldots, \mathcal{C}_{x_{n-2}}).$$

Since the B_n action preserves the set of $\vec{o}$-spherical collections, it follows that

$$\mathrm{Hom}(\mathcal{C}_u, \mathfrak{C}_t\langle k\rangle) = 0 \text{ for } k \neq 0,$$

giving (5).

To prove the converse statement (5) $\implies$ (1), for concreteness we will assume that $t = \sigma_1$, so that $\mathcal{C}_t = P_1$ and $\mathfrak{C}_t = \mathfrak{P}_1$. (The case of more general t can be reduced to this one using the Hurwitz action, although one can also give a more direct proof along the lines indicated below for the case $\mathcal{C}_t = P_1$.) So we now suppose that $\mathrm{Hom}(\mathcal{C}_u, \mathfrak{P}_1\langle k\rangle) = 0$ for $k \neq 0$. We claim that $\mathcal{C}_u$ is in the full triangulated subcategory generated by $\{P_2, \ldots, P_n\}$; that is, we claim that no shifts of P_1 appear in the chain groups of the minimal complex of $\mathcal{C}_u$.

For suppose that a shift of P_1 appears somewhere in the minimal complex of $\mathcal{C}_u$; let us denote such a summand of the chain group of the minimal complex of $\mathcal{C}_u$ by N_1, which for convenience we assume appears in homological degree 0. The boundary map of the minimal complex of $\mathcal{C}_u$ does not involve any non-zero scalar multiples of the map $e_1 : P_1 \to P_1$, as any appearance of e_1 would result in a contractible summand of $\mathcal{C}_u$. But the the only $\vec{o}$-degree 0 maps out of N_1 are scalar multiples of e_1, and all components of the differential in the minimal complex of $\mathcal{C}_u$ are of $\vec{o}$-degree 0. From this it follows that N_1 must be in the kernel of the boundary map of the minimal complex of $\mathcal{C}_u$. Now, let

$$\iota : P_1 \to \mathcal{C}_u$$

be the map which sends P_1 to N_1 isomorphically via e_1. Since N_1 is in the kernel of the boundary map of the minimal complex of $\mathcal{C}_u$, ι is a chain map and defines an element of $\mathrm{Hom}(P_1, \mathcal{C}_u)$. Let

$$z_{N_1} : \mathcal{C}_u \to P_1\langle 1\rangle$$

be the map which restricts to multiplication by the loop $z : N_1 \to P_1\langle 1\rangle$ and which is 0 on all other summands of the chain group of the minimal complex of $\mathcal{C}_u$. This, too, is a chain map, and composition

$$z_{N_1} \circ \iota = z : P_1 \to P_1\langle 1\rangle$$

is clearly non-zero in the homotopy category; hence z_{N_1} is non-zero in the homotopy category, too. Thus $\mathrm{Hom}(\mathcal{C}_u, \mathfrak{P}_1\langle 1\rangle) \neq 0$, contradicting the assumption of (5).

We conclude that if $\mathrm{Hom}(\mathcal{C}_u, \mathfrak{P}_1\langle k\rangle) = 0$ for all $k \neq 0$, then $\mathcal{C}_u$ must be in the full triangulated subcategory generated by $P_2, \ldots, P_n$. Now an inductive argument using the Hurwitz action implies that if a reflection complex $\mathcal{C}_u$ lies in the full triangulated subcategory generated by $P_2, \ldots, P_n$, then in fact the reflection u is a divisor of $\gamma' = \sigma_2 \ldots \sigma_n$. Thus $\sigma_1 u$ divides $\sigma_1\gamma' = \gamma$, so that $\sigma_1 u = tu$ is a simple Bessis element, giving (1).

The equivalence of (6) with (1) is given analogously. The equivalence of each of (7) – (10) with (5), (6) is obtained directly using the definition of the auto-equivalences $\Psi_t^{\pm 1}, \Psi_u^{\pm 1}$. □

The transitivity Hurwitz action of B_n on $\vec{\sigma}$-spherical collections in the orbit of $(P_1, \ldots, P_n)$ has some further important consequences for morphism spaces between reflection complexes. For example, we have the following lemma, whose proof we leave as an exercise for the interested reader. (One such proof is by induction on the length of the braid $\beta \in B_2 \cong \mathbb{Z}$ needed to obtain r as a component of $\beta(t, u)$ under the Hurwitz action.)

LEMMA 6.5. *Let t, u be γ-reflections with $tu \in \mathcal{B}e^+$ a Bessis simple element. Suppose that*

$$\mathrm{Hom}(\mathfrak{C}_t, Y) = \mathrm{Hom}(\mathfrak{C}_u, Y) = 0.$$

Then for any γ-reflection r dividing tu,

$$\mathrm{Hom}(\mathfrak{C}_r, Y) = 0.$$

Similarly, if

$$\mathrm{Hom}(Y, \mathfrak{C}_t) = \mathrm{Hom}(Y, \mathfrak{C}_u) = 0,$$

then for any γ-reflection r dividing tu,

$$\mathrm{Hom}(Y, \mathfrak{C}_r) = 0.$$

6.3. Dual ping pong. In this section we define the sets required for dual ping pong, as formulated in Lemma 3.2. We remind the reader that in this section $\langle \bullet \rangle$ denotes a shift in the $\vec{\sigma}$-grading, while $\mathcal{D}_k, \mathcal{D}_{\geq k}, \mathcal{D}_{\leq k}$, etc. denote the subcategories of $\mathcal{D}$ defined in Section 4.6 using the $\vec{\sigma}$-baric structure.

For $w \in \mathcal{B}e^+$ a simple Bessis element, set

$$X_w = \{Y \in \mathcal{D}_{\geq 0} : \mathrm{Hom}(Y, \mathfrak{C}_t) = 0 \iff t \text{ divides } w\}$$

We now show that the X_w satisfy the requirements for dual ping pong, as formulated in Lemma 3.3.

PROPOSITION 6.6. *The sets $\{X_w\}_{w\in \mathcal{B}e^+}$ are all non-empty and pairwise disjoint. For u a γ-reflection and $w \in \mathcal{B}e^+$,*

$$\Psi_u(X_w) \subset X_{lf(uw)}.$$

PROOF. We first show the sets X_w are all non-empty. For $w \neq \gamma \in \mathcal{B}e^+$, write $\gamma = (w^\vee)w$, and let

$$Y_w = \bigoplus_{x \text{ a } \gamma\text{-reflection dividing } w^\vee} \mathcal{C}_x.$$

A basic property of the Bessis monoid F_n^+ is that, for t a γ-reflection, $xt \in \mathcal{B}e^+$ for all x dividing $w^\vee$ if and only if t divides w. Thus it follows from from Proposition 6.4 that $\mathrm{Hom}(Y_w, \mathfrak{C}_t) = 0$ if and only if $xt \in \mathcal{B}e^+$ if and only if t divides w; this shows that $Y_w \in X_w$, so that $X_w \neq \emptyset$. Since $\mathcal{D}_{>0} \subset X_\gamma$, the set X_γ is clearly non-empty, too.

The fact that $X_w \cap X_y = \emptyset$ when $w \neq y$ is clear from the definition, since if $Y \in X_w$, then w is the least common multiple of the set $\{t \mid \mathrm{Hom}(Y, \mathfrak{C}_t) = 0\}$, and thus Y determines w.

We now show that $\Psi_u(X_w) \subset X_{lf(uw)}$. Let $a = lf(uw)$, and write $uw = ab = ua'b$, with $a'b = w$. Let $Y \in X_w$. We must now show that $\Psi_u Y \in X_a$, that is, we must show that

$$\mathrm{Hom}(\Psi_u Y, \mathfrak{C}_t) = 0 \iff t \text{ divides } a.$$

Suppose first that t divides a. One possibility is that $t = u$, in which case

$$\mathrm{Hom}(\Psi_u Y, \mathfrak{C}_u) \cong \mathrm{Hom}(Y, \Psi_u^{-1}\mathfrak{C}_u) \cong \mathrm{Hom}(Y, \mathfrak{C}_u\langle -1\rangle).$$

As $Y \in \mathcal{D}_{\geq 0}$, the summands of $\mathfrak{C}_u\langle -1\rangle$ are in $\mathcal{D}_{<0}$, and

$$\mathrm{Hom}(\mathcal{D}_{\geq 0}, \mathcal{D}_{<0}) = 0,$$

we have that $\mathrm{Hom}(Y, \mathfrak{C}_u\langle -1\rangle) = 0$.

Next, suppose that t divides a'. Since t divides w and $Y \in X_w$, we know that

$$\mathrm{Hom}(Y, \mathfrak{C}_t) = 0.$$

Also, as $ua' \in \mathcal{B}e^+$ is a simple Bessis element and ut divides ua', we know that $ut \in \mathcal{B}e^+$, too. Now, by Proposition 6.4 (10), we have that $\Psi_u^{-1}\mathcal{C}_t$ is isomorphic to a mapping cone of a direct sum of $\mathcal{C}_u\langle -1\rangle$'s mapping to $\mathcal{C}_t$. Since the $\mathcal{C}_u\langle -1\rangle$ terms of $\Psi_u^{-1}\mathcal{C}_t$ live in $\mathcal{D}_{<0}$, it follows that for $Z \in \mathcal{D}_{\geq 0}$,

$$\text{if } \mathrm{Hom}(Z, \mathfrak{C}_t) = 0 \text{ then } \mathrm{Hom}(Z, \Psi_u^{-1}\mathfrak{C}_t) = 0.$$

Taking $Z = Y$, we then have

$$\mathrm{Hom}(Y, \mathfrak{C}_t) = 0 \implies \mathrm{Hom}(\Psi_u Y, \mathfrak{C}_t) \cong \mathrm{Hom}(Y, \Psi_u^{-1}\mathfrak{C}_t) = 0.$$

Thus $\mathrm{Hom}(\Psi_u Y, \mathfrak{C}_t) = 0$ for $t = u$ and also for t dividing a'. Now Lemma 6.5 shows that $\mathrm{Hom}(\Psi_u Y, \mathfrak{C}_t) = 0$ for any t dividing $ua' = a = lf(uw)$.

We now show the converse statement:

$$\mathrm{Hom}(\Psi_u Y, \mathfrak{C}_t) = 0 \implies t \text{ divides } a.$$

Let a'' be the least common multiple of $\{t \mid \mathrm{Hom}(\Psi_u Y, \mathfrak{C}_t) = 0\}$. Since u is in this set, u divides a'', and we may write $a'' = ua'''$ for a simple Bessis element a'''. If we show that a''' divides w, then since $ua''' = a''$, it will follow that a'' divides $a = lf(uw)$.

To see that a''' divides w, suppose that t is a γ-reflection dividing a'''. We claim that $\mathrm{Hom}(Y, \mathfrak{C}_t) = 0$. As before, we know that

$$\text{if } \mathrm{Hom}(Y, \Psi_u^{-1}\mathfrak{C}_t) = 0 \text{ then } \mathrm{Hom}(Y, \mathfrak{C}_t) = 0.$$

Since

$$\mathrm{Hom}(\Psi_u Y, \mathfrak{C}_t) \cong \mathrm{Hom}(Y, \Psi_u^{-1}\mathfrak{C}_t) = 0,$$

and $Y \in X_w$, we conclude then t divides w. Since t was an arbitrary factor of a''', it follows that a''' divides w, and $a'' = ua'''$ divides $a = lf(uw)$, as desired. □

REMARK 2. *The statements Proposition* 6.6 *above together with the dual ping pong Lemma* 3.3 *together imply the faithfulness of the F_n action on $\mathcal{D}$, so it might appear as though we have given a second independent proof of Theorem* 1.1. *However, in preparation for Proposition* 6.6, *we used Theorem* 1.1 *to prove that the Hurwitz action of B_n on $\vec{o}$-spherical collections is free. Nevertheless, one can give a proof of Proposition* 6.6 *which does not use Theorem* 1.1 *(by, for example, proving the freeness of the Hurwitz action on $\vec{o}$-spherical collections directly) and so in principle one can prove Theorem* 1.1 *using either ping pong or dual ping pong. We give a dual ping pong faithfulness proof in setting of categorical actions of ADE spherical Artin-Tits groups in* [**21**].

7. Metrics on F_n from homological algebra

Let o be an orientation of Q, and $\{\mathcal{D}_k\}$ the associated o-baric structure of $\mathcal{D}$, as defined in Section 4.6. For any auto-equivalence β of $\mathcal{D}$ we may define a closed interval

$$[\phi_-(\beta), \phi_+(\beta)] \subset \mathbb{Z}$$

as the smallest interval such that

$$\Psi_\beta(\mathcal{D}_0) \subset \mathcal{D}_{[\phi_-(\beta), \phi_+(\beta)]}. \tag{1}$$

There is no a priori reason to consider only o-baric structure on $\mathcal{D}$ for such a definition, and one could (and should) consider instead t-structures, co-t structures, Bridgeland slicings or co-slicings, etc. In favourable situations, the length of this interval will give a metric on the auto-equivalence group of $\mathcal{D}$. It would be interesting to study the resulting metrics on the free group coming from categorical structures on $\mathcal{D}$ more systematically, and we certainly have not done that here. However, the ping pong and dual ping pong constructions considered in Sections 5 and 6 allow us to at least give a combinatorial description of the metrics arising from the $\widetilde{o}$- and $\vec{o}$-baric structures on $\mathcal{D}$. We collect the relevant statements in two theorems below. The proofs of these theorems are very similar to the proofs of Propositions 5.1 and 6.6, and as a result we omit the details. (In fact, the easiest way to prove these statements is via an examination of the ping pong and dual ping pong games constructed in Propositions 5.1 and 6.6. We also refer also [**21**],[**20**], where we prove analogous metric statements for spherical and right-angled Artin-Tits groups.)

For the first statement, we consider the symmetric orientation $\widetilde{o}$, which was used in Section 5 for our construction of ping pong and proof of Theorem 1.1. Thus the integers $\phi_+(\beta)$ and $\phi_-(\beta)$ are determined by the interval in (1), where $\mathcal{D}_k$ are given by the $\widetilde{o}$-baric structure on $\mathcal{D}$.

THEOREM 7.1. *For $n \geq 2$ and all $\beta \in F_n$, $\phi_-(\beta) \leq 0$ and $\phi_+(\beta) \geq 0$. Furthermore,*

- *$\phi_+(\beta)$ is equal to the number of positive generators from the set $\{\sigma_1, \dots, \sigma_n\}$ in a reduced expression*

$$\beta = \sigma_{i_1}^{\epsilon_1} \dots \sigma_{i_r}^{\epsilon_r}, \quad \epsilon_i \in \pm 1;$$

- *$-\phi_-(\beta)$ is equal to the number of negative generators from the set $\{\sigma_1^{-1}, \dots, \sigma_n^{-1}\}$ in a reduced expression*

$$\beta = \sigma_{i_1}^{\epsilon_1} \dots \sigma_{i_r}^{\epsilon_r}, \quad \epsilon_i \in \pm 1;$$

- *the $\widetilde{o}$-baric spread $l_{\widetilde{o}}(\beta) = \phi_+(\beta) - \phi_-(\beta)$ is equal to the word length of β in the standard generators $\{\sigma_i^{\pm 1}\}$ of F_n.*

We now replace the symmetric orientation $\widetilde{o}$ on Q by the ordered orientation $\vec{o}$ on Q; the $\vec{o}$-grading on $\widehat{A}$ was used in Section 6 for dual ping pong. Thus in the following theorem, the integers $\phi_+(\beta)$ and $\phi_-(\beta)$ are defined by the interval in (1), where now $\mathcal{D}_k$ are given by the $\vec{o}$-baric structure on $\mathcal{D}$.

THEOREM 7.2. *For any $\beta \in F_n$, we have the following:*

- *If $\phi_+(\beta) \geq 0$, then $\phi_+(\beta)$ is equal to the number of positive simple Bessis elements $w \in \mathcal{B}e^+$ in any minimal length expression*

$$\beta = w_{i_1}^{\epsilon_1} \dots w_{i_r}^{\epsilon_r}, \quad \epsilon_i \in \pm 1$$

  for β as a word in the generating set $(\mathcal{B}e^+)^{\pm 1}$.
- *If $\phi_+(\beta) \leq 0$, then $\beta \in (F_n^+)^{-1}$ is in the negative Bessis monoid. Moreover, in this case $-\phi_+(\beta)$ is equal to the number of γ^{-1}'s in any minimal length expression*

$$\beta = w_{i_1}^{\epsilon_1} \dots w_{i_r}^{\epsilon_r}, \quad \epsilon_i \in \pm 1$$

  for β as a word in the generating set $(\mathcal{B}e^+)^{\pm 1}$.
- *If $\phi_-(\beta) \leq 0$, then $-\phi_-(\beta)$ is equal to the number of negative simple Bessis elements w^{-1} in any minimal length expression*

$$\beta = w_{i_1}^{\epsilon_1} \dots w_{i_r}^{\epsilon_r}, \quad \epsilon_i \in \pm 1$$

  for β as a word in the generating set $(\mathcal{B}e^+)^{\pm 1}$.
- *If $\phi_-(\beta) \geq 0$, then $\beta \in F_n^+$ is in the positive Bessis monoid. Moreover, in this case $\phi_-(\beta)$ is equal to the number of γ's in any minimal length expression*

$$\beta = w_{i_1}^{\epsilon_1} \dots w_{i_r}^{\epsilon_r}, \quad \epsilon_i \in \pm 1$$

  for β as a word in the generating set $(\mathcal{B}e^+)^{\pm 1}$.
- *Let*

$$\phi_+^*(\beta) = max\{\phi_+(\beta), 0\} \text{ and } \phi_-^*(\beta) = min\{\phi_-(\beta), 0\},$$

  so that $\phi_+^*(\beta) \geq 0$ and $\phi_-^*(\beta) \leq 0$. The $\vec{o}$-baric spread

$$l_{\vec{o}}(\beta) = \phi_+^*(\beta) - \phi_-^*(\beta)$$

  is equal to the word length of β in the generating set $(\mathcal{B}e^+)^{\pm 1}$ of simple Bessis elements and their inverses.

Thus Theorems 7.1 and 7.2 give homological constructions of the standard and dual word-length metrics on F_n.

7.1. The exotic metric on F_n. We close by defining a distinguished homological metric on the free group arising from the faithful 2-representation Ψ. Recall that the triangulated category $\mathcal{D}$ has a distinguished t-structure $\{\mathcal{D}^k\}$ whose heart $\mathcal{D}^0$ is the abelian category of linear complexes of graded projective $\widehat{A}$-modules, where *here the grading on $\widehat{A}$ we consider is the grading by path-length* (see Section 4.3). Just as for the baric structures considered in the previous section, this t-structure gives rise to a metric on F_n, and a basic problem is to describe it combinatorially.

In more detail, for $\beta \in F_n$, define $[\phi_-(\beta), \phi_+(\beta)] \subset \mathbb{Z}$ to be the smallest interval such that

$$\Psi_\beta(\mathcal{D}^0) \subset \mathcal{D}^{[\phi_-(\beta),\phi_+(\beta)]}.$$

It is not difficult to see that for $n \geq 2$ and any $\beta \in F_n$, $-\infty < \psi_-(\beta) \leq 0$ and $0 \leq \phi_+(\beta) < \infty$. Using the injectivity of the 2-representation Ψ, one can also show the following.

PROPOSITION 7.3. *The function $d : F_n \times F_n \longrightarrow \mathbb{Z}_{\geq 0}$ defined by*

$$d_{exotic}(\alpha, \beta) = \phi_+(\beta^{-1}\alpha) - \phi_-(\beta^{-1}\alpha)$$

is a (left-invariant) metric on F_n, $n \geq 2$.

The triangle inequality and the symmetry of d_{exotic} in the arguments α and β both follow immediately from the definitions; the non-trivial point is to prove that

$$\text{if } d_{exotic}(\alpha, \beta) = 0 \text{ then } \alpha = \beta.$$

Thus, to prove Proposition 7.1, one must show that

$$\beta(\mathcal{D}^0) \subset \mathcal{D}^0 \iff \beta = 1. \tag{2}$$

This is in principle a slightly stronger condition than the faithfulness of the action of F_n on $\mathcal{D}$. However, the faithfulness of the action of F_n on $\mathcal{D}$ implies that for $\beta \neq 1$, the complex of $(\widehat{A}, \widehat{A})$ bimodules defining Ψ_β is not homotopic to the $(\widehat{A}, \widehat{A})$ bimodule $\widehat{A}$. Now, by examining the chain groups of the (minimal complexes of) the top and bottom slices of Ψ_β, one can show that there is either a projective module P_j such that the top t-slice of $\Psi_\beta(P_j)$ is in $\mathcal{D}^{>0}$ or a projective module P_j such that the bottom t-slice of $\Psi_\beta(P_j)$ is in $\mathcal{D}^{<0}$. From this it follows that

$$\Psi_\beta(\oplus_{j=1}^n P_j) \in \mathcal{D}^0 \iff \beta = 1,$$

which implies statement (2).

We refer to the metric d_{exotic} as the *exotic metric* on F_n, and pose the following problem.

PROBLEM 7.4. *Give a combinatorial description of the exotic metric on F_n.*

If β is an element of the standard positive monoid, so that a minimal expression for β in the generators $\{\sigma_i^{\pm 1}\}$ uses only positive generators $\{\sigma_i\}$, then one can show that $d(\beta, 1)$ is equal to the length of β with respect to the generating set of canonical positive lifts from the associated universal Coxeter group

$$W_n = F_n / \langle \sigma_i^2 \rangle.$$

We denote by d_{Cox} the word-length metric on F_n whose length one elements are these positive lifts and their inverses. Thus, for $n = 2$, the free group elements α with $d_{Cox}(\alpha) = 1$ are

$$\sigma_1,\ \sigma_2,\ \sigma_1\sigma_2,\ \sigma_2\sigma_1,\ \sigma_1\sigma_2\sigma_1,\ \sigma_2\sigma_1\sigma_2,\ \sigma_1\sigma_2\sigma_1\sigma_2, \ldots$$

and their inverses. In fact, when $n = 2$ one can show that for all $\alpha, \beta \in F_2$,

$$d_{exotic}(\alpha, \beta) = d_{Cox}(\alpha, \beta).$$

This solves Problem 7.4 when $n = 2$.

When $n > 2$, the situation appears to be more subtle. While we always have

$$d_{exotic}(\alpha, \beta) \leq d_{Cox}(\alpha, \beta),$$

typically we will not have equality (though we will have equality when $\beta^{-1}\alpha$ is in the standard positive monoid.) For example, in F_3, if

$$\alpha = \sigma_2\sigma_1 \text{ and } \beta = \sigma_1\sigma_3\sigma_1^{-1},$$

then $d_{Cox}(\alpha, \beta) = 3$, but $d_{exotic}(\alpha, \beta) = 2$. In fact we have no reason to expect that d_{exotic} is a word-length metric when $n > 2$.

References

[1] P. N. Achar and D. Treumann, *Baric structures on triangulated categories and coherent sheaves*, Int. Math. Res. Not. IMRN **16** (2011), 3688–3743, DOI 10.1093/imrn/rnq226. MR2824842

[2] E. Artin, *Theory of braids*, Ann. of Math. (2) **48** (1947), 101–126. MR0019087

[3] J. A. Baldwin and J. E. Grigsby, *Categorified invariants and the braid group*, Proc. Amer. Math. Soc. **143** (2015), no. 7, 2801–2814, DOI 10.1090/S0002-9939-2015-12482-3. MR3336606

[4] D. Bessis, *A dual braid monoid for the free group*, J. Algebra **302** (2006), no. 1, 55–69, DOI 10.1016/j.jalgebra.2005.10.025. MR2236594

[5] R. Bezrukavnikov and I. Losev, *Etingof conjecture for quantized quiver varieties*, arXiv:1309.1716.

[6] S. J. Bigelow, *Braid groups are linear*, J. Amer. Math. Soc. **14** (2001), no. 2, 471–486 (electronic), DOI 10.1090/S0894-0347-00-00361-1. MR1815219

[7] T. Braden, A. Licata, N. Proudfoot, and B. Webster, *Quantizations of conical symplectic resolutions ii: category o and symplectic duality*, arXiv:1407.0964.

[8] T. Braden, N. Proudfoot, and B. Webster, *Quantizations of conical symplectic resolutions i: local and global structure*, arXiv:1208.3863.

[9] C. Brav and H. Thomas, *Braid groups and Kleinian singularities*, Math. Ann. **351** (2011), no. 4, 1005–1017, DOI 10.1007/s00208-010-0627-y. MR2854121

[10] T. Bridgeland, *t-structures on some local Calabi-Yau varieties*, J. Algebra **289** (2005), no. 2, 453–483, DOI 10.1016/j.jalgebra.2005.03.016. MR2142382

[11] W. Crawley-Boevey, *Exceptional sequences of representations of quivers [MR1206935 (94c:16017)]*, Representations of algebras (Ottawa, ON, 1992), CMS Conf. Proc., vol. 14, Amer. Math. Soc., Providence, RI, 1993, pp. 117–124. MR1265279

[12] I. Frenkel, M. Khovanov, and O. Schiffmann, *Homological realization of Nakajima varieties and Weyl group actions*, Compos. Math. **141** (2005), no. 6, 1479–1503, DOI 10.1112/S0010437X05001727. MR2188446

[13] S. I. Gelfand and Y. I. Manin, *Methods of homological algebra*, 2nd ed., Springer Monographs in Mathematics, Springer-Verlag, Berlin, 2003. MR1950475

[14] R. S. Huerfano and M. Khovanov, *A category for the adjoint representation*, J. Algebra **246** (2001), no. 2, 514–542, DOI 10.1006/jabr.2001.8962. MR1872113

[15] A. Ishii, K. Ueda, and H. Uehara, *Stability conditions on A_n-singularities*, J. Differential Geom. **84** (2010), no. 1, 87–126. MR2629510

[16] Lars T. Jensen, *The 2-braid group and garside normal form*, arXiv:1505.05353.

[17] A. M. Keating, *Dehn twists and free subgroups of symplectic mapping class groups*, J. Topol. **7** (2014), no. 2, 436–474, DOI 10.1112/jtopol/jtt033. MR3217627

[18] M. Khovanov and P. Seidel, *Quivers, Floer cohomology, and braid group actions*, J. Amer. Math. Soc. **15** (2002), no. 1, 203–271, DOI 10.1090/S0894-0347-01-00374-5. MR1862802

[19] D. Krammer, *Braid groups are linear*, Ann. of Math. (2) **155** (2002), no. 1, 131–156, DOI 10.2307/3062152. MR1888796

[20] A. Licata, *2-representations of right-angled Artin groups*, in preparation.

[21] A. Licata and H. Queffelec, *2-representations of ADE braid groups and Garside structures*, in preparation.

[22] R. Lipshitz, P. S. Ozsváth, and D. P. Thurston, *A faithful linear-categorical action of the mapping class group of a surface with boundary*, J. Eur. Math. Soc. (JEMS) **15** (2013), no. 4, 1279–1307, DOI 10.4171/JEMS/392. MR3055762

[23] I. Losev, *On categories o for quantized symplectic resolutions*, arXiv:1502.00595.

[24] V. Mazorchuk, S. Ovsienko, and C. Stroppel, *Quadratic duals, Koszul dual functors, and applications*, Trans. Amer. Math. Soc. **361** (2009), no. 3, 1129–1172, DOI 10.1090/S0002-9947-08-04539-X. MR2457393

[25] R. Rouquier, *Categorification of $\mathfrak{sl}_2$ and braid groups*, Trends in representation theory of algebras and related topics, Contemp. Math., vol. 406, Amer. Math. Soc., Providence, RI, 2006, pp. 137–167, DOI 10.1090/conm/406/07657. MR2258045

[26] R. Rouquier and A. Zimmermann, *Picard groups for derived module categories*, Proc. London Math. Soc. (3) **87** (2003), no. 1, 197–225, DOI 10.1112/S0024611503014059. MR1978574

[27] P. Seidel and R. Thomas, *Braid group actions on derived categories of coherent sheaves*, Duke Math. J. **108** (2001), no. 1, 37–108, DOI 10.1215/S0012-7094-01-10812-0. MR1831820

[28] R. P. Thomas, *Stability conditions and the braid group*, Comm. Anal. Geom. **14** (2006), no. 1, 135–161. MR2230573

Mathematical Sciences Institute; Australian National University; Canberra, Australia

E-mail address: amlicata@gmail.com

Contemporary Mathematics
Volume **683**, 2017
http://dx.doi.org/10.1090/conm/683/13707

The Blanchet-Khovanov algebras

Michael Ehrig, Catharina Stroppel, and Daniel Tubbenhauer

Dedicated to Christian Blanchet's sixtieth birthday

ABSTRACT. Blanchet introduced certain singular cobordisms to fix the functoriality of Khovanov homology. In this paper we introduce graded algebras consisting of such singular cobordisms à la Blanchet. As the main result we explicitly describe these algebras in algebraic terms using the combinatorics of arc diagrams.

CONTENTS

1. Introduction

For an arbitrary field $\mathbb{K}$ we consider the $\mathfrak{gl}_2$-*web algebra* $\mathfrak{W}$, which we call *web algebra* for short. (For the reason why we like to call it $\mathfrak{gl}_2$-web algebra instead of $\mathfrak{sl}_2$-web algebra see in the introduction of [**14**].) This is a graded $\mathbb{K}$-algebra

2010 *Mathematics Subject Classification.* 17B37, 17B10, 16G10, 57M27, 81R50.

which naturally appears as an algebra of singular cobordisms. In particular, it is of topological origin. The underlying category of singular cobordisms was used in [**2**] by Blanchet to fix the functoriality of Khovanov homology. Its objects are certain trivalent graphs and its morphisms are singular cobordisms whose boundary are such trivalent graphs. We call these singular cobordisms $\mathfrak{gl}_2$-*foams* (or *foams* for short). Note that Blanchet's category is a sign modified version of the original cobordism category which describes Khovanov homology and which was for instance used by Bar-Natan in his formulation of Khovanov homology, see [**1**]. The fact that such a twist in the definition of Khovanov homology solves the functoriality leaves the question whether this could also be fixed *algebraically* using the original construction of Khovanov involving his arc algebra, see [**18**].

We therefore suggest here to study a certain signed (with highly non-trivial sign modifications) version $\mathfrak{A}^{\mathfrak{F}}$ of Khovanov's original algebra, which we call the *Blanchet-Khovanov algebra.* This is a graded $\mathbb{K}$-algebra defined diagrammatically via explicit multiplication rules on a distinguished set of basis vectors similar to the family of algebras from [**4**] or [**12**].

The main result of the paper is then that $\mathfrak{A}^{\mathfrak{F}}$ is an algebraic counterpart of $\mathfrak{W}$:

THEOREM. There is an equivalence of graded, $\mathbb{K}$-linear 2-categories

$$\mathbf{\Phi}: \mathfrak{W}\text{-}\mathbf{biMod}^p_{\mathrm{gr}} \xrightarrow{\cong} \mathfrak{A}^{\mathfrak{F}}\text{-}\mathbf{biMod}^p_{\mathrm{gr}}$$

induced by an isomorphism of graded algebras

$$\Phi: \mathfrak{W}^{\circ} \to \mathfrak{A}^{\mathfrak{F}}.$$

(Where $\mathfrak{W}^{\circ}$ is a certain subalgebra of $\mathfrak{W}$.) ∎

This provides a direct link between the topological and the algebraic point of view. As a consequence, computations (which are hard to do in practice on the topological side) can be done on the algebraic side, whereas the associativity (a non-trivial fact on the algebraic side) is clear from the topological point of view.

The set-up in more details. In his pioneering work [**18**], Khovanov introduced the so-called *arc algebra* H_m. One of his main purposes was to extend his celebrated categorification of the Jones polynomial [**17**] to tangles. To a given tangle with $2m$ bottom boundary points and $2m'$ top boundary points one associates a certain complex of graded H_m-$H_{m'}$-bimodules. He showed that the chain homotopy equivalence class of this complex is an invariant of the tangle. Moreover, taking the tensor product with a certain H_m-module from the left and a certain $H_{m'}$-module from the right produces a complex which is still an invariant. On the level of Grothendieck groups this invariant descends to the Kauffman bracket of the tangle.

In this set-up it makes sense to ask if cobordisms between tangles correspond to natural transformations between bimodules. Or said in other words, whether there is a 2-functor from the 2-category of tangles to a certain 2-category of $H = \bigoplus_{m\in\mathbb{Z}_{\geq 0}} H_m$-bimodules. This is often called *functoriality.*

In a series of papers [**4**], [**5**], [**6**], [**7**] and [**8**] a *generalization* $\mathfrak{A}_\Lambda$ of the arc algebra was studied revealing that Khovanov's arc algebra has, left aside its knot theoretical origin, interesting representation theoretical, algebraic geometrical and combinatorial properties. These algebras $\mathfrak{A}_\Lambda$ were defined using an algebraic approach via the *combinatorics of arc diagrams*, i.e. certain diagrams consisting of embedded lines in $\mathbb{R}^2$ inspired by the diagrams for Temperley-Lieb algebras.

This series of results has led to several variations and generalizations of Khovanov's original formulation, utilized in a large body of work by several researchers (including the authors of this paper), e.g. an $\mathfrak{sl}_3$-variation considered in [**27**], [**32**], [**31**] and [**36**], and an $\mathfrak{sl}_n$-variation studied in [**26**] and [**37**], all of them having relations to (cyclotomic) KL-R algebras as in [**20**] or [**33**], and link homologies in the sense of Khovanov and Rozansky [**21**]. There is also the $\mathfrak{gl}_{1|1}$-variation developed in [**34**] with relations to the Alexander polynomial as well as a type **D**-version introduced in [**11**] and [**12**] with connections to the representation theory of Brauer's centralizer algebras and orthosymplectic Lie superalgebras, see e.g. [**13**].

A fact we like to stress about the $\mathfrak{sl}_3/\mathfrak{sl}_n$-variations is that their *graded 2-categories of biprojective, finite-dimensional modules* are equivalent to certain *graded 2-categories of $\mathfrak{sl}_3/\mathfrak{sl}_n$-foams*, the analogues of Bar-Natan's cobordism category [**1**] studied e.g. in [**19**], [**24**], [**28**] and [**29**] from the viewpoint of link homologies. (We note hereby that such a topological description for the type **D**-version was recently found in [**15**], providing, in some sense, the first 'foamy description' outside of type **A**.)

We like to stress that Khovanov's original construction as well as Bar-Natan's reformulation from [**1**] are *not* functorial, but are *functorial up to signs*, see [**1**], [**16**], [**30**] or [**35**]. It became clear that Bar-Natan cobordisms miss some subtle extra signs (see for example [**10**] for the first fix of functoriality using "disoriented" cobordisms).

A solution to this problem, that is of key interest for us, was provided by Blanchet in [**2**]. He formulated Khovanov's link homology using certain singular cobordisms, that we call ($\mathfrak{gl}_2$-)*foams*, which, by construction, include highly non-trivial signs fixing the functoriality of Khovanov's link homology. Moreover, Blanchet's formulation fits neatly into the framework of graded 2-representations of the categorified quantum group in the sense of [**20**], as it was shown in [**24**].

Using Blanchet's construction it makes sense to define "foamy" versions $\mathfrak{W}_{\vec{k}}$ of Khovanov's arc algebra, which we call ($\mathfrak{gl}_2$-)*web algebras*. We set $\mathfrak{W} = \bigoplus_{\vec{k}\in\mathtt{bl}^\diamond}\mathfrak{W}_{\vec{k}}$. The web algebras $\mathfrak{W}_{\vec{k}}$ and $\mathfrak{W}$ are graded $\mathbb{K}$-algebras defined using Blanchet's singular cobordisms and the multiplication is given by composition of singular cobordisms (for our conventions see Section 2). The signs within this multiplication are quite sophisticated, e.g. even merges (which are quite easy in the formulations of [**4**] and [**12**]) can come with a sign.

Unfortunately calculating in $\mathfrak{W}_{\vec{k}}$ and $\mathfrak{W}$ is very hard. Indeed, it is not even clear what a basis of $\mathfrak{W}_{\vec{k}}$ or $\mathfrak{W}$ is - left aside the question how to rewrite an arbitrary foam in terms of some basis. Thus, the main purpose of this paper is to give *algebraic counterparts* of $\mathfrak{W}_{\vec{k}}$ and $\mathfrak{W}$, denoted by $\mathfrak{A}^{\mathfrak{F}}_{\Lambda}$ and $\mathfrak{A}^{\mathfrak{F}} = \bigoplus_{\Lambda\in\mathtt{bl}^\diamond}\mathfrak{A}^{\mathfrak{F}}_{\Lambda}$ where these questions about bases are easy. We call the algebraic counterparts, which are built up using certain combinatorics of arc diagrams, *Blanchet-Khovanov algebras*.

The proof of our main theorem relies on the rather subtle Theorem 4.18 which needs careful treatment of all involved signs. The whole Subsection 4.5 is devoted to its proof. Our main theorem clarifies algebraically the deficiency in the original theory. For brevity, we stop our investigation here although several natural questions remain open, e.g. a direct representation theoretic construction of Blanchet-Khovanov algebras, see Remark 4.20.

To keep the paper self-contained we start by a rather detailed exposition of the main ingredients and players adapted to the main purpose of the paper.

Conventions used throughout.

CONVENTION 1.1. Let $\mathbb{K}$ denote a field of arbitrary characteristic. (We sometimes need to work with $\mathbb{K}(q)$ for a formal parameter q. All notions below are similarly defined in this case.) An *algebra* always means a non-necessarily finite-dimensional, non-necessarily unital $\mathbb{K}$-algebra A. We do not assume that such A's are associative and it will be a non-trivial fact that all A's which we consider are actually associative. Given two algebras A and B, then an *A-B-bimodule* is a $\mathbb{K}$-vector space M with a left action of A and a right action of B in the usual compatible sense. If $A = B$, then we also write *A-bimodule* for short. We call an A-B-bimodule M *biprojective*, if it is projective as a left A-module as well as a right B-module (such finitely generated bimodules are called sweet in [**18**, Subsection 2.6]). We denote the category of locally finite-dimensional A-bimodules by A-**biMod**, i.e. the category of bimodules M such that eMe' is finite-dimensional for any two primitive idempotents e and e'. Diagrammatic left (or right) actions will be given by acting on the bottom (or top). ▲

CONVENTION 1.2. By a *graded algebra* we mean an algebra A which decomposes into graded pieces $A = \bigoplus_{i\in\mathbb{Z}} A_i$ such that $A_iA_j \subset A_{i+j}$ for all $i,j \in \mathbb{Z}$. Given two graded algebras A and B, we study (and only consider) *graded* A-B-bimodules, i.e. A-B-bimodules $M = \bigoplus_{i\in\mathbb{Z}} M_i$ such that $A_iM_jB_k \subset M_{i+j+k}$ for all $i,j,k \in \mathbb{Z}$. We also set $M_i\{s\} = M_{i-s}$ for $s \in \mathbb{Z}$ (thus, positive integers shift up).

If A is a graded algebra and M is a graded A-bimodule, then $\overline{M}$ obtained from M by forgetting the grading is in A-**biMod**.

Given such A-bimodules $\overline{M}, \overline{N}$, then

$$\mathrm{Hom}_{A\text{-}\mathbf{biMod}}(\overline{M}, \overline{N}) = \bigoplus_{s\in\mathbb{Z}} \mathrm{Hom}_0(M, N\{s\}). \tag{1}$$

Here Hom_0 means all degree-preserving A-homomorphisms, i.e. $f(M_i) \subset N_i$. ▲

CONVENTION 1.3. We consider three diagrammatic calculi in this paper: $\mathfrak{gl}_2$-webs (webs for short) in the sense of [**9**] and [**23**], foams whose definition is motivated from [**2**] and [**19**], and arc diagrams in the sense of [**4**] and [**6**]. Our reading convention for all of these is from bottom to top and from left to right. We often illustrate local pieces only; the diagram then is meant to be the identity or arbitrary outside of the displayed part (which one will be clear from the context). ▲

REMARK 1.4. We use colors in this paper. It is only necessary to distinguish colors for webs and foams. For the readers with a black-and-white version: we illustrate colored web edges using dashed lines, while colored foam facets appear shaded. ▲

2. $\mathfrak{gl}_2$-foams and $\mathfrak{gl}_2$-web algebras

In this section we introduce the foam 2-category $\mathfrak{F}$ and the web algebra $\mathfrak{W}$ in the spirit of Khovanov [**18**], but using foams à la Blanchet [**2**].

2.1. Webs, foams and TQFTs. We start by recalling the definition of a web. For this purpose, we denote by $\mathbb{bl}$ the set of all vectors $\vec{k} = (k_i)_{i\in\mathbb{Z}} \in \{0,1,\text{-}1,2,\text{-}2\}^{\mathbb{Z}}$ with $k_i = 0$ for $|i| \gg 0$. Abusing notation, we also sometimes write $\vec{k} = (k_a,\dots,k_b)$ for some fixed part of $\vec{k}$ (with $a < b \in \mathbb{Z}$) where it is to be understood that all non-displayed entries are zero. By convention, the *empty vector* is the unique vector containing only zeros. We consider $\vec{k} \in \overline{\mathbb{bl}}$ as a set of discrete labeled points in $\mathbb{R} \times \{\pm 1\}$ (or in $\mathbb{R} \times \{0\}$) by putting the symbols k_i at position $(i, \pm 1)$ (or $(i, 0)$). We denote by $\mathbb{bl} \subset \overline{\mathbb{bl}}$ the subset of all vectors with entries from $\{0, 1, 2\}$ only.

DEFINITION 2.1. A *web* is an embedded labeled, oriented, trivalent graph which can be obtained by gluing (whenever this makes sense and the labels fit) or juxtaposition of finitely many (possibly zero) of the following pieces:

(2) 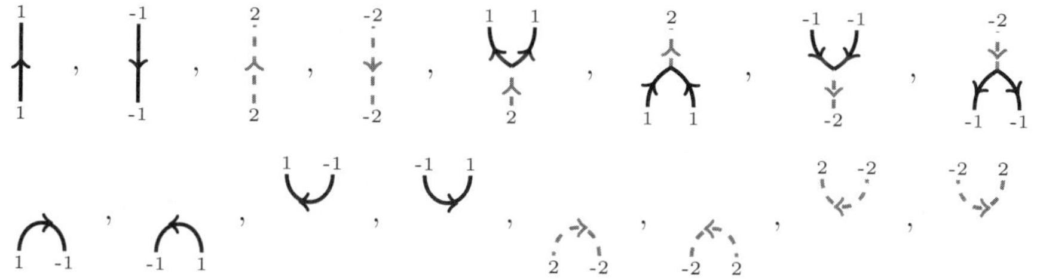

(Hence, this includes the *empty web*.) We assume that webs are embedded in $\mathbb{R} \times [-1, 1]$ such that each edge starts/ends either in a trivalent vertex or at the boundary of the strip at the points $(i, \pm 1)$. We assume that the points at $(i, \pm 1)$ are labeled 1, -1, 2 or -2. In particular, these webs have distinguished bottom $\vec{k}$ and top $\vec{l}$ boundary which we will throughout denote from left to right by $\vec{k} = (k_a, \ldots, k_b)$ and $\vec{l} = (l_{a'}, \ldots, l_{b'})$ where k_i is the label at $(i, -1)$ and l_i is the label at $(i, 1)$.

Edges come in two different version, i.e. *ordinary edges* which are only allowed to have boundary points labeled 1 or -1, and *phantom edges* which are only allowed to have boundary points labeled 2 or -2. As in (2), we draw phantom edges dashed (and colored); one should think of them as "non-existing".

We denote the set consisting of all webs with bottom boundary $\vec{k}$ and top boundary $\vec{l}$ by $\mathrm{Hom}_{\overline{\mathfrak{F}}}(\vec{k}, \vec{l})$ (for a reason that will become clear later). Given $\vec{k} \in \overline{\mathbb{bl}}$, we denote by $\mathbf{1}_{\vec{k}} \in \mathrm{Hom}_{\overline{\mathfrak{F}}}(\vec{k}, \vec{k})$ the identity web on $\vec{k}$. ▲

REMARK 2.2. For our purposes it will be mostly sufficient to consider webs in a "highest weight setup", i.e. only upwards pointing webs (see also Subsection 2.3). In particular, we do not need the labels -1 and -2 much in this paper. Still, all construction from this and the next subsection can be done in a more flexible setup using topological webs, which are however not necessary for our purposes. ▲

By a *surface* we mean a marked, orientable, compact surface with possible finitely many boundary components and with finitely many connected components. Additionally, by a *trivalent surface* we understand the same as in [**19**, Subsection 3.1], i.e. certain embedded, marked, singular cobordisms whose boundaries are webs.

Precisely, fix the following data denoted by $\boldsymbol{S}$:

(I) A surface S with connected components divided into two sets $\{S_1^{\mathrm{o}}, \ldots, S_r^{\mathrm{o}}\}$ and $\{S_1^{\mathrm{p}}, \ldots, S_{r'}^{\mathrm{p}}\}$. (The former are called *ordinary surfaces* and the latter are called *phantom surfaces*.)

(II) The boundary components of S are partitioned into triples $(C_i^{\mathrm{o}}, C_j^{\mathrm{o}}, C_k^{\mathrm{p}})$ such that each triple contains precisely one boundary component C_k^{p} of a phantom surface.

(III) The three circles $C_i^{\mathrm{o}}, C_j^{\mathrm{o}}$ and C_k^{p} in each triple are identified via diffeomorphisms $\varphi_{ij} \colon C_i^{\mathrm{o}} \to C_j^{\mathrm{o}}$ and $\varphi_{jk} \colon C_j^{\mathrm{o}} \to C_k^{\mathrm{p}}$.

(IV) A finite (possible empty) set of markers per connected components.

DEFINITION 2.3. Let $\boldsymbol{S}$ be as above. The *closed, singular trivalent surface* $f_c = f_c^{\boldsymbol{S}}$ attached to $\boldsymbol{S}$ is the CW-complex obtained as the quotient of S by the identifications φ_{ij} and φ_{jk}. We call all such f_c's *closed pre-foams* (following [**19**]) and their markers *dots*. A triple $(C_i^{\mathrm{o}}, C_j^{\mathrm{o}}, C_k^{\mathrm{p}})$ becomes one circle in f_c which we call a *singular seam*, while the interior of the connected components $S_1^{\mathrm{o}}, \dots, S_r^{\mathrm{o}}$ and $S_1^{\mathrm{p}}, \dots, S_{r'}^{\mathrm{p}}$ are facets of f_c, called *ordinary facets* respectively *phantom facets*. We embed these pre-foams into $\mathbb{R}^3$ in such a way that the three annuli glued to a singular seam can be oriented. We additionally choose orientations on the singular seams, compare to (3). ▲

EXAMPLE 2.4. Consider two spheres with two punctures respectively one puncture. The sphere with one puncture is assumed to be a phantom sphere (we color phantom facets in what follows).

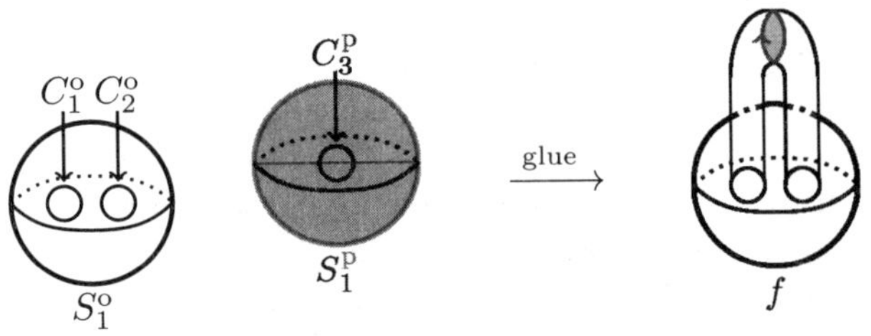

Then the pre-foam on the right is obtained by identifying the three boundary circles. (We have also chosen an orientation of the singular seam.) Another example of a closed pre-foam is given below in the proof of Lemma 2.13 (where we leave it to the reader to identify the precise labels). ▲

The pre-foams we have constructed so far are all closed. We will need *non-closed* pre-foams as well. To this end, we follow [**19**, Subsection 3.3] and consider a plane $P \cong \mathbb{R}^2 \subset \mathbb{R}^3$. We say that P intersects a closed pre-foam f_c *generically*, if $P \cap f_c$ is a non-oriented web (seen as a topological space).

DEFINITION 2.5. Let $P_{xy}^{\pm 1}$ be the xy-plane in $\mathbb{R}^3$ (embedded such that the third coordinate is ± 1). A *(non-closed) pre-foam* f is the intersection of $\mathbb{R}^2 \times [-1,1]$ with some closed pre-foam f_c such that $P_{xy}^{\pm 1}$ intersects f_c generically, and the intersection is a web as in Definition 2.1. We see such pre-foam f as a singular cobordism between $P_{xy}^{-1} \cap f_c$ (bottom, source) and $P_{xy}^{+1} \cap f_c$ (top, target) embedded in $\mathbb{R}^2 \times [-1,1]$. Moreover, there is an evident composition $g \circ f$ via gluing and rescaling. Similarly, we construct pre-foams embedded in $\mathbb{R} \times [-1,1] \times [-1,1]$ with vertical boundary components. These vertical boundary components should be the boundary of the webs at the bottom/top times $[-1,1]$. We consider such pre-foams modulo isotopies in $\mathbb{R} \times [-1,1] \times [-1,1]$ which fix the horizontal boundary as well as the vertical boundary, and the condition that generic slices are webs. ▲

We call pre-foam parts *ordinary*, if they do not contain singular seams or phantom facets, and we call pre-foam parts *ghostly*, if they only contain phantom facets.

EXAMPLE 2.6. Pre-foams can be seen as singular surfaces (with oriented, singular seams) in $\mathbb{R} \times [-1,1] \times [-1,1]$ such that the bottom boundary and the top boundary are webs with facets colored as follows:

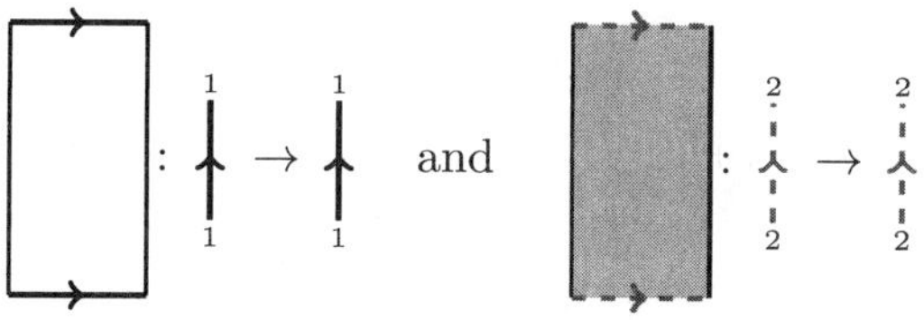

The leftmost facet is an ordinary facet. Whereas the rightmost facet is a phantom facet, and the reader might think of it as "non-existing" (similar to a phantom edge) - they only encode signs. The singularities of f are all locally of the following form (where the other orientations of the seams are also allowed)

(3) 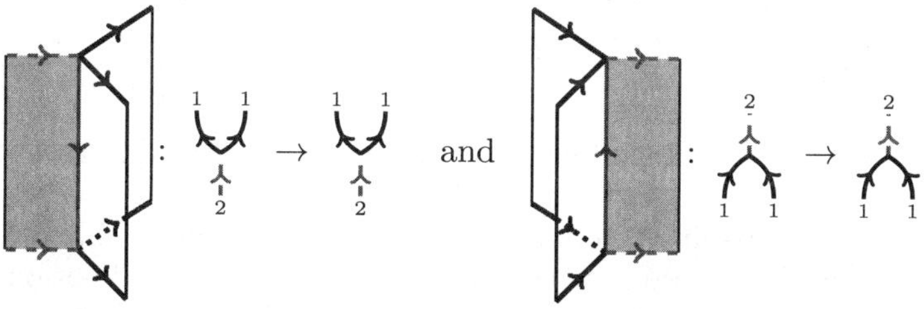

(Note that we consider webs in a "monoidal" way. Thus, we do not have to relate the orientations of facets/seam to the orientations of webs as e.g. in [**2**, Section 1].) Such pre-foams can carry dots that freely move around its facets:

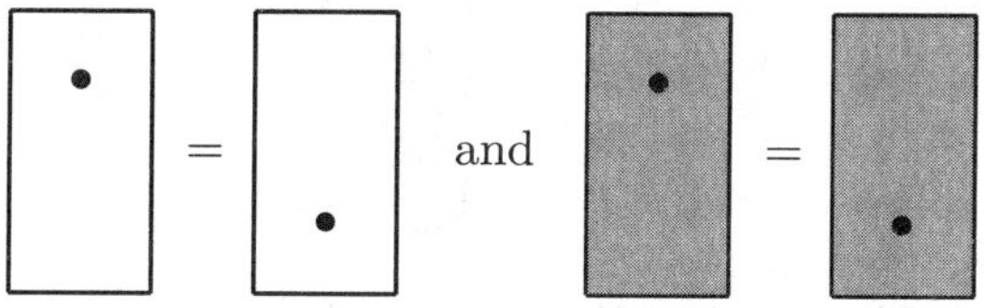

▲

REMARK 2.7. Pre-foams are considered modulo boundary preserving isotopies that do preserve the condition that each generic slice is a web. These isotopies form a finite list: isotopies coming from the two cobordism theories associated to the two different types of facets (see for example [**22**, Section 1.4]) and isotopies coming from isotopies of the singular seams seen as tangles in $\mathbb{R}^2 \times [-1, 1]$. ▲

To work with the 2-category of foams it will be enough (for our purposes) to consider its image under a certain (singular) TQFT functor defined by Blanchet [**2**]. Recall that equivalence classes of TQFTs for surfaces (i.e. a monoidal functor $\mathcal{Z}$ from the category of two-dimensional cobordisms to the category of $\mathbb{K}$-vector spaces) are in one-to-one correspondence with isomorphism classes of (finite-dimensional, associative) commutative Frobenius algebras. The reader unfamiliar with these notions might want to consult Kock's book [**22**] for a detailed account. Given a Frobenius algebra $\mathcal{A}$ corresponding to a TQFT $\mathcal{Z}_{\mathcal{A}}$, then the association is as follows. To a disjoint union of m circles one associates the m-fold tensor product $\mathcal{A}^{\otimes m}$. To a cobordism Σ with distinguished incoming and outgoing boundary components consisting of, let us say, m and m' circles, we assign a $\mathbb{K}$-linear map from $\mathcal{A}^{\otimes m}$ to $\mathcal{A}^{\otimes m'}$. Hereby the usual cup/cap respectively pants cobordisms correspond to the unit, counit, multiplication and comultiplication maps. These are the basic pieces of every cobordism. Then the TQFT assigns to Σ a $\mathbb{K}$-linear map

$$\mathcal{Z}_{\mathcal{A}}(\Sigma)\colon \mathcal{A}^{\otimes m} \to \mathcal{A}^{\otimes m'},$$

which is obtained by decomposing Σ into basic pieces.

The commutative Frobenius algebras we need are

$$(4) \qquad \mathcal{A}_{\mathrm{o}} = \mathbb{K}[X]/(X^2), \qquad \mathcal{A}_{\mathrm{p}} = \mathbb{K}$$

with induced multiplications, counits $\varepsilon_{\mathrm{o,p}}(\cdot)$ and comultiplications $\Delta_{\mathrm{o,p}}(\cdot)$ given via

$$\varepsilon_{\mathrm{o}}(1) = 0, \qquad \varepsilon_{\mathrm{o}}(X) = 1, \qquad \varepsilon_{\mathrm{p}}(1) = -1,$$
$$\Delta_{\mathrm{o}}(1) = 1 \otimes X + X \otimes 1, \qquad \Delta_{\mathrm{o}}(X) = X \otimes X, \qquad \Delta_{\mathrm{p}}(1) = -1 \otimes 1.$$

Thus, we have the traces

$$(5) \quad \mathrm{tr}_{\mathrm{o}}(1 \otimes 1) = \mathrm{tr}_{\mathrm{o}}(X \otimes X) = 0,\ \mathrm{tr}_{\mathrm{o}}(1 \otimes X) = \mathrm{tr}_{\mathrm{o}}(X \otimes 1) = 1,\ \mathrm{tr}_{\mathrm{p}}(1 \otimes 1) = -1.$$

We associate the Frobenius algebra $\mathcal{A}_{\mathrm{o}}$ to the ordinary parts, and the Frobenius algebra $\mathcal{A}_{\mathrm{p}}$ to the phantom parts of a pre-foam f using the usual notion of a TQFT functor, but extended to surfaces marked with dots via multiplication by X or -1:

(6) 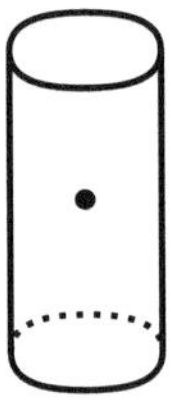 $\overset{\mathcal{Z}_{\mathcal{A}_{\mathrm{o}}}}{\longmapsto} \cdot X\colon \mathcal{A}_{\mathrm{o}} \to \mathcal{A}_{\mathrm{o}},$ 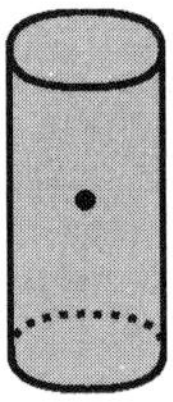 $\overset{\mathcal{Z}_{\mathcal{A}_{\mathrm{p}}}}{\longmapsto} \cdot(-1)\colon \mathcal{A}_{\mathrm{p}} \to \mathcal{A}_{\mathrm{p}}.$

(Note that pre-foams without singular seams and markers are surfaces in the usual sense.) By the universal construction given in [**3**], it is no problem to extend the usual construction of TQFTs to the marked setup using (6). We leave the details to the reader.

EXAMPLE 2.8. If we view a $\mathbb{K}$-linear map $\phi\colon \mathbb{K} \to \mathcal{A}_{\mathrm{o,p}}^{\otimes m}$ as $\phi(1) \in \mathcal{A}_{\mathrm{o,p}}^{\otimes m}$, then

(7)  $\overset{\mathcal{Z}_{\mathcal{A}_{\mathrm{o}}}}{\longmapsto} 1 \in \mathcal{A}_{\mathrm{o}},$  $\overset{\mathcal{Z}_{\mathcal{A}_{\mathrm{o}}}}{\longmapsto} X \in \mathcal{A}_{\mathrm{o}},$  $\overset{\mathcal{Z}_{\mathcal{A}_{\mathrm{p}}}}{\longmapsto} 1 \in \mathcal{A}_{\mathrm{p}}.$

Here we have from left to right ι_{o}, $(\cdot X) \circ \iota_{\mathrm{o}}$ and ι_{p} as maps. These are sometimes called (marked) *units*. The *counits* $\varepsilon_{\mathrm{o,p}}$ are obtained by flipping the pictures. ▲

Note that the values of the non-closed surfaces can be determined by closing them in all possible ways using (7) and its dual. Here and throughout, we say for short that a relation $a = b$ (of formal $\mathbb{K}$-linear combinations of marked surfaces) *lies in the kernel of a TQFT functor* $\mathcal{Z}$, if $\mathcal{Z}(a) = \mathcal{Z}(b)$ as $\mathbb{K}$-linear maps. (Similarly later on for singular TQFT functors as defined below.)

LEMMA 2.9. The following *ordinary sphere relations*, the *ghostly sphere relation* and the *cyclotomic relations*

(8) 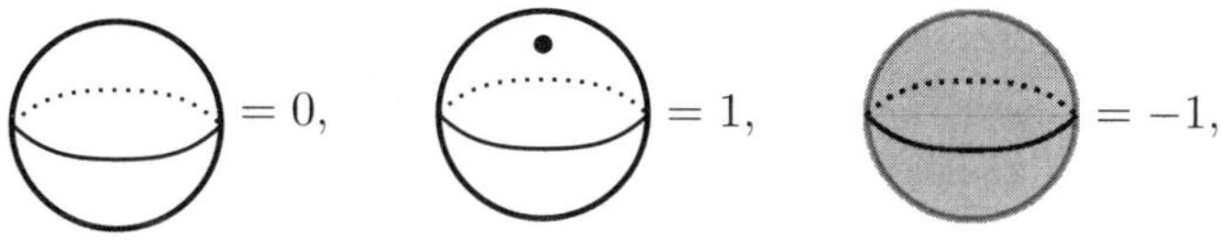

(9) 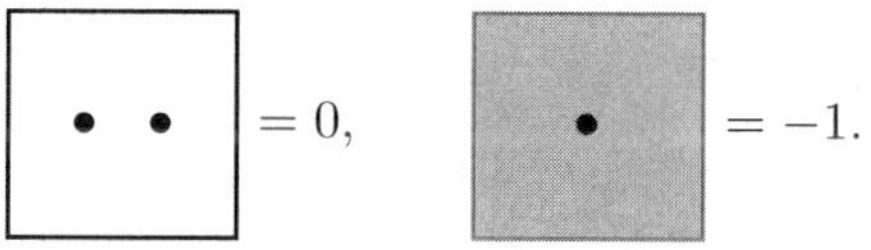

as well as the following *ordinary and ghostly neck cutting relations*

(10) 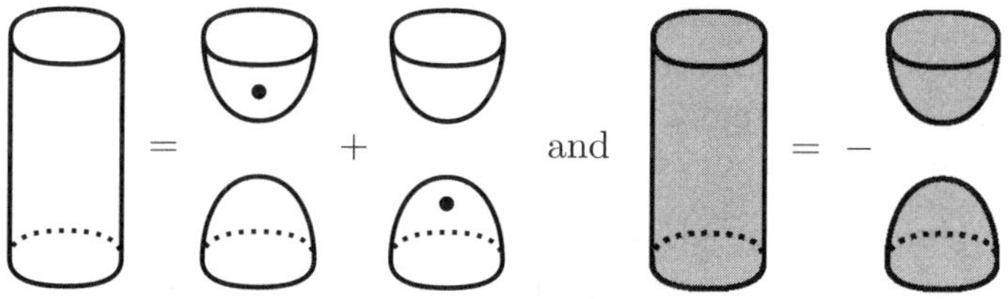

are in the kernel of $\mathcal{T}_{\mathcal{A}_o}$ (ordinary) respectively of $\mathcal{T}_{\mathcal{A}_p}$ (ghostly). ■

PROOF. Via direct calculation. For instance, the ordinary respectively ghostly neck cutting relations decompose the identity map as $\mathrm{id} = (\cdot X)\circ\iota_o\circ\varepsilon_o + \iota_o\circ\varepsilon_o\circ(\cdot X)$ respectively $\mathrm{id} = -\iota_p \circ \varepsilon_p$ with units $\iota_{o,p}$ and counits $\varepsilon_{o,p}$ as in (7). □

The neck cutting relations (10) give a topological interpretation of a dot as a shorthand notation for ($\frac{1}{2}$-times) a handle, see also [**1**, (4)].

2.2. Blanchet's singular TQFT construction. The following definition follows the construction given by Blanchet, see [**2**, Subsection 1.5].

We want to construct a monoidal functor $\mathcal{T}$ on the category whose objects are webs with values in finite-dimensional $\mathbb{K}$-vector spaces. To this end, let $p\mathbf{F}$ denote the category whose objects are webs and whose morphisms are pre-foams (composition is gluing of pre-foams). We view $p\mathbf{F}$ as a monoidal category by juxtaposition of webs and pre-foams. Moreover, we define for $a, b, c, d \in \mathbb{K}$ two maps:

$$(11)\qquad \begin{aligned} \alpha_{\mathcal{A}_o} &: \mathcal{A}_o \otimes \mathcal{A}_o \to \mathcal{A}_o,\ (a+bX)\otimes(c+dX) \mapsto (a+bX)(c-dX), \\ \alpha_{\mathcal{A}_p} &: \mathcal{A}_p \to \mathcal{A}_o,\ 1 \mapsto 1. \end{aligned}$$

DEFINITION 2.10. Let $\mathcal{Z}_{\mathcal{A}_o}$ and $\mathcal{Z}_{\mathcal{A}_p}$ denote the TQFTs associated to $\mathcal{A}_o$ and $\mathcal{A}_p$ from (4). Given a closed pre-foam f_c, let $\dot{f}_c = f_o \dot{\cup} f_p$ be the pre-foam obtained by cutting f_c along the singular seams (of which we assume to have m in total). Here f_o is the surface which in f_c is attached to the ordinary parts and f_p is the surface which in f_c is attached to phantom parts. Note that the boundary of f_o splits into σ_i^+ and σ_i^- for each $i \in \{1, \ldots, m\}$. Which one is which depends on the orientation of the singular seam: use the right hand rule with the index finger pointing in the direction of the singular seam and the middle finger pointing in direction of the attached phantom facet, then the thumb points in direction of σ_i^+. In contrast, f_p has only boundary components σ_i for each $i \in \{1, \ldots, m\}$. Now

$$\mathcal{Z}_{\mathcal{A}_o}(f_o) \in \bigotimes_{i=1}^{m}(\mathcal{Z}_{\mathcal{A}_o}(\sigma_i^+)\otimes\mathcal{Z}_{\mathcal{A}_o}(\sigma_i^-)), \qquad \mathcal{Z}_{\mathcal{A}_p}(f_p) \in \bigotimes_{i=1}^{m}\mathcal{Z}_{\mathcal{A}_p}(\sigma_i).$$

Let $\mathrm{tr}_{\mathrm{o}}\colon \mathcal{A}_{\mathrm{o}} \to \mathbb{K}$ be as in (5), and let $\alpha_{\mathcal{A}_{\mathrm{o}}}, \alpha_{\mathcal{A}_{\mathrm{p}}}$ be as in (11). Then we set

$$\mathcal{T}(f_c) = (\mathrm{tr}_{\mathrm{o}})^{\otimes 2m}(\alpha_{\mathcal{A}_{\mathrm{o}}}^{\otimes m}(\mathcal{Z}_{\mathcal{A}_{\mathrm{o}}}(f_{\mathrm{o}})) \otimes \alpha_{\mathcal{A}_{\mathrm{p}}}^{\otimes m}(\mathcal{Z}_{\mathcal{A}_{\mathrm{p}}}(f_{\mathrm{p}}))) \in \mathbb{K}^{\otimes 2m} \cong \mathbb{K}.$$

This gives a well-defined value $\mathcal{T}(f_c) \in \mathbb{K}$ for all closed pre-foams f_c. ▲

A crucial insight of Blanchet is that this extends to pre-foams:

THEOREM 2.11. The construction from Definition 2.10 can be extended to a monoidal functor $\mathcal{T}\colon p\mathbf{F} \to \mathbb{K}\text{-}\mathbf{Vect}$. (We call such a functor a *singular TQFT.*) ■

PROOF. This follows from the universal construction from [**3**]. That is, the only thing one really needs to check for this is that the $\mathbb{K}$-vector spaces constructed via the universal construction are finite-dimensional. This is not a priori clear, but also not hard to show. First observe that the evaluation given above ensures the case for the $\mathbb{K}$-vector space associated to the empty web. By using the relations found below, one can show an analog of Lemma 4.3, which in turn provides recursively that the $\mathbb{K}$-vector space associated to any web is finite-dimensional. □

Note the following properties of pre-foams f, which follow by construction.

(I) The *topological reduction* $\hat{f}$ obtained by removing all phantom facets of f is the cobordism theory corresponding to $\mathcal{A}_{\mathrm{o}}$ from (4).
(II) The *phantom* $\check{f}$ obtained by removing all 1-labeled facets of f is the cobordism theory corresponding to $\mathcal{A}_{\mathrm{p}}$ from (4).

Hence, the relations from (8), (9) and (10) are also in the kernel of the functor $\mathcal{T}$ (for all possible orientations of the boundary webs).

LEMMA 2.12. Let $\tilde{f}$ be the pre-foam obtained from a pre-foam f by reversing the orientation of a singular seam. Then $f + \tilde{f} = 0$ is in the kernel of $\mathcal{T}$. ■

PROOF. Switching the orientation of a singular seam swaps the attached parts of σ_1^+ and σ_1^-. In particular, it swaps the two copies of $\mathcal{A}_{\mathrm{o}}$ in the source of $\alpha_{\mathcal{A}_{\mathrm{o}}}$ from (11) and hence, produces an extra sign (we note that the case $b = d = 0$ is killed by applying the trace ε_{o} in the formula for $\mathcal{T}(f_c)$). □

LEMMA 2.13. Let $a, b \in \mathbb{Z}_{\geq 0}$. The *sphere relations*, i.e.

(12)
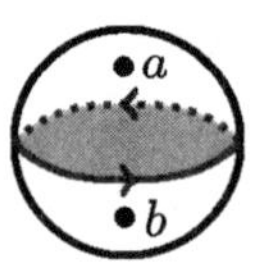

$$= \begin{cases} 1, & \text{if } a = 1, b = 0, \\ -1, & \text{if } a = 0, b = 1, \\ 0, & \text{otherwise,} \end{cases}$$

are in the kernel of $\mathcal{T}$. (We call such pre-foams *spheres.*) ■

PROOF. We prove the case $a = 0, b = 1$. The others are similar and omitted for brevity. Decompose f_c into (t=thumb, i=index finger, m=middle finger)

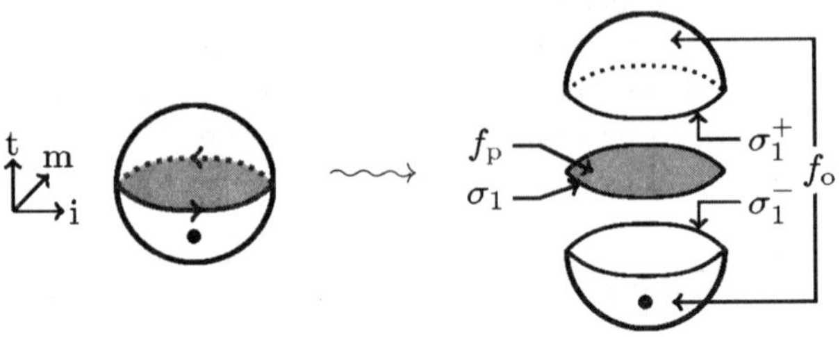

Now, because of the assignment in (7), we have $\mathcal{T}_{\mathcal{A}_{\text{o}}}(f_{\text{o}}) = 1 \otimes X$ and $\mathcal{T}_{\mathcal{A}_{\text{p}}}(f_{\text{p}}) = 1$. Thus, $\alpha_{\mathcal{A}_{\text{o}}}(\mathcal{T}_{\mathcal{A}_{\text{o}}}(f_{\text{o}})) = -X$ and $\alpha_{\mathcal{A}_{\text{p}}}(\mathcal{T}_{\mathcal{A}_{\text{p}}}(f_{\text{p}})) = 1$, both considered in $\mathcal{A}_{\text{o}}$. Applying the trace tr_{o} to $-X \otimes 1$ gives -1 as in (12). □

LEMMA 2.14. The *bubble removals* (a "sphere" in a phantom plane; the top dots are meant to be on the front facets and the bottom dots on the back facets)

(13) 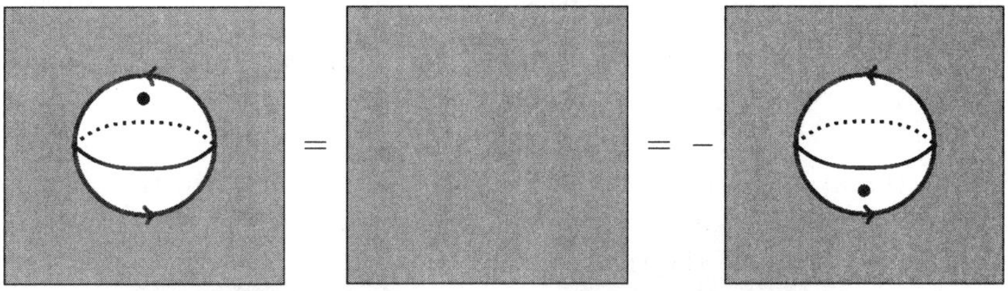

(14) 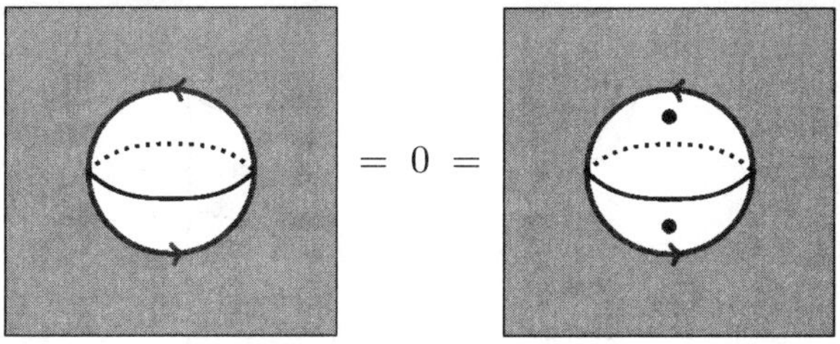

are in the kernel of $\mathcal{T}$. The *neck cutting relation*

(15) 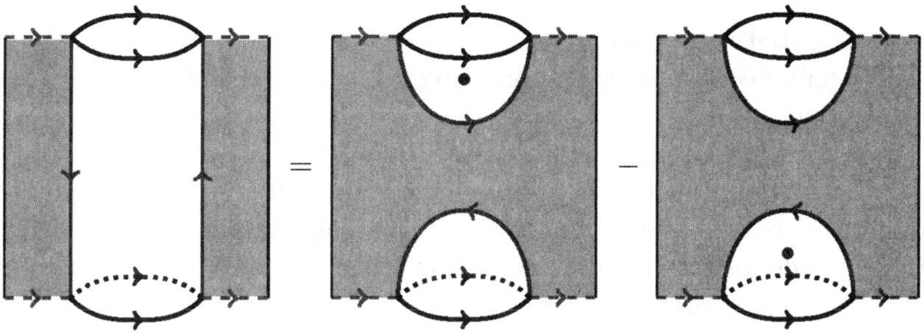

(with top dot on the front facet and bottom dot on the back facet) is also in the kernel of $\mathcal{T}$. Furthermore, the (left) *squeezing relation*

(16) 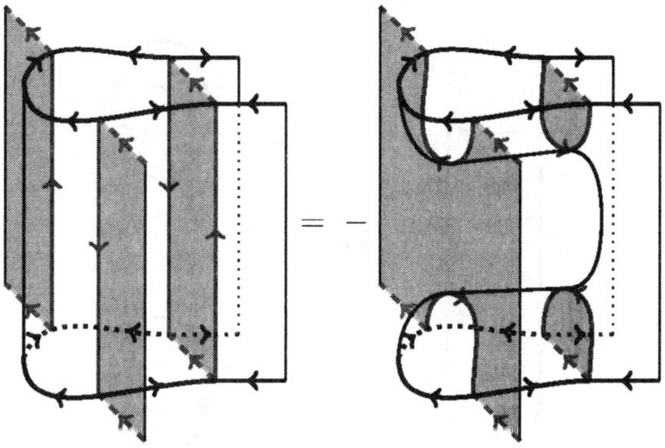

(there is also a similar right squeezing relation) and the *dot migrations*

(17) 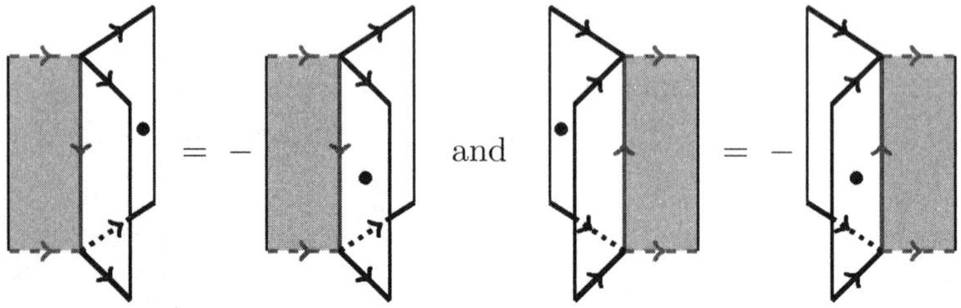

as well as the *ordinary-to-phantom neck cutting relations* (in the leftmost picture the upper closed circle is an ordinary facet, while the lower closed circle is a phantom facet, and vice versa for the rightmost picture)

(18) 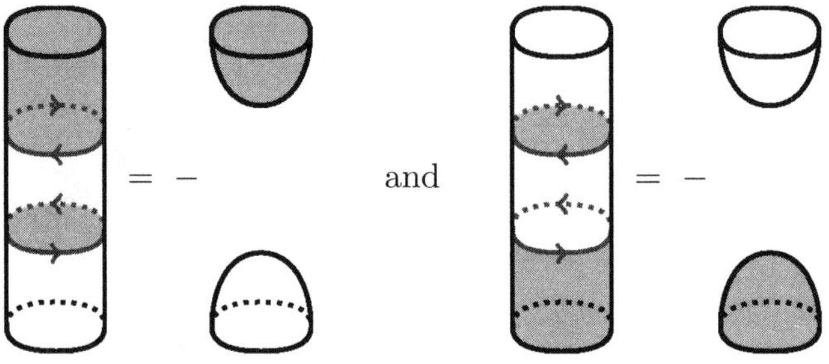

(we have omitted the orientations of the circle-shaped webs) are also in the kernel of $\mathcal{T}$. (We only need these in the following. But there are also similar relations with different orientations of the webs.) ∎

The leftmost situation in (15) is called a *cylinder* - as all local parts of pre-foams f such that the corresponding part in $\hat{f}$ is a cylinder. Note that the squeezing relation (16) enables us to use the neck cutting (15) on any such cylinders.

PROOF. We only prove the left equation in (18). First note that we have to consider all possible ways to close the non-closed pre-foam on the left-hand and on the right-hand side of the equation. We consider the closing

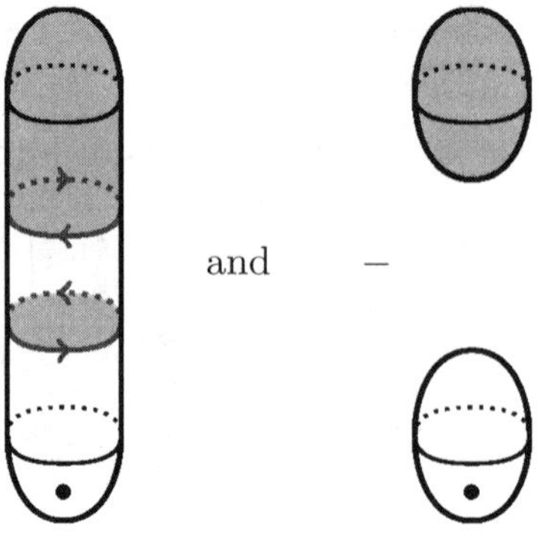

since all other possibilities give zero (as the reader might want to check). By (8), the right-hand closed pre-foams evaluate to $-(1 \cdot (-1))$. Now:

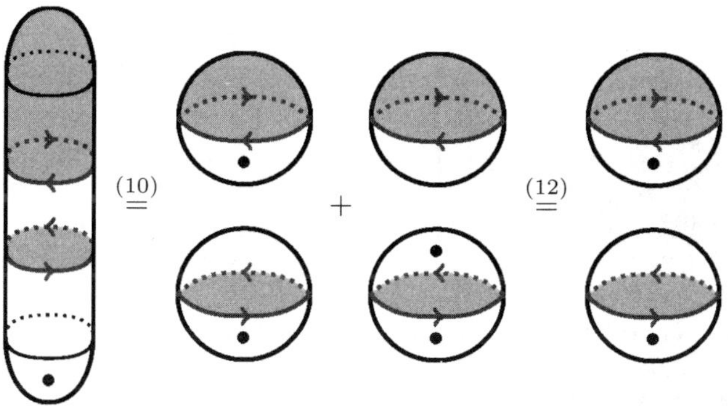

The bottom sphere evaluates to -1 because of (12). Moreover, performing the same steps as in the proof of Lemma 2.13, we see that the top sphere also evaluates to -1. Thus, the left-hand and the right-hand side evaluate to the same value. This shows that the first equation in (18) is in the kernel of $\mathcal{T}$. The other relations are verified similarly, see also [**2**, Lemma 1.3] or [**24**, Subsection 3.1]. $\square$

If we define a grading on the TQFT-modules by setting $\deg(1) = -1$ and $\deg(X) = 1$, then the TQFT $\mathcal{Z}_{\mathcal{A}_{\mathrm{o}}}$ respects the grading, where the degree of a cobordism Σ is given by $\deg(\Sigma) = -\chi(\Sigma) + 2 \cdot \text{dots}$. Here $\chi(\Sigma)$ is the topological Euler characteristic of Σ, that is, the number of vertices minus the number of edges plus the number of faces of Σ seen as a CW complex, and "dots" is the total number of dots. Additionally, we can see the TQFT $\mathcal{Z}_{\mathcal{A}_{\mathrm{p}}}$ as being trivially graded. Motivated by this we define the following.

DEFINITION 2.15. Given a pre-foam f, we define its *degree*

$$\deg(f) = -\chi(\hat{f}) + 2 \cdot \text{dots} + \tfrac{1}{2}\text{vbound},$$

where vbound is the total number of vertical boundary components. If $\hat{f}$ is the empty cobordism, then, by convention, $\chi(\hat{f}) = 0$. ▲

EXAMPLE 2.16. For example,

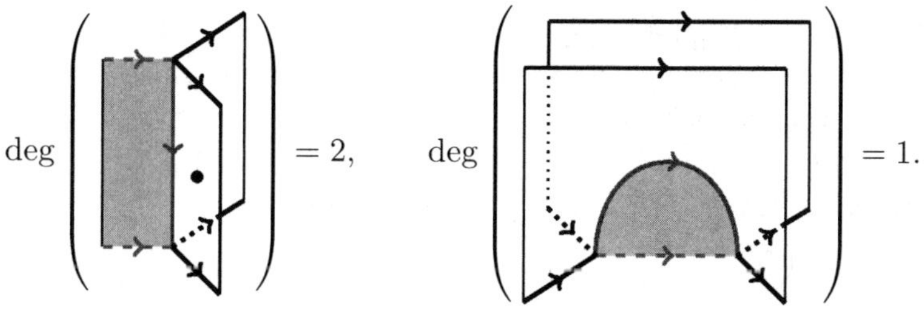

The leftmost pre-foam is called a *dotted cup* (the name will become clear in Lemma 2.27), while the rightmost pre-foam is called a *saddle* (there are also saddles obtained by flipping the picture upside down). Furthermore, we have

deg 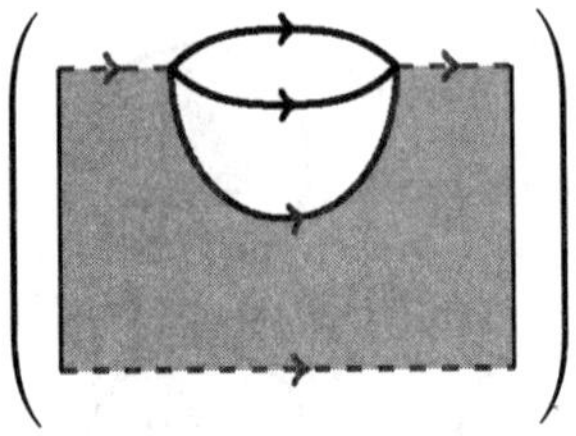 $= -1 =$ deg 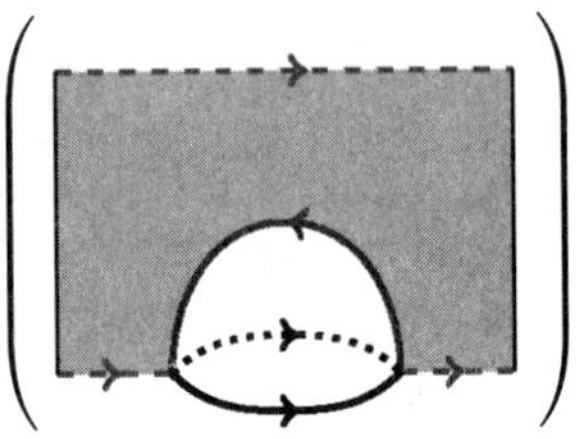

for the pre-foams called *cup* respectively *cap*. ▲

The 2-category we like to study is cooked up from $\mathcal{T}$ as follows.

DEFINITION 2.17. Let $\overline{\mathfrak{F}}$ be the $\mathbb{K}$-linear 2-category given by:

- The objects are all $\vec{k} \in \overline{\mathbb{bl}}$.
- The morphisms spaces $\mathrm{Hom}_{\overline{\mathfrak{F}}}(\vec{k}, \vec{l})$ are as in Definition 2.1.
- The 2-morphisms spaces $2\mathrm{Hom}_{\overline{\mathfrak{F}}}(u, v)$ for two webs u, v is the $\mathbb{K}$-linear span of all pre-foams with bottom boundary u and top boundary v.
- Vertical compositions $g \circ f$ of pre-foams by stacking g on top of f, horizontal composition $g \otimes f$ by putting g to the right of f (whenever those operations make sense).
- We take everything modulo the relations from (8), (9) and (10), as well as the relations found in Lemmas 2.12, 2.13 and 2.14.

Since the relations are degree preserving, $\overline{\mathfrak{F}}$ is a graded, $\mathbb{K}$-linear 2-category by taking the degree from Definition 2.15. Similarly, the "highest weight" 2-subcategory $\mathfrak{F}$ of $\overline{\mathfrak{F}}$ is the full 2-subcategory consisting of only webs with upwards pointing edges. (We will only consider $\mathfrak{F}$ in the following.) ▲

We call the 2-morphisms in $\mathfrak{F}$ (or in $\overline{\mathfrak{F}}$) *foams*. Moreover, all notions we had for pre-foams can be adapted to the setting of foams and we do so in the following. Note that the objects and the morphisms of $\mathfrak{F}$ can be seen as a $\mathbb{K}(q)$-linear category (by considering the spaces $\mathrm{Hom}_{\mathfrak{F}}(\vec{k}, \vec{l})$ as $\mathbb{K}(q)$-linear vector spaces whose basis are the webs in $\mathrm{Hom}_{\mathfrak{F}}(\vec{k}, \vec{l})$) which we denote by **Web**.

2.3. An action of the quantum group $\dot{\mathrm{U}}_q(\mathfrak{gl}_\infty)$. We denote by $\dot{\mathrm{U}}_q(\mathfrak{gl}_\infty)$ the $\mathbb{K}(q)$-linear category whose objects are given by $\vec{k} \in \mathbb{bl}$ and whose morphisms are generated by pairwise orthogonal idempotents $\mathbf{1}_{\vec{k}}$, and by $E_i^{(r)}\mathbf{1}_{\vec{k}}$ and $F_i^{(r)}\mathbf{1}_{\vec{k}}$ for $\vec{k} \in \mathbb{bl}$, $i \in \mathbb{Z}$ and $r \in \mathbb{Z}_{>0}$ (the generators $E_i^{(r)}\mathbf{1}_{\vec{k}}$ and $F_i^{(r)}\mathbf{1}_{\vec{k}}$ are called the *divided powers*) modulo some relations which are analogs of the relations in the quantum group $\mathrm{U}_q(\mathfrak{gl}_\infty)$ (see [**25**, Chapter 23]). We note that there exists a unique $\mathbf{1}_{\vec{l}}$ such that $\mathbf{1}_{\vec{l}}E_i^{(r)}\mathbf{1}_{\vec{k}} \neq 0$ and $\mathbf{1}_{\vec{l}}F_i^{(r)}\mathbf{1}_{\vec{k}} \neq 0$. This enable us to write $\mathbf{1}_{\vec{k}}$ only on one side of any expression. Now, there is a $\mathbb{K}(q)$-linear functor

$$\Phi_{\mathrm{Howe}}^{\mathbf{Web}} \colon \dot{\mathrm{U}}_q(\mathfrak{gl}_\infty) \to \mathbf{Web}$$

given on objects by $\Phi^{\mathbf{Web}}_{\text{Howe}}(\vec{k}) = \vec{k}$, and on morphisms by $\Phi^{\mathbf{Web}}_{\text{Howe}}(\mathbf{1}_{\vec{k}}) = \mathbf{1}_{\vec{k}}$ and (where we use a simplified, "rectangular", notation for webs):

(19)

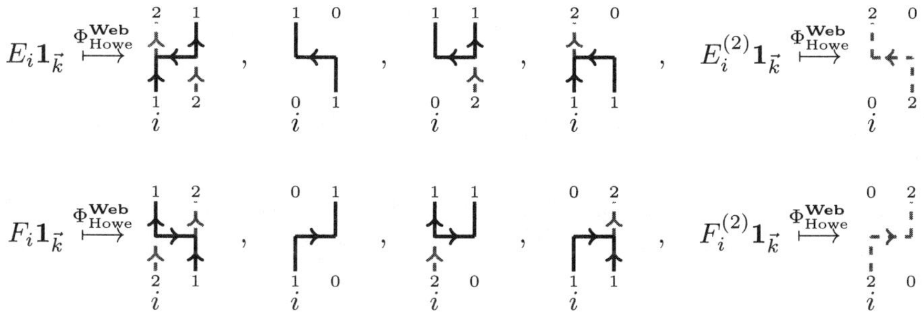

Here the generators $E_i\mathbf{1}_{\vec{k}}$, $F_i\mathbf{1}_{\vec{k}}$, $E_i^{(2)}\mathbf{1}_{\vec{k}}$ and $F_i^{(2)}\mathbf{1}_{\vec{k}}$ are sent to the local (between strand i and $i+1$) pictures above (we have displayed all possibilities depending on $\vec{k}$ at position k_i and k_{i+1}). Moreover, all higher divided powers $E_i^{(r)}\mathbf{1}_{\vec{k}}, F_i^{(r)}\mathbf{1}_{\vec{k}}$ for $r > 2$ are sent to zero. We call webs that arise as $\Phi^{\mathbf{Web}}_{\text{Howe}}(X)$, for X being any composition of the $\mathbf{1}_{\vec{k}}, E_i^{(r)}\mathbf{1}_{\vec{k}}, F_i^{(r)}\mathbf{1}_{\vec{k}}$ generators, *EF-generated*, and, on the other hand, webs *F-generated*, if X is any composition of only $\mathbf{1}_{\vec{k}}, F_i^{(r)}\mathbf{1}_{\vec{k}}$ generators (in both cases no coefficients from $\mathbb{K}(q)$ are allowed to occur). For details about the functor $\Phi^{\mathbf{Web}}_{\text{Howe}}$ we refer to [**9**, Section 5].

EXAMPLE 2.18. If $\vec{k} = (0,0,1,0,2,0,1,2,0,0)$, then $E_1E_0E_{-1}E_2F_0\mathbf{1}_{\vec{k}}$ is sent to

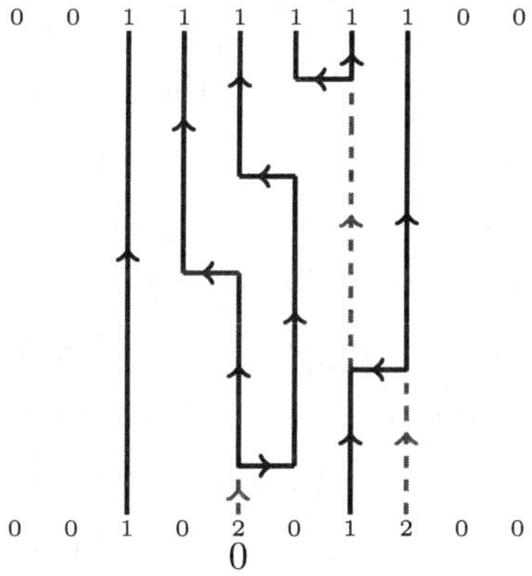

Here the first entry 2 is assumed to be at position 0. ▲

2.4. $\mathfrak{gl}_2$-web algebras. Now we define the following "algebraic" version of $\mathfrak{F}$. To this end, let $\ell \in \mathbb{Z}_{\geq 0}$ and let $\omega_\ell = (1,\dots,1,0,\dots,0)$ with ℓ numbers equal to 1. Fix $\vec{k}$ and let $\mathrm{Cup}(\vec{k}) = \mathrm{Hom}_{\mathfrak{F}}(2\omega_\ell, \vec{k})$ and $\mathrm{Cap}(\vec{k}) = \mathrm{Hom}_{\mathfrak{F}}(\vec{k}, 2\omega_\ell)$. (Thus, $\mathrm{Cup}(\vec{k}) = \emptyset$ and $\mathrm{Cap}(\vec{k}) = \emptyset$ if $\sum_{i\in\mathbb{Z}} k_i \neq 2\ell$.) Elements of these are called *cup webs* respectively *cap webs*. For diagrams in the multiplication process of $\mathfrak{W}$ described below, we need *cup-ray webs* as well, i.e. elements of $\mathrm{CupRay}(\vec{k}) = \mathrm{Hom}_{\mathfrak{F}}(\omega_{\ell+\ell'} + \omega_\ell, \vec{k})$ for $\vec{k} \in \mathbb{bl}$ (and similarly defined *cap-ray webs*).

DEFINITION 2.19. Let $u, v \in \mathrm{Cup}(\vec{k}), \vec{k} \in \mathbb{bl}$. We denote by ${}_u(\mathfrak{W}^{\natural}_{\vec{k}})_v$ the space $2\mathrm{Hom}_{\mathfrak{F}}(u, v)$. The *web algebra* $\mathfrak{W}^{\natural}_{\vec{k}}$ *for* $\vec{k} \in \mathbb{bl}$ and the *(full) web algebra* $\mathfrak{W}^{\natural}$ are the graded $\mathbb{K}$-vector spaces

$$\mathfrak{W}^{\natural}_{\vec{k}} = \bigoplus_{u,v \in \mathrm{Cup}(\vec{k})} {}_u(\mathfrak{W}^{\natural}_{\vec{k}})_v, \qquad \mathfrak{W}^{\natural} = \bigoplus_{\vec{k} \in \mathbb{bl}} \mathfrak{W}^{\natural}_{\vec{k}},$$

whose grading is induced by the grading in $\mathfrak{F}$. We consider these as graded algebras with multiplication given by composition in $\mathfrak{F}$. ▲

REMARK 2.20. Note that $\mathfrak{W}^{\natural}_{\vec{k}}$ is defined via composition of foams and thus, forms a graded, associative, unital algebra. Similarly for (the locally unital) algebra $\mathfrak{W}^{\natural}$. ▲

REMARK 2.21. Although new in this form, the algebras from Definition 2.19 are of course inspired by Khovanov's original arc algebras from [**18**]. Consequently, we obtain that $\mathfrak{W}_{\vec{k}}$ is a graded Frobenius algebra (by copying [**27**, Theorem 3.9]). ▲

DEFINITION 2.22. Denote by $\mathbb{bl}^{\diamond} \subset \mathbb{bl}$ the set of all $\vec{k} \in \mathbb{bl}$ which have an even number of entries 1. We call elements of $\mathbb{bl}^{\diamond}$ *balanced.* ▲

REMARK 2.23. Clearly there are no cups respectively caps if $\vec{k} \in \mathbb{bl} - \mathbb{bl}^{\diamond}$. Hence, $\mathfrak{W}^{\natural}_{\vec{k}} = 0$ iff $\vec{k} \in \mathbb{bl} - \mathbb{bl}^{\diamond}$ or $\vec{k} = \emptyset$. Consequently, we restrict ourselves to balanced $\vec{k}$ in what follows. We note that the full set $\mathbb{bl}$ would be needed if one wants to study *generalized* web algebras in the sense of [**4**] or [**6**]. ▲

There is an alternative way to define the web algebras which is the one we will use later on. Thus, we make the following definition (and show below that it agrees with the one from Definition 2.19). We denote by * the involution on webs which flips the diagrams upside down and reverses their orientations.

DEFINITION 2.24. Let $u, v \in \mathrm{Cup}(\vec{k}), \vec{k} \in \mathbb{bl}^{\diamond}$. We denote by ${}_u(\mathfrak{W}_{\vec{k}})_v$ the space $2\mathrm{Hom}_{\mathfrak{F}}(\mathbf{1}_{2\omega_\ell}, uv^*)\{d(\vec{k})\}$, where $d(\vec{k}) = \ell - \sum_{i \in \mathbb{Z}} k_i(k_i - 1)$. The *web algebra* $\mathfrak{W}_{\vec{k}}$ *for* $\vec{k} \in \mathbb{bl}^{\diamond}$ and the *(full) web algebra* $\mathfrak{W}$ are the graded $\mathbb{K}$-vector spaces

$$\mathfrak{W}_{\vec{k}} = \bigoplus_{u,v \in \mathrm{Cup}(\vec{k})} {}_u(\mathfrak{W}_{\vec{k}})_v, \qquad \mathfrak{W} = \bigoplus_{\vec{k} \in \mathbb{bl}^{\diamond}} \mathfrak{W}_{\vec{k}}.$$

We consider these as graded algebras with multiplication

$$\mathbf{Mult}\colon \mathfrak{W}_{\vec{k}} \otimes \mathfrak{W}_{\vec{k}} \to \mathfrak{W}_{\vec{k}},\ f \otimes g \mapsto \mathbf{Mult}(f, g) = fg \tag{20}$$

using multiplication foams as follows. To multiply $f \in {}_u(\mathfrak{W}_{\vec{k}})_v$ with $g \in {}_{\tilde{v}}(\mathfrak{W}_{\vec{k}})_w$ stack the diagram $\tilde{v}w^*$ on top of uv^* and obtain $uv^*\tilde{v}w^*$. Then $fg = 0$ if $v \neq \tilde{v}$. Otherwise, pick the leftmost cup-cap pair indicated in the picture to the left below and perform a "surgery"

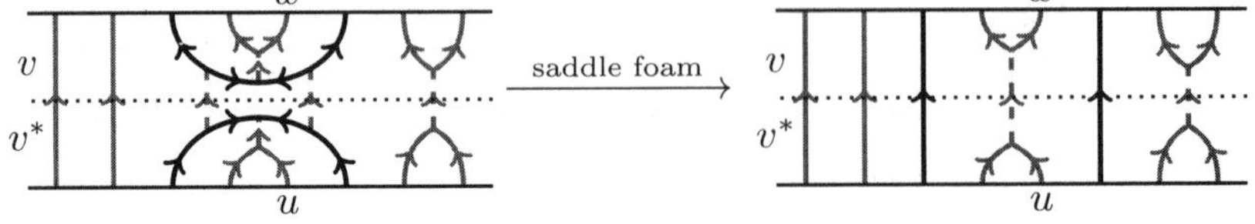

where the saddle foam is locally of the following form (and the identity elsewhere)

(21)

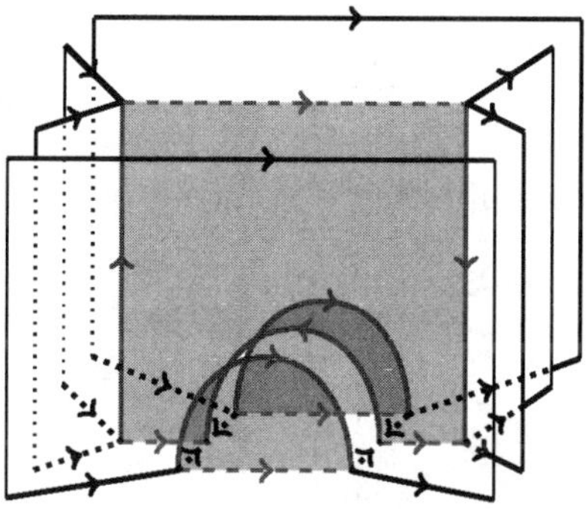

This foam should be read as follows: start with $f \in 2\mathrm{Hom}_{\mathfrak{F}}(\mathbf{1}_{2\omega_\ell}, uv^*vw)$ and stack on top of it a foam which is the identity at the bottom (u part) and top (w part) of the web and the saddle in between. Repeat until no cup-cap pair as above remains. This gives inductively rise to a multiplication foam (after the last surgery step we collapse the webs and foams). Compare also to [**27**, Definition 3.3]. ▲

Note that each intermediate step in the multiplication from (20) is a web of the form v^*v with $v \in \mathrm{CapRay}(\vec{k})$ and the multiplication foam is zero or a foam in $\mathrm{Hom}_{\mathfrak{F}}(uv^*vw^*, uw^*)$ (and thus, locally a foam in $\mathrm{Hom}_{\mathfrak{F}}(v^*v, \mathbf{1}_{\vec{k}})$). As a convention, we consider $u \in \mathrm{Cup}(\vec{k})$ as a web in $\mathbb{R} \times [-1, 0]$, $v^* \in \mathrm{Cap}(\vec{k})$ as a web in $\mathbb{R} \times [0, 1]$ such that $\vec{k} \in \mathbb{R} \times \{0\}$ whenever we use this viewpoint on $\mathfrak{W}$. Similarly for cup-ray and cap-ray webs.

LEMMA 2.25. The multiplication is degree preserving. ■

PROOF. The saddles as in (21) are always of degree 1. These have two boundary components labeled 1. Thus, the degree of all saddles within the multiplication procedure and the shift by $d(\vec{k})$ sum up to zero. □

EXAMPLE 2.26. An easy example illustrating the multiplication is

(22)

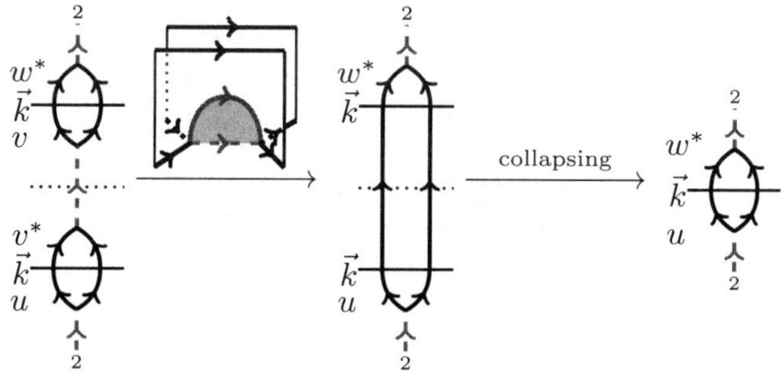

where the reader should think about any foam $f\colon \mathbf{1}_{2\omega_1} \to uv^*vw^*$ sitting underneath. The saddle is of degree 1 and thus, taking the shift $d(\vec{k})$ into account for $\vec{k} = (1, 1)$, the multiplication foam is of degree zero. ▲

The following shows that Definitions 2.19 and 2.24 agree.

LEMMA 2.27. We have $\mathfrak{W}^{\natural}_{\vec{k}} \cong \mathfrak{W}_{\vec{k}}$ and $\mathfrak{W}^{\natural} \cong \mathfrak{W}$ as graded algebras. ■

PROOF. Recall that the multiplication in $\mathfrak{W}^{\natural}_{\vec{k}}$ is composition, while the multiplication in $\mathfrak{W}_{\vec{k}}$ is given by multiplication foams. Thus, for the former we take foams $f \in 2\mathrm{Hom}_{\mathfrak{F}}(u,v)$ and $g \in 2\mathrm{Hom}_{\mathfrak{F}}(v,w)$ and obtain a foam $g \circ f \in 2\mathrm{Hom}_{\mathfrak{F}}(u,w)$, while for the latter we take foams $f \in 2\mathrm{Hom}_{\mathfrak{F}}(\mathbf{1}_{2\omega_\ell}, uv^*)$ and $g \in 2\mathrm{Hom}_{\mathfrak{F}}(\mathbf{1}_{2\omega_\ell}, vw^*)$ and obtain a foam $fg \in 2\mathrm{Hom}_{\mathfrak{F}}(\mathbf{1}_{2\omega_\ell}, uw^*)$ (in case the multiplication is non-zero). Now, the following "clapping of pictures" (as indicated by the arrows)

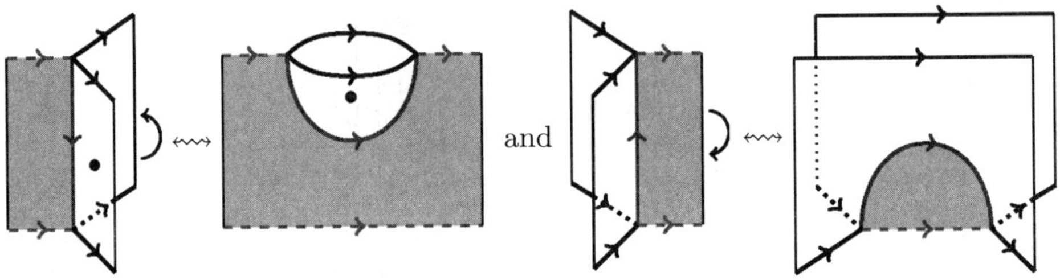

induces isomorphisms of $\mathbb{K}$-vector spaces

$$2\mathrm{Hom}_{\mathfrak{F}}(u,w) \cong 2\mathrm{Hom}_{\mathfrak{F}}(\mathbf{1}_{2\omega_\ell}, uw^*)\{d(\vec{k})\},$$
$$2\mathrm{Hom}_{\mathfrak{F}}(v,v) \cong 2\mathrm{Hom}_{\mathfrak{F}}(v^*v, \mathbf{1}_{2\omega_\ell})\{d(\vec{k})\}.$$

These are isomorphisms of graded $\mathbb{K}$-vector spaces since the shift by $d(\vec{k})$ encodes the vertical boundary components which are "lost" by the "clapping". Moreover, as indicated in the rightmost picture above, the multiplications in $\mathfrak{W}^{\natural}_{\vec{k}}$ and $\mathfrak{W}_{\vec{k}}$ are identified under this "clapping procedure". This shows the isomorphism of graded $\mathbb{K}$-algebras. For more details the reader might also consult [**27**, Lemma 3.7]. □

As a direct consequence of Remark 2.20 and Lemma 2.27 we obtain in particular the associativity of $\mathfrak{W}_{\vec{k}}$:

COROLLARY 2.28. The map **Mult**: $\mathfrak{W}_{\vec{k}} \otimes \mathfrak{W}_{\vec{k}} \to \mathfrak{W}_{\vec{k}}$ from Definition 2.24 is independent of the order in which the surgeries are performed. This turns $\mathfrak{W}_{\vec{k}}$ into a graded, associative, unital algebra. Similar for (the locally unital) algebra $\mathfrak{W}$. □

2.5. Web bimodules. We still consider only balanced $\vec{k}, \vec{l} \in \mathrm{bl}^{\diamond}$ in this subsection.

DEFINITION 2.29. Given any web $u \in \mathrm{Hom}_{\mathfrak{F}}(\vec{k}, \vec{l})$ (with boundaries $\vec{k}$ and $\vec{l}$ summing up to 2ℓ), we consider the $\mathfrak{W}$-bimodule

$$\mathfrak{W}(u) = \bigoplus_{\substack{v \in \mathrm{Cup}(\vec{k}),\\ w \in \mathrm{Cup}(\vec{l})}} 2\mathrm{Hom}_{\mathfrak{F}}(\mathbf{1}_{2\omega_\ell}, vuw^*)$$

with left (bottom) and right (top) action of $\mathfrak{W}$ as in Definition 2.24. We call such $\mathfrak{W}$-bimodules $\mathfrak{W}(u)$ *web bimodules*. ▲

PROPOSITION 2.30. Let $u \in \mathrm{Hom}_{\mathfrak{F}}(\vec{k}, \vec{l})$ be a web. Then the left (bottom) action of $\mathfrak{W}_{\vec{k}}$ and the right (top) action of $\mathfrak{W}_{\vec{l}}$ on $\mathfrak{W}(u)$ are well-defined and commute. Hence, $\mathfrak{W}(u)$ is a $\mathfrak{W}_{\vec{k}}$-$\mathfrak{W}_{\vec{l}}$-bimodule (and thus, a $\mathfrak{W}$-bimodule). ■

PROOF. Let $u \in \mathrm{Hom}_{\mathfrak{F}}(\vec{k}, \vec{l})$. Then, by construction, the left (bottom) action of $\mathfrak{W}_{\vec{k}}$ and the right (top) action of $\mathfrak{W}_{\vec{l}}$ commute since they are topologically "far apart". Hence, $\mathfrak{W}(u)$ is indeed a $\mathfrak{W}_{\vec{k}}$-$\mathfrak{W}_{\vec{l}}$-bimodule (and thus, a $\mathfrak{W}$-bimodule). □

Note that, given two webs $u, v \in \mathrm{Hom}_{\mathfrak{F}}(\vec{k}, \vec{l})$, then $\mathfrak{W}(u)$ and $\mathfrak{W}(v)$ could be isomorphic even though u and v are different, see for example (35).

PROPOSITION 2.31. The $\mathfrak{W}$-bimodules $\mathfrak{W}(u)$ are graded biprojective $\mathfrak{W}$-bimodules, with finite-dimensional subspaces for all pairs v, w. ■

PROOF. Clearly, they are graded, $\mathfrak{W}$-bimodules, but finite-dimensionality is not a priori clear. It follows from the existence of a cup foam basis as in Subsection 4.3. (More precisely, from Lemma 4.14.) They are also biprojective, because they are direct summands of some $\mathfrak{W}_{\vec{k}}$ (of some $\mathfrak{W}_{\vec{l}}$) as left (right) modules and for suitable $\vec{k} \in \mathbb{bl}^{\diamond}$ (or $\vec{l} \in \mathbb{bl}^{\diamond}$). See also [**27**, Proposition 5.11]. □

This proposition motivates the definition of the following 2-category which is one of the main objects that we are going to study.

DEFINITION 2.32. Given $\mathfrak{W}$ as above, let $\mathfrak{W}$-$\mathbf{biMod}^{p}_{\mathrm{gr}}$ be the following 2-category:

- Objects are the various $\vec{k} \in \mathbb{bl}^{\diamond}$.
- Morphisms are finite sums and tensor products (taken over the algebra $\mathfrak{W}$) of $\mathfrak{W}$-bimodules $\mathfrak{W}(u)$.
- The composition of $\mathfrak{W}$-bimodules is given by tensoring (over $\mathfrak{W}$).
- 2-morphisms are $\mathfrak{W}$-bimodule homomorphisms.
- The vertical composition of $\mathfrak{W}$-bimodule homomorphisms is the usual composition and the horizontal composition is given by tensoring (over $\mathfrak{W}$).

We consider $\mathfrak{W}$-$\mathbf{biMod}^{p}_{\mathrm{gr}}$ as a graded 2-category with 2-hom-spaces as in (1). ▲

3. Blanchet-Khovanov algebras

In this section we define the *Blanchet-Khovanov algebra*, following the framework of [**18**] and [**4**], but with signs differing at a number of crucial places.

3.1. Combinatorics of arc diagrams. We start with the notion of weights and blocks. These definitions are the same as in [**4**, Section 2] and, apart from the exact definition of blocks, as in [**12**, Sections 2 and 3].

DEFINITION 3.1. A *(diagrammatical) weight* is a sequence $\lambda = (\lambda_i)_{i\in\mathbb{Z}}$ with entries $\lambda_i \in \{\circ, \times, \vee, \wedge\}$, such that $\lambda_i = \circ$ for $|i| \gg 0$. Two weights λ and μ are said to be equivalent if one can obtain μ from λ by permuting some symbols $\wedge$ and $\vee$ in λ. The equivalence classes are called *blocks*. We denote by $\mathtt{bl}$ the set of blocks. ▲

To a block we assign a number of invariants.

DEFINITION 3.2. Let $\Lambda \in \mathtt{bl}$ be a block. To Λ we associate its (well-defined) *block sequence* $\mathrm{seq}(\Lambda) = (\mathrm{seq}(\Lambda)_i)_{i\in\mathbb{Z}}$ by taking any $\lambda \in \Lambda$ and replacing the symbols $\wedge, \vee$ by $\bigstar$. Moreover, we define $\mathrm{up}(\Lambda)$ respectively $\mathrm{down}(\Lambda)$ to be the total number of $\wedge$'s respectively $\vee$'s in Λ where we count $\times$ as both, $\wedge$ and $\vee$. ▲

Important for us is the following subset of blocks.

DEFINITION 3.3. A block $\Lambda \in \mathtt{bl}$ is called *balanced*, if $\mathrm{up}(\Lambda) = \mathrm{down}(\Lambda)$. We denote by $\mathtt{bl}^{\diamond} \subset \mathtt{bl}$ the set of balanced blocks. ▲

REMARK 3.4. The Blanchet-Khovanov algebras will only be defined for balanced blocks, while general blocks can be used to define a *generalized* version of these algebras in the spirit of [**4**]. ▲

A *cup diagram* c is a finite collection of non-intersecting arcs inside $\mathbb{R} \times [-1, 0]$ such that each arc intersects the boundary exactly in its endpoints, and either connecting two distinct points $(i, 0)$ and $(j, 0)$ with $i, j \in \mathbb{Z}$ (called a *cup*), or connecting one point $(i, 0)$ with $i \in \mathbb{Z}$ with a point on the lower boundary of $\mathbb{R} \times [-1, 0]$ (called a *ray*). Furthermore, each point in the boundary is endpoint of at most one arc. Two cup diagrams are equal if the arcs contained in them connect the same points. Similarly, a *cap diagram* d^* is defined inside $\mathbb{R} \times [0, 1]$. By construction, one can reflect a cup diagram c along the axis $\mathbb{R} \times \{0\}$, denote this operation by *, to obtain a cap diagram c^*. Clearly, $(c^*)^* = c$.

A cup diagram c (and similarly a cap diagram d^*) is *compatible* with a block $\Lambda \in \texttt{bl}$ if $\{(i, 0) \mid \mathrm{seq}(\Lambda)_i = \bigstar\} = (\mathbb{R} \times \{0\}) \cap c$.

We will view a weight λ as labeling integral points, called vertices, of the horizontal line $\mathbb{R} \times \{0\}$ inside $\mathbb{R} \times [-1, 0]$ and $\mathbb{R} \times [0, 1]$, putting the symbol λ_i at position $(i, 0)$. Together with a cup diagram c this forms a new diagram $c\lambda$.

DEFINITION 3.5. We say that $c\lambda$ is *oriented* if:

(I) An arc in c only contains vertices labeled by $\wedge$ or $\vee$.
(II) The two vertices of a cup are labeled by exactly one $\wedge$ and one $\vee$.
(III) Every vertex labeled $\wedge$ or $\vee$ is contained in an arc.
(IV) It is not possible to find $i < j$ such that $\lambda_i = \vee$, $\lambda_j = \wedge$, and each are contained in a ray. ▲

In the following, when depicting a composite diagram like $c\lambda$, we will omit the line and only draw the labels obtained from λ.

EXAMPLE 3.6. Consider the following diagrams.

(i)  , (ii)  , (iii)  , (iv) 

The diagrams (i) and (iii) are oriented. Diagram (ii) is not oriented since condition (II) is violated, while (iv) is not oriented because condition (IV) is not fulfilled. ▲

Similarly, a cap diagram d^* together with a weight λ forms a diagram λd^*, which is called *oriented* if $d\lambda$ is oriented. A cup respectively a cap in such diagrams is called *anticlockwise* (or *clockwise*), if its rightmost vertex is labeled $\wedge$ (or $\vee$).

Putting a cap diagram d^* on top of a cup diagram c such that they are connected to the line $\mathbb{R} \times \{0\}$ at the same points creates a *circle diagram*, denoted by cd^*. All connected component of this diagram that do not touch the boundary of $\mathbb{R} \times [-1, 1]$ are called *circles*, all others are called *lines*. Together with $\lambda \in \Lambda$ such that $c\lambda$ and λd^* are oriented this forms an *oriented circle diagram* $c\lambda d^*$.

DEFINITION 3.7. We define the *degree* of an oriented cup diagram $c\lambda$, of an oriented cap diagram λd^* and of an oriented circle diagram $c\lambda d^*$ as follows.

$$\begin{aligned} \deg(c\lambda) &= \text{number of clockwise cups in } c\lambda, \\ \deg(\lambda d^*) &= \text{number of clockwise caps in } \lambda d^*, \\ \deg(c\lambda d^*) &= \deg(c\lambda) + \deg(\lambda d^*). \end{aligned} \tag{23}$$

Note that the degree is always non-negative. ▲

EXAMPLE 3.8. In Example 3.6 above, diagram (i) has degree 1 (due to one clockwise cup) and diagram (iii) has degree 0. ▲

Finally, we associate to each $\lambda \in \Lambda$ a unique cup diagram, denoted by $\underline{\lambda}$, via:

(I) Connect neighboring pairs $\vee\wedge$ with a cup, ignoring symbols of the type $\circ$ and $\times$ as well as symbols already connected. Repeat this process until there are no more $\vee$'s to the left of any $\wedge$.

(II) Put a ray under any remaining symbols $\vee$ or $\wedge$.

It is an easy observation that $\underline{\lambda}$ always exists for a fixed λ. Furthermore, λ is the (unique) orientation of $\underline{\lambda}$, such that $\underline{\lambda}\lambda$ has minimal degree. Each cup diagram c is of the form $\underline{\lambda}$ for $\lambda \in \Lambda$, a block compatible with c.

Similarly we can define $\overline{\lambda} = \underline{\lambda}^*$, and, as before, in an oriented circle diagram $\underline{\lambda}\nu\overline{\mu}$ a circle C is said to be oriented *anticlockwise* if the rightmost vertex contained in the circle is $\wedge$ and *clockwise* in case its $\vee$. Two helpful facts about the degree of oriented circle diagrams are summarized below.

LEMMA 3.9. Fix a block Λ and $\lambda, \mu, \nu \in \Lambda$.

(a) The contribution to the degree of the arcs contained in a given circle C inside an oriented circle diagram $\underline{\lambda}\nu\overline{\mu}$ is equal to

$$\deg(C) = (\text{number of cups in } C) \pm 1,$$

with $+1$, if the circle C is oriented clockwise and -1 otherwise.

(b) If, in an oriented circle diagram $\underline{\lambda}\nu\overline{\mu}$, one changes the orientation such that all vertices contained in exactly one circle C are changed, then the degree increases by 2, if C was oriented anticlockwise, and decreases by 2, if C was oriented clockwise. ■

PROOF. (a) is a special case of [**12**, Proposition 4.9] while (b) follows from (a). □

We conclude this part with the notions of *distance* and *saddle width*, which will be important for spreading signs in the multiplication given below.

DEFINITION 3.10. For $i \in \mathbb{Z}$ and a block Λ define the *position of* i as

$$\mathrm{p}_\Lambda(i) = |\{j \mid j \leq i, \mathrm{seq}(\Lambda)_j = \star\}| + 2\,|\{j \mid j \leq i, \mathrm{seq}(\Lambda)_j = \times\}|\,.$$

For a cup or cap γ in a diagram connecting vertices $(i, 0)$ and $(j, 0)$ we define its *distance* $\mathrm{d}_\Lambda(\gamma)$ and *saddle width* $s_\Lambda(\gamma)$ by

$$\mathrm{d}_\Lambda(\gamma) = |\mathrm{p}_\Lambda(i) - \mathrm{p}_\Lambda(j)| \quad \text{respectively} \quad s_\Lambda(\gamma) = \tfrac{1}{2}\left(\mathrm{d}_\Lambda(\gamma) + 1\right).$$

For a ray γ set $\mathrm{d}_\Lambda(\gamma) = 0$. For a collection $M = \{\gamma_1, \dots, \gamma_r\}$ of distinct arcs (e.g. a circle or sequence of arcs connecting two vertices) set

$$\mathrm{d}_\Lambda(M) = \sum_{1 \leq k \leq r} \mathrm{d}_\Lambda(\gamma_k).$$

The saddle width will be interpreted in Subsection 4.4 as the number of phantom facets at the bottom of a saddle (e.g. $s = 2$ for the saddle from (21)).

We omit the subscript Λ, if no confusion can arise. ▲

3.2. The Blanchet-Khovanov algebras as graded $\mathbb{K}$-vector spaces. Fix a block $\Lambda \in \mathtt{bl}$, and consider the *basis set of oriented circle diagrams*

$$\mathbb{B}(\Lambda) = \{\underline{\lambda}\nu\overline{\mu} \mid \underline{\lambda}\nu\overline{\mu} \text{ is oriented and } \lambda, \mu, \nu \in \Lambda\}\,.$$

This set is subdivided into smaller sets of the form ${}_\lambda\mathbb{B}(\Lambda)_\mu$ which are those diagrams in $\mathbb{B}(\Lambda)$ which have $\underline{\lambda}$ as cup part and $\overline{\mu}$ as cap part.

From now on, we restrict to circle diagrams that only contain cups and caps. Formally this is done as follows: for a block $\Lambda \in \mathtt{bl}$ denote by Λ° the set of weights λ such that $\underline{\lambda}$ only contains cups. Note that $\Lambda^\circ \neq \emptyset$ iff Λ is balanced. Define

$$(24)\qquad \mathbb{B}^\circ(\Lambda) = \{\underline{\lambda}\nu\overline{\mu} \mid \underline{\lambda}\nu\overline{\mu} \text{ is oriented and } \lambda,\mu \in \Lambda^\circ, \nu \in \Lambda\} = \bigcup_{\lambda,\mu\in\Lambda^\circ} {}_\lambda\mathbb{B}(\Lambda)_\mu.$$

DEFINITION 3.11. The *Blanchet-Khovanov algebra* $\mathfrak{A}^{\mathfrak{F}}_\Lambda$ attached to a block $\Lambda \in \mathtt{bl}^\diamond$ and the (full) *Blanchet-Khovanov algebra* $\mathfrak{A}^{\mathfrak{F}}$ are the graded $\mathbb{K}$-vector space

$$\mathfrak{A}^{\mathfrak{F}}_\Lambda = \langle \mathbb{B}^\circ(\Lambda)\rangle_{\mathbb{K}} = \bigoplus_{(\underline{\lambda}\nu\overline{\mu})\in\mathbb{B}^\circ(\Lambda)} \mathbb{K}(\underline{\lambda}\nu\overline{\mu}), \qquad \mathfrak{A}^{\mathfrak{F}} = \bigoplus_{\Lambda\in\mathtt{bl}^\diamond} \mathfrak{A}^{\mathfrak{F}}_\Lambda,$$

with multiplication **mult** given in Subsection 3.3. Denote also by ${}_\lambda(\mathfrak{A}^{\mathfrak{F}}_\Lambda)_\mu$ the span of the basis vectors inside ${}_\lambda\mathbb{B}(\Lambda)_\mu$. ▲

PROPOSITION 3.12. The map $\mathbf{mult}\colon \mathfrak{A}^{\mathfrak{F}}_\Lambda \otimes \mathfrak{A}^{\mathfrak{F}}_\Lambda \to \mathfrak{A}^{\mathfrak{F}}_\Lambda$ given in Subsection 3.3 endows $\mathfrak{A}^{\mathfrak{F}}_\Lambda$ with the structure of a graded, unital algebra with pairwise orthogonal, primitive idempotents ${}_\lambda\mathbb{1}_\lambda = \underline{\lambda}\lambda\overline{\lambda}$ for $\lambda \in \Lambda$ and unit $\mathbb{1} = \sum_{\lambda\in\Lambda} {}_\lambda\mathbb{1}_\lambda$. Similar for (the locally unital) algebra $\mathfrak{A}^{\mathfrak{F}}$. ■

PROOF. The maps $\mathbf{mult}_{D_l,D_{l+1}}$ are homogeneous of degree 0 by [**12**, Proposition 5.19], since the proof is diagrammatic and independent of any signs or coefficients. The proof that the ${}_\lambda\mathbb{1}_\lambda$ are idempotents is the same as in [**12**, Theorem 6.2], since multiplying them only involves merges of non-nested circles, in which case the map $\mathbf{mult}_{D_l,D_{l+1}}$ agrees with the one defined in [**12**, Section 5]. That they are pairwise orthogonal and primitive is clear by definition. □

REMARK 3.13. Note that so far we do not know whether $\mathfrak{A}^{\mathfrak{F}}_\Lambda$ is associative. It will follow from the identification of $\mathfrak{A}^{\mathfrak{F}}_\Lambda$ with $\mathfrak{W}_{\vec{k}}$ that **mult** is independent of the chosen order in which the surgeries are performed and that $\mathfrak{A}^{\mathfrak{F}}_\Lambda$ is associative, see Corollary 4.19. Alternatively, the independence from the chosen order and associativity can be shown in the same spirit as [**12**, Theorem 5.34]. ▲

3.3. Multiplication of the Blanchet-Khovanov algebra. The multiplication on $\mathfrak{A}^{\mathfrak{F}}_\Lambda$ is based on the one defined in [**18**] and used in [**4**]. We will first recall the maps used in each step, which are the same as in [**4**] and afterwards go into details about how we modify these maps with different sign choices.

For $\lambda,\mu,\mu',\eta \in \Lambda^\circ$ we define a map $\mathbf{mult}\colon {}_\lambda(\mathfrak{A}^{\mathfrak{F}}_\Lambda)_\mu \otimes {}_{\mu'}(\mathfrak{A}^{\mathfrak{F}}_\Lambda)_\eta \to {}_\lambda(\mathfrak{A}^{\mathfrak{F}}_\Lambda)_\eta$ as follows. If $\mu \neq \mu'$ we declare the map to be identically zero. Thus, assume that $\mu = \mu'$, and stack the diagram, without orientations, $\underline{\mu}\overline{\eta}$ on top of the diagram $\underline{\lambda}\overline{\mu}$, creating a diagram $D_0 = \underline{\lambda}\overline{\mu}\underline{\mu}\overline{\eta}$. In [**12**, Definition 5.1] such a diagram is called a *stacked circle diagram*. Given such a diagram D_l, starting with $l = 0$, we construct below a new diagram D_{l+1} by choosing a certain symmetric pair of a cup and a cap in the middle section. If r is the number of cups in $\underline{\mu}$, then this can be done a total number of r times. We call this procedure a *surgery* at the corresponding cup-cap pair. For each such step we define below a map $\mathbf{mult}_{D_l,D_{l+1}}$. Observing that the space of orientations of the final diagram D_r is equal to the space of orientations of the diagram $\underline{\lambda}\overline{\eta}$, we define

$$\mathbf{mult} = \mathbf{mult}_{D_{r-1},D_r} \circ \dots \circ \mathbf{mult}_{D_0,D_1}\colon {}_\lambda(\mathfrak{A}^{\mathfrak{F}}_\Lambda)_\mu \otimes {}_{\mu'}(\mathfrak{A}^{\mathfrak{F}}_\Lambda)_\eta \to {}_\lambda(\mathfrak{A}^{\mathfrak{F}}_\Lambda)_\eta.$$

The global map **mult** is defined as the direct sum of all the ones defined here. In order to make **mult** a priori well-defined, we always pick the *leftmost available cup-cap pair*. Corollary 4.19 will finally ensure that this fixed choice is irrelevant.

3.3.1. *The surgery procedure.* To obtain D_{l+1} from $D_l = \underline{\lambda}c^*c\overline{\eta}$ (for some cup diagram c) choose the cup-cap pair with the leftmost endpoint in c^*c that can be connected without crossing any arcs (this means that the cup and cap are not nested inside any other arcs). Cut open the cup and the cap and stitch the loose ends together to form a pair of vertical line segments, call this diagram D_{l+1}:

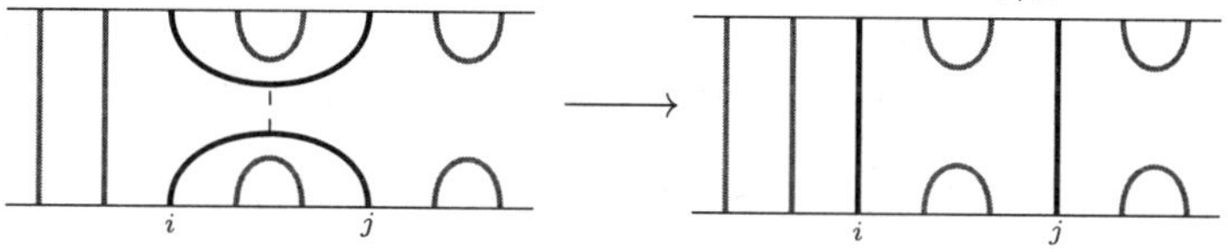

3.3.2. *The map without signs.* As in [**18**] and [**4**] the map $\mathbf{mult}_{D_{l+1},D_l}$, without any additional signs only depends on how the components change when going from D_l to D_{l+1}. The image of an orientation of D_l is constructed as follows in these cases (where we always leave the orientations on non-interacting arcs fixed).

Merge. If two circles, say C_1 and C_2, are merged into a circle C proceed as follows.
► *If both circles are oriented anticlockwise*, then orient C anticlockwise.
► *If one circle is oriented clockwise, one is oriented anticlockwise*, then orient C clockwise.
► *If both circles are oriented clockwise*, then the map is zero.

Split. If one circle C is replaced by two circles, say C_1 and C_2, proceed as follows.
► *If C is oriented anticlockwise*, then take two copies of D_{l+1}. In one copy orient C_1 clockwise and C_2 anticlockwise, in the other vice versa.
► *If C is oriented clockwise*, then orient both, C_1 and C_2, clockwise.

REMARK 3.14. These are the same rules as in [**18**] and [**4**]. They can also be given by a TQFT, but the direct definition simplifies the introduction of signs. ▲

3.3.3. *The map with signs.* In general, the formulas below include two types of signs. One type, which we call *dot moving signs*, also appear in [**12**], while the second, called *topological signs* and *saddle signs*, are topological in nature and more involved. These second types of signs will be given an interesting meaning in Subsection 4.5. The main difference to [**12**] will be that we distinguish whether the two circles, that are merged together or split into, are nested in each other or not. Define:

$$\mathrm{t}(C) = \text{(a choice of) a rightmost point in the circle } C.$$

Let γ denote the cup in the cup-cap pair we use to perform the surgery procedure in this step connecting vertices $i < j$.

Non-nested Merge. The non-nested circles C_1 and C_2 are merged into C. The only case that is modified here is:
► *One circle oriented clockwise, one oriented anticlockwise.* Let C_k (for $k \in \{1,2\}$) be the clockwise oriented circle and let γ^{dot} be a sequence of arcs in C connecting $\mathrm{t}(C_k)$ and $\mathrm{t}(C)$. (Neither $\mathrm{t}(C_k), \mathrm{t}(C)$ nor γ^{dot} are unique, but possible choices differ in distance by 2, making the sign well-defined, see also [**12**, Lemma 5.7]. Thus, the

reader may choose any of these.) Proceed as in Subsection 3.3.2 and multiply with the *dot moving sign*

$$(-1)^{\mathrm{d}_\Lambda(\gamma^{\mathrm{dot}})}. \tag{25}$$

Nested Merge. The nested circles C_1 and C_2 are merged into C. Denote by C_{in} the inner of the two original circles. The cases are modified as follows.

▶ *Both circles oriented anticlockwise.* Proceed as above, but multiply with

$$-(-1)^{\frac{1}{4}(\mathrm{d}_\Lambda(C_{\mathrm{in}})-2)} \cdot (-1)^{s_\Lambda(\gamma)}. \tag{26}$$

▶ *One circle oriented clockwise, one oriented anticlockwise.* Again perform the surgery procedure as described in Subsection 3.3.2 and multiply with

$$(-1)^{\mathrm{d}_\Lambda(\gamma^{\mathrm{dot}})} \cdot (-(-1)^{\frac{1}{4}(\mathrm{d}_\Lambda(C_{\mathrm{in}})-2)}) \cdot (-1)^{s_\Lambda(\gamma)},$$

where γ^{dot} is defined as in (25).

Non-nested Split. The circle C splits into the non-nested circles C_i, containing the vertices at position i, and C_j, containing the vertices at position j. For both orientations we introduce a *dot moving sign* as well as a *saddle sign* as follows.

▶ *C oriented anticlockwise.* Use the map as in Subsection 3.3.2, but the copy where C_i is oriented clockwise is multiplied with

$$(-1)^{\mathrm{d}_\Lambda(\gamma_i^{\mathrm{ndot}})} \cdot (-1)^{s_\Lambda(\gamma)},$$

while the one where C_j is oriented clockwise is multiplied with

$$-(-1)^{\mathrm{d}_\Lambda(\gamma_j^{\mathrm{ndot}})} \cdot (-1)^{s_\Lambda(\gamma)}.$$

Here γ_i^{ndot} and γ_j^{ndot} are sequences of arcs connecting $(i,0)$ and $\mathrm{t}(C_i)$ inside C_i respectively $(j,0)$ and $\mathrm{t}(C_j)$ in C_j (ndot can be read as "newly created dot").

▶ *C oriented clockwise.* In this case multiply the result with

$$(-1)^{\mathrm{d}_\Lambda(\gamma^{\mathrm{dot}})} \cdot (-1)^{\mathrm{d}_\Lambda(\gamma_i^{\mathrm{ndot}})} \cdot (-1)^{s_\Lambda(\gamma)}.$$

Here γ^{dot} is a sequence of arcs connecting $\mathrm{t}(C)$ and $\mathrm{t}(C_j)$ in C and γ_i^{ndot} is as before.

Nested Split. We use here the same notations as in the non-nested split case above and furthermore denote by C_{in} the inner of the two circles C_i and C_j. The difference to the non-nested case is that we substitute the *saddle sign* with a *topological sign*.

▶ *C oriented anticlockwise.* We use the map as defined in Subsection 3.3.2, but the copy where C_i is oriented clockwise is multiplied with

$$(-1)^{\mathrm{d}_\Lambda(\gamma_i^{\mathrm{ndot}})} \cdot (-1)^{\frac{1}{4}(\mathrm{d}_\Lambda(C_{\mathrm{in}})-2)},$$

while the copy where C_j is oriented clockwise is multiplied with

$$-(-1)^{\mathrm{d}_\Lambda(\gamma_j^{\mathrm{ndot}})} \cdot (-1)^{\frac{1}{4}(\mathrm{d}_\Lambda(C_{\mathrm{in}})-2)}.$$

Here γ_i^{ndot} and γ_j^{ndot} are as before.

▶ *C oriented clockwise.* We multiply with

$$(-1)^{\mathrm{d}_\Lambda(\gamma^{\mathrm{dot}})} \cdot (-1)^{\mathrm{d}_\Lambda(\gamma_i^{\mathrm{ndot}})} \cdot (-1)^{\frac{1}{4}(\mathrm{d}_\Lambda(C_{\mathrm{in}})-2)},$$

again with the same notations as above.

3.3.4. *Examples for the multiplication.* We give below examples for some of the shapes that can occur during the surgery procedure and determine the signs. In all examples assume that outside of the shown strip all entries of the weights are ∘.

EXAMPLE 3.15. This example illustrates the merge situation. First we look at a simple merge of *two anticlockwise, non-nested* circles. In this case no signs appear at all that means we have

(27)

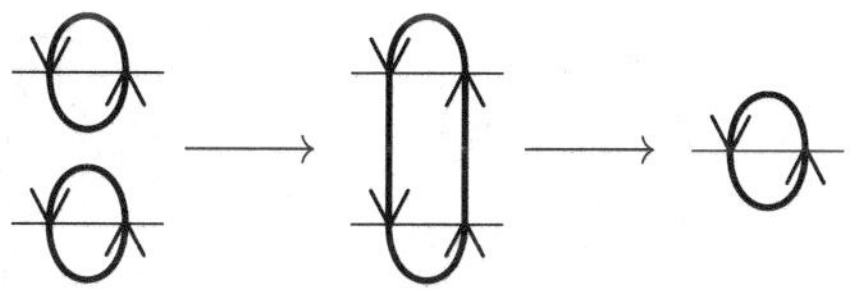

The rightmost step above, called *collapsing*, is always performed at the end of a multiplication procedure and is omitted in what follows.

Secondly, we consider a merge of *two anticlockwise, nested* circles. Depending on the concrete shape of the diagram it can produce different signs:

(28)

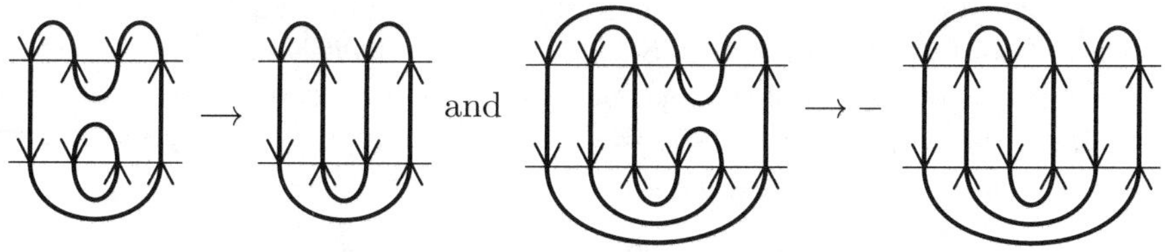

Note that, in contrast to [**18**], [**4**] or [**12**], nested merges can come with a sign. ▲

EXAMPLE 3.16. This example illustrated the two versions of a split. In both cases a non-nested merge is performed, followed by a split into *two non-nested* respectively *nested* circles. First, the H-shape:

(29)

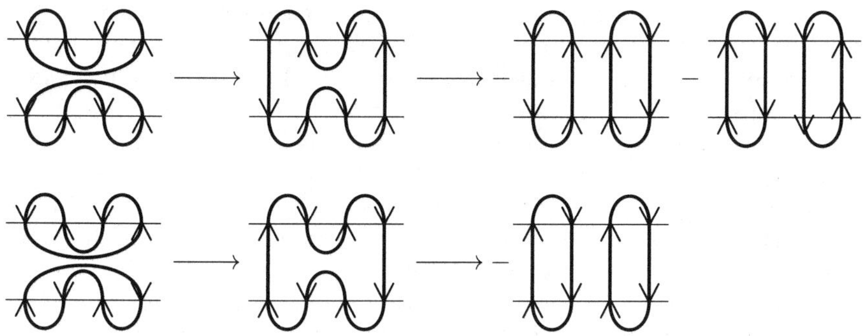

Next, the C shape:

(30)

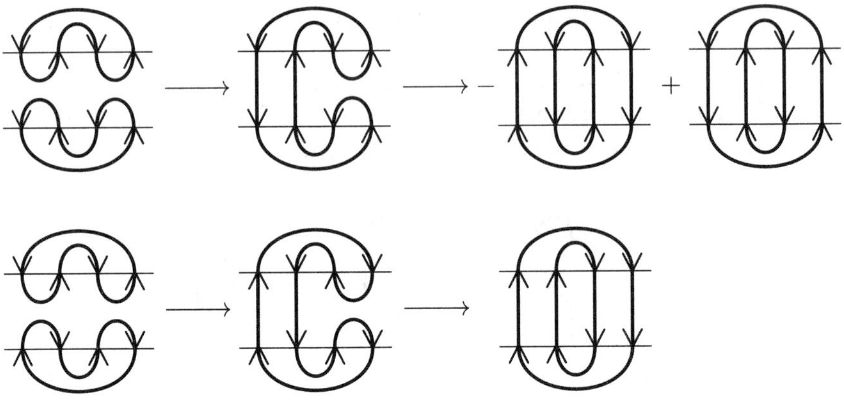

▲

REMARK 3.17. The Ↄ shape cannot appear as long as we impose the choice of the order of cup-cap pairs from left to right in the surgery procedure and it will not be needed in the proof of Theorem 4.18. ▲

3.4. Bimodules for Blanchet-Khovanov algebras. To define $\mathfrak{A}^{\mathfrak{F}}$-bimodules we need further diagrams moving from one block Λ to another block Γ.

Fix two blocks $\Lambda, \Gamma \in \mathtt{bl}^{\diamond}$ such that $\mathrm{seq}(\Lambda)$ and $\mathrm{seq}(\Gamma)$ coincide except at positions i and $i+1$. Following [**5**], a (Λ, Γ)-*admissible matching (of type* $\pm\alpha_i$*)* is a diagram t inside $\mathbb{R} \times [0,1]$ consisting of vertical lines connecting $(k,0)$ with $(k,1)$ if we have that $\mathrm{seq}(\Lambda)_k = \mathrm{seq}(\Gamma)_k = \star$ and, depending on the sign of α_i, an arc at positions i and $i+1$ of the form

(31)

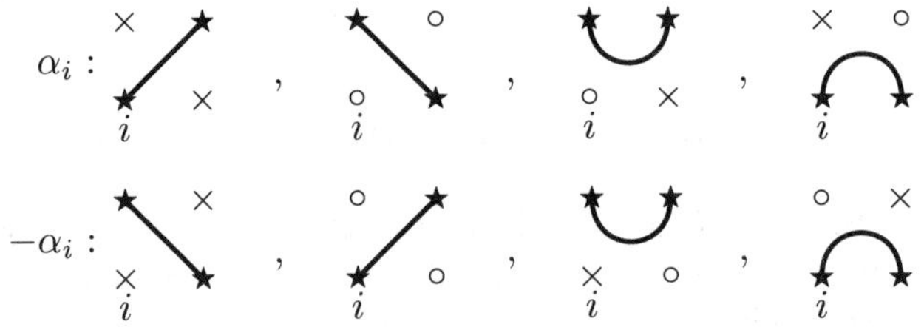

where we view $\mathrm{seq}(\Lambda)$ as decorating the integral points of $\mathbb{R} \times \{0\}$ and $\mathrm{seq}(\Gamma)$ as decorating the integral points of $\mathbb{R} \times \{1\}$. Again, the first two moves in each row are called *rays*, the third ones *cups* and the last ones *caps*.

For t a (Λ, Γ)-admissible matching, $\lambda \in \Lambda$, and $\mu \in \Gamma$ we say that $\lambda t \mu$ is *oriented* if cups respectively caps connect one $\wedge$ and one $\vee$ in λ respectively μ, and rays connect the same symbols in λ and μ. For $\mathbf{\Lambda} = (\Lambda_0, \ldots, \Lambda_r)$ a sequence of blocks a $\mathbf{\Lambda}$-*admissible composite matching* is a sequence of diagrams $\mathbf{t} = (t_1, \ldots, t_r)$ such that t_k is a $(\Lambda_{k-1}, \Lambda_k)$-admissible matching of some type. We view the sequence of matchings as being stacked on top of each other. A sequence of weights $\lambda_i \in \Lambda_i$ such that $\lambda_{k-1} t_k \lambda_k$ is oriented for all k is an *orientation* of the $\mathbf{\Lambda}$-admissible composite matching $\mathbf{t}$. For short, we tend to drop the word admissible, since the only matchings we consider are admissible. Moreover, if only start $\Lambda = \Lambda_0$ and end $\Gamma = \Lambda_r$ matter, then we call $\mathbf{t}$ short a (Λ, Γ)-*composite matching.*

We stress that $\mathbf{\Lambda}$-composite matching can contain lines, in contrast to the diagrams we consider for $\mathfrak{A}^{\mathfrak{F}}_{\Lambda}$ and $\mathfrak{A}^{\mathfrak{F}}$.

EXAMPLE 3.18. Below is a (Λ_0, Λ_5)-composite matching. Assume that outside of the indicated areas all symbols of the block sequences are equal to $\circ$.

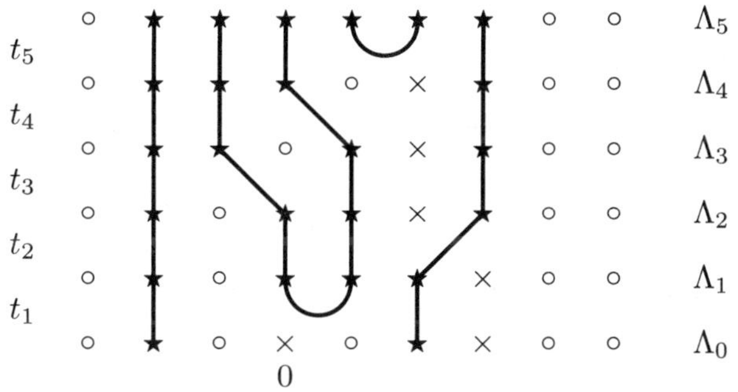

The types of the matchings are $-\alpha_0, \alpha_2, \alpha_{-1}, \alpha_0, \alpha_1$ (read from bottom to top). ▲

We now want to consider bimodules between Blanchet-Khovanov algebras for different blocks, or said differently, bimodules for the algebra $\mathfrak{A}^{\mathfrak{F}}$.

We start by defining a basis of the underlying $\mathbb{K}$-vector space. To a $\mathbf{\Lambda}$-composite matching $\mathbf{t}$ we again associate a set of diagrams from which to create a $\mathbb{K}$-vector space (its elements are called *stretched circle diagrams*)

$$\mathbb{B}^\circ(\mathbf{\Lambda}, \mathbf{t}) = \left\{ \underline{\lambda}(\mathbf{t}, \boldsymbol{\nu})\overline{\mu} \;\middle|\; \begin{array}{l} \lambda \in \Lambda_0^\circ,\ \mu \in \Lambda_r^\circ,\ \boldsymbol{\nu} = (\nu_0, \ldots, \nu_r) \text{ with } \nu_i \in \Lambda_i, \\ \underline{\lambda}\nu_0 \text{ oriented, } \nu_r\overline{\mu} \text{ oriented,} \\ \nu_{i-1}t_i\nu_i \text{ oriented for all } 1 \leq i \leq r. \end{array} \right\} \tag{32}$$

As before we obtain the set $\mathbb{B}(\mathbf{\Lambda}, \mathbf{t})$ by allowing $\lambda \in \Lambda_0$ and $\mu \in \Lambda_r$ in (32).

EXAMPLE 3.19. Let Λ be the block with block sequence $\star\ \star\ \circ\ \times$ at positions 0, 1, 2, 3 and Γ the block with sequence $\star\ \star\ \star\ \star$ at the same positions (both with $\circ$ everywhere else) and assume both blocks are balanced. Then an example for a (Λ, Γ)-matching of type α_1 is the third diagram in the first row of (31) denoted here by t_1. Taking this as our composite matching we obtain a $\mathbb{K}$-vector space of dimension 6 with basis consisting of

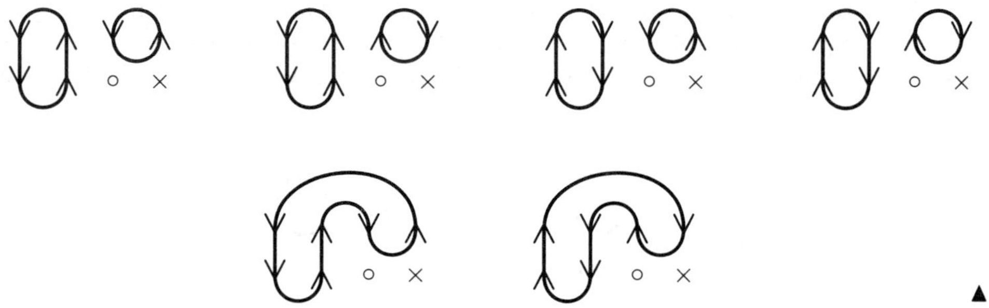

▲

For a basis vector $\underline{\mu}(\mathbf{t}, \boldsymbol{\nu})\overline{\eta}$ denote by $\overline{\eta}^{\downarrow}$ its *downwards reduction*. This is the cap diagram obtained by stacking the diagrams $t_1, \ldots, t_r, \overline{\eta}$ on top of each other from left to right, removing any components in this stacked diagram that are not connected to the bottom line of t_1, and replacing all components that are connected to the bottom in two vertices by a cap connecting the vertices. It is clearly independent of $\underline{\mu}$. Analogously, define its *upward reduction* $\underline{\mu}^{\uparrow}$, a cup diagram independent of $\overline{\eta}$.

DEFINITION 3.20. Let $\mathbf{t}$ be a $\mathbf{\Lambda}$-composite matching for $\mathbf{\Lambda} = (\Lambda_0, \dots, \Lambda_r)$. Set

$$\mathfrak{A}^{\mathfrak{F}}(\mathbf{\Lambda}, \mathbf{t}) = \langle \mathbb{B}^{\circ}(\mathbf{\Lambda}, \mathbf{t}) \rangle_{\mathbb{K}}$$

as a graded $\mathbb{K}$-vector space. Hereby the *degree* of a basis element itself is, by definition, minus the number of its anticlockwise circles plus the number of its clockwise circles.

For a basis element $\underline{\lambda}\nu\overline{\mu} \in \mathfrak{A}^{\mathfrak{F}}_{\Lambda_0}$ define a left (bottom) action via

$$(\underline{\lambda}\nu\overline{\mu})(\underline{\mu}(\mathbf{t}, \boldsymbol{\nu})\overline{\eta}) = \sum_{\boldsymbol{\nu}'} a_{\boldsymbol{\nu}'} \underline{\lambda}(\mathbf{t}, \boldsymbol{\nu}')\overline{\eta},$$

with the coefficients $a_{\nu'}$ given by the multiplication

$$(\underline{\lambda}\nu\overline{\mu})(\underline{\mu}\nu_0\overline{\eta}^{\downarrow}) = \sum_{\nu'} a_{\nu'} \underline{\lambda}\nu'\overline{\eta}^{\downarrow}.$$

Then $a_{\boldsymbol{\nu}'} = a_{\nu'}$ for $\boldsymbol{\nu}'$ the unique orientation with $\nu_0' = \nu'$ and all components not connected to the bottom line of t_1 have the same orientation as in $\underline{\mu}(\mathbf{t}, \boldsymbol{\nu})\overline{\eta}$. Analogously define a right (top) action of $\mathfrak{A}^{\mathfrak{F}}_{\Lambda_r}$ using the upwards reduction $\underline{\mu}^{\uparrow}$. ▲

EXAMPLE 3.21. The basis elements (read from left to right) from Example 3.19 have degrees $-2, 0, 0, 2$ (top row) and $-1, 1$ (bottom row). ▲

It is not clear that the above actions are well-defined and commute and we need the translation between $\mathfrak{W}$ and $\mathfrak{A}^{\mathfrak{F}}$ from Section 4 to prove it.

PROPOSITION 3.22. Let $\mathbf{t}$ be a $\mathbf{\Lambda}$-composite matching for $\mathbf{\Lambda} = (\Lambda_0, \dots, \Lambda_r)$. Then the left action of $\mathfrak{A}^{\mathfrak{F}}_{\Lambda_0}$ and the right action of $\mathfrak{A}^{\mathfrak{F}}_{\Lambda_r}$ on $\mathfrak{A}^{\mathfrak{F}}(\mathbf{\Lambda}, \mathbf{t})$ are well-defined and commute. Hence, $\mathfrak{A}^{\mathfrak{F}}(\mathbf{\Lambda}, \mathbf{t})$ is a $\mathfrak{A}^{\mathfrak{F}}_{\Lambda_0}$-$\mathfrak{A}^{\mathfrak{F}}_{\Lambda_r}$-bimodule (and thus, a $\mathfrak{A}^{\mathfrak{F}}$-bimodule). ■

PROOF. Using Theorem 4.18, we obtain an isomorphism of graded algebras of $\mathfrak{A}^{\mathfrak{F}}_{\Lambda}$ with $\mathfrak{W}^{\circ}_{\vec{k}}$. An isomorphism of graded $\mathbb{K}$-vector spaces of $\mathfrak{A}^{\mathfrak{F}}(\mathbf{\Lambda}, \mathbf{t})$ with $\mathfrak{W}(\mathbf{w}(\mathbf{\Lambda}, \mathbf{t}))$ is obtained by using Lemma 4.16. This isomorphism intertwines the actions of the two algebras on the bimodules by construction and hence, proves the claim. □

We introduce a slight generalization of the notion of an admissible matching, the so-called *empty moves* (of which the reader should think of switching neighboring ∘ and ×). This means that in the list of local moves (31) we also allow the following:

(33)

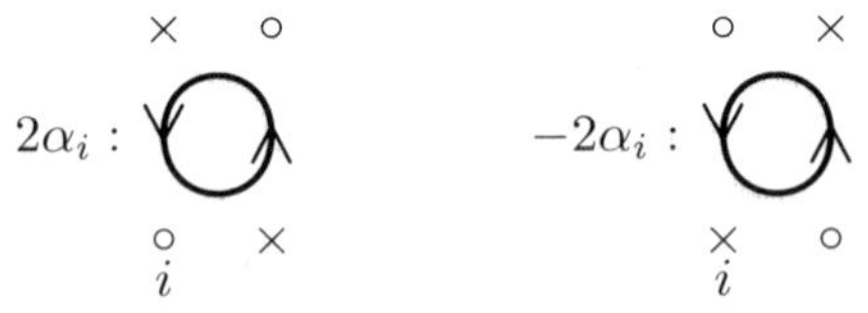

If a composite matching $\mathbf{t}$ contains empty moves, $\mathfrak{A}^{\mathfrak{F}}(\mathbf{\Lambda}, \mathbf{t})$ is constructed as follows. Take the composite matching $\mathbf{t}'$ that is obtained by substituting each empty move by the composition of a cup and cap local move such that its fits with the two blocks. Then take the submodule spanned by those basis elements such that the internal circles are all oriented anticlockwise. It is evident that this will be a submodule of the full bimodule where all orientations are allowed. Finally, shift the module up by the total number of empty moves in $\mathbf{t}$.

EXAMPLE 3.23. The pictures below are wildcards for the $\mathfrak{A}^{\mathfrak{F}}$-bimodules defined via the illustrated matchings.

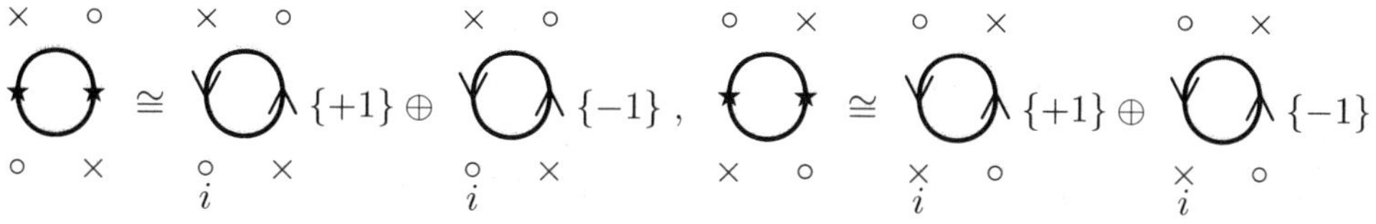

The isomorphisms of $\mathfrak{A}^{\mathfrak{F}}$-bimodules are evident by the construction above. ▲

PROPOSITION 3.24. The $\mathfrak{A}^{\mathfrak{F}}$-bimodules $\mathfrak{A}^{\mathfrak{F}}(\mathbf{\Lambda}, \mathbf{t})$ are finite-dimensional, graded biprojective $\mathfrak{A}^{\mathfrak{F}}$-bimodules. ■

PROOF. Clearly, $\mathfrak{A}^{\mathfrak{F}}(\mathbf{\Lambda}, \mathbf{t})$ are finite-dimensional $\mathfrak{A}^{\mathfrak{F}}$-bimodules.

That they are graded as $\mathfrak{A}^{\mathfrak{F}}$-bimodules follows by the identification with web bimodules from Lemma 4.16. We only prove here projectivity for the left action, the right action is done analogously. Denote by Λ and Γ the first and last block in $\mathbf{\Lambda}$. For any $\mu \in \Gamma$ denote by ${}_{\mu^{\downarrow}}\mathbb{1}_{\mu^{\downarrow}}$ the idempotent in $\mathfrak{A}^{\mathfrak{F}}_{\Lambda}$ corresponding to the downwards reduction of μ. Then, as an $\mathfrak{A}^{\mathfrak{F}}_{\Lambda}$-module, $\mathfrak{A}^{\mathfrak{F}}(\mathbf{\Lambda}, \mathbf{t})$ decomposes as a direct sum of modules of the form $\mathfrak{A}^{\mathfrak{F}}_{\Lambda} \cdot ({}_{\mu^{\downarrow}}\mathbb{1}_{\mu^{\downarrow}})$, which are projective $\mathfrak{A}^{\mathfrak{F}}_{\Lambda}$-modules, proving the claim. □

This proposition again motivates the definition of the following 2-category which is, as before, one of the main objects that we are going to study.

DEFINITION 3.25. Given $\mathfrak{A}^{\mathfrak{F}}$ as above, let $\mathfrak{A}^{\mathfrak{F}}$-$\mathbf{biMod}^{p}_{\mathrm{gr}}$ be the following 2-category:

- Objects are the various $\Lambda \in \mathtt{bl}^{\diamond}$.
- Morphisms are finite sums and tensor products (taken over $\mathfrak{A}^{\mathfrak{F}}$) of the $\mathfrak{A}^{\mathfrak{F}}$-bimodules $\mathfrak{A}^{\mathfrak{F}}(\mathbf{\Lambda}, \mathbf{t})$.
- The composition of $\mathfrak{A}^{\mathfrak{F}}$-bimodules is given by tensoring (over $\mathfrak{A}^{\mathfrak{F}}$).
- 2-morphisms are $\mathfrak{A}^{\mathfrak{F}}$-bimodule homomorphisms.
- The vertical composition of $\mathfrak{A}^{\mathfrak{F}}$-bimodule homomorphisms is the usual composition and the horizontal composition is given by tensoring (over $\mathfrak{A}^{\mathfrak{F}}$).

We consider $\mathfrak{A}^{\mathfrak{F}}$-$\mathbf{biMod}^{p}_{\mathrm{gr}}$ as a graded 2-category with 2-hom-spaces as in (1). ▲

In analogy to **Web** from the end of Subsection 2.2, the objects and morphisms (when seen as composite matchings) in $\mathfrak{A}^{\mathfrak{F}}$-$\mathbf{biMod}^{p}_{\mathrm{gr}}$ can be seen as a $\mathbb{K}(q)$-linear category, which we denote by **CM**.

4. Equivalences

In this section we assume that all appearing $\vec{k}$'s and Λ's are balanced. Our goal now is to construct an isomorphism of graded algebras $\Phi\colon \mathfrak{W}^{\circ} \to \mathfrak{A}^{\mathfrak{F}}$ (where $\mathfrak{W}^{\circ}$ is a certain subalgebra of $\mathfrak{W}$ defined in (40)). This isomorphism Φ induces the following:

THEOREM 4.1. There is an equivalence of graded, $\mathbb{K}$-linear 2-categories

$$\mathbf{\Phi}\colon \mathfrak{W}\text{-}\mathbf{biMod}^{p}_{\mathrm{gr}} \xrightarrow{\cong} \mathfrak{A}^{\mathfrak{F}}\text{-}\mathbf{biMod}^{p}_{\mathrm{gr}}$$

under which the $\mathfrak{W}$-bimodules $\mathfrak{W}(\mathbf{w}(\mathbf{\Lambda}, \mathbf{t}))$ (with $\mathbf{w}(\mathbf{\Lambda}, \mathbf{t})$ defined in Definition 4.9 below) are identified with the $\mathfrak{A}^{\mathfrak{F}}$-bimodules $\mathfrak{A}^{\mathfrak{F}}(\mathbf{\Lambda}, \mathbf{t})$. ■

4.1. Some useful lemmas. In order to prove Theorem 4.1, we need some lemmas.

LEMMA 4.2. Let $\ell \in \mathbb{Z}_{\geq 0}$. Then $\dim_{\mathbb{K}}(2\mathrm{End}_{\mathfrak{F}}(\mathbf{1}_{2\omega_\ell})) = 1$. ■

PROOF. Note that the identity foam on $\mathbf{1}_{2\omega_\ell}$ (i.e. ℓ parallel phantom facets) is an element of $2\mathrm{End}_{\mathfrak{F}}(\mathbf{1}_{2\omega_\ell})$ which shows that $\dim_{\mathbb{K}}(2\mathrm{End}_{\mathfrak{F}}(\mathbf{1}_{2\omega_\ell})) \geq 1$. Now, given any $f \in 2\mathrm{End}_{\mathfrak{F}}(\mathbf{1}_{2\omega_\ell})$, denote by $g \in 2\mathrm{End}_{\overline{\mathfrak{F}}}(\emptyset)$ (where $\emptyset$ denotes the empty web) the closed foam obtained from f by closings the ℓ bottom and top phantom facets to phantom spheres. Note that the question whether there are relations to reduce f to a scalar multiple of the identity foam on $\mathbf{1}_{2\omega_\ell}$ is the same as the question whether g can be evaluated to a scalar. But the latter can be achieved using the relations in $\overline{\mathfrak{F}}$ (which can be shown similar as in [**19**, Proposition 5]). Thus, every such f is a scalar multiple of the identity foam on $\mathbf{1}_{2\omega_\ell}$ which shows that $\dim_{\mathbb{K}}(2\mathrm{End}_{\mathfrak{F}}(\mathbf{1}_{2\omega_\ell})) \leq 1$. □

LEMMA 4.3. The following foams are locally one-sided invertible in $\mathfrak{F}$ (the ones in the first column have a right inverse obtained by gluing from the bottom, the ones in the second column a left inverse obtained by gluing from the top):

(34)

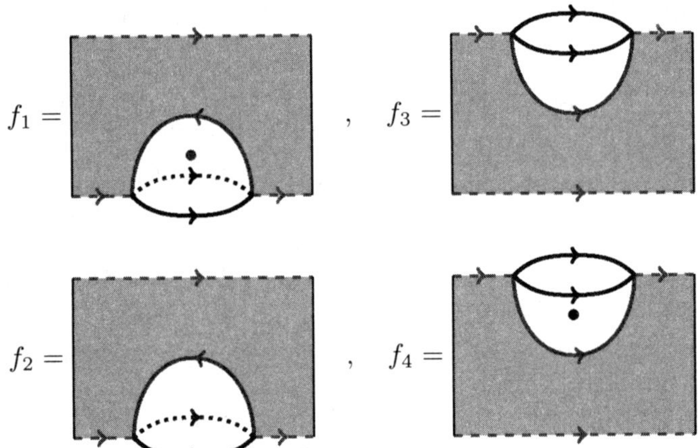

Similarly with the dots moved to the opposite facets. This implies locally that

(35)

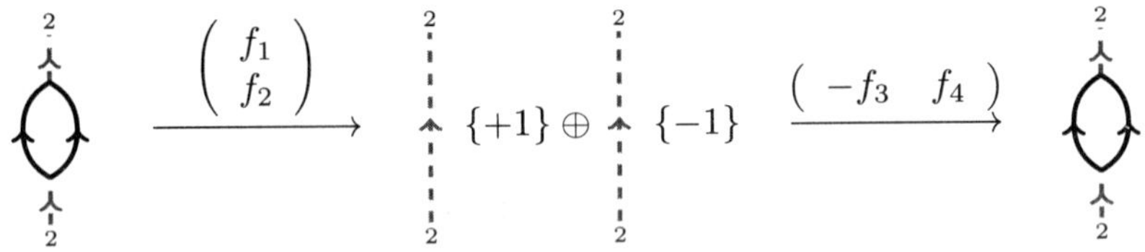

induce isomorphisms in $\mathfrak{W}\text{-}\mathbf{biMod}^p_{\mathrm{gr}}$ between web bimodules. ■

PROOF. The statement of one-sided invertibility follows from the top bubble removals (13) (by stacking the foams in the first column on top of the foams in the second column). The invertibility of the foams with a different dot placement follows from the above and the dot migrations (17). The claim that the given morphisms are isomorphisms follows from composing them and using both bubble removals (13) and (14) as well as the neck cutting relation (15). □

The next lemmas say that isotopic webs give isomorphic web modules.

LEMMA 4.4. We have locally

(36) 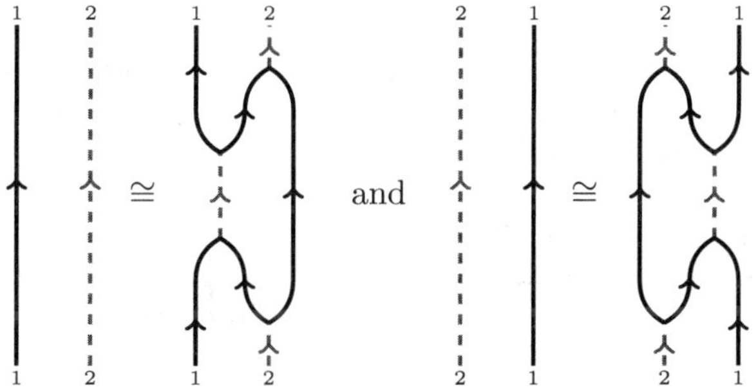

which are isomorphisms in $\mathfrak{W}$-$\mathbf{biMod}^p_{\mathrm{gr}}$ between web bimodules. ∎

PROOF. The proof is similar to the one of Lemma 4.3: we can cap the bulge part respectively cup the two straight lines using the evident (undotted) foams. Then the ordinary-to-phantom neck cutting relations (18) provide the isomorphisms. □

LEMMA 4.5. We have locally (where we use "rectangular" diagrams as in (19))

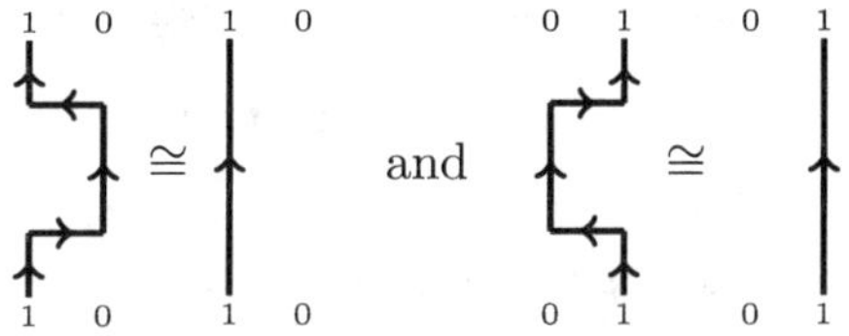

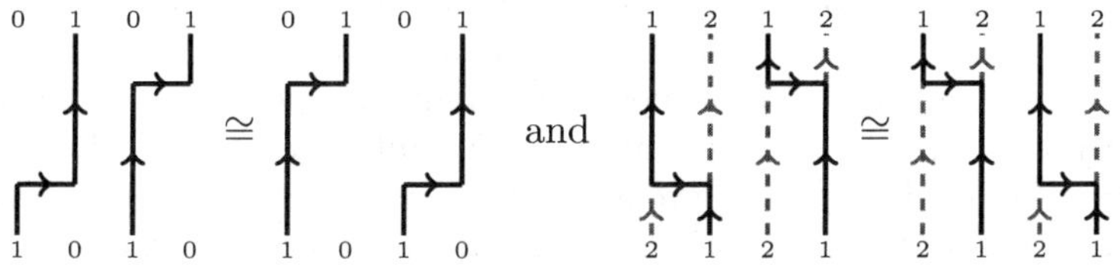 etc.

which are isomorphisms in $\mathfrak{W}$-$\mathbf{biMod}^p_{\mathrm{gr}}$ between web bimodules. Analogously for other isotopies of webs. ∎

PROOF. This is evident. □

Given a web u, then we denote by $\hat{u}$ the *topological web* obtained from u by forgetting orientations, labels and phantom edges, i.e. we have locally

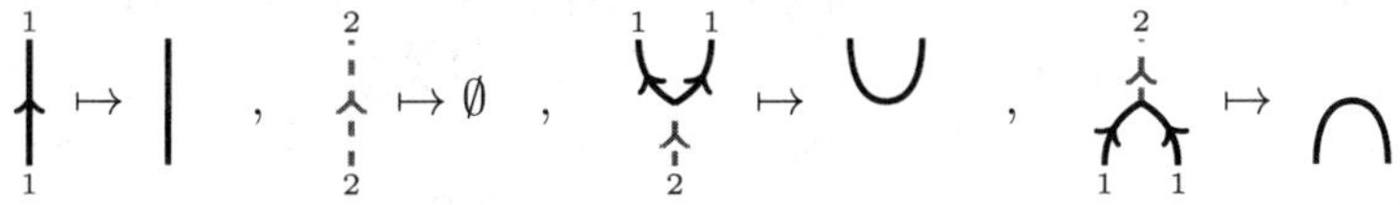

The topological webs are just non-crossing arcs and (closed) circles. We call any web u such that $\hat{u}$ is topologically a circle also a *circle*. Similarly for webs u such that $\hat{u}$ is topologically a cup, cap or a line. Moreover, we assume that the leftmost non-zero entries of $\Lambda \in \mathtt{bl}^\diamond$ and $\vec{k} \in \mathbb{bl}^\diamond$ are at the same position in what follows.

LEMMA 4.6. Let $\lambda \in \Lambda$ with j entries equal to $\times$. Consider the cup diagram $\underline{\lambda}$. Assume that $\underline{\lambda}$ does have ℓ' rays and in total $\ell + \ell' - j$ components. Then there exists a cup-ray web $u \in \mathrm{CupRay}(\vec{k})$ such that the entries of $\vec{k}$ sum up to $2\ell + \ell'$, and such that the topological web $\hat{u}$ is $\underline{\lambda}$. ∎

The sequences $\Lambda \in \mathtt{bl}^{\diamond}$ and $\vec{k} \in \mathbb{bl}^{\diamond}$ are related by the bijection from (38) below.

PROOF. Given a cup diagram $\underline{\lambda}$ with $\ell+\ell'-j$ components nested in any order, we can generate it using $E_i^{(r)}$'s and $F_i^{(r)}$'s acting on the sequence $\omega_{\ell+\ell'}+\omega_{\ell}$. It is clear from (19) that we can open one cup for each entry 2 of $\omega_{\ell+\ell'}+\omega_{\ell}$. By using

(37) 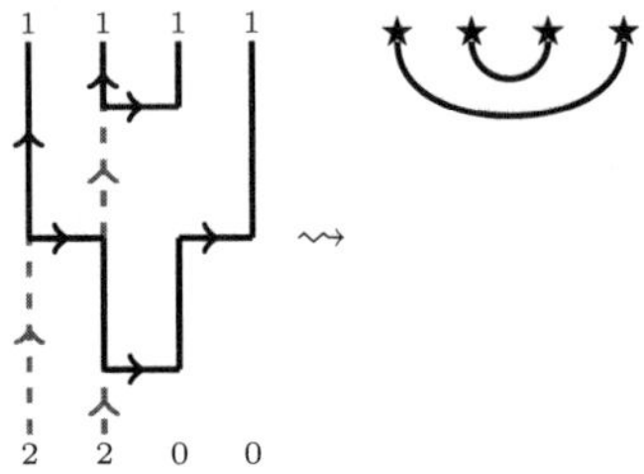

and 

we can nest them and place them next to each other in any order we like (by locally shifting everything in place as above). Last, by using the shifts from (19), we can move the remaining entries 1 to form the rays, and move the entries 2 to the positions of the $\times$ symbols while building the cup parts of $\underline{\lambda}$. □

LEMMA 4.7. For each $\mathbf{\Lambda}$-composite matching $\mathbf{t}$, there exists a web $u \in \mathrm{Hom}_{\mathfrak{F}}(\vec{k},\vec{l})$ such that the topological web $\hat{u}$ is $\mathbf{t}$. ■

PROOF. Mutatis mutandis as in the proof of Lemma 4.6. □

Even if we require a web u to be F-generated, there are still several ways to built $\underline{\lambda}$ or $\mathbf{t}$. In order to fix one, we say an F-generated web u *prefers right to left* if, inductively, the component with rightmost right boundary point of $\underline{\lambda}$ or $\mathbf{t}$ is created from the rightmost available 2, if its a cup, or the rightmost available 1, if its a ray (using a minimal number of possible moves). We create the right boundary point of a cup before the left. In the whole procedure we avoid creating circles.

LEMMA 4.8. In the set-up of Lemma 4.6 or of Lemma 4.7 we can make u unique by requiring it to be F-generated and preferring right to left. ■

PROOF. An easy observation shows that we do not need E_i's and $E_i^{(2)}$'s in order to built $\underline{\lambda}$ or $\mathbf{t}$ (this holds more generally, see [**37**, Lemma 4.9]). Moreover, we can always avoid creating circles. Thus, we obtain a set F_{gen} of F-generated webs such that $\hat{u}$ gives $\underline{\lambda}$ or $\mathbf{t}$. All of these differ by distant commutation relations as in Lemma 4.5 (bottom moves) or Serre relations $\Phi^{\mathbf{Web}}_{\mathrm{Howe}}((F_{i+1}F_i^{(2)} - F_iF_{i+1}F_i + F_i^{(2)}F_{i+1})\mathbf{1}_{\vec{k}}) = 0$. One checks that, for fixed $\vec{k} \in \mathbb{bl}^{\diamond}$, all Serre relations have only two non-zero terms and that the corresponding pictures are as in Lemmas 4.4 and 4.5. Hence, there is a unique web in F_{gen} that prefers right to left which can be shown by induction on the number m of components of $\underline{\lambda}$ or $\mathbf{t}$. This is clear if $m=0$ or $m=1$. For $m>1$ remove the leftmost connected component from $\underline{\lambda}$ or $\mathbf{t}$ and obtain $\underline{\lambda}'$ or $\mathbf{t}'$ with one connected component less. We can then apply the induction hypothesis and we get a unique web u' such that $\hat{u}'$ is $\underline{\lambda}'$ or $\mathbf{t}'$. Since we removed the leftmost component of $\underline{\lambda}$ or of $\mathbf{t}$, we can now just construct u from u' (the result is unique due to our conventions for such webs). □

Hence, the following definition makes sense.

DEFINITION 4.9. Let $\Lambda \in \mathtt{bl}^\diamond$. Given $\lambda \in \Lambda$, we denote by $\mathbf{w}(\lambda)$ the unique web as in Lemma 4.8, and given a $\mathbf{\Lambda}$-composite matching $\mathbf{t}$, we denote the unique web as in Lemma 4.8 by $\mathbf{w}(\mathbf{\Lambda}, \mathbf{t})$. ▲

Examples of such webs are given in (37). Moreover, the following assignment

$$(38) \qquad \vec{k} \in \mathbb{bl}^\diamond \mapsto \Lambda \in \mathtt{bl}^\diamond, \quad \text{via } 0 \mapsto \circ, \quad 1 \mapsto \bigstar, \quad 2 \mapsto \times,$$

clearly defines a bijection. Here $\circ, \bigstar, \times$ are entries of $\mathrm{seq}(\Lambda)$ and Λ is determined demanding that Λ is balanced.

The next lemma is important for the calculation of the signs that turn up in the multiplication procedure. For this purpose, fix a circle C in a web $\mathbf{w}(\lambda)\mathbf{w}(\mu)^*$ with corresponding circle C' in $\underline{\lambda}\overline{\mu}$. For such a circle let $\mathrm{nest}(C)$ be the total number of circles C_i^{in} nested in C. We denote by $\mathrm{ipe}(C)$ the total number of *internal phantom edges* of C (all such edges that lie in the interior of C, but not in the interior of any circle nested in C) and more generally by

$$\mathrm{ipe}\left(C - \textstyle\bigcup_{i=1}^{\mathrm{nest}(C)} C_i^{\mathrm{in}}\right)$$

the total number of internal phantom edges of C after removing C_i^{in} (by using simplifications as isotopies, (36) and (35)). Recalling $\mathrm{d}(\cdot)$ from Definition 3.10, we have the following lemma:

LEMMA 4.10. Given a circle diagram $\underline{\lambda}\overline{\mu}$ (for suitable $\lambda, \mu \in \Lambda$) and its associated web $\mathbf{w}(\lambda)\mathbf{w}(\mu)^*$. Fix any circle C' in $\underline{\lambda}\overline{\mu}$ and denote the associated circle in $\mathbf{w}(\lambda)\mathbf{w}(\mu)^*$ by C. Then

$$\mathrm{ipe}(C) = \tfrac{1}{4}\left(\mathrm{d}(C') + \textstyle\sum_{i=1}^{\mathrm{nest}(C)} \mathrm{d}(C_{\mathrm{in}}^i) - 2 + 2 \cdot \mathrm{nest}(C)\right),$$

$$\mathrm{ipe}\left(C - \textstyle\bigcup_{i=1}^{\mathrm{nest}(C)} C_i^{\mathrm{in}}\right) = \tfrac{1}{4}\left(\mathrm{d}(C') - 2\right),$$

where C_{in}^i denotes the counterparts in $\underline{\lambda}\overline{\mu}$ of the circles C_{in}^i of $\mathbf{w}(\lambda)\mathbf{w}(\mu)^*$.

PROOF. We prove this by induction on the total length of all components in question. Circles C in their easiest form as on the left in (42) below have zero internal phantom edges and $\mathrm{d}(C) = 2$, which is the start of our induction. The main observation now is that the move in (36) increases the number of internal phantom edges by one and the length by four (no matter which side of the diagram is the internal part of the circle). This shows the formulas in case that $\mathrm{ipe}(C)$ has no nested circles. The general formulas follow similarly, where we note that each $\times$ within C' increases its length by 2 and creates a internal phantom edge in C. □

EXAMPLE 4.11. The circles C' and C_{in}^1 as on the right in (42) have $\mathrm{d}(C') = 6$ and $\mathrm{d}(C_{\mathrm{in}}^1) = 2$, and its counterpart C has one internal phantom edge after removing the nested circle, while is has two internal phantom edges in total. Circles C' of H or C shape as in (43) have $\mathrm{d}(C') = 6$ respectively $\mathrm{d}(C') = 10$ and their counterparts have two respectively one internal phantom edge. ▲

4.2. An action of the quantum group $\dot{\mathrm{U}}_q(\mathfrak{gl}_\infty)$ and arc diagrams. We obtain now an action of the quantum group on $\mathbf{CM}$ (with $\mathbf{CM}$ being as at the end of Subsection 3.4) via the functor

$$\Phi_{\mathrm{Howe}}^{\mathbf{CM}} \colon \dot{\mathrm{U}}_q(\mathfrak{gl}_\infty) \to \mathbf{CM},$$

given on objects by $\Phi^{\mathbf{CM}}_{\mathrm{Howe}}(\vec{k}) = \Lambda$ (with associated $\Lambda \in \mathtt{bl}^{\diamond}$ as in (38)) and on identity morphisms by $\Phi^{\mathbf{CM}}_{\mathrm{Howe}}(\mathbf{1}_{\vec{k}}) = \mathbf{1}_{\Lambda}$. For the generating morphism $E_i\mathbf{1}_{\vec{k}}$, $\Phi^{\mathbf{CM}}_{\mathrm{Howe}}(E_i\mathbf{1}_{\vec{k}})$ is the bimodule corresponding to the unique local picture in the first row of (31) that fits with Λ at position i and $i+1$, or zero if none of them fits, $\Phi^{\mathbf{CM}}_{\mathrm{Howe}}(F_i\mathbf{1}_{\vec{k}})$ is defined analogously using the local moves in the second row of (31). The divided powers $E_i^{(2)}\mathbf{1}_{\vec{k}}$ respectively $F_i^{(2)}\mathbf{1}_{\vec{k}}$ are sent to the bimodules for the empty moves in (33) for $2\alpha_i$ respectively $-2\alpha_i$, again only if the block fits at position i and $i+1$, and to zero otherwise. As in Subsection 2.3, all higher divided powers $E_i^{(r)}\mathbf{1}_{\vec{k}}, F_i^{(r)}\mathbf{1}_{\vec{k}}$ for $r > 2$ are sent to zero. By our discussion in the previous subsection, we see that $\Phi^{\mathbf{CM}}_{\mathrm{Howe}}$ is a functor of $\mathbb{K}$-linear categories.

4.3. The cup basis. The goal of this subsection is to define a basis of the morphism spaces on the side of foams which we call the *cup basis*.

DEFINITION 4.12. Let ${}_u(\mathfrak{W}_{\vec{k}})_v$ be as in Definition 2.24. Perform the following steps.

(I) Label each circle in uv^* by either "no dot" or "dot". Consider all possibilities of labeling the circles in such a way.

(II) For each such possibility we construct a foam $f\colon \mathbf{1}_{2\omega_\ell} \to uv^*$ via:
- If a circle is in its *easiest form* (i.e. only one incoming phantom edge and only one outgoing phantom edge), then we locally use the foam at the top right in (34) for circles with label "no dot" and the foam at the bottom right in (34) for circles with label "dot".
- If a circle is more complicated, then we first apply the isomorphisms described in Lemma 4.4 to reduce the circle to its easiest form, then we proceed as before, and finally we rebuild the circle using the inverses as in Lemma 4.4. Then we move the dot to the rightmost facet using (17) (the meticulous reader will note that this is ill-defined since there could be more than one rightmost facet, but one can choose any of them: dots passing from one rightmost facet to another always have to move an even number of times across phantom facets, see (36)).

From this we obtain a set of foams $\mathrm{Cup}(uv^*)$ which we call the *cup basis* of ${}_u(\mathfrak{W}_{\vec{k}})_v$. Basis elements of this basis are called *basis cup foams*. This name is justified: by construction, the foams in ${}_u(\mathfrak{W}_{\vec{k}})_v$ are topological obtained by "cupping off circles in the evident way". Similarly, given $u \in \mathrm{Hom}_{\mathfrak{F}}(\vec{k}, \vec{l})$ we can define a *cup basis* $\mathrm{Cup}(u)$ *of* $\mathfrak{W}(u)$ as above with the extra condition that we first close the web u in any possible way. ▲

LEMMA 4.13. Let $u, v \in \mathrm{Cup}(\vec{k})$. The set $\mathrm{Cup}(uv^*)$ is a homogeneous, $\mathbb{K}$-linear basis of the space ${}_u(\mathfrak{W}_{\vec{k}})_v$. ■

PROOF. That $\mathrm{Cup}(uv^*)$ is homogeneous is evident. To show that we get a basis, we use induction on the number m of circles in uv^*. If $m = 0$, then the statement is clear by Lemma 4.2. So let $m > 0$. Choose now any maximally nested circle C (a circle that does not have any nested components) in uv^* (we have to choose such a maximally nested circle in order use locally the foams from (34)). Remove it from uv^* and obtain a new web $\tilde{u}\tilde{v}^*$ with fewer number of circles. By induction, $\mathrm{Cup}(\tilde{u}\tilde{v}^*)$ is a basis of ${}_{\tilde{u}}(\mathfrak{W}_{\vec{k}})_{\tilde{v}}$. Now apply the isomorphism from (35) and create a new circle by taking two copies of the basis elements from $\mathrm{Cup}(\tilde{u}\tilde{v}^*)$ (shifted up by one respectively down by one). The claim of the lemma follows if C was in its easiest form. Otherwise, we rebuild the chosen circle C from the newly created circle using isomorphisms as

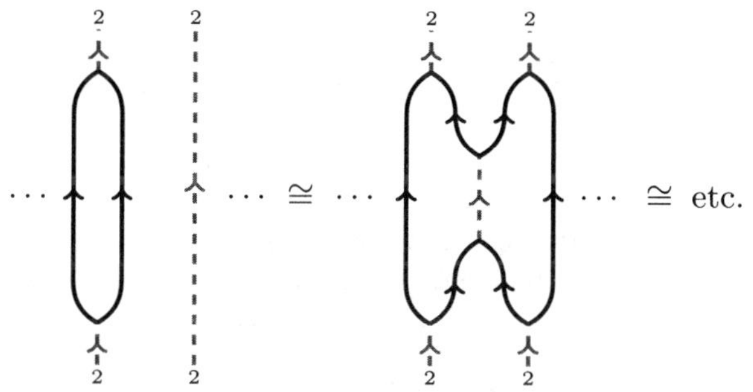

from Lemma 4.4. Finally move dots to the rightmost facets using the dot migrations (17). Thus, the lemma also follows in this case. □

LEMMA 4.14. Let $u \in \mathrm{Hom}_{\mathfrak{F}}(\vec{k}, \vec{l})$. The set $\mathrm{Cup}(u)$ is a homogeneous, $\mathbb{K}$-linear basis of the web bimodule $\mathfrak{W}(u)$. ■

PROOF. Similar to the proof of Lemma 4.13: the induction start is the same and uses Lemma 4.2. Note that $\mathfrak{W}(u)$ is obtained from u (which has possibly only cups, caps and rays and no circles) by closing the bottom and top in any possible way. Thus, for each such "closure" of $\mathfrak{W}(u)$ we can use the argument from above. □

We now match the cup basis $\mathrm{Cup}(uv^*)$ with the basis ${}_{\lambda}\mathbb{B}^{\circ}(\Lambda)_{\mu}$ defined in (24).

LEMMA 4.15. Let u, v be webs such that $u = \mathbf{w}(\lambda)$ and $v = \mathbf{w}(\mu)$. There is an isomorphism of graded $\mathbb{K}$-vector spaces

$$\Phi_{uv}^{\lambda\mu} \colon {}_{u}(\mathfrak{W}_{\vec{k}})_{v} \to {}_{\lambda}(\mathfrak{A}_{\Lambda}^{\mathfrak{F}})_{\mu} \tag{39}$$

which sends $\mathrm{Cup}(uv^*)$ to ${}_{\lambda}\mathbb{B}^{\circ}(\Lambda)_{\mu}$ by identifying the cup foams without dots with anticlockwise circles and the foams with dots with clockwise circles. ■

PROOF. The sets $\mathrm{Cup}(uv^*)$ and ${}_{\lambda}\mathbb{B}^{\circ}(\Lambda)_{\mu}$ are clearly in bijective correspondence. Moreover, recalling Lemma 3.9 and the shift as in Definition 2.24, we obtain that $\Phi_{uv}^{\lambda\mu}$ is homogeneous, which proves the lemma. □

Similarly, we match the cup basis $\mathrm{Cup}(u)$ with the basis $\mathbb{B}^{\circ}(\mathbf{\Lambda}, \mathbf{t})$ from (32):

LEMMA 4.16. Let u be a web such that $u = \mathbf{w}(\mathbf{\Lambda}, \mathbf{t})$. There is a surjection of graded $\mathbb{K}$-vector spaces

$$\phi_{uv}^{\lambda\mu} \colon \mathfrak{W}(u) \to \mathfrak{A}^{\mathfrak{F}}(\mathbf{\Lambda}, \mathbf{t})$$

which sends $\mathrm{Cup}(u)$ to $\mathbb{B}^{\circ}(\mathbf{\Lambda}, \mathbf{t})$ by identifying the basis cup foams without dots with anticlockwise circles and the basis cup foams with dots with clockwise circles. ■

The statement of Lemma 4.16 can be easily strengthened using Lemma 4.22. (Morally, the web bimodules are infinite-dimensional in a "stupid way".)

PROOF. To show that (41) is indeed a homogeneous, $\mathbb{K}$-linear surjection we can proceed as in the proof of Lemma 4.15 since both bases, $\mathrm{Cup}(u)$ and $\mathbb{B}^{\circ}(\mathbf{\Lambda}, \mathbf{t})$, are in the end defined by closing u respectively $\mathbf{t}$ in all possible ways. □

Let still $\Lambda \in \mathtt{bl}^{\diamond}$. Given $\lambda, \mu \in \Lambda$, let us denote for $u = \mathbf{w}(\lambda)$ and $v = \mathbf{w}(\mu)$ (and only these)

$$\mathfrak{W}^{\circ}_{\vec{k}} = \bigoplus_{u,v \in \mathrm{Cup}(\vec{k})} {}_u(\mathfrak{W}_{\vec{k}})_v \quad \text{and} \quad \mathfrak{W}^{\circ} = \bigoplus_{\vec{k} \in \mathbb{b}\mathbb{l}} \mathfrak{W}_{\vec{k}}. \tag{40}$$

The following is a direct consequence of Lemma 4.15.

COROLLARY 4.17. For any $\lambda, \mu \in \Lambda$ and $u = \mathbf{w}(\lambda), v = \mathbf{w}(\mu)$: the isomorphisms $\Phi^{\lambda\mu}_{uv}$ from (39) extend to isomorphisms of graded, $\mathbb{K}$-vector spaces

$$\Phi^{\Lambda}_{\vec{k}} \colon \mathfrak{W}^{\circ}_{\vec{k}} \to \mathfrak{A}^{\mathfrak{F}}_{\Lambda}, \qquad \Phi \colon \mathfrak{W}^{\circ} \to \mathfrak{A}^{\mathfrak{F}}. \tag{41}$$

where we identify $\vec{k}$ and Λ as in (38). □

4.4. Proof of the main result. We deduce now Theorem 4.1 from the following.

THEOREM 4.18. The maps from (41) are isomorphisms of graded algebras. ■

Before we prove Theorem 4.18 in Subsection 4.5, we deduce some consequences, e.g. our main result. Moreover, the identification from Theorem 4.18 allows us to use topological arguments (i.e. foams) to deduce algebraic properties. In particular, we obtain the associativity of the Blanchet-Khovanov algebras.

COROLLARY 4.19. The multiplication rule from Subsection 3.3 is independent of the order in which the surgeries are performed. This turns $\mathfrak{A}^{\mathfrak{F}}_{\Lambda}$ into a graded, associative, unital algebra. Similar for (the locally unital) algebra $\mathfrak{A}^{\mathfrak{F}}$. ■

PROOF. This follows directly from Theorem 4.18 (note that the stated properties are clear if we work with the web algebra, see Corollary 2.28). □

REMARK 4.20. Theorem 4.18 leaves the question how the Blanchet-Khovanov algebra $\mathfrak{A}^{\mathfrak{F}}_{\Lambda}$ and Khovanov's original arc algebra H_m are related (and thus, how the Blanchet foam construction relates to the Khovanov [**17**] and Bar-Natan [**1**] theory using cobordisms). To answer this question, note that the action of $\dot{\mathrm{U}}_q(\mathfrak{gl}_n)$ from Subsection 4.2 extends to a 2-representation of Khovanov-Lauda's categorification of $\dot{\mathrm{U}}_q(\mathfrak{gl}_n)$, see [**24**, Proposition 3.3]. The same holds on the side of Khovanov's arc algebra, see [**6**, Remark 5.7]. Hence, it follows, for suitable choices of Λ and m, that $\mathfrak{A}^{\mathfrak{F}}_{\Lambda}$ and H_m are (graded) Morita equivalent. This can be deduced from Rouquier's universality theorem, see [**33**, Proposition 5.6 and Corollary 5.7], compare also to [**27**, Proposition 5.18]. Since $\mathfrak{A}^{\mathfrak{F}}_{\Lambda}$ and H_m are basic algebras, it follows by abstract nonsense that $\mathfrak{A}^{\mathfrak{F}}_{\Lambda}$ and H_m are isomorphic algebras (in fact, as graded algebras). This approach however does not provide an explicit isomorphism. Such an isomorphism is constructed in [**14**, Section 4] using a slightly more general framework. ▲

REMARK 4.21. Theorem 4.18 gives a way to compute the functorial chain complex (which is a link invariant) defined by Blanchet, see [**2**], and all its involved maps. Indeed, in our framework $\mathfrak{A}^{\mathfrak{F}}$ comes equipped with an easy to handle basis and all appearing $\mathfrak{A}^{\mathfrak{F}}$-module homomorphisms can explicitly be computed in this basis. This is in contrast to the local action used in [**24**, Proposition 3.3] to define

Blanchet's link homologies, see [**24**, Subsection 4.1], because it is not a priori clear in their formulation how to do explicit computations (since globally a significant number of non-trivial signs come into play). ▲

Note that we consider webs in $u \in \mathrm{Cup}(\vec{k})$ without imposing any relations. On the "uncategorified level" in the sense of Kuperberg [**23**], this has to be modified: denote by $\mathrm{Cup}(\vec{k})^{\mathbb{K}(q)}_{\mathrm{rel}} = \langle \mathrm{Hom}(\mathbf{1}_{2\omega}, \vec{k}) \rangle_{\mathbb{K}(q)}$ the $\mathbb{K}(q)$-linear vector space obtained from $\mathrm{Cup}(\vec{k})$ by linearization and modding out by the circle removal relation

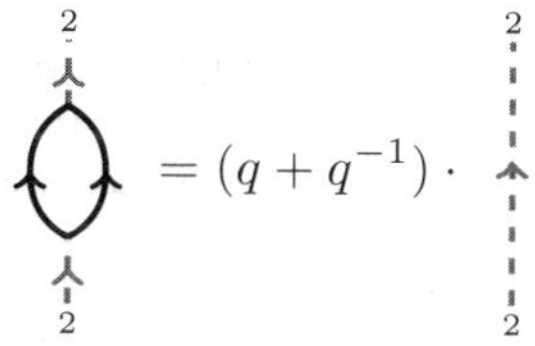

and isotopy relations as in the Lemmas 4.4 and 4.5. Note that, on the "categorified level" in which we work in the rest of the paper, we do not need to impose these relations since we "lift" them to the isomorphisms from Lemmas 4.3, 4.4 and 4.5.

LEMMA 4.22. *Let $\Lambda \in \mathtt{bl}^{\diamond}$ and $\vec{k}$ be its associated element in $\mathbb{bl}^{\diamond}$ (see (38)). Then*

$$\{u \in \mathrm{Cup}(\vec{k}) \mid u = \mathbf{w}(\lambda), \lambda \in \Lambda\}$$

is a $\mathbb{K}(q)$-linear basis of $\mathrm{Cup}(\vec{k})^{\mathbb{K}(q)}_{\mathrm{rel}}$.

PROOF. This is clear by the relations imposed on $\mathrm{Cup}(\vec{k})^{\mathbb{K}(q)}_{\mathrm{rel}}$. □

We are finally able to prove our main result.

PROOF OF THEOREM 4.1. Instead of taking all webs $u \in \mathrm{Cup}(\vec{k})$, it suffices to take a basis of $\mathrm{Cup}(\vec{k})^{\mathbb{K}(q)}_{\mathrm{rel}}$ and the webs $\mathbf{w}(\lambda)$ form such a basis by Lemma 4.22. Concretely, the algebras $\mathfrak{W}_{\vec{k}}$ (all webs) and $\mathfrak{W}^{\circ}_{\vec{k}}$ (only basis webs) are graded Morita equivalent (this can be seen similar to [**26**, Lemma 7.5]) and the statement follows from Theorem 4.18. The identification of the bimodules as graded $\mathbb{K}$-vector spaces is clear by Lemma 4.16, while the actions agree by Theorem 4.18 and construction of the actions. □

4.5. The proof of the graded isomorphism. We now prove Theorem 4.18.

PROOF OF THEOREM 4.18. By Lemma 4.15, it suffices to show that $\Phi^{\Lambda}_{\vec{k}}$ is a homomorphism of algebras (since then so is Φ as well). For this purpose, fix $\lambda, \mu, \nu \in \Lambda$ and set $u = \mathbf{w}(\lambda)$, $v = \mathbf{w}(\mu)$, and $w = \mathbf{w}(\nu)$. We show that any product of two basis cup foams $f \in \mathrm{Cup}(uv^*)$ and $g \in \mathrm{Cup}(vw^*)$ satisfies

$$\Phi^{\lambda\nu}_{uw}(fg) = \Phi^{\lambda\mu}_{uv}(f)\Phi^{\mu'\nu}_{v'w}(g), \quad v = v', \mu = \mu'$$

(This is enough since all non-zero multiplications on the side of $\mathfrak{A}^{\mathfrak{F}}_{\Lambda}$ as well as on the side of $\mathfrak{W}_{\vec{k}}$ satisfy $v = v'$ and $\mu = \mu'$ by definition.)

In order to do so, we show that each step in the multiplication procedure from Definition 2.24 locally agrees with the one from Subsection 3.3. There are four topological different situations to check (compare with the cases in Subsection 3.3 introducing the different signs):

(i) **Non-nested merge.** Two non-nested circles are replaced by one circle.

(ii) **Nested merge.** Two nested circles are replaced by one circle.
(iii) **Non-nested split.** One circle is replaced by two non-nested circles.
(iv) **Nested split.** One circle is replaced by two nested circles.

We will consider these four cases step-by-step and compare the corresponding multiplications rules. In addition to the four cases we will further distinguish the following shapes of the involved underlying webs:

(A) **Basic shape.** The involved components are as small as possible with the minimal number of phantom facets.
(B) **Minimal saddle.** While the components themselves are allowed to be of any shape, the involved saddle only includes a single phantom facet.
(C) **General case.** Both, the shape as well as the saddle, are arbitrary.

We start by comparing the two multiplication rules for the basic shapes first. This will only involve simplified versions of the *dot moving signs*. The basic shapes for (i), (ii), (iii) and (iv) are the following (with the multiplication step taking place in the marked region)

(42)

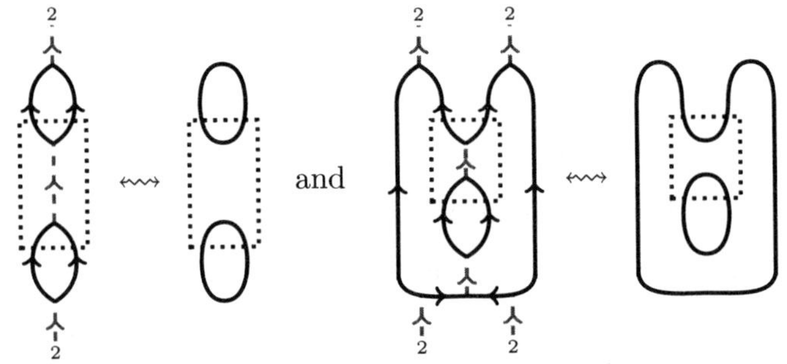

in cases (i) and (ii), and the following H-shape and C-shape

(43)

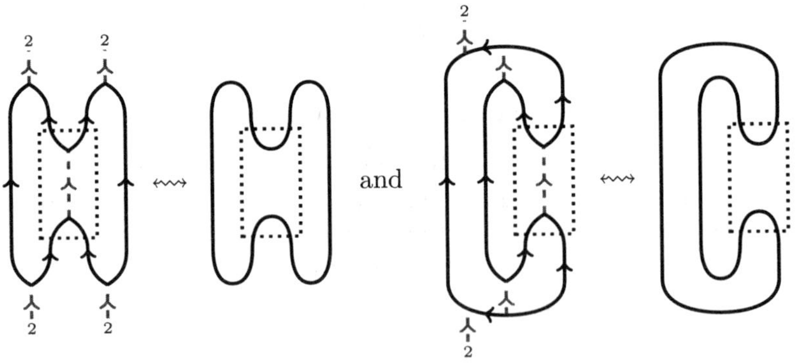

in cases (iii) and (iv). Here we always display both, the web as well as its corresponding arc diagram.

▶ **Non-nested merge - basic shape.** In this case, we merge two simple cup webs on the web side and two oriented circles on the arc diagram side. The following table gives the multiplication results in the four possible orientation combinations.

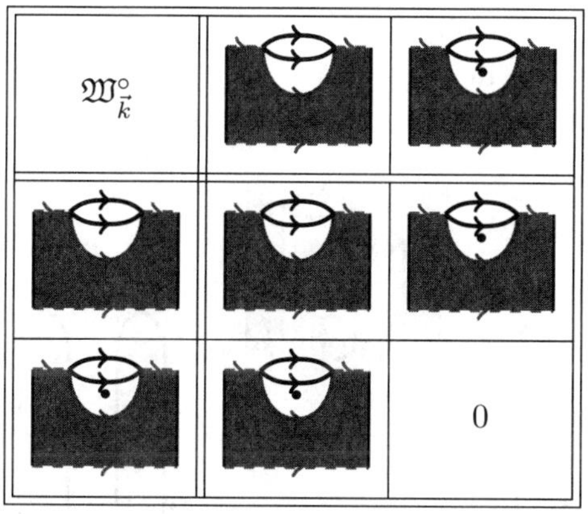

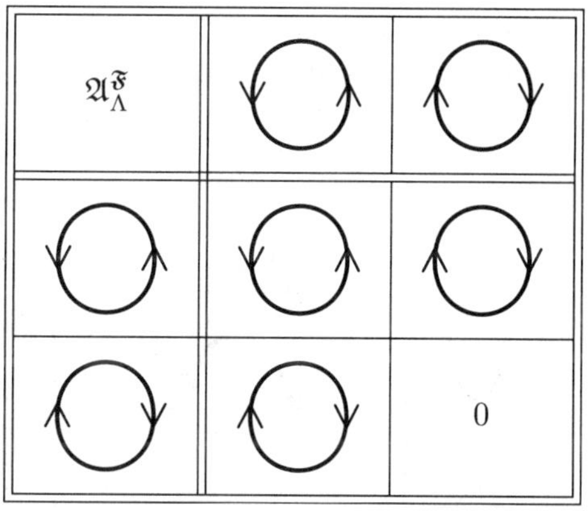

To obtain this table one argues as follows: merging two basis cup foams via a saddle creates a new basis cup foam. Thus, only the position of the dot matters if we rewrite the result in the cup basis. If there is no dot or there are two dots on the new cup foam, then we are done (the latter follows from (8)). If there is only one dot note that it is automatically on the rightmost facet and we are done as well (no signs). This is precisely as in (27). □

► **Nested merge - basic shape.** This step in case of $\mathfrak{A}^{\mathfrak{F}}_{\Lambda}$ was calculated in (28). In case of $\mathfrak{W}^{\circ}_{\vec{k}}$ this is given by stacking the saddle displayed below on top of a given foam (the second foam displayed below is shown to illustrate the cylinder we want to cut).

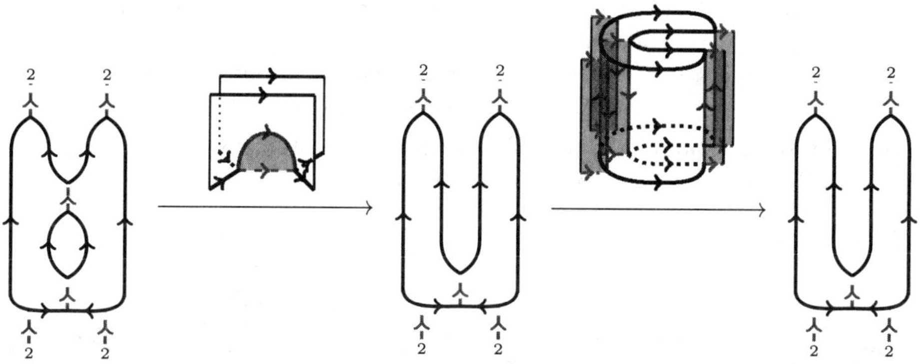

Note now that the difference to the non-nested merge above is that, if a basis cup foam is sitting underneath the leftmost picture, then the end result is topological not a basis cup foam. In order to turn the result into a basis cup foam, we apply (15) to the cylinder illustrated above. Here we have to use (16) first, which gives an overall sign. After neck cutting the cylinder we create a "bubble" (recalling that a basis cup foam is sitting underneath) with two internal phantom facets in the bottom part of the picture. By (18), we can remove the phantom facets with the cost of a sign and create an "honest" bubble instead. Thus, by (14), only the term in (15) with the dot on the bottom survives (with a sign). By (13) the remaining bubble evaluates to -1. Hence, we get in total four overall signs which is the same as no extra sign. The dots behave as in the table above, since before the neck cut we can move any of them to the top and thus they do not interfere with the above procedure. Thus, using (17), we get the same result as in (28). □

▶ **Non-nested split - basic shape.** This step in case of $\mathfrak{A}_\Lambda^{\mathfrak{F}}$ was calculated in (29). For $\mathfrak{W}^{\circ}_{\vec{k}}$ the multiplication is given by the composition of the following foams.

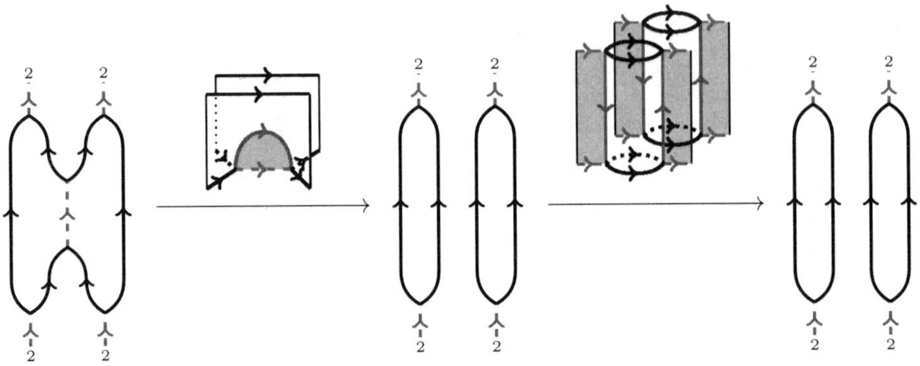

As before, we can use the neck cutting (15) on the cylinder that corresponds to the right circle (we could also choose the other one and slightly change the steps below). The result is precisely as in (29):

- If the original basis cup foams sitting underneath has no dots, then the one with the dot on the rightmost circle gets no sign (the dot is automatically on the rightmost facet), and the other does not as well (the dot from the neck cutting needs to pass one phantom facets to move to the right).
- If the original basis cup foams sitting underneath has already a dot, then only the positive term in (15) survives and the dots are already in the rightmost positions.
- In both cases, the resulting foam is topological not a basis cup foam, but using (18) once reduces it to a basis cup foam, giving an overall sign. □

▶ **Nested split - basic shape.** This step in case of $\mathfrak{A}_\Lambda^{\mathfrak{F}}$ was calculated in (30). In case of $\mathfrak{W}^{\circ}_{\vec{k}}$ we again give the composite of foams that we stack on top of each other for the multiplication.

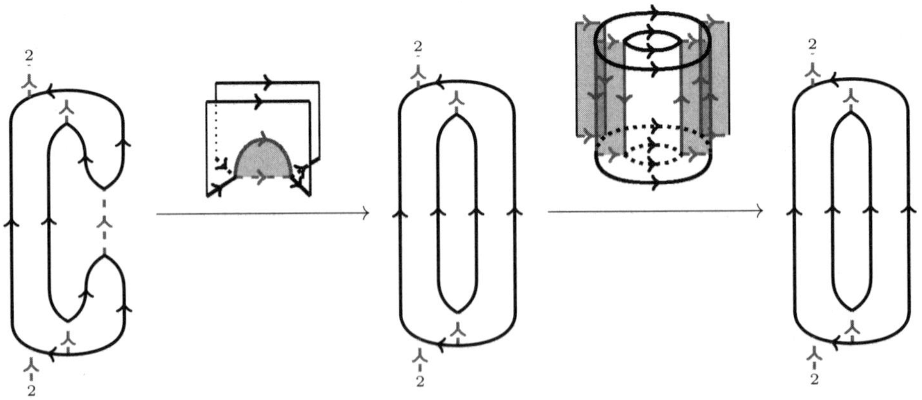

Again we can apply neck cutting. This time to the internal cylinder in the second foam between the middle web and the rightmost web connecting the two nested circles that we can cut using (15). The result is precisely as in (30):

- If the original basis cup foams sitting underneath has no dots, then the one with the dot on the non-nested circle gets a sign (the dot is automatically on the rightmost facet), while the other does not (the dot created in the neck cutting needs to pass two phantom facets to get to the right).
- If the original basis cup foams sitting underneath has already a dot, then only the positive term in (15) survives and the dots are already in the rightmost positions.
- In both cases, the resulting foam is already topological a cup basis foam and nothing needs to be done anymore. □

All other situations (i.e. general shapes) are similar to the ones discussed, but with the main difference that dots need to be shifted to the rightmost facets, some (intermediate) results might not be in the topological form of a basis cup foam and one needs to take care of saddles with a possible number of additional internal phantom facets as in (21). These three facts together explain the signs turning up in the multiplication described in Subsection 3.3 with the first one corresponding to the dot moving sign as in (25), the second corresponding to the topological signs as in (26) and the latter to the saddle sign as in (26).

Thus, we next deal with the minimal saddle situation as for example in (22). In those cases $s(\gamma) = 1$ on the side of $\mathfrak{A}_{\Lambda}^{\mathfrak{F}}$. This will introduce the *topological signs* and simplified versions (for $s(\gamma) = 1$) of the *saddle signs*.

► **Non-nested merge - minimal saddle.** This is topologically the same as in *non-nested merge - basic shape*, since the resulting foam will already be a basis cup foam. Thus, the only sign comes from moving dots to the right which matches the dot moving sign turning up in the multiplication on the side of $\mathfrak{A}_{\Lambda}^{\mathfrak{F}}$ in (25). □

► **Nested merge - minimal saddle.** Assume now that we merge a circle C^{out} with some nested circle C^{in} inside of it.

The dot moving is as above in *non-nested merge - minimal saddle* and gives the same sign as for $\mathfrak{A}_{\Lambda}^{\mathfrak{F}}$. The difference to the nested merge in the basic shape is that we have to bring the resulting foam in the topological form of a basis cup foam. To this end, we can proceed as in *nested merge - basic shape* and cut the same cylinder as there. We first note that neither the dots which are already on the foam sitting underneath nor the internal circles in C^{out} which are different from C^{in} matter: we can topologically move them "away from the local picture".

Thus, to simplify a little bit, assume that C^{in} is the only circle nested in C^{out} and there are no dots. Following the procedure given as in *nested merge - basic shape* above, we see that the only things of importance are signs that come from cutting the cylinder with possible internal phantom facets and evaluating the "bubble" with possible more than one internal phantom facets. Indeed, what matters is the number of times we need to use (16) in the cutting procedure of the cylinder and the number of times we need to use (18) in the bursting of the "bubble" (the rest stays the same as before in *nested merge - basic shape*). Now, the number of times we need to apply (16) is $\mathrm{ipe}(C^{\mathrm{out}}) - 1$ (the -1 comes in because we apply a saddle which removes one of the internal phantom edges of the starting picture) and the number of times we need to apply (18) is $\mathrm{ipe}(C^{\mathrm{out}} - C^{\mathrm{in}})$. By Lemma 4.10, we obtain that

$$(-1)^{\mathrm{ipe}(C^{\mathrm{out}})-1+\mathrm{ipe}(C^{\mathrm{out}}-C^{\mathrm{in}})} = (-1)^{\frac{1}{4}(2(\mathrm{d}(C_{\mathrm{out}})-2)+\mathrm{d}(C_{\mathrm{in}})-2)} \tag{44}$$
$$= -(-1)^{\frac{1}{4}(\mathrm{d}(C_{\mathrm{in}})-2)} \cdot (-1)^1,$$

where C_{out} and C_{in} are the cup diagram counterparts of C^{out} and C^{in}. This is precisely the same sign turning up on the side of $\mathfrak{A}_\Lambda^{\mathfrak{F}}$ (compare to (26) with $s_\Lambda(\gamma) = 1$).

The case where C^{out} has several nested components, is similar since all nested components of C^{out}, which are not C^{in}, increase the number of times we need to use (16) in the same way as the number of times we need to use (18) (hence, no change modulo 2). Again, this matches the side of $\mathfrak{A}_\Lambda^{\mathfrak{F}}$ in (26). □

▶ **Non-nested split - minimal saddle.** The dot moving is as before. Observe now that we do not have an extra sign turning up although the resulting foam is not in the topological shape of a basis cup foam. To see this, we first simplify by assuming that the circle C which is split does not contain any nested components. Using the same cutting as in *non-nested split - basic shape*, we have signs coming from squeezing cylinders and simplifying "bubbles" (similar as above in *nested merge - minimal saddle*). But the number of times we need to apply (16) in this case is now the same as the number of times one has to apply (18), namely $\mathrm{ipe}(C)-1$. Thus, again no change modulo 2. The case with nested components in C is now analogously as above in *nested merge - minimal saddle* since we can move dots and nested circles "away". As before, this increases the number of times we need to use (16) in the same way as the number of times we need to use (18) (hence, no change mod 2). Thus, we do not get an extra overall sign as in case of $\mathfrak{A}_\Lambda^{\mathfrak{F}}$ (see Subsection 3.3.3 non-nested split case). □

▶ **Nested split - minimal saddle.** Moving dots is again as before. Furthermore, again, as in *nested merge - minimal saddle*, we need to topologically manipulate the resulting foam until it is in basis cup foam shape. We can proceed as before in *nested split - basic shape* and, similar as above in *nested merge - minimal saddle*, we pick up signs coming from cylinder cuts and bubble removals. In fact, the total sign can be calculated analogously as in *nested merge - minimal saddle* (and is the same as there). Again, this matches the side of $\mathfrak{A}_\Lambda^{\mathfrak{F}}$ (see Subsection 3.3.3 nested split case). □

▶ **Non-nested merge - general case.** In fact, nothing changes compared to the discussion in *non-nested merge - minimal saddle*, since dots passing a saddle always pass an odd number of phantom facets (compare to (21)) and the resulting foams are topological already basis cup foams. □

▶ **Nested merge - general case.** The dot moving stays as before. The only thing that changes in contrast to *nested merge - minimal saddle* is that we obtain

$$(-1)^{\mathrm{ipe}(C^{\mathrm{out}})-s+\mathrm{ipe}(C^{\mathrm{out}}-C^{\mathrm{in}})} = (-1)^{\frac{1}{4}(2(\mathrm{d}(C_{\mathrm{out}})-2)+\mathrm{d}(C_{\mathrm{in}})-2)} \cdot (-1)^{s(\gamma)-1}$$
$$= -(-1)^{\frac{1}{4}(\mathrm{d}(C_{\mathrm{in}})-2)} \cdot (-1)^{s(\gamma)}.$$

within the topological re-writing procedure instead of the formula from (44). Thus, this matches the side of $\mathfrak{A}_\Lambda^{\mathfrak{F}}$ (see (26)). □

► **Non-nested split - general case.** Again, the dot moving stays as before. The difference to *non-nested split - minimal saddle* is that we have to apply (18) s-times instead of once, which gives the sign turning up for $\mathfrak{A}_\Lambda^{\mathfrak{F}}$ (see Subsection 3.3.3 non-nested split case). □

► **Nested split - general case.** There is no difference to the arguments given in *nested split - minimal saddle*. Again, this matches the side of $\mathfrak{A}_\Lambda^{\mathfrak{F}}$ (see Subsection 3.3.3 nested split case). □

Hence, in each case the topological multiplication agrees with the algebraically defined one. This concludes the proof. □

Acknowledgements

The authors would like to thank David Rose and Nathalie Wahl for helpful conversations, and Paul Wedrich and the referee for helpful comments. M.E. and D.T. thank the whiteboard in their office for many helpful illustrations.

References

[1] D. Bar-Natan, *Khovanov's homology for tangles and cobordisms*, Geom. Topol. **9** (2005), 1443–1499, DOI 10.2140/gt.2005.9.1443. MR2174270

[2] C. Blanchet, *An oriented model for Khovanov homology*, J. Knot Theory Ramifications **19** (2010), no. 2, 291–312, DOI 10.1142/S0218216510007863. MR2647055

[3] C. Blanchet, N. Habegger, G. Masbaum, and P. Vogel, *Topological quantum field theories derived from the Kauffman bracket*, Topology **34** (1995), no. 4, 883–927, DOI 10.1016/0040-9383(94)00051-4. MR1362791

[4] J. Brundan and C. Stroppel, *Highest weight categories arising from Khovanov's diagram algebra I: cellularity* (English, with English and Russian summaries), Mosc. Math. J. **11** (2011), no. 4, 685–722, 821–822. MR2918294

[5] J. Brundan and C. Stroppel, *Highest weight categories arising from Khovanov's diagram algebra. II. Koszulity*, Transform. Groups **15** (2010), no. 1, 1–45, DOI 10.1007/s00031-010-9079-4. MR2600694

[6] J. Brundan and C. Stroppel, *Highest weight categories arising from Khovanov's diagram algebra III: category $\mathcal{O}$*, Represent. Theory **15** (2011), 170–243, DOI 10.1090/S1088-4165-2011-00389-7. MR2781018

[7] J. Brundan and C. Stroppel, *Highest weight categories arising from Khovanov's diagram algebra IV: the general linear supergroup*, J. Eur. Math. Soc. (JEMS) **14** (2012), no. 2, 373–419, DOI 10.4171/JEMS/306. MR2881300

[8] J. Brundan and C. Stroppel, *Gradings on walled Brauer algebras and Khovanov's arc algebra*, Adv. Math. **231** (2012), no. 2, 709–773, DOI 10.1016/j.aim.2012.05.016. MR2955190

[9] S. Cautis, J. Kamnitzer, and S. Morrison, *Webs and quantum skew Howe duality*, Math. Ann. **360** (2014), no. 1-2, 351–390, DOI 10.1007/s00208-013-0984-4. MR3263166

[10] D. Clark, S. Morrison, and K. Walker, *Fixing the functoriality of Khovanov homology*, Geom. Topol. **13** (2009), no. 3, 1499–1582, DOI 10.2140/gt.2009.13.1499. MR2496052

[11] M. Ehrig and C. Stroppel, *2-row Springer fibres and Khovanov diagram algebras for type D*, Canad. J. Math. **68** (2016), no. 6, 1285–1333, DOI 10.4153/CJM-2015-051-4. MR3563723

[12] M. Ehrig and C. Stroppel, *Diagrammatic description for the categories of perverse sheaves on isotropic Grassmannians*, Selecta Math. (N.S.) **22** (2016), no. 3, 1455–1536, DOI 10.1007/s00029-015-0215-9. MR3518556

[13] M. Ehrig and C. Stroppel, *On the category of finite-dimensional representations of* $\mathrm{OSp}(r|2n)$*: Part I*, Representation theory–Current Trends and Perspectives, EMS Ser. Congr. Rep. (2016), pp. 109–170, Eur. Math. Soc., Zürich

[14] M. Ehrig, C. Stroppel, and D. Tubbenhauer, *Generic* $\mathfrak{gl}_2$*-foams, web and arc algebras*, URL: `http://arxiv.org/abs/1601.08010`.

[15] M. Ehrig, D. Tubbenhauer, and A. Wilbert, *Singular* TQFTs*, foams and type D arc algebras*, URL: `http://arxiv.org/abs/1611.07444`

[16] M. Jacobsson, *An invariant of link cobordisms from Khovanov homology*, Algebr. Geom. Topol. **4** (2004), 1211–1251 (electronic), DOI 10.2140/agt.2004.4.1211. MR2113903

[17] M. Khovanov, *A categorification of the Jones polynomial*, Duke Math. J. **101** (2000), no. 3, 359–426, DOI 10.1215/S0012-7094-00-10131-7. MR1740682

[18] M. Khovanov, *A functor-valued invariant of tangles*, Algebr. Geom. Topol. **2** (2002), 665–741, DOI 10.2140/agt.2002.2.665. MR1928174

[19] M. Khovanov, *sl(3) link homology*, Algebr. Geom. Topol. **4** (2004), 1045–1081, DOI 10.2140/agt.2004.4.1045. MR2100691

[20] M. Khovanov and A. D. Lauda, *A categorification of quantum* sl(n), Quantum Topol. **1** (2010), no. 1, 1–92, DOI 10.4171/QT/1. MR2628852

[21] M. Khovanov and L. Rozansky, *Matrix factorizations and link homology*, Fund. Math. **199** (2008), no. 1, 1–91, DOI 10.4064/fm199-1-1. MR2391017

[22] J. Kock, *Frobenius algebras and 2D topological quantum field theories*, London Mathematical Society Student Texts, vol. 59, Cambridge University Press, Cambridge, 2004. MR2037238

[23] G. Kuperberg, *Spiders for rank* 2 *Lie algebras*, Comm. Math. Phys. **180** (1996), no. 1, 109–151. MR1403861

[24] A. D. Lauda, H. Queffelec, and D. E. V. Rose, *Khovanov homology is a skew Howe 2-representation of categorified quantum* $\mathfrak{sl}_m$, Algebr. Geom. Topol. **15** (2015), no. 5, 2517–2608, DOI 10.2140/agt.2015.15.2517. MR3426687

[25] G. Lusztig, *Introduction to quantum groups*, Modern Birkhäuser Classics, Birkhäuser/Springer, New York, 2010. Reprint of the 1994 edition. MR2759715

[26] M. Mackaay, *The* $\mathfrak{sl}_N$*-web algebras and dual canonical bases*, J. Algebra **409** (2014), 54–100, DOI 10.1016/j.jalgebra.2014.02.036. MR3198835

[27] M. Mackaay, W. Pan, and D. Tubbenhauer, *The* $\mathfrak{sl}_3$*-web algebra*, Math. Z. **277** (2014), no. 1-2, 401–479, DOI 10.1007/s00209-013-1262-6. MR3205780

[28] M. Mackaay, M. Stošić, and P. Vaz, $\mathfrak{sl}(N)$*-link homology* ($N \geq 4$) *using foams and the Kapustin-Li formula*, Geom. Topol. **13** (2009), no. 2, 1075–1128, DOI 10.2140/gt.2009.13.1075. MR2491657

[29] H. Queffelec and D. E. V. Rose, *The* $\mathfrak{sl}_n$ *foam 2-category: a combinatorial formulation of Khovanov-Rozansky homology via categorical skew Howe duality*, Adv. Math. **302** (2016), 1251–1339, DOI 10.1016/j.aim.2016.07.027. MR3545951

[30] J. Rasmussen, *Khovanov's invariant for closed surfaces*, URL: http://arxiv.org/abs/math/0502527.

[31] L.-H. Robert, *A characterization of indecomposable web modules over Khovanov-Kuperberg algebras*, Algebr. Geom. Topol. **15** (2015), no. 3, 1303–1362, DOI 10.2140/agt.2015.15.1303. MR3361138

[32] L.-H. Robert, *A large family of indecomposable projective modules for the Khovanov-Kuperberg algebras of* sl_3*-webs*, J. Knot Theory Ramifications **22** (2013), no. 11, 1350062, 25, DOI 10.1142/S0218216513500624. MR3143579

[33] R. Rouquier, 2*-Kac-Moody algebras*, URL: http://arxiv.org/abs/0812.5023.

[34] A. Sartori, *Categorification of tensor powers of the vector representation of* $U_q(\mathfrak{gl}(1|1))$, Selecta Math. (N.S.) **22** (2016), no. 2, 669–734, DOI 10.1007/s00029-015-0202-1. MR3477333

[35] C. Stroppel, *TQFT with corners and tilting functors in the Kac-Moody case*, URL: http://arxiv.org/abs/math/0605103.

[36] D. Tubbenhauer, $\mathfrak{sl}_3$*-web bases, intermediate crystal bases and categorification*, J. Algebraic Combin. **40** (2014), no. 4, 1001–1076, DOI 10.1007/s10801-014-0518-5. MR3273401

[37] D. Tubbenhauer, $\mathfrak{sl}_n$*-webs, categorification and Khovanov-Rozansky homologies*, URL: http://arxiv.org/abs/1404.5752.

School of Mathematics & Statistics, Carslaw Building, University of Sydney, NSW 2006, Australia

E-mail address: michael.ehrig@sydney.edu.au

Mathematisches Institut, Universität Bonn, Endenicher Allee 60, Room 4.007, 53115 Bonn, Germany

E-mail address: stroppel@math.uni-bonn.de

Mathematisches Institut, Universität Bonn, Endenicher Allee 60, Room 1.003, 53115 Bonn, Germany

E-mail address: dtubben@math.uni-bonn.de

Contemporary Mathematics
Volume **683**, 2017
http://dx.doi.org/10.1090/conm/683/13705

Generic character sheaves on groups over $\mathbf{k}[\epsilon]/(\epsilon^r)$

G. Lusztig

ABSTRACT. Let G be a connected reductive group over k, an algebraic closure of a finite field F_q. For an integer $r \geq 1$ let $G_r = G(k[\epsilon]/(\epsilon^r))$ viewed as an algebraic group of dimension $r \dim G$ over k. We show that the character of the generic principal series representation of $G_r(F_q)$ can be realized by a simple perverse sheaf on G_r provided that $r = 2$ or $r = 4$ and we give a strategy to prove the same statement for any r. The case where $r = 1$ is already known.

Introduction

0.1. Let $\mathbf{k}$ be an algebraic closure of the finite field $\mathbf{F}_q$ with q elements where q is a power of a prime number p. Let G be a connected reductive group over $\mathbf{k}$ with a fixed split $\mathbf{F}_q$-rational structure, a fixed Borel subgroup B defined over $\mathbf{F}_q$, with unipotent radical U and a fixed maximal torus T of B defined over $\mathbf{F}_q$. Let $\mathfrak{g}, \mathfrak{b}, \mathfrak{t}, \mathfrak{n}$ be the Lie algebras of G, B, T, U. We fix a prime number $l \neq p$. If $\lambda : T(\mathbf{F}_q) \to \bar{\mathbf{Q}}_l^*$ is a character, we can lift λ to a character $\tilde{\lambda} : B(\mathbf{F}_q) \to \bar{\mathbf{Q}}_l^*$ trivial on $U(\mathbf{F}_q)$ and we can form the induced representation $\mathrm{ind}_{B(\mathbf{F}_q)}^{G(\mathbf{F}_q)} \tilde{\lambda}$ of $G(\mathbf{F}_q)$. Its character is the class function $G(\mathbf{F}_q) \to \bar{\mathbf{Q}}_l$ given by

(a) $$y \mapsto \sum_{\substack{B(\mathbf{F}_q)x \in B(\mathbf{F}_q)\backslash G(\mathbf{F}_q);\\ xyx^{-1} \in B(\mathbf{F}_q)}} \tilde{\lambda}(xyx^{-1}).$$

This class function has a geometric analogue. Namely, we consider the diagram $T \xleftarrow{h} \tilde{G} \xrightarrow{\pi} G$ where $\tilde{G} = \{(Bx, y) \in (B\backslash G) \times G; xyx^{-1} \in B\}$, $\pi(Bx, y) = y$ is the Springer map and $h(Bx, y) = d(xyx^{-1})$; here $d : B \to T$ is the obvious homomorphism with kernel U. Let $\mathcal{E}$ be a fixed $\bar{\mathbf{Q}}_l$-local system of rank 1 on T such that $\mathcal{E}^{\otimes m} \cong \bar{\mathbf{Q}}_l$ for some $m \geq 1$ prime to p. The geometric analogue of (a) is the complex $L_1 = \pi_! h^* \mathcal{E} \in \mathcal{D}(G)$. (For any algebraic variety X over $\mathbf{k}$, $\mathcal{D}(X)$ denotes the bounded derived category of constructible $\bar{\mathbf{Q}}_l$-sheaves on X.) When $\mathcal{E}$ is defined over $\mathbf{F}_q$ and has characteristic function λ then L_1 is defined over $\mathbf{F}_q$ and its characteristic function is (up to a nonzero scalar factor) the function (a). Thus L_1 can be viewed as a categorified version of the function (a). More precisely, L_1 is (up to shift) a perverse sheaf on G; indeed, one of the main observations of [**4**] was that the (proper) map π is small, which implies that L_1 is an intersection cohomology

2010 *Mathematics Subject Classification.* Primary 20G99.
Key words and phrases. Perverse sheaf, reductive group, Fourier-Deligne transform.
Supported in part by National Science Foundation grant 1303060.

complex; this was the starting point of the theory of character sheaves on G, see [**5**]. This point of view is useful since the complexes L_1 are defined independently of the $\mathbf{F}_q$-structure and from them one can extract not only the characters (a) for any q but even their twisted versions defined in [**1**].

0.2. For any integer $r \geq 1$ we consider the ring $\mathbf{k}_r = \mathbf{k}[\epsilon]/(\epsilon^r)$ (ϵ is an indeterminate). Let $G_r = G(\mathbf{k}_r)$ be the group of points of G with values in $\mathbf{k}_r$, viewed as an algebraic group over $\mathbf{k}$ of dimension $r\Delta$ where $\Delta = \dim G$. For example, if $G = GL_n(\mathbf{k})$, then $G(\mathbf{k}_r)$ is the group of invertible $n \times n$ matrices with entries in $\mathbf{k}_r$, viewed as an affine subvariety of $\mathbf{k}^{n^2 r}$ (the complement of a hypersurface). Let $B_r = B(\mathbf{k}_r)$, $T_r = T(\mathbf{k}_r)$, $U_r = U(\mathbf{k}_r)$. Note that G_r inherits from G a natural $\mathbf{F}_q$-structure and that B_r, T_r, U_r are defined over $\mathbf{F}_q$. For $r = 1$, G_r reduces to G; we would like to extend as much as possible the results in 0.1 from $r = 1$ to a general r. If $\lambda : T_r(\mathbf{F}_q) \to \bar{\mathbf{Q}}_l^*$ is a character, we can lift λ to a character $\tilde{\lambda} : B_r(\mathbf{F}_q) \to \bar{\mathbf{Q}}_l^*$ trivial on $U_r(\mathbf{F}_q)$ and we can form the induced representation $\mathrm{ind}_{B(\mathbf{F}_q)}^{G(\mathbf{F}_q)} \tilde{\lambda}$ of $G_r(\mathbf{F}_q)$. Its character is the class function $G_r(\mathbf{F}_q) \to \bar{\mathbf{Q}}_l$ given by

(a)
$$g' \mapsto \sum_{\substack{B_r(\mathbf{F}_q)g \in B_r(\mathbf{F}_q)\backslash G_r(\mathbf{F}_q);\\ gg'g^{-1} \in B_r(\mathbf{F}_q)}} \tilde{\lambda}(gg'g^{-1}).$$

It generalizes the function 0.1(a). Again this class function has a geometric analogue. Namely, we consider the diagram $T_r \xleftarrow{h_r} \tilde{G}_r \xrightarrow{\pi_r} G_r$ where

$$\tilde{G}_r = \{(B_r g, g') \in (B_r\backslash G_r) \times G_r; gg'g^{-1} \in B_r\},$$
$$\pi_r(B_r g, g') = g', h_r(B_r g, g') = d_r(gg'g^{-1});$$

here $d_r : B_r \to T_r$ is the obvious homomorphism with kernel U_r. We can identify $T = \mathcal{Y} \otimes \mathbf{k}^*$ where $\mathcal{Y}$ is the lattice of one parameter subgroups of T and $T(\mathbf{k}_r) = \mathcal{Y} \otimes \mathbf{k}_r^*$ where $\mathbf{k}_r^*$ is the group of units of $\mathbf{k}_r$. The isomorphism $\mathbf{k}^* \times \mathbf{k}^{r-1} \xrightarrow{\sim} \mathbf{k}_r^*$,

$$(a_0, a_1, \dots, a_{r-1}) \mapsto a_0 + a_1\epsilon + \cdots + a_{r-1}\epsilon^{r-1},$$

identifies $T(\mathbf{k}_r)$ with $T \times \mathfrak{t}^{r-1}$. Let $f_1, \dots, f_{r-1}$ be linear functions $\mathfrak{t} \to \mathbf{k}$ and let $\mathcal{E}$ be as in 0.1. We can form the local system $\mathcal{E} \boxtimes \mathcal{L}_{f_1} \boxtimes \dots \boxtimes \mathcal{L}_{f_{r-1}}$ on $T \times \mathfrak{t}^{r-1} = T(\mathbf{k}_r)$ (for the notation $\mathcal{L}_{f_i}$ see 0.3). The geometric analogue of (a) is the complex $L = \pi_{r!} h_r^*(\mathcal{E} \boxtimes \mathcal{L}_{f_1} \boxtimes \dots \boxtimes \mathcal{L}_{f_{r-1}}) \in \mathcal{D}(G_r)$. Again from the complexes L one can extract the characters (a) for any q. In this paper we are interested in the conjecture in [**6**, 8(a)] according to which, L is (up to shift) an intersection cohomology complex on G_r, provided that f_{r-1} is sufficiently general. This would imply that there is a theory of generic character sheaves on G_r. In the case where $r = 1$, $f_1, \dots, f_{r-1}$ are missing, the hypothesis that f_{r-1} is sufficiently general is vacuous and the conjecture is true by [**4**]. The conjecture was proved in [**6**, no.12] in the case where $G = GL_2$ and $r = 2$. In this paper we give a method to attack the conjecture for any G and even r (but with some restriction on p) in which case π_r is neither small nor proper in general; we carry out the method in detail in the cases where $r = 2$ and $r = 4$ and we prove the conjecture in these cases (with some restriction on p). We also prove a weak form of the conjecture assuming that $r = 3$ (see Theorem 4.7). I believe that the method of this paper should be applicable with any $r \geq 2$.

Our method is to first replace L by another complex K which is a geometric (categorified) form of the character of a representation constructed by Gérardin [**2**] in 1975, then to try to describe explicitly the Fourier-Deligne transform of K on

G_r (viewed as a vector bundle over G). For $r = 2$ and $r = 4$ we show that this is a simple perverse sheaf of a very special kind, namely one associated to a local system of rank 1 on a closed smooth irreducible subvariety of G_r; by a result of Laumon [**3**], this implies that K is itself a simple perverse sheaf, up to twist. Recall that Laumon's theorem states that an l-adic complex of sheaves on a vector bundle is a perverse sheaf if and only if its Fourier-Deligne transform (a complex of sheaves on the dual vector bundle) is a perverse sheaf.

Finally, we show that L is a shift of K for the values of r that we consider and this gives the desired result.

0.3. Notation. In this paper all algebraic varieties are over $\mathbf{k}$. We fix a nontrivial homomorphism $\psi : \mathbf{F}_p \to \bar{\mathbf{Q}}_l^*$. For any morphism $f : X \to \mathbf{k}$ let $X_f = \{(x, \lambda) \in X \times \mathbf{k}; \lambda^q - \lambda = f(x)\}$ and let $\iota : X_f \to X$ be the Artin-Schreier covering $(x, \lambda) \mapsto x$. Then $\iota_! \bar{\mathbf{Q}}_l$ is a local system with a natural action of $\mathbf{F}_p$ (coming from the $\mathbf{F}_p$-action $\zeta : (x, \lambda) \mapsto (x, \lambda + \zeta)$ on $\tilde{X}$); we denote by $\mathcal{L}_f$ the ψ-eigenspace of this action (a local system of rank 1 on X).

Let $\delta = \dim T$.

For $x \in G$ if X is an element of $\mathfrak{g}$ or a subset of $\mathfrak{g}$ we write ${}^xX$ instead of $\mathrm{Ad}(x)X$ and ${}_xX$ instead of $\mathrm{Ad}(x^{-1})X$.

1. The complex K

1.1. Let $X_1, X_2, \ldots$ and $Y_1, Y_2, \ldots$ be two sequences of noncommuting indeterminates. From the Campbell-Baker-Hausdorff formula we deduce the equality

$$(e^{\epsilon X_1} e^{\epsilon^2 X_2} \ldots)(e^{\epsilon Y_1} e^{\epsilon^2 Y_2} \ldots) = e^{\epsilon z_1} e^{\epsilon^2 z_2} \ldots$$

where $z_i = z_i(X_1, X_2, \ldots, X_i, Y_1, Y_2, \ldots, Y_i)$, $(i \geq 1)$ are universal Lie polynomials with coefficients in $\mathbf{Z}[(i!)^{-1}]$. (Here ϵ commutes with each X_i, Y_i.) For example,

$z_1(X_1, Y_1) = X_1 + Y_1$,

$z_2(X_1, X_2, Y_1, Y_2) = X_2 + Y_2 + [X_1, Y_1]/2$,

$z_3(X_1, X_2, X_3, Y_1, Y_2, Y_3) = X_3 + Y_3 + [X_2, Y_1] - [X_1, [X_1, Y_1]]/6 - [Y_1, [X_1, Y_1]]/3$.

We deduce that if $X_1, X_2, \ldots,$ $X'_1, X'_2, \ldots$ and $Y_1, Y_2, \ldots$ are three sequences of noncommuting indeterminates then we have the equality

$$(e^{\epsilon X'_1} e^{\epsilon^2 X'_2} \ldots)(e^{\epsilon Y_1} e^{\epsilon^2 Y_2} \ldots)(e^{\epsilon X_1} e^{\epsilon^2 X_2} \ldots)^{-1} = e^{\epsilon u_1} e^{\epsilon^2 u_2} \ldots$$

where $u_i = u_i(X'_1, \ldots, X'_i, Y_1, \ldots, Y_i, X_1, \ldots, X_i)$, $(i \geq 1)$ are universal Lie polynomials with coefficients in $\mathbf{Z}[(i!)^{-1}]$. For example,

$$u_1(X_1, Y_1, X'_1) = X'_1 - X_1 + Y_1,$$

$$u_2(X_1, X_2, Y_1, Y_2, X'_1, X'_2) = X'_2 - X_2 + Y_2 + [X'_1, Y_1]/2 - [X'_1, X_1]/2 - [Y_1, X_1]/2,$$

$$\begin{aligned}
&u_3(X_1, X_2, X_3, Y_1, Y_2, Y_3, X'_1, X'_2, X'_3)\\
&= X'_3 - X_3 + Y_3 + [X'_2, Y_1] + [X_2, X_1] - [X'_2, X_1] - [Y_2, X_1] - [X'_1, [X'_1, Y_1]]/6\\
&- [Y_1, [X'_1, Y_1]]/3 + [X_1, [X'_1, Y_1]]/2 + [X'_1, [X'_1, X_1]]/6 + [X'_1, [Y_1, X_1]]/6\\
&+ [Y_1, [X'_1, X_1]]/6 + [Y_1, [Y_1, X_1]]/6 - [X_1, [X'_1, X_1]]/3 - [X_1, [Y_1, X_1]]/3.
\end{aligned}$$

Note that

(a) $u_i(X'_1, \ldots, X'_i, Y_1, \ldots, Y_i, X_1, \ldots, X_i) = X'_i - X_i + Y_i + u'_i$ where
$u'_i = u'_i(X'_1, \ldots, X'_{i-1}, Y_1, \ldots, Y_{i-1}, X_1, \ldots, X_{i-1})$
is a Lie polynomial in $X'_1, \ldots, X'_{i-1}, Y_1, \ldots, Y_{i-1}, X_1, \ldots, X_{i-1}$.

1.2. We now fix $r \geq 2$. We write $r = 2r'$ if r is even and $r = 2r'+1$ if r is odd. We always assume that $p \geq r$. Then for any $X \in \mathfrak{g}$ and any $m \geq 1$, the exponential $e^{\epsilon^m X} \in G_r$ is well defined. For any $X_1, X_2, \dots, X_{r-1}$ in $\mathfrak{g}$ we set

$$|X_1, X_2, \dots, X_{r-1}| = e^{\epsilon X_1} e^{\epsilon^2 X_2} \dots e^{\epsilon^{r-1} X_{r-1}} \in G_r.$$

We have an isomorphism of algebraic varieties

$$G \times \mathfrak{g}^{r-1} \xrightarrow{\sim} G_r$$

given by

$$(x, X_1, X_2, \dots, X_{r-1}) \mapsto x|X_1, X_2, \dots, X_{r-1}| = |{}^x X_1, {}^x X_2, \dots, {}^x X_{r-1}|x.$$

This restricts to isomorphisms of algebraic varieties $B \times \mathfrak{b}^{r-1} \xrightarrow{\sim} B_r$, $U \times \mathfrak{n}^{r-1} \xrightarrow{\sim} U_r$, $T \times \mathfrak{t}^{r-1} \xrightarrow{\sim} T_r$. (The last isomorphism is the same as one in 0.2.)

Let $X_1, X_2, \dots, X_{r-1}$ and $Y_1, Y_2, \dots, Y_{r-1}$ be two sequences in $\mathfrak{g}$ and let x, y be in G. We have

$$(x|X_1, \dots, X_{r-1}|)(y|Y_1, \dots, Y_{r-1}|) = xy|Z_1, \dots, Z_{r-1}|$$

where $Z_i = z_i({}_yX_1, \dots, {}_yX_i, Y_1, \dots, Y_i) \in \mathfrak{g}$ $(i = 1, \dots, r-1)$ with notation of 1.1 and where $[,]$ becomes the Lie bracket in $\mathfrak{g}$; note that Z_i are well defined since $p \geq r$. Moreover, we have

$$(x|X_1, \dots, X_{r-1}|)(y|Y_1, \dots, Y_{r-1}|)(x|X_1, \dots, X_{r-1}|)^{-1} = xyx^{-1}|U_1, \dots, U_{r-1}|$$

where $U_i = {}^x u_i({}_yX_1, \dots, {}_yX_i, Y_1, \dots, Y_i, X_1, \dots, X_i) \in \mathfrak{g}$ $(i = 1, \dots, r-1)$ with notation of 1.1); note that U_i are well defined since $p \geq r$.

1.3. Let $\phi : E \to X$ be an algebraic vector bundle with fibres of constant dimension N. Let $f : E \to \mathbf{k}$ be a morphism such that for any $x \in X$ the restriction $f^x : \phi^{-1}(x) \to \mathbf{k}$ is affine-linear. Let X_0 be the set of all $x \in X$ such that f^x is a constant (depending of x) and let $f_0 : X_0 \to \mathbf{k}$ be such that $f(e) = f_0(\phi(e))$ for all $e \in \phi^{-1}(X_0)$. Let $j : X_0 \to X$ be the (closed) imbedding. We show:

(a) $\phi_! \mathcal{L}_f \cong j_! \mathcal{L}_{f_0}[-2N]$.

For any $x \in X - X_0$ we have $H^i_c(\phi^{-1}(x), \mathcal{L}_f) = 0$ for all i. Hence $\phi_! \mathcal{L}_f|_{X - X_0} = 0$. We are reduced to the case where $X = X_0$. In this case we have $\mathcal{L}_f = \phi^* j_! \mathcal{L}_{f_0}$ hence

$$\phi_! \mathcal{L}_f = \phi_! \phi^* j_! \mathcal{L}_{f_0} = j_! \mathcal{L}_{f_0} \otimes \phi_! \phi^* \bar{\mathbf{Q}}_l \cong j_! \mathcal{L}_{f_0}[-2N],$$

as required. (We ignore Tate twists.) This completes the proof of (a).

If in addition we are given a local system $\mathcal{F}$ on X and we denote $\phi^* \mathcal{F}$ and $j^* \mathcal{F}$ again by $\mathcal{F}$, then from (a) we have immediately

(b) $\phi_!(\mathcal{F} \otimes \mathcal{L}_f) \cong j_!(\mathcal{F} \otimes \mathcal{L}_{f_0})[-2N]$.

1.4. In the rest of this paper we assume that a nondegenerate symmetric bilinear invariant form $\langle , \rangle : \mathfrak{g} \times \mathfrak{g} \to \mathbf{k}$ is given and that a sequence $A_1, A_2, \dots, A_{r-1}$ of elements of $\mathfrak{t}$ is given such that A_{r-1} is regular semisimple. This requires a further restriction on p in addition to the restriction $p \geq r$.

For a subspace E of $\mathfrak{g}$ we set $E^\perp = \{\xi \in \mathfrak{g}; \langle \xi, E \rangle = 0\}$.

Let $\mathcal{X}$ be the variety of all

$$(Tx, y, X_1, X_2, \dots, X_{r-1}, Y_1, Y_2, \dots, Y_{r-1}) \in (T \backslash G) \times G \times \mathfrak{g}^{2r-2}$$

such that $xyx^{-1} \in T$ and

$$u_j({}_yX_1, \dots, {}_yX_j, Y_1, \dots, Y_j, X_1, \dots, X_j) \in {}_x\mathfrak{t} \text{ for } 1 \le j \le r' - 1,$$
$$u_j({}_yX_1, \dots, {}_yX_j, Y_1, \dots, Y_j, X_1, \dots, X_j) \in {}_x\mathfrak{b} \text{ if } j = r' \text{ and } r \text{ is odd.}$$

We have a diagram

$$G_r \xleftarrow{\pi} \mathcal{X} \xrightarrow{h} \mathbf{k}$$

where $\pi(Tx, y, X_1, \dots, X_{r-1}, Y_1, \dots, Y_{r-1}) = y|Y_1, \dots, Y_{r-1}|$,

$$h(Tx, y, X_1, \dots, X_{r-1}, Y_1, \dots, Y_{r-1})$$
$$= \sum_{j \in [1, r-1]} \langle {}_xA_j, u_j({}_yX_1, \dots, {}_yX_j, Y_1, \dots, Y_j, X_1, \dots, X_j)\rangle.$$

(Note that if x is replaced by tx, $(t \in T)$ in the last sum, the sum remains unchanged since ${}_tA_j = A_j$ for all j.) We define $\iota : \mathcal{X} \to T$ by

$$\iota(Tx, y, X_1, X_2, \dots, X_{r-1}, Y_1, Y_2, \dots, Y_{r-1}) = xyx^{-1}$$

and we set $\tilde{\mathcal{E}} = \iota^*\mathcal{E}$. Let $K = \pi_!(\tilde{\mathcal{E}} \otimes \mathcal{L}_h) \in \mathcal{D}(G_r)$. Via the identification $G_r = G \times \mathfrak{g}^{r-1}$ (see 1.2) we can regard G_r as a vector bundle over G with fibre $\mathfrak{g}^{r-1}$ endowed with a nondegenerate symmetric bilinear form. Hence the Fourier-Deligne transform $\hat{K} \in \mathcal{D}(G_r)$ along these fibres is well defined. More explicitly, for $i = 1, 2$ we have the diagram $G_r \xleftarrow{\rho_i} G_r \times_G G_r \xrightarrow{h'} \mathbf{k}$ where ρ_i is the projection to the i-th factor and

$$h'(x|Y_1, \dots, Y_{r-1}|, x|R_1, R_2, \dots, R_{r-1}|) = \sum_{j \in [1, r-1]} \langle Y_j, R_j\rangle.$$

Then $\hat{K} = \rho_{2!}(\rho_1^*K \otimes \mathcal{L}_{h'})[(r-1)\Delta]$ that is,

$$\hat{K} = \rho_{2!}(\rho_1^*\pi_!(\tilde{\mathcal{E}} \otimes \mathcal{L}_h) \otimes \mathcal{L}_{h'})[(r-1)\Delta].$$

Let $\tilde{\mathcal{X}}$ be the variety of all

$$(Tx, y, X_1, X_2, \dots, X_{r-1}, Y_1, Y_2, \dots, Y_{r-1}, R_1, R_2, \dots, R_{r-1}) \in (T\backslash G) \times G \times \mathfrak{g}^{3r-3}$$

such that $xyx^{-1} \in T$ and

$$u_j({}_yX_1, \dots, {}_yX_j, Y_1, \dots, Y_j, X_1, \dots, X_j) \in {}_x\mathfrak{t} \text{ for } 1 \le j \le r' - 1,$$
$$u_j({}_yX_1, \dots, {}_yX_j, Y_1, \dots, Y_j, X_1, \dots, X_j) \in {}_x\mathfrak{b} \text{ if } j = r' \text{ and } r \text{ is odd .}$$

We have a cartesian diagram

$$\begin{array}{ccc} \tilde{\mathcal{X}} & \xrightarrow{\tilde{\rho}_1} & \mathcal{X} \\ {\scriptstyle\sigma}\downarrow & & \downarrow{\scriptstyle\pi} \\ G_r \times_G G_r & \xrightarrow{\rho_1} & G_r \end{array}$$

where

$$\tilde{\rho}_1(Tx, y, X_1, \dots, X_{r-1}, Y_1, \dots, Y_{r-1}, R_1, \dots, R_{r-1})$$
$$= (Tx, y, X_1, \dots, X_{r-1}, Y_1, \dots, Y_{r-1}),$$
$$\sigma(Tx, y, X_1, \dots, X_{r-1}, Y_1, \dots, Y_{r-1}, R_1, \dots, R_{r-1})$$
$$= (y|Y_1, \dots, Y_{r-1}|, y|R_1, \dots, R_{r-1}|).$$

It follows that

$$\begin{aligned}\hat{K} &= \rho_{2!}(\sigma_!(\tilde{\mathcal{E}} \otimes \tilde{\rho}_1^* \mathcal{L}_h \otimes \mathcal{L}_{h'})[(r-1)\Delta] \\ &= \rho_{2!}\sigma_!(\tilde{\mathcal{E}} \otimes \mathcal{L}_{\tilde{h}'} \otimes \mathcal{L}_{\tilde{h}''}) = (\rho_2\sigma)_!(\tilde{\mathcal{E}} \otimes \mathcal{L}_{\tilde{h}})[(r-1)\Delta]\end{aligned}$$

where $\tilde{h}'' = h\mathrm{tr}_1 : \tilde{\mathcal{X}} \to \mathbf{k}$, $\tilde{h}' = h'\sigma : \tilde{\mathcal{X}} \to \mathbf{k}$, $\tilde{h} = \tilde{h}' + \tilde{h}'' : \tilde{\mathcal{X}} \to \mathbf{k}$ and the inverse image of $\tilde{\mathcal{E}}$ under $\tilde{\mathcal{X}} \to T$,

$$(Tx, y, X_1, \dots, X_{r-1}, Y_1, \dots, Y_{r-1}, R_1, \dots, R_{r-1}) \mapsto xyx^{-1}$$

is denoted again by $\tilde{\mathcal{E}}$. Thus,

$$\hat{K} = \pi'_!(\tilde{\mathcal{E}} \otimes \mathcal{L}_{\tilde{h}})[(r-1)\Delta]$$

where $\pi' : \tilde{\mathcal{X}} \to G_r$ and $\tilde{h} : \tilde{\mathcal{X}} \to \mathbf{k}$ are given by

$$\pi'(Tx, y, X_1, \dots, X_{r-1}, Y_1, \dots, Y_{r-1}, R_1, \dots, R_{r-1}) = y|R_1, \dots, R_{r-1}|,$$

$$\begin{aligned}&\tilde{h}(Tx, y, X_1, \dots, X_{r-1}, Y_1, \dots, Y_{r-1}, R_1, \dots, R_{r-1}) \\ &= \sum_{j \in [1, r-1]} \langle Y_j, R_j \rangle + \sum_{j \in [1, r-1]} \langle {}_xA_j, u_j({}_yX_1, \dots, {}_yX_j, Y_1, \dots, Y_j, X_1, \dots, X_j) \rangle.\end{aligned}$$

1.5. Let $\tilde{\mathcal{X}}''$ be the variety of all

$$\begin{aligned}&(Tx, y, X_1, X_2, \dots, X_{r-2}, Y_1, Y_2, \dots, Y_{r-2}, R_1, R_2, \dots, R_{r-1}) \\ &\in (T\backslash G) \times G \times \mathfrak{g}^{(r-2)+(r-2)+(r-1)}\end{aligned}$$

such that $xyx^{-1} \in T$ and

$$u_j({}_yX_1, \dots, {}_yX_j, Y_1, \dots, Y_j, X_1, \dots, X_j) \in {}_x\mathfrak{t} \text{ for } 1 \le j \le r' - 1,$$

$$u_j({}_yX_1, \dots, {}_yX_j, Y_1, \dots, Y_j, X_1, \dots, X_j) \in {}_x\mathfrak{b} \text{ if } j = r' \text{ and } r \text{ is odd }.$$

(The equations make sense since if $1 \le j \le r' - 1$ then $j \le r - 2$ and since when r is odd we have $r' = r - r' - 1 \le r - 2$.) We define $\mu : \tilde{\mathcal{X}} \to \tilde{\mathcal{X}}''$ by

$$\begin{aligned}&(Tx, y, X_1, X_2, \dots, X_{r-1}, Y_1, Y_2, \dots, Y_{r-1}, R_1, R_2, \dots, R_{r-1}) \mapsto \\ &(Tx, y, X_1, X_2, \dots, X_{r-2}, Y_1, Y_2, \dots, Y_{r-2}, R_1, R_2, \dots, R_{r-1}).\end{aligned}$$

This is a vector bundle; for a fixed

$$s = (Tx, y, X_1, X_2, \dots, X_{r-2}, Y_1, Y_2, \dots, Y_{r-2}, R_1, R_2, \dots, R_{r-1}) \in \tilde{\mathcal{X}}'',$$

the fibre $\mu^{-1}(s)$ can be identified with $\mathfrak{g}^2$ with coordinates X_{r-1}, Y_{r-1}. The restriction of $\tilde{h}$ to $\mu^{-1}(s)$ is of the form

$$(X_{r-1}, Y_{r-1}) \mapsto \langle Y_{r-1}, R_{r-1} + {}_xA_{r-1} \rangle + c$$

where c is a constant depending on s. We use that $\langle {}_xA_j, {}_yX_{r-1} - X_{r-1} \rangle = 0$; this holds since ${}^{yx^{-1}}A_{r-1} = {}^{x^{-1}}A_{r-1}$ (recall that $xyx^{-1} \in T$). Thus this restriction is affine-linear and is constant precisely when $R_{r-1} = -{}_xA_{r-1}$. Hence the results in 1.3 are applicable. Let $\bar{\mathcal{X}}$ be the variety of all

$$(Tx, y, X_1, X_2, \dots, X_{r-2}, Y_1, Y_2, \dots, Y_{r-2}, R_1, R_2, \dots, R_{r-1}) \in \tilde{\mathcal{X}}''$$

such that $R_{r-1} = -{}_xA_{r-1}$. We define $\bar{\pi} : \bar{\mathcal{X}} \to G_r$, $\bar{h} : \bar{\mathcal{X}} \to \mathbf{k}$ by

$$\bar{\pi}(Tx, y, X_1, X_2, \dots, X_{r-2}, Y_1, Y_2, \dots, Y_{r-2}, R_1, R_2, \dots, R_{r-1})$$
$$= y|R_1, R_2, \dots, R_{r-1}|,$$
$$\bar{h}(Tx, y, X_1, X_2, \dots, X_{r-2}, Y_1, Y_2, \dots, Y_{r-2}, R_1, R_2, \dots, R_{r-1})$$
$$= \sum_{j\in[1,r-2]} \langle Y_j, R_j\rangle + \sum_{j\in[1,r-2]} \langle {}_xA_j, u_j({}_yX_1, \dots, {}_yX_j, Y_1, \dots, Y_j, X_1, \dots, X_j)\rangle$$
$$+ \langle {}_xA_{r-1}, u'_{r-1}({}_yX_1, \dots, {}_yX_{r-2}, Y_1, \dots, Y_{r-2}, X_1, \dots, X_{r-2})\rangle,$$

with notation of 1.1(a). The inverse image of $\mathcal{E}$ under $\bar{\mathcal{X}} \to T$,

$$(Tx, y, X_1, X_2, \dots, X_{r-2}, Y_1, Y_2, \dots, Y_{r-2}, R_1, R_2, \dots, R_{r-1}) \mapsto xyx^{-1}$$

is denoted again by $\tilde{\mathcal{E}}$. Then from 1.3(b) we deduce

(a) $$\hat{K} = \bar{\pi}_!(\tilde{\mathcal{E}} \otimes \mathcal{L}_{\bar{h}})[(r-5)\Delta].$$

Let $\mathcal{C} \subset \mathfrak{g}$ be the G-orbit of $-A_{r-1}$ for the adjoint action, a regular semisimple orbit. Let $V = \{(y, R) \in G \times \mathcal{C}; {}^yR = R\}$. Let Σ be the support of $\hat{K}$ (a closed subset of G_r). From (a) we see that

$$\Sigma \subset \{y|R_1, R_2, \dots, R_{r-1}| \in G_r; (y, R_{r-1}) \in V\}.$$

It is likely that Σ is a smooth subvariety of G_r, isomorphic to a vector bundle over V with fibres isomorphic to $(\mathfrak{t}^{\perp})^{r-2}$. We will we show that this is the case at least when $r \in \{2, 3, 4\}$. Moreover, it is likely that when r is even, $\hat{K}$ is up to shift the intersection cohomology complex associated to a local system of rank 1 on the smooth closed subvariety Σ. We will show that this is the case when $r \in \{2, 4\}$ and that the analogous statement is not true when $r = 3$.

1.6. The method used in 1.5 to eliminate the variables X_{r-1}, Y_{r-1} can be used to eliminate all variables $X_{r-r'}, \dots, X_{r-1}, Y_{r-r'}, \dots, Y_{r-1}$. Let $\tilde{\mathcal{X}}_1''$ be the variety of all

$$(Tx, y, X_1, X_2, \dots, X_{r-r'-1}, Y_1, Y_2, \dots, Y_{r-r'-1}, R_1, R_2, \dots, R_{r-1})$$
$$\in (T\backslash G) \times G \times \mathfrak{g}^{r-1+2(r-r'-1)}$$

such that $xyx^{-1} \in T$ and

$$u_j({}_yX_1, \dots, {}_yX_j, Y_1, \dots, Y_j, X_1, \dots, X_j) \in {}_x\mathfrak{t} \text{ for } 1 \le j \le r'-1,$$
$$u_j({}_yX_1, \dots, {}_yX_j, Y_1, \dots, Y_j, X_1, \dots, X_j) \in {}_x\mathfrak{b} \text{ if } j = r' \text{ and } r \text{ is odd}.$$

(The equations make sense since if $1 \le j \le r'-1$ then $j \le r-r'-1$ and since when r is odd we have $r' = r-r'-1$.) We define $\mu_1 : \tilde{\mathcal{X}} \to \tilde{\mathcal{X}}_1''$ by

$$(Tx, y, X_1, X_2, \dots, X_{r-1}, Y_1, Y_2, \dots, Y_{r-1}, R_1, R_2, \dots, R_{r-1}) \mapsto$$
$$(Tx, y, X_1, X_2, \dots, X_{r-r'-1}, Y_1, Y_2, \dots, Y_{r-r'-1}, R_1, R_2, \dots, R_{r-1}).$$

This is a vector bundle; for a fixed

$$s = (Tx, y, X_1, X_2, \dots, X_{r-r'-1}, Y_1, Y_2, \dots, Y_{r-r'-1}, R_1, R_2, \dots, R_{r-1}) \in \bar{\mathcal{X}},$$

the fibre $\mu^{-1}(s)$ can be identified with $\mathfrak{g}^{2r'}$ with coordinates

$$X_{r-r'}, do, X_{r-1}, Y_{r-r'}, \dots, Y_{r-1}.$$

The restriction of $\tilde{h}$ to $\mu^{-1}(s)$ in an affine linear function. This follows from the fact that for $j \in [1, r-1]$, the Lie polynomial

$$u_j(X'_1, \dots, X'_j, Y_1, \dots, Y_j, X_1, \dots, X_j)$$

is a linear combination of terms which are iterated brackets of indeterminates X'_h, Y_h, X_h with sum of indices equal to j (hence $\le r-1$) hence containing at most one X'_h, Y_h or X_h with $h \ge r - r'$. (If they contained more than one, we would have $2(r-r') \le r-1$ hence $r \le 2r'-1$, a contradiction.) Hence the results in 1.3 are applicable and they result in a description of $\hat{K}$ which does not involve $X_{r-r'}, \dots, X_{r-1}, Y_{r-r'}, \dots, Y_{r-1}$. But even after this method is applied, one needs further arguments to analyze $\hat{K}$, as we will see in Sections 2 and 3.

2. The cases $r = 2$ and $r = 4$

2.1. In this subsection we assume that $r = 2$. Now

$$\bar{\mathcal{X}} = \{(Tx, y, R_1) \in (T\backslash G) \times G \times \mathfrak{g}; xyx^{-1} \in T, R_1 = -{}_xA_1.\}$$

We have $\bar{\pi}(Tx, y, R_1) = y|R_1|$ and $\bar{h} : \bar{\mathcal{X}} \to \mathbf{k}$ is identically 0. Using 1.5(a) we have

(a) $$\hat{K} = \bar{\pi}_! \tilde{\mathcal{E}}[-3\Delta]$$

Note that $\bar{\pi}$ defines an isomorphism of $\bar{\mathcal{X}}$ with

$$\mathcal{Z} = \{y|R_1| \in G_2; R_1 \in \mathcal{C}, {}^yR_1 = R_1\}$$

and that $\mathcal{Z}$ is closed in G_2 (we use that $\mathcal{C}$ is closed in $\mathfrak{g}$). Moreover $\mathcal{Z}$ is a smooth subvariety of G_2 and $\bar{\pi}$ can be viewed as the imbedding $\mathcal{Z} \to G_2$. Since $\mathcal{Z}$ is closed in G_2 and smooth, irreducible of dimension Δ we see that $\bar{\pi}\tilde{\mathcal{E}}[\Delta]$ is a simple perverse sheaf on G_2. Hence $\hat{K}[4\Delta]$ is a simple perverse sheaf on G_2 with support $\Sigma = \mathcal{Z}$. Using Laumon's theorem (see 0.2), it follows that

(b) $K[4\Delta]$ *is a simple perverse sheaf on* G_2.

2.2. We now assume (until the end of 2.5) that $r = 4$. Now $\bar{\mathcal{X}}$ is the variety of all

$$(Tx, y, X_1, X_2, Y_1, Y_2, R_1, R_2, R_3) \in (T\backslash G) \times G \times \mathfrak{g}^7$$

such that $xyx^{-1} \in T$, ${}_yX_1 - X_1 + Y_1 \in {}_x\mathfrak{t}$ and $R_3 = -{}_xA_3$. We have

$$\bar{\pi}(Tx, y, X_1, X_2, Y_1, Y_2, R_1, R_2, R_3) = y|R_1, R_2, R_3|,$$

$$\begin{aligned}
&\bar{h}(Tx, y, X_1, X_2, Y_1, Y_2, R_1, R_2, R_3) \\
&= \langle Y_1, R_1\rangle + \langle Y_2, R_2\rangle + \langle {}_xA_1, {}_yX_1 - X_1 + Y_1\rangle \\
&+ \langle {}_xA_2, {}_yX_2 - X_2 + Y_2 + [{}_yX_1, Y_1]/2 - [{}_yX_1, X_1]/2 - [Y_1, X_1]/2\rangle \\
&+ \langle {}_xA_3, [{}_yX_2, Y_1] + [X_2, X_1] - [{}_yX_2, X_1] - [Y_2, X_1] - [{}_yX_1, [{}_yX_1, Y_1]]/6 \\
&- [Y_1, [{}_yX_1, Y_1]]/3 + [X_1, [{}_yX_1, Y_1]]/2 + [{}_yX_1, [{}_yX_1, X_1]]/6 + [{}_yX_1, [Y_1, X_1]]/6 \\
&+ [Y_1, [{}_yX_1, X_1]]/6 + [Y_1, [Y_1, X_1]]/6 - [X_1, [{}_yX_1, X_1]]/3 - [X_1, [Y_1, X_1]]/3\rangle.
\end{aligned}$$

We make a change of variable $Y_1 = X_1 - {}_yX_1 + {}_x\tau$ where $\tau \in \mathfrak{t}$. Then $\bar{\mathcal{X}}$ becomes the variety of all

$$(Tx, y, \tau, X_1, X_2, Y_2, R_1, R_2, R_3) \in (T\backslash G) \times G \times \mathfrak{t} \times \mathfrak{g}^6$$

such that $xyx^{-1} \in T$ and $R_3 = -{}_xA_3$. Now $\bar{\pi} : \bar{\mathcal{X}} \to G_4$ and $\bar{h} : \bar{\mathcal{X}} \to \mathbf{k}$ become

$$\bar{\pi}(Tx, y, \tau, X_1, X_2, Y_2, R_1, R_2, R_3) = y|R_1, R_2, R_3|,$$

$$\begin{aligned}
&\bar{h}(Tx, y, \tau, X_1, X_2, Y_2, R_1, R_2, R_3) \\
&= \langle X_1 - {}_yX_1 + {}_x\tau, R_1\rangle + \langle Y_2, R_2\rangle + \langle {}_xA_1, {}_x\tau\rangle \\
&+ \langle {}_xA_2, {}_yX_2 - X_2 + Y_2 + [{}_yX_1, X_1 + {}_x\tau]/2 - [{}_yX_1, X_1]/2 - [-{}_yX_1 + {}_x\tau, X_1]/2\rangle \\
&+ \langle {}_xA_3, [{}_yX_2, X_1 - {}_yX_1 + {}_x\tau] + [X_2, X_1] - [{}_yX_2, X_1] - [Y_2, X_1] \\
&- [{}_yX_1, [{}_yX_1, X_1 + {}_x\tau]]/6 - [X_1 - {}_yX_1 + {}_x\tau, [{}_yX_1, X_1 + {}_x\tau]]/3 \\
&+ [X_1, [{}_yX_1, X_1 + {}_x\tau]]/2 + [{}_yX_1, [{}_yX_1, X_1]]/6 + [{}_yX_1, [-{}_yX_1 + {}_x\tau, X_1]]/6 \\
&+ [X_1 - {}_yX_1 + {}_x\tau, [{}_yX_1, X_1]]/6 + [X_1 - {}_yX_1 + {}_x\tau, [-{}_yX_1 + {}_x\tau, X_1]]/6 \\
&- [X_1, [{}_yX_1, X_1]]/3 - [X_1, [-{}_yX_1 + {}_x\tau, X_1]]/3\rangle.
\end{aligned}$$

For $i = 1, 2, 3$ we have $[A_i, \tau] = 0$ since $\mathfrak{t}$ is abelian; it follows that $\langle A_i, [\xi, \tau]\rangle = 0$ for any $\xi \in \mathfrak{g}$. We also have $\langle {}_xA_i, {}_yX_j - X_j\rangle = 0$; indeed the left hand side is $\langle {}^{yx^{-1}}A_i - {}^{x^{-1}}A_i, X_j\rangle$ and this is zero since ${}^{xyx^{-1}}A_i = A_i$ (recall that $xyx^{-1} \in T$). Similarly we have $\langle {}_xA_3, [{}_yX_2, -{}_yX_1] + [X_2, X_1]\rangle = 0$; indeed, the left hand side is $\langle {}^{yx^{-1}}A_3 - {}^{x^{-1}}A_3, [X_1, X_2]\rangle = 0$. We see that

$$\begin{aligned}
&\bar{h}(Tx, y, \tau, X_1, X_2, Y_2, R_1, R_2, R_3) = \langle X_1 - {}_yX_1 + {}_x\tau, R_1\rangle \\
&+ \langle Y_2, R_2\rangle + \langle {}_xA_1, {}_x\tau\rangle + \langle {}_xA_2, Y_2 - [-{}_yX_1, X_1]/2\rangle \\
&+ \langle {}_xA_3, -[Y_2, X_1] + [{}_yX_1, [{}_yX_1, {}_x\tau]]/6 + [X_1, [X_1, {}_x\tau]]/6 \\
&+ [X_1, [{}_yX_1, {}_x\tau]]/6 + [{}_yX_1, [{}_yX_1, X_1]]/6 + [X_1, [{}_yX_1, X_1]]/6\rangle.
\end{aligned}$$

Next we use the identity

$$\langle {}_xA_3, [Z, [Z', {}_x\tau]]\rangle = \langle {}_x\tau, [Z, [Z', {}_xA_3]]\rangle$$

for any Z, Z' in $\mathfrak{g}$. (This follows from $[{}_xA_3, {}_x\tau] = 0$.) We also use the equality

$$\langle {}_xA_3, [{}_yX_1, [{}_yX_1, {}_x\tau]]\rangle = \langle {}_xA_3, [X_1, [X_1, {}_x\tau]]\rangle.$$

(Since ${}^{yx^{-1}}\tau = {}^{x^{-1}}\tau$, ${}^{yx^{-1}}A_3 = {}^{x^{-1}}A_3$, the left hand side is

$$\langle {}_xA_3, {}_y[X_1, [X_1, {}_x\tau]]\rangle = \langle {}^{yx^{-1}}A_3, [X_1, [X_1, {}_x\tau]]\rangle = \langle {}_xA_3, [X_1, [X_1, {}_x\tau]]\rangle,$$

as required.) We see that

$$\begin{aligned}
&\bar{h}(Tx, y, \tau, X_1, X_2, Y_2, R_1, R_2, R_3) = \langle Y_2, R_2 + {}_xA_2 - [X_1, {}_xA_3]\rangle \\
&+ \langle {}_x\tau, R_1 + {}_xA_1 + [X_1, [X_1, {}_xA_3]]/6 + [X_1, [{}_yX_1, {}_xA_3]]/3\rangle + \langle X_1 - {}_yX_1, R_1\rangle \\
&+ \langle {}_xA_2, [{}_yX_1, X_1]/2\rangle + \langle {}_xA_3, [{}_yX_1, [{}_yX_1, X_1]]/6 + [X_1, [{}_yX_1, X_1]]/6\rangle.
\end{aligned}$$

2.3. Let $\mathcal{T} = \{(Tx, y, X_1, R_1, R_2, R_3) \in (T\backslash G) \times G \times \mathfrak{g}^4; xyx^{-1} \in T, R_3 + {}_xA_3 = 0\}$. Let $\mathcal{T}_0$ be the closed subset of $\mathcal{T}$ consisting of all $(Tx, y, X_1, R_1, R_2, R_3)$ such that

$$R_2 + {}_xA_2 - [X_1, {}_xA_3] = 0,$$
$$R_1 + {}_xA_1 + [X_1, [X_1, {}_xA_3]]/3 + [X_1, [{}_yX_1, {}_xA_3]]/6 \in ({}_x\mathfrak{t})^{\perp}.$$

Define $\tilde{h}_0 : \mathcal{T}_0 \to \mathbf{k}$ by

$$\begin{aligned}
&\tilde{h}_0(Tx, y, X_1, R_1, R_2, R_3) = \langle X_1 - {}_yX_1, R_1\rangle + \langle {}_xA_2, [{}_yX_1, X_1]/2\rangle \\
&+ \langle {}_xA_3, [{}_yX_1, [{}_yX_1, X_1]]/6 + [X_1, [{}_yX_1, X_1]]/6\rangle.
\end{aligned}$$

Define $\bar{\mathcal{X}} \xrightarrow{\phi} \mathcal{T} \xrightarrow{\phi'} G_4$ by

$$\phi(Tx, y, \tau, X_1, X_2, Y_2, R_1, R_2, R_3) = (Tx, y, X_1, R_1, R_2, R_3),$$

$$\phi'(Tx, y, X_1, R_1, R_2, R_3) = y|R_1, R_2, R_3|$$

so that $\pi' = \phi'\phi$. Now ϕ is a vector bundle with fibres of dimension $N = 2\Delta + \delta$. Note that the restriction of $\tilde{h} : \mathcal{X} \to \mathbf{k}$ to any fibre of ϕ is affine linear and this restriction is constant precisely at the fibres over points in $\mathcal{T}_0$; moreover the constant is given by the value of $\tilde{h}_0$. Using 1.3(b), we see that

(a) $$\hat{K} = j(\tilde{\mathcal{E}} \otimes \mathcal{L}_{\tilde{h}_0})[-5\Delta - 2\delta]$$

where $j : \mathcal{T}_0 \to G_4$ is the restriction of ϕ'.

2.4. Let R be a regular semisimple element in $\mathfrak{g}$. Let $\mathfrak{t}_R$ be the centralizer of R in $\mathfrak{g}$; let T_R be the centralizer of R in T. For any $z \in T_R$ we define a linear map $\Xi_{R,z} : \mathfrak{t}_R^\perp \to \mathfrak{g}/\mathfrak{t}_R^\perp$ by $\Xi_{R,z}(\xi) = [X, {}^z\xi]$ mod $\mathfrak{t}_R^\perp$ where X is any element of $\mathfrak{g}$ such that $\xi = [X, R]$. Note that such X exists; if X' is a another element such that $\xi = [X', R]$, then ${}^{z^{-1}}(X' - X) \in \mathfrak{t}_R$ hence $X' = X + \rho$ for some $\rho \in \mathfrak{t}_R$ and $[X', {}^z\xi] = [X, {}^z\xi] + [\rho, {}^z\xi]$. Since $[\rho, {}^z\xi] \in \mathfrak{t}_R^\perp$ we see that our map $\Xi_{R,z}$ is well defined.

2.5. Let $\mathcal{C} \subset \mathfrak{g}$ be the G-orbit of $-A_3$ for the adjoint action, a regular semisimple orbit. Let $\mathcal{Z}$ be the subset of G_4 consisting of all $y|R_1, R_2, R_3|$ such that

$R_3 \in \mathcal{C}$;

$y \in T_{R_3}$;

$R_2 + {}_xA_2 \in \mathfrak{t}_{R_3}^\perp$ where $Tx \in T\backslash G$ is uniquely determined by $R_3 = -{}_xA_3$;

$R_1 + {}_xA_1 + \Xi_{R_3,1}(R_2')/3 + \Xi_{R_3,y^{-1}}(R_2')/6 = 0$ in $\mathfrak{g}/\mathfrak{t}_{R_3}^\perp$ where $R_2' = R_2 + {}_xA_2$.

Note that $\mathcal{Z}$ is closed in G_4 (we use that $\mathcal{C}$ is closed in $\mathfrak{g}$). Moreover $\mathcal{Z}$ is a smooth subvariety of G_4. Indeed, $V = \{(y, R) \in G \times \mathcal{C}; y \in T_R\}$ is clearly smooth and $\mathcal{Z}$ is a fibration over V with fibres isomorphic to $\mathfrak{t}^\perp \times \mathfrak{t}^\perp$.

From the definitions we see that $\mathcal{T}_0 = \phi'^{-1}\mathcal{Z}$ and that the restriction of ϕ' defines a morphism $\underline{\phi}' : \mathcal{T}_0 \to \mathcal{Z}$ whose fibres are exactly the orbits of the free $\mathfrak{t}$-action on $\mathcal{T}_0$ given by

$$\tau : (Tx, y, X_1, R_1, R_2, R_3) \mapsto (Tx, y, X_1 + {}_x\tau, R_1, R_2, R_3).$$

Clearly, the local system $\tilde{\mathcal{E}}$ on $\mathcal{T}_0$ is the inverse image under $\underline{\phi}'$ of a local system on $\mathcal{Z}$ denoted again by $\tilde{\mathcal{E}}$. Next we show that

(a) *the function $\tilde{h}_0 : \mathcal{T}_0 \to \mathbf{k}$ is constant on each orbit of the $\mathfrak{t}$-action on $\mathcal{T}_0$* that is, if $\tau \in \mathfrak{t}$ and $(Tx, y, X_1, R_1, R_2, R_3) \in \mathcal{T}_0$, then

$$\tilde{h}_0(Tx, y, X_1 + {}_x\tau, R_1, R_2, R_3) = \tilde{h}_0(Tx, y, X_1, R_1, R_2, R_3).$$

Thus, we must show that

$$\begin{aligned}
&\langle X_1 + {}_x\tau - {}_yX_1 - {}_x\tau, R_1\rangle + \langle {}_xA_2, [{}_yX_1 + {}_x\tau, X_1 + {}_x\tau]/2\rangle \\
&+ \langle {}_xA_3, [{}_yX_1 + {}_x\tau, [{}_yX_1 + {}_x\tau, X_1 + {}_x\tau]]/6 + [X_1 + {}_x\tau, [{}_yX_1 + {}_x\tau, X_1 + {}_x\tau]]/6\rangle \\
&= \langle X_1 - {}_yX_1, R_1\rangle + \langle {}_xA_2, [{}_yX_1, X_1]/2\rangle \\
&+ \langle {}_xA_3, [{}_yX_1, [{}_yX_1, X_1]]/6 + [X_1, [{}_yX_1, X_1]]/6\rangle.
\end{aligned}$$

(We have used that ${}_{xy}\tau = {}_x\tau$.) It is enough to show that

$$\langle {}_xA_2, [{}_x\tau, X_1]/2\rangle + \langle {}_xA_2, [{}_yX_1, {}_x\tau]/2\rangle$$
$$+ \langle {}_xA_3, [{}_yX_1, [{}_yX_1, {}_x\tau]]/6 + [{}_yX_1, [{}_x\tau, X_1]]/6 + [{}_x\tau, [{}_yX_1, {}_x\tau]]/6$$
$$+ [{}_x\tau, [{}_yX_1, X_1]]/6 + [{}_x\tau, [{}_x\tau, X_1]]/6 + [X_1, [{}_yX_1, {}_x\tau]]/6 + [X_1, [{}_x\tau, X_1]]/6$$
$$+ [{}_x\tau, [{}_yX_1, X_1]]/6 + [{}_x\tau, [{}_yX_1, {}_x\tau]]/6 + [{}_x\tau, [{}_x\tau, X_1]]/6\rangle = 0.$$

Since $\langle {}_xA_i, [{}_x\tau, \xi]\rangle = 0$ for any $\xi \in \mathfrak{g}$, we see that it is enough to show

$$\langle {}_xA_3, [{}_yX_1, [{}_yX_1, {}_x\tau]]/6 + [{}_yX_1, [{}_x\tau, X_1]]/6 + [X_1, [{}_yX_1, {}_x\tau]]/6$$
$$+ [X_1, [{}_x\tau, X_1]]/6\rangle = 0.$$

It is enough to show the following two equalities:

(b) $$\langle {}_xA_3, [{}_yX_1, [{}_yX_1, {}_x\tau]] + [X_1, [{}_x\tau, X_1]]\rangle = 0,$$

(c) $$\langle {}_xA_3, [{}_yX_1, [{}_x\tau, X_1]] + [X_1, [{}_yX_1, {}_x\tau]]\rangle = 0.$$

The left hand side of (b) is

$$\langle {}_xA_3, {}_y[X_1, [X_1, {}_x\tau]] - [X_1, [X_1, {}_x\tau]]\rangle = \langle {}^{yx^{-1}}A_3 - {}^{x^{-1}}A_3, [X_1, [X_1, {}_x\tau]]\rangle$$

and this is zero since ${}^{yx^{-1}}A_3 = {}^{x^{-1}}A_3$. The left hand side of (c) is

$$\langle {}_xA_3, [{}_x\tau, [{}_yX_1, X_1]]\rangle$$

and this is zero since $\langle {}_xA_3, [{}_x\tau, \xi]\rangle = 0$ for any $\xi \in \mathfrak{g}$. This completes the proof of (a).

From (a) we see that there is a unique morphism $\hat{h} : \mathcal{Z} \to \mathbf{k}$ such that $\tilde{h}_0(s) = \hat{h}(\underline{\phi}'(s))$ for any $s \in \mathcal{T}_0$. It follows that $\mathcal{L}_{\tilde{h}_0} = \underline{\phi}'^*\mathcal{L}_{\hat{h}}$. Now $j : \mathcal{T}_0 \to G_4$ in 2.3 is a composition $\underline{j}\underline{\phi}'$ where $\underline{j} : \mathcal{Z} \to G_4$ is the imbedding. It follows that $j(\tilde{\mathcal{E}} \otimes \mathcal{L}_{\tilde{h}_0}) = \underline{j}(\tilde{\mathcal{E}} \otimes \underline{\phi}'\underline{\phi}'^*\mathcal{L}_{\hat{h}}) \cong \underline{j}(\tilde{\mathcal{E}} \otimes \mathcal{L}_{\hat{h}})[-2\delta]$. Combining with 2.3(a) we see that

$$\hat{K} \cong \underline{j}(\tilde{\mathcal{E}} \otimes \mathcal{L}_{\hat{h}})[-5\Delta - 4\delta].$$

Since $\mathcal{Z}$ is closed in G_4 and smooth, irreducible of dimension $3\Delta - 2\delta$ we see that $\underline{j}(\tilde{\mathcal{E}} \otimes \mathcal{L}_{\hat{h}})[3\Delta - 2\delta]$ is a simple perverse sheaf on G_4. Hence $\hat{K}[8\Delta + 2\delta]$ is a simple perverse sheaf on G_4 with support $\Sigma = \mathcal{Z}$. Using Laumon's theorem (see 0.2), it follows that

(b) *$K[8\Delta + 2\delta]$ is a simple perverse sheaf on G_4.*

3. The case $r = 3$

3.1. In this section we assume that $r = 3$. Now $\bar{\mathcal{X}}$ is the variety of all

$$(Tx, y, X_1, Y_1, R_1, R_2) \in (T\backslash G) \times G \times \mathfrak{g}^4$$

such that $xyx^{-1} \in T$, ${}_yX_1 - X_1 + Y_1 \in {}_x\mathfrak{b}$ and $R_2 = -{}_xA_2$. In our case we have

$$\hat{K} = \bar{\pi}(\tilde{\mathcal{E}} \otimes \mathcal{L}_{\overline{h}})[-2\Delta]$$

where $\bar{\pi} : \bar{\mathcal{X}} \to G_3$ and $\overline{h} : \bar{\mathcal{X}} \to \mathbf{k}$ are given by

$$\bar{\pi}(Tx, y, X_1, Y_1, R_1, R_2) = y|R_1, R_2|,$$

$$\overline{h}(Tx, y, X_1, Y_1, R_1, R_2) = \langle Y_1, R_1\rangle + \langle {}_xA_1, {}_yX_1 - X_1 + Y_1\rangle$$
$$+ \langle {}_xA_2, [{}_yX_1, Y_1]/2 - [{}_yX_1, X_1]/2 - [Y_1, X_1]/2\rangle.$$

Let

$$\bar{\mathcal{X}}' = \{(Tx, y, X_1, R_1, R_2, \beta); (Tx, y) \in (T\backslash G) \times G, (X_1, R_1, R_2) \in \mathfrak{g}^3, \beta \in {}_x\mathfrak{b},$$
$$xyx^{-1} \in T, R_2 = -{}_xA_2\}.$$

We define an isomorphism $\bar{\mathcal{X}} \xrightarrow{\sim} \bar{\mathcal{X}}'$ by

$$(Tx, y, X_1, Y_1, R_1, R_2) \mapsto (Tx, y, X_1, R_1, R_2, \beta)$$

where $\beta \in {}_x\mathfrak{b}$ is given by $\beta = {}_yX_1 - X_1 + Y_1$. We identify $\bar{\mathcal{X}} = \bar{\mathcal{X}}'$ via this isomorphism. Then $\bar{\pi}, \bar{h}$ become

$$\bar{\pi}(Tx, y, X_1, R_1, R_2, \beta) = y|R_1, R_2|,$$

$$\bar{h}(Tx, y, X_1, R_1, R_2, \beta) = \langle X_1 - {}_yX_1 + \beta, R_1\rangle + \langle {}_xA_1, \beta\rangle$$
$$+ \langle {}_xA_2, [{}_yX_1, X_1 - {}_yX_1 + \beta]/2 - [{}_yX_1, X_1]/2 - [X_1 - {}_yX_1 + \beta, X_1]/2\rangle$$
$$= \langle X_1 - {}_yX_1, R_1\rangle + \langle {}_xA_1 + R_1 + [{}_xA_2, X_1 + {}_yX_1]/2, \beta\rangle + \langle {}_xA_2, [{}_yX_1, X_1]/2\rangle.$$

Let

$$Z = \{(Tx, y, X_1, R_1, R_2) \in (T\backslash G) \times G \times \mathfrak{g}^3; xyx^{-1} \in T, R_2 = -{}_xA_2\},$$
$$Z_0 = \{(Tx, y, X_1, R_1, R_2) \in Z; {}_xA_1 + R_1 + [{}_xA_2, X_1 + {}_yX_1]/2 \in {}_x\mathfrak{n}\}$$

Define $\pi'_0 : Z_0 \to G_3$, $\tilde{h}_0 : Z_0 \to \mathbf{k}$ by

$$\pi'_0(Tx, y, X_1, R_1, R_2) = y|R_1, R_2|,$$

$$\tilde{h}_0(Tx, y, X_1, R_1, R_2) = \langle X_1 - {}_yX_1, R_1\rangle + \langle {}_xA_2, [{}_yX_1, X_1]/2\rangle.$$

The map $\bar{\mathcal{X}}' \to Z$ given by $(Tx, y, X_1, R_1, R_2, \beta) \mapsto (Tx, y, X_1, R_1, R_2)$ is a vector bundle with fibres isomorphic to $\mathfrak{b}$. Applying 1.3(b) to this vector bundle we see that

$$\hat{K} = \pi'_{0!}(\tilde{\mathcal{E}} \otimes \mathcal{L}_{\tilde{h}_0})[-3\Delta - \delta]$$

where the inverse image of $\mathcal{E}$ under $Z_0 \to T$, $(Tx, y, X_1, R_1, R_2) \mapsto xyx^{-1}$ is denoted again by $\tilde{\mathcal{E}}$. (We have used that $2\dim \mathfrak{b} = \Delta + \delta$.)

For any $R \in \mathcal{C}$ (see 1.5) let T_R be the centralizer of R in G and let $\mathfrak{t}_R$ be the centralizer of R in $\mathfrak{g}$. Let $\mathcal{R} \subset \mathrm{Hom}(\mathfrak{t}, \mathbf{k}^*)$ be the set of roots of $\mathfrak{g}$ with respect to $\mathfrak{t}$; for any $\alpha \in \mathcal{R}$ let $\mathfrak{g}^\alpha$ be the corresponding (1-dimensional) root subspace and let $e^\alpha : T \to \mathbf{k}^*$ be the correspondings root homomorphism.

Let $\mathcal{R}^+ = \{\alpha \in \mathcal{R}; \mathfrak{g}^\alpha \subset \mathfrak{n}\}$, $\mathcal{R}^- = \mathcal{R} - \mathcal{R}^+$. For $R \in \mathcal{C}$ let $\mathfrak{g}^-_R = \oplus_{a\in\mathcal{R}^-}{}_x\mathfrak{g}^\alpha$, $\mathfrak{g}^+{-}_R = \oplus_{a\in\mathcal{R}^+}{}_x\mathfrak{g}^\alpha$ (where $R = -{}_xA_2$); we have a direct sum decomposition $\mathfrak{g} = \mathfrak{g}^-_R \oplus \mathfrak{t}_R \oplus \mathfrak{g}^+_R$. Hence for any $X \in \mathfrak{g}$ we can write uniquely $X = X^-_R + X^0_R + X^+_R$ with $X^-_R \in \mathfrak{g}^-_R$, $X^0_R \in \mathfrak{t}_R$, $X^+_R \in \mathfrak{g}^+_R$. Let $\hat{Z}$ be the variety of all (y, X, R_1, R) where $R \in \mathcal{C}$, $R_1 \in \mathfrak{g}$, $y \in T_R$, $X \in \mathfrak{g}^-_R$ such that ${}_xA_1 + R_1 + [{}_xA_2, X + {}_yX]/2 \in {}_x\mathfrak{n}$ for some/any $x \in G$ such that $R = -{}_xA_2$. Define $\hat{\pi} : \hat{Z} \to G_3$, $\hat{h} : \hat{Z} \to \mathbf{k}$ by $\hat{\pi}(y, X, R_1, R) = y|R_1, R|$,

$$\hat{h}(y, X, R_1, R) = \langle X - {}_yX, R_1\rangle.$$

We define $\zeta : Z_0 \to \hat{Z}$ by $(Tx, y, X, R_1, R) \mapsto (y, X^-_R, R_1, R)$. This is well defined since, if $\beta \in \mathfrak{b}$, $R = -{}_xA_2$ and $y \in T_R$, then $[{}_xA_2, {}_x\beta + {}_y({}_x\beta)] \in {}_x\mathfrak{n}$. Now ζ is a vector bundle with fibres isomorphic to $\mathfrak{b}$. Note also that $\pi'_0 = \hat{\pi}\zeta$. We show that $\tilde{h}_0 = \hat{h}\zeta$.

For a fixed $(Tx, y, X, R_1, R) \in Z_0$ we have ${}_xA_1 + R_1 + [{}_xA_2, X + {}_yX]/2 \in {}_x\mathfrak{n}$ and in particular

$$R_1^0 + {}_xA_1 = 0,$$

(a) $$R_1^- = -[{}_xA_2, X^- + {}_y(X^-)]/2$$

where we write X^+, X^-, X^0 instead of X_R^+, X_R^-, X_R^0. We must show that

$$\langle X - {}_yX, R_1\rangle + \langle {}_xA_2, [{}_yX, X]/2\rangle = \langle X^- - {}_y(X^-), R_1\rangle$$

or equivalently

$$\langle X^+ + X^0 - {}_y(X^+ + X^0), R_1\rangle + \langle {}_xA_2, [{}_y(X^+ + X^0), X^-]/2$$
$$+ [{}_y(X^-), X^+ + X^0]/2 + [{}_y(X^+ + X^0), X^+ + X^0]/2\rangle = 0,$$

that is,

$$\langle X^+ - {}_y(X^+), R_1^-\rangle + \langle {}_xA_2, [{}_y(X^+), X^-]/2 + [{}_y(X^-), X^+]/2\rangle = 0.$$

In the left hand side we replace R_1^- by the expression (a) and we obtain

$$\langle X^+ - {}_y(X^+), -[{}_xA_2, X^- + {}_y(X^-)]/2\rangle + \langle {}_xA_2, [{}_y(X^+), X^-]/2 + [{}_y(X^-), X^+]/2\rangle$$
$$= \langle {}_xA_2, [X^- + {}_y(X^-), X^+ - {}_y(X^+)]/2 + [{}_y(X^+), X^-]/2 + [{}_y(X^-), X^+]/2\rangle$$
$$= \langle {}_xA_2, [X^-, X^+] - [{}_y(X^-), {}_y(X^+)]/2\rangle = \langle {}^{yx^{-1}}A_2 - {}^{x^{-1}}A_2, [X^-, X^+]\rangle = 0$$

since ${}^{yx^{-1}}A_2 - {}^{x^{-1}}A_2 = 0$. Thus our claim is proved. Applying 1.3(b) to the vector bundle ζ we deduce

$$\hat{K} = K'[-4\Delta - 2\delta], \quad K' = \hat{\pi}(\tilde{\mathcal{E}} \otimes \mathcal{L}_{\hat{h}}),$$

where the inverse image of $\mathcal{E}$ under $\hat{Z} \to T$, $(y, X, R_1, R) \mapsto xyx^{-1}$ (where $R = -{}_xA_2$) is denoted again by $\tilde{\mathcal{E}}$. (We have used that $2\dim\mathfrak{b} = \Delta + \delta$.)

3.2. Let $\mathcal{Z}$ be the set of all $y|R_1, R| \in G_3$ such that $R \in \mathcal{C}$, $y \in T_R$ and $(R_1)_R^0 = -{}_xA_1$ where $R = -{}_xA_2$. This is clearly a closed, smooth subvariety of G_3; it is irreducible of dimension $2\Delta - \delta$. For any $R \in \mathcal{C}$ let $\mathcal{Z}_R$ be the inverse image of R under the map $G_3 \to \mathcal{C}$, $y|R_1, R| \mapsto R$.

Let $\mathcal{H}^i_{y,R_1,R}$ be the stalk at $y|R_1, R| \in G_r$ of the i-th cohomology sheaf of K', see 3.1. We want to describe the vector spaces $\mathcal{H}^i_{y,R_1,R}$. Note that $\mathcal{H}^i_{y,R_1,R} = 0$ unless $y|R_1R| \in \mathcal{Z}$; we now assume that this condition is satisfied. Using G-equivariance and the G-homogeneity of $\mathcal{C}$, we see that we may also assume that $R = -A_2$ and we write $\mathcal{H}^i_{y,R_1}$ instead of $\mathcal{H}^i_{y,R_1,R}$. We have

$$\mathcal{H}^i_{y,R_1} = H^i_c(\hat{\pi}^{-1}(y|R_1, R|), \tilde{\mathcal{E}} \otimes \mathcal{L}_{\hat{h}}).$$

For any $X \in \mathfrak{g}$ we can write uniquely $X = X^0 + \sum_{\alpha\in\mathcal{R}} X^\alpha$ where $X^0 \in \mathfrak{t}$, $X^\alpha \in \mathfrak{g}^\alpha$. Note that we have ${}_yX = X^0 + \sum_\alpha e^\alpha(y^{-1})X^\alpha$.

Then $\hat{\pi}^{-1}(y|R_1R|)$ can be identified with the affine space

(a) $$\{(X^{-\alpha})_{\alpha\in\mathcal{R}^+}; \alpha(A_2)(1 + e^\alpha(y))X^{-\alpha}/2 = R_1^{-\alpha}\}.$$

The restriction of $\hat{h}$ to $\hat{\pi}^{-1}(y|R_1R|)$ becomes the affine-linear function

(b) $$(X^{-\alpha})_{\alpha\in\mathcal{R}^+} \mapsto \sum_{a\in\mathcal{R}^+} (1 - e^\alpha(y))\langle X^{-\alpha}, R_1^\alpha\rangle.$$

We consider several cases.

(1) for some $\alpha \in \mathcal{R}^+$ we have $1 + e^\alpha(y) = 0$ and $R_1^{-\alpha} \neq 0$;

(2) for any $\alpha \in \mathcal{R}^+$ such that $1+e^\alpha(y)=0$ we have $R_1^{-\alpha}=0$ but for some such α we have $R_1^\alpha \neq 0$;

(3) for any $\alpha \in \mathcal{R}^+$ such that $1+e^\alpha(y)=0$ we have $R_1^{-\alpha}=0$ and $R_1^\alpha=0$;

In case (1), the affine space (a) is empty and $\mathcal{H}^i_{y,R_1}=0$.

In case (2), the affine space (a) is nonempty and (b) is non-constant hence $\mathcal{H}^i_{y,R_1}=0$.

For any $y \in T$ we set $\Xi_y=\{\alpha \in \mathcal{R}^+; 1+e^\alpha(y)=0\}$. In case (3), the affine space (a) is nonempty of dimension equal to $\sharp(\Xi_y)$ and (b) is constant, hence $\mathcal{H}^i_{y,R_1}$ is 1-dimensional if $i=2\sharp(\Xi_y)$ and is 0 $i \neq 2\sharp(\Xi_y)$.

3.3. For any subset Ξ of $\mathcal{R}^+$ let $T^\Xi=\{y \in T; \Xi_y=\Xi\}$ (the sets T^Ξ form a partition of T). Note that $T^\emptyset$ is an open dense subset of T. For $\Xi \subset \mathcal{R}^+$ let $\mathcal{Z}_R^\Xi$ be the set of all $y|R_1R| \in \mathcal{Z}_R$ such that $y \in T^\Xi$ and $R_1^\alpha=0$, $R_1^{-\alpha}=0$ for all $\alpha \in \Xi$. The subsets $\mathcal{Z}_R^\Xi$ are clearly disjoint. Let $\mathcal{Z}'_R=\mathcal{Z}_R-\cup_{\Xi\subset\mathcal{R}^+}\mathcal{Z}_R^\Xi$. Note that for $y|R_1,R| \in \mathcal{Z}_R^\Xi$, $\mathcal{H}^i_{y,R_1}$ is 1-dimensional if $i=2\sharp(\Xi)$ and is 0 if $i \neq 2\sharp(\Xi)$. Moreover, for $y|R_1,R| \in \mathcal{Z}'_R$, we have $\mathcal{H}^i_{y,R_1}=0$ for all i. We show that for any $\Xi \subset \mathcal{R}^+$ we have

(a) $\dim \mathcal{X}_R^\Xi + 2\sharp(\Xi) \leq \dim \mathcal{Z}_R$ *with strict inequality unless* $\Xi=\emptyset$.

Indeed, we have $\dim \mathcal{X}_R^\Xi = \dim T^\Xi + 2\sharp(\mathcal{R}^+-\Xi)$. On the other hand, $\dim \mathcal{Z}_R = \delta + 2\sharp(\mathcal{R}^+)$. Thus (a) is equivalent to $\dim T^\Xi \leq \delta$, with strict inequality unless $\Xi=\emptyset$; this is obvious.

From (a) we see that $K'|_{\mathcal{Z}_R}$ satisfies half of the defining properties of an intersection cohomology complex (the ones not involving Verdier duality). It follows that $K'|_{\mathcal{Z}}$ itself satisfies the same half of the defining properties of an intersection cohomology complex; moreover Σ (the support of K') is equal to $\mathcal{Z}$. Hence the perverse cohomology sheaves of $K'[2\Delta-\delta]$ satisfy ${}^pH^i(K'[2\Delta-d])=0$ for $i>0$ and ${}^pH^0(K'[2\Delta-d])$ is a simple perverse sheaf on G_3. Since $\hat{K}=K'[-4\Delta-2\delta]$, it follows that ${}^pH^i(\hat{K}[6\Delta+\delta])=0$ for $i>0$ and ${}^pH^0(\hat{K}[6\Delta+\delta])$ is a simple perverse sheaf on G_3. Using Laumon's theorem (see 0.2) we deduce:

(b) ${}^pH^i(K[6\Delta+\delta])=0$ *for* $i>0$ *and* ${}^pH^0(K[6\Delta+\delta])$ *is a simple perverse sheaf on* G_3.

3.4. Let $\mathcal{Z}^\emptyset$ be the set of all $y|R_1,R| \in \mathcal{Z}$ such that for any $\alpha \in \mathcal{R}^+$ we have $e^\alpha(xyx^{-1}) \neq -1$ (where $R=-_xA_2$); this is an open dense subset of $\mathcal{Z}$. We define $f:\mathcal{Z}^\emptyset \to \mathbf{k}$ by

(a) $$f(y|R_1,R|)=\sum_{\alpha\in\mathcal{R}^+}\frac{2}{\alpha(A_2)}\frac{1-e^\alpha(xyx^{-1})}{1+e^\alpha(xyx^{-1})}\langle({}_xR_1)^\alpha,({}_xR_1)^{-\alpha}\rangle.$$

For $(y,X,R_1,R) \in \hat{\pi}^{-1}(\mathcal{Z}^\emptyset)$ we have

(b) $$f(\hat{\pi}(y,X,R_1,R))=\hat{h}(y,X,R_1,R).$$

To prove (b) we can assume that $R=-A_2$. We then have

$$\hat{h}(y,X,R_1,R)=\sum_{a\in\mathcal{R}^+}(1-e^\alpha(y))\langle X^{-\alpha},R_1^\alpha\rangle.$$

Replacing here $X^{-\alpha}$ by $\frac{2}{\alpha(A_2)(1+e^\alpha(y))}R_1^{-\alpha}$ we obtain

$$\hat{h}(y,X,R_1,R)=\sum_{a\in\mathcal{R}^+}(1-e^\alpha(y))\frac{2}{\alpha(A_2)(1+e^\alpha(y))}\langle R_1^{-\alpha},R_1^\alpha\rangle=f(y|R_1,R|),$$

as required. Since $\hat{h}$ is an isomorphism $\hat{\pi}^{-1}(\mathcal{Z}^\emptyset) \xrightarrow{\sim} \mathcal{Z}^\emptyset$ (by results in 3.3), we see that $K'|_{\mathcal{Z}^\emptyset}$ is the rank 1 local system $\tilde{\mathcal{E}} \otimes \mathcal{L}_f$ on $\mathcal{Z}^\emptyset$ where the inverse image of $\mathcal{E}$ under $\mathcal{Z}^\emptyset \to T$, $y|R_1, R| \mapsto xyx^{-1}$ (where $R = -{}_xA_2$) is denoted again by $\tilde{\mathcal{E}}$. It follows that the simple perverse sheaf ${}^pH^0(\hat{K}[6\Delta + \delta])$ on G_3 is associated to the local system $\tilde{\mathcal{E}} \otimes \mathcal{L}_f$ on the locally closed smooth irreducible subvariety $\mathcal{Z}^\emptyset$ of G_3.

3.5. It is likely that $K'[2\Delta - d]$ is a simple perverse sheaf on G_3. This would imply that $K[6\Delta + \delta]$ is a simple perverse sheaf on G_3.

4. A comparison of two complexes

4.1. We preserve the assumptions in 1.4. Let L be as in 0.2 where $f_i : \mathfrak{t} \to \mathbf{k}$ is $\tau \mapsto \langle A_i, \tau\rangle$ for $i = 1, \dots, r-1$. In this section we describe a strategy for showing that a shift of L is isomorphic to K in 1.4.

We define a sequence of algebraic varieties $\mathcal{X}_r, \mathcal{X}_{r-1}, \dots, \mathcal{X}_{r-2r'}$ as follows. For $i \in \{r-r', r-r'+1, \dots, r\}$ let $\mathcal{X}_i$ be the variety consisting of all

$(Tx, y, X_1, \dots, X_{r-1}, Y_1, \dots, Y_{r-1}) \in (T\backslash G) \times G \times \mathfrak{g}^{2r-2}$

such that $xyx^{-1} \in B$ and

$u_j({}_yX_1, \dots, {}_yX_j, Y_1, \dots, Y_j, X_1, \dots, X_j) \in {}_x\mathfrak{b}$ for $1 \le j \le i-1$.

For $i \in \{r-2r', r-2r'+1, \dots, r-r'-1\}$ let $\mathcal{X}_i$ be the variety of all

$(Tx, y, X_1, \dots, X_{r-1}, Y_1, \dots, Y_{r-1}) \in (T\backslash G) \times G \times \mathfrak{g}^{2r-2}$

such that $xyx^{-1} \in T$ and

$u_j({}_yX_1, \dots, {}_yX_j, Y_1, \dots, Y_j, X_1, \dots, X_j) \in {}_x\mathfrak{t}$ for $1 \le j \le r-r'-i-1$,

$u_j({}_yX_1, \dots, {}_yX_j, Y_1, \dots, Y_j, X_1, \dots, X_j) \in {}_x\mathfrak{b}$ for $r-r'-i \le j \le r-r'-1$.

For $i = r, r-1, \dots, r-2r'$ we have a diagram

$$G_r \xleftarrow{\pi_i} \mathcal{X}_i \xrightarrow{h_i} \mathbf{k}$$

where $\pi_i(Tx, y, X_1, \dots, X_{r-1}, Y_1, \dots, Y_{r-1}) = y|Y_1, Y_2, \dots, Y_{r-1}|$,

$$h_i(Tx, y, X_1, \dots, X_{r-1}, Y_1, \dots, Y_{r-1}) = \sum_{j\in[1,r-1]} \langle {}_xA_j, u_j({}_yX_1, \dots, {}_yX_j, Y_1, \dots, Y_j, X_1, \dots, X_j)\rangle.$$

We define $\iota : \mathcal{X}_i \to T$ by $\iota(Tx, y, X_1, \dots, X_{r-1}, Y_1, \dots, Y_{r-1}) = d(xyx^{-1})$ where $d : B \to T$ is as in 0.1. Let $\tilde{\mathcal{E}} = \iota^*\mathcal{E}$. The inverse image of $\tilde{\mathcal{E}}$ under various maps to $\mathcal{X}_i$ is denoted again by $\tilde{\mathcal{E}}$.

Let $L_i = \pi_{i!}(\tilde{\mathcal{E}} \otimes \mathcal{L}_{h_i}) \in \mathcal{D}(G_r)$. Let H be the kernel of the obvious map $B_r \to T$, a connected unipotent group of dimension $r(\Delta+\delta)/2 - \delta$. Note that $\mathcal{X}_r$ is a principal H-bundle over $\tilde{G}_r$ in 0.2. It follows that $L_r \cong K[r(\Delta+\delta) - 2\delta]$. On the other hand we have $L = L_{r-2r'}$. We would like to show that $L \cong K[r(\Delta+\delta) - 2\delta]$. It is enough to show that

(a) $L_r = L_{r-1} = \dots = L_{r-2r'}$.

Note that $\mathcal{X}_r \subset \mathcal{X}_{r-r'} \subset \dots \subset \mathcal{X}_{r'} \supset \mathcal{X}_{r'-1} \supset \dots \supset \mathcal{X}_{r-2r'}$. For $r' \le i \le r-1$ let $\pi'_i : \mathcal{X}_i - \mathcal{X}_{i+1} \to G_r$, $h'_i : \mathcal{X}_i - \mathcal{X}_{i+1} \to \mathbf{k}$ be the restrictions of π_i, h_i to $\mathcal{X}_i - \mathcal{X}_{i+1}$; let $L'_i = p'_{i!}(\tilde{\mathcal{E}} \otimes \mathcal{L}_{h'_i}) \in \mathcal{D}(G_r)$.

For $r-2r'+1 \le i \le r'$ let $\pi''_i : \mathcal{X}_i - \mathcal{X}_{i-1} \to G_r$, $h''_i : \mathcal{X}_i - \mathcal{X}_{i-1} \to \mathbf{k}$ be the restrictions of π_i, h_i to $\mathcal{X}_i - \mathcal{X}_{i-1}$; let $L''_i = p''_{i!}(\tilde{\mathcal{E}} \otimes \mathcal{L}_{h''_i}) \in \mathcal{D}(G_r)$.

From the definitions we have distinguished triangles (L'_i, L_i, L_{i+1}) (for $i = r', r'+1, \dots, r-1$) and (L''_i, L_i, L_{i-1}) (for $r-2r'+1 \le i \le r'$). Hence (a) would follow from statements (b),(c) below.

(b) $L''_{r'} = L''_{r'-1} = \dots = L''_{r-2r'+1} = 0$.

(c) $L'_{r'} = L'_{r'+1} = \dots = L'_{r-1} = 0$.

Here is a strategy to prove (b),(c).

For $r - 2r' + 1 \le i \le r'$ one should partition $\mathcal{X}_i - \mathcal{X}_{i-1}$ into pieces isomorphic to $\mathfrak{g}$ so that the restriction of h''_i to each piece is a nonconstant affine linear function and the restriction of $\tilde{\mathcal{E}}$ is $\bar{\mathbf{Q}}_l$. For $r' \le i \le r-1$ one should partition $\mathcal{X}_i - \mathcal{X}_{i+1}$ into pieces isomorphic to an affine space so that the restriction of h'_i to each piece is a nonconstant affine linear function and the restriction of $\tilde{\mathcal{E}}$ is $\bar{\mathbf{Q}}_l$. This should give the desired result. In 4.2-4.5 we carry out this strategy in several cases which are sufficient to deal with the cases where $r \in \{2, 3, 4\}$.

4.2. In this subsection we show that

(a) $L''_{r'} = 0$.

Note that $\mathcal{X}_{r'} - \mathcal{X}_{r'-1}$ is the set of all

$(Tx, y, X_1, \dots, X_{r-1}, Y_1, \dots, Y_{r-1}) \in (T\backslash G) \times G \times \mathfrak{g}^{2r-2}$

such that $xyx^{-1} \in B - T$ and

$u_j({}_yX_1, \dots, {}_yX_j, Y_1, \dots, Y_j, X_1, \dots, X_j) \in {}_x\mathfrak{b}$ for $1 \le j \le r - r' - 1$.

Let Z be the set of all

$(Tx, y, X_1, \dots, X_{r-2}, Y_1, \dots, Y_{r-1}) \in (T\backslash G) \times G \times \mathfrak{g}^{2r-3}$

such that $xyx^{-1} \in B - T$ and

$u_j({}_yX_1, \dots, {}_yX_j, Y_1, \dots, Y_j, X_1, \dots, X_j) \in {}_x\mathfrak{b}$ for $1 \le j \le r - r' - 1$.

Now $\pi''_{r'}$ is a composition $\mathcal{X}_{r'} - \mathcal{X}_{r'-1} \xrightarrow{a} Z \xrightarrow{a'} G_r$ where

$a(Tx, y, X_1, \dots, X_{r-1}, Y_1, \dots, Y_{r-1}) = (Tx, y, X_1, \dots, X_{r-2}, Y_1, \dots, Y_{r-1})$,

$a'(Tx, y, X_1, \dots, X_{r-2}, Y_1, \dots, Y_{r-1}) = y|Y_1, \dots, Y_{r-1}|$.

It is enough to show that $a(\tilde{\mathcal{E}} \otimes \mathcal{L}_{h''_{r'}}) = 0$. Clearly, $\tilde{\mathcal{E}}$ is the inverse image under a of a local system on Z denoted again by $\tilde{\mathcal{E}}$. Hence

$$a(\tilde{\mathcal{E}} \otimes \mathcal{L}_{h''_{r'}}) = \tilde{\mathcal{E}} \otimes a(\mathcal{L}_{h''_{r'}})$$

and it is enough to show that $a(\mathcal{L}_{h''_{r'}}) = 0$. It is also enough to show that for any $s = (Tx, y, X_1, \dots, X_{r-2}, Y_1, \dots, Y_{r-1}) \in Z$ we have $H^*_c(a^{-1}(s), \mathcal{L}_{h''_{r'}}) = 0$. Now $a^{-1}(s)$ may be identified with the affine space $\mathfrak{g}$ with coordinate X_{r-1} and $h''_{r'}$ is of the form $X_{r-1} \mapsto \langle {}_xA_{r-1}, {}_yX_{r-1} - X_{r-1} \rangle + c$ where c is a constant (for fixed s). It is enough to show that the linear form

$$X_{r-1} \mapsto \langle {}_xA_{r-1}, {}_yX_{r-1} - X_{r-1} \rangle = \langle {}^{yx^{-1}}A_{r-1} - {}^{x^{-1}}A_{r-1}, X_{r-1} \rangle$$

on $\mathfrak{g}$ is not identically zero. If it was identically zero, we would have ${}^{yx^{-1}}A_{r-1} = {}^{x^{-1}}A_{r-1}$ hence xyx^{-1} centralizes A_{r-1} hence $xyx^{-1} \in T$ contradicting $xyx^{-1} \in B - T$. This completes the proof of (a).

4.3. In this subsection we show that

(a) $L''_{r'-1} = 0$ (assuming that $r = 4$).

Note that $\mathcal{X}_1 - \mathcal{X}_0$ is the set of all $(Tx, y, X_1, X_2, X_3, Y_1, Y_2, Y_3) \in (T\backslash G) \times G \times \mathfrak{g}^6$ such that $xyx^{-1} \in T$ and ${}_yX_1 - X_1 + Y_1 \in {}_x\mathfrak{b} - {}_x\mathfrak{t}$.

We have a free action of $\mathfrak{g}$ on $\mathcal{X}_1 - \mathcal{X}_0$:

$$E : (Tx, y, X_1, X_2, X_3, Y_1, Y_2, Y_3) \mapsto (Tx, y, X_1, X_2 + E, X_3 + [E, X_1], Y_1, Y_2, Y_3).$$

The orbit of $(Tx, y, X_1, 0, X_3, Y_1, Y_2, Y_3)$ is

$$\mathcal{O} = \{(Tx, y, X_1, E, X_3 + [E, X_1], Y_1, Y_2, Y_3), E \in \mathfrak{g}\}.$$

It is enough to show that $H_c^*(\mathcal{O}, \tilde{\mathcal{E}} \otimes \mathcal{L}_{h_1''}) = 0$ for any such $\mathcal{O}$. Clearly $\tilde{\mathcal{E}} \cong \bar{\mathbf{Q}}_l$ on $\mathcal{O}$. Hence it is enough to show that $H_c^*(\mathcal{O}, \mathcal{L}_{h_1''}) = 0$. We can identify $\mathcal{O}$ with the affine space $\mathfrak{g}$ with coordinate E. On this affine space h_1'' is of the form

$$E \mapsto \langle {}_xA_2, {}_yE - E\rangle + \langle {}_xA_3, {}_y[E, X_1] - [E, X_1] + [{}_yE, Y_1] + [E, X_1] - [{}_yE, X_1]\rangle + c$$

where c is a constant (for our fixed $\mathcal{O}$). We have

$$\langle {}_xA_2, {}_yE - E\rangle = \langle {}^{yx^{-1}}A_2 - {}^{x^{-1}}A_2, E\rangle = 0$$

since ${}^{yx^{-1}}A_2 = {}^{x^{-1}}A_2$ (recall that $xyx^{-1} \in T$). Hence h_1'' is of the form

$$E \mapsto \langle {}_xA_3, [{}_yE, {}_yX_1] + [{}_yE, Y_1] - [{}_yE, X_1]\rangle + c = \langle {}_yE, [\xi, {}_xA_3]\rangle + c$$

where $\xi = {}_yX - X_1 + Y_1$. It is enough to show that the linear form $E \mapsto \langle {}_yE, [\xi, {}_xA_3]\rangle$ on $\mathfrak{g}$ is not identically zero. If it is identically zero we would have $[\xi, {}_xA_3] = 0$ that is, ξ is in the centralizer of ${}_xA_3$ so that $\xi \in {}_x\mathfrak{t}$, contradicting $\xi \in {}_x\mathfrak{b} - {}_x\mathfrak{t}$. This completes the proof of (a).

4.4. In this subsection we show that

(a) $L'_{r-1} = 0$.

Note that $\mathcal{X}_{r-1} - \mathcal{X}_r$ is the set of all

$(Tx, y, X_1, X_2, \dots, X_{r-1}, Y_1, Y_2, \dots, Y_{r-1}) \in (T\backslash G) \times G \times \mathfrak{g}^{2r-2}$

such that $xyx^{-1} \in B$ and

$u_j({}_yX_1, \dots, {}_yX_j, Y_1, \dots, Y_j, X_1, \dots, X_j) \in {}_x\mathfrak{b}$ for $j = 1, 2, \dots, r-2$,

$u_j({}_yX_1, \dots, {}_yX_j, Y_1, \dots, Y_j, X_1, \dots, X_j) \notin {}_x\mathfrak{b}$ for $j = r-1$.

Now π'_{r-1} is a composition $\mathcal{X}_{r-1} - \mathcal{X}_r \xrightarrow{a} Z \xrightarrow{a'} G_r$ where Z is the set of all

$$(Bx, y, X_1, X_2, \dots, X_{r-1}, Y_1, Y_2, \dots, Y_{r-1}) \in (B\backslash G) \times G \times \mathfrak{g}^{2r-2}$$

satisfying the same conditions as the points of $\mathcal{X}_{r-1} - \mathcal{X}_r$ and a is the obvious map. It is enough to show that $a(\tilde{\mathcal{E}} \otimes \mathcal{L}_{h'_{r-1}}) = 0$. Clearly $\tilde{\mathcal{E}}$ is the inverse image under a of a local system on Z denoted again by $\tilde{\mathcal{E}}$. Hence $a(\tilde{\mathcal{E}} \otimes \mathcal{L}_{h'_{r-1}}) = \tilde{\mathcal{E}} \otimes a(\mathcal{L}_{h'_{r-1}})$ and it is enough to show that $a(\mathcal{L}_{h'_{r-1}}) = 0$. It is also enough to show that for any $s = (Bx, y, X_1, X_2, \dots, X_{r-2}, Y_1, Y_2, \dots, Y_{r-1}) \in Z$ we have $H_c^*(a^{-1}(s), \mathcal{L}_{h'_{r-1}}) = 0$. Now $a^{-1}(s)$ may be identified with U by

$$u \mapsto (Tux, y, X_1, X_2, \dots, X_{r-1}, Y_1, Y_2, \dots, Y_{r-1})$$

where x is a fixed representative of Bx. For $j \in [1, r-1]$ we set

$$\xi_j = {}^xu_j({}_yX_1, \dots, {}_yX_j, Y_1, \dots, Y_j, X_1, \dots, X_j) \in \mathfrak{g}.$$

Then h'_{r-1} becomes the function $U \to \mathbf{k}$ given by

$$u \mapsto \sum_{j\in[1,r-1]} \langle {}_{ux}A_j, {}_x\xi_j\rangle$$
$$= \sum_{j\in[1,r-2]} (\langle A_j, {}^u\xi_j - \xi_j\rangle + \langle A_j, \xi_j\rangle) + \langle A_{r-1}, \xi_{r-1}\rangle + \langle {}_uA_{r-1} - A_{r-1}, \xi_{r-1}\rangle.$$

For $j \in [1, r-2]$ we have $\xi_j \in \mathfrak{b}$ hence ${}^u\xi_j - \xi_j \in \mathfrak{n}$ so that $\langle A_j, {}^u\xi_j - \xi_j\rangle = 0$. Thus h'_{r-1} becomes the function $U \to \mathbf{k}$ given by

$$u \mapsto \langle {}_uA_{r-1} - A_{r-1}, \xi_{r-1}\rangle + c$$

where c is a constant (for fixed s). We identify U with $\mathfrak{n}$ by $u \mapsto {}_uA_{r-1} - A_{r-1}$. Then h'_{r-1} becomes the function $\mathfrak{n} \to \mathbf{k}$ given by $\zeta \mapsto \langle \zeta, \xi_{r-1}\rangle + c$. This function is affine linear and nonconstant since $\xi_{r-1} \notin \mathfrak{b} = \mathfrak{n}^\perp$. It follows that $H_c^*(a^{-1}(s), \mathcal{L}_{h'_{r-1}}) = 0$ and (a) is proved.

4.5. In this subsection we show that

(a) $L'_{2r'-2} = 0$ (assuming that $r = 4$).

Note that $\mathcal{X}_2 - \mathcal{X}_3$ is the set of all

$(Tx, y, X_1, X_2, X_3, Y_1, Y_2, Y_3) \in (T\backslash G) \times G \times \mathfrak{g}^6$

such that $xyx^{-1} \in B$,

(b) ${}_yX_1 - X_1 + Y_1 \in {}_x\mathfrak{b}$,

(c) ${}_yX_2 - X_2 + Y_2 + [{}_yX_1, Y_1]/2 - [{}_yX_1, X_1]/2 - [Y_1, X_1]/2 \notin {}_x\mathfrak{b}$.

Let $Z = \{(Tx, y, Y_1, Y_2, Y_3) \in (T\backslash G) \times G \times \mathfrak{g}^3; xyx^{-1} \in B\}$. The inverse image of $\mathcal{E}$ under $Z \to T$, $(Tx, y, Y_1, Y_2, Y_3) \mapsto d(xyx^{-1})$ is denoted by $\tilde{\mathcal{E}}_0$.

Now π'_2 is a composition $\mathcal{X}_2 - \mathcal{X}_3 \xrightarrow{a} Z \xrightarrow{a'} G_r$ where

$$a(Tx, y, X_1, X_2, X_3, Y_1, Y_2, Y_3) = (Tx, y, Y_1, Y_2, Y_3),$$
$$a'(Tx, y, Y_1, Y_2, Y_3) = y|Y_1, Y_2, Y_3|$$

We have $a^*\tilde{\mathcal{E}}_0 = \tilde{\mathcal{E}}$. It is enough to show that $a(\tilde{\mathcal{E}} \otimes \mathcal{L}_{h'_2}) = 0$ that is, $\tilde{\mathcal{E}}_0 \otimes a(\mathcal{L}_{h'_2}) = 0$. Thus it is enough to prove that $a(\mathcal{L}_{h'_2}) = 0$. Hence it is enough to show that for any $s = (Tx, y, Y_1, Y_2, Y_3) \in (T\backslash G) \times G \times \mathfrak{g}^3$ we have $H_c^*(a^{-1}(s), \mathcal{L}_{h'_2}) = 0$.

Let $\mathcal{G} = \{|E, E', E''| \in G_4; E \in \mathfrak{n}, E' \in \mathfrak{n}, E'' \in \mathfrak{n}\}$; this is a closed subgroup of G_4.

We fix a representative x in Tx and we define a free $\mathcal{G}$-action on $a^{-1}(s)$ by

$$|E, E', E''| : (Tx, y, X_1, X_2, X_3, Y_1, Y_2, Y_3) \mapsto (Tx, y, X_1 + {}_xE,$$
$$X_2 + {}_xE' + [{}_xE, X_1]/2, X_3 + {}_xE''$$
$$+ [{}_xE', X_1] - [{}_xE, [{}_xE, X_1]]/6 - [X_1, [{}_xE, X_1]]/3, Y_1, Y_2, Y_3).$$

We verify that this action is well defined (that is, the equations (b),(c) are preserved). To show that (b) is preserved it is enough to verify that ${}_{xy}E - {}_xE \in {}_x\mathfrak{b}$ or that ${}^{xy^{-1}x^{-1}}E - E \in \mathfrak{b}$; this follows from $E \in \mathfrak{n}$, $xyx^{-1} \in B$. To show that (c) is preserved it is enough to verify that

$$ {}_{xy}E' + [{}_{xy}E, {}_yX_1]/2 - {}_xE' - [{}_xE, X_1]/2 + [{}_{xy}E, Y_1]/2$$
$$- [{}_yX_1, {}_xE]/2 - [{}_{xy}E, X_1]/2 - [{}_{xy}E, {}_xE]/2 - [Y_1, {}_xE]/2 \in {}_x\mathfrak{b}$$

(when (b) holds) or that

$$[{}_{xy}E, {}_yX_1 - X_1 + Y_1]/2 + [{}_xE, {}_yX_1 - X_1 + Y_1]/2 - [{}_{xy}E, {}_xE]/2 + {}_{xy}E' - {}_xE' \in {}_x\mathfrak{b}$$

and this follows from (b) and from ${}_{xy}E \in {}_x\mathfrak{b}$, ${}_xE \in {}_x\mathfrak{b}$, ${}_{xy}E' \in {}_x\mathfrak{b}$, ${}_xE' \in {}_x\mathfrak{b}$.

It is enough to show that for any $\mathcal{G}$-orbit $\mathcal{O}$ in $a^{-1}(s)$ we have $H_c^*(\mathcal{O}, \mathcal{L}_{h'_2}) = 0$. We may identify $\mathcal{O} = \mathcal{G}$ using a base point $(Tx, y, X_1, X_2, X_3, Y_1, Y_2, Y_3) \in \mathcal{O}$ (with a fixed representative x for Tx) and we identify $\mathcal{G} = \mathfrak{n}^3$ using $[E, E', E''] \leftrightarrow (E, E', E'')$. Then h'_2 becomes a function $h'' : \mathfrak{n}^3 \to \mathbf{k}$ of the following form (we have substituted $Y_1 = X_1 - {}_yX_1 + {}_x\beta$ where $\beta \in \mathfrak{b}$):

$$(E, E', E'') \mapsto h''(E, E', E'') = \langle {}_xA_1, \xi_1\rangle + \langle {}_xA_2, \xi_2 + \xi'_2\rangle + \langle {}_xA_3, \xi_3 + \xi'_3 + \xi''_3\rangle$$

where

$$\xi_1 = {}_x\beta + {}_{xy}E - {}_xE,$$

$$\begin{aligned}\xi_2 &= {}_yX_2 + [{}_{xy}E, {}_yX_1]/2 - X_2 - [{}_xE, X_1]/2 + Y_2\\ &+ [{}_yX_1 + {}_{xy}E, X_1 - {}_yX_1 + {}_x\beta]/2 - [{}_yX_1 + {}_{xy}E, X_1 + {}_xE]/2\\ &- [X_1 - {}_yX_1 + {}_x\beta, X_1 + {}_xE]/2,\end{aligned}$$

$$\xi_2' = {}_{xy}E' - {}_xE'$$

$$\begin{aligned}\xi_3 &= {}_yX_3 - X_3 + Y_3 - [{}_{xy}E, [{}_{xy}E, {}_yX_1]]/6 - [{}_yX_1, [{}_{xy}E, {}_yX_1]]/3\\ &+ [{}_xE, [{}_xE, X_1]]/6 + [X_1, [{}_xE, X_1]]/3\\ &+ [{}_yX_2 + [{}_{xy}E, {}_yX_1]/2, X_1 - {}_yX_1 + {}_x\beta] + [X_2 + [{}_xE, X_1]/2, X_1 + {}_xE]\\ &- [{}_yX_2 + [{}_{xy}E, {}_yX_1]/2, X_1 + {}_xE] - [Y_2, X_1 + {}_xE]\\ &- [{}_yX_1 + {}_{xy}E, [{}_yX_1 + {}_{xy}E, X_1 - {}_yX_1 + {}_x\beta]]/6\\ &- [X_1 - {}_yX_1 + {}_x\beta, [{}_yX_1 + {}_{xy}E, X_1 - {}_yX_1 + {}_x\beta]]/3\\ &+ [X_1 + {}_xE, [{}_yX_1 + {}_{xy}E, X_1 - {}_yX_1 + {}_x\beta]]/2\\ &+ [{}_yX_1 + {}_{xy}E, [{}_yX_1 + {}_{xy}E, X_1 + {}_xE]]/6\\ &+ [{}_yX_1 + {}_{xy}E, [X_1 - {}_yX_1 + {}_x\beta, X_1 + {}_xE]]/6\\ &+ [X_1 - {}_yX_1 + {}_x\beta, [{}_yX_1 + {}_{xy}E, X_1 + {}_xE]]/6\\ &+ [X_1 - {}_yX_1 + {}_x\beta, [X_1 - {}_yX_1 + {}_x\beta, X_1 + {}_xE]]/6\\ &- [X_1 + {}_xE, [{}_yX_1 + {}_{xy}E, X_1 + {}_xE]]/3\\ &- [X_1 + {}_xE, [X_1 - {}_yX_1 + {}_x\beta, X_1 + {}_xE]]/3\rangle,\end{aligned}$$

$$\begin{aligned}\xi_3' &= {}_{xy}E'' + [{}_{xy}E', {}_yX_1] - {}_xE'' - [{}_xE', X_1]\\ &+ [{}_{yx}E', X_1 - {}_yX_1 + {}_x\beta] + [{}_xE', X_1] - [{}_{xy}E', X_1],\end{aligned}$$

$$\xi_3'' = [{}_xE', {}_xE] - [{}_{xy}E', {}_xE].$$

It is enough to show that for any fixed E', E'' in $\mathfrak{n}$, the function $E \mapsto h_1''(E) = h''(E, E', E'')$ is affine linear and nonconstant. Let

$$S = {}_yX_2 - X_2 + Y_2 + [{}_yX_1, Y_1]/2 - [{}_yX_1, X_1]/2 - [Y_1, X_1]/2.$$

A computation shows that

$$\xi_1 - C_1 \in {}_x\mathfrak{n}, \xi_2 - C_2 \in {}_x\mathfrak{n}, \xi_3 - [{}_xE, S] - C_3 \in {}_x\mathfrak{n}, \xi_3' = C_4$$

where C_1, C_2, C_3, C_4 are vectors in $\mathfrak{g}$ independent of E. Moreover, $\xi_2' \in {}_x\mathfrak{n}$, $\xi_3'' \in {}_x\mathfrak{n}$. Since $\langle {}_xA_i, {}_x\mathfrak{n}\rangle = 0$, for some constant $c \in \mathbf{k}$ we have

$$h_1''(E) = \langle {}_xA_3, [{}_xE, S]\rangle + c = \langle S, [{}_xA_3, {}_xE]\rangle + c.$$

In particular, $E \mapsto h_1''(E)$ is affine linear on $\mathfrak{n}$. To show that it is nonconstant it is enough to show that $E \mapsto \langle S, [{}_xA_3, {}_xE]\rangle$ is not identically zero. Assume that it is identically zero. Since $E \mapsto [A_3, E]$ is a vector space isomorphism $\mathfrak{n} \xrightarrow{\sim} \mathfrak{n}$ it would follow that $\langle S, {}_x\tilde{E}\rangle = 0$ for any $\tilde{E} \in \mathfrak{n}$ hence $S \in {}_x(\mathfrak{n}^\perp)$ that is, $S \in {}_x\mathfrak{b}$. This contradicts the definition of $\mathcal{X}_2 - \mathcal{X}_3$ and proves (a).

4.6. In this subsection we assume that $r \in \{2, 3, 4\}$. From 4.2, 4.3, 4.4, 4.5 we see that 4.1(b),(c) hold. Hence 4.1(a) holds. Hence $L \cong K[2\delta]$ if $r = 2$, $L \cong K[3\Delta + \delta]$ if $r = 3$ and $L \cong K[4\Delta + 2\delta]$ if $r = 4$.

Using now 2.1(b), 2.5(b), 3.3(b) we deduce the folowing result.

THEOREM 4.7. *(a) $L[r\delta]$ is a simple perverse sheaf on G_r provided that $r = 2$ or $r = 4$.*

(b) If $r = 3$ we have ${}^pH^i(L[r\delta]) = 0$ for $i > 0$ and ${}^pH^0(L[r\delta]) = 0$ is a simple perverse sheaf on G_r.

It is likely that in fact $L[r\delta]$ is a simple perverse sheaf on G_r for any $r \geq 2$. For $r = 3$ this would follow if the truth of the statements in 3.5 could be established.

Acknowledgement

The author wishes to thank Dongkwan Kim for pointing out an inaccuracy in an earlier version of this paper.

References

[1] P. Deligne and G. Lusztig, *Representations of reductive groups over finite fields*, Ann. of Math. (2) **103** (1976), no. 1, 103–161. MR0393266

[2] P. Gérardin, *Construction de séries discrètes p-adiques*, Lecture Notes in Mathematics, Vol. 462, Springer-Verlag, Berlin-New York, 1975. Sur les séries discrètes non ramifiées des groupes réductifs déployés p-adiques. MR0396859

[3] G. Laumon, *Transformation de Fourier, constantes d'équations fonctionnelles et conjecture de Weil* (French), Inst. Hautes Études Sci. Publ. Math. **65** (1987), 131–210. MR908218

[4] G. Lusztig, *Green polynomials and singularities of unipotent classes*, Adv. in Math. **42** (1981), no. 2, 169–178, DOI 10.1016/0001-8708(81)90038-4. MR641425

[5] G. Lusztig, *Character sheaves. I*, Adv. in Math. **56** (1985), no. 3, 193–237, DOI 10.1016/0001-8708(85)90034-9. MR792706

[6] G. Lusztig, *Character sheaves and generalizations*, The unity of mathematics, Progr. Math., vol. 244, Birkhäuser Boston, Boston, MA, 2006, pp. 443–455, DOI 10.1007/0-8176-4467-9_12. MR2181813

DEPARTMENT OF MATHEMATICS, M.I.T., CAMBRIDGE, MASSACHUSETTS 02139

Contemporary Mathematics
Volume **683**, 2017
http://dx.doi.org/10.1090/conm/683/13706

Integral presentations of quantum lattice Heisenberg algebras

Diego Berdeja Suárez

ABSTRACT. We give integral presentations of quantum lattice Heisenberg algebras by viewing them as Heisenberg doubles. Our presentations generalize those appearing previously in the literature.

CONTENTS

1. Introduction

The Heisenberg algebra has its origins in the description of the relationship between the momentum and position operators of quantum mechanics, as well as the creation and annihilation operators employed in the description of the quantum harmonic oscillator. It also plays a vital role in Lie theory, especially in the field of vertex operators.

More recently, mathematicians have been developing so called *categorifications* of the Heisenberg algebra and its quantum analogues (see [**Kho14**, **CL12**, **LS13**, **SV12**, **LS12**, **SY15**, **RS15a**, **RS**]). In categorification, generators with nice integrality properties are often an important ingredient. The usual presentation of the Heisenberg algebra is in terms of a generating set that does not possess these properties. Thus, the categorification literature often works in a different presentation.

The goal of the current paper is to develop explicit formulas for integral presentations of quantum lattice Heisenberg algebras (the Heisenberg algebras appearing most often in categorification). While some special cases of these formulas have

---

2010 *Mathematics Subject Classification.* Primary 16T05, 17B35.
Key words and phrases. Heisenberg algebra, Hopf algebra, Heisenberg double, symmetric functions, quantum algebra.
This research was supported by the MITACS Globalink program.

appeared before in the literature, the description in the current paper is more complete. We obtain the integral presentations by adopting the point of the view of the Heisenberg double. In particular, the Heisenberg algebra is the Heisenberg double of the Hopf algebra of symmetric functions. Then our presentations arise from considering the complete and elementary symmetric functions as generators and performing computations in the Heisenberg double. This approach towards quantum lattice Heisenberg algebras, followed, for example, by Rosso and Savage in [**RS15b, RS**], simplifies proofs.

We begin in Section 2 and Section 3 by reviewing the concepts of Hopf algebras, Hopf pairings, and the ring of symmetric functions. In particular we show, in Proposition 3.1, that a Hopf pairing of finitely many copies of the ring of symmetric functions with itself is uniquely determined by its values on the power sum symmetric functions. In Section 4, we discuss the graded dimensions of the symmetric and exterior algebras of $\mathbb{Z}$-graded vector spaces, since these algebras appear in our presentations. Finally, in Section 5, we state our main results. Namely, we give two integral presentations of quantum lattice Heisenberg algebras in Theorems 5.3 and 5.4. We conclude with pointing out the connection between our results and those appearing in the literature.

Note on the arXiv version. For the interested reader, the tex file of the arXiv version of this paper includes hidden details of some straightforward computations and arguments that are omitted in the pdf file. These details can be displayed by switching the `details` toggle to true in the tex file and recompiling.

2. Hopf algebras, Hopf pairings, and the Heisenberg double

In this section we recall the definition of the Heisenberg double. We begin with the concepts of Hopf algebras and Hopf pairings. For the remainder of the paper, we fix a field $\Bbbk$. Unless otherwise indicated, vector spaces, tensor products, associative algebras, Hopf algebras, etc. are over this field.

DEFINITION 2.1 (Graded Hopf algebra). A *$\mathbb{Z}$-graded Hopf algebra* is a tuple $(H, \nabla, \epsilon, \Delta, \eta, S)$, where $H = \bigoplus_{c\in\mathbb{Z}} H_c$, with each H_c a finite-dimensional vector space, such that (H, ∇, ϵ) is a $\mathbb{Z}$-graded algebra, (H, Δ, η) is a $\mathbb{Z}$-graded coalgebra, and $S\colon H \to H$ is an *antipode*. This means, in particular, that the following conditions are satisfied:

$$\nabla(H_c \otimes H_d) \subseteq H_{c+d}, \quad \Delta(H_c) \subseteq \bigoplus_{d\in\mathbb{Z}} H_d \otimes H_{c-d},$$
$$\eta(\Bbbk) \subseteq H_0, \quad \epsilon(H_c) = 0 \text{ for } c \neq 0,$$
$$\nabla \circ (S \otimes 1_H) \circ \Delta = \nabla \circ (1_H \otimes S) \circ \Delta = \epsilon \circ \eta.$$

DEFINITION 2.2 (Hopf pairing, dual Hopf pair). Fix two $\mathbb{Z}$-graded Hopf algebras H^+ and H^-. Then a bilinear map

$$\langle -, - \rangle \colon H^+ \times H^- \longrightarrow \Bbbk$$

is a *Hopf pairing* if

$$\langle -, - \rangle|_{H_c^+ \times H_d^-} = 0$$

for all $c, d \in \mathbb{Z}$ and $c \neq d$, and we have

$$\langle xy, a \rangle = \langle x \otimes y, \Delta(a) \rangle, \tag{2.1a}$$

$$\langle x, ab\rangle = \langle \Delta(x), a\otimes b\rangle, \tag{2.1b}$$

$$\langle 1_{H^+}, a\rangle = \eta(a), \tag{2.1c}$$

$$\langle x, 1_{H^-}\rangle = \eta(x), \tag{2.1d}$$

for all $x, y \in H^+$ and $a, b \in H^-$. Given such a Hopf pairing, we define

$$\langle -,-\rangle\colon (H^+\otimes H^+)\otimes(H^-\otimes H^-)\longrightarrow \Bbbk, \quad \langle x\otimes y, a\otimes b\rangle \mapsto \langle x,a\rangle\langle y,b\rangle.$$

If the Hopf pairing is nondegenerate, we say that (H^+, H^-) are a *dual Hopf pair.*

When dealing with Hopf algebras, it is often useful to use *Sweedler notation.* For any Hopf algebra H, and $a \in H$, we write

$$\Delta(a) = \sum_i a_{1,i}\otimes a_{2,i} = \sum_{(a)} a_1\otimes a_2 \in H\otimes H,$$

where $a_{j,i} \in H$ for all i and for $j = 1, 2$.

DEFINITION 2.3 (Heisenberg double). The *Heisenberg double* $\mathfrak{h}(H^+, H^-)$ associated to a dual Hopf pair (H^+, H^-) is defined as follows. We set $\mathfrak{h}(H^+, H^-) = H^+\otimes H^-$ as $\Bbbk$-modules, and write $a\#x$ for $a\otimes x$ as an element of $\mathfrak{h}(H^+, H^-)$, with $a \in H^+$ and $x \in H^-$. We define the multiplication of two elements of $\mathfrak{h}(H^+, H^-)$ by

$$(a\#x)(b\#y) = \sum_{(x),(b)} \langle x_1, b_2\rangle ab_1\#x_2y, \tag{2.2}$$

extended linearly. It can be shown that this endows $\mathfrak{h}(H^+, H^-)$ with the structure of an associative algebra.

3. The ring of symmetric functions

In this section, we recall some basic facts about symmetric functions. We refer the reader to [**Mac95**, §I.1 and §I.2] for details.

We let $\mathcal{P}$ denote the set of partitions. We reserve Greek symbols to denote partitions. For $\lambda \in \mathcal{P}$, we let λ_i denote the i-th part of the partition λ. We let $|\lambda|$ denote the sum of the parts of λ, i.e. $|\lambda| = \sum \lambda_i$. We may also write $\lambda \vdash n$ for $|\lambda| = n$. We define $\ell(\lambda)$ to be the length of λ, i.e., the number of nonzero parts of λ. For $k \in \mathbb{N}$, we let $\mu \ominus k$ be the partition μ with a part k removed. If μ has no part k, then we set $\mu\ominus k$ to be the empty partition. In an analogous manner, we define $\mu\oplus k$ to be the partition μ with a part k added. We then let $\mu\oplus\lambda = (\cdots(\mu\oplus\lambda_1)\oplus\cdots)\oplus\lambda_{\ell(\lambda)})$. Similarly $\mu\ominus\lambda = (\cdots(\mu\ominus\lambda_1)\ominus\cdots)\ominus\lambda_{\ell(\lambda)})$. We let $m_i(\lambda)$ denote the number of parts of λ equal to i.

Now fix a finite set $I = \{1, 2, \dots, |I|\}$. We let bold Greek symbols denote elements of $\mathcal{P}^I$. For $\boldsymbol{\lambda} \in \mathcal{P}^I$, we let λ^i denote the i-th partition, $i \in I$. We set $|\boldsymbol{\lambda}| = \sum_{i,k} \lambda^i_k$ and $\ell(\boldsymbol{\lambda}) = \sum_i \ell(\lambda^i)$. We interpret $\boldsymbol{\lambda}\oplus\boldsymbol{\mu}$ to be the element of $\mathcal{P}^I$ whose i-th partition is $\lambda^i\oplus\mu^i$ for all $i \in I$. Finally, we let $\boldsymbol{\lambda}\oplus(a)^j$ be the element of $\mathcal{P}^I$ whose i-th partition is $\lambda^i\oplus\delta_{ij}a$. Note that $\lambda\oplus 0 = \lambda$.

Following [**RS15b**, p. 1068], we define $\underline{\boldsymbol{\lambda}}$ to be the sequence of elements of $\mathbb{N}_{>0}\times I$, in lexicographical order, such that the element (k, i) appears $m_k(\lambda^i)$ times. In other words $\underline{\boldsymbol{\lambda}}$ consists of all the parts of the λ^i, $i \in I$, with their "color" i recorded.

We will write the n-th term of the sequence $\underline{\boldsymbol{\lambda}}$ as

$$\underline{\boldsymbol{\lambda}}_n = (\operatorname{part}(\underline{\boldsymbol{\lambda}}_n), \operatorname{color}(\underline{\boldsymbol{\lambda}}_n)) \in \mathbb{N}_{>0}\times I, \quad \text{for } n = 1, \dots, \ell(\boldsymbol{\lambda}).$$

Let Sym denote the ring of symmetric functions in countably many variables $x_1, x_2, \ldots$ with coeffcients in $\mathbb{Z}$ and define $\mathrm{Sym}_{\mathbb{Q}} = \mathbb{Q} \otimes_{\mathbb{Z}} \mathrm{Sym}$. It is known that $\mathrm{Sym}_{\mathbb{Q}}$ is generated as a $\mathbb{Q}$-algebra by the *power sums*:

$$\mathrm{Sym}_{\mathbb{Q}} = \mathbb{Q} \otimes_{\mathbb{Z}} \mathrm{Sym} = \mathbb{Q}[p_1, p_2, \ldots], \quad p_n = \sum_{i=1}^{\infty} x_i^n, \quad n \in \mathbb{N}.$$

Consider the tensor product Hopf Algebra $\mathrm{Sym}_{\mathbb{Q}}^{\otimes I}$. We have that $\mathrm{Sym}_{\mathbb{Q}}^{\otimes I}$ is generated by $\{p_{n,i} \mid n \in \mathbb{N}_{>0},\ i \in I\}$, where here, and in what follows, the second subscript indicates the factor in which an element of Sym lives. For example, $p_{n,i}$ denotes the simple tensor in $\mathrm{Sym}_{Q}^{\otimes I}$ with i-th component equal to p_n and all other components equal to 1.

The coproduct on $\mathrm{Sym}_{\mathbb{Q}}^{\otimes I}$ is determined by

$$\Delta(p_{n,i}) = 1 \otimes p_{n,i} + p_{n,i} \otimes 1. \tag{3.1}$$

For $\lambda \in \mathcal{P}$ and $\boldsymbol{\lambda} \in \mathcal{P}^I$, define

$$p_\lambda = \prod_{i=1}^{\ell(\lambda)} p_{\lambda_i}, \quad p_{\boldsymbol{\lambda}} = \prod_{i \in I} p_{\lambda^i, i}.$$

Thus, the $p_{\boldsymbol{\lambda}}$, $\boldsymbol{\lambda} \in \mathcal{P}^I$, form a basis for $\mathrm{Sym}_{\mathbb{Q}}^{\otimes I}$.

The *complete symmetric functions* h_r, $r \in \mathbb{N}$, and *elementary symmetric functions* e_n, $n \in \mathbb{N}$, are defined by

$$h_r = \sum_{|\lambda| = r} \frac{p_\lambda}{z_\lambda}, \qquad e_n = \sum_{\lambda \vdash n} (-1)^{|\lambda| - \ell(\lambda)} \frac{p_\lambda}{z_\lambda}, \tag{3.2}$$

where

$$z_\lambda = \prod_{i \geq 1} i^{m_i(\lambda)} m_i(\lambda)!. \tag{3.3}$$

By convention, we set $e_0 = p_0 = h_0 = 1$. We have

$$\mathrm{Sym} = \mathbb{Z}[h_1, h_2, \ldots] = \mathbb{Z}[e_1, e_2, \ldots,].$$

In other words, the h_n and e_n generated Sym over $\mathbb{Z}$, as opposed to the p_n, which only generated $\mathrm{Sym}_{\mathbb{Q}}$ over $\mathbb{Q}$. We have (see, for example, [**Zab15**, Prop. 3.2])

$$\Delta(h_n) = \sum_{r=0}^{n} h_r \otimes h_{n-r}, \qquad \Delta(e_n) = \sum_{r=0}^{n} e_r \otimes e_{n-r}. \tag{3.4}$$

Let us now grab two copies of $\mathrm{Sym}_{\mathbb{Q}}^{\otimes I}$, rename them H^+ and H^-, and consider the associated Heisenberg double $\mathfrak{h}(H^+, H^-)$. For the rest of the paper, we shall consider bilinear forms on $\mathrm{Sym}_{\mathbb{Q}}^{\otimes I} \otimes \mathrm{Sym}_{\mathbb{Q}}^{\otimes I}$ of the type described in the following proposition.

PROPOSITION 3.1. *Let I be a finite set. Fix a $C_{j,i}^n \in \Bbbk$ for every $i, j \in I$ and $n \in \mathbb{N}$. Then there exists a unique Hopf pairing $\langle -, - \rangle \colon \mathrm{Sym}_{\mathbb{Q}}^{\otimes I} \otimes \mathrm{Sym}_{\mathbb{Q}}^{\otimes I} \longrightarrow \Bbbk$ such that*

$$\langle p_{n,j}, p_{m,i} \rangle = \delta_{nm} C_{j,i}^n. \tag{3.5}$$

This pairing is given by

$$\langle p_{\boldsymbol{\lambda}}, p_{\boldsymbol{\mu}} \rangle = \delta_{\ell(\boldsymbol{\lambda}),\ell(\boldsymbol{\mu})} \sum_{\sigma \in S_{\ell(\boldsymbol{\lambda})}} \prod_{r=1}^{\ell(\boldsymbol{\lambda})} \delta_{\mathrm{part}(\underline{\boldsymbol{\lambda}}_r),\mathrm{part}(\underline{\boldsymbol{\mu}}_{\sigma(r)})} C^{\mathrm{part}(\underline{\boldsymbol{\lambda}}_r)}_{\mathrm{color}(\underline{\boldsymbol{\lambda}}_r),\mathrm{color}(\underline{\boldsymbol{\mu}}_{\sigma(r)})}, \tag{3.6}$$

for all $\boldsymbol{\lambda}, \boldsymbol{\mu} \in \mathcal{P}^I$. *In particular*

$$\langle p_{\lambda,j}, p_{\mu,i} \rangle = \delta_{\lambda\mu} \prod_{k\geq 1} (C^k_{j,i})^{m_k(\lambda)} m_k(\lambda)!. \tag{3.7}$$

PROOF. The coproduct of the p_λ is given by (cf. [**Mac95**, Eq. 3.2])

$$\Delta(p_\lambda) = \sum_{\mu\oplus\nu=\lambda} p_\mu \otimes p_\nu \prod_{i=1}^{\lambda_1} \binom{m_i(\lambda)}{m_i(\mu)}.$$

Thus, we have

$$\begin{aligned}
&\langle p_{\boldsymbol{\lambda}} \otimes p_{\boldsymbol{\nu}}, \Delta(p_{\boldsymbol{\mu}}) \rangle \\
&= \sum_{\boldsymbol{\alpha}\oplus\boldsymbol{\beta}=\boldsymbol{\mu}} \langle p_{\boldsymbol{\lambda}}, p_{\boldsymbol{\alpha}} \rangle \langle p_{\boldsymbol{\nu}}, p_{\boldsymbol{\beta}} \rangle \prod_{j\in I} \prod_{k\geq 1} \binom{m_k(\mu^j)}{m_k(\alpha^j)} \\
&= \sum_{\boldsymbol{\alpha}\oplus\boldsymbol{\beta}=\boldsymbol{\mu}} \delta_{\ell(\boldsymbol{\lambda}),\ell(\boldsymbol{\alpha})} \delta_{\ell(\boldsymbol{\nu}),\ell(\boldsymbol{\beta})} \prod_{j\in I} \prod_{k\geq 1} \binom{m_k(\mu^j)}{m_k(\alpha^j)} \\
&\qquad \sum_{\substack{\sigma\in S_{\ell(\boldsymbol{\lambda})} \\ \tau \in S_{\ell(\boldsymbol{\nu})}}} \prod_{r=1}^{\ell(\boldsymbol{\lambda})} \prod_{s=1}^{\ell(\boldsymbol{\nu})} \delta_{\mathrm{part}(\underline{\boldsymbol{\lambda}}_r),\mathrm{part}(\underline{\boldsymbol{\alpha}}_{\sigma(r)})} \delta_{\mathrm{part}(\underline{\boldsymbol{\nu}}_s),\mathrm{part}(\underline{\boldsymbol{\beta}}_{\tau(s)})} \\
&\qquad\qquad C^{\mathrm{part}(\underline{\boldsymbol{\lambda}}_r)}_{\mathrm{color}(\underline{\boldsymbol{\lambda}}_r),\mathrm{color}(\underline{\boldsymbol{\alpha}}_{\sigma(r)})} C^{\mathrm{part}(\underline{\boldsymbol{\nu}}_s)}_{\mathrm{color}(\underline{\boldsymbol{\nu}}_s),\mathrm{color}(\underline{\boldsymbol{\beta}}_{\tau(s)})} \\
&= \sum_{\boldsymbol{\alpha}\oplus\boldsymbol{\beta}=\boldsymbol{\mu}} \delta_{\ell(\boldsymbol{\lambda}),\ell(\boldsymbol{\alpha})} \delta_{\ell(\boldsymbol{\nu}),\ell(\boldsymbol{\beta})} \prod_{j\in I} \prod_{k\geq 1} \binom{m_k(\mu^j)}{m_k(\alpha^j)} \\
&\qquad \sum_{\sigma\in S_{\ell(\boldsymbol{\lambda})}\times S_{\ell(\boldsymbol{\nu})}} \prod_{r=1}^{\ell(\boldsymbol{\lambda}\oplus\boldsymbol{\nu})} \delta_{\mathrm{part}(\underline{\boldsymbol{\lambda}\oplus\boldsymbol{\nu}}_r),\mathrm{part}(\underline{\boldsymbol{\mu}}_{\sigma(r)})} C^{\mathrm{part}(\underline{\boldsymbol{\lambda}\oplus\boldsymbol{\nu}}_r)}_{\mathrm{color}(\underline{\boldsymbol{\lambda}\oplus\boldsymbol{\nu}}_r),\mathrm{color}(\underline{\boldsymbol{\mu}}_{\sigma(r)})} \\
&= \delta_{\ell(\boldsymbol{\lambda}\oplus\boldsymbol{\nu}),\ell(\boldsymbol{\mu})} \sum_{\sigma\in S_{\ell(\boldsymbol{\mu})}} \prod_{r=1}^{\ell(\boldsymbol{\lambda}\oplus\boldsymbol{\nu})} \delta_{\mathrm{part}(\underline{\boldsymbol{\lambda}\oplus\boldsymbol{\nu}}_r),\mathrm{part}(\underline{\boldsymbol{\mu}}_{\sigma(r)})} C^{\mathrm{part}(\underline{\boldsymbol{\lambda}\oplus\boldsymbol{\nu}}_r)}_{\mathrm{color}(\underline{\boldsymbol{\lambda}\oplus\boldsymbol{\nu}}_r),\mathrm{color}(\underline{\boldsymbol{\mu}}_{\sigma(r)})} \\
&= \langle p_{\boldsymbol{\lambda}\oplus\boldsymbol{\mu}}, p_{\boldsymbol{\nu}} \rangle = \langle p_{\boldsymbol{\lambda}} p_{\boldsymbol{\mu}}, p_{\boldsymbol{\nu}} \rangle.
\end{aligned}$$

Thus, (3.6) is a Hopf pairing. It is clear that (3.6) reduces to (3.7) and (3.5).

Now, the relation $\langle p_{\boldsymbol{\lambda}} p_{\boldsymbol{\mu}}, p_{\boldsymbol{\nu}} \rangle = \langle p_{\boldsymbol{\lambda}} \otimes p_{\boldsymbol{\nu}}, \Delta(p_\mu) \rangle$ implies that the pairing between any $p_{\boldsymbol{\lambda}}$ and $p_{\boldsymbol{\mu}}$, where $\boldsymbol{\lambda}$ and $\boldsymbol{\mu}$ have length greater than one, is determined by the pairing between $p_{\boldsymbol{\nu}}$ of strictly smaller length. It follows by induction that the pairing is uniquely determined by (3.5). □

For $a \in \mathrm{Sym}_{\mathbb{Q}}$, we will write $a^\pm$ for the corresponding element of $H^\pm$. We regard H^+ and H^- as subalgebras of $\mathfrak{h}(H^+, H^-)$, via the natural maps $a^+ \mapsto a\#1$ and $a^- \mapsto 1\#a$ for all $a \in \mathrm{Sym}_{\mathbb{Q}}^{\otimes I}$.

REMARK 3.2. The ring isomorphism $\Omega\colon \mathrm{Sym}_{\mathbb{Q}} \longrightarrow \mathrm{Sym}_{\mathbb{Q}}$, $p_n \mapsto (-1)^{n-1}p_n$ for all $n \geq 1$, is an algebra involution. This follows from the fact that Ω simply rescales the variables p_n of the polynomial algebra $\mathrm{Sym}_{\mathbb{Q}}$, and any such rescaling is an algebra isomorphism. We denote the induced isomorphisms of $\mathrm{Sym}_{\mathbb{Q}}^{\otimes I}$ and $\mathrm{Sym}_{\mathbb{Q}}^{\otimes I} \otimes \mathrm{Sym}_{\mathbb{Q}}^{\otimes I}$ again by Ω.

LEMMA 3.3. *We have* $\Omega(h_n) = e_n$ *and* $\Omega(e_n) = h_n$ *for all* $n \in \mathbb{N}$.

PROOF. The proof follows directly from (3.2). □

LEMMA 3.4. *The map* Ω *preserves any bilinear form* $\langle -,- \rangle\colon H^+ \times H^- \longrightarrow \Bbbk$ *for which the* $p_{\boldsymbol{\lambda}}$ *are orthogonal.*

PROOF. We have

$$\Omega(p_{\boldsymbol{\lambda}}) = \Omega\left(\prod_{i\in I}\prod_{k=1}^{\ell(\lambda^i)} p_{\lambda^i_k}\right) = \prod_{i\in I}\prod_{k=1}^{\ell(\lambda^i)} (-1)^{\lambda^i_k - 1} p_{\lambda^i_k} = (-1)^{|\boldsymbol{\lambda}|-\ell(\boldsymbol{\lambda})} p_{\boldsymbol{\lambda}}.$$

Thus, since $\langle p_{\boldsymbol{\lambda}}, p_{\boldsymbol{\mu}}\rangle \in \delta_{\boldsymbol{\lambda}\boldsymbol{\mu}}\mathbb{Q}(q)$,

$$\langle \Omega(p_{\boldsymbol{\lambda}}), \Omega(p_{\boldsymbol{\mu}})\rangle = (-1)^{2(|\boldsymbol{\lambda}|-\ell(\boldsymbol{\lambda}))}\langle p_{\boldsymbol{\lambda}}, p_{\boldsymbol{\mu}}\rangle = \langle p_{\boldsymbol{\lambda}}, p_{\boldsymbol{\mu}}\rangle.$$ □

LEMMA 3.5. *The map* Ω *is compatible with the coproduct; i.e.*

$$\Delta(\Omega(a)) = \Omega(\Delta(a)), \quad a \in \mathrm{Sym}_{\mathbb{Q}}^{\otimes I}.$$

PROOF. Since Ω and Δ are both algebra homomorphisms, we need only prove the claim on a generating set, and we choose the h_n. Since the coproduct of a bialgebra (in this case a Hopf algebra) is an algebra homomorphism, we have

$$\Omega\left(\Delta(h_n)\right) \stackrel{(3.4)}{=} \Omega\left(\sum_{r=0}^{n} h_r \otimes h_{n-r}\right) = \sum_{r=0}^{n} e_r \otimes e_{n-r} = \Delta(e_n) = \Delta\left(\Omega(h_n)\right).$$ □

PROPOSITION 3.6. *The map* Ω *is an algebra endomorphism of the Heisenberg double* $\mathfrak{h}(H^+, H^-)$, *i.e.* $\Omega(a\#b)\Omega(c\#d) = \Omega((a\#b)(c\#d))$ *for all* $a,b,c,d \in \mathrm{Sym}_{\mathbb{Q}}$.

PROOF. Since $\{p_{\boldsymbol{\lambda}}\}_{\boldsymbol{\lambda}\in\mathcal{P}^I}$ is a basis for $\mathrm{Sym}_{\mathbb{Q}}$, a basis for $\mathfrak{h}(H^+, H^-)$ is $\{p_{\boldsymbol{\lambda}}\#p_{\boldsymbol{\mu}}\}_{\boldsymbol{\lambda},\boldsymbol{\mu}\in\mathcal{P}^I}$, and we only need to prove the proposition for the elements of the latter. We have

$$\begin{aligned}
\Omega((p_{\boldsymbol{\lambda}}\#p_{\boldsymbol{\mu}})(p_{\boldsymbol{\nu}}\#p_{\boldsymbol{\omega}})) &= \Omega\left(\sum_{(p_{\boldsymbol{\mu}}),(p_{\boldsymbol{\nu}})} \langle (p_{\boldsymbol{\mu}})_1, (p_{\boldsymbol{\nu}})_2\rangle p_{\boldsymbol{\lambda}}(p_{\boldsymbol{\nu}})_1 \# p_{\boldsymbol{\omega}}(p_{\boldsymbol{\mu}})_2\right)\\
&= \sum_{(p_{\boldsymbol{\mu}}),(p_{\boldsymbol{\nu}})} \langle \Omega((p_{\boldsymbol{\mu}})_1), \Omega((p_{\boldsymbol{\nu}})_2)\rangle \Omega(p_{\boldsymbol{\lambda}})\Omega((p_{\boldsymbol{\nu}})_1) \# \Omega(p_{\boldsymbol{\omega}})\Omega((p_{\boldsymbol{\mu}})_2)\\
&= \sum_{(p_{\boldsymbol{\mu}}),(p_{\boldsymbol{\nu}})} \langle (\Omega(p_{\boldsymbol{\mu}}))_1, (\Omega(p_{\boldsymbol{\nu}}))_2\rangle \Omega(p_{\boldsymbol{\lambda}})(\Omega(p_{\boldsymbol{\nu}}))_1 \# \Omega(p_{\boldsymbol{\omega}})(\Omega(p_{\boldsymbol{\mu}}))_2\\
&= \Omega(p_{\boldsymbol{\lambda}}\#p_{\boldsymbol{\mu}})\Omega(p_{\boldsymbol{\nu}}\#p_{\boldsymbol{\omega}}),
\end{aligned}$$

where we have used Lemma 3.4 in the second equality and Lemma 3.5 in the third equality. □

The following generating functions for the complete symmetric functions, the elementary symmetric functions, and the power sums can be found in [**Mac95**, p. 19–23]:

$$E(t) = \sum_{r\geq 0} e_r t^r = \prod_{i\geq 1}(1 + x_i t), \tag{3.8a}$$

$$H(t) = \sum_{r\geq 0} h_r t^r = \prod_{i\geq 1}(1 - x_i t)^{-1}, \tag{3.8b}$$

$$P(t) = \sum_{r\geq 1} p_r t^{r-1} = \sum_{i\geq 1} \frac{\mathrm{d}}{\mathrm{d}t} \log\left(\frac{1}{1 - x_i t}\right). \tag{3.8c}$$

LEMMA 3.7. *We have*

$$H(t) = \exp\left(\sum_{r\geq 1} p_r \frac{t^r}{r}\right). \tag{3.9}$$

PROOF. We have, by (3.8b) and (3.8c),

$$\int P(t)\mathrm{d}t = \log\left(\prod_{i\geq 1}\frac{1}{1 - x_i t}\right) = \log(H(t)),$$

$$H(t) = \exp\left(\int P(t)\mathrm{d}t\right) = \exp\left(\sum_{r\geq 1} p_r \frac{t^r}{r}\right).$$ □

LEMMA 3.8. *We have*

$$E(t) = \exp\left(-\sum_{r\geq 1} p_r \frac{(-t)^r}{r}\right). \tag{3.10}$$

PROOF. The proof follows from the fact that $E(-t)H(t) = 1$, and Lemma 3.7. □

4. Symmetric and exterior algebras

To properly describe our approach to the presentation of quantum lattice Heisenberg algebras, we first introduce some ingredients used in the construction. For the remainder of the paper, we let q denote an indeterminate.

DEFINITION 4.1 (Graded dimension). Let $V = \bigoplus_{n\in\mathbb{Z}} V_n$ be a $\mathbb{Z}$-graded vector space over $\Bbbk$. The *graded dimension* of V is

$$\mathrm{grdim}(V) = \sum_{n\in\mathbb{Z}} \dim(V_n) q^n \in \mathbb{N}[q, q^{-1}].$$

DEFINITION 4.2 (Symmetric algebra). Let W be a vector space over $\Bbbk$. The *symmetric algebra* of W is the quotient algebra of the tensor algebra of W by the ideal $\mathcal{I}$ generated by the elements $a \otimes b - b \otimes a$, with $a, b \in W \subseteq \mathcal{T}(W)$. This is

$$S(W) = \frac{\bigoplus_{k\geq 0} \mathcal{T}^k(W)}{\langle a \otimes b - b \otimes a\rangle}.$$

The *k-th symmetric power* of W, $S^k(W)$, is the image of $\mathcal{T}^k(W)$ under the quotient map $\mathcal{T}(W) \longrightarrow \mathcal{T}(W)/\mathcal{I}$.

The generating function for the graded dimension of the symmetric powers of a $\mathbb{Z}$-graded vector space V is

$$\sum_{k\in\mathbb{N}} t^k \operatorname{grdim}\left(S^k(V)\right) = \prod_{n\in\mathbb{Z}} \left(\frac{1}{1-q^n t}\right)^{\dim(V_n)}. \tag{4.1}$$

DEFINITION 4.3 (Exterior algebra). Let W be a vector space over $\Bbbk$. The *exterior algebra* or *antisymmetric algebra* of W is the quotient algebra of the tensor algebra of W by the ideal $\mathcal{I}$ generated by the elements $a\otimes a$, with $a \in W \subseteq \mathcal{T}(W)$. This is

$$\Lambda(W) = \frac{\bigoplus_{k\geq 0} \mathcal{T}^k(W)}{\langle a\otimes a\rangle}.$$

The *k-th exterior power* of W, $\Lambda^k(W)$, is the image of $\mathcal{T}^k(W)$ under the quotient map $\mathcal{T}(W) \longrightarrow \mathcal{T}(W)/\mathcal{I}$.

The generating function for the graded dimension of the exterior powers of a $\mathbb{Z}$-graded vector space V is

$$\sum_{k\in\mathbb{N}} t^k \operatorname{grdim}\left(\Lambda^k(V)\right) = \prod_{n\in\mathbb{Z}} (1+q^n t)^{\dim(V_n)}. \tag{4.2}$$

Since our main theorem will involve graded dimensions of symmetric and exterior powers, it is useful to have explicit expressions for these.

Consider the set of functions from $\mathbb{Z}$ to $\mathbb{N}$, $\mathbb{N}^{\mathbb{Z}}$. Let bold Latin characters denote functions $\mathbf{m} \in \mathbb{N}^{\mathbb{Z}}$ with $\mathbf{m}(n) = 0$ for all but finitely many $n \in \mathbb{Z}$. We then set, for all such $\mathbf{m} \in \mathbb{N}^{\mathbb{Z}}$, $|\mathbf{m}| = \sum_{n\in\mathbb{Z}} \mathbf{m}(n)$ and $\|\mathbf{m}\| = \sum_{n\in\mathbb{Z}} n\mathbf{m}(n)$, as well as

$$s^{\mathbf{m}}V = \prod_{n\in\mathbb{Z}} \binom{\dim(V_n)+\mathbf{m}(n)-1}{\mathbf{m}(n)}, \qquad \ell^{\mathbf{m}}V = \prod_{n\in\mathbb{Z}} \binom{\dim(V_n)}{\mathbf{m}(n)},$$

with V a $\mathbb{Z}$-graded vector space of finite dimension.

PROPOSITION 4.4. *Suppose V is a finite-dimensional $\mathbb{Z}$-graded vector space. Then*

$$\operatorname{grdim}(S^k(V)) = \sum_{|\mathbf{m}|=k} q^{\|\mathbf{m}\|} s^{\mathbf{m}}V \quad \text{and} \quad \operatorname{grdim}(\Lambda^k(V)) = \sum_{|\mathbf{m}|=k} q^{\|\mathbf{m}\|} \ell^{\mathbf{m}}V. \tag{4.3}$$

PROOF. The Taylor series of $(1-q^n t)^{-\beta}$ about $t=0$, with $n\in\mathbb{Z}$, $\beta\in\mathbb{N}$, is

$$(1-q^n t)^{-\beta} = \sum_{m\geq 0} t^m q^{nm} \binom{\beta+m-1}{m}. \tag{4.4}$$

Substitution of (4.4) into (4.1) yields

$$\sum_{k\in\mathbb{N}} t^k \operatorname{grdim}(S^k(V)) = \prod_{n\in\mathbb{Z}} \sum_{m_n\in\mathbb{N}} t^{m_n} q^{nm_n} \binom{\dim(V_n)+m_n-1}{m_n} = \sum_{\mathbf{m}\in\mathbb{N}^{\mathbb{Z}}} t^{|\mathbf{m}|} q^{\|\mathbf{m}\|} s^{\mathbf{m}}V.$$

Comparing coefficients of powers of t gives the first equation in (4.3). The proof of the second is analogous. $\square$

5. Presentations of quantum lattice Heisenberg algebras

In this section we present our main result on presentations of quantum lattice Heisenberg algebras. For $n \in \mathbb{N}$, we let $[n]$ denote the quantum integer

$$[n] = \frac{q^{-n} - q^n}{q^{-1} - q} = q^{-n+1} + q^{-n+3} + \cdots + q^{n-3} + q^{n-1} \in \mathbb{N}[q, q^{-1}].$$

Note that if we set $q = 1$ in the last expression above, we obtain $[n] = n$. We define $[-n] = [n]$ for $n \in \mathbb{N}_{>0}$.

Let us consider the pairing

$$\langle p_{n,i}, p_{m,j} \rangle = \delta_{nm} [n \langle i, j \rangle] \frac{n}{[n]}, \tag{5.1}$$

which, by Proposition 3.1, extends to a unique Hopf pairing $\langle -, - \rangle \colon H^+ \times H^- \to \mathbb{Z}[q, q^{-1}]$.

DEFINITION 5.1 (Quantum lattice Heisenberg algebra). Let L be the free $\mathbb{Z}$-module on the the set $\{v_i\}_{i \in I}$. Let $\langle -, - \rangle_L \colon L \times L \longrightarrow \mathbb{Z}$ be a symmetric bilinear form. The *quantum lattice Heisenberg algebra* $\mathfrak{h}_q^L$ associated to L is the unital $\mathbb{Z}$-algebra generated by $q_{i,n}$, $i \in I$, $n \in \mathbb{Z} \setminus \{0\}$, and commutation relations

$$q_{m,i} q_{n,j} = q_{n,j} q_{m,i} + \delta_{m,-n} [n \langle v_i, v_j \rangle_L] \frac{n}{[n]}, \quad i, j \in I, \quad n, m \in \mathbb{Z} \setminus \{0\}.$$

LEMMA 5.2. *With the pairing defined in* (5.1), *the Heisenberg double* $\mathfrak{h}(H^+, H^-)$ *is isomorphic to the quantum lattice Heisenberg algebra via the map*

$$p^-_{m,i} \mapsto q_{-m,i}, \quad p^+_{m,i} \mapsto q_{m,i}, \quad m \in \mathbb{N}_{>0}, \ i \in I.$$

In particular,

$$p^-_{m,i} p^+_{n,j} = p^+_{n,j} p^-_{m,i} + \delta_{nm} [n \langle i, j \rangle] \frac{n}{[n]}, \quad n, m \in \mathbb{N}_{>0}, \ i, j \in I. \tag{5.2}$$

PROOF. We have, recalling (2.2) and (3.1),

$$\begin{aligned}
p^-_{m,i} p^+_{n,j} &= (1 \# p_{m,i})(p_{n,j} \# 1), \\
&= \sum_{\substack{(p_{m,i}) \\ (p_{n,j})}} \langle (p_{m,i})_1, (p_{n,j})_2 \rangle (p_{n,j})_1 \# (p_{m,i})_2, \\
&= \langle p_{m,i}, 1 \rangle p_{n,j} \# 1 + \langle 1, p_{n,j} \rangle 1 \# p_{m,i} + \langle 1, 1 \rangle p_{n,j} \# p_{m,i} + \langle p_{m,i}, p_{n,j} \rangle 1 \# 1, \\
&\overset{(5.1)}{=} p^+_{n,j} p^-_{m,i} + p_{n,j} \# \delta_{m0} + \delta_{0n} \# p_{m,i} + \delta_{nm} [n \langle i, j \rangle] \frac{n}{[n]}, \\
&= p^+_{n,j} p^-_{m,i} + \delta_{nm} [n \langle i, j \rangle] \frac{n}{[n]}.
\end{aligned}$$

□

The following two theorems give presentations of the quantum lattice Heisenberg algebra in terms of the complete and elementary symmetric functions.

THEOREM 5.3. *The Heisenberg double* $\mathfrak{h}(H^+, H^-)$ *is generated by the complete symmetric functions* $\{h^+_n, h^-_n\}_{n \in \mathbb{N}}$, *with relations*

$$h^-_{n,j} h^+_{m,i} = \sum_{r=0}^{\min(m,n)} \operatorname{grdim}(S^r(V)) h^+_{m-r,i} h^-_{n-r,j} \quad \text{if } \langle i, j \rangle \in \mathbb{Z}_{\geq 0},$$

$$h^-_{n,j}h^+_{m,i} = \sum_{r=0}^{\min(m,n)} \operatorname{grdim}(\Lambda^r(V))h^+_{m-r,i}h^-_{n-r,j} \quad \textit{if } \langle i,j\rangle \in \mathbb{Z}_{<0},$$

where V is a $\mathbb{Z}$-graded vector space such that $\operatorname{grdim}(V) = [\langle i,j\rangle]$.

PROOF. For notational convenience, let $\chi := \langle i,j\rangle$. First set

$$A(\zeta) := \sum_{k\geq 1} p^+_{k,i}\frac{\zeta^k}{k}, \qquad B(\omega) := \sum_{k\geq 1} p^-_{k,j}\frac{\omega^k}{k}.$$

Now, by Lemma 3.7, we have

$$\sum_{n,m\geq 0} h^-_{n,j}h^+_{m,i}\omega^n\zeta^m = \exp(B)\exp(A).$$

Let us calculate

$$\begin{aligned}
[B,A] &= \left(\sum_{k\geq 1} p^-_{k,j}\frac{\omega^k}{k}\right)\left(\sum_{r\geq 1} p^+_{r,i}\frac{\zeta^r}{r}\right) - \left(\sum_{r\geq 1} p^+_{r,i}\frac{\zeta^r}{r}\right)\left(\sum_{k\geq 1} p^-_{k,j}\frac{\omega^k}{k}\right)\\
&= \sum_{k,r\geq 1}(p^-_{k,j}p^+_{r,i} - p^+_{r,i}p^-_{k,j})\frac{\omega^k\zeta^r}{kr}\\
&\overset{(5.2)}{=} \sum_{k,r\geq 1} \delta_{kr}[k\chi]\frac{k}{[k]r}\omega^k\zeta^r = \sum_{k\geq 1}\frac{[k\chi]}{[k]k}(\omega\zeta)^k\\
&= \begin{cases} \sum_{k\geq 1} \frac{(\omega\zeta)^k}{k}\frac{q^{-k\chi}-q^{k\chi}}{q^{-k}-q^k} & \text{if } \chi\in\mathbb{Z}_{\geq 0},\\ \sum_{k\geq 1}\frac{(\omega\zeta)^k(-1)^{-k\chi+1}}{k}\frac{(-q)^{-k(-\chi)}-(-q)^{k(-\chi)}}{(-q)^{-k}-(-q)^k} & \text{if } \chi\in\mathbb{Z}_{<0}.\end{cases}
\end{aligned}$$

For any two operators A, B, it is routine to prove (see, for example, [**Nak99**, Lem. 9.43]) that as long as $[A,B]$ commutes with both A and B, then

$$\exp(B)\exp(A) = \exp([B,A])\exp(A)\exp(B). \tag{5.3}$$

We have

$$\begin{aligned}
[B,A] &= \begin{cases} \sum_{k\geq 1}\sum_{g=0}^{\chi-1}\frac{1}{k}(q^{\chi-(2g+1)}\omega\zeta)^k & \text{if } \chi\in\mathbb{Z}_{\geq 0},\\ -\sum_{k\geq 1}\sum_{g=0}^{(-\chi)-1}\frac{1}{k}(-q^{(-\chi)-(2g+1)}\omega\zeta)^k & \text{if } \chi\in\mathbb{Z}_{<0},\end{cases}\\
&= \begin{cases} \log\left(\prod_{g=0}^{\chi-1}(1-q^{\chi-(2g+1)}\omega\zeta)^{-1}\right) & \text{if } \chi\in\mathbb{Z}_{\geq 0},\\ \log\left(\prod_{g=0}^{-\chi-1}(1+q^{(-\chi)-(2g+1)}\omega\zeta)\right) & \text{if } \chi\in\mathbb{Z}_{<0}.\end{cases}
\end{aligned}$$

By (5.3), we have

$$\sum_{n,m\geq 0} h^-_{n,j}h^+_{m,i}\omega^n\zeta^m = \begin{cases}\sum_{s,t\geq 0} h^+_{t,i}h^-_{s,j}\omega^s\zeta^t\prod_{g=0}^{\chi-1}(1-q^{\chi-(2g+1)}\omega\zeta)^{-1} & \text{if } \chi\in\mathbb{Z}_{\geq 0},\\ \sum_{s,t\geq 0} h^+_{t,i}h^-_{s,j}\omega^s\zeta^t\prod_{g=0}^{-\chi-1}(1+q^{(-\chi)-(2g+1)}\omega\zeta) & \text{if } \chi\in\mathbb{Z}_{<0}.\end{cases}$$

Now, since $\operatorname{grdim}(V) = [\chi] = \sum_{g=0}^{\chi-1} q^{\chi-(2g+1)}$, we recall (4.1) and (4.2) and thus write

$$\sum_{n,m\geq 0} h^-_{n,j}h^+_{m,i}\omega^n\zeta^m = \begin{cases}\sum_{s,t\geq 0} h^+_{t,i}h^-_{s,j}\omega^s\zeta^t\sum_{k\geq 0}(\omega\zeta)^k\operatorname{grdim}\left(S^k(V)\right) & \text{if } \chi\in\mathbb{Z}_{\geq 0},\\ \sum_{s,t\geq 0} h^+_{t,i}h^-_{s,j}\omega^s\zeta^t\sum_{k\geq 0}(\omega\zeta)^k\operatorname{grdim}\left(\Lambda^k(V)\right) & \text{if } \chi\in\mathbb{Z}_{<0}.\end{cases}$$

Upon comparison of the coefficients of $\omega\zeta$, we finally arrive at

$$h^-_{n,j}h^+_{m,i} = \begin{cases} \sum_{r=0}^{\min(m,n)} h^+_{m-r,i}h^-_{n-r,j}\,\mathrm{grdim}(S^r(V)) & \text{if } \chi \in \mathbb{Z}_{\geq 0}, \\ \sum_{r=0}^{\min(m,n)} h^+_{m-r,i}h^-_{n-r,j}\,\mathrm{grdim}(\Lambda^r(V)) & \text{if } \chi \in \mathbb{Z}_{<0}. \end{cases}$$ □

THEOREM 5.4. *The Heisenberg double $\mathfrak{h}(H^+, H^-)$ is generated by the complete and elementary symmetric functions $\{h^+_n, e^-_n\}_{n\in\mathbb{N}}$, with relations*

$$e^-_{n,j}h^+_{m,i} = \sum_{r=0}^{\min(m,n)} \mathrm{grdim}(\Lambda^r(V))h^+_{m-r,i}e^-_{n-r,j} \quad \textit{if } \langle i,j\rangle \in \mathbb{Z}_{\geq 0},$$

$$e^-_{n,j}h^+_{m,i} = \sum_{r=0}^{\min(m,n)} \mathrm{grdim}(S^r(V))h^+_{m-r,i}e^-_{n-r,j} \quad \textit{if } \langle i,j\rangle \in \mathbb{Z}_{<0},$$

where V is a $\mathbb{Z}$-graded vector space such that $\mathrm{grdim}(V) = [\langle i,j\rangle]$.

PROOF. The proof is the same as that of Theorem 5.3, but with

$$A(\zeta) := \sum_{k\geq 1} p^+_{k,i}\frac{\zeta^k}{k}, \qquad B(\omega) := -\sum_{k\geq 1} p^-_{k,j}\frac{(-\omega)^k}{k}.$$ □

REMARK 5.5. We can obtain presentations in terms of the generating sets $\{e^+_n, e^-_n\}_{n\in\mathbb{N}}$ and $\{e^+_n, h^-_n\}_{n\in\mathbb{N}}$ immediately by considering the algebra endomorphism Ω of Proposition 3.6.

REMARK 5.6. Since the presentations of Theorems 5.3 and 5.4 are in terms of the h_n and e_n which generate Sym over $\mathbb{Z}$, they in fact yield integral forms of the quantum lattice Heisenberg algebra. It is for this reason that these generating sets are important in categorification. In fact, the presentations given in Theorems 5.3 and 5.4 correspond to those of [**RS**, Prop. 5.5]. However, since the power sums do not seem to have a natural interpretation in the categorical setting of [**RS**], the connection to the presentation in terms of power sums is not given there.

Integral presentations of quantum lattice Heisenberg algebras have been considered previously. For example, in [**CL12**, §2.2.1], the authors considered such presentations, but with the restriciton that $(\langle i,j\rangle)_{i,j\in I}$ was an extended Cartan matrix of type ADE. In their paper, they define elements $a_i(n)$, $p_i^{(m)}$, and $q_j^{(m)}$ of the lattice Heisenberg algebra, related by the equations

$$[a_i(m), a_j(n)] = \delta_{m,-n}[n\langle i,j\rangle]\frac{[n]}{n},$$

$$\sum_{n\geq 0} p_i^{(n)}z^n = \exp\left(\sum_{m\geq 1}\frac{a_i(-m)}{[m]}\right), \quad \sum_{n\geq 0} q_i^{(n)}z^n = \exp\left(\sum_{m\geq 1}\frac{a_i(m)}{[m]}\right).$$

The connection to the notation of the current paper is given by

$$p_i^{(n)} = h^+_{n,i}, \quad q_i^{(n)} = h^-_{n,i}, \quad a_i(m)\frac{m}{[m]} = \begin{cases} p^-_{n,i} & \text{if } m \in \mathbb{Z}_{\geq 0}, \\ p^+_{m,i} & \text{if } m \in \mathbb{Z}_{<0}. \end{cases}$$

Thus, the results described in [**CL12**, §2.2] are specific cases of Theorem 5.3 and Theorem 5.4. However, because of the restricted setting, [**CL12**] treats only the case where $\langle i,j\rangle$ is equal to 0, 1, or 2, instead of an arbitrary integer as in the current paper. In addition, the authors of [**CL12**] do not consider the algebra from the Heisenberg double point of view, and so their proofs are different.

Setting $q = 1$ corresponds to the vector space V appearing in Theorems 5.3 and 5.4 being concentrated in degree zero. Under this specialization, Theorem 5.3 recovers [**Kru**, Lem. 2.1]. Note however, that the Heisenberg double approach of the current paper simplifies the proofs considerably. In particular, the second part of the proof of [**Kru**, Lem. 1.2], where the author shows that no further relations exists between the generators, is an immediate consequence of the definition of the Heisenberg double and Lemma 5.2.

The quantum lattice Heisenberg algebra and its nonquantum $q = 1$ specialization are treated from the Heisenberg double point of view in [**RS15b**, §7 and §8], respectively. However, as in [**CL12**], explicit presentations are only given when $\langle i, j \rangle$ is equal to 0, 1, or 2.

Acknowledgements

The author would like to thank the immensely generous and patient supervision of Prof. Alistair Savage, facilitated by the great opportunity to conduct a research internship at the University of Ottawa underwritten by the MITACS Globalink program.

If one can be allowed some sentimentality on their first paper, the author would also like to thank his family and friends: *ustedes son mi vida entera.*

References

[CL12] S. Cautis and A. Licata, *Heisenberg categorification and Hilbert schemes*, Duke Math. J. **161** (2012), no. 13, 2469–2547, DOI 10.1215/00127094-1812726. MR2988902

[Kho14] M. Khovanov, *Heisenberg algebra and a graphical calculus*, Fund. Math. **225** (2014), no. 1, 169–210, DOI 10.4064/fm225-1-8. MR3205569

[Kru] A. Krug. Symmetric quotient stacks and Heisenberg actions. `arXiv:1501.07253v1`.

[LS12] A. Licata and A. Savage, *A survey of Heisenberg categorification via graphical calculus*, Bull. Inst. Math. Acad. Sin. (N.S.) **7** (2012), no. 2, 291–321. MR3024895

[LS13] A. Licata and A. Savage, *Hecke algebras, finite general linear groups, and Heisenberg categorification*, Quantum Topol. **4** (2013), no. 2, 125–185, DOI 10.4171/QT/37. MR3032820

[Mac95] I. G. Macdonald, *Symmetric functions and Hall polynomials*, 2nd ed., Oxford Mathematical Monographs, The Clarendon Press, Oxford University Press, New York, 1995. With contributions by A. Zelevinsky; Oxford Science Publications. MR1354144

[Nak99] H. Nakajima, *Lectures on Hilbert schemes of points on surfaces*, University Lecture Series, vol. 18, American Mathematical Society, Providence, RI, 1999. MR1711344

[RS] D. Rosso and A. Savage. A general approach to Heisenberg categorification via wreath product algebras. `arXiv:1507.06298`.

[RS15a] D. Rosso and A. Savage, *Towers of graded superalgebras categorify the twisted Heisenberg double*, J. Pure Appl. Algebra **219** (2015), no. 11, 5040–5067, DOI 10.1016/j.jpaa.2015.03.016. MR3351572

[RS15b] D. Rosso and A. Savage, *Twisted Heisenberg doubles*, Comm. Math. Phys. **337** (2015), no. 3, 1053–1076, DOI 10.1007/s00220-015-2330-z. MR3339171

[SV12] P. Shan and E. Vasserot, *Heisenberg algebras and rational double affine Hecke algebras*, J. Amer. Math. Soc. **25** (2012), no. 4, 959–1031, DOI 10.1090/S0894-0347-2012-00738-3. MR2947944

[SY15] A. Savage and O. Yacobi, *Categorification and Heisenberg doubles arising from towers of algebras*, J. Combin. Theory Ser. A **129** (2015), 19–56, DOI 10.1016/j.jcta.2014.09.002. MR3275113

[Zab15] Mike Zabrocki. Introduction to symmetric functions. Available at: https://www.yumpu.com/en/document/view/19079932/introduction-to-symmetric-functions-mike-zabrocki, accessed May 2015

Department of Mathematics and Statistics, University of Ottawa
E-mail address: diego.berdejasuarez@merton.ox.ac.uk

Contemporary Mathematics
Volume **683**, 2017
http://dx.doi.org/10.1090/conm/683/13721

Categorification at prime roots of unity and hopfological finiteness

You Qi and Joshua Sussan

Dedicated to Christian Blanchet on the occasion of his sixtieth birthday

ABSTRACT. We survey some recent results in hopfological algebra and the program of categorification at prime roots of unity. A categorical Jones-Wenzl projector at prime roots of unity is studied, and it is shown that this projector is hopfologically finite in special characteristics, while generically it is infinite.

1. Introduction

The Witten-Reshetikhin-Turaev (WRT) invariants [**Wit89, RT90**] of links and tangles can be constructed from quantum groups at generic values of q. For the purposes of this note, we will only focus on the $\mathfrak{sl}_2$-case. When q equals a root of unity, these invariants specialized at q naturally arise, from a quantum physical point of view, as invariants assigned to certain decorated 3-spheres using the 3d-WRT topological quantum field theory (TQFT) Z_n. Here the 3d TQFT Z_n, parameterized by the order of q (which is related to the level of the TQFT), admits representation theoretical explanations from quantum groups at roots of unity [**RT90**].

The goal of the categorification program envisioned by Crane and Frenkel [**CF94**] was to lift the WRT 3-manifold invariant $Z_n(M^3)$, whose values live in the ring of cyclotomic integers $\mathbb{Z}[\zeta_n]$ (see, for instance, [**BCL14**]), to a homological invariant. If this hypothetical homological structure were truly a functorial theory, then one should obtain invariants of 4-manifolds with boundaries, which are viewed as cobordisms between 3-manifolds. This would in turn lift Z_n to a 4-dimensional theory, as conjectured in [**CF94**].

Strong evidence of the plausibility of Crane and Frenkel's proposal is provided by Khovanov homology [**Kho00**], which assigns a functorial homological invariant to any (framed) tangle. Since this ground-breaking work, there has been significant progress towards categorifying link invariants at generic q-values. We refer

---

2010 *Mathematics Subject Classification.* Primary 81R50; Secondary 16E20, 16E35.

Key words and phrases. Categorification, hopfological algebra, prime roots of unity.

The authors are very grateful to Mikhail Khovanov for his support and encouragement, as well as sharing with us his insights on the subject. The authors would also like to thank Catharina Stroppel for helpful discussions about categorifying Jones-Wenzl projectors.

The second author was supported by NSF grant DMS-1407394, PSC-CUNY Award 67144-00 45, and an Alfred P. Sloan Foundation CUNY Junior Faculty Award.

the reader to the work of Webster [**Web15**] (and the references therein) for more historical accounts, as well as a striking general construction of homological WRT tangle invariants.

In 2004, Khovanov proposed the first steps towards categorification at roots of unity [**Kho16**]. He introduced the subject of *hopfological algebra*, which combines theory of Hopf algebras with homological algebra. The key observation of [**Kho16**] is that, to categorify the ring of cyclotomic integers at a prime root of unity, one could utilize Mayer's homotopy category of p-complexes $\Bbbk[\partial]/(\partial^p)\text{–}\underline{\mathrm{gmod}}$ [**May42a**, **May42b**] over a field of characteristic $p > 0$:

$$K_0(\Bbbk[\partial]/(\partial^p)-\underline{\mathrm{gmod}}) \cong \mathbb{Z}[\zeta_p]. \tag{1.1}$$

Mayer's p-complexes can be understood as a special case of hopfological algebra, in the sense that it is the hopfological theory of the ground field $\Bbbk$ associated with the graded Hopf algebra $H = \Bbbk[\partial]/(\partial^p)$, where $\mathrm{char}(\Bbbk) = p$ and $\deg(\partial) = 1$. A characteristic-zero analogue of Mayer's construction, known as n-complexes, has been considered by Kapranov [**Kap96**] and Sakaria [**Sar95**]. This subject is, roughly speaking, a hopfological theory attached to the Taft algebra H_n at an nth root of unity. However, due to the lack of a braiding functor on $H_n-\mathrm{mod}$, it is relatively difficult to implement this theory in the categorification program. It remains an interesting open problem to devise a monoidal categorification of a general cyclotomic integer ring[1], and apply such a construction in categorification of the WRT 3d TQFT.

In order to apply the "categorical cyclotomic integers" $\Bbbk[\partial]/(\partial^p)\text{–}\underline{\mathrm{gmod}}$ to categorification, one needs a good supply of algebra objects in the monoidal category $\Bbbk[\partial]/(\partial^p)-\underline{\mathrm{gmod}}$. Such algebra objects arise naturally from p-differential graded (p-DG) algebras, an analogue of the usual DG-algebra over graded (super)vector spaces. Some basic technical machinery of hopfological algebra, and in particular, p-DG algebra, has been developed in [**Qi14**]. The theory parallels that of the usual DG theory, and in fact the usual DG theory is a special case of hopfological algebra for the Hopf superalgebra $\Bbbk[d]/(d^2)$. For instance, one has the notion of the abelian category[2] of p-DG modules over a p-DG algebra A, its homotopy category of p-DG modules $\mathcal{H}(A)$, and the derived category $\mathcal{D}(A)$ obtained from $\mathcal{H}(A)$ under localization at quasi-isomorphisms. Both categories $\mathcal{H}(A)$ and $\mathcal{D}(A)$ afford categorical module structures over $H\text{–}\underline{\mathrm{gmod}}$ under tensor product by p-complexes. Taking Grothendieck groups in the appropriate sense, the decategorification realizes $K_0(\mathcal{D}(A))$ (or $K_0(\mathcal{H}(A))$, but the former is more relevant in our story) as modules over $\mathbb{Z}[\zeta_p]$:

$$\begin{array}{ccc}
\Bbbk[\partial]/(\partial^p)-\underline{\mathrm{gmod}} \times \mathcal{D}(A) & \xrightarrow{\quad\otimes\quad} & \mathcal{D}(A) \\
\Big\Downarrow {\scriptstyle K_0} & & \Big\Downarrow {\scriptstyle K_0} \\
K_0(\Bbbk[\partial]/(\partial^p)-\underline{\mathrm{gmod}}) \times K_0(\mathcal{D}(A)) & \xrightarrow{\quad\times\quad} & K_0(\mathcal{D}(A)).
\end{array} \tag{1.2}$$

[1]See the recent preprint [**Mir15**] for an interesting new case.
[2]These notions will be recalled in Section 3.

One can also easily check that the usual associativity and distributivity axioms of a module over $\mathbb{Z}[\zeta_p]$ are lifted to the corresponding properties of direct sums and tensor products in the triangulated categories.

In this note we will sketch out what interesting p-DG algebras have been investigated in earlier works. A natural way to find nontrivial p-DG algebras that may be relevant to quantum topology is to combine hopfological algebra of p-differentials with the machineries that have successfully lifted integral quantum structures (for generic q). It appears to us that, so far, building on the beautiful works of [**KL09**, **KL11**, **KL10**, **Lau10**, **Rou08**, **Web15**] etc., working along this direction seems promising. A summary of what is already known will be carried out in the next few sections, and at the end we will point out one new interesting feature of a hopfological finiteness property and its dependence on the characteristic of the ground field.

In Section 2 we review quantum $\mathfrak{sl}_2$ at roots of unity and related structures. The results are by now classical, and are included to fix notation for the rest of the paper.

Starting from Section 3, we recall the definition of a p-DG algebra and some basic hopfological properties. Following [**QS13**], we illustrate the definitions with a simple example $A_n^!$, whose derived category carries a categorical braid group action. Via this example, we hope the reader is convinced that hopfological algebra is indeed connected to quantum topology at roots of unity.

Then we proceed to more representation theoretical aspects of the story. In Section 4, we review the p-DG nilHecke algebra defined by Khovanov and Qi [**KQ15**] used to categorify the lower half of quantum $\mathfrak{sl}_2$ at prime roots of unity. This is further extended by Elias and Qi [**EQ16a**] to categorify the full small quantum group, following Lauda's [**Lau10**] work on the generic case. A categorification of the Beilinson-Lusztig-MacPherson (BLM) form of quantum $\mathfrak{sl}_2$ at prime roots of unity has been constructed in [**EQ15**] by equipping the thick calculus of Khovanov-Lauda-Mackaay-Stošić [**KLMS12**] with a compatible p-differential.

A categorified Jones-Wenzl projector is constructed in Section 5. When equipped with p-differentials, we find some interesting new phenomenon. In certain characteristics, this functor is quasi-isomorphic to a finite complex, although for almost all primes it is an infinite complex unbounded in one direction. This suggests to us that the homology of the colored unknot may be finite only in special prime characteristics. We hope to investigate this phenomenon more carefully in upcoming works.

Finally, we should mention a recent perspective on variants of WRT TQFTs. The original WRT TQFTs build upon semisimplification of certain tensor subcategories of representation theory of quantum groups at roots of unity. A very recent construction of Blanchet, Costantino, Geer and Patureau-Mirand [**BCGPM14**] utilizes only projective objects in the representation category to define a non semisimple TQFT. It is an exciting problem to categorify their story at a prime root of unity.

2. The small quantum group

2.1. The small quantum group. Let ζ_{2l} be a primitive $2l$th root of unity. The quantum group $u_{\mathbb{Q}[\zeta_{2l}]}(\mathfrak{sl}_2)$, which we will denote simply by $u_{\mathbb{Q}[\zeta_{2l}]}$, is the $\mathbb{Q}[\zeta_{2l}]$-algebra generated by $E, F, K^{\pm 1}$ subject to relations:

(1) $KK^{-1} = K^{-1}K = 1$,
(2) $K^{\pm 1}E = \zeta_{2l}^{\pm 2}EK^{\pm 1}, \quad K^{\pm 1}F = \zeta_{2l}^{\mp 2}FK^{\pm 1}$,
(3) $EF - FE = \frac{K-K^{-1}}{\zeta_{2l}-\zeta_{2l}^{-1}}$,
(4) $E^l = F^l = 0$.

The quantum group is a Hopf algebra whose comultiplication map

$$\Delta\colon u_{\mathbb{Q}[\zeta_{2l}]} \longrightarrow u_{\mathbb{Q}[\zeta_{2l}]} \otimes_{\mathbb{Q}[\zeta_{2l}]} u_{\mathbb{Q}[\zeta_{2l}]}$$

is given on generators by

$$\Delta(E) = E\otimes 1 + K \otimes E, \qquad \Delta(F) = 1 \otimes F + F \otimes K^{-1}, \qquad \Delta(K^{\pm 1}) = K^{\pm 1} \otimes K^{\pm 1}.$$

For the purpose of categorification, it is more convenient to use the idempotented quantum group $\dot{u}_{\mathbb{Q}[\zeta_{2l}]}(\mathfrak{sl}_2)$, which we will denote simply by $\dot{u}_{\mathbb{Q}[\zeta_{2l}]}$. It is a non-unital $\mathbb{Q}[\zeta_{2l}]$-algebra generated by E, F and idempotents 1_λ ($\lambda \in \mathbb{Z}$), subject to relations:

(1) $1_\lambda 1_\mu = \delta_{\lambda,\mu}1_\lambda$,
(2) $E1_\lambda = 1_{\lambda+2}E, \quad F1_{\lambda+2} = 1_\lambda F$,
(3) $EF1_\lambda - FE1_\lambda = [\lambda]1_\lambda$,
(4) $E^l = F^l = 0$.

Here $[\lambda] = \sum_{i=0}^{|\lambda|-1} \zeta_{2l}^{1-|\lambda|+2i}$ is the quantum integer specialized at ζ_{2l}.

The quantum group $\dot{u}_{\mathbb{Q}[\zeta_{2l}]}$ has an integral lattice subalgebra which we now recall. For any integer $n \in \{0, 1, \ldots, l-1\}$, let $E^{(n)} = \frac{E^n}{[n]!}$, and $F^{(n)} = \frac{F^n}{[n]!}$. The elements $E^{(n)}$, $F^{(n)}$ $(0 \le n \le l-1)$, and 1_λ ($\lambda \in \mathbb{Z}$) generate an algebra over the ring of cyclotomic integers $\mathcal{O}_{2l} = \mathbb{Z}[\zeta_{2l}]$. Denote this integral form by $\dot{u}_{\zeta_{2l}}$.

Now let $l = p$ be prime. Introduce the auxiliary ring

$$\mathbb{O}_p = \mathbb{Z}[q]/(\Psi_p(q^2)), \tag{2.1}$$

where $\Psi_p(q)$ is the p-th cyclotomic polynomial. We can define in a similar fashion an integral form $u_{\mathbb{O}_p}$ and its dotted version $\dot{u}_{\mathbb{O}_p}$ for the small quantum $\mathfrak{sl}_2$ over $\mathbb{O}_p$. For more details see [**KQ15**, Section 3.3]. Let the lower half of $u_{\mathbb{O}_p}$ be the subalgebra generated by the $F^{(n)}$ $(0 \le n \le p-1)$ and denote it by $u_{\mathbb{O}_p}^-$. Likewise, write the upper half as $u_{\mathbb{O}_p}^+$.

Let V_n be the unique (up to isomorphism) irreducible module for $\dot{u}_{\mathbb{O}_p}$ of rank $n+1$. It has a basis $\{v_0, v_1, \ldots, v_n\}$ such that

$$F1_\lambda v_i = \delta_{\lambda,-n+2i}[n-i+1]v_{i-1} \qquad E1_\lambda v_i = \delta_{\lambda,-n+2i}[i+1]v_{i+1}. \tag{2.2}$$

On $V_1^{\otimes n}$ there is an action of $u_{\mathbb{O}_p}$ given via the comultiplication map Δ. There is also a commuting action of the braid group that factors through the Hecke algebra. We mostly consider the second lowest weight space which we denote by $V_1^{\otimes n}[-n+2]$. This is spanned by vectors $v_{i_1} \otimes \cdots \otimes v_{i_n}$ where $i_r \in \{0,1\}$ and exactly one i_r is 1. More generally $V_1^{\otimes n}[-n+2k]$ is spanned by vectors $v_{i_1} \otimes \cdots \otimes v_{i_n}$ where $i_r \in \{0,1\}$ and k of the i_r are 1. It is convenient to use another basis of $V_1^{\otimes n}[-n+2]$ which is spanned by vectors l_r where

$$l_r = \begin{cases} v_0^{\otimes(r-1)} \otimes (v_1 \otimes v_0 - qv_0 \otimes v_1) \otimes v_0^{\otimes(n-r-1)} & r \neq n \\ v_0^{\otimes(n-1)} \otimes v_1 & r = n. \end{cases}$$

2.2. The Temperley-Lieb algebra. The *Temperley-Lieb algebra* TL_n is the $\mathbb{O}_p$-algebra generated by elements U_i for $i = 1, \ldots, n-1$ subject to relations

(1) $U_i^2 = -(q + q^{-1})U_i$,
(2) $U_iU_j = U_jU_i$ for $|i-j| > 1$,
(3) $U_iU_jU_i = U_i$ for $|i-j| = 1$.

Define an operator $\tilde{U}$ on $V_1^{\otimes 2}$ by

$$\tilde{U}(v_0 \otimes v_0) = 0$$
$$\tilde{U}(v_1 \otimes v_1) = 0$$
$$\tilde{U}(v_0 \otimes v_1) = v_1 \otimes v_0 - qv_0 \otimes v_1$$
$$\tilde{U}(v_1 \otimes v_0) = -q^{-1}(v_1 \otimes v_0 - qv_0 \otimes v_1).$$

Then there is an action of the Temperley-Lieb algebra on $V_1^{\otimes n}$ where the generator $U_i \colon V_1^{\otimes n} \to V_1^{\otimes n}$ acts by

$$U_i = \mathrm{Id}^{\otimes(i-1)} \otimes \tilde{U} \otimes \mathrm{Id}^{\otimes(n-i-1)}.$$

In terms of the basis $\{l_1, \ldots, l_n\}$ the action of the Temperley-Lieb algebra is given by

$$U_i(l_j) = \begin{cases} -(q+q^{-1})l_i & \text{if } i = j \\ l_i & \text{if } |i-j| = 1 \\ 0 & \text{if } |i-j| > 1. \end{cases} \tag{2.3}$$

2.3. The braid group. The braid group B_n is generated by elements t_i for $i = 1, \ldots, n-1$ subject to the relations that

(1) $t_it_j = t_jt_i$ if $|i-j| > 1$,
(2) $t_it_{i+1}t_i = t_{i+1}t_it_{i+1}$ for $i = 1, \ldots, n-2$.

For $i = 1, \ldots, n-1$, it will be convenient to introduce the inverse of t_i and denote it by t_i'.

Define an operator $\tilde{t}$ on $V_1^{\otimes 2}$ by

$$\tilde{t}(v_0 \otimes v_0) = v_0 \otimes v_0$$
$$\tilde{t}(v_1 \otimes v_1) = v_1 \otimes v_1$$
$$\tilde{t}(v_0 \otimes v_1) = qv_1 \otimes v_0 + (1-q^2)v_0 \otimes v_1$$
$$\tilde{t}(v_1 \otimes v_0) = qv_0 \otimes v_1.$$

Then there is an action of the braid group generator $t_i \colon V_1^{\otimes n} \to V_1^{\otimes n}$ given by

$$t_i = \mathrm{Id}^{\otimes(i-1)} \otimes \tilde{t} \otimes \mathrm{Id}^{\otimes(n-i-1)}.$$

Define an operator $\tilde{t}'$ on $V_1^{\otimes 2}$ by

$$\tilde{t}'(v_0 \otimes v_0) = v_0 \otimes v_0$$
$$\tilde{t}'(v_1 \otimes v_1) = v_1 \otimes v_1$$
$$\tilde{t}'(v_0 \otimes v_1) = q^{-1}v_1 \otimes v_0$$
$$\tilde{t}'(v_1 \otimes v_0) = (1-q^{-2})v_1 \otimes v_0 + q^{-1}v_0 \otimes v_1.$$

Then there is an action of the braid group generator $t_i' \colon V_1^{\otimes n} \to V_1^{\otimes n}$ given by

$$t_i' = \mathrm{Id}^{\otimes(i-1)} \otimes \tilde{t}' \otimes \mathrm{Id}^{\otimes(n-i-1)}.$$

It is easy to see that these operators preserve each weight space of $V_1^{\otimes n}$ and it is straightforward to check that they do indeed satisfy the braid relations. The action of B_n on $V_1^{\otimes n}[-n+2]$ is the Burau representation.

In terms of the basis $\{l_1, \ldots, l_n\}$ the action of the braid group is given by

$$t_i(l_j) = \begin{cases} -q^2 l_i & \text{if } i=j \\ q l_i + l_j & \text{if } |i-j|=1 \\ l_j & \text{if } |i-j|>1 \end{cases} \tag{2.4}$$

$$t_i'(l_j) = \begin{cases} -q^{-2} l_i & \text{if } i=j \\ q^{-1} l_i + l_j & \text{if } |i-j|=1 \\ l_j & \text{if } |i-j|>1. \end{cases} \tag{2.5}$$

2.4. The Jones-Wenzl projector. The Jones-Wenzl projector is the unique linear map

$$p_n \colon V_1^{\otimes n} \to V_n \to V_1^{\otimes n}$$

which commutes with the action of the quantum group and is idempotent. There are many formulas for this map, usually given in terms of Temperley-Lieb generators. When restricting to weight spaces we could write this map in terms of generators of the quantum group

$$\frac{1}{\begin{bmatrix} n \\ k \end{bmatrix}} E^{(k)}F^{(k)} \colon V_1^{\otimes n}[-n+2k] \to V_1^{\otimes n}[-n+2k].$$

For full details see [**SS**].

PROPOSITION 2.1. *The operator $\frac{1}{\begin{bmatrix} n \\ k \end{bmatrix}} E^{(k)}F^{(k)} \colon V_1^{\otimes n}[-n+2k] \to V_1^{\otimes n}[-n+2k]$ is equal to the restriction of the Jones-Wenzl projector $p_n[-n+2k]$.*

PROOF. One checks that $\frac{1}{\begin{bmatrix} n \\ k \end{bmatrix}} E^{(k)}F^{(k)}$ is idempotent, commutes with E and F and is annihilated by the elements U_i of the Temperley-Lieb algebra. For more details see [**SS**]. □

3. A categorical braid group action at a prime root of unity

3.1. Elements of p-DG algebras. As a matter of notation for the rest of the paper, the undecorated tensor product symbol $\otimes$ will always denote tensor product over the ground field $\Bbbk$. We first recall some basic notions.

DEFINITION 3.1. Let $\Bbbk$ be a field of positive characteristic p. A p-DG algebra A over $\Bbbk$ is a $\mathbb{Z}$-graded $\Bbbk$-algebra equipped with a degree-two[3] endomorphism ∂_A, such that, for any elements $a, b \in A$, we have

$$\partial_A^p(a) = 0, \qquad \partial_A(ab) = \partial_A(a)b + a\partial_A(b).$$

Compared with the usual DG case, the lack of the usual sign in the second equation above is because of the fact that the Hopf algebra $\Bbbk[\partial]/(\partial^p)$ is a genuine Hopf algebra, not a Hopf super-algebra.

As in the DG case, one has the notion of left and right p-DG modules.

[3]In general one should define the degree of ∂_A to be one. We adopt this degree only to match earlier grading conventions in categorification. One may adjust the gradings of the algebras we consider so as to make the degree of ∂_A to be one, but we choose not to do so.

DEFINITION 3.2. Let (A, ∂_A) be a p-DG algebra. A left p-DG module (M, ∂_M) is a $\mathbb{Z}$-graded A-module endowed with a degree-two endomorphism ∂_M, such that, for any elements $a \in A$ and $m \in M$, we have

$$\partial_M^p(m) = 0, \qquad \partial_M(am) = \partial_A(a)m + a\partial_M(m).$$

Similarly, one has the notion of a right p-DG module.

It is readily checked that the category of left (right) p-DG modules, denoted $A_\partial-\mathrm{mod}$ ($A_\partial^{op}-\mathrm{mod}$), is abelian, with morphisms grading preserving A-module maps that also commute with differentials. When no confusion can be caused, we will drop all subscripts in differentials.

DEFINITION 3.3. Let M and N be two p-DG modules. A morphism $f : M \longrightarrow N$ in A_∂–mod is called *null-homotopic* if there is an A-module map h of degree $2-2p$ such that

$$f = \sum_{i=0}^{p-1} \partial_N^i \circ h \circ \partial_M^{p-1-i}.$$

It is an easy exercise to check that null-homotopic morphisms form an ideal in $A_\partial-\mathrm{mod}$. The resulting quotient category, denoted $\mathcal{H}(A)$, is called the *homotopy category* of left p-DG modules over A, and it is a triangulated category.

The simplest p-DG algebra is the ground field equipped with the trivial differential, whose homotopy category is denoted $\Bbbk[\partial]/(\partial^p)-\underline{\mathrm{gmod}}$[4]. In general, given any p-DG algebra A, one has a forgetful functor

$$\mathrm{For} : \mathcal{H}(A) \longrightarrow \Bbbk[\partial]/(\partial^p)-\underline{\mathrm{gmod}} \tag{3.1}$$

by remembering only the underlying p-complex structure up to homotopy of any p-DG module over A. A morphism between two p-DG modules $f : M \longrightarrow N$ (or its image in the homotopy category) is called a *quasi-isomorphism* if $\mathrm{For}(f)$ is an isomorphism in $\Bbbk[\partial]/(\partial^p)-\underline{\mathrm{gmod}}$. Denoting the class of quasi-isomorphisms in $\mathcal{H}$ by $\mathcal{Q}$, we define the p-DG derived category of A to be

$$\mathcal{D}(A) := \mathcal{H}(A)[\mathcal{Q}^{-1}], \tag{3.2}$$

the localization of $\mathcal{H}(A)$ at quasi-isomorphisms. By construction, $\mathcal{D}(A)$ is triangulated.

Many constructions in the usual homological algebra of DG-algebras translate over into the p-DG context without any trouble. We will see a few illustrations of this phenomenon in what follows.

Generalizing some useful concepts from the usual DG theory [**Kel06**], we make the following definitions.

DEFINITION 3.4. Let A be a p-DG algebra, and K be a (left or right) p-DG module.

(i) The module K is said to satisfy *property (P)* if there exists an increasing, possibly infinite, exhaustive ∂_K-stable filtration $F^\bullet$, such that each subquotient $F^\bullet/F^{\bullet-1}$ is isomorphic to a direct sum of p-DG direct summands of A.

(ii) The module K is called a *finite cell module*, if it satisfies property (P), and as an A-module, it is finitely generated (necessarily projective by the property (P) requirement).

[4]This is usually known as the graded stable category of $\Bbbk[\partial]/(\partial^p)$.

Property-(P) modules are the analogues of projective modules in usual homological algebra. For instance, the morphism spaces from a property-(P) module to any p-DG module coincide in both the homotopy and derived categories.

It is a theorem [**Qi14**, Theorem 6.6] that there are always sufficiently many property-(P) modules: for any p-DG module M, there is a surjective quasi-isomorphism

$$\mathbf{p}(M) \longrightarrow M \tag{3.3}$$

of p-DG modules, with $\mathbf{p}(M)$ satisfying property (P). We will usually refer to such a property-(P) replacement $\mathbf{p}(M)$ for M as a *bar resolution*. The proof of its existence is similar to that of the usual (simplicial) bar resolution for DG modules over DG algebras. We omit the proof here, but an example of such a construction will be used in the proof of Theorem 5.5.

In a similar vein, finite cell modules play the role of finite projective modules in usual homological algebra. We give an example of a finite cell module whose construction mimics the notion of "one-sided twisted complexes" for DG modules.

EXAMPLE 3.5. If A is a p-DG algebra, then A itself is finite cell. Now, if $a_0 \in A$ is an element such that $\partial^{p-1}(a_0) = 0$, then one can easily check that the differential rule on the rank-two free module $Ax \oplus Ay$

$$\partial(ax) := \partial(a)x + aa_0y, \quad \partial(ay) := \partial(a)y$$

defines the structure of a left p-DG module which is finite cell. We will schematically depict this module as

$$Ax \xrightarrow{a_0} Ay,$$

even though a_0 does not necessarily commute with differentials. One should understand this diagram as a filtered p-DG module, with $F^0 = Ay$ and $F^1 = Ax \oplus Ay$.

A p-DG bimodule ${}_AM_B$ over two p-DG algebras A and B is a p-DG module over $A \otimes B^{\mathrm{op}}$. One has the associated derived tensor product functor

$$M \otimes^{\mathbf{L}}_B (-) : \mathcal{D}(B) \longrightarrow \mathcal{D}(A), \quad N \mapsto \mathbf{p}(M) \otimes_B N, \tag{3.4}$$

where $\mathbf{p}(M)$ is a bar resolution for M as a p-DG module over $A \otimes B^{\mathrm{op}}$

One useful fact about such functors is the following theorem, whose proof can be found in [**Qi14**, Section 8].

THEOREM 3.6. *Let $f : M_1 \longrightarrow M_2$ be a quasi-isomorphism of p-DG bimodules. Then f descends to an isomorphism of the induced derived tensor product functors.* □

As the last example of adapting the usual DG theory to the present situation, one may define for the triangulated category $\mathcal{D}(A)$ its *Grothendieck group* $K_0(A)$ as usual. What we gain as dividend in the current situation is that, since $\mathcal{D}(A)$ is a "categorical module" over $\Bbbk[\partial]/(\partial^p)-\underline{\mathrm{gmod}}$, the abelian group $K_0(A)$ naturally has a module structure over the auxiliary cyclotomic ring at a pth root of unity, which was defined in equation (2.1) of the previous Section:

$$\mathbb{O}_p := K_0(\Bbbk[\partial]/(\partial^p)-\underline{\mathrm{gmod}}). \tag{3.5}$$

This explains the diagram (1.2) in the Introduction.

The Grothendieck groups are the algebraic invariants that we will be interested in for certain p-DG algebras in this note. As in the DG case, in order to avoid trivial cancellations one should restrict the class of objects used to define $K_0(A)$. It turns out that a good condition is the *compactness* of a p-DG module. However, we will not recall this definition, but refer the reader to [**Qi14**, Section 7] for the details.

3.2. The p-DG algebra $A_n^!$. As an illustration of the general theory discussed above, we now present a p-DG algebra that is relevant to quantum topology and easy to define.

Let $\Bbbk$ be a field of any characteristic $p \geq 0$ (later we will only focus on the case $p > 0$). Let n be a natural number greater than or equal to two, and Q_n be the following quiver:

(3.6)

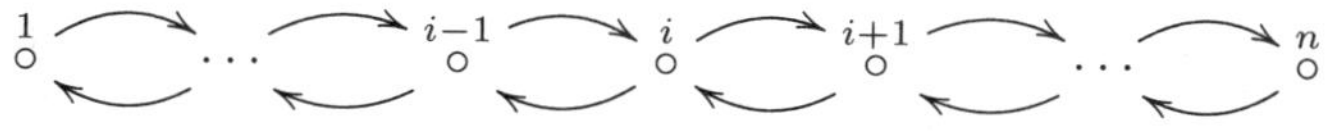

Let $\Bbbk Q_n$ be the path algebra associated to Q_n over the ground field. Following Khovanov-Seidel [**KS02**], we use, for instance, the symbol $(i|j|k)$, where i, j, k are vertices of the quiver Q_n, to denote the path which starts at a vertex i, then goes through j (necessarily $j = i \pm 1$) and ends at k. The composition of paths is given by

$$(i_i|i_2|\cdots|i_r)\cdot(j_1|j_2|\cdots|j_s) = \begin{cases} (i_i|i_2|\cdots|i_r|j_2|\cdots|j_s) & \text{if } i_r = j_1, \\ 0 & \text{otherwise,} \end{cases} \tag{3.7}$$

where $i_1, \ldots, i_r$ and $j_1, \ldots, j_s$ are sequences of neighboring vertices in Q_n.

DEFINITION 3.7. The algebra $A_n^!$ is the quotient of the path algebra $\Bbbk Q_n$ by the relations

$$(1|2|1) = 0 \quad \text{and} \quad (i|i-1|i) = (i|i+1|i) \quad \text{for} \quad i = 2, \ldots, n-1.$$

It should be mentioned that $A_n^!$ is isomorphic to the endomorphism algebra of a minimal projective generator of the category $\mathcal{O}_{1,n-1}(\mathfrak{gl}_n)$ considered in [**BFK99**]. The sum of categories $\oplus_{i=0}^n \mathcal{O}_{i,n-i}(\mathfrak{gl}_n)$ has been used to categorify $V_1^{\otimes n}$ in [**BFK99**]. Moreover, the algebra $A_n^!$ is Koszul, whose quadratic dual is isomorphic to the algebra A_n considered in [**KS02**].

This algebra arises naturally from the truncated polynomial algebra $C_n := \Bbbk[x]/(x^n)$. For $i = 1, \ldots, n$, there are ideals[5]

$$G^i = x^{n-i} C_n\{i-n\},$$

where the curly bracket $\{i\}$ stands for a graded module with the grading shifted up by i. A routine calculation shows that

$$A_n^! \cong \mathrm{END}_{C_n}(\bigoplus_{i=1}^n G^i),$$

where the capital END_{C_n} stands for the space of all endomorphisms of a finite-dimensional C_n-module, not necessarily grading preserving. The path $(i|i+1)$ gets sent to the homomorphism $G^{i+1} \to G^i$ which maps x^{n-i-1} to x^{n-i}. The path $(i|i-1)$ gets sent to the homomorphism $G^i \to G^{i+1}$ which maps x^{n-i} to

[5] The ideals $x^{n-i}C_n$ $(i = 1, \ldots, n)$ are isomorphic to the quotient modules $C/(x^i)$, only as C-modules but not as p-DG modules. However, the results obtained below will not be affected by using the quotient modules instead: the differentials obtained in different interpretations are conjugate to each other by the involution on $A_n^!$ which reverses arrow directions.

x^{n-i}. The endomorphism algebra inherits a differential determined, for any $\phi \in \mathrm{END}_{C_n}(\bigoplus_{i=1}^n G^i)$ and $z \in \bigoplus_{i=1}^n G^i$, by

$$\partial(\phi)(z) = \partial(\phi(z)) - \phi(\partial(z)).$$

This realization of $A_n^!$ endows the algebra with a p-DG structure. On C_n there is a differential ∂ given by $\partial(x) = x^2$. It is clear that each G^i is a p-DG submodule of C_n. This gives $A_n^!$ the structure of a p-DG algebra where the differential is given by

$$\partial(i|i+1) = (i|i+1|i|i+1), \qquad \partial(i|i-1) = 0.$$

Since the algebra $A_n^!$ is a finite-dimensional Koszul algebra with p-differential acting trivially on the degree zero component, the computation of the Grothendieck group of the derived category of $A_n^!$ reduces to the computation of the Grothendieck without the presence of the differential calculated in Proposition 5.1. See [**EQ16a**, Corollary 2.18] for more details.

PROPOSITION 3.8. There is an isomorphism of $\mathbb{O}_p$-modules

$$K_0(\mathcal{D}(A_n^!)) \cong V_1^{\otimes n}[-n+2].$$

For each $i = 1, \ldots, n$ there is a simple left $A_n^!$-module L_i spanned by $1 \in \Bbbk$ in degree zero such that $(j)1 = \delta_{i,j}1$. It is a p-DG module with trivial differential. Similarly one may define the simple right $A_n^!$-module ${}_iL$. In the Grothendieck group the image of the module L_i gets identified with $l_i \in V_1^{\otimes n}[-n+2]$ defined earlier.

3.3. A braid group action. For $i = 1, \ldots, n-1$ define derived functors $\mathfrak{U}_i$ as follows:

$$\mathfrak{U}_i \colon \mathcal{D}(A_n^!) \to \mathcal{D}(A_n^!) \qquad \mathfrak{U}_i(M) = (L_i \otimes {}_iL \otimes^{\mathbf{L}}_{A_n^!} M)[-1]\{-1\}.$$

The following result shows that these functors define a (weak) categorical Temperley-Lieb algebra action on the derived category $\mathcal{D}(A_n^!)$.

THEOREM 3.9 ([**QS13**]). For $i = 1, \ldots, n-1$, the functors $\mathfrak{U}_i$ are self-biadjoint, and they enjoy the following functor-isomorphism relations.

(i) $\mathfrak{U}_i \circ \mathfrak{U}_i \cong \mathfrak{U}_i[-1]\{-1\} \oplus \mathfrak{U}_i[1]\{1\}$,
(ii) $\mathfrak{U}_i \circ \mathfrak{U}_j \cong \mathfrak{U}_j \circ \mathfrak{U}_i$ for $|i-j| > 1$,
(iii) $\mathfrak{U}_i \circ \mathfrak{U}_j \circ \mathfrak{U}_i \cong \mathfrak{U}_i$ for $|i-j| = 1$.

These categorical equations can be roughly explained by the fact that the p-DG modules L_i are *spherical objects* in the p-DG derived category $\mathcal{D}(A_n^!)$. Spherical objects in the usual DG case were studied in [**AL13, ST01, KS02**].

To convert spherical objects into braid group generators, we mimic the construction in [**KS02**] in the p-DG context. Consider the following canonical maps:

$$\epsilon_1 \colon L_i \otimes {}_iL[-2]\{-2\} \cong L_i \otimes \mathbf{RHOM}_{A_n^!}(L_i, A_n^!) \xrightarrow{\lambda_1} A_n^!, \tag{3.8}$$

$$\eta_2 \colon A_n^! \xrightarrow{\lambda_2} \mathbf{RHOM}_{\Bbbk}(\mathbf{RHOM}_{A_n^!}(A_n^!, L_i), L_i) \cong L_i \otimes {}_iL. \tag{3.9}$$

We use them to define *filtered p-DG bimodules* and derived functors.

DEFINITION 3.10. For each $i \in \{1, \ldots, n-1\}$, we define the functors $\mathfrak{T}_i$ and $\mathfrak{T}_i'$ as follows.

(i) The functor $\mathfrak{T}_i : \mathcal{D}(A_n^!) \longrightarrow \mathcal{D}(A_n^!)$ is given by the derived tensor product with the cocone of the bimodule adjunction map (3.9)

$$A_n^! \xrightarrow{\lambda_2[-1]} \mathbf{RHOM}_{\Bbbk}(\mathbf{RHOM}_{A_n^!}(A_n^!, L_i), L_i)[-1] \cong \mathfrak{U}_i\{1\},$$

(ii) The functor $\mathfrak{T}_i' : \mathcal{D}(A_n^!) \longrightarrow \mathcal{D}(A_n^!)$ is given by the derived tensor product with the cone of the bimodule adjunction map (3.8):

$$\mathfrak{U}_i\{-1\} \cong L_i \otimes \mathbf{RHOM}_{A_n^!}(L_i, A_n^!)[1] \xrightarrow{\lambda_1[-1]} A_n^!,$$

THEOREM 3.11 ([**QS13**]). *The functors $\mathfrak{T}_i, \mathfrak{T}_i'$ for $i = 1, \ldots, n-1$ satisfy braid relations, i.e., there are isomorphisms of functors*

(1) $\mathfrak{T}_i \circ \mathfrak{T}_i' \cong \mathrm{Id} \cong \mathfrak{T}_i' \circ \mathfrak{T}_i$
(2) $\mathfrak{T}_i \circ \mathfrak{T}_j \cong \mathfrak{T}_j \circ \mathfrak{T}_i$ if $|i-j| > 1$
(3) $\mathfrak{T}_i \circ \mathfrak{T}_{i+1} \circ \mathfrak{T}_i \cong \mathfrak{T}_{i+1} \circ \mathfrak{T}_i \circ \mathfrak{T}_{i+1}$.

The proofs of both Theorems 3.9 and 3.11 are applications of Theorem 3.6 plus explicit comparisons of the bimodules involved.

4. A categorification of quantum $\mathfrak{sl}_2$ at prime roots of unity

4.1. The p-DG nilHecke category. The ring C_n considered in the previous section has a natural topological origin as $C_n \cong \mathrm{H}^*(\mathbb{P}^{n-1}, \Bbbk)$. Next, we will consider other similarly defined rings from topology, and equip them with p-differentials over a field of characteristic p.

Let Fl_n be the flag variety of complete flags in $\mathbb{C}^n$. It is well known that the $GL_n(\mathbb{C})$-equivariant cohomology of Fl_n is identified with

$$\mathrm{H}^*_{GL_n(\mathbb{C})}(Fl_n, \Bbbk) \cong \mathrm{H}^*_{B_n}(\mathrm{pt}, \Bbbk) \cong \mathrm{Pol}_n, \tag{4.1}$$

where B_n stands for a Borel subgroup of $GL_n(\mathbb{C})$, and Pol_n denotes the algebra of polynomial functions on n-letters. As such, Pol_n is naturally a module over

$$\mathrm{H}^*_{GL_n(\mathbb{C})}(\mathrm{pt}) \cong \Lambda_n. \tag{4.2}$$

The algebra of symmetric functions Λ_n is nothing but the polynomial algebra on the usual elementary symmetric functions e_i ($i = 1, \ldots, n$) on n-letters. We define a p-DG algebra structure on Λ_n by setting

$$\partial(e_i) := e_1 e_i - (i+1)e_{i+1}, \tag{4.3}$$

with it understood that $e_{n+1} = 0$. Write the resulting p-DG algebra as (Λ_n, ∂).

The *nilHecke algebra on n-letters* NH_n, by definition, is the endomorphism algebra of the module Pol_n over Λ_n:

$$\mathrm{NH}_n \cong \mathrm{END}_{\Lambda_n}(\mathrm{Pol}_n). \tag{4.4}$$

It is useful to give a generators-and-relations presentation of NH_n as follows. Let x_i ($i = 1, \ldots, n$) be the operator of multiplication on Pol_n by the ith variable, and define τ_i ($i = 1, \ldots, n-1$) to be the ith *divided difference operator* given by

$$\tau_i(f) = \frac{f - {}^i f}{x_i - x_{i+1}}.$$

Here f is any polynomial in Pol_n, and ${}^i f$ stands for f with the ith and $i+1$st variables exchanged. A full set of relations satisfied by the generators is given by

$$(4.5)\qquad \begin{gathered} x_i x_j = x_j x_i, \qquad x_i \tau_j = \tau_j x_i \ (|i-j|>1), \qquad \tau_i\tau_j = \tau_j\tau_i \ (|i-j|>1), \\ \tau_i^2 = 0, \quad x_i\tau_i - \tau_i x_{i+1} = 1 = \tau_i x_i - x_{i+1}\tau_i, \quad \tau_i\tau_{i+1}\tau_1 = \tau_{i+1}\tau_i\tau_{i+1}. \end{gathered}$$

Define

$$\mathrm{NH} := \bigoplus_{n\in\mathbb{N}} \mathrm{NH}_n\text{-gmod}.$$

A key property of NH is that it is a *monoidal category*, with the monoidal structure coming from a natural inclusion map

$$(4.6)\qquad \mathrm{NH}_n \otimes \mathrm{NH}_m \subset \mathrm{NH}_{n+m}.$$

It is quite surprising to us that, not until the recent work of Khovanov, Lauda and Rouquier [**KL09, Lau10, Rou08**], the following simple relationship between quantum $\mathfrak{sl}_2$ and nilHecke algebras, the latter dating back to the work of Bernstein-Gelfand on Schubert calculus, was not stated anywhere in the literature.

THEOREM 4.1 ([**KL09, Lau10, Rou08**]). There is an isomorphism of $\mathbb{Z}[q,q^{-1}]$-bialgebras

$$U_q^-(\mathfrak{sl}_2) \cong K_0(\mathrm{NH}),$$

with the symbols of induction and restriction functors on the right hand side corresponding to the multiplication and comultiplication on $U_q^-(\mathfrak{sl}_2)$.

To categorify half of the small quantum $\mathfrak{sl}_2$, we combine this construction with hopfological algebra of p-differentials. To do so, define Pol_n to be the p-DG algebra with the p-differential $\partial(x_i) = x_i^2$ $(i=1,\ldots,n)$. Following [**KQ15**, Section 3], define a p-DG module structure on the rank-one Pol_n module $\mathcal{P}_n$ as follows:

$$(4.7)\qquad \mathcal{P}_n := \mathrm{Pol}_n \cdot v_0, \qquad \partial(v_0) = -\sum_{i=1}^{n}(n-i)x_i v_0.$$

Upon restriction, we regard $\mathcal{P}_n$ as a p-DG module over (Λ_n, ∂). It is shown in [**KQ15**, Section 3] that, up to an isomorphism, this is the unique p-DG structure on Pol_n over (Λ_n,∂) that makes the module finite cell in all characteristic $p>0$. Furthermore, it also induces a monoidal-structure preserving (see equation (4.6)) differential on the NH category. A simple computation identifying $\mathrm{NH}_n \cong \mathrm{END}_{\Lambda_n}(\mathcal{P}_n)$ as p-DG algebras yields the following definition.

DEFINITION 4.2. The p-DG monoidal category is the direct sum of p-DG algebras

$$(\mathrm{NH},\partial) := \oplus_{n\in\mathbb{N}}(\mathrm{NH}_n,\partial),$$

where each NH_n is equipped with the p-differential defined on the generators (see equation (4.5)) by

$$\partial(x_i) = x_i^2, \qquad \partial(\tau_i) = -x_i\tau_i - \tau_i x_{i+1}.$$

THEOREM 4.3 ([**KQ15**]). There is an isomorphism of $\mathbb{O}_p$-bialgebras

$$u_{\mathbb{O}_p}^- \cong K_0(\mathcal{D}(\mathrm{NH})),$$

under which $F^{(n)}$ is identified with the symbol $[\mathcal{P}_n]$.

To see that this is indeed the case, we justify the relation $F^n = 0$ if $n \geq p$. One can readily see that, as a p-DG module over Λ_n, $\mathcal{P}_n$ has a ∂-stable basis consisting of elements in the set

$$B_n := \{x_1^{a_1} \dots x_n^{a_n} | 0 \leq a_i \leq n - i\}.$$

The $\Bbbk$-span of B_n forms an $n!$-dimensional p-complex, which is acyclic whenever $n \geq p$. Therefore, since $\mathrm{NH}_n \cong \mathrm{END}_{\Lambda_n}(\mathcal{P}_n)$, NH_n is an acyclic p-DG algebra whenever $n \geq p$. It follows that $\mathcal{D}(\mathrm{NH}_n) \cong 0$ if $n \geq p$, and this categorical vanishing result lifts the above relation $F^n = 0$ $(n \geq p)$ for the small quantum group.

4.2. The p-DG 2-category $\mathcal{U}$. We seek to "Drinfeld-double" the one-half construction of the categorified small quantum group following Khovanov-Lauda and Rouquier [**Lau10, KL10, Rou08**]. To do so for a generic q, one takes two copies of the NH-category, and "glue" them together with biadjuntions. More specifically, let us recall the diagrammatic definition of $\mathcal{U}$ introduced by Lauda in [**Lau10**].

DEFINITION 4.4. The 2-category $\mathcal{U}$ is an additive graded $\Bbbk$-linear category whose objects λ are elements of the weight lattice of $\mathfrak{sl}_2$. The 1-morphisms are (direct sums of grading shifts of) composites of the generating 1-morphisms $\mathbb{1}_{\lambda+2}\mathcal{E}\mathbb{1}_\lambda$ and $\mathbb{1}_\lambda\mathcal{F}\mathbb{1}_{\lambda+2}$, for each $\lambda \in \mathbb{Z}$. Each $\mathbb{1}_{\lambda+2}\mathcal{E}\mathbb{1}_\lambda$ will be drawn the same, regardless of the object λ.

1-morphism Generator	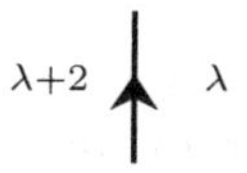	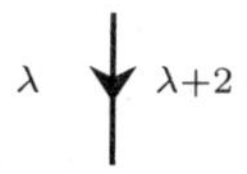
Name	$\mathbb{1}_{\lambda+2}\mathcal{E}\mathbb{1}_\lambda$	$\mathbb{1}_\lambda\mathcal{F}\mathbb{1}_{\lambda+2}$

The weight of any region in a diagram is determined by the weight of any single region. When no region is labeled, the ambient weight is irrelevant.

The 2-morphisms will be generated by the following pictures.

Generator	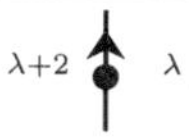			
Degree	2	2	-2	-2

Generator	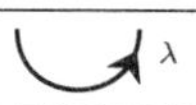			
Degree	$1+\lambda$	$1-\lambda$	$1+\lambda$	$1-\lambda$

We will give a list of relations shortly, after we discuss some notation. For a product of m dots on a single strand, we draw a single dot labeled by m. Here is the case when $m = 4$.

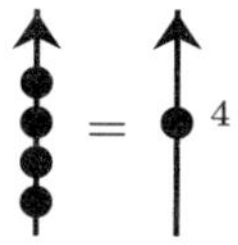

A *closed diagram* is a diagram without boundary, constructed from the generators above. The simplest non-trivial closed diagram is a *bubble*, which is a closed diagram without any other closed diagrams inside. Bubbles can be oriented clockwise or counter-clockwise.

A simple calculation shows that the degree of a bubble with m dots in a region labeled λ is $2(m+1-\lambda)$ if the bubble is clockwise, and $2(m+1+\lambda)$ if the bubble is counter-clockwise. Instead of keeping track of the number m of dots a bubble has, it will be much more illustrative to keep track of the degree of the bubble, which is in $2\mathbb{Z}$. We will use the following shorthand to refer to a bubble of degree $2k$.

This notation emphasizes the fact that bubbles have a life of their own, independent of their presentation in terms of caps, cups, and dots.

Note that λ can be any integer, but $m \geq 0$ because it counts dots. Therefore, we can only construct a clockwise (resp. counter-clockwise) bubble of degree k when $k \geq 1-\lambda$ (resp. $k \geq 1+\lambda$). These are called *real bubbles*. Following Lauda, we also allow bubbles drawn as above with arbitrary $k \in \mathbb{Z}$. Bubbles with k outside of the appropriate range are not yet defined in terms of the generating maps; we call these *fake bubbles*. One can express any fake bubble in terms of real bubbles (see Remark 4.5).

Now we list the relations. Whenever the region label is omitted, the relation applies to all ambient weights.

(i) **Biadjointness relations.**

(4.8a)

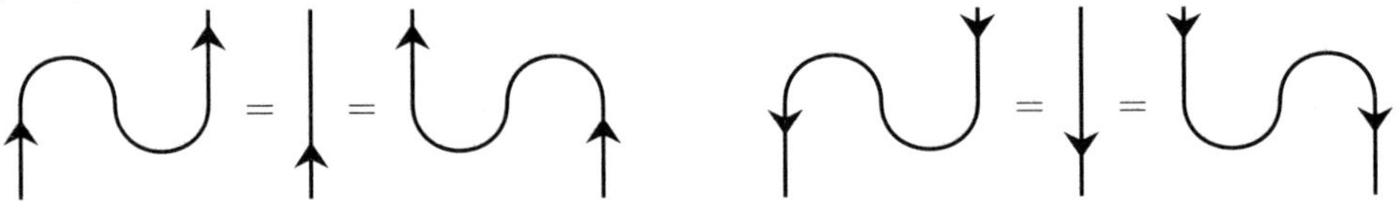

(4.8b) 

(4.8c)

(ii) **Positivity and Normalization of bubbles.** Positivity states that all bubbles (real or fake) of negative degree should be zero.

(4.9a) 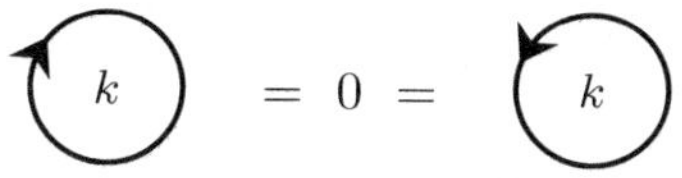

if $k < 0$,

Normalization states that degree 0 bubbles are equal to the empty diagram (i.e. the identity 2-morphism of the identity 1-morphism).

(4.9b) 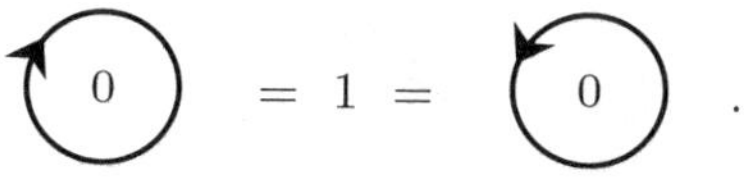

(iii) **NilHecke relations.** The upward pointing strands satisfy nilHecke relations as in (4.5). Note that, diagrammatically, far-away commuting elements become isotopy relations and are thus built in by default.

(4.10a) 

(4.10b) 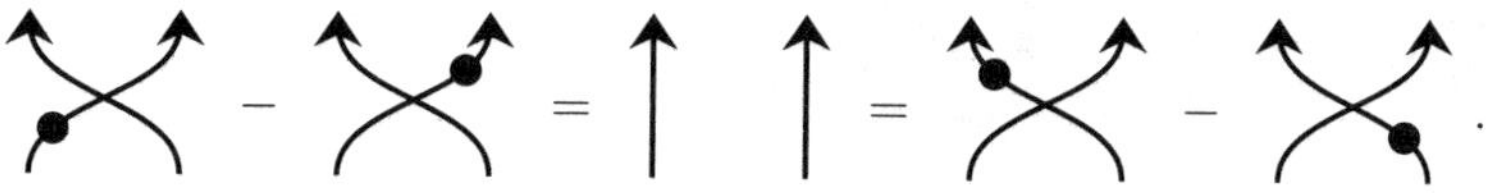

(iv) **Reduction to bubbles.** The following equalities hold for all $\lambda \in \mathbb{Z}$.

(4.11a) 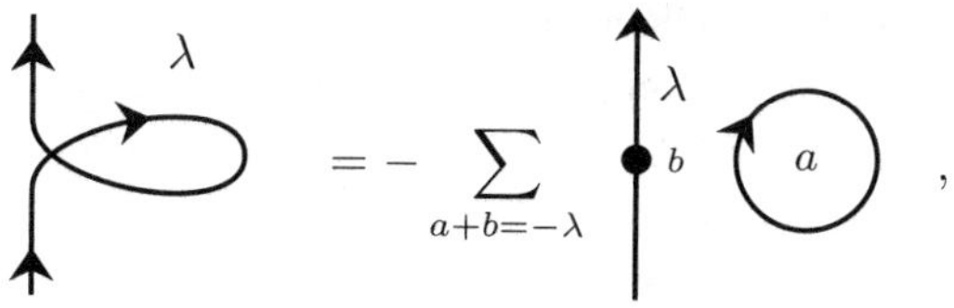

(4.11b) 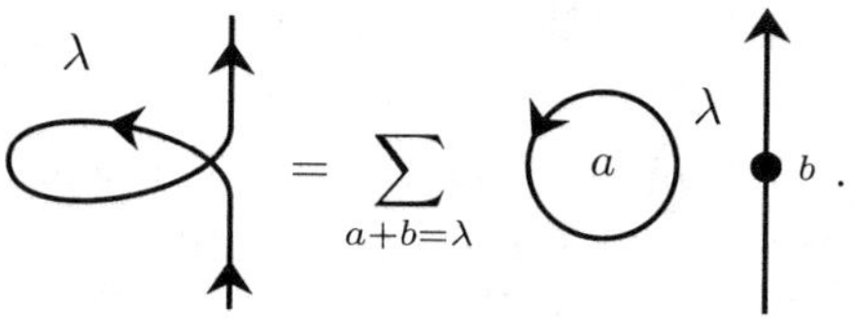

These sums only take values for $a, b \geq 0$. Therefore, when $\lambda \neq 0$, either the right curl or the left curl is zero.

(v) **Identity decomposition.** The following equations hold for all $\lambda \in \mathbb{Z}$.

(4.12a)

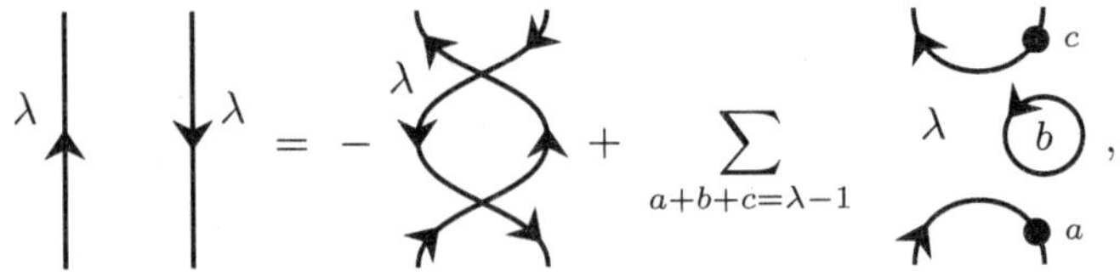

(4.12b)

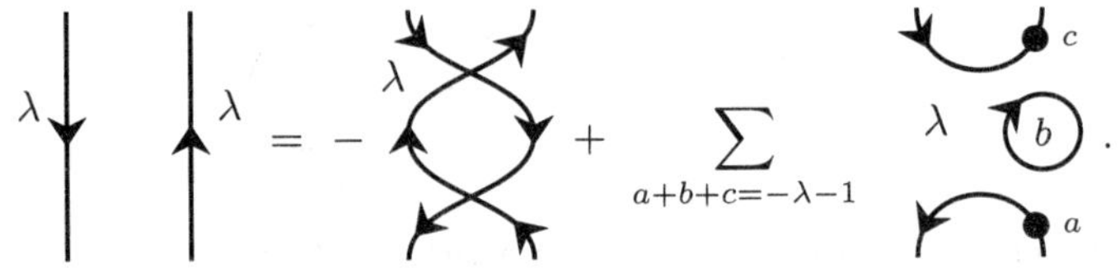

The sum in the first equality vanishes for $\lambda \leq 0$, and the sum in the second equality vanishes for $\lambda \geq 0$.

The terms on the right hand side form a collection of orthogonal idempotents.

REMARK 4.5 (Infinite Grassmannian relations). These relations, which follow from the above defining relations, can be expressed most succinctly in terms of generating functions.

(4.13)

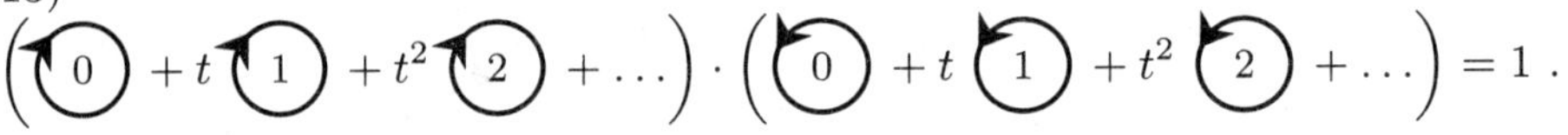

The cohomology ring of the "infinite dimensional Grassmannian" is the ring Λ of symmetric functions. Inside this ring, there is an analogous relation $e(t)h(t) = 1$, where $e(t) = \sum_{i \geq 0} (-1)^i e_i t^i$ is the total Chern class of the tautological bundle, and $h(t) = \sum_{i \geq 0} h_i t^i$ is the total Chern class of the dual bundle. Lauda has proved that the bubbles in a single region generate an algebra inside $\mathcal{U}$ isomorphic to Λ.

Looking at the homogeneous component of degree m, we have the following equation.

(4.14)

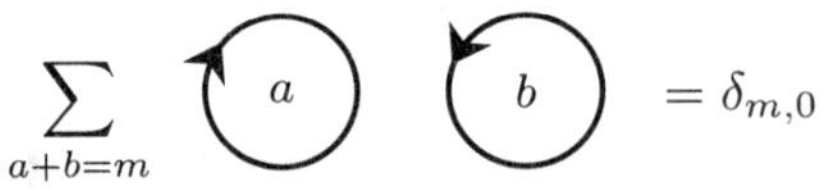

Because of the positivity of bubbles relation, this equation holds true for any $m \in \mathbb{Z}$, and the sum can be taken over all $a, b \in \mathbb{Z}$.

Using these equations one can express all (positive degree) counter-clockwise bubbles in terms of clockwise bubbles, and vice versa. Consequentially, all fake bubbles can be expressed in terms of real bubbles.

DEFINITION 4.6. Let ∂ be the derivation defined on the 2-morphism generators of $\mathcal{U}$ as follows.

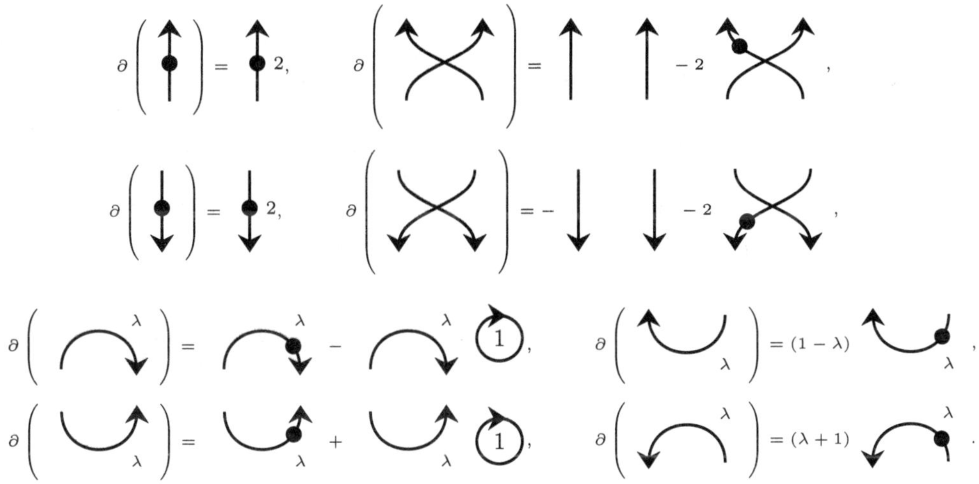

THEOREM 4.7 ([**EQ16a**]). There is an isomorphism of $\mathbb{O}_p$-algebras

$$\dot{u}_{\mathbb{O}_p} \longrightarrow K_0(\mathcal{D}(\mathcal{U})),$$

sending $1_{\lambda+2}E1_\lambda$ to $[\mathbb{1}_{\lambda+2}\mathcal{E}\mathbb{1}_\lambda]$ and $1_\lambda F 1_{\lambda+2}$ to $[\mathbb{1}_\lambda \mathcal{F}\mathbb{1}_{\lambda+2}]$ for any weight $\lambda \in \mathbb{Z}$.

Without the presence of the differential, Khovanov, Lauda, Mackaay and Stošić (KLMS) has given a diagrammatic presentation of the idempotent completion $\dot{\mathcal{U}}$ of $\mathcal{U}$. By construction, $\dot{\mathcal{U}}$ is Morita equivalent to $\mathcal{U}$, and they both categorify quantum $\mathfrak{sl}_2$ at generic q values.

The KLMS calculus $\dot{\mathcal{U}}$ has been equipped with a compatible differential derived from $\mathcal{U}$. We will refer the reader to [**EQ15**] for the details. However, unlike the abelian case, the p-DG derived categories are drastically different! There is a functor

$$\mathcal{J} : \mathcal{H}(\mathcal{U}) \longrightarrow \mathcal{H}(\dot{\mathcal{U}}),$$

which is given by tensor product with $\dot{\mathcal{U}}$ regarded as a $(\dot{\mathcal{U}}, \mathcal{U})$-bimodule with a compatible differential. This functor, not surprisingly, is an equivalence of homotopy categories. However, under localization, it is no longer an equivalence, but instead is a fully-faithful embedding of derived categories:

$$\mathcal{J} : \mathcal{D}(\mathcal{U}) \longrightarrow \mathcal{D}(\dot{\mathcal{U}}).$$

This derived embedding categorifies the embedding of $\dot{u}_{\mathbb{O}_p}$ into the BLM form $\dot{U}_{\mathbb{O}_p}$ for quantum $\mathfrak{sl}(2)$, the latter generated by all divided powers $E^{(n)}$, $F^{(n)}$ $(n \in \mathbb{N})$.

REMARK 4.8. When $p = 2$, the constructions in this section can be lifted to characteristic zero, using the usual DG theory. The nilHecke category should then be replaced by the odd nilHecke algebra of [**EKL14**]. We refer the reader to [**EQ16b**] for the details. But a natural question still remains: Can one lift the constructions in positive characteristics to characteristic zero? A fundamental problem one needs to tackle is to have a characteristic zero categorification of the ring of cyclotomic integers, and then build nilHecke-like objects in that category.

5. A categorification of the Jones-Wenzl projector

5.1. Categorification at a generic q. We present here a special case of a functor categorifying the Jones-Wenzl projector first considered in [**FSS12**] in the context of category $\mathcal{O}$. A further specialization has been studied in greater detail in [**SS14**], where the Koszul dual functor is computed and shown to be related to the independent constructions of Cooper and Krushkal [**CK12**] and Rozansky [**Roz14**]. What follows below is the construction for the category categorifying the second lowest weight space. Details for other weight spaces will appear in [**SS**].

First we ignore the differential thus categorifying the Jones-Wenzl projector for generic q. For ease of notation set

$$C := \Bbbk[x]/(x^n) \qquad\qquad A := A_n^!.$$

PROPOSITION 5.1. The Grothendieck groups of finitely generated graded modules over A and C categorify weight spaces of $V_1^{\otimes n}$ and V_n respectively.

(1) $K_0(A\text{-gmod}) \cong V_1^{\otimes n}[-n+2]$.
(2) $K_0(C\text{-gmod}) \cong V_n[-n+2]$.

PROOF. It is easy to see that the projective modules $A(i)$ for $i = 1, \ldots, n$ are mutually non-isomorphic and exhaust the list of indecomposable A-modules.

The second part is even easier since C is a graded local algebra. □

The connection between these categories is given by the Soergel functor and its adjoint. It is easy to see that $\mathrm{END}_A(A(n)) \cong C$.

Consider the following two functors. Let

$$\Pi\colon A\text{-gmod} \to C\text{-gmod}$$

be the exact functor given by

$$\Pi(M) = (n)A \otimes_A M,$$

which automatically extends to the derived categories. And let

$$\mathrm{I}\colon \mathcal{D}^<(C\text{-gmod}) \to \mathcal{D}^<(A\text{-gmod})$$

be the derived functor given by

$$\mathrm{I}(M) = A(n) \otimes_C^{\mathbf{L}} M.$$

Then define the categorified Jones-Wenzl projector to be

$$\mathfrak{P} = \mathrm{I} \circ \Pi\colon \mathcal{D}^<(A\text{-gmod}) \to \mathcal{D}^<(A\text{-gmod}).$$

In the next result, we construct a projective resolution of $(n)A$ as a graded (C, A)-bimodule (recall that unadorned tensor product $\otimes$ is always taken over the ground field).

PROPOSITION 5.2. The following complex of (C, A)-bimodules

$$\cdots \longrightarrow C\otimes(n)A\{2n+2\} \xrightarrow{h_3} C\otimes(n)A\{2n\} \xrightarrow{h_2} C\otimes(n)A\{2\} \xrightarrow{h_1} C\otimes(n)A \xrightarrow{h_0} (n)A$$

is exact, where the differentials are bimodule maps which are defined on the generators by

$$h_k(1\otimes(n)) = \begin{cases} (n) & \text{if } k=0 \\ x\otimes(n) - 1\otimes(n|n-1|n) & \text{if } k=1,3,\ldots \\ \sum_{j=0}^{n-1} x^j \otimes (n|n-1|n)^{n-1-j} & \text{if } k=2,4,\ldots. \end{cases}$$

PROOF. It is clear that the square of the differential is zero. Thus it suffices to show that the complex is exact.

To do so, notice that, as a left module over C, $(n)A$ decomposes, up to grading shifts, into a direct sum of ideals

$$(n)A \cong C \oplus xC \oplus \cdots \oplus x^{n-1}C. \tag{5.1}$$

Replacing $(n)A$ in each term of the complex by this decomposition, and we are reduced to showing that, for each $i \in \{0, 1, \ldots, n-1\}$, the complex

(5.2)

$$\cdots \longrightarrow C \otimes x^iC\{2n+2\} \xrightarrow{h_3} C \otimes x^iC\{2n\} \xrightarrow{h_2} C \otimes x^iC\{2\} \xrightarrow{h_1} C \otimes x^iC \xrightarrow{h_0} x^iC$$

is exact. We will first prove it for $i = 0$, in this case, the complex is isomorphic to

(5.3)

$$\cdots \longrightarrow \frac{\Bbbk[x,y]}{(x^n,y^n)}\{2n+2\} \xrightarrow{h_3} \frac{\Bbbk[x,y]}{(x^n,y^n)}\{2n\} \xrightarrow{h_2} \frac{\Bbbk[x,y]}{(x^n,y^n)}\{2\} \xrightarrow{h_1} \frac{\Bbbk[x,y]}{(x^n,y^n)} \xrightarrow{h_0} \frac{\Bbbk[z]}{(z^n)},$$

and the boundary maps are now identified with $h_0(x) = z$, $h_0(y) = z$, and

$$h_k(1) = \begin{cases} x - y & \text{if } k = 1, 3, \ldots \\ \sum_{j=0}^{n-1} x^j y^{n-1-j} & \text{if } k = 2, 4, \ldots. \end{cases}$$

The exactness of this complex is clear.

For $i > 0$, notice that $x^iC \cong C/(x^{n-i})$. If one chops off the right most term of equation (5.2), that complex has higher homology groups isomorphic to $\mathrm{Tor}_C^{\bullet}(C, C/(x^{n-i})) = 0$ $(\bullet > 0)$, and the zeroth homology isomorphic to $C/(x^{n-i})$. The result follows. □

Now tensor the projective resolution in Proposition 5.2 on the left by $A(n)$ over C to get a complex of (A, A)-bimodules quasi-isomorphic to the functor $\mathfrak{P}$.

THEOREM 5.3. (1) The Jones-Wenzl functor $\mathfrak{P}$ is quasi-isomorphic to the following complex of (A, A)-bimodules

$$\cdots \to A(n)\otimes(n)A\{2n+2\} \xrightarrow{g_3} A(n)\otimes(n)A\{2n\} \xrightarrow{g_2} A(n)\otimes(n)A\{2\} \xrightarrow{g_1} A(n)\otimes(n)A$$

where

$$g_i = \begin{cases} (n|n-1|n) \otimes 1 - 1 \otimes (n|n-1|n) & \text{if i is odd} \\ \sum_{j=0}^{n-1} (n|n-1|n)^j \otimes (n|n-1|n)^{n-1-j} & \text{if i is even.} \end{cases}$$

(2) The functor $\mathfrak{P}$ is idempotent: $\mathfrak{P} \circ \mathfrak{P} \cong \mathfrak{P}$.

(3) The Temperley-Lieb functors $\mathfrak{U}_i$, $i = 1, \ldots, n-1$, annhilate the Jones-Wenzl projector:

$$\mathfrak{U}_i \circ \mathfrak{P} \cong 0 \cong \mathfrak{P} \circ \mathfrak{U}_i.$$

PROOF. The first statement is clear from the discussion above.

The second statement follows almost directly from the definition of the categorified Jones-Wenzl projector and associativity of (derived) tensor products. The tensor product of the bimodules representing $\mathfrak{P}$ is given by

$$A(n) \otimes_C^{\mathbf{L}} (n)A \otimes_A A(n) \otimes_C^{\mathbf{L}} (n)A \cong A(n) \otimes_C^{\mathbf{L}} (n)A(n) \otimes_C^{\mathbf{L}} (n)A \cong A(n) \otimes_C^{\mathbf{L}} (n)A.$$

The result follows.

The third part is a direct consequence of the first part since

$${}_iL \otimes_A^{\mathbf{L}} A(n) \cong (n)A \otimes_A^{\mathbf{L}} L_i = 0$$

for $i = 1, \ldots, n-1$. □

As a 2-representation of Lauda's category this complex may be rewritten as

$$\left(\cdots \xrightarrow{f_4} \mathcal{E}\mathcal{F}\{2n+2\} \xrightarrow{f_3} \mathcal{E}\mathcal{F}\{2n\} \xrightarrow{f_2} \mathcal{E}\mathcal{F}\{2\} \xrightarrow{f_1} \mathcal{E}\mathcal{F}\right)\{n-1\}$$

where

$$f_i = \begin{cases} x \otimes 1 - 1 \otimes y & \text{if } i \text{ is odd} \\ \sum_{j=0}^{n-1} x^j \otimes y^{n-1-j} & \text{if } i \text{ is even} \end{cases}$$

and x and y represent the degree-two endomorphisms of $\mathcal{E}$ and $\mathcal{F}$ respectively. They were diagrammatically represented by a dot on an upward pointing and downward pointing strand respectively in $\mathcal{U}$ (see Section 4).

COROLLARY 5.4. The equality

$$[\mathfrak{P}] = \frac{q^{n-1}(1-q^2)}{1-q^{2n}} EF = \frac{1}{[n]} EF.$$

holds in the usual Grothendieck group of $\mathcal{U}$. □

5.2. Categorification at a prime root of unity. In the presence of ∂ we occasionally expect that the categorified Jones-Wenzl projector is quasi-isomorphic to a finite complex of (A, A)-bimodules. On the decategorified level this expectation comes from the observation that for certain values of p the quantity $\frac{1}{[n]}$ is actually quite simple. For instance, if p divides $n-1$, then we have $[n] = (-1)^k$ where $n - 1 = kp + r$ with $0 \leq r < p$. More generally, if p does not divde n, $[n]$ is always a unit in $\mathbb{O}_p$. On the level of categories, we will see that, whenever p divides $n-1$, one can construct a finite cell bimodule representing $\mathfrak{P}$. Furthermore, for a fixed n, these are precisely all the prime characteristics for which this happens. Indeed we obtain the following theorem:

THEOREM 5.5. Let n be a positive integer, and denote by $A = A_n^!$ and $C = C_n$ the p-DG algebras which depend on n.

(i) If $n-1$ is divisible by p, then the p-DG bimodule over (A, A) representing $\mathfrak{P}$ is quasi-isomorphic to a finite cell bimodule.
(ii) If $n-1$ is not divisible by p, then $\mathfrak{P}$ can not be represented by a finite cell p-DG module over (A, A).

PROOF OF PART (I). To prove the first statement we may explicitly construct a resolution of $\mathfrak{P}$. In this case, $n = kp + 1$ for some $k \geq 0$, and the algebra $C = \Bbbk[x]/(x^{kp+1})$ is quasi-isomorphic to $\Bbbk$ since

$$\Bbbk(x \to x^2 \to \cdots \to x^p) \oplus \Bbbk(x^{p+1} \to x^{p+2} \to \cdots \to x^{2p})$$
$$\oplus \cdots \oplus \Bbbk(x^{(k-1)p+1} \to x^{(k-1)p+2} \to \cdots \to x^{kp})$$

is an acyclic ideal inside C. Thus, as a p-DG bimodule over (C, A), we have a surjective quasi-isomorphism coming from the multiplication map of C on the left module $(n)A$:

$$C \otimes (n)A \xrightarrow{m} (n)A.$$

It follows that the functor $\mathfrak{P}$ is given by tensor product with the p-DG bimodule

$$A(n) \otimes_C^{\mathbf{L}} (n)A \cong A(n) \otimes_C (C \otimes (n)A) \cong A(n) \otimes (n)A.$$

Therefore

$$\mathfrak{P} \cong (A(n) \otimes (n)A) \otimes^{\mathbf{L}}_A (-) = \mathcal{E}\mathcal{F}.$$

□

It is a bit more difficult to prove the second statement in the Theorem. To do so we will make some preparations.

LEMMA 5.6. Let A be a p-DG algebra. Suppose

$$\cdots \xrightarrow{\phi_{m+4}} P_{m+3} \xrightarrow{\phi_{m+3}} P_{m+2} \xrightarrow{\phi_{m+2}} P_{m+1} \xrightarrow{\phi_{m+1}} P_m \longrightarrow 0$$

is a bounded-from-above exact sequence of p-DG modules over A. Then the filtered p-DG module below, with every other term repeated $(p-1)$-times, is acyclic:

$$\cdots \xrightarrow{\phi_{m+4}} P_{m+3} = \cdots = P_{m+3} \xrightarrow{\phi_{m+3}} P_{m+2} \xrightarrow{\phi_{m+2}} P_{m+1} = \cdots = P_{m+1} \xrightarrow{\phi_{m+1}} P_m \longrightarrow 0.$$

PROOF. The result reduces easily to the corresponding result for short exact sequences of p-DG modules. So let us assume that $P_k = 0$ when $k \geq m+3$. We will analyze the p-complex structure of the filtered p-DG module obtained.

As a p-complex, it is easy to see that the filtered module[6] has as a subcomplex

$$0 \longrightarrow \underbrace{\mathrm{Im}(\phi_{m+1}) = \cdots = \mathrm{Im}(\phi_{m+1})}_{(p-1)\text{ terms}} \xrightarrow{\cong} P_m \longrightarrow 0.$$

Modulo the subcomplex, the quotient is isomorphic to

$$0 \longrightarrow P_{m+2} \xrightarrow{\cong} \underbrace{\mathrm{Ker}(\phi_{m+1}) = \cdots = \mathrm{Ker}(\phi_{m+1})}_{(p-1)\text{ terms}} \longrightarrow 0.$$

Both complexes are obviously acyclic. The result follows. □

PROOF OF PART (II). To prove the statement, we will show that $\mathfrak{P}(L_n)$ has infinite p-cohomology. Since $(n)A \otimes_A L_n \cong \Bbbk$, we have $\mathfrak{P}(L_n) \cong A(n) \otimes^{\mathbf{L}}_C \Bbbk$. We are reduced to finding a property-(P) (Defnition 3.4) resolution for $\Bbbk$ as a left p-DG module over C.

Now we start constructing the resolution. The procedure is an augmented version of the construction of a bar resolution in [**Qi14**, Theorem 6.6], which will be done inductively.

Consider the short exact sequence of p-DG modules over C

$$0 \longrightarrow I \longrightarrow C \longrightarrow \Bbbk \longrightarrow 0. \tag{5.4}$$

Here I stands for the ideal generated by x, which is a p-DG submodule inside C.

The surjective multiplication map $C \otimes I \longrightarrow I$ is a map of p-DG modules. Denote its kernel by I_2, and it is easy to see that

$$I_2 = \{y \otimes z - 1 \otimes yz | y,\ z \in I\}.$$

[6]Here the filtration on the p-DG module is the natural one with respect to the differential: P_m is a sub p-DG module that is the starting term, and the higher terms in the filtration are obtained by adding more terms to the left of P_m.

As p-complexes, $C \otimes I \cong \Bbbk \otimes I \oplus I \otimes I$, and thus $I_2 \cong I \otimes I$. Gluing this map to to the above short exact sequence, we get an exact sequence of left p-DG modules

$$0 \longrightarrow I_2 \longrightarrow C \otimes I \longrightarrow C \longrightarrow \Bbbk \longrightarrow 0. \tag{5.5}$$

Inductively, suppose we have constructed an exact sequence of p-DG modules

$$0 \longrightarrow I_{k-1} \longrightarrow C \otimes I_{k-2} \longrightarrow \cdots \longrightarrow C \otimes I \longrightarrow C \longrightarrow \Bbbk \longrightarrow 0 \tag{5.6}$$

where the arrow between each $C \otimes I_r \longrightarrow C \otimes I_{r-1}$ $(r = 1, \ldots, k-2)$ is given by the multiplication of C on a p-DG module; the kernel $I_{r+1} = \{y \otimes w - 1 \otimes yw | y \in I,\ w \in I_r\}$ of the arrow, as a p-complex, is isomorphic to $I^{\otimes(r+1)}$.

Then glue to the left most end of the sequence above the C algebra action map

$$C \otimes I_{k-1} \longrightarrow I_{k-1}. \tag{5.7}$$

by the discussion similar to the $k = 2$ case, one sees that the kernel of the map is a p-DG submodule inside $C \otimes I_{k-1}$, which can be identified with

$$I_k = \{y \otimes w - 1 \otimes yw | y \in I,\ w \in I_{k-1}\},$$

whose underlying p-complex is isomorphic to $I \otimes I_{k-1} \cong I^{\otimes k}$. This finishes the induction step, and we obtain a bounded-above sequence of p-DG modules ($I_1 := I$).

$$\cdots \longrightarrow C \otimes I_k \longrightarrow C \otimes I_{k-1} \longrightarrow \cdots \longrightarrow C \otimes I_1 \longrightarrow C \longrightarrow \Bbbk \longrightarrow 0. \tag{5.8}$$

Now, applying Lemma 5.6 to the sequence (5.8), we obtain a desired filtered module which satisfies property (P) (Definition 3.4) and is quasi-isomorphic to $\Bbbk$:
(5.9)

$$\cdots \xrightarrow{m} C \otimes I_3 = = \cdots = = C \otimes I_3 \xrightarrow{m} C \otimes I_2 \xrightarrow{m} C \otimes I_1 = = \cdots = = C \otimes I_1 \xrightarrow{m} C \longrightarrow 0,$$

where the maps m indicate the algebra C multiplication acting on a C-module.

It then follows that $\mathfrak{P}(L_n)$ is quasi-isomorphic to the p-complex

$$\cdots \xrightarrow{m} A(n) \otimes I_3 = = \cdots = = A(n) \otimes I_3 \xrightarrow{m} A(n) \otimes I_2 \xrightarrow{m} A(n) \otimes I_1 = = \cdots = = A(n) \otimes I_1 \xrightarrow{m} A(n) \longrightarrow 0.$$

As the summand decomposition (5.1) is also a decomposition of p-DG modules, we have that $A(n)$ contains $x^{n-1}C \cong \Bbbk\{2n-2\}$ as a p-complex direct summand. Hence $\mathfrak{P}(L_n)$ contains the infinite p-complex (up to an overall shift of $2n-2$)

$$\cdots \xrightarrow{0} I^{\otimes 3} = = \cdots = = I^{\otimes 3} \xrightarrow{0} I^{\otimes 2} \xrightarrow{0} I = = \cdots = = I \xrightarrow{0} \Bbbk \longrightarrow 0,$$

where the zero-maps come from the fact that the element x^{n-1} kills the ideal I: Under the map m, we have, for any $y \otimes w - 1 \otimes yw \in I_k$, where $y \in I$ and $w \in I_{k-1}$,

$$x^{n-1} \otimes_C (y \otimes w - 1 \otimes yw) = -x^{n-1} \otimes yw \mapsto -x^{n-1}yw = 0.$$

The theorem now follows. □

5.3. Two examples. In Theorem 5.5, when $n-1$ is not divisible by p, the number of copies of C in each piece of the property-(P) replacement of $\Bbbk$ grows exponentially. However, there are two special cases where the bar resolution used can be replaced by a much smaller resolution: the number of copies of C in each degree is almost a constant.

EXAMPLE 5.7. We consider the case when $n=(k+1)p$ for some $k\in\mathbb{N}$. As before, first we take $C_0 := C\cdot 1_0$ and map it onto $\Bbbk$ by identifying $\Bbbk$ with the lowest degree subspace spanned by 1_0. Next, we form the filtered complex (c.f. Example 3.5)

$$C1_1 \xrightarrow{x^{kp+1}} C1_0 \ .$$

Since $\partial^{p-1}(x^{kp+1})=0$, this is a rank-two finite cell module.

We compute that

$$\begin{aligned}\partial^{p-1}(x^{kp+1}1_1) &= \sum_{i=0}^{p-1}\binom{p-1}{i}\partial^i(x^{kp+1})\partial^{p-i-1}(1_1)\\ &= \sum_{i=0}^{p-2}\frac{(p-1)!}{i!(p-1-i)!}i!x^{kp+1+i}\partial^{p-i-2}(x1_0)\\ &= \sum_{i=0}^{p-2}\frac{(p-1)!}{(p-1-i)}x^{(k+1)p}1_0.\end{aligned}$$

The second equality holds since $\partial^{p-1}(x^{kp+1})=0$. Now, as i ranges over $\{0,\dots,p-2\}$, $1/(p-1-i)$ ranges over all non-zero elements of $\mathbb{F}_p$, therefore the sum in the last term is zero.

It follows that we may repeat the construction, and we obtain an infinite filtered module satisfying property (P) as follows:

$$\cdots \xrightarrow{x^{kp+1}} C\{2(r+1)kp\} \xrightarrow{x^{kp+1}} C\{2rkp\} \xrightarrow{x^{kp+1}} \cdots \xrightarrow{x^{kp+1}} C\{2kp\} \xrightarrow{x^{kp+1}} C.$$

This resolution of $\Bbbk$ as a left p-DG module has an interesting implication in the Grothendieck group $K_0(C)$, which says that

$$[\Bbbk]=\sum_{r=0}^{\infty}q^{2rkp}[C].$$

Since $q^{2p}=1$ in $\mathbb{O}_p$, this equation reduces to

$$[\Bbbk]=\sum_{r=0}^{\infty}[C]. \tag{5.10}$$

Although the identity $\sum_{r=0}^{\infty}1 = 1/(1-1)$ no longer makes sense, yet equation (5.10) can be understood as

$$[C]=[\Bbbk]+[I]=[\Bbbk]+\sum_{i=1}^{p-1}q^{2i}[\Bbbk]=[\Bbbk]-[\Bbbk]=0.$$

EXAMPLE 5.8. In this example we consider Theorem 5.5 when $n=kp+2$ ($k\geq 0$), in which case the p-DG resolution of $\Bbbk$ over C may also be simplified significantly.

For $j = 0, \ldots, k-1$ there are contractible p-complexes of C of the form

$$(x^{jp+1} \to x^{jp+2} \to \cdots \to x^{(j+1)p}).$$

Contracting these p-complexes one establishes a quasi-isomorphism between C and $\Bbbk[x^{kp+1}]/(x^{kp+1})^2$ with trivial differential.

As a p-DG module over C, the trivial module $\Bbbk$ is quasi-isomorphic to

(5.11)

$$\cdots \xrightarrow{x^{kp+1}} C\{2(3kp-2p+4)\} = \cdots = C\{2(3kp-p+2)\} \xrightarrow{x^{kp+1}} C\{2(3kp-p+2)\} \xrightarrow{x^{kp+1}} C\{2(kp-p+2)\} = \cdots = C\{2kp\} \xrightarrow{x^{kp+1}} C \longrightarrow 0$$

Similar as above, this equation has an interesting Grothendieck group implication saying that

$$[\Bbbk] = \sum_{r=0}^{\infty} (-1)^r q^{2r} [C].$$

Although, as before, q^2 is a root of unity in $\mathbb{O}_p$, and it does not make sense for $\sum_{r=0}^{\infty}(-1)^r q^{2r}$ to converge, this equation can, however, be "explained" via the equality of symbols

$$[C] = [\Bbbk] + [I] = [\Bbbk] + q^{2(kp+1)}[\Bbbk] = (1+q^2)[\Bbbk].$$

Both examples are instances of why one wants to restrict to *compact objects* for defining the Grothendieck group of the triangulated category $\mathcal{D}(C)$ of a p-DG algebra. The module $\Bbbk$ in these cases are not compact in $\mathcal{D}(C)$. We refer the reader to [**EQ16a**] for a discussion about compact p-DG modules over a *positive p-DG algebra*. This notion includes both C and A in this paper as special instances.

References

[AL13] R. Anno and T. Logvinenko, *Spherical DG-functors*, arXiv:1309.5035.

[BCGPM14] Christian Blanchet, Francesco Costantino, Nathan Geer, and Bertrand Patureau-Mirand, *Non semi-simple $\mathfrak{sl}(2)$ quantum invariants, spin case*, Acta Math. Vietnam. **39** (2014), no. 4, 481–495. MR3319702

[BCL14] Anna Beliakova, Qi Chen, and Thang T. Q. Le, *On the integrality of the Witten-Reshetikhin-Turaev 3-manifold invariants*, Quantum Topol. **5** (2014), no. 1, 99–141, DOI 10.4171/QT/48. MR3176311

[BFK99] Joseph Bernstein, Igor Frenkel, and Mikhail Khovanov, *A categorification of the Temperley-Lieb algebra and Schur quotients of $U(\mathfrak{sl}_2)$ via projective and Zuckerman functors*, Selecta Math. (N.S.) **5** (1999), no. 2, 199–241, DOI 10.1007/s000290050047. MR1714141

[CF94] Louis Crane and Igor B. Frenkel, *Four-dimensional topological quantum field theory, Hopf categories, and the canonical bases*, J. Math. Phys. **35** (1994), no. 10, 5136–5154, DOI 10.1063/1.530746. Topology and physics. MR1295461

[CK12] Benjamin Cooper and Vyacheslav Krushkal, *Categorification of the Jones-Wenzl projectors*, Quantum Topol. **3** (2012), no. 2, 139–180, DOI 10.4171/QT/27. MR2901969

[EKL14] Alexander P. Ellis, Mikhail Khovanov, and Aaron D. Lauda, *The odd nilHecke algebra and its diagrammatics*, Int. Math. Res. Not. IMRN **4** (2014), 991–1062. MR3168401

[EQ15] B. Elias and Y. Qi, *A categorification of quantum sl(2) at prime roots of unity*, arXiv:1503.05114.

[EQ16a] Ben Elias and You Qi, *An approach to categorification of some small quantum groups II*, Adv. Math. **288** (2016), 81–151, DOI 10.1016/j.aim.2015.10.009. MR3436383

[EQ16b] Alexander P. Ellis and You Qi, *The Differential Graded Odd NilHecke Algebra*, Comm. Math. Phys. **344** (2016), no. 1, 275–331, DOI 10.1007/s00220-015-2569-4. MR3493144

[FSS12] Igor Frenkel, Catharina Stroppel, and Joshua Sussan, *Categorifying fractional Euler characteristics, Jones-Wenzl projectors and 3j-symbols*, Quantum Topol. **3** (2012), no. 2, 181–253, DOI 10.4171/QT/28. MR2901970

[Kap96] M. Kapranov, *On the q-analog of homological algebra*, arXiv:q-alg/9611005.

[Kel06] Bernhard Keller, *On differential graded categories*, International Congress of Mathematicians. Vol. II, Eur. Math. Soc., Zürich, 2006, pp. 151–190. MR2275593

[Kho00] Mikhail Khovanov, *A categorification of the Jones polynomial*, Duke Math. J. **101** (2000), no. 3, 359–426, DOI 10.1215/S0012-7094-00-10131-7. MR1740682

[Kho16] Mikhail Khovanov, *Hopfological algebra and categorification at a root of unity: The first steps*, J. Knot Theory Ramifications **25** (2016), no. 3, arXiv:math/0509083.

[KL09] Mikhail Khovanov and Aaron D. Lauda, *A diagrammatic approach to categorification of quantum groups. I*, Represent. Theory **13** (2009), 309–347, DOI 10.1090/S1088-4165-09-00346-X. MR2525917

[KL10] Mikhail Khovanov and Aaron D. Lauda, *A categorification of quantum* sl(n), Quantum Topol. **1** (2010), no. 1, 1–92, DOI 10.4171/QT/1. MR2628852

[KL11] Mikhail Khovanov and Aaron D. Lauda, *A diagrammatic approach to categorification of quantum groups II*, Trans. Amer. Math. Soc. **363** (2011), no. 5, 2685–2700, DOI 10.1090/S0002-9947-2010-05210-9. MR2763732

[KLMS12] Mikhail Khovanov, Aaron D. Lauda, Marco Mackaay, and Marko Stošić, *Extended graphical calculus for categorified quantum* sl(2), Mem. Amer. Math. Soc. **219** (2012), no. 1029, vi+87, DOI 10.1090/S0065-9266-2012-00665-4. MR2963085

[KQ15] Mikhail Khovanov and You Qi, *An approach to categorification of some small quantum groups*, Quantum Topol. **6** (2015), no. 2, 185–311, DOI 10.4171/QT/63. MR3354331

[KS02] Mikhail Khovanov and Paul Seidel, *Quivers, Floer cohomology, and braid group actions*, J. Amer. Math. Soc. **15** (2002), no. 1, 203–271, DOI 10.1090/S0894-0347-01-00374-5. MR1862802

[Lau10] Aaron D. Lauda, *A categorification of quantum* sl(2), Adv. Math. **225** (2010), no. 6, 3327–3424, DOI 10.1016/j.aim.2010.06.003. MR2729010

[May42a] W. Mayer, *A new homology theory. I, II*, Ann. of Math. (2) **43** (1942), 370–380, 594–605. MR0006514

[May42b] W. Mayer, *A new homology theory. II*, Ann. of Math. (2) **43** (1942), no. 3, 594–605.

[Mir15] D. Mirmohades, *Categorification of the ring of cyclotomic integers for products of two primes*, arXiv:1506.08755.

[Qi14] You Qi, *Hopfological algebra*, Compos. Math. **150** (2014), no. 1, 1–45, DOI 10.1112/S0010437X13007380. MR3164358

[QS13] Y. Qi and J. Sussan, *A categorification of the Burau representation at prime roots of unity*, arXiv:1312.7692.

[Rou08] R. Rouquier, *2-Kac-Moody algebras*, arXiv:0812.5023.

[Roz14] Lev Rozansky, *An infinite torus braid yields a categorified Jones-Wenzl projector*, Fund. Math. **225** (2014), no. 1, 305–326, DOI 10.4064/fm225-1-14. MR3205575

[RT90] N. Yu. Reshetikhin and V. G. Turaev, *Ribbon graphs and their invariants derived from quantum groups*, Comm. Math. Phys. **127** (1990), no. 1, 1–26. MR1036112

[Sar95] K. S. Sarkaria, *Combinatorial methods in topology*, Available at http://kssarkaria.org/.

[SS] C. Stroppel and J. Sussan, *A Lie theoretic categorification of the coloured Jones polynomial*, In preparation.

[SS14] ———, *Categorified Jones-Wenzl projectors: a comparison*, Perspectives in representation theory, Contemp. Math., vol. 610, Amer. Math. Soc., Providence, RI, 2014, arXiv:1105.3038, pp. 333–351.

[ST01] Paul Seidel and Richard Thomas, *Braid group actions on derived categories of coherent sheaves*, Duke Math. J. **108** (2001), no. 1, 37–108, DOI 10.1215/S0012-7094-01-10812-0. MR1831820

[Web15] B. Webster, *Knot invariants and higher representation theory*, arXiv:1309.3796.

[Wit89] Edward Witten, *Quantum field theory and the Jones polynomial*, Comm. Math. Phys. **121** (1989), no. 3, 351–399. MR990772

Department of Mathematics, Yale University, New Haven, Connecticut 06511
E-mail address: you.qi@yale.edu

Department of Mathematics, CUNY Medgar Evers, Brooklyn, New York 11225
E-mail address: jsussan@mec.cuny.edu

Contemporary Mathematics
Volume **683**, 2017
http://dx.doi.org/10.1090/conm/683/13704

Folding with Soergel bimodules

Ben Elias

ABSTRACT. We give a gentle introduction to the concept of folding. That is, we provide an elementary discussion of equivariant categories, their weighted Grothendieck groups, and the technical aspects of computing with them. We then perform the computations required to confirm that quasi-split Hecke algebras with unequal parameters are categorified by equivariant Soergel bimodules, in almost every case.

1. Introduction

Let σ be an automorphism of a Dynkin diagram Γ. Let W be the Weyl group corresponding to Γ, which inherits an action of σ. It is an old theorem of Steinberg [**25**] that the σ-fixed elements of W themselves form a Coxeter group W^σ. Similar statements can be made about the invariant parts of the corresponding Lie algebra or root system, under the restriction that the orbits of σ contain only mutually-distant vertices in Γ. The most familiar examples of this are the embeddings of B_{k+1} inside A_{2k+1}, and of C_{k+1} inside D_{k+2}, where in both examples the Dynkin diagrams of the larger group are visually "folded" by an involution σ to produce the Dynkin diagram of the subgroup. Thus, *folding* is the word chosen to describe the general phenomenon that constructions attached to W, when considered σ-invariantly or σ-equivariantly, can be related to constructions attached to W^σ.

Automorphisms of Dynkin diagrams and the induced automorphisms of Chevalley groups were studied by Chevalley in the 1950s [**1**]. Steinberg [**25**] introduced the twisted Chevalley groups which are obtained by taking invariants under these automorphisms. The modern, categorical approach was pioneered by Lusztig in his books [**15, 17**]: it studies σ-equivariant categories and their decategorifications. This will be the version of folding discussed in this paper.

A primary example of folding is the categorification of quantum groups due to Lusztig [**17**]. For a simply-laced Dynkin diagram Γ, an appropriate category of perverse sheaves on the quiver variety corresponding to Γ (with arbitrary orientation) will categorify the positive half of the quantum group. For non-simply-laced Dynkin diagrams, this construction will not work. However, every non-simply-laced Dynkin diagram (i.e. types B, C, F, and G) can be folded from a simply-laced Dynkin diagram. The automorphism σ lifts to an automorphism of the quiver variety, and one can consider σ-equivariant perverse sheaves. This equivariant category will have a Grothendieck group that is too large, but imposing some *trace relation* on this

2010 *Mathematics Subject Classification.* Primary 20C08.

Grothendieck group will yield the positive half of the quantum group for the folded Dynkin diagram. In this example, folding is crucial as a means of approaching non-simply-laced Dynkin diagrams.

Another example of folding is the categorification of Hecke algebras, due to Kazhdan-Lusztig (with the folding part due to Lusztig [**15**]). For any Weyl group W, the category of perverse sheaves on the flag variety of W, stratified with respect to the Bruhat decomposition, will categorify the Hecke algebra $\mathcal{H}(W)$ with equal parameters. Again, the automorphism σ lifts to an automorphism of the flag variety. The Grothendieck group of the category of σ-equivariant perverse sheaves, after imposing the trace relation, is isomorphic to the Hecke algebra $\mathcal{H}(W^\sigma, L)$ with unequal parameters. Here, L is the *weight function* on W^σ determined by the embedding, which sends $w \in W^\sigma$ to its length viewed as an element of W (as opposed to its length as an element of the Coxeter group W^σ). In this example, non-simply-laced Weyl groups can be approached in two ways, using their intrinsic flag varieties or using folding, and this categorifies two very different algebras! One should not confuse $\mathcal{H}(W^\sigma)$ with $\mathcal{H}(W^\sigma, L)$, but should think of them as being essentially unrelated.

In this paper, we refer to the Grothendieck group of an equivariant category with a trace relation imposed as a *weighted Grothendieck group*.

The main examples of folding have all been geometric, and for this reason it is traditionally limited to Weyl groups and affine Weyl groups, Dynkin diagrams and affine Dynkin diagrams. However, there is no essential reason for this limitation. Generalizing Steinberg's theorem, Lusztig [**18**, Appendix] and Hée [**9**] have shown the following: when a finite group G acts on a Coxeter group W by automorphisms, preserving the set S of simple reflections, and the simple reflections in the orbit of any $s \in S$ generate a finite subgroup, then the invariant subgroup W^G is itself a Coxeter group. An embedding of one Coxeter group inside another in this fashion is called a *quasi-split embedding*, and the corresponding weight function L is called *quasi-split*.

REMARK 1.1. The literature also occasionally makes assumptions about the action of G which are not truly necessary. It is sometimes assumed that the orbits of G consist of mutually-distant simple reflections. It is also sometimes assumed that G is cyclic, generated by σ; there are geometric reasons that this is reasonable, because one can often interpret σ as a modified Frobenius automorphism in finite characteristic. Regardless, even when G is not cyclic, a choice of a *weighting element* $\sigma \in G$ is required to impose a trace relation on the Grothendieck group.

Meanwhile, Soergel [**24**] has given an algebraic (not geometric) definition of a graded monoidal category $\mathbb{S}\mathrm{Bim}(W)$ attached to any Coxeter group W, whose objects are now called *Soergel bimodules*, and he proved under mild restrictions that the Grothendieck group of $\mathbb{S}\mathrm{Bim}(W)$ is isomorphic to $\mathcal{H}(W)$. For Weyl groups, with coefficients in a field of characteristic zero, Soergel proved that $\mathbb{S}\mathrm{Bim}(W)$ is equivalent to the aforementioned category of perverse sheaves.

Thus it is very natural to ask whether equivariant Soergel bimodules $\mathbb{S}\mathrm{Bim}(W)_G$ should categorify Hecke algebras with unequal parameters (in the quasi-split case). The goal of this paper is to give an elementary, computational proof of this fact. We give a precise statement in Theorem 5.4.

Continuing work of the author, Khovanov, and Libedinsky [**3, 4, 14**] in special cases, the author and Williamson have provided a presentation of $\mathbb{S}\mathrm{Bim}(W)$ by generators and relations, using the language of planar diagrammatics, for an arbitrary Coxeter group W [**5**]. The new mathematical content in this paper is a series of calculations performed using this diagrammatic calculus, which explicitly computes decompositions of certain tensor products in the equivariant category $\mathbb{S}\mathrm{Bim}(W)_G$. We hope this paper serves as an advertisement for the power of diagrammatics, and its ability to effectively address questions even in equivariant categories. Interesting results can be reduced to elaborate computational exercises.

REMARK 1.2. Suppose that G acts on the Coxeter system (W, S). As noted above, one should not confuse $\mathbb{S}\mathrm{Bim}(W)_G$, the category of equivariant Soergel bimodules for W, with $\mathbb{S}\mathrm{Bim}(W^G)$. Even though both categories have Grothendieck groups related in some way to W^G, the categories are extremely different and wholly unrelated. It should be possible to derive a diagrammatic calculus for the equivariant category itself, though we have not done this here.

The idea to examine equivariant Soergel bimodules, being as natural as it is, was independently pursued by multiple parties. Since the proof of the Soergel conjecture (in characteristic zero) by the author and Williamson [**6**], Soergel bimodules have become a more tractable and attractive place to work. Soon after [**6**] appeared, Lusztig updated his book [**18**] to account for Soergel bimodules, adding a section on equivariant Soergel bimodules and also giving a proof of the above result. Our approach is quite different from Lusztig's, being entirely computational in nature, and so we believe our independent proof will still be quite useful. While our proof as stated does rely on the Soergel conjecture, our computations do not, and thus this paper will be important in future analysis of equivariant Soergel bimodules in situations, such as finite characteristic, where the Soergel conjecture fails.

REMARK 1.3. The statement that $\mathbb{S}\mathrm{Bim}(W)_G$ has weighted Grothendieck group isomorphic to $\mathcal{H}(W^G, L)$ is false in general. It will be true when G is abelian and the weighting element $\sigma \in G$ acts transitively on the G-orbits in S; under these conditions, we reduce to the case of a cyclic group generated by σ in §3. When G is abelian and σ is the identity, then the weighted Grothendieck group is isomorphic to $\mathcal{H}(W)^G$ instead. For general G and $\sigma \in G$, the weighted Grothendieck group of $\mathbb{S}\mathrm{Bim}(W)_G$ is some genuinely new algebra, which should have a number of attractive properties (like cell theory and a canonical basis). It remains to see whether these "exotic Hecke algebras" are worth studying. Regardless, the computations performed in this paper will help one to compute the weighted Grothendieck group in general, such as for the action of the symmetric group S_3 on D_4.

When W is a Weyl group or affine Weyl group, our main theorem already followed from (highly nontrivial) geometric considerations. Allowing for general Coxeter groups does extend the generality of the notion of folding slightly. For example, one can now view the Coxeter group of type A_1 as the invariants inside the Coxeter group of type $I_2(m)$ (the dihedral group) under the non-trivial automorphism of $I_2(m)$. However, this is effectively the only new example amongst irreducible, finite Coxeter groups, as the Coxeter systems of type H_3 and H_4 have no automorphisms. One can also add new examples of folding amongst reducible

Coxeter groups, such as folding k copies of H_3 into a single copy under an automorphism which permutes the copies cyclically; such examples are not usually referred to as folding in the literature, but we do treat them here.

A similar question can be posed for quantum groups, where Khovanov-Lauda-Rouquier's quiver Hecke algebras [**12**, **22**] give an algebraic categorification of the positive half of the quantum group, and agree with perverse sheaves on quiver varieties in simply-laced type by work of Varagnolo-Vasserot [**26**]. It should be possible to perform the analogous computational analysis of the equivariant quiver Hecke category in simply-laced type. McNamara [**21**] has a recent paper to this effect, although with different methods (he exploits the crystal structure on simple objects to prove his main theorem). For non-simply-laced Dynkin diagrams, this would provide two distinct categorifications of the positive half of the quantum group: the equivariant, geometric, folding categorification which will categorify Lusztig's canonical basis, and the non-geometric categorification which categorifies a different positive basis. As with Remark 1.2, these two categories should not be confused.

There are some numerical artifacts which arise freely from categorification results. For example, the Soergel conjecture [**6**] demonstrated that the indecomposable objects in $\mathbb{S}\mathrm{Bim}(W)$ descend to the Kazhdan-Lusztig basis of $\mathcal{H}(W)$. One consequence is that the structure constants of $\mathcal{H}(W)$, with respect to this basis, are non-negative integers, being the dimensions of certain multiplicity spaces. The dimension is the trace of the identity map of the multiplicity space. Meanwhile, "multiplicity spaces" in the equivariant category are representations of a stabilizer group (this is a lie, but philosophically true), and the trace relation on the Grothendieck group implies that structure constant of $\mathcal{H}(W^G, L)$ are traces of other elements acting on multiplicity spaces. When $G = \mathbb{Z}/2\mathbb{Z}$, this statement is a form of *signed positivity*, that a (possibly negative) number is the difference of two positive numbers, the dimensions of the $+1$ and the -1 eigenspace.

There is another "folding-esque" scenario, due to Lusztig and Vogan [**19**]. Here, one begins with an involution σ of a Coxeter system (W, S), and considers the involution of W given by $w \mapsto \sigma(w^{-1})$. It is easy to lift this involution to an anti-monoidal involution on $\mathbb{S}\mathrm{Bim}(W)$, using adjunction to replace the inverse. In a follow-up paper, we will apply the same techniques to this equivariant category, and prove that the equivariant Grothendieck group (with a trace relation) is isomorphic to the Lusztig-Vogan representation. This proves a signed positivity conjecture from [**19**]. Just like for this paper, Lusztig and Vogan have already proven this result themselves [**20**] using Soergel bimodules and the results of [**6**], though we hope our direct computational approach will still be illuminating.

We now discuss the contents of this paper. This paper is purposely written for the reader who may be familiar with diagrammatic categorification, but less familiar with equivariant categories and Coxeter groups.

Chapter 2 contains an introduction to quasi-split embeddings of Coxeter groups, and to Hecke algebras with unequal parameters. The least well-known aspect of this introduction is a non-standard presentation of the Hecke algebra with unequal parameters in the quasi-split case, using the Kazhdan-Lusztig generators. The braid relations are transformed into certain unfamiliar relations, one for each finite dihedral group. These are the relations we categorify in §5.

Chapter 3 discusses the generalities of equivariant categories and introduces the idea of a *weighted Grothendieck group*. We believe that weighted Grothendieck groups are slightly easier to think about and work with than the trace relation mentioned above. It also allows for more general discussion, beyond the case of cyclic groups. Thus we have presented the ideas in a different way than is done in Lusztig's books; however, the main ideas are clearly indebted to him. We provide some general results that help to compute the weighted Grothendieck groups of mixed categories. In particular, for an abelian group G with an element σ which acts transitively on each G-orbit, computing the σ-weighted Grothendieck group of a mixed category will reduce to the trace relation of Lusztig. Most of the results in this chapter should be well-known to the experts, but we were not able to find a comprehensive reference, so we have tried to give an exposition for the apprentice.

Chapter 4 gives a terse summary of the category of Soergel bimodules, its diagrammatic presentation due to the author and Williamson, and the Soergel categorification theorem.

Finally, chapter 5 performs numerous computations in the category of equivariant Soergel bimodules, for group actions where W^G is dihedral. These computations are sufficient to categorify the braid-like relations of the Hecke algebra with unequal parameters. Then, using the Soergel conjecture to deduce that $\mathbb{S}\mathrm{Bim}(W)$ is a mixed category, and using the abstract results of §3, we deduce our main theorem.

REMARK 1.4. There is an action of $\mathbb{Z}/2\mathbb{Z}$ on the Weyl group of type F_4, with invariant subgroup $I_2(8)$. We have not even attempted to perform the computations here. This missing case should follow by analogous computations. Regardless, the result has already been proven by Lusztig.

Again, we do rely heavily on the truth of the Soergel conjecture for our categorification result. However, we do not use any of the Hodge-theoretic machinery required to prove the Soergel conjecture in [**6**]. Moreover, our computations do not rely on the Soergel conjecture, and apply in more generality.

2. Hecke algebras and unequal parameters

We assume the reader is familiar with the notion of a Coxeter group, as well as the notions of length, reduced expressions, and the Bruhat order. An excellent introduction can be found in [**11**]. We now give some background on weighted Coxeter groups and Hecke algebras with unequal parameters, following Lusztig's book [**15**].

2.1. Weighted Coxeter groups. Fix a Coxeter system (W, S). For a pair of simple reflections $s, t \in S$, we let m_{st} denote the order of (st). We let ℓ denote the length function. For simplicity, we assume S is finite, though everything can be adapted to the case where S is infinite.

DEFINITION 2.1. A *weight function* is a map $L\colon W \to \mathbb{Z}$ such that $L(vw) = L(v) + L(w)$ whenever $\ell(vw) = \ell(v) + \ell(w)$. A Coxeter system paired with a weight function (W, S, L) is called a *weighted Coxeter system*.

A length function is determined by the values of $L(s)$ for $s \in S$. It is easy to deduce that $L(s) = L(t)$ whenever m_{st} is odd. We say that L is *positive* if $L(s) > 0$ for all $s \in S$, which we assume henceforth.

EXAMPLE 2.2. The length function ℓ is also a weight function, and weighted Coxeter systems of the form (W, S, ℓ) are called *split*.

Non-split weight functions arise naturally when one Coxeter group is embedded inside another in a "Coxeter-esque" way.

DEFINITION 2.3. Let (W', S') be a Coxeter system. A partition of S' into subsets $I(s)$, parametrized by elements s of a set S, is called a *finitary partition* if the parabolic subgroup generated by $I(s)$ is finite for each $s \in S$. We identify the set S with a subset of W', letting $s = w_{I(s)}$, the longest element in the parabolic subgroup for $I(s)$.

DEFINITION 2.4. Let (W', S') be a Coxeter system, and I a finitary partition parametrized by S. Let W be the subgroup of W' generated by $s \in S$. It is possible that (W, S) is also a Coxeter system, in which case we call this setup a *Coxeter embedding*, and let $\varphi_I \colon W \to W'$ denote the inclusion map. It equips (W, S) with a positive weight function, called the *embedded weight function*, defined by $L(w) = \ell(\varphi_I(w))$.

DEFINITION 2.5. A Coxeter embedding where, for each $s \in S$, the set $I(s)$ consists only of commuting simple reflections will be called a *standard Coxeter embedding*. In this case, $w_{I(s)} = \prod_{t \in I(s)} t$, and $L(s) = \#I(s)$.

When discussing Coxeter embeddings we will often omit φ_I from our notation, thus letting $s \in S$ denote both an element in W and its image in W'.

EXAMPLE 2.6. Let (W', S') have type A_n, with $S' = \{s_1, s_2, \ldots, s_n\}$. Let (W, S) have type $I_2(n+1)$, a dihedral group with $S = \{t, u\}$ and $m_{tu} = n + 1$. Set $I(t) = \{s_i\}_{i \text{ even}}$ and $I(u) = \{s_i\}_{i \text{ odd}}$. This yields a standard Coxeter embedding. This embedding is related to the 2-sided cell in $I_2(n + 1)$ consisting of elements with a unique reduced expression, as shown by Lusztig [**16**].

Sometimes the partition of S' into subsets comes from the orbits of a group action. As the example above shows, this is not always the case.

DEFINITION 2.7. Let G be a finite group acting on a Coxeter system (W', S'). In other words, G acts on the set S', with $m_{st} = m_{g(s)g(t)}$. A Coxeter embedding where S is in bijection with the orbit space S'/G, and $I(s)$ is the corresponding orbit, will be called *quasi-split*. In this case, $W = (W')^G$ is the subgroup of G-invariants. If we want to emphasize the group, we will call the embedding *G-split*.

It is known that every finitary partition of S' which comes from the orbits of a group action gives rise to a Coxeter embedding. See [**7**] for an elementary proof. However, we do not know whether a general finitary partition always gives rise to a Coxeter embedding. We were unable to find this result, or a counterexample, in the literature.

There is a classification of quasi-split embeddings where the Coxeter groups involved are finite. The table on [**8**, p227] lists the possibilities when W' is a connected Weyl group. Allowing for connected non-crystallographic finite Coxeter groups only adds one family of examples (Example 2.13 below).

EXAMPLE 2.8. The group $\mathbb{Z}/2\mathbb{Z}$ acts on the Coxeter system of type A_{2k+1}, and the invariants have type B_{k+1}. This is a standard quasi-split embedding. If the simple reflections in type B_{k+1} are labeled t_i for $1 \leq i \leq k$ and u, where $\{t_i\}$ generates a subgroup of type A_k, then the embedded weight function satisfies $L(t_i) = 2$ and $L(u) = 1$.

EXAMPLE 2.9. The group $\mathbb{Z}/2\mathbb{Z}$ acts on the Coxeter system of type A_{2k+2}, and the invariants have type B_{k+1}. This is a non-standard quasi-split embedding, as the middle orbit has type A_2 instead of $A_1 \times A_1$. In this example, $L(t_i) = 2$ and $L(u) = 3$.

EXAMPLE 2.10. The group $\mathbb{Z}/2\mathbb{Z}$ acts on the Coxeter system of type D_{k+2}, and the invariants have type B_{k+1}. This is a standard quasi-split Coxeter embedding, with $L(t_i) = 1$ and $L(u) = 2$.

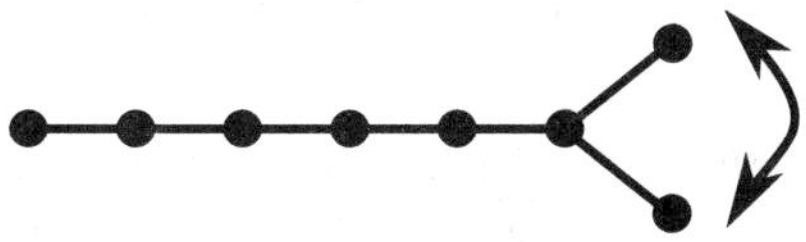

EXAMPLE 2.11. The group $\mathbb{Z}/2\mathbb{Z}$ acts on E_6 with invariant subgroup F_4.

EXAMPLE 2.12. Let (W', S') be a star-shaped Coxeter group. That is, $S' = \{s_1, s_2, \ldots, s_n, v\}$ where $m_{s_i s_j} = 2$ and $m_{s_i v} = 3$ for all i, j. The symmetric group S_n or its cyclic subgroup $\mathbb{Z}/n\mathbb{Z}$ will permute the spokes of the star, giving a standard quasi-split embedding whose invariant subgroup will be dihedral. If t corresponds to the vertex and u corresponds to the orbit containing the spokes, then $L(t) = 1$ and $L(u) = n$. Moreover, $m_{tu} = 3$ when $n = 1$, $m_{tu} = 4$ when $n = 2$, $m_{tu} = 6$ when $n = 3$, and $m_{tu} = \infty$ when $n \geq 4$. The embedding of G_2 inside D_4 is the special case $n = 3$.

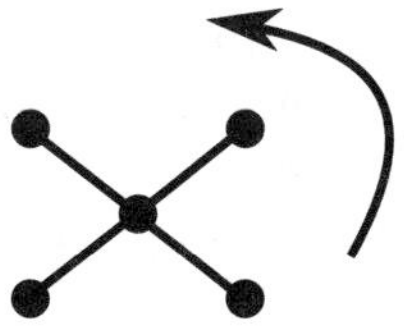

EXAMPLE 2.13. The group $\mathbb{Z}/2\mathbb{Z}$ acts on the dihedral group $I_2(m)$ for $m < \infty$, with invariant group of type A_1.

EXAMPLE 2.14. The group $\mathbb{Z}/2\mathbb{Z}$ acts on F_4 with quotient $I_2(8)$.

Let us list some other examples of quasi-split embeddings.

EXAMPLE 2.15. There are many examples of inclusions of affine Weyl groups, analogous to some of the embeddings of finite Weyl groups above.

EXAMPLE 2.16. Let (W, S) be a Coxeter group, and let (W', S') be the k-fold disjoint multiple of W. In order words, $S' = S \times \{1, \ldots, k\}$, with $m_{(s,i)(t,i)} = m_{st}$ and $m_{(s,i)(t,j)} = 2$ when $1 \leq i \neq j \leq k$. Let $I(s) = \{(s,i)\}_{1 \leq i \leq k}$. Clearly this gives a (diagonal) standard quasi-split embedding of Coxeter groups, with $L(s) = k$.

EXAMPLE 2.17. Suppose $k \geq 3$. The Dynkin diagram of the affine Weyl group $\widetilde{A}_{nk-1}$ looks like a circle with (nk) nodes. It has an action of $\mathbb{Z}/n\mathbb{Z}$, with invariant subgroup of type $\widetilde{A}_{k-1}$.

EXAMPLE 2.18. Let (W, S) be a universal Coxeter group (that is, $m_{st} = \infty$ for all $s, t \in S$). One can obtain any positive weight function L as an embedded weight function for a standard quasi-split embedding, in many ways. A simple such realization is to let S' be a complete multipartite graph, where an edge means $m_{uv} = \infty$, and the lack of an edge means $m_{uv} = 2$.

For fun and edification, we list some non-quasi-split embeddings.

EXAMPLE 2.19. There is an embedding of A_1 into any finite Coxeter group, sending the generator to the longest element. This is quasi-split only for (k-fold products of) dihedral groups.

EXAMPLE 2.20. There is an embedding of G_2 into B_3. It is standard but not quasi-split.

EXAMPLE 2.21. There is a standard, non-quasi-split embedding of H_4 into E_8, where $L(s) = 2$ for all simple reflections. There is a similar embedding of H_3 into D_6. These were shown to me by Lusztig, and like Example 2.6 above are associated to the 2-sided cell of elements with a unique reduced expression in H_3 and H_4. See [**16**, Remark 3.9].

EXAMPLE 2.22. Embed B_2 into A_n by letting $I(t) = \{s_1, s_n\}$ and $I(u) = \{s_i\}_{i=2}^{n-1}$. This is a non-standard, non-quasi-split Coxeter embedding, for which the longest element of B_2 is not sent to the longest element of A_n.

REMARK 2.23. Categorifying Hecke algebras with unequal parameters for non-quasi-split embeddings is a major open question. Examples 2.6 and 2.21 are particularly important, because they might provide an explanation for the apparent "geometric" behavior of non-crystallographic Coxeter groups like $I_2(m)$ and H_4.

2.2. Quasi-split embeddings and cyclic groups. When quasi-split embeddings are discussed in the literature, it is sometimes implicitly assumed that the group G which acts is cyclic, generated by an element σ. This is not much of an assumption, for the following reason. The embedded group (W, S, L) only depends on the orbits of the action, and not on the group action itself, and every quasi-split embedding can be realized for a cyclic group. This is evidenced by the comprehensive list of examples above. The reason that cyclic groups are important is that the original definition of quasi-split embeddings came from embeddings of algebraic groups in finite characteristic, and σ arises from the Frobenius automorphism. See [**15**, §16] for more details.

Later in this paper we will define a category $\mathbb{S}\mathrm{Bim}_G$ attached to any Coxeter system (W', S') with a group action by a group G, and a weighted equivariant Grothendieck group corresponding to an element $\sigma \in G$. Now, it is not just the orbits, but the group itself which plays a significant role. One hopes that this weighted

Grothendieck group is isomorphic to the Hecke algebra with unequal parameters corresponding to the invariant subgroup for the Coxeter embedding (to be defined soon). We will prove in §5.2 that this is only possible when σ acts transitively on each orbit, so that the orbits of the cyclic subgroup $\langle\sigma\rangle$ agree with the orbits of G. This is one reason why one may wish to assume that G is cyclic.

EXAMPLE 2.24. There is an embedding of B_2 into $A_3 \times A_3$, with $L(s) = 2$ and $L(t) = 4$. These orbits could arise from an action of $\mathbb{Z}/4\mathbb{Z}$ or an action of $\mathbb{Z}/2\mathbb{Z} \times \mathbb{Z}/2\mathbb{Z}$; this latter group does not contain any element of order 4, so no cyclic subgroup will have the same orbits.

However, even when an element $\sigma \in G$ exists which acts transitively on each orbit, the G-equivariant σ-weighted Grothendieck group and the $\langle\sigma\rangle$-equivariant σ-weighted Grothendieck group need not be equal.

The equivariant Grothendieck groups attached to actions of non-cyclic groups will not be Hecke algebras in the traditional sense. However, they may still be algebras with similarly attractive features, such as an interesting cell theory. It remains to be seen whether these algebras merit further investigation.

2.3. Dihedral quasi-split embeddings. Let us make a quick note about the possibilities for finite dihedral subgroups in quasi-split embeddings, which is also evident from the classification above.

PROPOSITION 2.25. *Suppose one has a quasi-split embedding $W \to W'$, and $s, t \in S$ with m_{st} finite. The following list encapsulates all possibilities for $I(s) \cup I(t)$.*

- *m_{st} is arbitrary, and $I(s) \cup I(t)$ is a disjoint union of k copies of $I_2(m_{st})$. One has $L(s) = L(t) = k$.*
- *$m_{st} = 2$, and $I(s) \cup I(t)$ is a disjoint union of $k + l$ copies of A_1. One has $L(s) = k$ and $L(t) = l$.*
- *$m_{st} = 4$, and $I(s) \cup I(t)$ is a disjoint union of k copies of A_3. One has $L(s) = k$ and $L(t) = 2k$.*
- *$m_{st} = 4$, and $I(s) \cup I(t)$ is a disjoint union of k copies of A_4. One has $L(s) = 3k$ and $L(t) = 2k$.*
- *$m_{st} = 6$, and $I(s) \cup I(t)$ is a disjoint union of k copies of D_4. One has $L(s) = k$ and $L(t) = 3k$.*
- *$m_{st} = 8$, and $I(s) \cup I(t)$ is a disjoint union of k copies of F_4. One has $L(s) = 4k$ and $L(t) = 2k$.*

In particular, $L(s) = L(t)$ unless $m_{st} = 2, 4, 6, 8$.

PROOF. Without loss of generality, W is dihedral, and $S' = I(s) \cup I(t)$. It is easy to classify the finite Coxeter systems (W', S') with a group action having only two orbits, and they correspond to those listed above. It remains to show that W' must be finite in order that $m_{st} < \infty$.

When the embedding is standard, the Coxeter element $h = st$ of W is sent under the embedding to a Coxeter element h' of W'. Since the order of h is finite, so is the order of h'. Howlett [**10**] has shown that the order of a Coxeter element of an infinite group is infinite; therefore, W' must be finite.

In general, one can make the following argument, explained to me by Meinolf Geck. The longest element of W goes to some element $w \in (W')^G$. We claim that w has every simple reflection in its (left) descent set, which forces it to be a longest element of W'. If some $u \in I(s)$ is not in the left descent set of w, then neither is any element in the entire orbit $I(s)$. Standard observations about cosets of parabolic subgroups indicate that w must be a minimal coset representative for the parabolic subgroup generated by $I(s)$, and thus $sw > w$. But then sw is a longer element in W, a contradiction. See [**7**, Lemma 2] for a similar proof. □

2.4. Hecke algebras with unequal parameters. The *Hecke algebra (with unequal parameters)* $\mathcal{H} = \mathcal{H}(W, S, L)$ is a $\mathbb{Z}[v, v^{-1}]$-algebra with *standard generators* H_s for $s \in S$, satisfying relations

$$(H_s + v^{L(s)})(H_s - v^{-L(s)}) = 0, \tag{2.1a}$$

$$H_s H_t \cdots = H_t H_s \cdots \tag{2.1b}$$

where both terms in (2.1b) are alternating of length $m_{s,t}$. Relation (2.1b) is omitted when $m_{s,t} = \infty$. We assume that L is a positive weight function henceforth.

An *expression* $\underline{x} = s_1 s_2 \dots s_d$ is a sequence of elements $s_i \in S$, and is denoted with an underlined letter. Removing the underline, one obtains the product $x \in W$. An expression as above is reduced if $\ell(x) = d$. We write $H_{\underline{x}} \stackrel{\text{def}}{=} H_{s_1} \cdots H_{s_d}$. When $\underline{x}$ is reduced, let $H_x = H_{\underline{x}}$, a product which does not depend on the chosen reduced expression. By convention $H_1 = 1$. The collection $\{H_x\}_{x \in W}$ is the *standard basis* of $\mathcal{H}$ over the ring $\mathbb{Z}[v, v^{-1}]$.

Given a polynomial $P \in \mathbb{Z}[v, v^{-1}]$ we write P^k for the coefficient of v^k in P. We write $\mathcal{H}_{\leq n}$ for those linear combinations $\sum_{w \in W} P_w H_w$ where $P_w^k = 0$ for $k > n$. We define $\mathcal{H}_{\geq n}$, $\mathcal{H}_{<n}$ and $\mathcal{H}_{>n}$ in the same way.

The *bar involution* is the $\mathbb{Z}$-linear map $\overline{(\cdot)} \colon \mathcal{H} \to \mathcal{H}$ satisfying $\overline{v} = v^{-1}$ and $\overline{H_x} = H_{x^{-1}}^{-1}$. It is an algebra involution. An element $b \in \mathcal{H}$ satisfying $\overline{b} = b$ is called *self-dual*. For each $x \in W$ there is a unique element $b_x \in \mathcal{H}_{\geq 0}$ which is self-dual, and for which $b_x = H_x$ modulo $\mathcal{H}_{>0}$. Together, these elements $\{b_x\}_{x \in W}$ are called the *Kazhdan-Lusztig basis* or *KL basis*. The *KL polynomials* $P_{w,x}$ are defined by $b_x = \sum_w P_{w,x} H_w$. One can show that $P_{w,x} = 0$ unless $w \leq x$, and that $P_{x,x} = 1$.

Our notation for the Hecke algebra differs from Lusztig's. To compare this definition with the definition in [**15**], one must switch v and v^{-1}. Then our H_x agrees with Lusztig's T_x, and our b_x agrees with his c_x.

In the split case (W, S, ℓ), there is a *standard pairing* on $\mathcal{H}$ with a variety of nice properties. This is a map $\mathcal{H} \times \mathcal{H} \to \mathbb{Z}[v, v^{-1}]$ satisfying

$$(b_w x, y) = (x, b_{w^{-1}} y) \tag{2.2a}$$

$$(x b_w, y) = (x, y b_{w^{-1}}) \tag{2.2b}$$

$$(b_w, b_v) = \delta_{wv} + v\mathbb{Z}[v] \tag{2.2c}$$

for all $v, w \in W$ and all $x, y \in \mathcal{H}$. The property (2.2c) states that the KL basis is *graded orthonormal.*

2.5. Another presentation. One special KL basis element is $b_s = H_s + v^{L(s)}$, which we call a *KL generator*. Just as the standard generators generate the Hecke algebra over $\mathbb{Z}[v, v^{-1}]$, so too do the KL generators. The presentation for $\mathcal{H}$ in terms of the KL generators is less well-known; it does not appear to be in the literature.

It is easy to observe that the quadratic relation on standard generators (2.1a) is equivalent to the following quadratic relation.

$$b_s b_s = (v^{L(s)} + v^{-L(s)}) b_s. \tag{2.3}$$

Given a sequence $\underline{x}$, we write $b_{\underline{x}}$ for the product $b_{s_1} b_{s_2} \cdots b_{s_d}$. Note that typically $b_{\underline{x}} \neq b_x$, even for reduced expressions. However, $b_{\underline{x}}$ is self-dual, and one can deduce that for a reduced expression $\underline{x}$,

$$b_{\underline{x}} = b_x + \sum_{y<x} Q_{y,\underline{x}} b_y \tag{2.4}$$

for some polynomials $Q_{y,\underline{x}} \in \mathbb{Z}[(v+v^{-1})]$. For the moment, we will call these terms $Q_{y,\underline{x}} b_y$ by the name "lower terms."

Pick $s, t \in S$ with $m_{st} < \infty$. Let w_0 denote the longest element of the dihedral parabolic subgroup $W_{s,t}$ generated by s and t. It has two reduced expressions, and therefore one can find two separate equations for b_{w_0}, one as $b_s b_t b_s \ldots$ plus certain lower terms, and one as $b_t b_s b_t \ldots$ plus certain other lower terms. The equality of these two expressions is an algebraic relation on the generators b_s and b_t, and this relation will be equivalent to the braid relation on standard generators (2.1b).

Therefore, the Hecke algebra has a presentation with KL generators, the relation (2.3), and some dihedral relation for each pair s, t with $m_{st} < \infty$, given by equating two expressions for b_{w_0}. It remains to find these two expressions for b_{w_0}, which we do here for the quasi-split case (i.e. using the restrictions on L coming from Proposition 2.25).

PROPOSITION 2.26. *Suppose that (W, S, L) is quasi-split, and let $s, t \in S$ with $m_{st} < \infty$. We give two descriptions of b_{w_0}, where w_0 is the longest element of $W_{s,t}$. Equivalently, given either reduced expression $\underline{x}$ for $w_0 \in W_{s,t}$, we decompose the product $b_{\underline{x}}$ into the KL basis.*

- *Suppose that $L(s) = L(t)$, and m_{st} is arbitrary. For $1 \leq k \leq m$ let $\underline{k}_s$ denote the reduced expression which alternates between s and t, has length k, and ends in s. Then*

$$b_{w_0} = b_{\underline{m}_s} + \sum_{k<m} \rho_{k,m} b_{\underline{k}_s}. \tag{2.5}$$

In this equation, the coefficients $\rho_{k,m}$ are all integers, and the same integers give a relation

$$V_{m+1} \cong V^{\otimes(m+1)} + \sum_{k<m} \rho_{k,m} V^{\otimes(k+1)}$$

in the Grothendieck group of $\mathfrak{sl}_2$ representations. The same equation holds with s and t switched. Equivalently, one has

$$b_{\underline{m}_s} = b_{w_0} + \sum_{k<m} \kappa_{k,m} b_{k_s}. \tag{2.6}$$

The positive integers $\kappa_{k,m}$ satisfy

$$V^{\otimes(m+1)} \cong V_{m+1} \oplus \bigoplus_{k<m} V_{k+1}^{\oplus \kappa_{k,m}}.$$

- *Suppose that $m_{st} = 2$, $L(s) = k$ and $L(t) = l$. Then*

$$b_{w_0} = b_s b_t = b_t b_s. \tag{2.7}$$

- *Suppose that* $m_{st} = 4$, $L(s) = k$ *and* $L(t) = 2k$. *Then*

$$b_{w_0} = b_s b_t b_s b_t - (v^k + v^{-k}) b_s b_t = b_t b_s b_t b_s - (v^k + v^{-k}) b_t b_s. \tag{2.8}$$

Equivalently, one has

$$b_s b_t b_s b_t = b_{stst} + (v^k + v^{-k}) b_{st}. \tag{2.9}$$

In fact, the same computation holds for $L(s) = 3k$ *and* $L(t) = 2k$.

- *Suppose that* $m_{st} = 6$, $L(s) = k$ *and* $L(t) = 3k$. *Then*

$$\begin{aligned} b_{w_0} &= b_s b_t b_s b_t b_s b_t - (2v^{2k} + 2v^{-2k}) b_s b_t b_s b_t + (v^{4k} + 1 + v^{-4k}) b_s b_t \\ &= b_t b_s b_t b_s b_t b_s - (2v^{2k} + 2v^{-2k}) b_t b_s b_t b_s + (v^{4k} + 1 + v^{-4k}) b_t b_s. \end{aligned} \tag{2.10}$$

Equivalently, one has

$$b_s b_t b_s b_t b_s b_t = b_{ststst} + (2v^{2k} + 2v^{-2k}) b_{stst} + (v^{-4k} + 3 + v^{4k}) b_{st}. \tag{2.11}$$

- *The case when* $m_{st} = 8$ *is not difficult, but we set a precedent here and ignore it.*

The proof of the case when $L(s) = L(t) = 1$ is discussed at length in [**3**], and implies the case $L(s) = L(t) = k$. The remaining cases are just straightforward computations in the dihedral group. which we leave as an exercise to the reader. For practice computing with dihedral groups in the unequal parameter case, see chapter 7 of [**15**].

The general method is to find a recursion formula for multiplying a KL basis element with a KL generator. For equal parameters, this recursion is exemplified by the formula

$$b_s b_{tststs} = b_{ststststs} + b_{ststs}.$$

The "closed form" of (2.6) comes from the observation that this recursion agrees with plethyism formulas for representations of $\mathfrak{sl}_2$, where $V_1 \otimes V_k \cong V_{k+1} \oplus V_{k-1}$ for $k \geq 1$. This agreement is actually a shadow of the geometric Satake equivalence. See [**3**] for further explanation.

Meanwhile, the recursion formulas for arbitrary unequal parameters are not difficult to derive. For example, when $0 < L(s) < L(t)$ one has the exemplary formulas

$$b_s b_{tststs} = b_{stststs}$$

and

$$b_t b_{stststs} = b_{tstststs} + (v^{L(t)-L(s)} + v^{L(s)-L(t)}) b_{tststs} + b_{tsts}.$$

A "closed form" formula for b_{w_0} for arbitrary dihedral groups would require some elementary combinatorics. However, it would be far more interesting to find a category with similar plethyism rules.

2.6. Invariant subalgebras of Hecke algebras. In this chapter, we fix a G-split embedding $(W, S, L) \to (W', S', \ell)$. One mystery this paper seeks to unravel is the connection between $\mathcal{H}(W, S, L)$ and $\mathcal{H}(W', S', \ell)$. For an element $x \in W$, we write c_x for the KL basis element in $\mathcal{H}(W, S, L)$. For an element $y \in W'$ (resp. $y \in W$) we write b_y for the KL basis element of y (resp. of $\varphi_I(y)$) in $\mathcal{H}(W', S', \ell)$.

The group G acts on W' and on $\mathcal{H}(W', S', \ell)$. Though one has $W = (W')^G \subset W'$, the size of the subalgebra $\mathcal{H}(W', S', \ell)^G$ is much larger than the size of W. After all, $\mathcal{H}(W', S', \ell)^G$ has a basis parametrized by orbits of G on W', given by sums $\sum_y b_y$ over the elements y of each orbit. Considering only the singleton orbits, one obtains a sub-basis $\{b_x\}_{x \in (W')^G}$, though the span of this sub-basis is not closed under multiplication.

We now give examples which compare, for various $x \in W$, the multiplication of b_x in $\mathcal{H}(W', S', \ell)^G$ with the multiplication of c_x in $\mathcal{H}(W, S, L)$. These examples will cover (almost) all the cases in Proposition 2.26.

EXAMPLE 2.27. Let W' have type $A_1 \times A_1$, with G acting transitively on $S' = \{t, u\}$. Then $s = tu$ generates $(W')^G = W$. One has

$$b_s b_s = b_t b_u b_t b_u = (v + v^{-1})^2 b_t b_u = (v^2 + 2 + v^{-2}) b_s. \tag{2.12}$$

In $\mathcal{H}(W, S, L)$, however, (2.3) implies that

$$c_s c_s = (v^2 + v^{-2}) c_s. \tag{2.13}$$

More generally, let W' be any finite Coxeter group, with A_1 embedded by sending s to the longest element of W'. Then one has $b_s^2 = [W'] b_s$ while $c_s^2 = (v^{L(s)} + v^{-L(s)}) c_s$. Here, $[W']$ denotes the balanced Poincaré polynomial of W', which has highest degree term $v^{L(s)}$ and lowest degree term $v^{-L(s)}$, but has various other terms in between.

EXAMPLE 2.28. Let W' have type $A_1^{\times k} \times A_1^{\times l}$, with G acting transitively on the first k factors and the last l factors. Let s denote the product of the first k simple reflections, and t denote the product of the last l. Then

$$b_s b_t = b_{st}, \tag{2.14}$$

just as

$$c_s c_t = c_{st}. \tag{2.15}$$

EXAMPLE 2.29. Let W' have type $I_2(m)^{\times k}$, with G acting transitively on these copies of $I_2(m)$. Let s and t be the generators of W, and let w_0 denote the longest element of W', also the longest element of W. Then

$$b_{\underline{m}_s} = b_{w_0} + \sum_{k<m} \kappa_{k,m} b_{k_s}, \tag{2.16}$$

just as

$$c_{\underline{m}_s} = c_{w_0} + \sum_{k<m} \kappa_{k,m} c_{k_s}. \tag{2.17}$$

EXAMPLE 2.30. Let W' have type A_3, with $G = \mathbb{Z}/2\mathbb{Z}$ acting on $S' = \{x, y, z\}$ to switch x and z. Then $s = y$ and $t = xz$ generate $(W')^G = W$. A computation yields

$$\begin{aligned} b_s b_t b_s b_t &= b_y b_x b_z b_y b_x b_z = b_{yxzyxz} + b_{yxyz} + b_{yzyx} + (v + v^{-1}) b_{yxz} \\ &= b_{stst} + (b_{yxyz} + b_{yzyx}) + (v + v^{-1}) b_{st}. \end{aligned} \tag{2.18}$$

In $\mathcal{H}(W, S, L)$, however, (2.9) implies that

$$c_s c_t c_s c_t = c_{stst} + (v + v^{-1}) c_{st}. \tag{2.19}$$

The same calculation suffices when W' has type $A_3^{\times k}$, and G also permutes these k factors. However, one must replace v with v^k. This is a general principle.

EXAMPLE 2.31. Let W' have type A_4, with $G = \mathbb{Z}/2\mathbb{Z}$ acting on $S' = \{s_1, s_2, s_3, s_4\}$ in the usual way. Then $t = s_1s_4$ and $u = s_2s_3s_2$ generate $(W')^G = W$ of type B_2. Let w_{123} and w_{234} be the longest elements of the corresponding parabolic subgroups of type A_3. A computation yields

(2.20)
$$b_tb_ub_tb_u = b_{tutu}+(b_{s_2s_1w_{234}} + b_{s_3s_4w_{123}})+(v + v^{-1})(b_{s_1w_{234}} + b_{s_4w_{123}})+(v+v^{-1})b_{tu}.$$

In $\mathcal{H}(W, S, L)$, however, (2.9) implies that

$$c_tc_uc_tc_u = c_{tutu} + (v + v^{-1})c_{tu}. \tag{2.21}$$

EXAMPLE 2.32. Let W' have type D_4, where $S' = \{u_1, u_2, u_3, v\}$, and let G act transitively on the vertices u_i. Let $s = v$ and $t = u_1u_2u_3$ denote the generators of W. Let $w_0 = ststst$ denote the longest element in W' and W. A computation yields

(2.22)
$$\begin{aligned} b_sb_tb_sb_tb_sb_t &= b_{ststst} + (b_{u_1u_2stst} + b_{u_1u_3stst} + b_{u_2u_3stst})+(v + v^{-1})(b_{u_1stst}+b_{u_2stst}+b_{u_3stst}) \\ &\quad + (2v^2 + 6 + 2v^{-2})b_{stst} + (v + v^{-1})(b_{vu_1u_2st} + b_{vu_1u_3st} + b_{vu_2u_3st}) \\ &\quad + (v^4 + 6v^2 + (3 + 9) + 6v^{-2} + v^{-4})b_{st}. \end{aligned}$$

In $\mathcal{H}(W, S, L)$, however, (2.11) implies that

$$c_sc_tc_sc_tc_sc_t = c_{ststst} + (2v^2 + 2v^{-2})c_{stst} + (v^{-4} + 3 + v^4)c_{st}. \tag{2.23}$$

We continue our precedent and ignore the case where W' has type F_4.

As evidenced by the examples above, mutliplication of the sub-basis $\{b_x\}$ for $x \in W \subset W'$ seems to behave approximately like the multiplication of $\{c_x\}$ in $\mathcal{H}(W, S, L)$, although with extraneous terms. Some of these extraneous terms have the form $\sum_y b_y$ over a non-trivial orbit of G in W', while others are multiples nb_x for $x \in W$. If only there were some way to eliminate these extraneous terms...

2.7. Some motivation. Our goal in the remainder of this paper is to give a categorical explanation for the similarities and differences between these formulas. Let us briefly motivate this, as in the introduction. One important result in the split case is called *KL positivity*.

THEOREM 2.33. *When $L = \ell$ (i.e. in the split case), the following properties hold.*

- $P_{w,x} \in \mathbb{N}[v]$.
- $b_xb_y = \sum_z m^z_{x,y}b_z$, *and* $m^z_{x,y} \in \mathbb{N}[v, v^{-1}]$.

This theorem was proven using categorification-theoretic methods by Kazhdan and Lusztig for Weyl groups, and by the author and Williamson for general Coxeter groups [**6**]. One proves that each coefficient of these polynomials is in fact the dimension of a certain vector space. On the other hand, KL positivity as stated will fail for Hecke algebras with unequal parameters. Various KL polynomials $P_{w,x}$ and structure coefficients $m^z_{x,y}$ have negative coefficients.

In this paper, for a G-split embedding, we construct a category $\mathbb{S}\mathrm{Bim}_G$ from which one can extract vector spaces equipped with an action of G. Taking the dimension of these vector spaces, one can obtain the (positive) coefficients for the subalgebra $\mathcal{H}(W', S', \ell)^G$. The dimension is the trace of the identity element in G.

If instead one takes the trace of a different element of G in special cases, one can obtain the coefficients for the algebra $\mathcal{H}(W, S, L)$. Traces of other elements need not be positive integers. In this sense does one cancel out the extraneous terms above: they correspond to sub-vector spaces with zero trace. In other words, the monoidal structure on $\mathbb{S}\mathrm{Bim}_G$ categorifies the formulas involving b_s above, while some trace cancellation reduces to the formulas involving c_s. The precise way in which $\mathbb{S}\mathrm{Bim}_G$ is said to categorify $\mathcal{H}(W, S, L)$ involves weighted Grothendieck groups, and will be elaborated in §3.

3. Equivariant categories and weighted Grothendieck groups

This section contains general background information on equivariant categories and the weighted Grothendieck groups associated to them. Definitions like these were made by Lusztig in his description of folding for quantum groups [**17**]. The results of this chapter are entirely folklore, though we were unable to find an exposition with the same level of generality.

3.1. Definitions.

DEFINITION 3.1. A *strict action* of a group G on a (graded) category $\mathcal{C}$ is the following data. For each $g \in G$ one has an autoequivalence of $\mathcal{C}$, also denoted g, such that the identity $e \in G$ is assigned the identity functor. When $\mathcal{C}$ is graded, these autoequivalences must commute naturally with the grading shift. For each $g, h \in G$, there is an invertible natural transformation $a_{g,h} \colon g \circ h \to gh$, satisfying an *associativity* condition for each triple $g, h, k \in G$:

$$a_{gh,k} \circ (a_{g,h} \otimes 1) = a_{g,hk} \circ (1 \otimes a_{h,k}) \colon g \circ h \circ k \to ghk. \tag{3.1}$$

REMARK 3.2. Another definition of a strict action often found in the literature will allow the identity $e \in G$ to be assigned to a functor naturally isomorphic to the identity functor. One fixes this natural isomorphism, and places additional conditions upon it. It is hardly less general to assert that e is sent to the identity functor.

We will often abuse notation and identify the functors $g \circ h$ and gh, leaving the natural transformation $a_{g,h}$ understood. The associativity condition allows us to identify the functors $g \circ h \circ k$ and ghk canonically.

DEFINITION 3.3. Let $\mathcal{D}$ be an additive (graded) category with a strict action of a group K, and let $\eta \in K^*$. The *η-weighted Grothendieck group* $[\mathcal{D}]_\eta$ of $\mathcal{D}$ is the free $\mathbb{C}$-module generated by the symbols $[M]$ for objects $M \in \mathcal{D}$, modulo the usual relation that $[M \oplus N] = [M] + [N]$, and the new relation that $[kM] = \eta(k)[M]$ for $k \in K$. When $\mathcal{D}$ is graded, this is also a $\mathbb{C}[v^{\pm 1}]$-module via $v[M] = [M(1)]$.

REMARK 3.4. One can generalize this definition to define a V-weighted Grothendieck group, for any representation V of K. A typical element is $[M] \otimes v$ for $v \in V$, and the new relation is $[kM] \otimes v = [M] \otimes k^{-1}v$.

We will apply this construction not to the action of G on $\mathcal{C}$, but to the action of G^* on the equivariant category $\mathcal{C}_G$.

DEFINITION 3.5. Suppose one has a strict action of G on $\mathcal{C}$. Given an object $B \in \mathcal{C}$, an *equivariant structure* on B is a *compatible* system of isomorphisms

$\varphi_g \colon B \xrightarrow{\sim} gB$ for each $g \in G$. Here, the system φ is compatible if $g(\varphi_h) \circ \varphi_g = \varphi_{gh} \colon B \to ghB$ for all $g, h \in G$.

$$\begin{array}{ccc} B & \xrightarrow{\varphi_g} & gB \\ & \searrow^{\varphi_{gh}} & \downarrow{\scriptstyle g(\varphi_h)} \\ & & ghB \end{array}$$

A *morphism* of equivariant structures from (B, φ) to (B', φ') is a morphism $f \in \mathrm{Hom}_{\mathcal{C}}(B, B')$ such that $\varphi'_g \circ f = g(f) \circ \varphi_g \colon B \to g(B')$ for all $g \in G$.

$$\begin{array}{ccc} B & \xrightarrow{\varphi_g} & gB \\ {\scriptstyle f}\downarrow & & \downarrow{\scriptstyle g(f)} \\ B' & \xrightarrow{\varphi'_g} & gB' \end{array}$$

The collection of equivariant structures (B, φ), and the morphisms between them, form the *equivariant category* $\mathcal{C}_G$.

Let $\mathcal{C}$ be an additive (graded) $\mathbb{C}$-linear category. Then $\mathcal{C}_G$ is also an additive (graded) $\mathbb{C}$-linear category. Let $G^* = \mathrm{Hom}(G, \mathbb{C}^*)$ denote the Pontrjagin dual of G, which is an abelian group.

Definition 3.6. We define a natural strict action of G^* on $\mathcal{C}_G$. For any character $\xi \in G^*$ and any compatible system of isomorphisms $\varphi = (\varphi_g)_{g \in G}$ for an object $B \in \mathcal{C}$, there is a new compatible system of isomorphisms $\xi \cdot \varphi$ given by the rescaling $(\xi \cdot \varphi)_g = \xi(g)\varphi_g$. Define $\xi(B, \varphi) = (B, \xi \cdot \varphi)$. If $f \in \mathrm{Hom}(B, B')$ gives rise to a map $(B, \varphi) \to (B', \varphi')$, then f also gives rise to a map $\xi(B, \varphi) \to \xi(B', \varphi')$. Thus $\xi \colon \mathcal{C}_G \to \mathcal{C}_G$ is a functor. Moreover, for $\xi, \eta \in G^*$, the composition of functors $\xi \circ \eta$ is actually equal (not just isomorphic) to the functor $\xi\eta$. Thus we can let $a_{\xi,\eta}$ be the identity natural transformation.

When we apply Definition 3.3 for $K = G^*$ and $\mathcal{D} = \mathcal{C}_G$, the character η is actually an element of the Pontrjagin double dual G^{**}, which is canonically isomorphic to the abelianization $G/[G, G]$.

Definition 3.7. Given a strict action of G on $\mathcal{C}$ and an element $g \in G$, the *g-weighted Grothendieck group* $[\mathcal{C}]_g$ of $\mathcal{C}$ will be the $\overline{g}$-weighted Grothendieck group of $\mathcal{C}_G$, where $\overline{g}$ is the class of g in $G/[G, G]$.

Remark 3.8. One can generalize these constructions slightly, either by changing the base ring of the category, or by changing the base ring of the decategorification. If $\mathcal{C}$ is $\Bbbk$-linear for some complete commutative local ring $\Bbbk$, one has an action of the abelian group $\mathrm{Hom}(G, \Bbbk^*)$ on $\mathcal{C}_G$. Similarly, if K acts on $\mathcal{D}$ and $\eta \in K^*$, then one can define the η-weighted Grothendieck group to be linear over any subring of $\mathbb{C}$ containing the image of η.

The most common such situation is when $G = \mathbb{Z}/2\mathbb{Z}$, as all characters are defined over $\mathbb{Z}$. There is an action of G^* on $\mathcal{C}_G$ even when $\mathcal{C}$ is only $\mathbb{Z}$-linear, and one can consider a $\mathbb{Z}$-form of the weighted Grothendieck group.

Suppose that $\mathcal{C}$ is a monoidal category, and that the action of G is by monoidal functors. Then $\mathcal{C}_G$ is also a monoidal category, with $(B, \varphi) \otimes (B', \varphi') = (B \otimes B', \varphi \otimes$

$\varphi')$. Moreover, for $\eta, \xi \in G^*$ one has $\eta(B,\varphi) \otimes \xi(B',\varphi') = (\eta\xi)(B \otimes B', \varphi \otimes \varphi')$, so that the action of G^* on $\mathcal{C}_G$ respects the monoidal structure in the appropriate way. Therefore, for any $g \in G$, the g-weighted Grothendieck group of $\mathcal{C}$ will inherit the structure of a ring.

Similarly, suppose that $\mathcal{C}$ is equipped with a contravariant duality functor $\mathbb{D}$, satisfying $\mathbb{D}^2 \cong \mathbb{1}$, and that the action of G commutes with $\mathbb{D}$ up to isomorphism. Then $\mathcal{C}_G$ can also be equipped with a duality functor, via $\mathbb{D}(B,\varphi) = (\mathbb{D}B, \mathbb{D}\varphi)$. The structure map is given by $(\mathbb{D}\varphi)_g = \mathbb{D}(g\varphi_{g^{-1}})\colon \mathbb{D}B \to \mathbb{D}gB$. The action of $\mathbb{D}$ is conjugate-linear with respect to the action of G^*, in the sense that $\mathbb{D}(\xi(B,\varphi)) = \xi^{-1}\mathbb{D}(B,\varphi)$. In particular, the action of $\mathbb{D}$ descends to the g-weighted Grothendieck group to be $\mathbb{C}$-conjugate-linear. When $\mathcal{C}$ is graded and $\mathbb{D}(M(1)) \cong \mathbb{D}(M)(-1)$, the action of $\mathbb{D}$ on the Grothendieck group is also v-antilinear.

Let $\mathcal{F}\mathrm{or}\colon \mathcal{C}_G \to \mathcal{C}$ denote the forgetful functor which only remembers the underlying object/morphism. It is faithful. $\mathcal{F}\mathrm{or}$ induces a map of Grothendieck groups $[\mathcal{C}_G] \to [\mathcal{C}]$, but it need not descend to $[\mathcal{C}]_g \to [\mathcal{C}]$. However, $\mathcal{F}\mathrm{or}$ clearly induces a map $[\mathcal{C}]_e \to [\mathcal{C}]$ for the identity element $e \in G$ (or for any element $g \in [G,G]$).

3.2. Equivariant mixed categories and indecomposable equivariant objects. For a graded category $\mathcal{C}$, linear over $\mathbb{C}$, we write

$$\mathrm{Hom}^k(X,Y) = \mathrm{Hom}^0(X, Y(k))$$

to denote the space of maps of degree k, and we write

$$\mathrm{Hom}(X,Y) = \oplus_{k\in\mathbb{Z}} \mathrm{Hom}^k(X,Y)$$

for the graded vector space of maps of all degrees. Assume that $\mathrm{Hom}^k(X,Y)$ is finite dimensional for all k and all X, Y. The graded dimension $\mathrm{gdim}\,\mathrm{Hom}(X,Y)$ will be a polynomial $P \in \mathbb{N}[v,v^{-1}]$, satisfying $P^k = \dim \mathrm{Hom}^k(X,Y)$.

DEFINITION 3.9. Consider an additive graded Karoubian $\Bbbk$-linear category $\mathcal{C}$ (for some base field $\Bbbk$), equipped with a duality functor $\mathbb{D}$ satisfying $\mathbb{D}(M(1)) = \mathbb{D}(M)(-1)$. Let $\mathbb{X}$ index the set of indecomposable objects, up to isomorphism and grading shift. Then $\mathcal{C}$ is called *mixed* if for each $x \in \mathbb{X}$ there is a choice of representative (i.e. choice of grading shift) $M_x \in \mathcal{C}$ for which:

- $\mathbb{D}(M_x) = M_{x^*}$ for some $x^* \in \mathbb{X}$;
- $\mathrm{gdim}\,\mathrm{Hom}(M_x, M_y) = \delta_{xy} + v\mathbb{N}[v]$.

In particular, no object is isomorphic to a grading shift of itself, and $\mathcal{C}$ has the graded Krull-Schmidt property. We say the mixed category is *self-dual* if $x^* = x$ for all $x \in \mathbb{X}$.

This definition (or a similar one) is found, for instance, in [**27**], which also contains many examples and a great deal of discussion.

Suppose that a finite group G acts on a self-dual mixed category $\mathcal{C}$, commuting with the grading shift and the duality functor. Clearly G will preserve the set of self-dual indecomposable objects in $\mathcal{C}$. Thus G will act on the set $\mathbb{X}$, by $gM_x \cong M_{gx}$. Let

$$M_{Gx} \stackrel{\text{def}}{=} \oplus_{y \in G\cdot x} M_y,$$

a direct sum of the indecomposables in the orbit of x. By the Krull-Schmidt property, for any $Y \in \mathcal{C}$ one has

(3.2)
$$Y \cong gY \text{ for all } g \in G \qquad \Longleftrightarrow \qquad Y \cong \bigoplus M_{Gx}(n) \text{ for various } x \in \mathbb{X}, n \in \mathbb{Z}.$$

Only an object Y satisfying (3.2) can be equipped with a G-equivariant structure.

Let us consider the possible equivariant structures which can be placed on $M_{Gx}^{\oplus n}$.

PROPOSITION 3.10. *Fixing an element x in the orbit Gx, and fixing an arbitrary compatible system on M_{Gx}, one has an equivalence of categories between compatible systems on $M_{Gx}^{\oplus n}$ for various n, and representations of the stabilizer G_x. Thus the indecomposable objects in $\mathcal{C}_G$ are classified (after fixing some data), up to isomorphism and grading shift, by an orbit $Gx \subset \mathbb{X}$ and an irreducible representation of the stabilizer G_x of x.*

First we look at some examples.

EXAMPLE 3.11. Suppose that $G \cong \mathbb{Z}/4\mathbb{Z}$, generated by the element σ. For an object $M \in \mathcal{C}$, a compatible system of isomorphisms φ on M would be determined by an isomorphism $\varphi_\sigma \colon M \to \sigma M$. The compatibility of the system amounts to the fact that

$$\sigma^3(\varphi_\sigma) \circ \sigma^2(\varphi_\sigma) \circ \sigma(\varphi_\sigma) \circ \varphi_\sigma = \varphi_{\sigma^4} = 1_M \in \mathrm{End}^0(M). \tag{3.3}$$

An orbit of G on $\mathbb{X}$ has size 1, 2, or 4. Consider first an orbit $G \cdot x$ of size 1, yielding an indecomposable object $M = M_x$ such that $M \cong \sigma M$. Fix an arbitrary compatible system, determined by an isomorphism $\alpha \colon M \to \sigma M$. This yields an identification of the one-dimensional spaces $\mathrm{Hom}^0(M, \sigma M)$ and $\mathrm{End}^0(M)$, sending $\varphi \in \mathrm{Hom}^0(M, \sigma M)$ to $\psi = \alpha^{-1}\varphi \in \mathrm{End}^0(M)$. Moreover, any element $\psi \in \mathrm{End}^0(M)$ is a scalar multiple of 1_M, so that $\sigma(\psi)$ is that same scalar multiple of $1_{\sigma M}$. From this we deduce that $\alpha\psi = \sigma(\psi)\alpha$ and $\alpha\psi^{-1} = \sigma(\psi^{-1})\alpha$. Using this, it is not hard to show that the relation (3.3) on a map $\varphi_\sigma \in \mathrm{Hom}^0(M, \sigma M)$ is equivalent to the relation

$$\psi^4 = 1_M.$$

Therefore, there are four equivariant structures on M, given by $\psi = \zeta_4^k 1_M$ for $k = 0, 1, 2, 3$. It is easy to see that these four structures are non-isomorphic in $\mathcal{C}_G$. The chosen compatible system α corresponds to $k = 0$.

Now consider an orbit of size 4, yielding a sum $M = M_x \oplus M_{\sigma x} \oplus M_{\sigma^2 x} \oplus M_{\sigma^3 x}$ of four distinct indecomposables, such that $M \cong \sigma M$. Fix an arbitrary compatible system, giving an identification of $\sigma(M_x)$ with $M_{\sigma x}$. A map $\varphi_\sigma \colon M \to \sigma M$ is determined by a quadruple of scalars $a, b, c, d \in \mathbb{C}$, corresponding to the maps $a 1_{M_x} \in \mathrm{End}^0(M_x) = \mathrm{Hom}^0(\sigma(M_x), M_{\sigma x})$, $b 1_{M_{\sigma x}} \in \mathrm{End}^0(M_{\sigma x})$, etcetera. The compatibility condition states that $abcd = 1$. There is one equivariant structure on M for each such quadruple. However, which quadruple a, b, c, d is chosen is irrelevant up to change of basis, or in other words, $(M, a, b, c, d) \cong (M, a', b', c', d')$ in $\mathcal{C}_G$, for any two quadruples. Therefore, there is only one isomorphism class of equivariant structure on M.

For an orbit of size 2, the argument is a mix of the two cases above. After choosing an isomorphism $\alpha \colon M \to \sigma^2 M$, the map φ_σ is determined by a pair of scalars $a, b \in \mathbb{C}$; compatibility states that $(ab)^2 = 1$; changing basis can alter the individual values of a and b, but will not change the product ab. Therefore, there are two isomorphism classes of equivariant structure on M, given by $ab = \zeta_2^k$ for $k = 0, 1$.

Note that these arguments rely heavily on the fact that $\mathrm{End}^0(M_x) \cong \mathbb{C}$ for any indecomposable object M_x. Without this assumption, equivariant objects can be more complex. This is one of the crucial features of mixed categories.

Now we sketch the proof of the proposition.

PROOF. Fix $x \in \mathbb{X}$ and consider an equivariant object $(M_{Gx} \boxtimes V, \varphi)$, for some multiplicity space V. By fixing an arbitrary G_x-equivariant structure on M_x, it is easy to see that φ induces a representation of G_x on V, in similar fashion to the examples above (where α determined the G_x-equivariant structure). It is an exercise to see that there is a morphism between equivariant objects precisely when there is a morphism of G_x representations, and that these morphisms compose appropriately. □

Note that this classification depended on an arbitrary choice of x and a G_x-equivariant structure on M_x. In the example above for an orbit of size 1, the indecomposable attached to the trivial representation of G depended upon a choice of isomorphism α. Once one chooses an equivariant structure to correspond to the trivial representation, the equivariant structures attached to other representations are determined. After making this choice, we denote the indecomposable objects in $\mathcal{C}_G$ by $(M_{Gx} \boxtimes V)$, for a representation V of G_x. We let $(M_{Gx} \boxtimes \mathbb{C})$ or $(M_{Gx}, +1)$ denote the irreducible corresponding to the trivial representation.

Because $\mathcal{F}$or is faithful, it is easy to see that morphism spaces between various $(M_{Gx} \boxtimes V)$ are concentrated in non-negative degree, with all degree zero morphisms being isomorphisms. Moreover, $\mathbb{D}(M_{Gx} \boxtimes V) \cong (M_{Gx} \boxtimes V^*)$. Therefore $\mathcal{C}_G$ is a mixed category (though not necessarily self-dual), and in particular it has the Krull-Schmidt property.

In future sections, we will write $[M_{Gx}, V]$ for the image of $(M_{Gx} \boxtimes V)$ in a Grothendieck group or weighted Grothendieck group.

3.3. Nonabelian examples. Let G be an arbitrary group acting on a self-dual mixed category $\mathcal{C}$, and let $g \in G$. The g-weighted Grothendieck group is (by the above definition) the quotient of $[\mathcal{C}_G]$ by the relation

$$[M_{Gx}, \xi \otimes V] = \xi(g)[M_{Gx}, V] \tag{3.4}$$

for any $\xi \in G^*$ and $V \in \mathrm{Rep}(G_x)$. We provide some non-abelian examples.

EXAMPLE 3.12. Let G be nonabelian, let $H = [G, G]$ be the commutator subgroup, and let K be the abelianization G/H. Suppose that $\mathbb{X}$ is identified with K as a G-set. Then $\mathbb{X}/G$ has one orbit, with stablizer H, and any character of G will fix any representation of H. There are two possibilities for the weighted Grothendieck group. If $g \in H$ then any relation of the form (3.4) will be trivial, as $\xi(g) = 1$ and $\xi \otimes V = V$ for any $\xi \in G^*$ and $V \in \mathrm{Rep}(H)$. Therefore, $[\mathcal{C}]_g = [\mathcal{C}_G] = [\mathrm{Rep}(H)]$. On the other hand, if $g \notin H$ then there exists some character $\xi \in G^*$ such that $\xi(g) \neq 1$. Then the relation $[V] = \xi(g)[V]$ will force $[V] = 0$ for all $V \in \mathrm{Rep}(H)$. Therefore $[\mathcal{C}]_g = 0$.

EXAMPLE 3.13. Let $G = S_n$, acting by the standard action on a set $\mathbb{X}$ of size n. Therefore, a given stabilizer is isomorphic to S_{n-1}. The only non-trivial character of S_n is the sign representation, which descends to the sign representation of any stabilizer. The irreducible representations V_λ of S_{n-1} are parametrized by partitions λ of $n-1$, and the action of the sign representation switches λ with its transpose partition λ^t. If g is even then $[\mathcal{C}]_g$ will have a basis given by partitions modulo transpose, with $[V_\lambda] = [V_{\lambda^t}]$. If g is odd, then $[\mathcal{C}]_g$ has a basis given by non-self-transpose partitions modulo transpose, with $[V_\lambda] = -[V_{\lambda^t}]$.

3.4. Decomposition in equivariant mixed categories. Let M be an arbitrary object in a self-dual mixed category $\mathcal{C}$. Within $\mathcal{C}$ there is a direct sum decomposition

$$M \cong \bigoplus_{x \in \mathbb{X}, k \in \mathbb{Z}} M_x(k)^{\oplus d_{x,k}}$$

for some multiplicities $d_{x,k}$. When M is self-dual one has $d_{x,k} = d_{x,-k}$. To compute the numbers $d_{x,k}$, one uses the *local intersection pairing* or *LIP*, which is the pairing given by composition

$$\mathrm{Hom}^{-k}(M_x, M) \times \mathrm{Hom}^k(M, M_x) \to \mathrm{End}^0(M_x) = \mathbb{C}.$$

Then $d_{x,k}$ will be the rank of this pairing. We let $V^k(M, M_x)$ denote the quotient of $\mathrm{Hom}^k(M, M_x)$ by the kernel of this pairing, a vector space of dimension $d_{x,k}$.

REMARK 3.14. It is not difficult to observe that the kernel of the local intersection pairing is contained in the Jacobson radical of $\mathcal{C}$. In fact, the union of all such kernels will generate the Jacobson radical, and modulo the Jacobson radical all morphisms will split. Any family of endomorphisms which become orthogonal idempotents modulo the Jacobson radical can be lifted to a family of orthogonal idempotents, so that questions of decomposition can be answered modulo the Jacobson radical, where $V^k(M, M_x)$ is a genuine Hom space. We will not pursue this style of algebra in this paper.

REMARK 3.15. Suppose that $\mathcal{C}$ is monoidal and that the monoidal identity $\mathbb{1}$ is a self-dual indecomposable object. Its endomorphism ring is a non-negatively-graded commutative ring $\mathrm{End}(\mathbb{1}) = R$, with only scalars in degree 0. This ring will act on the right and the left of any morphism space $\mathrm{Hom}(M, N)$, and all composition maps are R-bimodule maps. The action of the positive part R_+ will send any morphism into the kernel of the LIP. The reason is that the image of (either the right or left action of) R_+ in $\mathrm{End}(M_x)$ does not intersect $\mathrm{End}^0(M_x)$.

Now let (M, φ) be an arbitrary object in $\mathcal{C}_G$. Decomposing (M, φ) in $\mathcal{C}_G$ is more work than decomposing M in $\mathcal{C}$: a decomposition of representations, rather than a dimension count. For any $x \in \mathbb{X}$, let ψ be the chosen equivariant structure map for $(M_{Gx}, +1)$. The vector space $V^k(M, M_x)$ admits an action of the stabilizer G_x as follows: for $g \in G_x$ and $f \in V^k(M, M_x)$, let

$$g \cdot f = g(\psi_{g^{-1}}) \circ g(f) \circ \varphi_g \colon M \to gM \to g(M_x) \to M_x. \tag{3.5}$$

CLAIM 3.16. *For a representation V of G_x, the multiplicity of $(M_{Gx} \boxtimes V)$ in (M, φ) is equal to the multiplicity of V in $V^k(M, M_x)$. We call this multiplicity $d_{x,V,k}$.*

PROOF. This is a fairly straightforward argument, which we leave to the reader. The key point is that a morphism in a V-component of $V^k(M, M_x)$ will help produce an equivariant morphism from (M, φ) to $(M_{Gx} \boxtimes V)$. □

Our reason for computing decompositions in $\mathcal{C}_G$ will mainly be to compute the ring structure on $[\mathcal{C}]_g$.

3.5. The abelian case. Abelian groups have several advantages which contribute to the ease in understanding of their weighted Grothendieck groups. These advantages are

- For any subgroup $H \subset G$ (such as a stabilizer subgroup), the action of G^* on H^* is transitive.
- All representations are characters, so that the action of H^* on the objects in $\mathrm{Rep}(H)$ is transitive.
- For any subgroup H, G/H is abelian. Therefore, for any $g \notin H$, there is a character $\xi \in G^*$ such that $\xi(g) \neq 1$ but $\xi(H) = 1$.

When we write $x \in \mathbb{X}/G$, we mean that we choose a representative for an orbit. Because G is abelian, the stabilizer G_x is independent of the chosen orbit representative x. For a character $\xi \in G^*$ we write (M_{Gx}, ξ) for the corresponding indecomposable object, viewing ξ as a character of G_x. We write $[M, \varphi]$ for the class of an object (M, φ) in some weighted Grothendieck group of $\mathcal{C}$.

PROPOSITION 3.17. *Let G be an abelian group and let $g \in G$. The g-weighted Grothendieck group $[\mathcal{C}]_g$ has a basis $\{[M_{Gx}, +1]\}$ in bijection with the orbits Gx which g fixes pointwise, that is, for which $g \in G_x$.*

PROOF. Certainly the ordinary Grothendieck group $[\mathcal{C}_G]$ has a basis given by

$$\{[M_{Gx}, \xi]\}_{x \in \mathbb{X}/G, \xi \in G_x^*},$$

which splits as a vector space into subspaces spanned by $[M_{Gx}, \xi]$ for each fixed x. The relations imposed by the g-weighting preserve this decomposition. One has $[M_{Gx}, \xi] = \xi(g)[M_{Gx}, +1]$, so that certainly $[M_{Gx}, +1]$ spans its corresponding subspace, which is therefore at most one-dimensional. Additional relations can only be imposed by characters in $(G/G_x)^*$.

If $g \notin G_x$ then there exists some character $\xi \in G^*$ with $\xi(g) \neq 1$ and $\xi(G_x) = 1$. Therefore ξ preserves $(M_{Gx}, +1)$, so $[M_{Gx}, +1] = \xi(g)[M_{Gx}, +1]$, implying that $[M_{Gx}, +1] = 0$. On the other hand, if $g \in G_x$ then any character which preserves $(M_{Gx}, +1)$ will also send g to 1, so that no non-trivial relations exist on $[M_{Gx}, +1]$. □

COROLLARY 3.18. *The forgetful functor $\mathcal{F}\mathrm{or} \colon \mathcal{C}_G \to \mathcal{C}$ induces an isomorphism $[\mathcal{C}]_e \cong [\mathcal{C}]^G$.*

PROOF. By the previous proposition, $[\mathcal{C}]_e$ has a basis given by $\{[M_{Gx}, 1]\}$ ranging over all orbits $x \in \mathbb{X}/G$. The basis element $[M_{Gx}, +1]$ is sent to $\sum_{y \in Gx}[M_y]$, and together these images form a basis for $[\mathcal{C}]^G$. □

The following corollary will help to quickly compute the ring structure on $[\mathcal{C}]_g$.

COROLLARY 3.19. *For any $(M, \varphi) \in \mathcal{C}_G$, one has the following equality in $[\mathcal{C}]_g$:*

$$[M, \varphi] = \sum_{k \in \mathbb{Z}} \sum_{x \in \mathbb{X}/G \text{ such that } gx = x} \mathrm{Tr}_g V^k(M, M_x) \cdot v^k [M_{Gx}, +1]. \tag{3.6}$$

PROOF. We have already seen in Claim 3.16 how (M, φ) splits into indecomposables in $\mathcal{C}_G$. Let V be a representation of G for which the character ξ appears with multiplicity d_ξ. Then $\sum d_\xi \xi(g) = \mathrm{Tr}_g V$. The corollary follows easily. □

3.6. Reductions in the abelian case. Consider the situation where G is an abelian group acting on $\mathcal{C}$ trivially. The category $\mathcal{C}_G$ is much larger than $\mathcal{C}$, having an object (M,ξ) for each object $M \in \mathcal{C}$ and character $\xi \in G^*$. However, Proposition 3.17 implies that, for any $g \in G$, the g-weighted Grothendieck group $[\mathcal{C}]_g$ is isomorphic to the usual Grothendieck group $[\mathcal{C}]$ as a vector space; Corollary 3.19 implies that they are isomorphic as rings when $\mathcal{C}$ is monoidal. For purposes of the weighted Grothendieck group, we may as well assume that G is trivial. In this section, we consider several similar reductions.

Let G be an abelian group acting on a self-dual equivariant mixed category $\mathcal{C}$. Let $\mathcal{K} \subset \mathcal{C}$ be a monoidal Karoubian subcategory, closed under the action of G. Then $\mathcal{K}_G \subset \mathcal{C}_G$ is fully faithful, and this inclusion induces an injection of weighted Grothendieck groups which preserves the ring structure.

Let $K \subset G$, and suppose that it acts trivially (on both morphisms and objects) on a subcategory $\mathcal{K}$ as above. Let $G' = G/K$. The equivariant categories $\mathcal{K}_G$ and $\mathcal{K}_{G'}$ are certainly not equivalent. However, the results above imply that they have the same g-weighted Grothendieck groups, for an element $g \in G$ viewed as an element of either G or G'.

Combining these previous two observations allows us to simplify slightly when computing the ring structure on $[\mathcal{C}]_g$. For instance, if we want to compute the product $[M_{Gx}, +1][M_{Gy}, +1]$, we may work within the smallest G-closed monoidal Karoubian subcategory $\mathcal{K}$ which contains both M_{Gx} and M_{Gy}. Then we may assume that G acts faithfully on $\mathcal{K}$, because we may replace G with the quotient by the kernel of the action.

REMARK 3.20. More general statements can be made, because the condition that K acts trivially on objects is not a categorical one, and can be weakened. However, this reduction will help us in our discussion of folding below. In this case, one has a mixed category $\mathcal{D}$ attached to any Coxeter graph, with a monoidally-closed subcategory attached to any subgraph. The group G will act by automorphisms of a Coxeter graph, and for any union of orbits in the Coxeter graph there will be a subcategory like $\mathcal{K}$ above. Any subgroup K acting trivially on this union of orbits will act trivially on the corresponding subcategory. Thus when computing in the weighted Grothendieck group, we can replace G with its image in the permutation group of these orbits.

We must clarify this statement slightly. This comment may not make sense until one reads the chapter on Soergel diagrammatics. If a subgroup K acts trivially on a certain union of orbits in a Coxeter graph, but non-trivially somewhere, then K does not actually act trivially on morphism spaces in $\mathcal{K} \subset \mathcal{D}$. This is because $\mathbb{1} \in \mathcal{K}$ and $\mathrm{End}(\mathbb{1})$ is a graded polynomial ring R, whose positive part depends on the entire Coxeter graph and not just the union of orbits. However, as noted in Remark 3.15, the image of the action of R_+ on any Hom space is in the kernel of the LIP, and therefore will not contribute to any decomposition computations.

Here is another simplication we can use. Suppose that G is abelian, and that some element $g \in G$ acts transitively on each orbit in $\mathbb{X}$. Let $H \subset G$ be the cyclic subgroup generated by g. We now compare the g-weighted Grothendieck group of $\mathcal{C}$, when acted on by G or by H. The orbit spaces $\mathbb{X}/G$ and $\mathbb{X}/H$ are the same, and only singleton orbits will contribute to the g-weighted Grothendieck group in either case. The trace of g on $V^k(M, M_x)$ also is unaffected by the choice G versus H. Therefore, $[\mathcal{C}]_g$ is independent of whether $\mathcal{C}$ is acted on by G or by H. So if weighted by such an element, we may as well assume that G is cyclic. However, once again, the actual equivariant categories $\mathcal{C}_G$ and $\mathcal{C}_H$ are quite different.

REMARK 3.21. The combined assumptions that G is abelian, $g \in G$ acts transitively on each orbit, and G acts faithfully, are not sufficient to imply that $G = H$. One could have $G = \mathbb{Z}/3\mathbb{Z} \times \mathbb{Z}/3\mathbb{Z}$ and $\sigma = (1,1)$, acting on a disjoint union of two copies of $\mathbb{Z}/3\mathbb{Z}$.

As seen in the non-abelian examples above, one can not make any of these reductions when G is not abelian. For example, the standard action of S_n has a cyclic subgroup which acts transitively, but the weighted Grothendieck groups of S_n and this cyclic subgroup are very different.

Suppose G is non-abelian and $K \subset G$ acts on $\mathcal{D}$ trivially. Consider a tensor product M of indecomposables $(M_{Gx} \boxtimes V)$ in $\mathcal{D}_G$. Suppose that K acts trivially on each representation V of G_x. Then K will act trivially on each space $V^k(M, M_{Gx})$, and on every representation appearing in the decomposition of M. Therefore, the full subcategory $\mathcal{D}_G^K \subset \mathcal{D}_G$ whose indecomposable summands are $(M_{Gx} \boxtimes V)$, where V is a representation of G_x on which K acts trivially, is actually a monoidal subcategory. Thus, we may assume that G acts faithfully for some computations, such as computing decompositions and the monoidal structure within $\mathcal{D}_G^K$. Unfortunately, this will not help when it comes time to compute the ring structure on the weighted Grothendieck group, as the weighted relation will not "preserve" $\mathcal{D}_G^K$.

4. Soergel bimodules

In 1990, Soergel [**23**] defined a graded monoidal category of bimodules over a coinvariant ring of a Weyl group, known now as Soergel bimodules. This monoidal category has Grothendieck ring isomorphic to the Hecke algebra. Since then, Soergel has modified his construction to yield bimodules over a polynomial ring associated to any Coxeter group, which categorify the Hecke algebra $\mathcal{H}(W, S, \ell)$ of that Coxeter group. See [**24**] for more details.

However, in [**5**], the authors have given another description of the same monoidal category, in a fashion which does not reference bimodules at all. This description is by generators and relations, and uses the language of planar diagrammatics. We refer the reader to [**5**] for additional background information and references. Definition 4.2 is almost copied verbatim from [**5**]. For experts who do not need the review, it is important to note Claim 4.4, and to recall some relevant features of the category outlined in §4.3.

In this chapter we fix a Coxeter group (W, S). (This will be the Coxeter group (W', S') into which another Coxeter group will be embedded, but there is no need for apostrophes at the moment.)

4.1. Realizations.

The input to Soergel's construction (as modified in [**5**]) is a *realization* of (W, S), which is roughly a generalization of the reflection representation of W. A realization is the data of a representation $\mathfrak{h}$ of W over a base ring $\Bbbk$, together with a choice of simple roots $\alpha_s \in \mathfrak{h}^*$ and simple coroots $\alpha_s^\vee \in \mathfrak{h}$, satisfying certain nondegeneracy conditions. One requires that the action of a simple reflection on $\mathfrak{h}$ and $\mathfrak{h}^*$ is given by "reflection," so that for $v \in \mathfrak{h}^*$ one has $s(v) = v - \alpha_s^\vee(v)\alpha_s$.

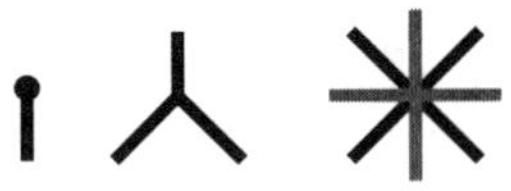

FIGURE 1. The vertices in an S-graph

Let $R = \mathcal{O}(\mathfrak{h}) = \mathrm{Sym}[\mathfrak{h}^*]$ denote the polynomial ring over $\Bbbk$ with linear terms equal to $\mathfrak{h}^*$. The ring R inherits an action of W. We equip R with a grading such that linear terms have degree 2.

4.2. Diagrammatics for Soergel bimodules.

DEFINITION 4.1. An *S-graph* is a certain kind of colored planar graph with boundary, properly embedded in the strip $\mathbb{R} \times [0, 1]$. In other words, all vertices of the graph lie on the interior of the strip, and edges may terminate either at a vertex or on the boundary of the strip. Additionally, an edge may form a closed loop. The number of vertices and the number of components are required to be finite. The edges of this graph are unoriented, and colored by elements $s \in S$. We call the place where an edge meets the boundary a *boundary point*; boundary points are not vertices. The boundary points on $\mathbb{R} \times \{0\}$ (resp. $\{1\}$) give a finite sequence of colored points, known as the *bottom boundary* (resp. *top boundary*) of the graph.

The only vertices allowed in an S-graph are of 3 types (see Figure 4.2):

- Univalent vertices (dots). These have degree $+1$.
- Trivalent vertices, where all three adjoining edges have the same color. These have degree -1.
- $2m$-valent vertices, where the adjoining edges alternate in color between two elements $s \neq t \in S$, and $m_{st} = m < \infty$. These have degree 0.

We also allow decorations called *polynomials* to float in the regions of the graph. These will be labeled by homogeneous polynomials $f \in R$. The *degree* of an S-graph is the sum of the degree of each vertex, and the degree of each polynomial.

We say that two Soergel graphs are *isotopic* if there is an isotopy of Soergel graph embeddings between them, relative to the boundary. Such an isotopy can alter the precise location of the boundary points on $\mathbb{R} \times \{0, 1\}$, but can not change the bottom or the top boundary sequence.

DEFINITION 4.2. Let $\mathcal{D}$ (or if there is ambiguity, $\mathcal{D}(W, S)$) denote the $\Bbbk$-linear monoidal category defined as follows. Objects are sequences $\underline{w}$, sometimes denoted $B_{\underline{w}}$, with monoidal structure given by concatenation. The space $\mathrm{Hom}_{\mathcal{D}}(\underline{w}, \underline{y})$ is the free $\Bbbk$-module generated by S-graphs with bottom boundary $\underline{w}$ and top boundary $\underline{y}$, modulo the relations below. Hom spaces will be graded by the degree of the graphs, and all the relations below are homogeneous.

The first (unwritten) relation is that two isotopic S-graphs are equal.

The polynomial relations:

(4.1)

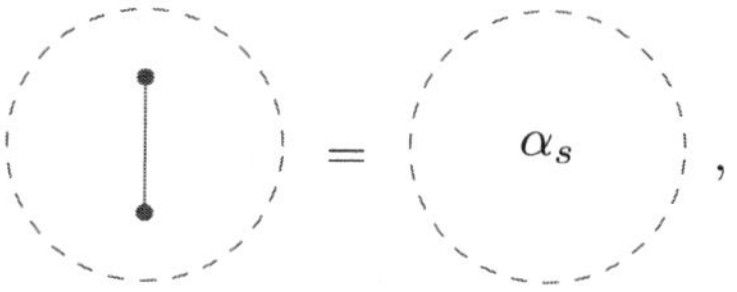

(4.2)

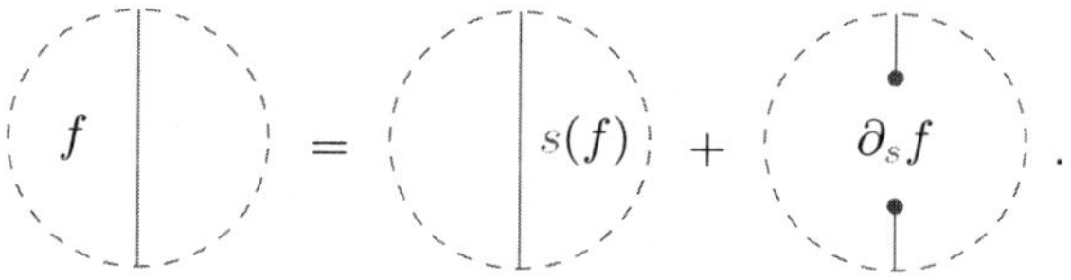

The one color relations:

(4.3)

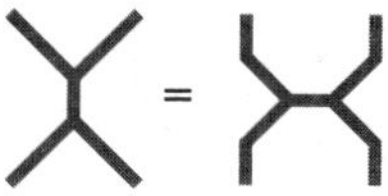

(4.4)

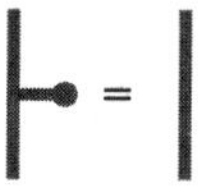

(4.5)

The two color relations: The color scheme depends slightly on the parity of $m = m_{st} < \infty$. We give one example of each relation for each parity; the reader can guess the general form.

(4.6)

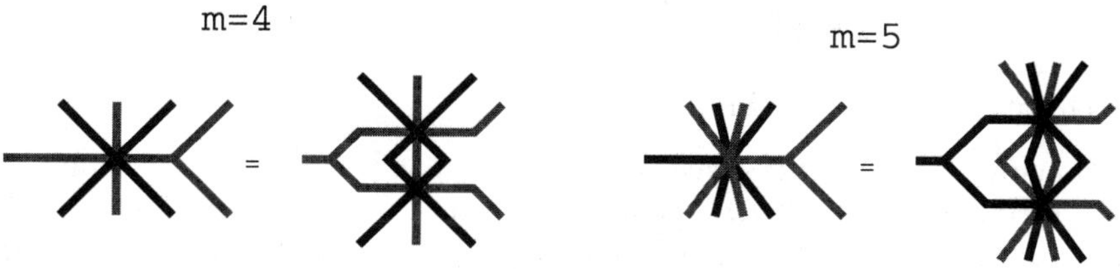

(4.7)

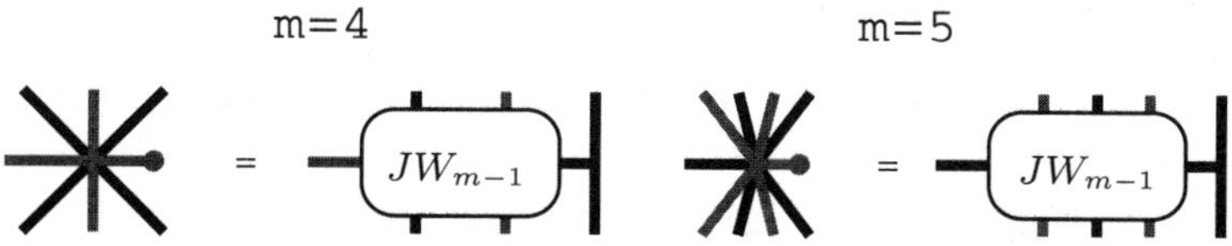

In equation (4.7) above, the *Jones-Wenzl morphism* JW_{m-1} is a $\Bbbk$-linear combination of graphs constructed only out of dots and trivalent vertices. For details, see [**5**].

The three color relations: It will be clear from the graphs which colors represent which indices.

For a triplet of colors forming a sub-Coxeter system of type $A_1 \times I_2(m)$ for $m < \infty$, we have

(4.8) $(A_1 \times I_2(m))$

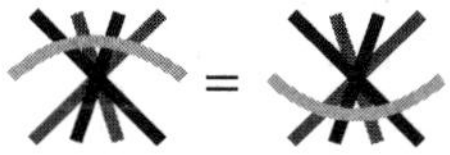

A specific example, when $m = 2$, is the case $A_1 \times A_1 \times A_1$:

(4.9) $(A_1 \times A_1 \times A_1)$ 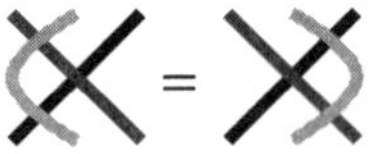

The last three relations are for types A_3, B_3, and H_3 respectively, and are known as the *Zamolodzhikov relations.*

(4.10) (A_3) 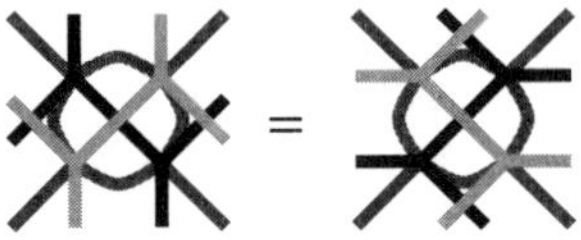

(4.11) (B_3) 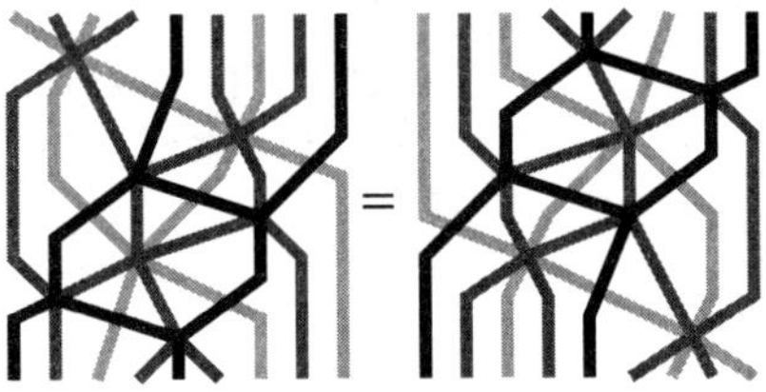

(4.12) (H_3) 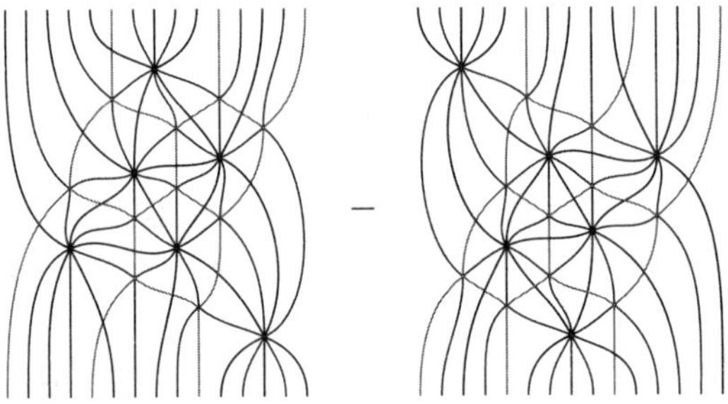 = lower terms

This concludes the definition of $\mathcal{D}$.

For any $w \in W$, the space of all (linear combinations of) morphisms which factor through objects $B_y(n)$ for $y < w$ is an ideal in $\mathcal{D}$, called $\mathcal{D}_{<w}$. When the element w is understood, elements of this ideal are called *lower terms.* For instance, in (4.12) above, $w = w_0$ is meant to be the longest element of H_3. Unfortunately, the relation (4.12) is only in general form, as the coefficients of the lower terms which appear are not known. Thankfully, the precise form of (4.12) will not be relevant for this paper.

Given an element $w \in W$, one can construct the (oriented) graph Γ_w whose vertices are the reduced expressions of w. There is a (bidirectional) edge in Γ_w between two expressions if they differ by a single application of a braid relation. There is a functor from this graph to $\mathcal{D}$, which sends a reduced expression $\underline{w}$ to the object $B_{\underline{w}}$, and sends an edge to the appropriate $2m$-valent vertex. To a path in Γ_w one associates a morphism in $\mathcal{D}$ known as a *rex move.* Two distinct rex moves with the same source and target need not be equal as morphisms in $\mathcal{D}$, but they are equal modulo lower terms. For example, all the three color relations can be thought of as equalities between two different rex moves in Γ_{w_0} (or in the case of H_3, equality modulo lower terms).

Note that (graded) Hom spaces are enriched in graded R-bimodules, since one can place a polynomial in the leftmost or rightmost region.

REMARK 4.3. Let us observe some of the consequences of these relations for *distant colors* $s, t \in S$ with $m_{st} = 2$. In this case, the two color relations are

(4.13)

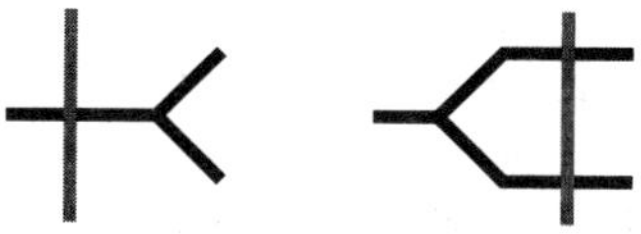

(4.14)

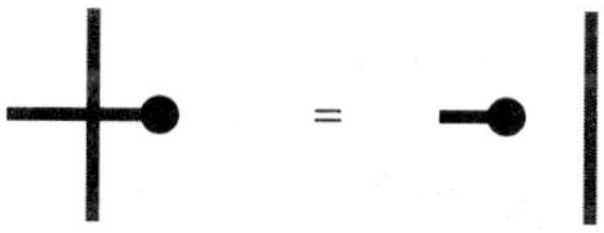

A consequence of these two relations is a "pulling apart" relation, which implies that the objects $B_s B_t$ and $B_t B_s$ are isomorphic in $\mathcal{D}$.

(4.15)

Combining these relations with the three color relations of type $A_1 \times I_2(m)$, one can conclude the following.

CLAIM 4.4. *Consider S-graphs whose colors lie in $S_1 \cup S_2 \subset S$, and contain no polynomials. Assume that every color in S_1 is distant from every color in S_2. We say that two such S-graphs with the same boundary are* distant-isotopic *if their S_1-colored subgraphs are isotopic, and their S_2-colored subgraphs are isotopic. Then any two distant-isotopic S-graphs represent equal morphisms in $\mathcal{D}$.*

4.3. Categorification results. Let $\mathbf{Kar}(\mathcal{D})$ denote the Karoubi envelope of $\mathcal{D}$, a graded monoidal category. The following result, known as the Soergel Categorification Theorem, was originally proven by Soergel. An easier proof was given by the authors in slightly more generality in [**5**].

THEOREM 4.5. *The category $\mathbf{Kar}(\mathcal{D})$ is Krull-Schmidt. The indecomposable objects are $\{B_w(n)\}$ for $n \in \mathbb{Z}$, parametrized up to shift by $w \in W$. Moreover, B_w is characterized as the unique summand of $B_{\underline{w}}$, for a reduced expression of w, which is not also a summand of $B_{\underline{y}}$ for any shorter sequence $\underline{y}$. The map*

$$b_s \mapsto [B_s]$$

gives an isomorphism of $\mathbb{Z}[v, v^{-1}]$-algebras from $\mathcal{H}(W, S, \ell)$ to $[\mathbf{Kar}(\mathcal{D})]$. The character map (defined below) gives an inverse isomorphism. Morphism spaces in $\mathbf{Kar}(\mathcal{D})$ are free as left or right R-modules, and counting the graded rank of morphism spaces gives the standard pairing on $\mathcal{H}$:

$$([M], [N]) = \operatorname{grank} \operatorname{Hom}_{\mathcal{D}}(M, N).$$

Let us record a number of significant features of these indecomposable bimodules.

- The category $\mathcal{D}$ also admits a duality functor, which we do not discuss. This duality descends in the Grothendieck group to the bar involution on $\mathcal{H}(W, S, \ell)$. The indecomposables B_w are self-dual.

- Suppose one has two reduced expressions $\underline{x}$ and $\underline{y}$ for $w \in W$. Then $B_{\underline{x}}$ and $B_{\underline{y}}$ each have a unique summand which does not factor through lower terms. However, these two summands are canonically isomorphic, and this isomorphism is induced by any rex move from $\underline{x}$ to $\underline{y}$.
- In the Grothendieck group, one has $[B_{\underline{w}}] = [B_w] + \sum_{y<w} m_{\underline{w},y}[B_y]$, which is an upper triangular decomposition. In particular, this implies that the Grothendieck group is generated as an algebra by $[B_s]$ for $s \in S$.
- Suppose that $I \subset S$ is a set of mutually commuting reflections. Let $\underline{x}$ be a sequence involving each element of I once; it is a reduced expression for the longest element w_I of the parabolic subgroup W_I. Then $B_{w_I} \cong B_{\underline{x}}$.

The following theorem, originally known as the Soergel conjecture, was proven in generality by the authors in [**6**].

THEOREM 4.6. *Suppose that the realization is obtained by base change from a realization defined over* $\mathbb{R}$. *Under the isomorphism* $\mathcal{H}(W,S,\ell) \to [\mathbf{Kar}(\mathcal{D})]$, *one has* $b_w \mapsto [B_w]$ *for all* $w \in W$.

One of the most significant implications of this theorem is the graded orthonormality of Hom spaces (c.f. section 2.4). In other words, $\mathrm{End}(\oplus_{w \in W} B_w)$ is concentrated in non-negative degree, and the only degree zero maps are (scalar multiples of) identity maps. Therefore, $\mathbf{Kar}(\mathcal{D})$ is a mixed category.

Given any object $M \in \mathbf{Kar}(\mathcal{D})$, let $P_{w,M} \in \mathbb{N}[v,v^{-1}]$ denote the graded dimension of the morphism space $\mathrm{Hom}(M, B_w)$ modulo $\mathcal{D}_{<w}$. For any choice of reduced expression $\underline{w}$, this morphism space is isomorphic to $\mathrm{Hom}(M, B_{\underline{w}})$ modulo lower terms. The sum $\sum_{w \in W} P_{w,M} H_w \in \mathcal{H}(W,S,\ell)$ is called the *character* of M. Character is obviously additive under direct sums, and thus yields a $\mathbb{Z}[v,v^{-1}]$-module map $[\mathbf{Kar}(\mathcal{D})] \to \mathcal{H}(W,S,\ell)$. As Theorem 4.5 states, this gives the inverse isomorphism to the map $b_s \mapsto [B_s]$.

Let us fix a reduced expression $\underline{w}$. For any sequence $\underline{x}$, Libedinsky [**13**] gives a (non-canonical) combinatorial construction of morphisms in $\mathrm{Hom}(B_{\underline{x}}, B_{\underline{w}})$, called *light leaves*, which descend to a (homogeneous) basis of $\mathrm{Hom}(B_{\underline{x}}, B_w)$ modulo $\mathcal{D}_{<w}$. We give light leaves a diagrammatic interpretation in [**5**]. For an idempotent $e \in \mathrm{End}(B_{\underline{x}})$, one can compute the character of the image of e by computing the graded dimension of $\mathrm{Hom}(B_{\underline{x}}, B_{\underline{w}})e$ modulo $\mathcal{D}_{<w}$. This is not easy to do in general, but light leaves make the computation of individual cases fairly explicit.

Libedinsky's light leaves can be used to construct a cellular basis of $\mathcal{D}$, known as the basis of *double leaves*. For an reduced expression $\underline{w}$, the identity of $\mathrm{End}(B_{\underline{w}})$ can be considered to span the highest cell of this endomorphism ring. The idempotent which projects from $B_{\underline{w}}$ to B_w is the unique idempotent which, expressed in the double leaves basis, has a non-zero coefficient of the identity; moreover, this coefficient is 1.

5. Folding

Fix a G-split embedding $(W,S,L) \to (W',S',\ell)$. Let $\mathcal{D}$ denote the category defined above for (W',S'). Then there is a strict action of G on $\mathcal{D}$ given by permuting colors. In other words, G acts on the generating objects by sending B_s to $B_{g(s)}$; G acts on S-graphs by permuting the colors of edges. We assume henceforth that the realization of (W',S') used to define $\mathcal{D}$ is the base change of a realization defined over $\mathbb{R}$.

As discussed in section 4.3, the Soergel conjecture implies that $\mathbf{Kar}(\mathcal{D})$ is mixed. This in turn allows us to apply the results of section 3.5, and to state a great deal about the equivariant category $\mathcal{D}_G$. For instance, Corollary 3.18 already implies the following result.

PROPOSITION 5.1. *If G is abelian, then there is an isomorphism of $\mathbb{Z}[v, v^{-1}]$-algebras from $[\mathcal{D}]_e$ to $\mathcal{H}(W', S', \ell)^G$.*

Let us be precise in naming the indecomposable objects of $\mathcal{D}_G$. Let $\underline{w}$ be a reduced expression for $w \in W'$, and let us fix an inclusion $B_w \overset{\oplus}{\subset} B_{\underline{w}}$ (this is unique, up to scalar). For any $g \in G$, this also fixes an inclusion $B_{g(w)} \overset{\oplus}{\subset} B_{g(\underline{w})}$. Suppose that $g(w) = w$, so that $\underline{w}$ and $g(\underline{w})$ are two different reduced expressions for the same element. Then any rex move $B_{\underline{w}} \to B_{g(\underline{w})}$ induces the canonical isomorphism $B_w \to g(B_w)$. Together, the collection of these canonical isomorphisms places the *standard* equivariant structure on B_w, which we denote as $(B_w, +1)$. Any other equivariant structure has the form $(B_w \boxtimes V)$ for some representation V of the stabilizer G_w.

One immediate consequence is the following.

CLAIM 5.2. *Let $\underline{w} = s_1 s_2 \ldots$ denote a reduced expression in W, for an element $w \in W \subset W'$. Then $(B_w, +1)$ is the top summand of $(B_{s_1}, +1) \otimes (B_{s_2}, +1) \otimes \cdots$ in $\mathcal{D}_G$.*

Warning: s_i is not a simple reflection in W', but instead a product of commuting simple reflections $I(s_i)$. The indecomposable B_{s_i} is in fact isomorphic to $B_{t_1} B_{t_2} \cdots B_{t_{L(s_i)}}$ for any order on the simple reflections $t_j \in R(s_i)$.

PROOF. We know that B_w is the top summand of $B_{s_1} \otimes B_{s_2} \otimes \cdots$ inside $\mathcal{D}$, and the space of possible idempotents is one-dimensional. The equivariant structure map of $(B_{s_1}, +1) \otimes (B_{s_2}, +1) \otimes \ldots$ is a tensor product of rex moves, and is therefore itself a rex move. This induces the standard equivariant structure on B_w. □

Our goal is to compute the weighted Grothendieck ring $[\mathcal{D}]_\sigma$ for $\sigma \in G$. We will not be able to do this in general, though we can do it when G is abelian and $\sigma \in G$ acts transitively on each orbit. In this case, the reductions of section 3.6 allow us to assume that G is cyclic, generated by σ. The results of section 3.5 imply that $[\mathcal{D}]_\sigma$ is a free $\mathbb{Z}[v, v^{-1}]$-module with basis given by $[B_w, +1]$ for $w \in (W')^G = W$.

COROLLARY 5.3. *Suppose that G is abelian, and $\sigma \in G$ is such that $\{[B_w, +1]\}_{w \in W \subset W'}$ span $[\mathcal{D}]_\sigma$. Then the elements $\{[B_s, +1]\}_{s \in S}$ generate $[\mathcal{D}]_\sigma$ as a ring.*

PROOF. By the previous claim, $(B_w, +1)$ is a summand of a tensor product of various $(B_s, +1)$, having multiplicity one. The remaining summands are $(B_v \boxtimes V)$ for $v < w$, whose images in the Grothendieck group are in the span of $[B_v, +1]$. Therefore, the fact that each $[B_w, +1]$ is in the subring generated by $[B_s, +1]$ is a simple upper-triangularity argument. □

THEOREM 5.4. *Let G act on (W', S'), and let (W, S, L) be the corresponding quasi-split embedded Coxeter subgroup. Suppose G is abelian, and $\sigma \in G$ acts transitively on each G-orbit in S'. Then there is an isomorphism of $\mathbb{Z}[v, v^{-1}]$-algebras $\mathcal{H}(W, S, L) \to [\mathcal{D}]_\sigma$, sending $b_w \mapsto [B_w, +1]$. The weighted character map (defined in §5.8) provides an inverse isomorphism.*

PROOF. The proof will take place throughout the remainder of this chapter, but we outline it here.

First we show the existence of an algebra map $\psi\colon \mathcal{H}(W,S,L) \to [\mathcal{D}]_\sigma$, sending $b_s \mapsto [B_s, +1]$. To show this, we must check the relations of the presentation given in section 2.5. This is really the crux of the proof, and follows from the explorations into decomposition of the following sections.

The remainder of the proof is an abstract and mostly tautological unwinding of the consequences of the categorification results in section 4.3. By the previous corollary, there is a unipotent upper triangular change of basis matrix between certain monomials in $[B_s, +1]$, and $\{[B_w, +1]\}$. Since $[B_s, +1]$ generates $[\mathcal{D}]_\sigma$ as a ring, so the map ψ is surjective. As another consequence, the change of basis matrix between $\psi(b_w)$ and $[B_w, +1]$ must also be unipotent upper triangular, and thus ψ is injective.

In the final section we define the weighted character map, and claim that it expresses $[M, \varphi]$ in terms of the standard basis of $\mathcal{H}(W,S,L)$. Using this, one can show that $[B_w, +1]$ satisfies the defining properties of the KL basis element b_w. However, we only sketch this result, as our favorite proof requires a much more in-depth discussion of localization and light leaves in order to state rigorously. □

Let G be abelian, and suppose that $\sigma \in G$ does not act transitively on each orbit. It is easy to find an element $x \in W' \setminus W$ fixed by σ, and the direct sum B_{Gx} will contribute to the σ-weighted Grothendieck group of $\mathcal{D}$. Therefore, it is not possible that $[\mathcal{D}]_\sigma$ is the Hecke algebra with unequal parameters, because it has a basis strictly larger than W. Nonetheless, when we perform our decompositions below, we will not assume that σ acts transitively on each orbit. We will still be able to compute certain formulas in the Grothendieck ring, though this will not be enough to determine the entire ring structure. Some non-abelian examples are also computable in this partial sense.

For standard Coxeter embeddings, the computations below are done mostly in full. For the embedding of A_1 in $I_2(m)$, or B_2 in A_4, the computations are not hard but use the *thick calculus* of [**2**] and [**3**]. In order to keep this paper shorter, we just sketch these computations. We have not attempted the computation for the embedding of $I_2(8)$ inside F_4.

5.1. Checking the quadratic relation I. We begin by checking the quadratic relation (2.3). This calculation will provide an excellent illustration of the methods involved.

First, we remind the reader why (2.3) holds in the split case. Suppose that s is a simple reflection. Then

$$B_s B_s \cong B_s(1) \oplus B_s(-1) \tag{5.1}$$

in $\mathcal{D}$. To prove this, one constructs inclusion maps $i_1, i_2\colon B_s \to B_s B_s$ of degrees $+1, -1$ respectively, and projection maps $p_1, p_2\colon B_s B_s \to B_s$ of degrees $-1, +1$. These maps are pictured below, and they satisfy:

$$1_{B_s B_s} = i_1 p_1 + i_2 p_2, \tag{5.2a}$$

$$1_{B_s} = p_1 i_1 = p_2 i_2, \tag{5.2b}$$

$$0 = p_1 i_2 = p_2 i_1. \tag{5.2c}$$

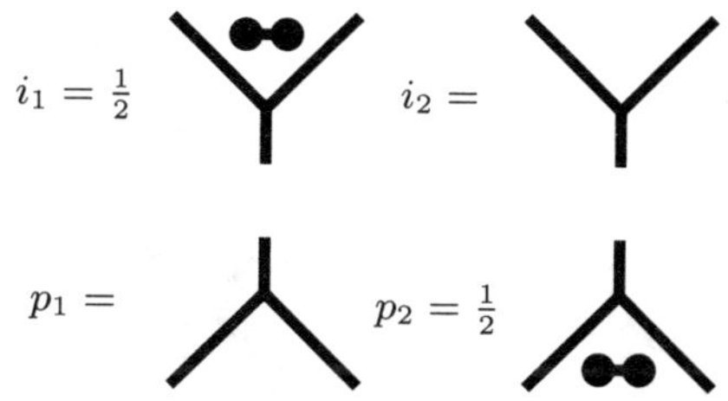

As an extended warmup, let us consider the diagonal quasi-split embedding of A_1 into $A_1 \times A_1$, with $G = \mathbb{Z}/2\mathbb{Z}$, so that σ switches the two colors in $A_1 \times A_1$. The reader should keep Example 2.27 in mind. Let $S' = \{t, u\}$ and let $S = \{s\}$, with the embedding $s = tu$. Viewing s as an element of W' one has $B_s = B_t B_u$. By the pulling apart relation (see Remark 4.3), one has $B_t B_u \cong B_u B_t$. Therefore, in $\mathcal{D}$ one has

$$B_s B_s \cong B_t B_u B_t B_u \cong B_t B_t B_u B_u \cong B_s(2) \oplus B_s \oplus B_s \oplus B_s(-2). \tag{5.3}$$

Thus we can write the identity of $B_t B_u B_t B_u$ as a sum of four orthogonal idempotents, as below. These factor with inclusion and projection maps i_k, p_k, for $1 \leq k \leq 4$.

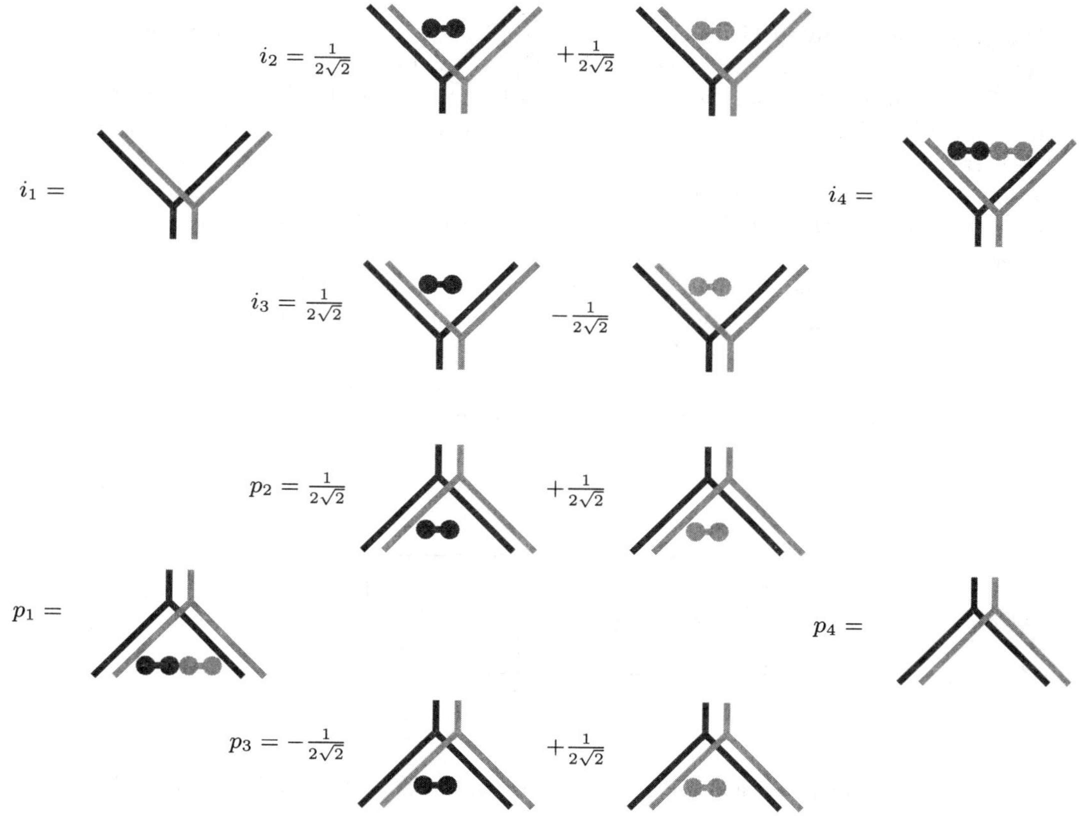

Now we ask what happens in the equivariant category $\mathcal{D}_G$. The equivariant structure map φ_σ of $(B_t B_u, +1)$ is the 4-valent vertex . Then $(B_t B_u, +1) \otimes$

$(B_tB_u, +1)$ has equivariant structure map 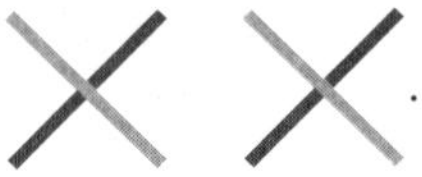 . Using distant-isotopy one can observe that

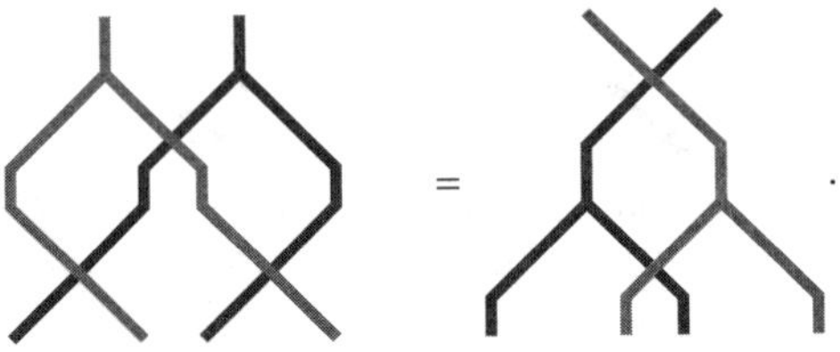

Symbolically, this says that $\sigma(p_4) \circ \varphi = \varphi \circ p_4$, which is precisely the condition that p_4 is a map from $(B_tB_u, +1) \otimes (B_tB_u, +1)$ to $(B_tB_u, +1)$. However, the reader can observe that p_3 and i_3 are maps between $(B_tB_u, +1) \otimes (B_tB_u, +1)$ and $(B_tB_u, -1)$, as $\sigma(p_3) \circ \varphi = -\varphi \circ p_3$.

Together, these maps imply the direct sum decomposition

$$\begin{aligned}(5.4)\quad &(B_tB_u, +1) \otimes (B_tB_u, +1) \cong \\ &\quad (B_tB_u(2), +1) \oplus (B_tB_u(0), +1) \oplus (B_tB_u(0), -1) \oplus (B_tB_u(-2), +1)\end{aligned}$$

in $\mathcal{D}_G$. In the weighted Grothendieck group $[\mathcal{D}]_\sigma$, the two terms with no degree shift cancel, leaving $[B_s, +1]^2 = (v^2 + v^{-2})[B_s, +1]$ as desired.

The decomposition (5.4) holds true for any G acting transitively on $A_1 \times A_1$, not just when $G = \mathbb{Z}/2\mathbb{Z}$. Let G be a group acting transitively on $A_1 \times A_1$, and let K be the kernel of the action, so that $G/K \cong \mathbb{Z}/2\mathbb{Z}$. Let -1 denote the representation of G which factors through the sign representation of G/K. Then G is the stabilizer of the element $s = tu \in W'$, and it is clear that (5.4) holds, interpreting $+1$ and -1 as representations of G.

When G is abelian, one can compute directly in the weighted Grothendieck group without needing to construct the specific inclusion and projection maps, instead using Corollary 3.19. One can examine the graded vector space V^k of (linear combinations of) morphisms $B_tB_uB_tB_u \to B_tB_u$ in degree k, modulo the kernel of the LIP. The G action of (3.5) is given by:

(5.5) 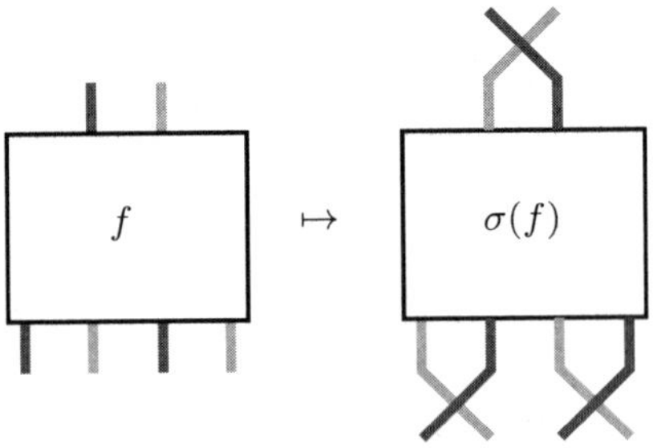

Computing the trace of σ will give multiplicities in the Grothendieck group $[\mathcal{D}]_\sigma$. For instance, V^{-2} one-dimensional, spanned by p_4. Since σ acts by the identity on this space, $[B_tB_u(2), +1]$ appears with multiplicity 1. On the other hand, V^0 is spanned by the two pictures in either p_2 or p_3, and these two pictures are switched by σ, so the trace is zero. Therefore $[B_tB_u(0), +1]$ appears with multiplicity zero.

REMARK 5.5. The actual space of degree 0 morphisms is larger than two-dimensional, also allowing for positive degree polynomials to be placed next to i_1. However, all such maps are in the kernel of the LIP, as is any map in the image of the left (or right) action of positive degree polynomials.

5.2. Checking the quadratic relation II. More generally, A_1 can embed in a quasi-split embedding as the longest element of A_1^k, or as the longest element of $I_2(m)^k$.

First we treat the case of A_1^k, where $k = L(s)$. Let $s \in S$ be the product of the commuting reflections $I(s) = \{t_1, \ldots, t_{L(s)}\} \subset S'$. One has $B_s = B_{t_1} \cdots B_{t_{L(s)}}$. Our goal is to decompose $(B_s, +1) \otimes (B_s, +1)$ into indecomposables in $\mathcal{D}_G$. Because of the reductions in section 3.6, we may work within the subcategory of $\mathcal{D}$ which only involves the colors in $I(s)$, and we may assume that G acts on $I(s)$ faithfully, so that we identify G with a subgroup of the symmetric group $S_{L(s)}$. We can make this assumption even when G is not abelian; however, translating this decomposition into a statement about the weighted Grothendieck ring will be difficult unless G is abelian.

In $\mathcal{D}$, the only summands of B_sB_s are grading shifts of B_s. We already know what the graded vector space $V^\bullet(B_sB_s, B_s)$ of projection maps is: it has a basis given by maps of the following form.

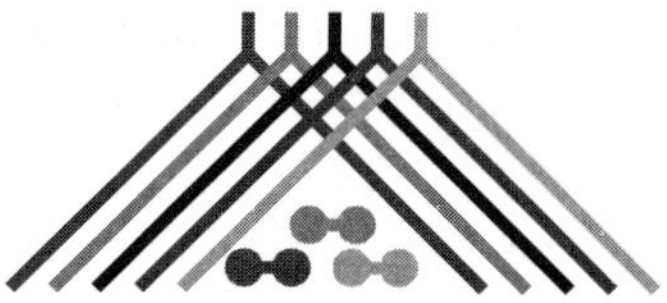

There is one such map for each subset $J \subset I(s)$ indicating which colors of barbells appear, and the map has degree $-L(s) + 2|J|$. The group G acts on this basis precisely as it acts on the subsets of $I(s)$, so that we need only decompose this subset representation into irreducibles for G.

If G is abelian, then computing the product in the σ-weighted Grothendieck group is easy. One has

$$[B_s, +1]^2 = \sum_{i=0}^{L(s)} v^{-L(s)+2i} c_i [B_s, +1], \tag{5.6}$$

where c_i is the number of subsets $J \subset I(s)$ which are fixed by σ, with $|J| = i$. Thinking of σ as an element in the symmetric group $S_{L(s)}$, the cycle decomposition of σ will determine these multiplicities c_i. For instance, if $L(s) = 3$ and σ is a 2-cycle, then $[B_s, +1]^2 = (v^3 + v^1 + v^{-1} + v^{-3})[B_s, +1]$. Similarly, if $L(s) = 4$ and σ is a product of disjoint 2-cycles then $[B_s, +1]^2 = (v^4 + 2 + v^{-4})[B_s, +1]$. In particular, one has

$$[B_s, +1]^2 = (v^{L(s)} + v^{-L(s)})[B_s, +1] \tag{5.7}$$

if and only if σ is a cycle with no fixed points. This was the quadratic relation in $\mathcal{H}(W, S, L)$.

When σ is a cycle, only the idempotents where each color behaved the same will contribute to the weighted Grothendieck group. *Mixed idempotents*, where some colors behave in one way, and others in another way, may appear as summands in

$\mathcal{D}_G$ but will not contribute to the weighted Grothendieck group. This is because mixed idempotents are permuted by σ with no fixed points.

Unraveling our reductions, one has the following statement: let G be abelian, acting (possibly non-faithfully) on (W', S', ℓ), and choose $\sigma \in G$. Suppose the G-split embedding is standard. Then the quadratic relation in $\mathcal{H}(W, S, L)$ holds for $[B_s, +1]$ in $[\mathcal{D}]_\sigma$ if and only if σ acts transitively on the orbit $I(s)$.

Now we sketch what happens for the embedding of $W = A_1$ into $W' = I_2(m)$, sending s to the longest element w_0 of the dihedral group. Let $\sigma \in \mathbb{Z}/2\mathbb{Z}$ swap the two colors of the dihedral group. We know that $B_{w_0} \otimes B_{w_0} \cong [W']B_{w_0}$, where again $[W']$ denotes the balanced Poincaré polynomial of the dihedral group. A choice of inclusion and projection maps can be made by choosing dual bases for the Frobenius extension $R^{W'} \subset R$, relative to the Demazure operator ∂_{w_0}. For more information on this decomposition, see the discussion of thick calculus in [**3**, §6.3], and perhaps also the diagrammatic discussion surrounding [**2**, Equation (3.43)].

Finding dual bases is an exercise in Schubert calculus; observing that σ is traceless on all but the highest and lowest degrees, where it acts as the identity, is also an exercise. Thus one has

$$[B_s, +1]^2 = (v^m + v^{-m})[B_s, +1] \tag{5.8}$$

inside the σ-weighted Grothendieck group.

The embedding of A_1 into $I_2(m)^k$ merely combines these different observations, but is no more difficult, and we leave it to the reader.

5.3. Type $A_1 \times A_1$ inside type $A_1^{\times k} \times A_1^{\times l}$. Now we begin checking the dihedral relations, as enumerated in Proposition (2.26). We wish to show that the expression given for $b_{w_0} \in \mathcal{H}(W, S, L)$ in terms of the KL generators b_s is categorified by the appropriate decomposition of certain tensor products of $(B_s, +1)$.

We begin with the easiest case, the embedding of $A_1 \times A_1$ inside $A_1^{\times k} \times A_1^{\times l}$. The reader may consult example 2.28 and (2.7). Let $S = \{s, t\}$. We seek to show that

$$(B_s, +1) \otimes (B_t, +1) \cong (B_{st}, +1), \tag{5.9}$$

regardless of the group G. Certainly, one already knows that $B_s B_t \cong B_{st}$ in $\mathcal{D}$.

For any $g \in G$, the corresponding equivariant structure map $\varphi_g \colon B_s \to g(B_s)$ inside $(B_s, +1)$ is just a rex move, as is the map $\varphi_g \colon B_t \to g(B_t)$. Their tensor product is also a rex move. Therefore, the induced equivariant structure on the unique summand B_{st} is $(B_{st}, +1)$.

As an example, let $R(s) = \{u_1, \ldots, u_k\}$ and $R(t) = \{v_1, \ldots, v_l\}$, and suppose that g acts on each orbit cyclically and transitively. We may assume that g sends u_i to u_{i+1} modulo k, and sends v_i to v_{i+1} modulo l. Choosing an isomorphism $B_s \cong B_{u_1} B_{u_2} \cdots B_{u_k}$, the equivariant structure map φ_g for $(B_s, +1)$ is

The map for $(B_t, +1)$ is similar. Therefore, the equivariant structure map on B_sB_t induced by $(B_s, +1) \otimes (B_t, +1)$ is

This is a rex move, and therefore it agrees with the standard equivariant structure map for B_{st}.

5.4. Type B_2 inside type A_3. Now we consider the embedding of B_2 inside A_3. The reader may consult Example 2.30, and the $k = 1$ case of (2.8). For purposes of computing decompositions we can assume that G acts on A_3 faithfully with the appropriate orbits, so that $G = \mathbb{Z}/2\mathbb{Z}$. Let $\sigma \in G$ be the generator.

Let $S = \{s, t\}$ and $S' = \{x, y, z\}$, and let $\sigma \in G$ switch x and z. Let $s = y$ and $t = xz$. Inside $\mathcal{D}$ one has the decomposition

$$B_yB_xB_zB_yB_xB_z \cong B_{yxzyxz} \oplus (B_{yxyz} \oplus B_{yzyx}) \oplus B_{yxz}(1) \oplus B_{yxz}(-1), \tag{5.10}$$

as can be easily computed using the Hecke algebra $\mathcal{H}(W', S', \ell)$. It remains to compute what happens in $\mathcal{D}_G$.

The equivariant structure map of σ on $(B_s, +1)\otimes(B_t, +1)\otimes(B_s, +1)\otimes(B_t, +1)$ is

As discussed previously, this induces the equivariant structure $(B_{stst}, +1)$ on the top summand. It induces the only possible equivariant structure on the summand $B_{yxyz} \oplus B_{yzyx}$. We wish to show that the equivariant structures induced on $B_{st}(1)$ and $B_{st}(-1)$ are the standard ones.

The projection to $B_{st}(-1)$ is (up to scalar) the map

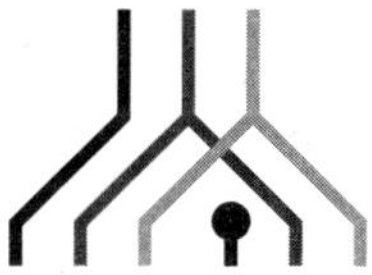

and its possible inclusions have the form

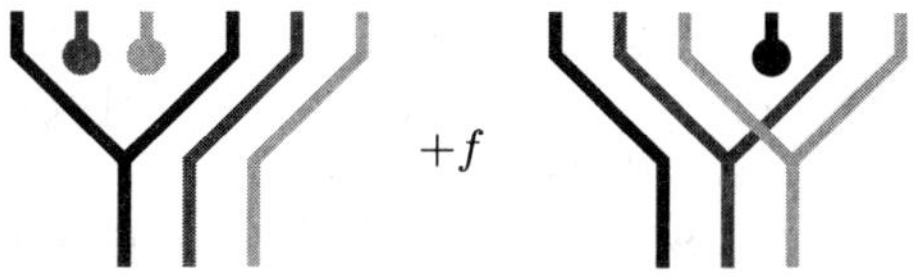

for some linear polynomial f. Placing these maps upside down one obtains the possible projections and inclusions for $B_{st}(1)$, except with some polynomial g instead of f. In order for the corresponding idempotents to be orthogonal, the required condition is that $f + g = -\partial_y(\alpha_x + \alpha_z)$. For instance, we may let $f = 0$ for the inclusion of $B_{st}(-1)$, and $g = -\partial_y(\alpha_x + \alpha_z)$ for the projection to $B_{st}(1)$. Both polynomials are invariant under the action of σ. It is easy to see that these idempotents are acted on trivially by σ, in the sense of (5.5).

REMARK 5.6. One need not actually find the two orthogonal idempotents to perform this computation. After all, the terms with polynomials above are in the kernel of the LIP, and can effectively be ignored. Also, by duality, one only need consider the trace of σ on projection maps of degree -1, and the trace on projection maps of degree $+1$ will be determined.

Therefore, in $\mathcal{D}_G$ one has the decomposition

$$(B_s, +1) \otimes (B_t, +1) \otimes (B_s, +1) \otimes (B_t, +1) \cong (B_{stst}, +1) \oplus (B_{yxyz} \oplus B_{yzyx}, \varphi) \oplus (B_{st}(1), +1) \oplus (B_{st}(-1), +1). \tag{5.11}$$

This is true for all G, not just $\mathbb{Z}/2\mathbb{Z}$. When G is abelian and σ descends to a generator of $\mathbb{Z}/2\mathbb{Z}$, this categorifies (2.9) in the σ-weighted Grothendieck group.

5.5. Type B_2 inside type A_4. The reader should consult Example 2.31, and we continue to use its notation. Let $S = \{t, u\}$ have type B_2 and $S' = \{s_1, s_2, s_3, s_4\}$ have type A_4, and set $t = s_1 s_4$ and $u = s_2 s_3 s_2$. We already know from Example 2.31 that

$$B_t \otimes B_u \otimes B_t \otimes B_u \cong B_{tutu} \oplus B_{tu}(1) \oplus B_{tu}(-1) \oplus B_{s_2 s_1 w_{234}} \oplus B_{s_3 s_4 w_{123}} \oplus B_{s_1 w_{234}}(1) \oplus B_{s_1 w_{234}}(-1) \oplus B_{s_4 w_{123}}(1) \oplus B_{s_4 w_{123}}(-1). \tag{5.12}$$

If σ is the usual automorphism of S', then the blue terms have a unique equivariant structure for σ, and will not contribute to the σ-weighted Grothendieck group when σ lives in an abelian group G. As in the previous section, we need only check that the projection map from $B_t B_u B_t B_u$ to $B_{tu}(-1)$ is acted on trivially by σ.

In the thick calculus of [2], this projection map can be drawn as follows.

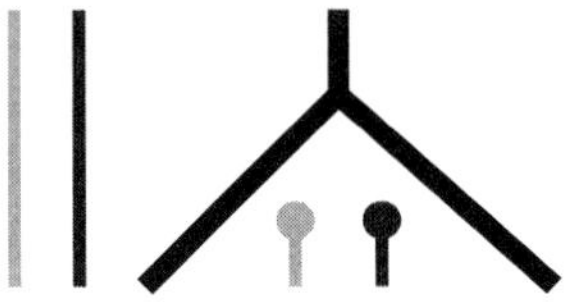

The thick trivalent vertex has degree -3, so the overall diagram has degree -1. It is evidently fixed by σ.

5.6. Diagonal embeddings. In section 5.4 we treated the quasi-split embedding of B_2 inside A_3, though we did not treat the embedding of B_2 inside $A_3^{\times k}$. Similarly, we have not yet treated the case of the diagonal embedding of $I_2(m)$ inside $I_2(m)^{\times k}$. In this section, we demonstrate by example that such multiplicities do not further complicate the computations in the σ-weighted Grothendieck group, when G is abelian and σ acts transitively on the orbits; the essential difficulties which arise have already been addressed in section 5.3. The reader should be able to formulate the general argument from this example.

Consider the embedding of B_2 inside $A_3^{\times 3}$. Let $S = \{s, t\}$, with $L(s) = 3$ and $L(t) = 6$. Color the three elements of $R(s)$ by different shades of blue: light, medium and dark. Do the same for the elements of $R(t)$, and different shades of red and green. The dark colors commute with the light and medium colors, and so forth. The equivariant structure map on $(B_s,+1)\otimes(B_t,+1)\otimes(B_s,+1)\otimes(B_t,+1)$ is

Given any Soergel diagram for A_3, one can obtain a Soergel diagram for $A_3^{\times 3}$ by replacing each blue strand with three shades of blue strands, and so forth. This operation will triple the degree of the map, but will not change how the maps compose thanks to the principle of distant-isotopy (see Remark 4.3). Thus we still have idempotents inside $(B_s,+1)\otimes(B_t,+1)\otimes(B_s,+1)\otimes(B_t,+1)$ for each summand mentioned in (5.11), except that the degree shifts are tripled.

However, there will also be a host of mixed idempotents. For example, one may project to $B_{st}(1)$ in the dark shade, while projecting to $B_{st}(-1)$ in the medium and light shades. Here is such a projection map:

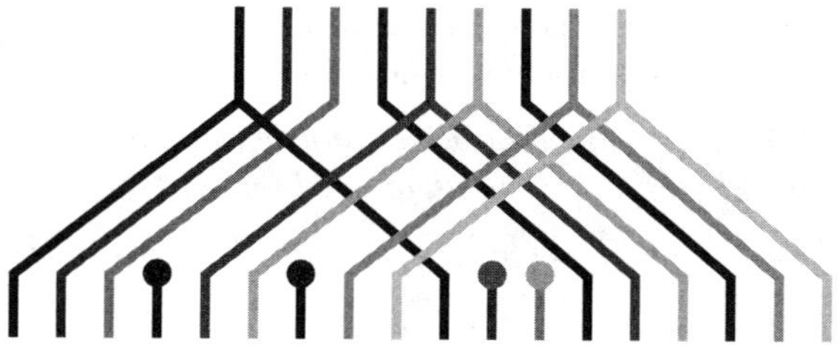

Just as in section 5.3, such mixed idempotents are permuted by σ with no fixed points, so that they do not contribute to the σ-weighted Grothendieck group.

Computing the full decomposition of $(B_s,+1)\otimes(B_t,+1)\otimes(B_s,+1)\otimes(B_t,+1)$ in this and more general situations seems to be a tedious problem.

5.7. Type G_2 inside type D_4. Finally, we treat the embedding of G_2 inside D_4. The embedding of G_2 inside $D_4^{\times k}$ is treated in similar fashion to the preceding section. The reader may consult Example 2.32, and the $k = 1$ case of (2.10). Let $S = \{s,t\}$ and $S' = \{u_1, u_2, u_3, v\}$. As above, we assume that G acts faithfully on D_4, thus embedding into the symmetric group on $\{u_1, u_2, u_3\}$. Since G acts transitively, it is either $\mathbb{Z}/3\mathbb{Z}$ or S_3. Either way there is some element $\sigma \in G$ which cyclically permutes the u_i. Let $s = v$ and $t = u_1u_2u_3$.

To shorten our notation, we denote the bimodule $B_sB_tB_sB_tB_sB_t$ as X. In this case, the indecomposables B_s and B_t are no more than products of the generators, so that X corresponds to the sequence $\underline{vu_1u_2u_3vu_1u_2u_3vu_1u_2u_3}$, an object of $\mathcal{D}$, and not some more complicated object in the Karoubi envelope. We write $(X,+1)$ for the equivariant object induced as a tensor product. We also give names to several of the representations of G or stabilizers inside G. Let $V_{\rm std}$ denote the standard 3-dimensional representation of S_3, and let $V_{\rm reg}$ denote the 6-dimensional regular representation; when $G = \mathbb{Z}/3\mathbb{Z}$ we also let these represent the restriction of these S_3-representations to the subgroup G. Let $+1$ denote the trivial representation of the stabilizer of u_1 in G, whether this stabilizer be S_2 or the trivial group.

We already know the decomposition of X in $\mathcal{D}$ for D_4, given on the decategorified level in Example 2.32. We reprint the equality here.

$$\begin{aligned}
&b_s b_t b_s b_t b_s b_t \\
&\quad = b_{ststst} + (b_{u_1u_2stst} + b_{u_1u_3stst} + b_{u_2u_3stst}) \\
&\qquad + (v+v^{-1})(b_{u_1stst} + b_{u_2stst} + b_{u_3stst}) \\
&\qquad + (2v^2 + 6 + 2v^{-2})b_{stst} \\
&\qquad + (v+v^{-1})(b_{vu_1u_2st} + b_{vu_1u_3st} + b_{vu_2u_3st}) \\
&\qquad + (v^4 + 6v^2 + (3+3+6) + 6v^{-2} + v^{-4})b_{st}.
\end{aligned}$$

Terms which are blue correspond to summands whose orbits have size 3; there is a unique equivariant structure when $G = \mathbb{Z}/3\mathbb{Z}$, and two possible equivariant structures when $G = S_3$. Such terms can not possibly contribute the σ-weighted Grothendieck group when G is abelian, because the indecomposables are not σ-fixed. We seek to show that the terms which are black have the standard equivariant structure, and the terms which are red correspond to summands with equivariant structure V_{std}, and the green term corresponds to a summand with equivariant structure V_{reg}. The red and green terms would then cancel in the σ-weighted Grothendieck group when G is abelian, since the trace of σ on V_{std} or V_{reg} is zero.

PROPOSITION 5.7. *The following direct sum decomposition holds in $\mathcal{D}_G$. We write polynomials to indicate direct sums of grading shifts.*

$$\begin{aligned}
(5.13)\qquad (X,+1) \cong\ & (B_{ststst},+1) \oplus (B_{u_1u_2stst} \oplus B_{u_1u_3stst} \oplus B_{u_2u_3stst},+1) \\
& \oplus (v+v^{-1})(B_{u_1stst} \oplus B_{u_2stst} \oplus B_{u_3stst},+1) \\
& \oplus (2v^2+2v^{-2})(B_{stst},+1) \oplus 2(B_{stst} \boxtimes V_{\text{std}}) \\
& \oplus (v+v^{-1})(B_{vu_1u_2st} + B_{vu_1u_3st} + B_{vu_2u_3st},+1) \\
& \oplus (v^4+v^{-4})(B_{st},+1) \oplus (2v^2+2v^{-2})(B_{st} \boxtimes V_{\text{std}}) \\
& \oplus 3(B_{st},+1) \oplus (B_{st} \boxtimes V_{\text{std}}) \oplus (B_{st} \boxtimes V_{\text{reg}}).
\end{aligned}$$

Let us discuss the algorithm by which we compute the spaces $V^k(X, B_w)$, for various k and w, in order to determine the action of the stabilizer G_w on them. By duality, we need only investigate the case $k \le 0$. The algorithm will proceed degreewise, from the most negative k up to zero, for a reason which will become clear soon.

Clearly $\beta\colon \operatorname{Hom}^k(X, \underline{w}) \to V^k(X, B_w)$ is a surjection. First, we find a set of linearly independent elements of $\operatorname{Hom}^k(X, \underline{w})$, for a reduced expression of w, which descend to a spanning set of $V^k(X, B_w)$. The light leaves of degree exactly k serve this purpose. This is because double leaves form a basis of $\operatorname{Hom}(X, \underline{w})$ as a right R-module; any map in the image of R_+ acting on the right will be in the kernel of β, and any double leaf which is not a light leaf followed by the identity map will factor through lower terms, and thus be in the kernel as well. Thus we actually think of β as a map from $Y^k(X, w)$, the span of light leaves of degree k, to $V^k(X, B_w)$.

Next, we compute the kernel of β, i.e. the kernel of the Local Intersection Pairing. It is an extrapolation of the Soergel conjecture that any morphism of non-positive degree must have a "reason" to be in the kernel; that the kernel is generated by maps from X to B_x of even more negative degree $k' < k$, followed by a map $B_x \to B_w$ of positive degree. We have already restricted to light leaves to ignore lower terms, so that we may assume $x > w$. This map from X to B_x of degree k' indicates that $B_x(k')$ has already occurred as a summand of X (i.e. we have already computed $V^{k'}(X, B_x)$ in some sense).

Working with objects in $\mathcal{D}$, rather than indecomposables in its Karoubi envelope, we find the kernel of β by looking at light leaves maps from X to $\underline{x}$ of degree $< k$, and composing with a map $\underline{x} \to \underline{w}$ of positive degree which is itself in the kernel of the LIP (i.e. it has nothing of negative degree to pair against). This process will become clear in practice below.

Consider $V^{-4}(X, B_{st})$. The space of light leaves $Y^{-4}(X, st)$ is one-dimensional, spanned by

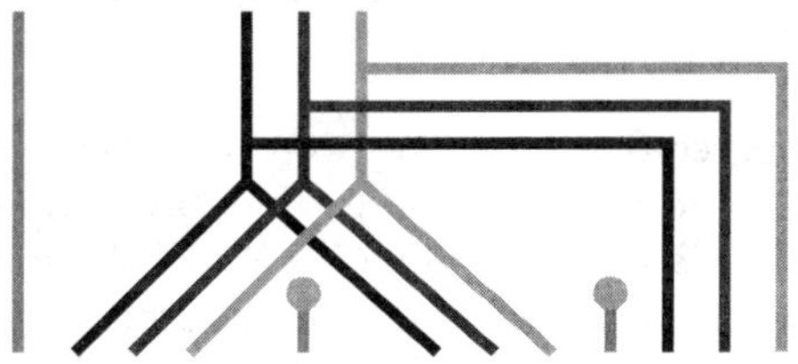

There is no kernel for β, because there are no maps of degree < -4 to any B_w with $w > st$; if there were such a map, it must lead to a summand, and there are no such summands. Thus $Y^{-4}(X, B_{st}) = V^{-4}(X, B_{st})$, or in other words, the map pictured above is a split morphism, a projection. This diagram is fixed by the action of G, so that $(B_{st}(-4), +1) \oplus (B_{st}(+4), +1)$ appear as equivariant summands of X. This explains the black term $(v^4 + v^{-4})b_{st}$.

Consider $V^{-2}(X, B_{stst})$. The space $Y^{-2}(X, stst)$ is spanned by the two pictures below.

(5.14)

There is no kernel for β, because there are no summands $B_x(k)$ with $k < -2$ and $x > stst$. Thus the pictures below provide two orthogonal projection morphisms. The action of G fixes both diagrams, so that one has two equivariant summands of X of the form $(B_{stst}(2), +1)$, and two of the form $(B_{stst}(-2), +1)$. This yields the black term $(2v^2 + 2v^{-2})b_{stst}$ above.

Consider $V^{-2}(X, B_{st})$. The space $Y^{-2}(X, st)$ is six-dimensional, spanned by the σ-conjugates of these two light leaves.

Again, there is no kernel for β, because there are no summands $B_x(k)$ with $k < -2$ and $x > st$. The action of G permutes the σ-conjugates of each picture, so that this six-dimensional representation is isomorphic to $V_{\mathrm{std}} \oplus V_{\mathrm{std}}$. This yields the red term $6(v^2 + v^{-2})b_{st}$.

Consider $V^{-1}(X, B_{u_1stst})$. The space $Y^{-1}(X, u_1stst)$ is one-dimensional, spanned by this picture.

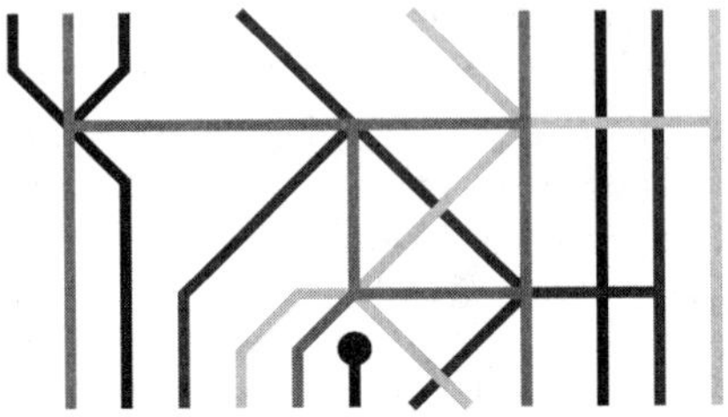

There is no kernel for degree reasons. The two other σ-conjugates of this map live in different Hom spaces, and together these three projections yield the blue term $(v+v^{-1})(b_{u_1stst}+b_{u_2stst}+b_{u_3stst})$. When $G=\mathbb{Z}/3\mathbb{Z}$ there is only one equivariant structure possible. When $G=S_3$ we need to determine the action of the stabilizer (switching u_2 and u_3) on this one-dimensional space. The reader can verify that, because of the Zamolodzhikov relation (4.10) applied to the central green square, the action of this stabilizer is trivial.

Consider $V^{-1}(X, B_{vu_1u_2st})$. The space $Y^{-1}(X, vu_1u_2st)$ is three-dimensional. Two of the three light leaves spanning this space are

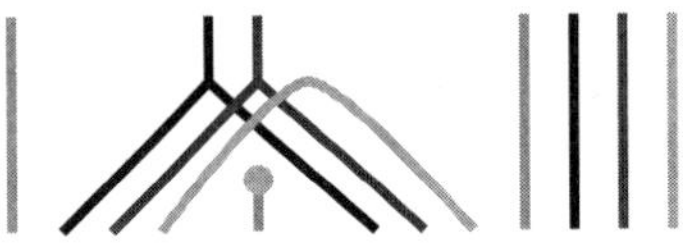

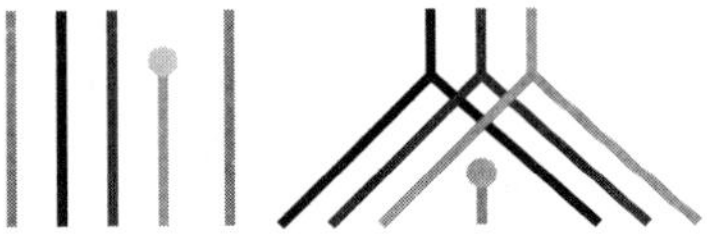

Any contribution to the kernel of β must come from $V^{-2}(X, B_{stst})$ composed with a degree $+1$ map in the kernel of the LIP. Considering the two maps in (5.14), and applying a dot to the first u_3 strand (which must be in the kernel of the LIP, as there are no negative degree maps to balance it out), one obtains the two maps pictured here. Therefore, $V^{-1}(X, B_{vu_1u_2st})$ is one-dimensional, spanned by the remaining light leaf map, which is

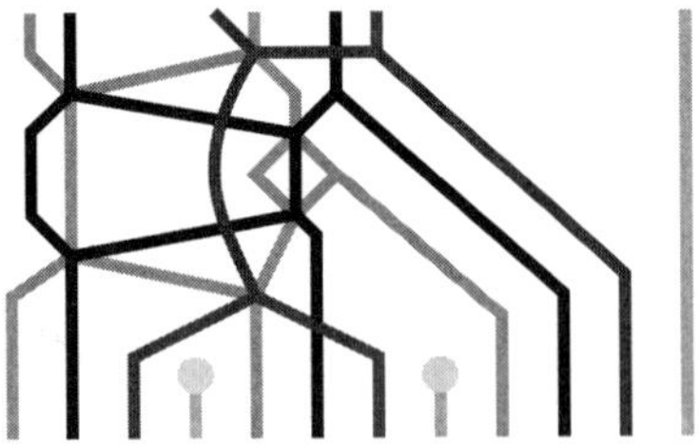

This yields the blue term $(v+v^{-1})(b_{vu_1u_2st}+b_{vu_1u_3st}+b_{vu_2u_3st})$.When $G=S_3$ we need to determine the action of the stabilizer (switching u_1 and u_2) on this one-dimensional space. Applying (4.10) and (4.6) several times one can see that this action is trivial.

Consider $V^0(X, B_{stst})$. The space $Y^0(X, stst)$ is nine-dimensional. We give three maps below; the other 6 are the σ-conjugates of these.

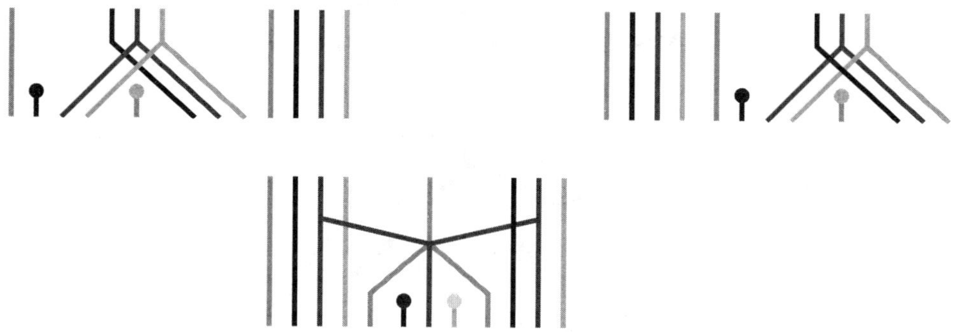

However, there is a three-dimensional subspace in the kernel of β. One element in the kernel is obtained by taking the degree -1 map to $\underline{u_1 stst}$, and applying a dot to the first strand. This dot is in the kernel, as no maps of negative degree exist to balance it out.

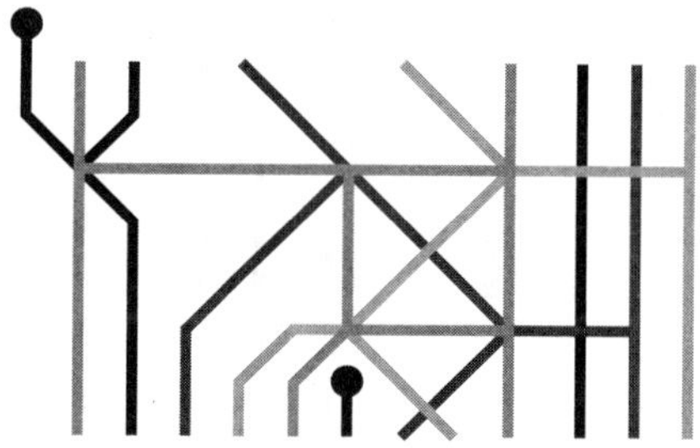

The other two generators of the kernel are the σ-conjugates of this. Expanding this diagram in the nine-dimensional basis above is an exercise in the relations of $\mathcal{D}$, but is not relevant here. What is obvious is that the original nine-dimensional space gives three copies of the standard representation of G, and the three-dimensional kernel is a copy of the standard representation inside. What remains is a sum of two copies of the standard representation. This yields the red term $6b_{stst}$.

Consider $V^0(X, B_{st})$. The space $Y^0(X, st)$ is eighteen-dimensional, and the kernel of β is six-dimensional. We will not bother to write down the basis of either space, which we leave as a combinatorial exercise to the intrepid reader, but we will at least record the basis of $V^0(X, B_{st})$.

Here are the three projection maps which are G-fixed, giving the black term $3b_{st}$.

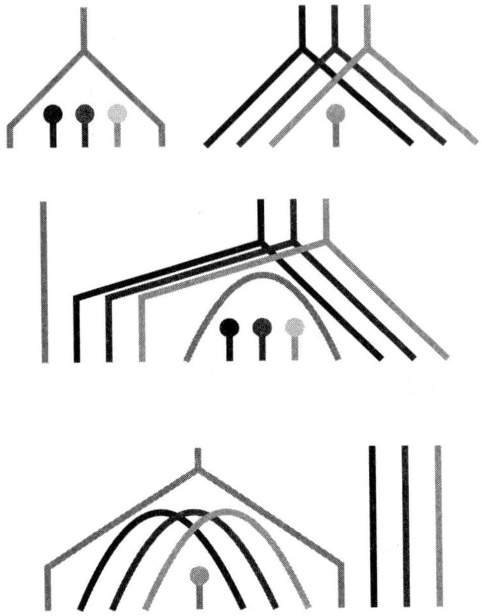

The cautious reader may worry if two of these maps are in the kernel, as they appear to factor through the degree -2 maps of (5.14), followed by this degree $+2$ map.

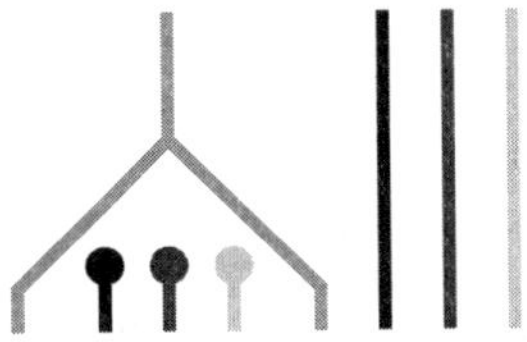

However, this degree $+2$ map is not in the kernel, as it pairs against a map in $\mathrm{Hom}^{-2}(B_sB_tB_sB_t, B_{st})$. This "appearance" of being in the kernel was an artifact of working with objects of $\mathcal{D}$ rather than indecomposables in the Karoubi envelope.

Here is a projection map which, along with its σ-conjugates, forms a subrepresentation isomorphic to V_{std}, yielding the red term $3b_{st}$.

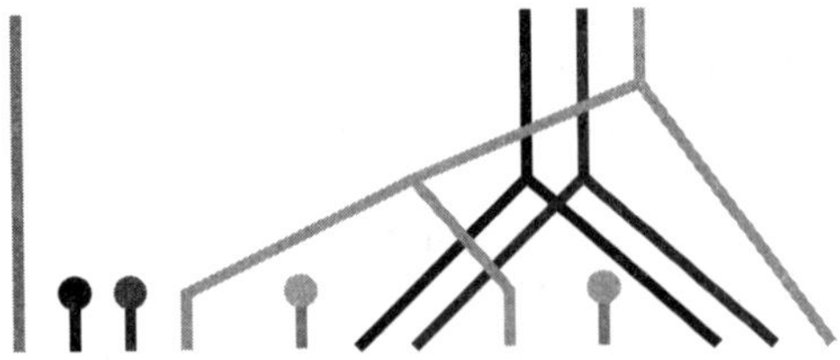

Here is a projection map which, along with its S_3-conjugates, forms a subrepresentation isomorphic to V_{reg}, yielding the green term $6b_{st}$.

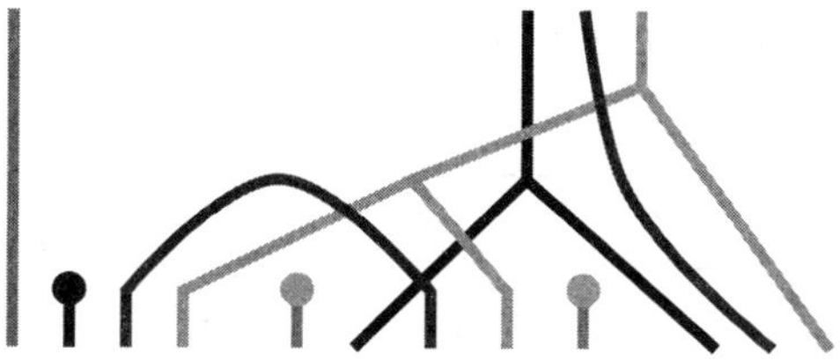

5.8. The weighted character map. We continue to assume that G is abelian and that σ acts transitively on the orbits of G.

We have finally proven that $\mathcal{H}(W,S,L) \cong [\mathcal{D}]_\sigma$. Let $(M,\varphi) \in \mathcal{D}_G$. There is some expression $[M,\varphi] = \sum_{x\in W} P_{(M,\varphi),x} H_x$ for some polynomials $P_x \in \mathbb{Z}[v,v^{-1}]$. We now reinterpret P_x as the trace of σ on a certain graded vector space.

The usual character map was discussed in §4.3. That is, for an object M in the Karoubi envelope of $\mathcal{D}$, its character in $\mathcal{H}(W')$ is given as $\sum P_{w,M} H_w$, where $P_{w,M}$ is the graded dimension of $\mathrm{Hom}(M,B_w)$ modulo $\mathcal{D}_{<w}$. As discussed in that section, one can compute the characters of objects in $\mathcal{D}$ using light leaves.

Now let $w \in W \subset W'$ be G-fixed, and let $(M,\varphi) \in \mathcal{D}_G$. There is an action of G on $\mathrm{Hom}(M,B_w)$, analogous to (3.5), which sends $f\colon M \to B_w$ to the composition

$$\psi_{B_w}^{-1} \circ g(f) \circ \varphi\colon M \to gM \to gB_w \to B_w.$$

CLAIM 5.8. *$P_{(M,\varphi),x}$ is the trace of σ acting on the graded vector space $\mathrm{Hom}(M,B_w)$ modulo $\mathcal{D}_{<w}$. In other words, define the* weighted character map *as the map $[\mathcal{D}]_\sigma \to \mathcal{H}(W,S,L)$, which sends $[M,\varphi]$ to $\sum_{x\in W} P_{(M,\varphi),x} H_x$ where P is the trace just defined. Then it is the inverse isomorphism to the isomorphism defined by sending the Kazhdan-Lusztig generator $c_s \in \mathcal{H}(W,S,L)$ to $[B_s,+1]$.*

If Claim 5.8 is true then, assuming that $[B_w] = b_w$ under the isomorphism $[\mathcal{D}] \cong \mathcal{H}(W')$, it is easy to deduce that $[B_w,+1] = c_w$ under the isomorphism $[\mathcal{D}]_\sigma \cong \mathcal{H}(W,L)$. After all, we need only show that $[B_w,+1]$ is self-dual, and that the polynomials $P_{(B_w,+1),x}$ are concentrated in strictly positive degrees (except when $x = w$). The self-duality follows immediately from the self-duality of B_w (see also the discussion of duality in §3.2). The morphism spaces $\mathrm{Hom}(B_w,B_x)$ are known to be concentrated in strictly positive degrees by the Soergel Hom formula, and taking the trace under any element will not change this fact.

We sketch a proof of Claim 5.8, which imitates the non-equivariant proof (using light leaves) that the character map from $[\mathcal{D}]$ to $\mathcal{H}(W')$ is the inverse isomorphism. Both sides (the weighted character map and the weighted Grothendieck group) are compatible with direct sum decompositions and with twisting by a character in G^*. Using this, we can reduce to the case where (M,φ) is a tensor product of the generating objects $(B_s,+1)$ for $s \in S$. The weighted character map clearly sends $[\mathbb{1},+1]$ to 1. One need only check an induction step: that the weighted character map intertwines the tensor product by $(B_s,+1)$ and multiplication by c_s.

Let us temporarily assume that the Coxeter embedding is standard, so that B_s is actually an object $\underline{s}$ in $\mathcal{D}$ instead of its Karoubi envelope (i.e. it is a sequence of distant reflections). Then the induction step follows from the "light leaf philosophy." Essentially, the claim is that, for an arbitrary sequence $\underline{x}$ in $\mathcal{D}$,

$\text{Hom}(M \otimes \underline{x}, B_w)$ has a basis modulo $\mathcal{D}_{<w}$ given by diagrams of the following form.

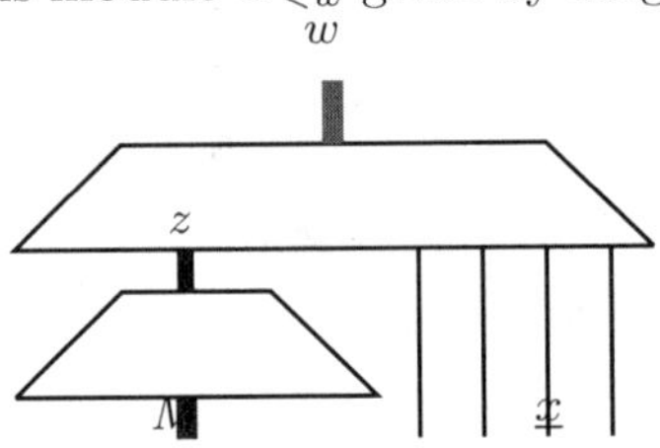

Here, the bottom trapezoid is a map in a basis for $\text{Hom}(M, B_z)$ modulo $\mathcal{D}_{<z}$, for some other element z. Then, one chooses a subexpression $\mathbf{e}$ of $\underline{x}$, expressing an element y with $zy = w$. The light leaf algorithm states how to construct the top trapezoid, a map from $B_z \otimes \underline{x} \to B_w$. This algorithm works one simple reflection at a time. For the first simple reflection in $\underline{x}$, the subexpression $\mathbf{e}$ has determined whether we take it or leave it, and the element z determines whether that reflection would go up or down in the Bruhat order. For each of the four possibilities, there is a diagram which is the first tier of the trapezoid. This handles the special case where $\underline{x}$ has length 1; the general case just iterates this procedure.

Now suppose that $\underline{x}$ is the sequence of commuting simple reflections expressing $s \in S$. The subexpression $\mathbf{e}$ of $\underline{x}$ is determined by a subset $J \subset I(s)$, specifying which simple reflections are included and which excluded. If $z \in W$ is G-invariant, then z is either a minimal or a maximal coset representative for the parabolic subgroup determined by $I(s)$. Thus, for each simple reflection in $\underline{x}$, whether it goes up or down is determined by z, not by the subset J. In particular, the action of G on the upper trapezoids is the same as the action of G on the formal span of subsets $J \subset I(s)$.

In general, the space of morphisms $\text{Hom}(M\underline{x}, B_w)$ modulo $\mathcal{D}_{<w}$ is the direct sum, over all z, of the tensor product of $\text{Hom}(M, B_z)$ with the formal span of subsets $J \subset I(s)$ for which $zy = w$. The action of G will permute the various z and the various J accordingly. As can easily be seen, a large portion of this representation is automatically traceless for σ. Only the terms where $z \in W$ and $w \in W$, and $J = \emptyset$ or $J = I(s)$ will contribute to the trace. These two terms, the "multiplication by 1" and the "multiplication by s" terms, agree with the two terms that appear when you multiply a standard basis element by c_s.

When the Coxeter embedding is not standard, a similar argument can be made. There are *generalized light leaves* (due to the author and Williamson) which describe morphisms between tensor products of longest elements of parabolic subgroups, not just tensor products of simple reflections, and they follow an analogous light leaf philosophy, constructed by a tiered algorithm. They have not yet appeared in the literature, though hopefully this will be remedied soon. The proof using generalized light leaves is very similar, only instead of considering the action of G on subset $J \subset I(s)$, one considers the action of G on the parabolic subgroup generated by $I(s)$. This is very similar to the sketched argument made in §5.2.

Acknowledgments

The author would like to thank George Lusztig and Meinolf Geck for useful conversations,as well as Geordie Williamson, whose mellifluous influence is hardly superfluous. This research was performed while the author was supported by an NSF postdoctoral fellowship, DMS-1103862.

References

[1] C. Chevalley, *Classification des groupes algébriques semi-simples* (French), Springer-Verlag, Berlin, 2005. Collected works. Vol. 3; Edited and with a preface by P. Cartier; With the collaboration of Cartier, A. Grothendieck and M. Lazard. MR2124841

[2] B. Elias, *Thicker Soergel calculus in type A*, Proc. Lond. Math. Soc. (3) **112** (2016), no. 5, 924–978. MR3502025

[3] B. Elias, *The two-color Soergel calculus*, Compos. Math. **152** (2016), no. 2, 327–398. MR3462556

[4] B. Elias and M. Khovanov, *Diagrammatics for Soergel categories*, Int. J. Math. Math. Sci. (2010), Art. ID 978635, 58. MR3095655

[5] B. Elias and G. Williamson, *Soergel calculus*, Represent. Theory **20** (2016), 295–374. MR3555156

[6] B. Elias and G. Williamson, *The Hodge theory of Soergel bimodules*, Ann. of Math. (2) **180** (2014), no. 3, 1089–1136, DOI 10.4007/annals.2014.180.3.6. MR3245013

[7] M. Geck and L. Iancu, *Coxeter groups and automorphisms*, Preprint, 2014, arXiv:1412.5428.

[8] M. Geck and N. Jacon, *Representations of Hecke algebras at roots of unity*, Algebra and Applications, vol. 15, Springer-Verlag London, Ltd., London, 2011. MR2799052

[9] J.-Y. Hée, *Système de racines sur un anneau commutatif totalement ordonné* (French, with English summary), Geom. Dedicata **37** (1991), no. 1, 65–102, DOI 10.1007/BF00150405. MR1094228

[10] R. B. Howlett, *Coxeter groups and M-matrices*, Bull. London Math. Soc. **14** (1982), no. 2, 137–141, DOI 10.1112/blms/14.2.137. MR647197

[11] J. E. Humphreys, *Reflection groups and Coxeter groups*, Cambridge Studies in Advanced Mathematics, vol. 29, Cambridge University Press, Cambridge, 1990. MR1066460

[12] M. Khovanov and A. D. Lauda, *A diagrammatic approach to categorification of quantum groups. I*, Represent. Theory **13** (2009), 309–347, DOI 10.1090/S1088-4165-09-00346-X. MR2525917

[13] N. Libedinsky, *Sur la catégorie des bimodules de Soergel* (French, with French summary), J. Algebra **320** (2008), no. 7, 2675–2694, DOI 10.1016/j.jalgebra.2008.05.027. MR2441994

[14] N. Libedinsky, *Presentation of right-angled Soergel categories by generators and relations*, J. Pure Appl. Algebra **214** (2010), no. 12, 2265–2278, DOI 10.1016/j.jpaa.2010.02.026. MR2660912

[15] G. Lusztig, *Hecke algebras with unequal parameters*, CRM Monograph Series, vol. 18, American Mathematical Society, Providence, RI, 2003. MR1974442

[16] G. Lusztig, *Some examples of square integrable representations of semisimple p-adic groups*, Trans. Amer. Math. Soc. **277** (1983), no. 2, 623–653, DOI 10.2307/1999228. MR694380

[17] G. Lusztig, *Introduction to quantum groups*, Modern Birkhäuser Classics, Birkhäuser/Springer, New York, 2010. Reprint of the 1994 edition. MR2759715

[18] G. Lusztig, *Hecke algebras with unequal parameters*, Revision, 2014, arXiv:0208154v2.

[19] G. Lusztig and D. A. Vogan Jr., *Hecke algebras and involutions in Weyl groups*, Bull. Inst. Math. Acad. Sin. (N.S.) **7** (2012), no. 3, 323–354. MR3051317

[20] G. Lusztig and D. A. Vogan Jr., *Hecke algebras and involutions in Coxeter groups*, Prog. in Math. **312** (2015), 365–398, arXiv:1312.3237.

[21] P. J. McNamara, *Folding KLR algebras*, Preprint, 2016, arXiv 1603.05768.

[22] R. Rouquier, *2-Kac-Moody algebras*, Preprint, 2008, arXiv:0812.5023.

[23] W. Soergel, *Kategorie $\mathcal{O}$, perverse Garben und Moduln über den Koinvarianten zur Weylgruppe* (German, with English summary), J. Amer. Math. Soc. **3** (1990), no. 2, 421–445, DOI 10.2307/1990960. MR1029692

[24] W. Soergel, *Kazhdan-Lusztig-Polynome und unzerlegbare Bimoduln über Polynomringen* (German, with English and German summaries), J. Inst. Math. Jussieu **6** (2007), no. 3, 501–525, DOI 10.1017/S1474748007000023. MR2329762

[25] R. Steinberg, *Lectures on Chevalley groups*, Yale University, New Haven, Conn., 1968, Notes prepared by John Faulkner and Robert Wilson. MR0466335 (57 #6215)

[26] M. Varagnolo and E. Vasserot, *Canonical bases and KLR-algebras*, J. Reine Angew. Math. **659** (2011), 67–100, DOI 10.1515/CRELLE.2011.068. MR2837011

[27] B. Webster, *Canonical bases and higher representation theory*, Compos. Math. **151** (2015), no. 1, 121–166, DOI 10.1112/S0010437X1400760X. MR3305310

Department of Mathematics, University of Oregon, Eugene, OR 97403-1222
E-mail address: `belias@uorgegon.edu`

Contemporary Mathematics
Volume **683**, 2017
http://dx.doi.org/10.1090/conm/683/13719

The p-canonical basis for Hecke algebras

Lars Thorge Jensen and Geordie Williamson

ABSTRACT. We describe a positive characteristic analogue of the Kazhdan-Lusztig basis of the Hecke algebra of a crystallographic Coxeter system and investigate some of its properties. Using Soergel calculus we describe an algorithm to calculate this basis. We outline some known or expected applications in modular representation theory. We conclude by giving several examples.

1. Introduction

One consequence of categorification is that it often leads to a "canonical" basis in the structure being categorified. Canonical bases have numerous remarkable applications in representation theory and beyond.

The first examples of canonical bases are the Kazhdan-Lusztig basis of Hecke algebras and the canonical basis of quantum groups. These bases have geometric origins, arising as the shadow of a sheaf (an intersection cohomology sheaf) on a variety. That one should obtain interesting bases in this way is, in some sense, predicted by Grothendieck's function-sheaf correspondence.[1]

More recently, considerable progress has been made via algebraic approaches to categorification. Here the canonical basis arises as the shadow of a simple or indecomposable projective module over an algebra. For example, the canonical basis of the quantum group is realized in simply laced type as the classes in the Grothendieck group of indecomposable projective modules for KLR algebras (see [**VV11**]). Similarly, the Kazhdan-Lusztig basis arises as the classes of indecomposable Soergel bimodules (see [**EW14**]). The passage between algebra and geometry is made by realizing an algebra as an extension algebra of geometric origin [**Rou12, VV11, Soe01**].

The canonical basis (or rather, the base change matrix between the canonical basis and a standard basis) usually records multiplicities in categories of interest in representation theory, which are defined over a field of characteristic 0. This is reflected by the choice of coefficients of characteristic zero (both in the geometric and algebraic settings). Usually it is only via appeal to powerful geometric theorems

2010 *Mathematics Subject Classification.* Primary 20C20; Secondary 20C08, 17B10, 14M15.

[1]Grothendieck's correspondence can be seen as an early encouragement to pursue categorification, by replacing a function (an object of a set) by a sheaf or D-module (an object of a category). Despite many more examples illustrating the power of this correspondence (e.g. Lusztig's theory of character sheaves, the geometric Langlands program, ...), it still seems remarkable that Grothendieck's prophecy really works!

(e.g. [**Del80, BBD82, Sai90, dCM09**]) that one has an algorithm to compute the canonical basis (however see [**EW14**]).

There are several questions in modular representation theory where certain multiplicities are expected to be given in terms of a canonical basis. Famous examples include Lusztig's conjecture for simple rational modules for algebraic groups [**Lus80**], the LLT conjecture for representations of Hecke algebras at roots of unity [**LLT96, Ari96**], and the James conjecture on representations of the symmetric group [**Jam90**]. These conjectures predict that the situation in characteristic 0 agrees with the one in characteristic $p \gg 0$, sometimes state an explicit bound on the "bad" primes to exclude, but usually aren't so bold as to make any statement about what happens for small p.

However recent work by several authors suggests a different way of thinking about this problem. Namely, in important examples there exists a "p-canonical basis" for each choice of prime number p. The p-canonical basis has the same indexing set as the usual canonical basis, and each of its elements agrees with the corresponding element of the canonical basis for large enough p, but may differ from it for small p. One hopes that the p-canonical basis gives the correct answer in representation theory in characteristic p. Examples of this phenomenon include:

(1) The work of Grojnowski, Ariki and others [**Gro99, Ari96**] identifying the Grothendieck group of the category of representations of all symmetric groups with the basic representation of an affine Lie algebra. Here the p-canonical basis is defined via the classes of indecomposable projective modules, and hence the p-canonical basis contains deep information in modular representation essentially by definition. (It is from [**Gro99**] that we first learnt the term p-canonical basis.)
(2) The work of Soergel [**Soe00**], connecting certain multiplicities in the rational representation theory of algebraic groups with indecomposable summands of Bott-Samelson sheaves. Subsequent work (e.g. [**JMW14a, WB12**]) shows that Soergel's results may be restated as giving these multiplicities in terms of coefficients of the p-canonical basis of the Hecke algebra of the finite Weyl group.
(3) A recent conjecture of Riche and the second author, which predicts that the characters of indecomposable tilting modules for algebraic groups should be given in terms of coefficients of the p-canonical basis in the anti-spherical module of the affine Weyl group. By Schur-Weyl duality this conjecture implies that all decomposition numbers for symmetric groups occur as coefficients in this basis. In contrast to (1), this equivalence is not "by definition", and might lead to more efficient algorithms. In recent work, Riche and the second author have confirmed their conjecture for the general linear group.

The above approach breaks problems in modular representation theory into two subproblems. First one should connect the problem in representation theory to the p-canonical basis, and then one should calculate the p-canonical basis. In general the second step seems to be extremely difficult, and it is possibly too optimistic to expect an answer in general [**Wil13**]. However at the very least this approach has the potential to unify questions in modular representation theory (just as different questions may have answers given by the same Kazhdan-Lusztig polynomials).

Also, there are combinatorial constraints on the p-canonical basis which make the situation very rigid, and might merit further investigation.

1.1. In this paper we recall the definition of and study the p-canonical basis of the Hecke algebra of a crystallographic Coxeter system (W, S) (see [**Wil12**] where the definition appears in print for the first time). To motivate its definition, we first recall how the Kazhdan-Lusztig basis arises from categorification. In the introduction, let us assume that W is the Weyl group of a complex reductive group G with a maximal torus T. Let $B \subset G$ denote the Borel subgroup corresponding to the simple reflections $S \subset W$.

The Hecke algebra $\mathcal{H}$ of (W, S) has two (essentially equivalent) categorifications, often referred to loosely as the Hecke category:

(1) *Geometric:* The additive, monoidal (under convolution) category of B-biequivariant semi-simple complexes on G: the full subcategory of the equivariant derived category $D^b_{B\times B}(G, \mathbb{Q})$ consisting of direct sums of shifts of equivariant intersection cohomology complexes.
(2) *Algebraic:* The monoidal category of Soergel bimodules: a certain full subcategory of the monoidal category of graded R-bimodules, where R denotes the regular functions on the Lie algebra of the maximal torus T.

In the first setting, the Kazhdan-Lusztig basis arises as the graded dimensions of stalks of the intersection cohomology complexes (see [**KL80**]). In the second case, the Kazhdan-Lusztig basis is realized as the characters of the indecomposable self-dual Soergel bimodules (see [**Soe98, EW14**]).

In [**EW13**] the monoidal category of Soergel bimodules is described by generators and relations, following earlier work by Elias [**Eli13**], Elias-Khovanov [**EK10**] and Libedinsky [**Lib10**] (we recall this description in detail below). The upshot is that there exists a graded monoidal category $\mathbf{H}$ which is defined over the integers, and whose extension of scalars to $\mathbb{Q}$ is equivalent to Soergel bimodules. Hence one can think of $\mathbf{H}$ as an integral form of the Hecke category. For any field k we can consider the extension of scalars ${}^k\mathbf{H}$ to k and it is proved in [**EW13**] that one has an canonical "character" isomorphism of $\mathbb{Z}[v, v^{-1}]$-algebras

$$\mathrm{ch} : [{}^k\mathbf{H}] \xrightarrow{\sim} \mathcal{H}$$

between the split Grothendieck group of ${}^k\mathbf{H}$ and the Hecke algebra. Hence for any field ${}^k\mathbf{H}$ provides a categorification of the Hecke algebra.[2] (Note that while the coefficients of $\mathbf{H}$ change, the Grothendieck group is always the same Hecke algebra over $\mathbb{Z}[v, v^{-1}]$.)

In [**EW13**] the indecomposable objects of ${}^k\mathbf{H}$ are classified, following Soergel's classification of the indecomposable Soergel bimodules in [**Soe07**]. It turns out that for all $w \in W$ there exists an indecomposable object ${}^kB_w \in {}^k\mathbf{H}$, and that any indecomposable object is isomorphic to a grading shift of ${}^kB_w$ for some $w \in W$. The p-canonical basis is defined as the character of this indecomposable object:

$${}^p\underline{H}_w := \mathrm{ch}({}^kB_w)$$

where p denotes the characteristic of k. From basic properties of ${}^k\mathbf{H}$ it is easy to see that $\{{}^p\underline{H}_w \mid w \in W\}$ is a basis for the Hecke algebra which only depends on

[2]There is a minor additional technical assumption if the characteristic of k is 2. We ignore this point in the introduction.

the characteristic of k, and that its structure constants are positive (see Proposition 4.2). Moreover, because ${}^{\mathbb{Q}}\mathbf{H}$ is equivalent to Soergel bimodules with $\mathbb{Q}$-coefficients, $\{{}^{0}\underline{H}_w \mid w \in W\}$ is the Kazhdan-Lusztig basis.

1.2. In this subsection we outline the connection between the p-canonical basis and parity sheaves on (affine) flag varieties. The reader unfamiliar with affine flag varieties may keep the important case of a (finite) flag variety in mind.

To any based root datum we can associate a connected, reductive algebraic group scheme G over $\mathbb{Z}$ with Borel subgroup $B \subseteq G$ and maximal torus $T \subseteq B \subseteq G$. In the finite case, let X denote the $\mathbb{C}$-points of the flag variety $\mathcal{F}l := G/B$ with its classical metric topology.

In the affine setting, define the loop group LG (resp. positive loop group L^+G) of G as the $\mathbb{Z}$-functor given by $R \mapsto G(R((t)))$ (resp. $R \mapsto G(R[[t]])$). Denote by I the Iwahori subgroup given by the inverse image of B under the morphism $L^+G \to G$ induced by $t \mapsto 0$. In this case we define X to be the $\mathbb{C}$-points of the affine flag variety $\mathcal{F}l_a := LG/I$ viewed as ind-variety (see [**Gör10**] for more information).

In both settings we have an (Iwahori)-Bruhat decomposition expressing the corresponding flag variety as a disjoint union of left $B(\mathbb{C})$ (resp. $I(\mathbb{C})$)-orbits indexed by the (extended affine) Weyl group. The closure relation is given by the Bruhat order. Note that the affine flag variety is isomorphic as an ind-variety to a suitable disjoint union of Kac-Moody flag varieties (see e.g. [**Kum02**]).

Fix a field k of coefficients. We may consider $D^b_H(X)$ the H-equivariant bounded derived category of k-sheaves where H is either $B(\mathbb{C})$ or $I(\mathbb{C})$ depending on the setting (see [**BL94**] for more information about equivariant derived categories). In [**JMW14a**, §4.1] Juteau, Mautner and the second author introduce and prove the existence of parity sheaves on generalized flag varieties, a class of objects in $D^b_H(X)$ whose stalks satisfy a cohomological parity vanishing condition (for the trivial pariversity function). Their work was motivated by Soergel's idea to consider another class of objects as "replacements" for intersection cohomology complexes with positive characteristic coefficients (see [**Soe00**]). Observe that while X is still a variety (resp. an ind-variety) over $\mathbb{C}$ equipped with its classical topology, the coefficients of the sheaves we are studying may lie in a field of positive characteristic.

The theory of parity sheaves parallels the theory of perverse sheaves in the following points:

(1) indecomposable parity sheaves are classified analogously to simple perverse sheaves, being up to isomorphism the unique extension of an irreducible local system on a stratum;
(2) in our setting, if the coefficients are a field of characteristic zero the intersection cohomology sheaves are parity sheaves.

But while the decomposition theorem for perverse sheaves fails in positive characteristic, the pushforward along a proper, even, stratified map preserves parity sheaves and it is possible to calculate the multiplicities of the occurring indecomposable parity sheaves via intersection forms (see [**JMW14a**, §3.3.]). Thus parity sheaves are particularly interesting in the case of positive characteristic coefficients.

Parity sheaves on various varieties have also been used for categorification. In [**Mak15**] Maksimau realizes Lusztig's integral form of the positive half of the quantum group associated to a Dynkin quiver as a coalgebra geometrically, by considering parity sheaves on quiver moduli spaces. In our setting, parity sheaves also give canonical bases of the Hecke algebra. If the coefficients are a field of characteristic zero, then the graded dimensions of stalks of parity sheaves give the Kazhdan-Lusztig basis (as mentioned above, see [**KL80**, **Spr82**]). For a field of positive characteristic the indecomposable parity sheaves realize the p-canonical basis in this way (see [**WB12**] in the setting of the flag variety)[3].

From this perspective, analyzing when the p-canonical basis differs from the Kazhdan-Lusztig basis has several interesting geometric implications. First of all, it allows one to study the failure of the decomposition theorem in the modular setting (see [**JMW14a**, §3]). Secondly, there are close connections between the decomposition matrix for intersection cohomology complexes and the base change matrix between the Kazhdan-Lusztig and the p-canonical basis. In [**AR14**, Theorem 2.6] Achar and Riche show that the base change matrix gives a q-refinement of the decomposition matrix on the Langlands dual flag variety. Moreover, in [**Wil14**] the second author proves that certain base change coefficients and decomposition numbers coincide and uses this to give an example of a reducible characteristic variety in type A. Non-trivial decomposition numbers for an intersection cohomology complex can only occur when the characteristic variety of the corresponding regular holonomic D_X-module is reducible (see [**VW13**]).

1.3. This article gives a survey on known results about the p-canonical basis and its main purpose is to discuss an algorithm to calculate the p-canonical basis. This algorithm relies heavily on the diagrammatic description of the Hecke category $\mathbf{H}$. Motivated by de Cataldo and Migliorini's proof of the decomposition theorem (see [**dCM09**]), the second author showed in [**JMW14a**, §3.3] together with Juteau and Mautner that certain intersection forms govern the decomposition behaviour of the pushforward of the constant sheaf along a surjective, stratified, even resolution of singularities. In the diagrammatic framework this algorithm becomes more feasible, thanks to an explicit description of bases of Hom spaces. This leads to the algorithm described in this paper. We conclude the paper by giving many examples to show that the p-canonical basis may depend in a subtle way on p.

1.4. Structure of the Paper:

Section 2: We introduce notation and recall important results about the Hecke algebra and Soergel calculus.

Section 3: After recalling the definition of the p-canonical basis, we explain how to calculate it using intersection forms.

Section 4: The elementary properties of the p-canonical basis are stated and proved.

Section 5: We give several new and interesting examples of the p-canonical basis.

Acknowledgements. We would like to thank Ben Elias for very valuable comments and some minor corrections. We would also like to thank Nicolas Libedinsky for his feedback and the Max Planck Institute of Mathematics for financial support.

[3]In many settings this statement follows from [**Soe07**, **FW14**, **AR13**]. In the generality discussed in this paper this result will appear in [**RW15**].

2. Background

2.1. Coxeter Systems and Based Root Data. Let S be a finite set and $(m_{s,t})_{s,t\in S}$ be a matrix with entries in $\mathbb{N}\cup\{\infty\}$ such that $m_{s,s}=1$ and $m_{s,t}=m_{t,s}\geqslant 2$ for all $s\neq t\in S$. Denote by W the group generated by S subject to the relations $(st)^{m_{s,t}}=1$ for $s,t\in S$ with $m_{s,t}<\infty$. We say that (W,S) is a *Coxeter system* and W a *Coxeter group*. The Coxeter group W comes equipped with a length function $l: W\to\mathbb{N}$ and the Bruhat order $\leqslant$ (see [**Hum90**] for more details). A Coxeter system (W,S) is called *crystallographic* if $m_{s,t}\in\{2,3,4,6,\infty\}$ for all $s,t\in S$.

Define an *expression* to be a finite sequence of elements in S. We denote by

$$\mathrm{Ex}(S) \coloneqq \{\varnothing\}\cup\bigcup_{i\in\mathbb{N}\setminus\{0\}}\underbrace{S\times\cdots\times S}_{i\text{-times}}$$

the set of all expressions in S. For an expression $\underline{w}=(s_1,s_2,\dots,s_n)$ denote its *length* by $l(\underline{w})=n$. The multiplication gives a canonical map $\mathrm{Ex}(S)\to W$, $\underline{w}\mapsto \underline{w}_\bullet$. An expression $\underline{w}$ in S is called *reduced* if $l(\underline{w})=l(\underline{w}_\bullet)$. For an expression $\underline{w}=(s_1,s_2,\dots,s_n)$ in S a *subexpression* of $\underline{w}$ is a sequence $\underline{w}^{\underline{e}}=(s_1^{e_1},s_2^{e_2},\dots s_n^{e_n})$ where $e_i\in\{0,1\}$ for all i. The sequence $\underline{e}=(e_1,e_2,\dots,e_k)$ is called the *associated* 01*-sequence*. We usually decorate $\underline{e}$ as follows: For $1\leqslant k\leqslant n$ denote by $\underline{w}_{\leqslant k}\coloneqq(s_1,s_2,\dots,s_k)$ the first k terms and set $w_k\coloneqq(\underline{w}_{\leqslant k}^{\underline{e}_{\leqslant k}})_\bullet$. Assign to e_i a decoration $d_i\in\{U,D\}$ where U stands for *Up* and D for *Down* as follows:

$$d_i \coloneqq \begin{cases} U & \text{if } w_{i-1}s_i > w_{i-1},\\ D & \text{if } w_{i-1}s_i < w_{i-1}.\end{cases}$$

We often write the decorated sequence as $(d_1e_1,d_2e_2,\dots,d_ne_n)$. The *defect* of $\underline{e}$ is defined to be

$$\mathrm{df}(e)\coloneqq |\{i\mid d_ie_i=U0\}|-|\{i\mid d_ie_i=D0\}|.$$

To illustrate the definitions, consider for example the case $S=\{s,t\}$ and $m_{s,t}=m_{t,s}=3$ (i.e. type A_2). The reduced expression (s,t,s) admits two decorated 01-sequences expressing s:

$$\begin{aligned}&(U1,U0,D0) \quad \text{of defect } 0\\ &(U0,U0,U1) \quad \text{of defect } 2\end{aligned}$$

Recall that given an abstract root datum $\Psi=(X,\Phi,X^\vee,\Phi^\vee)$ and a basis $\Delta\subseteq\Phi$ the quadruple $\Psi_0=(X,\Delta,X^\vee,\Delta^\vee)$ is called a *based root datum* where $\Delta^\vee$ is the set of simple coroots (see [**Spr89**, §7.4] for the definition of a root datum). From now on, fix a based root datum Ψ_0. The matrix $(\langle\alpha_s,\alpha_t^\vee\rangle)_{s,t\in S}$ is called the *Cartan matrix* associated to this based root datum. By the existence theorem (see [**SGA70**, Exposé XXV, Théorème 1.1]), starting from Ψ_0 we get G, a split connected reductive algebraic group scheme over $\mathbb{Z}$, together with a Borel subgroup $B\subseteq G$ and a maximal torus $T\subseteq B\subseteq G$ such that the root datum determined by (G,T) and the basis given by the simple roots whose root groups are contained in B give the corresponding based root datum Ψ_0. For a root $\alpha\in\Phi$ we define the corresponding reflection on X via

$$s_\alpha\lambda\coloneqq\lambda-\langle\lambda,\alpha^\vee\rangle\alpha \qquad \text{for all } \lambda\in X.$$

We denote the Weyl group of Φ by $W\coloneqq\langle s_\alpha\mid\alpha\in\Phi\rangle$.

For $S := \{s_\alpha \mid \alpha \in \Delta\}$ the triple $(X^\vee, \Delta, \Delta^\vee)$ gives a (not necessarily symmetric) faithful realization of the Coxeter system (W, S) over $\mathbb{Z}$ (as defined in the appendix of [**Eli13**]). Fix a commutative ring k. Then ${}^kV := X^\vee \otimes_{\mathbb{Z}} k$ yields a (potentially non-faithful) realization of (W, S) over k. Set ${}^kV^* := \operatorname{Hom}_k({}^kV, k)$ and note that ${}^kV^*$ is isomorphic to $X \otimes_{\mathbb{Z}} k$. Throughout we will assume the following:

Assumption 2.1 (Demazure Surjectivity). *The maps* $\alpha_s : {}^kV \to k$ *and* $\alpha_s^\vee : {}^kV^* \to k$ *are surjective for all* $s \in S$.

This is automatically satisfied if 2 is invertible in k or if the Coxeter system (W, S) is of simply-laced type and of rank $|S| \geqslant 2$.

We denote by $R = S({}^kV^*)$ the symmetric algebra of ${}^kV^*$ over k and view it as a graded ring with ${}^kV^*$ in degree 2. The action of W on ${}^kV$ induces an action on R by functoriality. For any $s \in S$ we define the *Demazure operator* $\partial_s : R \to R(-2)$ via

$$\partial_s(f) := \frac{f - s(f)}{\alpha_s}$$

where (1) denotes the grading shift down by one: Given a graded R-bimodule $B = \bigoplus_{i \in \mathbb{Z}} B^i$, we denote by $B(1)$ the shifted bimodule with $B(1)^i = B^{i+1}$. Observe that ∂_s is a well-defined graded R^s-bimodule homomorphism (see [**EW13**, §3.3] for more details).

2.2. The Hecke Algebra. The Hecke algebra is the free $\mathbb{Z}[v, v^{-1}]$-algebra with $\{H_w \mid w \in W\}$ as basis and multiplication determined by:

$$\begin{aligned} H_s^2 &= (v^{-1} - v)H_s + 1 && \text{for all } s \in S, \\ H_x H_y &= H_{xy} && \text{if } l(x) + l(y) = l(xy). \end{aligned}$$

There is a unique $\mathbb{Z}$-linear involution $\overline{(-)}$ on $\mathcal{H}$ satisfying $\overline{v} = v^{-1}$ and $\overline{H_x} = H_{x^{-1}}^{-1}$. The Kazhdan-Lusztig basis element $\underline{H}_x$ is the unique element in $H_x + \sum_{y<x} v\mathbb{Z}[v] H_y$ that is invariant under $\overline{(-)}$. This is Soergel's normalization from [**Soe97**] of a basis introduced originally in [**KL79**].

2.3. Soergel Calculus. We define an *S-graph* to be a finite, decorated, planar graph with boundary properly embedded in the planar strip $\mathbb{R} \times [0, 1]$ whose edges are coloured by S and all of whose vertices are of the following types:

univalent vertices ("dots") of degree 1,

trivalent vertices of degree -1,

$2m_{s,t}$-valent vertices of degree 0,

where we require the $2m_{s,t}$-valent vertex to have exactly $2m_{s,t}$ edges, coloured alternately by s and t around the vertex.

The regions of an S-graph (i.e. the connected components of the complement of the graph in $\mathbb{R} \times [0, 1]$) may be *decorated* by homogeneous elements of R. The degree of a decorated S-graph is defined as the sum of the degrees of its vertices and of the degrees of the polynomials decorating its regions.

Next, we introduce the diagrammatic category of Soergel bimodules. The main reference for this is [**EW13**] (see also [**Eli13**] in the dihedral case and [**EK10**] in type A).

Let $\mathbf{H}_{BS}$ be the strict monoidal category with $\mathbb{Z}$-graded Hom-spaces which is monoidally generated by the elements in S. Thus the objects of $\mathbf{H}_{BS}$ are given by $\mathrm{Ex}(S)$ and the monoidal structure on the level of objects is given by concatenation of sequences in S. For any $\underline{x}, \underline{y} \in \mathrm{Ex}(S)$, $\mathrm{Hom}_{\mathbf{H}_{BS}}(\underline{x}, \underline{y})$ is defined to be the free R-module generated by isotopy classes of decorated S-graphs with bottom boundary $\underline{x}$ and top boundary $\underline{y}$ modulo the local relations below. The composition (resp. tensor product) of two morphisms is given by vertical (resp. horizontal) concatenation of diagrams.

We now recall the relations defining $\mathbf{H}_{BS}$:

2.3.1. *One-colour Relations.* For all $s \in S$ we have:

- Frobenius Unit:

(1) 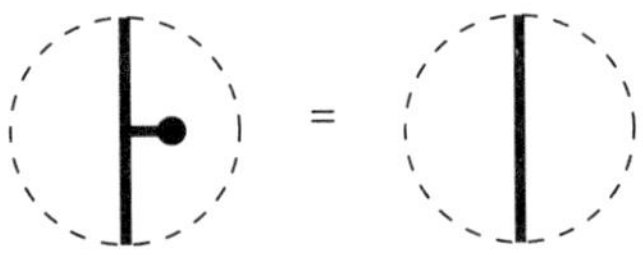

- Frobenius Associativity:

(2) 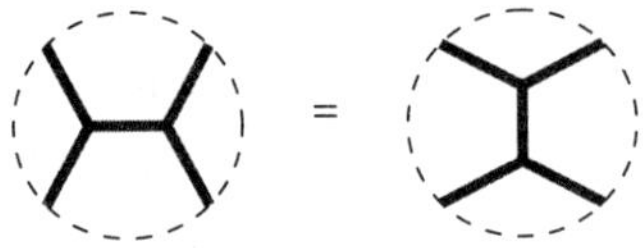

- Needle Relation:

(3) 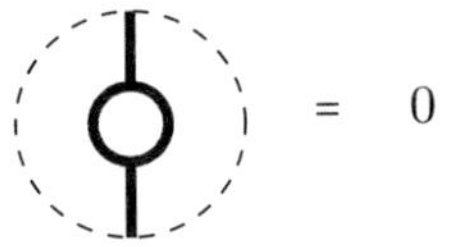

- Barbell Relation:

(4) 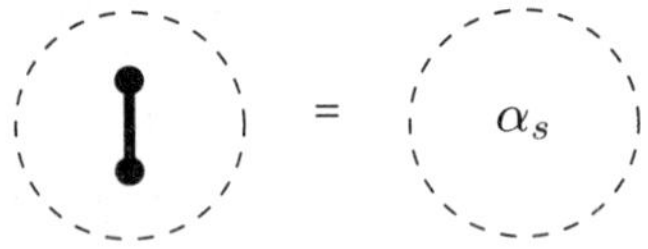

- Nil Hecke Relation:

(5) 

2.3.2. *Two-colour Relations.* There are two colour relations for all pairs $s, t \in S$ such that $m_{s,t} < \infty$ (so that the $2m_{st}$-valent vertex is defined).

The first two-colour relation is called *Two-colour Associativity* and describes what happens what happens when we pull a trivalent vertex through a $2m_{s,t}$-valent vertex. We give it for $m_{s,t} \in \{2, 3, 4\}$ and let the reader guess the general form (see

[**Eli13**, (6.12)]):

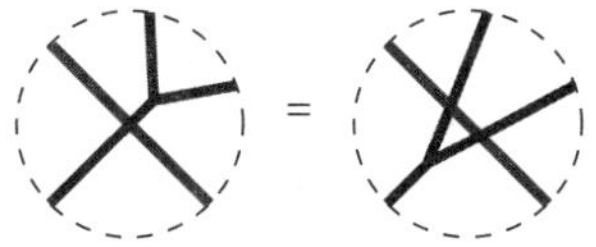

if $m_{s,t} = 2$ (type $A_1 \times A_1$),

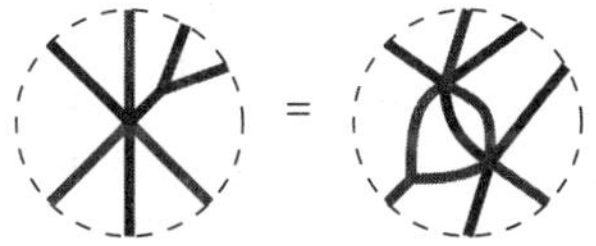

if $m_{s,t} = 3$ (type A_2),

if $m_{s,t} = 4$ (type B_2).

The next two-colour relation is called *Jones-Wenzl Relation* and expresses a $2m_{s,t}$-valent with a dot on one strand as a linear combination of diagrams in which only dots and trivalent vertices appear. We state it only for $m_{s,t} \in \{2, 3, 4\}$ and refer the reader to [**Eli13**] for more detail:

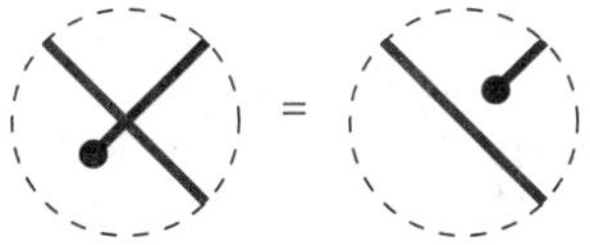

if $m_{s,t} = 2$,

if $m_{s,t} = 3$,

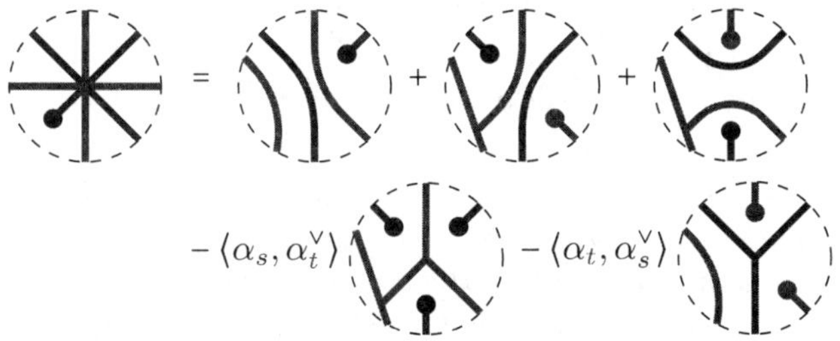

if $m_{s,t} = 4$.

2.3.3. *Three-Colour Relations.* We do not repeat the definition of the Zamolodchikov relations or "higher braid relations" here. The reader can find them in [**EW13**, §1.4.3] and is referred to [**EW12**] for more detail on the topological origins of the Zamolodchikov relations.

2.4. Light Leaves and Double Leaves. In this section we briefly discuss how to describe bases for morphism spaces in $\mathbf{H}_{BS}$. Fix an expression $\underline{w}$ and a reduced expression $\underline{x}$. In [**EW13**, §6.1] it is described how one may associate a "light leaves morphism" $\mathrm{LL}_{\underline{w},\underline{e}} \in \mathrm{Hom}_{\mathbf{H}_{BS}}(\underline{w}, \underline{x})$ to each subexpression $\underline{e}$ of $\underline{w}$ such that $(\underline{w}^{\underline{e}})_\bullet = x$. We will not need the explicit construction here, but the reader is encouraged to consult [**EW13**, §6.1] to follow our examples. The construction of light leaves follows a construction of Libedinsky for Soergel bimodules [**Lib08**] and depends on certain additional non-canonical choices.

In the special case of $x = e$, the identity of W, we get (see [**EW13**, Proposition 6.12]):

Proposition 2.2. *The set of all light leaves indexed by subsequences $\underline{e}$ of $\underline{w}$ expressing the identity of W gives an R-basis of* $\mathrm{Hom}_{\mathbf{H}_{BS}}(\underline{w}, \varnothing)$.

For an S-graph D denote by $\overline{D}$ the S-graph obtained by flipping the diagram upside down. This induces a contravariant equivalence on the monoidal category $\mathbf{H}_{BS}$ fixing all objects.

Out of light leaves one can construct double leaves as follows. Let $\underline{x}$ and $\underline{y}$ be arbitrary expressions in S. For any subsequences $\underline{e}$ (resp. $\underline{f}$) of $\underline{x}$ (resp. $\underline{y}$) both expressing $w \in W$ define $\mathbb{LL}_{w,\underline{e},\underline{f}} \coloneqq \overline{\mathrm{LL}_{\underline{y},\underline{f}}} \circ \mathrm{LL}_{\underline{x},\underline{e}}$. The following result can be found in [**EW13**, Theorem 6.11] (and was proved earlier in the setting of Soergel bimodules by Libedinsky in [**Lib15**]):

Theorem 2.3. *The set of all double leaves ranging over all $w \in W$ and pairs of subsequences $\underline{e}$ (resp. $\underline{f}$) of $\underline{x}$ (resp. $\underline{y}$) both expressing w gives an R-basis of* $\mathrm{Hom}_{\mathbf{H}_{BS}}(\underline{x}, \underline{y})$.

2.5. The Diagrammatic Category: Properties. Note that all relations in $\mathbf{H}_{BS}$ are homogeneous for our grading on S-graphs and thus $\mathbf{H}_{BS}$ is a category enriched in graded left R-modules; multiplying an S-graph D with a homogeneous polynomial $f \in R$ from the left is defined by decorating the leftmost region of D with f.

Let $\mathbf{H}$ be the Karoubian envelope of the graded version of the additive closure of $\mathbf{H}_{BS}$, in symbols $\mathbf{H} = \mathcal{K}ar(\mathbf{H}_{BS})$. We call $\mathbf{H}$ the *diagrammatic category of Soergel bimodules.* In other words, in the passage from $\mathbf{H}_{BS}$ to $\mathbf{H}$ we first allow direct sums and grading shifts (restricting to degree preserving homomorphisms) and then the taking of direct summands. The following properties can be found in [**EW13**, Lemma 6.24, Theorem 6.25 and Corollary 6.26]:

Theorem 2.4 (Properties of $\mathbf{H}$).
Let k be a complete local ring (e.g. a field or the p-adic integers $\mathbb{Z}_p$).

(1) $\mathbf{H}$ *is a Krull-Schmidt category.*
(2) *For all $w \in W$ there exists a unique, indecomposable object $B_w \in \mathbf{H}$ which is a direct summand of $\underline{w}$ for any reduced expression $\underline{w}$ of w and which is not isomorphic to a grading shift of any direct summand of any expression $\underline{v}$ for $v < w$. The object B_w does not depend up to isomorphism on the reduced expression $\underline{w}$ of w.*
(3) *The set $\{B_w \mid w \in W\}$ gives a complete set of representatives of the isomorphism classes of indecomposable objects in $\mathbf{H}$ up to grading shift.*
(4) *There exists a unique isomorphism of $\mathbb{Z}[v, v^{-1}]$-algebras*

$$\mathrm{ch} : [\mathbf{H}] \longrightarrow \mathcal{H}$$

sending $[B_s]$ to $\underline{H}_s$ for all $s \in S$, where $[\mathbf{H}]$ denotes the split Grothendieck group of $\mathbf{H}$. (We view $[\mathbf{H}]$ as a $\mathbb{Z}[v, v^{-1}]$-algebra as follows: the monoidal structure on $\mathbf{H}$ induces a unital, associative multiplication and v acts via $v[B] \coloneqq [B(1)]$ for an object B of $\mathbf{H}$.)

It should be noted that we do not have a diagrammatic presentation of $\mathbf{H}$ as determining the idempotents in $\mathbf{H}_{BS}$ is usually extremely difficult.

Observe that $\overline{(-)}$ extends to a contravariant equivalence of the graded, R-linear, additive, monoidal category $\mathbf{H}$ sending $B_w(n)$ to $B_w(-n)$ for all $n \in \mathbb{Z}$ and $w \in W$.

In order to explicitly give the isomorphism in the last part of Theorem 2.4, we need to introduce some more notation. For $x \in W$, let $\mathbf{H}^{\not< x}$ be the quotient category of $\mathbf{H}$ by the 2-sided ideal of morphisms factoring through any grading shift of a reduced expression $\underline{y}$ for some $y < x$. Write $\mathrm{Hom}_{\not< x}(-,-)$ for homomorphism spaces in $\mathbf{H}^{\not< x}$. In $\mathbf{H}^{\not< x}$ any two reduced expressions for x become canonically isomorphic. We denote the image of any reduced expression for x in $\mathbf{H}^{\not< x}$ by x as well. Under the assumptions of Theorem 2.4 we can define the *diagrammatic character* on an object B of $\mathbf{H}$ as follows:

$$\begin{aligned} \mathrm{ch} : [\mathbf{H}] &\longrightarrow \mathcal{H}, \\ [B] &\longmapsto \sum_{w \in W} \mathrm{grk}\, \mathrm{Hom}^{\bullet}_{\not< w}(B, w) H_w \end{aligned}$$

and extend $\mathbb{Z}[v, v^{-1}]$-linearly. In the last definition grk denotes the graded rank of the free R-module of homomorphisms of all degrees

$$\mathrm{Hom}^{\bullet}_{\not< w}(B_x, w) := \bigoplus_{n \in \mathbb{Z}} \mathrm{Hom}_{\not< w}(B_x, w(n)).$$

3. The p-Canonical Basis and Intersection Forms

In this section we recall the definition of the p-canonical basis of the Hecke algebra (see [**Wil12**]) and explain how to calculate it using intersection forms. Fix a field k of characteristic $p \geqslant 0$ and the realization ${}^kV$ of (W, S). We use this realization to define $\mathbf{H}$.

It is an interesting question what basis of the Hecke algebra the classes of the self-dual indecomposable objects in $\mathbf{H}$ correspond to. The answer is given for $k = \mathbb{R}$ by *Soergel's conjecture* which Elias and the second author proved in [**EW14**].

Theorem 3.1 (Elias-W. 2013).
$\mathrm{ch}([B_w]) = \underline{H}_w$ *for all* $w \in W$.

This illustrates that the basis of self-dual indecomposable objects in $\mathbf{H}$ gives an extremely interesting basis of $\mathcal{H}$ for $k = \mathbb{R}$ and motivates our definition of the p-canonical basis.

DEFINITION 3.2. Define ${}^p\underline{H}_w = \mathrm{ch}([B_w])$ for all $w \in W$ where $\mathrm{ch} : [\mathbf{H}] \xrightarrow{\cong} \mathcal{H}$ is the isomorphism of $\mathbb{Z}[v, v^{-1}]$-algebras introduced earlier, and p denotes the characteristic of k as above.

Theorem 2.4 implies that $\{{}^p\underline{H}_w \mid w \in W\}$ gives a basis of $\mathcal{H}$, called the *p-canonical basis.* As will become clearer later, the p-canonical basis depends only on the type of the root system chosen and on the characteristic p of the field k.

Next, we are going to explain how to use intersection forms to explicitly calculate the p-canonical basis. In order to calculate ${}^p\underline{H}_w$ we proceed by induction on $l(w)$. The induction start is given by ${}^p\underline{H}_e = \underline{H}_e = H_e$. Assume that we have already calculated ${}^p\underline{H}_v$ for all $v < w$. Let $\underline{w}$ be an arbitrary reduced expression for w. According to Theorem 2.4, we need to decompose $\underline{w}$ into indecomposable objects $B_x(n)$ for $x \in W$ and $n \in \mathbb{Z}$ in $\mathbf{H}$. For this we need local intersection forms:

Write $\mathrm{Hom}_{\nless x,k}(-,-)$ for the homomorphism spaces in $k \otimes_R \mathbf{H}^{\nless x}$ where we kill the action of the unique maximal ideal of all polynomials of positive degree. Since $\mathrm{Hom}^{\bullet}_{\nless x}(\underline{w}, x)$ is a graded free R-module with basis (see Section 2.4):

$$\{\mathrm{LL}_{\underline{w},\underline{e}} \mid \underline{e} \text{ is a subexpression of } \underline{w} \text{ expressing } x\},$$

$\mathrm{Hom}^{\bullet}_{\nless x,k}(\underline{w}, x)$ is a graded k-vector space on the same basis.

For an arbitrary expression $\underline{w}$ in S and $x \in W$, consider the k-bilinear Hom-pairing

$$\begin{aligned}\mathrm{Hom}^{\bullet}_{\nless x,k}(x,\underline{w}) \times \mathrm{Hom}^{\bullet}_{\nless x,k}(\underline{w},x) &\longrightarrow \mathrm{End}^{\bullet}_{\nless x,k}(x) = k,\\ (f,g) &\longmapsto g \circ f.\end{aligned}$$

Observe that $\mathrm{End}^{\bullet}_{\nless x,k}(x)$ is concentrated in degree 0 and that the duality $\overline{(-)}$ on $\mathbf{H}$ gives an isomorphism between $\mathrm{Hom}^{\bullet}_{\nless x,k}(x,\underline{w})$ and $\overline{\mathrm{Hom}^{\bullet}_{\nless x,k}(\underline{w},x)}$. This allows us to define:

DEFINITION 3.3. The *local intersection form of $\underline{w}$ at x* is the k-bilinear pairing on the graded free k-vector space $\mathrm{Hom}^{\bullet}_{\nless x,k}(\underline{w},x)$ given by

$$\begin{aligned}I_{\underline{w},x} : \mathrm{Hom}^{\bullet}_{\nless x,k}(\underline{w},x) \times \mathrm{Hom}^{\bullet}_{\nless x,k}(\underline{w},x) &\longrightarrow \mathrm{End}_{\nless x,k}(x) = k,\\ (f,g) &\longmapsto g \circ \overline{f}.\end{aligned}$$

The local intersection form of $\underline{w}$ at x can be split up into degree pieces as follows: Since $\mathrm{End}^{\bullet}_{\nless x,k}(x)$ is concentrated in degree 0, a homomorphism in $\mathrm{Hom}^{\bullet}_{\nless x,k}(\underline{w},x(d))$ for some $d \in \mathbb{Z}$ can only pair non-trivially with elements of $\mathrm{Hom}^{\bullet}_{\nless x,k}(\underline{w},x(-d))$. The *$d$-th grading piece of the intersection form* can thus be defined as:

$$I^d_{\underline{w},x} : \mathrm{Hom}_{\nless x,k}(\underline{w},x(-d)) \times \mathrm{Hom}_{\nless x,k}(\underline{w},x(d)) \longrightarrow \mathrm{End}_{\nless x,k}(x(d)) = k$$

Finally, the *graded rank* of $I_{\underline{w},x}$ is denoted by $n_{x,w} \in \mathbb{Z}_{\geqslant 0}[v,v^{-1}]$ and defined as

$$n_{x,w} := \sum_{d\in\mathbb{Z}} \mathrm{rk}(I^d_{\underline{w},x})v^d.$$

The following lemma illustrates the importance of intersection forms for the calculation of the p-canonical basis and follows from an argument similar to [**JMW14a**, Lemma 3.1]:

Lemma 3.4. *The multiplicity of B_x in $\underline{w}$ in* $\mathbf{H}$ *is given by the graded rank of $I_{\underline{w},x}$.*

After calculating the graded ranks of all $I_{\underline{w},x}$ for $x < w$, we can write for $\underline{w} = s_1 s_2 \dots s_n$:

$$\underline{H}_{s_1}\underline{H}_{s_2}\cdots\underline{H}_{s_n} = {}^p\underline{H}_w + \sum_{x<w} n_{x,w}\,{}^p\underline{H}_x.$$

REMARK 3.5. By comparing the intersection forms over $\mathbb{Q}$ and k, one may deduce that one only needs to calculate the graded ranks of $I_{\underline{w},x}$ for those x such that $\underline{H}_x$ occurs with a non-trivial coefficient when expressing $\underline{H}_{s_1}\underline{H}_{s_2}\cdots\underline{H}_{s_n}$ in terms of the Kazhdan-Lusztig basis.

In order to determine ${}^p\underline{H}_w$ we have to invert the matrix $(n_{x,y})_{x,y\leqslant w}$ which is upper triangular with ones on the diagonal in any total order refining the Bruhat order.

Finally, it should be noted that in practice one calculates the intersection form once over $\mathbb{Z}$ and reduces modulo different primes.

REMARK 3.6. There are other ways to calculate the p-canonical basis (which, however, are much more difficult in practice).

(1) If one can describe the geometry of the corresponding Schubert varieties quite explicitly, one can do calculations using parity sheaves (see [**WB12**, appendix]).
(2) In [**FW14**] Fiebig and the second author show that for a field k of characteristic p (or more generally a complete local PID), the Braden-MacPherson algorithm on the Bruhat graph allows one to compute the p-canonical basis.

3.1. Calculations in the nil Hecke Ring. In [**HW15**] Xuhua He and the second author reduce the calculation of certain entries in the intersection form to a simple formula in the nil Hecke ring. Instead of going into too much detail, we will try to give a survey of these results.

First, recall the definition of the nil Hecke ring. Let Q be the field of fractions of R. Denote by $Q * W$ the smash product. In other words, $Q * W$ is the free left Q-module with basis $\{\delta_w \mid w \in W\}$ and multiplication given by

$$(f\delta_x)(g\delta_y) = f(xg)\delta_{xy}$$

for $f, g \in Q$ and $x, y \in W$. Inside $Q * W$, we consider the elements

$$D_s = \frac{1}{\alpha_s}(\delta_e - \delta_s) = (\delta_e + \delta_s)\frac{1}{\alpha_s}$$

which satisfy the following relations:

$$\begin{aligned} D_s^2 &= 0, \\ D_s f &= (sf)D_s + \partial_s(f) \quad \text{for all } f \in Q, \\ \underbrace{D_s D_t D_s \dots}_{m_{s,t} \text{ terms}} &= \underbrace{D_t D_s D_t \dots}_{m_{s,t} \text{ terms}} . \end{aligned}$$

The last relation ensures that for $y \in W$ and any reduced expression $\underline{y} = (s_1, s_2, \dots, s_n)$ of y we get well-defined elements

$$D_y = D_{s_1} D_{s_2} \dots D_{s_n} \in Q * W.$$

The nil Hecke ring $\mathcal{NH}$ is the left R-submodule of $Q*W$ generated by $\{D_y \mid v \in W\}$.

Next, we briefly introduce gobbling morphisms. For any expression $\underline{w} = (s_1, s_2, \dots, s_n)$ in S, consider the following 01-sequence $\underline{e}$ with:

$$e_i = \begin{cases} 1 & \text{if } w_{i-1}s_i > w_{i-1}, \\ 0 & \text{otherwise,} \end{cases}$$

where at each step w_i is defined as in Section 2.1. Note that $(\underline{w}^{\underline{e}})_\bullet$ is the maximal element in W expressible as a subexpression of $\underline{w}$, and that the decoration of $\underline{e}$ consists entirely of $U1$'s and $D0$'s. Therefore any choice of light leaf morphism $\mathrm{LL}_{\underline{w},\underline{e}}$ has degree $l((\underline{w}^{\underline{e}})_\bullet) - l(\underline{w})$ and consists only of $2m_{st}$-valent and trivalent vertices. Denote by $G_{\underline{w}}$ the image of $\mathrm{LL}_{\underline{w},\underline{e}}$ in $\mathbf{H}^{\not<(\underline{w}^{\underline{e}})_\bullet}$. The morphism $G_{\underline{w}}$ is called a *gobbling morphism* and can be characterized as follows (see [**HW15**, Proposition 3.4]):

Proposition 3.7. *Let $\underline{w}$, $\underline{e}$ be as above. Any morphism $\underline{w} \to (\underline{w}^{\underline{e}})_\bullet$ in $\mathbf{H}^{\not<(\underline{w}^{\underline{e}})_\bullet}$ given by diagrams consisting only of $2m_{st}$-valent vertices and $l(\underline{w}) - l((\underline{w}^{\underline{e}})_\bullet)$ trivalent vertices is equal to $G_{\underline{w}}$.*

From this they deduce the canonicity of any light leaf morphism $\mathrm{LL}_{\underline{w},\underline{f}}$ in $\mathbf{H}^{\nless(\underline{w}^{\underline{f}})_\bullet}$ indexed by a 01-sequence $\underline{f}$ without $D1$'s in its decoration. (This follows because the morphism is given as the composition of a sequence of dots on strands corresponding to $U0$'s followed by a gobbling morphism.)

Finally, we come to their formula in the nil Hecke ring for certain entries of the intersection form. Let $\underline{e}^1$ and $\underline{e}^2$ be two subexpressions of $\underline{w}$. Assume that $\underline{e}^1$ and $\underline{e}^2$ both express the same element $x \in W$ (i.e. $x = (\underline{w}^{\underline{e}^1})_\bullet = (\underline{w}^{\underline{e}^2})_\bullet$) and that their decorations do not contain any $D1$. Define an element of the nil Hecke ring as the product $f(\underline{e}^1, \underline{e}^2) = f_1 f_2 \dots f_m$ where

$$f_i = \begin{cases} \alpha_{s_i} & \text{if } e_i^1 = e_i^2 = U0, \\ 1 & \text{if exactly one of } e_i^1 \text{ and } e_i^2 \text{ is } U0, \\ D_{s_i} & \text{otherwise.} \end{cases}$$

Denote by $d(\underline{e}^1, \underline{e}^2) \in R$ the coefficient of $D_{(\underline{w}^{\underline{e}^1})_\bullet}$ in $f(\underline{e}^1, \underline{e}^2)$. The main result is [**HW15**, Theorem 5.1]:

Theorem 3.8. *For $\underline{e}^1$ and $\underline{e}^2$ as above, we have*

$$I_{\underline{w},x}(\mathrm{LL}_{\underline{w},\underline{e}^1}, \mathrm{LL}_{\underline{w},\underline{e}^2}) = d(\underline{e}^1, \underline{e}^2).$$

This theorem gives a combinatorial formula for some entries in the intersection form. Sometimes one is lucky, and it can be used to calculate the complete intersection form, as we will see in examples below.

4. First Properties of the p-Canonical Basis

The goal of this section is to prove elementary properties of the p-canonical basis and to compare it to the Kazhdan-Lusztig basis. For this we need a p-modular system. Let $\mathbb{O}$ be a complete local ring with residue field k of characteristic $p > 0$ and quotient field $\mathbb{K}$ of characteristic 0. Fix the realization ${}^{\mathbb{O}}V$ of (W, S) and use it to define $\mathbf{H}$. For $x \in W$ we will denote by B_x (resp. ${}^kB_x$ or ${}^{\mathbb{K}}B_x$) the indecomposable object in $\mathbf{H}$ (resp. ${}^k\mathbf{H} := \mathbf{H} \otimes_{\mathbb{O}} k$ or ${}^{\mathbb{K}}\mathbf{H} := \mathbf{H} \otimes_{\mathbb{O}} \mathbb{K}$).

The following lemma shows that indecomposable objects in $\mathbf{H}$ remain indecomposable when passing to ${}^k\mathbf{H}$.

Lemma 4.1. *We have for all $w \in W$: $B_w \otimes_{\mathbb{O}} k \cong {}^kB_w$.*

Proof. Assume $B_w \otimes_{\mathbb{O}} k$ is not indecomposable in ${}^k\mathbf{H}$. Then there exists a non-trivial idempotent $e \in \mathrm{End}_{{}^k\mathbf{H}}(B_w \otimes_{\mathbb{O}} k)$. Since $\mathrm{End}_{\mathbf{H}}(B_w)$ is a finitely generated $\mathbb{O}$-module, we can use idempotent lifting techniques for complete local rings (see [**Lam01**, Theorem 21.31]) and find an idempotent $\tilde{e} \in \mathrm{End}_{\mathbf{H}}(B_w)$ mapping to e in $\mathrm{End}_{{}^k\mathbf{H}}(B_w \otimes_{\mathbb{O}} k) \cong \mathrm{End}_{\mathbf{H}}(B_w) \otimes_{\mathbb{O}} k$. Since B_w is indecomposable, this idempotent has to be trivial, a contradiction. □

Some of the following properties can also be found in [**WB12**] and [**Wil14**]:

Proposition 4.2. *For all $x, y \in W$ we have:*

(1) $\overline{{}^p\underline{H}_x} = {}^p\underline{H}_x$, *i.e. ${}^p\underline{H}_x$ is self-dual,*
(2) ${}^p\underline{H}_x = H_x + \sum_{y<x} {}^ph_{y,x} H_y$ *with ${}^ph_{y,x} \in \mathbb{Z}_{\geqslant 0}[v, v^{-1}]$,*
(3) ${}^p\underline{H}_x = \underline{H}_x + \sum_{y<x} {}^pm_{y,x} \underline{H}_y$ *with self-dual ${}^pm_{y,x} \in \mathbb{Z}_{\geqslant 0}[v, v^{-1}]$,*
(4) ${}^pm_{y,x} = {}^pm_{y^{-1},x^{-1}}$ *for all $x, y \in W$,*

(5) ${}^p m_{y,x} = 0$ *unless* $\mathcal{L}(x) \subseteq \mathcal{L}(y)$ *and* $\mathcal{R}(x) \subseteq \mathcal{R}(y)$ *where* $\mathcal{L}$ *and* $\mathcal{R}$ *denote the left and right descent sets,*
(6) ${}^p\underline{H}_x {}^p\underline{H}_y = \sum_{z\in W} {}^p\mu_{x,y}^z {}^p\underline{H}_z$ *with self-dual* ${}^p\mu_{x,y}^z \in \mathbb{Z}_{\geqslant 0}[v, v^{-1}]$,
(7) ${}^p\underline{H}_x = \underline{H}_x$ *for* $p \gg 0$ *(i.e. there are only finitely many primes for which* ${}^p\underline{H}_x \neq \underline{H}_x$*).*

PROOF. (1) We proceed by induction on $l(x)$. For small x the statement is clear as ${}^p\underline{H}_e = \underline{H}_e$ and ${}^p\underline{H}_s = \underline{H}_s$ for all $s \in S$ and all primes p. Assume that we have shown that ${}^p\underline{H}_y$ is self-dual for $y < x$. Choose $s \in \mathcal{L}(x)$ and set $y = sx$. The characterization of ${}^kB_x$ in Theorem 2.4 implies that ${}^kB_x$ occurs with multiplicity one in ${}^kB_s {}^kB_y$ and that $\overline{{}^kB_x(n)} = {}^kB_x(-n)$. Thus we can write

$$ {}^kB_s {}^kB_y = {}^kB_x \oplus \bigoplus_{\substack{z<x \\ n\in\mathbb{Z}}} ({}^kB_z(n))^{\oplus a_{z,n}} $$

with $a_{y,n} \in \mathbb{Z}_{\geqslant 0}$ for all $y < x$ and $n \in \mathbb{Z}$ and all but finitely many of the $a_{y,n}$ are zero. Applying the duality $\overline{(-)}$ to both sides and using that the left hand side is self-dual yields $a_{z,n} = a_{z,-n}$ for all $z < x$ and $n \in \mathbb{Z}$. This implies

$$ {}^p\underline{H}_s {}^p\underline{H}_y = {}^p\underline{H}_x + \sum_{z<x} {}^p\mu_{z,y} {}^p\underline{H}_z $$

where ${}^p\mu_{z,y} = \sum_{n\in Z} a_{z,n}v^n \in \mathbb{Z}_{\geqslant 0}[v, v^{-1}]$ is self-dual for all $z < x$. Therefore the self-duality of the left-hand side and our induction hypothesis imply the self-duality of ${}^p\underline{H}_x$.

(2) The unicity in the characterization of ${}^kB_x$ in Theorem 2.4 implies that it occurs with multiplicity 1 in any reduced expression $\underline{x}$ of x. In the quotient $\mathbf{H}^{\not< x}$ all other direct summands of x are killed. Thus we get $\operatorname{grk}\operatorname{Hom}_{\not< x}(B_x, x) = 1$. Note that the Laurent polynomials ${}^ph_{y,x}$ have non-negative coefficients as they are given by graded ranks of free R-modules.

(3) According to (2), $\underline{H}_x$ occurs precisely with coefficient 1 in ${}^p\underline{H}_x$. The self-duality of the Laurent polynomials ${}^pm_{y,x}$ follows from (1) and the self-duality of the Kazhdan-Lusztig basis. Since $\operatorname{Hom}^\bullet_{\not< w}(B_x, w)$ is a free R-module, we get for $F \in \{k, \mathbb{K}\}$:

$$ \operatorname{Hom}^\bullet_{\not< w}(B_x, w) \otimes_{\mathbb{O}} F \cong \operatorname{Hom}^\bullet_{\mathbf{H}^{\not< w}\otimes_{\mathbb{O}} F}(B_x \otimes_{\mathbb{O}} F, w \otimes_{\mathbb{O}} F). $$

This implies in particular using Lemma 4.1:

$$ \operatorname{ch}({}^kB_x) = \operatorname{ch}(B_x) = \operatorname{ch}(B_x \otimes_{\mathbb{O}} \mathbb{K}). $$

Thus the ${}^pm_{y,x}$ have non-negative coefficients as they come from decomposing $B_x \otimes_{\mathbb{O}} \mathbb{K}$ into indecomposable objects of the form ${}^{\mathbb{K}}B_x$ in ${}^{\mathbb{K}}\mathbf{H}$ whose character is given by $\underline{H}_x$ by Theorem 3.1.

(4) There is an equivalence ι on ${}^k\mathbf{H}$ viewed as a k-linear category induced by the horizontal flip of Soergel graphs. It maps $(s_1, s_2, \dots, s_m) \in {}^k\mathbf{H}$ to $(s_m, \dots, s_2, s_1)$ and thus ${}^kB_x$ to ${}^kB_{x^{-1}}$ for all $x \in W$. It is easy to see that ι descends to a well-known $\mathbb{Z}[v, v^{-1}]$-linear anti-involution on $\mathcal{H}$ sending $\underline{H}_x$ to $\underline{H}_{x^{-1}}$ as well as ${}^p\underline{H}_x$ to ${}^p\underline{H}_{x^{-1}}$ for all $x \in W$. By slight abuse of notation we will denote this map also by ι. Thus we have: $\operatorname{ch}\circ\iota = \iota\circ\operatorname{ch}$. Finally expressing ${}^p\underline{H}_x$ in the Kazhdan-Lusztig basis and applying ι proves the claim after comparing coefficients.

(5) The statement for left descent sets follows from Lemma 4.3 below and the fact that the Laurent polynomials ${}^p m_{y,x}$ have non-negative coefficients. Using (4) we can reduce the statement about the right descent sets to the case we have just proven.
(6) This follows immediately from the analogue of Soergel's categorification theorem (part (4) of Theorem 2.4). Indeed, in order to express ${}^p\underline{H}_x {}^p\underline{H}_y$ in the p-canonical basis, we need to decompose ${}^k B_x {}^k B_y$ into indecomposable objects in ${}^k\mathbf{H}$ and thus the Laurent polynomial ${}^p\mu_{x,y}^z$ encodes the graded multiplicity of ${}^k B_z$ in this tensor product. Therefore ${}^p\mu_{x,y}^z$ has non-negative coefficients. The self-duality of these Laurent polynomials follows from (1).
(7) As explained in Section 3 we need to calculate the graded rank of finitely many local intersection forms in order to calculate ${}^p\underline{H}_x$. The rank of each of these intersection forms can only decrease for finitely many primes. (7) now follows.

□

The *multiplication formula* from [**KL79**, (2.3.a) and (2.3.c)]) reads for $x \in W$ and $s \in S$ as follows:

$$\underline{H}_s \underline{H}_x = \begin{cases} (v+v^{-1})\underline{H}_x & \text{if } sx < x, \\ \underline{H}_{sx} + \sum_{\substack{y<x \text{ s.t.} \\ sy<y}} \mu(y,x)\underline{H}_y & \text{otherwise.} \end{cases} \tag{6}$$

where $\mu(y,x)$ is the coefficient of v in the Kazhdan-Lusztig polynomial ${}^0h_{y,x}$. One remnant of this multiplication formula for the p-canonical basis is the following result:

Lemma 4.3. *For $x \in W$ and $s \in \mathcal{L}(x)$ we have:*

$${}^p\underline{H}_s {}^p\underline{H}_x = (v+v^{-1}) {}^p\underline{H}_x.$$

PROOF. Since $sx < x$, we can write using Proposition 4.2 (2) for $\mathrm{ch}({}^kB_{sx})$:

$$\mathrm{ch}({}^kB_s {}^kB_{sx}) = \underline{H}_s \mathrm{ch}({}^kB_{sx}) = H_x + vH_{sx} + \sum_{\substack{y<x \\ y\neq sx}} h_y H_y = \underline{H}_x + \sum_{\substack{y<x \\ y\neq sx}} m_y \underline{H}_y.$$

As ${}^kB_x$ is a summand of ${}^kB_s {}^kB_{sx}$ we can apply Proposition 4.2 (3) to deduce

$$\mathrm{ch}({}^kB_x) = \underline{H}_x + \sum_{\substack{y<x \\ y\neq sx}} {}^p m_{y,x} \underline{H}_y.$$

This implies ${}^p m_{sx,x} = 0$. Next, we calculate as follows:

$$\begin{aligned} {}^p\underline{H}_s {}^p\underline{H}_x &= \underline{H}_s {}^p\underline{H}_x \\ &= \underline{H}_s(\underline{H}_x + \sum_{\substack{y<x \\ y\neq sx}} {}^p m_{y,x} \underline{H}_y) \\ &\in (v+v^{-1}) {}^p\underline{H}_x + \sum_{y<x} \mathbb{Z}_{\geqslant 0}[v,v^{-1}] {}^p\underline{H}_y \end{aligned}$$

where in the last equality we used Equation (6) and the observation ${}^p m_{sx,x} = 0$ to determine the coefficient in front of ${}^p\underline{H}_x$. After evaluating at $v=1$ and acting on the trivial module we see that no other terms besides $(v+v^{-1}) {}^p\underline{H}_x$ can occur on the right hand side. □

4.1. The Geometric Satake and the p-canonical Basis. In this section, we will explain the consequences of results about parity sheaves on the affine Grassmannian for the p-canonical basis.

A very interesting question is under which hypothesis the indecomposable parity sheaves on the affine Grassmannian are perverse. This is equivalent to the p-canonical basis being a $\mathbb{Z}$-linear combination of Kazhdan-Lusztig basis elements.

To shorten our notation, write $\mathcal{K} := \mathbb{C}((t))$ and $\mathcal{O} := \mathbb{C}[[t]]$. Define the affine Grassmannian $\mathcal{G}r_a$ to be the $\mathbb{Z}$-functor given by $R \mapsto LG(R)/L^{+}G(R)$. Its complex points coincide with $G(\mathcal{K})/G(\mathcal{O})$. For $\lambda \in X^\vee$ denote by $t^\lambda \in G(\mathcal{K})$ the following composition

$$\operatorname{Spec}(\mathcal{K}) \longrightarrow \operatorname{Spec}(\mathcal{O}) = \mathbb{G}_m \xrightarrow{\lambda} T \hookrightarrow G$$

where the first morphism comes from the inclusion $\mathcal{O} \hookrightarrow \mathcal{K}$. The Cartan decomposition

$$\mathcal{G}r_a(\mathbb{C}) = \bigcup_{\lambda \in X_+^\vee} \underbrace{G(\mathcal{O})t^\lambda G(\mathcal{O})/G(\mathcal{O})}_{\mathcal{G}r_a^\lambda :=}$$

gives a stratification of $\mathcal{G}r_a(\mathbb{C})$ where each stratum $\mathcal{G}r_a^\lambda$ is a vector bundle over a partial flag variety $G(\mathbb{C})t^\lambda G(\mathcal{O})/G(\mathcal{O})$. Since the $G(\mathcal{O})$-orbits are all simply-connected, the indecomposable parity sheaves are all parametrized by $\lambda \in X_+^\vee$ (see [**JMW14a**, Theorem 4.6]):

Theorem 4.4. *Assume that the characteristic of k is not a torsion prime*[4] *for G. For each $\lambda \in X_+^\vee$ there exists up to isomorphism a unique indecomposable parity complex $\mathcal{E}(\lambda)$ such that* $\operatorname{supp}(\mathcal{E}(\lambda)) = \overline{\mathcal{G}r_a^\lambda}$ *and* $\mathcal{E}(\lambda)|_{\mathcal{G}r_a^\lambda} = \underline{k}_{\mathcal{G}r_a^\lambda}[\dim \mathcal{G}r_a^\lambda]$. *Every indecomposable parity complex is isomorphic to $\mathcal{E}(\lambda)$ for some $\lambda \in X_+^\vee$.*

Denote by $P_{G(\mathcal{O})}(\mathcal{G}r_a(\mathbb{C}), k)$ the $G(\mathcal{O})$-equivariant perverse sheaves on $\mathcal{G}r_a(\mathbb{C})$ with coefficients in k. It comes equipped with a monoidal structure induced by the convolution product $*$.

Recall that we denote by $\Psi = (X, \Phi, X^\vee, \Phi^\vee)$ the root datum of G. Define $G^\vee$ to be the split connected reductive group scheme over k with root datum (Langlands) dual to that of G. To each dominant coweight $\lambda \in X_+^\vee$ we can associate an induced representation $\nabla(\lambda) := \operatorname{ind}_{B^\vee}^{G^\vee} k_\lambda$, its dual $\Delta(\lambda)$ called the *Weyl module* and a simple module $L(\lambda)$ sitting in the following sequence $\Delta(\lambda) \twoheadrightarrow L(\lambda) \hookrightarrow \nabla(\lambda)$ given by projection to the head and inclusion of the socle. Moreover, the set $\{L(\lambda) \mid \lambda \in X_+^\vee\}$ gives a complete set of representatives for the isomorphism classes of simple, rational $G^\vee$-modules. The category $\operatorname{Rep}(G^\vee)$ of rational representations of $G^\vee$ forms a highest-weight category with Weyl modules as standard objects and induced modules as costandard objects. A rational representation of $G^\vee$ is called *tilting* if it admits two filtrations, one with successive quotients isomorphic to Weyl modules and the other one with successive quotients isomorphic to induced modules. In [**Don93**, Theorem 1.1] Donkin classifies indecomposable tilting modules for algebraic groups giving in our setting:

Theorem 4.5. *For each $\lambda \in X_+^\vee$ there exists up to isomorphism a unique indecomposable tilting module $T(\lambda)$ of highest weight λ. Moreover, λ occurs with multiplicity one as a weight of $T(\lambda)$. Every indecomposable tilting module is isomorphic to $T(\lambda)$ for some $\lambda \in X_+^\vee$.*

[4]See [**JMW14a**, §2.6] for the definition of torsion primes. This restriction can be removed by working in the non-equivariant setting.

The Geometric Satake equivalence (see [**MV07**]) gives a monoidal equivalence

$$(P_{G(\mathcal{O})}(\mathcal{G}r_a(\mathbb{C}),k),*) \xrightarrow{\cong} (\mathrm{Rep}(G^\vee),\otimes).$$

In [**JMW14b**] Juteau, Mautner and the second author show that if the characteristic p of k is larger than an explicit bound depending on the root system Φ of G, then the indecomposable k-parity sheaves on the affine Grassmannian are perverse. More precisely, they show that for $\lambda \in X_+^\vee$ the indecomposable tilting module $T(\lambda)$ is mapped to $\mathcal{E}(\lambda)$ under the geometric Satake equivalence for $p > b(\Phi)$ (see below).

From now on let us assume that G is adjoint, so $G^\vee$ is simply connected. In particular, the cocharacter lattice $X^\vee$ coincides with the coweight lattice. Denote by $W_a := W \ltimes \mathbb{Z}\Phi^\vee$ the *affine Weyl group* and by $W_{ext} := W \ltimes X^\vee$ the *extended affine Weyl group*. View W_a as a Coxeter group (W_a, S_a). We can write

$$W_{ext} = \Omega \ltimes W_a$$

where Ω is a finite subgroup of "length zero elements" which acts by automorphisms of the Coxeter system (W_a, S_a) (see [**Lus83**, §2]). The action of W_{ext} on $X^\vee$ gives bijections:

$$\begin{aligned} X^\vee &\overset{\sim}{\longleftrightarrow} W_{ext}/W, \\ w(0) &\longleftrightarrow w, \\ X_+^\vee &\overset{\sim}{\longleftrightarrow} W\backslash W_{ext}/W = \bigcup_{\sigma\in\Omega} \sigma(W)\backslash W_a/W. \end{aligned}$$

Given $\lambda \in X_+^\vee$ we denote by $w_\lambda \in W_a$ the maximal element in the double coset in $\bigcup_{\sigma\in\Omega}\sigma(W)\backslash W_a/W$ that corresponds to λ under the bijection above. In this section we will work with the Hecke algebra $\mathcal{H}_a$ associated to (W_a, S_a).

The next results follows because the projection from the affine flag variety to the affine Grassmannian is a stratified fiber bundle with fibers isomorphic to the finite flag variety $\mathcal{F}l$:

Theorem 4.6. *Suppose that all parity sheaves on the affine Grassmannian are perverse. Then* ${}^p m_{w_\lambda, w_\nu} \in \mathbb{Z}$ *for all* $\lambda, \nu \in X_+^\vee$.

The explicit bound has later been improved by Mautner and Riche in [**MR15**]. They show that the parity sheaves on the affine Grassmannian are perverse whenever the characteristic p of k is good for G, thus giving the following bounds $p > b(\Phi)$ if the root system Φ of G is irreducible:

Type of Φ	A_n	B_n, C_n, D_n	E_6, E_7, F_4, G_2	E_8
$b(\Phi)$	1	2	3	5

The following result gives an interpretation of the multiplicities in the p-canonical basis in terms of $\mathrm{Rep}(G^\vee)$ (see [**JMW14b**, Corollary 4.1] for the first part):

Lemma 4.7. *Suppose that the characteristic p of k satisfies $p > b(\Phi)$. Then we have for $\lambda, \mu \in X_+^\vee$:*

(1) ${}^p h_{w_\mu, w_\lambda}(1) = \dim T(\lambda)_\mu$,
(2) ${}^p m_{w_\mu, w_\lambda} = [T(\lambda) : \Delta(\mu)] = [T(\lambda) : \nabla(\mu)]$ *where* $[T(\lambda) : \Delta(\mu)]$ *denotes the multiplicity of* $\Delta(\mu)$ *in a* Δ*-flag on* $T(\lambda)$.

The first part of the last result shows that ${}^p h_{w_\mu, w_\lambda}$ gives a q-analogue of the weight multiplicity of μ in $T(\lambda)$. In characteristic 0 (where $T(\lambda) = L(\lambda)$) this result can be found in [**Lus83**, Theorem 6.1]. In particular, knowledge of the characters of indecomposable tilting modules for $G^\vee$ is equivalent to the knowledge of the p-canonical basis elements $\{{}^p\underline{H}_{w_\lambda} \mid \lambda \in X_+^\vee\}$.

5. Examples

According to the classification of root systems and connected semi-simple algebraic groups, a Dynkin diagram fixes a semi-simple, adjoint algebraic group G together with a maximal torus $T \subseteq G$ such that the root system determined by (G,T) corresponds to the given Dynkin diagram. In this section we will only give the Dynkin diagram and consider the corresponding root datum of this pair (G,T) together with an arbitrary basis labelled by the nodes of the Dynkin diagram as input.

5.1. Type B_2. We label the simple reflections as follows:

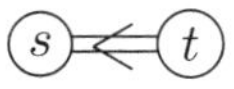

That means for the pairing between the simple coroots and roots:

$$\langle \alpha_t, \alpha_s^\vee \rangle = -2,$$
$$\langle \alpha_s, \alpha_t^\vee \rangle = -1.$$

Because the Schubert varieties associated to e, s, t, st, ts and $stst$ are smooth, we have ${}^p\underline{H}_x = \underline{H}_x$ for $x \in \{e, s, t, st, ts, stst\}$ and all primes p. (This can also be checked directly.) The remaining two elements are sts and tst. The two subsequences of (s,t,s) expressing s and corresponding light leaves are:

$(U1, U0, D0)$ of defect 0 $\rightsquigarrow$ $L_1 =$ ,

$(U0, U0, U1)$ of defect 2 $\rightsquigarrow$ $L_2 =$ .

Thus the pairing of the light leaves in $\mathbf{H}^{\nless s}$ is given by

$$(L_i \circ \overline{L_j})_{i,j \in \{1,2\}} = \begin{pmatrix} \langle \alpha_t, \alpha_s^\vee \rangle & \alpha_t \\ \alpha_t & \alpha_s \alpha_t \end{pmatrix}$$

where the top left entry comes from the following calculation:

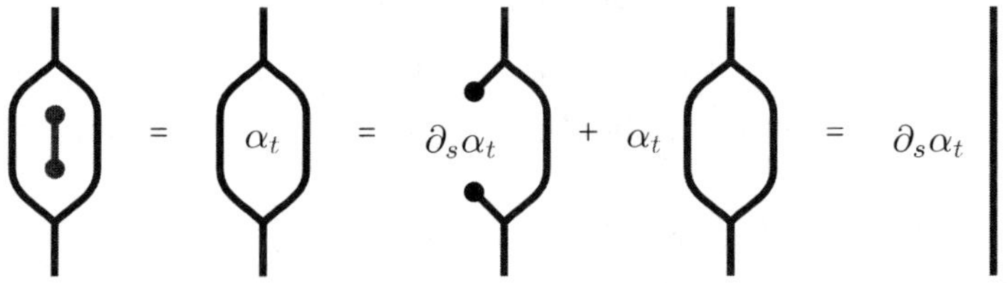

Therefore the 0-th degree piece of the local intersection form of (s,t,s) at s is $I^0_{sts,s} = (-2)$ (recall that $\langle \alpha_t, \alpha_s^\vee \rangle = \partial_s \alpha_t = -2$). This shows that if $p = 2$, then

${}^kB_s\,{}^kB_t\,{}^kB_s$ does not decompose as ${}^kB_{sts} \oplus {}^kB_s$, but remains indecomposable. We get

$$ {}^p\underline{H}_{sts} = \begin{cases} \underline{H}_{sts} + \underline{H}_s & \text{if } p = 2, \\ \underline{H}_{sts} & \text{otherwise.} \end{cases} $$

Swapping the roles of s and t, the same calculation yields

$$ {}^p\underline{H}_{tst} = \underline{H}_{tst} $$

for all primes p as $\langle \alpha_s, \alpha_t^\vee \rangle = -1$.

Observe that in this case the whole local intersection form of (s,t,s) at s can be calculated using the formula in the nil Hecke ring which we explained in Section 3.1.

5.2. Type G_2. We label the simple reflections as follows:

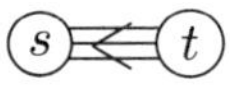

That means for the pairing between the simple coroots and roots:

$$ \langle \alpha_t, \alpha_s^\vee \rangle = -3, $$
$$ \langle \alpha_s, \alpha_t^\vee \rangle = -1. $$

For all primes $p > 3$ the Kazhdan-Lusztig basis coincides with the p-canonical basis. Since the Cartan matrix is symmetric modulo 2, the 2-canonical basis is stable under swapping s and t. Here is a summary of the results for $p \in \{2,3\}$:

$$ {}^2\underline{H}_{stst} = \underline{H}_{stst} + \underline{H}_{st} \qquad {}^3\underline{H}_{sts} = \underline{H}_{sts} + \underline{H}_s $$
$$ {}^2\underline{H}_{ststs} = \underline{H}_{ststs} + \underline{H}_s \qquad {}^3\underline{H}_{ststs} = \underline{H}_{ststs} + \underline{H}_{sts} $$

$$ {}^2\underline{H}_x = \underline{H}_x \text{ for } x \notin \{stst, tsts, ststs, tstst\} \qquad {}^3\underline{H}_x = \underline{H}_x \text{ for } x \notin \{sts, ststs\} $$

In this example, all the calculations needed to determine ${}^p\underline{H}_x$ for $x \notin \{ststs, tstst\}$ can be carried out using the formula in the nil Hecke ring from Section 3.1. For (s,t,s,t,s) (resp. (t,s,t,s,t)) there is a subexpression of defect 0 expressing s (resp. t) that contains a $D1$:

$$ (U1, U1, U0, D1, D0). $$

We will illustrate how useful the formula in the nil Hecke ring is by calculating the intersection form of (s,t,s,t) at st. For the Kazhdan-Lusztig basis we know:

$$ \underline{H}_s\underline{H}_t\underline{H}_s\underline{H}_t = \underline{H}_{stst} + 2\underline{H}_{st}. $$

There are two subexpressions of (s,t,s,t) of defect 0 expressing st:

$$ \underline{e}^1 := (U1, U0, D0, U1) $$
$$ \underline{e}^2 := (U1, U1, U0, D0) $$

We need to calculate the coefficient of D_{st} in the following elements of the nil Hecke ring:

$$ d(\underline{e}^1, \underline{e}^1): \quad D_s\alpha_t D_s D_t = \partial_s(\alpha_t) D_{st} $$
$$ d(\underline{e}^1, \underline{e}^2): \quad D_s 1 1 D_t = D_{st} $$
$$ d(\underline{e}^2, \underline{e}^2): \quad D_s D_t \alpha_s D_t = \partial_t(\alpha_s) D_{st} $$

Therefore the local intersection form of $stst$ at st is given by

$$\begin{pmatrix} -3 & 1 \\ 1 & 1 \end{pmatrix}$$

which implies the result stated above.

5.3. Type $\tilde{A}_1$. We label the simple reflections in S_a as follows:

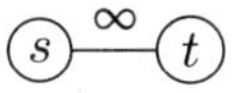

That means for the pairing between the simple coroots and roots:

$$\langle \alpha_t, \alpha_s^\vee \rangle = -2,$$
$$\langle \alpha_s, \alpha_t^\vee \rangle = -2.$$

As we discussed in Section 4.1, knowledge of the characters of the tilting modules for SL_2 in characteristic p gives us part of the p-canonical basis for the Weyl group of type $\widetilde{A}_1$. Actually in this case two miracles occur:

(1) Donkin's tilting tensor product theorem ([**Don93**, Proposition 2.1]) together with the knowledge of the characters of the fundamental tilting modules (e.g. see [**DH05**, Lemma 1.1]) allow us to determine the characters of all indecomposable tilting modules. (This is the only semi-simple group for which all tilting characters are known).
(2) The set $\{w_\lambda \mid \lambda \in X_+^\vee\}$ together with its image under the automorphism $s \leftrightarrow t$ yields the set $W_a \setminus \{\mathrm{id}\}$. In particular, knowledge of the tilting characters gives the whole p-canonical basis for SL_2.

To bypass the calculations, the following result (see [**EH02**, Lemma 6]) is useful because it gives a combinatorial description of the Δ-multiplicities of an indecomposable tilting module.

Lemma 5.1. *Let $n \in \mathbb{Z}_{\geqslant 0}$. Write n uniquely as $\sum_{i=0}^{l} n_i p^i$ with $p-1 \leqslant n_i \leqslant 2p-2$ for $i < l$ and $0 \leqslant n_l \leqslant p-1$.*

Then $\Delta(m)$ occurs in a Δ-flag of $T(n)$ if and only if $m = \sum_{i=0}^{l} m_i p^i$ where $m_j = n_j$ or $m_j = 2p-2-n_j$ for $j < l$ and $m_l = n_l$. Moreover, the multiplicity $[T(n) : \Delta(m)]$ is at most one.

Observe that in the last lemma for $m = \sum_{i=0}^{l} m_i p^i$ the coefficients are not required to satisfy $p-1 \leqslant m_i \leqslant 2p-2$ for $i < m$. For $n = \sum_{i=0}^{l} n_i p^i$ as in the lemma there are precisely 2^r natural numbers m such that $\Delta(m)$ occurs in a Δ-flag of $T(n)$ where $r = |\{0 \leqslant i < l \mid n_i \neq p-1\}|$.

To give an example how this works, consider $n = 15$ and $p = 3$. This is uniquely written as $15 = 3 \cdot 3^0 + 4 \cdot 3 + 0 \cdot 3^2$ (as described in the last lemma; the last 0 matters!) and thus 4 standard modules occur (with multiplicity one) in a Δ-flag of $T(15)$, namely $\Delta(m)$ for m among the following:

$$1 = 1 \cdot 3^0 + 0 \cdot 3^1 + 0 \cdot 3^2,$$
$$3 = 3 \cdot 3^0 + 0 \cdot 3^1 + 0 \cdot 3^2,$$
$$13 = 1 \cdot 3^0 + 4 \cdot 3^1 + 0 \cdot 3^2,$$
$$15 = 3 \cdot 3^0 + 4 \cdot 3^1 + 0 \cdot 3^2.$$

See Figure 1 at the end of the paper for the multiplicities of $\Delta(m)$ in $T(n)$ for $p = 3$ where each black box represents a one. Using Lemma 4.7 we get for example:

$$
\begin{array}{lllllll}
{}^3\underline{H}_s & = \underline{H}_s & & & & & \\
{}^3\underline{H}_{st} & = & \underline{H}_{st} & & & & \\
{}^3\underline{H}_{sts} & = & & \underline{H}_{sts} & & & \\
{}^3\underline{H}_{stst} & = & \underline{H}_{st} & + \ \underline{H}_{stst} & & & \\
{}^3\underline{H}_{ststs} & = \underline{H}_s & & + & \underline{H}_{ststs} & & \\
{}^3\underline{H}_{ststst} & = & & & & \underline{H}_{ststst} & \\
{}^3\underline{H}_{stststs} & = & & & \underline{H}_{ststs} & + & \underline{H}_{stststs} \\
{}^3\underline{H}_{stststst} & = & & \underline{H}_{stst} & & + & \underline{H}_{stststst}
\end{array}
$$

Note that even in this relatively simple example one sees a beautiful fractal-like structure emerging!

5.4. Types B_3 and C_3. In the Dynkin diagrams of types B_3 and C_3 we label the simple reflections as follows:

B_3 :
C_3 :
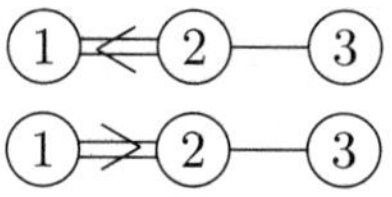

The only interesting case is $p = 2$. The following table gives an overview over all the Weyl group elements for which the 2-canonical basis differs from the Kazhdan-Lusztig basis. It illustrates the dependence of the 2-canonical basis on the type of the root system. Even though the combinatorics in types B_3 and C_3 are the same, the corresponding 2-canonical bases are quite different.

	B_3	C_3
${}^2\underline{H}_{212}$	$\underline{H}_{212}$	$\underline{H}_{212} + \underline{H}_2$
${}^2\underline{H}_{121}$	$\underline{H}_{121} + \underline{H}_1$	$\underline{H}_{121}$
${}^2\underline{H}_{3212}$	$\underline{H}_{3212}$	$\underline{H}_{3212} + \underline{H}_{32}$
${}^2\underline{H}_{2123}$	$\underline{H}_{2123}$	$\underline{H}_{2123} + \underline{H}_{23}$
${}^2\underline{H}_{1321}$	$\underline{H}_{1321} + \underline{H}_{13}$	$\underline{H}_{1321}$
${}^2\underline{H}_{1213}$	$\underline{H}_{1213} + \underline{H}_{13}$	$\underline{H}_{1213}$
${}^2\underline{H}_{32123}$	$\underline{H}_{32123}$	$\underline{H}_{32123} + \underline{H}_{232} + \underline{H}_3$
${}^2\underline{H}_{21232}$	$\underline{H}_{21232}$	$\underline{H}_{21232} + \underline{H}_{232}$
${}^2\underline{H}_{23212}$	$\underline{H}_{23212}$	$\underline{H}_{23212} + \underline{H}_{232}$
${}^2\underline{H}_{21321}$	$\underline{H}_{21321} + \underline{H}_{213}$	$\underline{H}_{21321}$
${}^2\underline{H}_{12132}$	$\underline{H}_{12132} + \underline{H}_{132}$	$\underline{H}_{12132}$
${}^2\underline{H}_{232123}$	$\underline{H}_{232123}$	$\underline{H}_{232123} + (v + v^{-1})\underline{H}_{232}$
${}^2\underline{H}_{212321}$	$\underline{H}_{212321}$	$\underline{H}_{212321} + \underline{H}_{2321}$
${}^2\underline{H}_{121321}$	$\underline{H}_{121321} + \underline{H}_{1212} + \underline{H}_{1321} + \underline{H}_{1213} + \underline{H}_{13}$	$\underline{H}_{121321}$
${}^2\underline{H}_{123212}$	$\underline{H}_{123212}$	$\underline{H}_{123212} + \underline{H}_{1232}$
${}^2\underline{H}_{2123212}$	$\underline{H}_{2123212}$	$\underline{H}_{2123212} + \underline{H}_{21232} + \underline{H}_{23212} + \underline{H}_{232}$
${}^2\underline{H}_{1212321}$	$\underline{H}_{1212321} + \underline{H}_{12123}$	$\underline{H}_{1212321}$
${}^2\underline{H}_{1213212}$	$\underline{H}_{1213212} + \underline{H}_{13212}$	$\underline{H}_{1213212}$
${}^2\underline{H}_{21232123}$	$\underline{H}_{21232123}$	$\underline{H}_{21232123} + \underline{H}_{232123}$
${}^2\underline{H}_{12123212}$	$\underline{H}_{12123212} + \underline{H}_{1212}$	$\underline{H}_{12123212}$
${}^2\underline{H}_{12132123}$	$\underline{H}_{12132123} + \underline{H}_{132123}$	$\underline{H}_{12132123}$

The most interesting entry in the whole table occurs for type C_3 and the element $232123 \in W$ where we have

$$ {}^2\underline{H}_{232123} = \underline{H}_{232123} + (v + v^{-1})\underline{H}_{232} $$

This means that, in the decomposition of $B_{232123} \otimes_{\mathbb{O}} \mathbb{K}$ into indecomposable objects in ${}^{\mathbb{K}}\mathbf{H}$, non-self-dual summands (i.e. with a non-trivial grading shift) occur.

5.5. Type D_4. We label the simple reflections as follows:

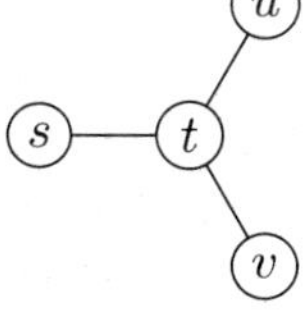

It turns out that the p-canonical basis and the Kazhdan-Lusztig basis coincide for all primes except for $p = 2$. There are four elements $x \in W$ with ${}^2\underline{H}_x \neq \underline{H}_x$. If $x = suvtsuv$ then we have

$$ {}^2\underline{H}_{t_1 x t_2} = \underline{H}_{t_1 x t_2} + \underline{H}_{t_1 suv t_2} $$

for $t_1, t_2 \in \langle t \rangle$. We will give some more details on how to calculate ${}^2\underline{H}_x$. We start out by decomposing the corresponding Bott-Samelson object into indecomposable objects in ${}^{\mathbb{K}}\mathbf{H}$ to get

$$ H_s H_u H_v H_t H_s H_u H_v = \underline{H}_x + (v^{-2} + 3 + v^2)\underline{H}_{suv}. $$

As subexpressions of (s,u,v,t,s,u,v) expressing suv we get

$$\begin{array}{ll}(U1,U1,U1,U0,D0,D0,D0) & \text{of defect } -2,\\ (U1,U1,U0,U0,D0,D0,U1) & \text{of defect } 0,\\ (U1,U0,U1,U0,D0,U1,D0) & \text{of defect } 0,\\ (U0,U1,U1,U0,U1,D0,D0) & \text{of defect } 0,\\ (U1,U0,U0,U0,D0,U1,U1) & \text{of defect } 2,\\ (U0,U1,U0,U0,U1,D0,U1) & \text{of defect } 2,\\ (U0,U0,U1,U0,U1,U1,D0) & \text{of defect } 2,\\ (U0,U0,U0,U0,U1,U1,U1) & \text{of defect } 4.\end{array}$$

The light leaf morphism of degree −2 pairs with the three light leaf morphisms of degree 2 to give the matrix

$$\begin{pmatrix}-1\\-1\\-1\end{pmatrix}.$$

The light leaf morphisms corresponding to the subexpressions of defect 0 are the following:

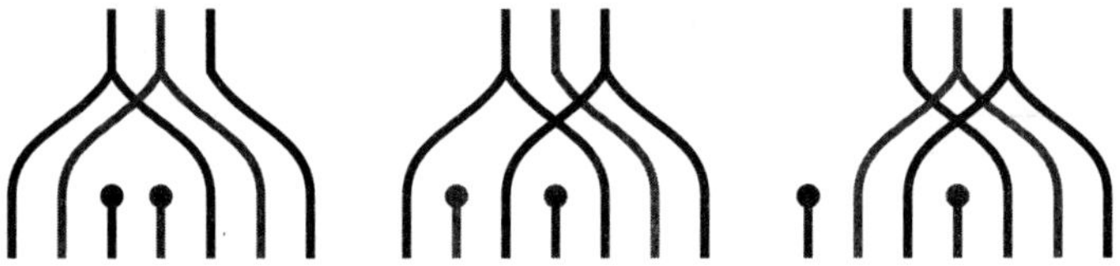

Pairing them gives the following degree 0 piece of the intersection form:

$$\begin{pmatrix}0 & -1 & -1\\ -1 & 0 & -1\\ -1 & -1 & 0\end{pmatrix}.$$

Note that the determinant of this matrix is −2 and its rank in characteristic 2 is 2. Therefore ${}^{k}B_{suvtsuv} \otimes_{\mathbb{O}} \mathbb{K}$ decomposes as

$${}^{\mathbb{K}}B_{suvtsuv} \oplus {}^{\mathbb{K}}B_{suv}$$

giving the result we stated above. The geometry of this example is discussed in the appendix of [**WB12**]. Note that all calculations presented in this section can also be carried out using the formula in the nil Hecke ring (see [**HW15**, §6.2]).

5.6. Type A_n. According to [**WB12**], the p-canonical basis and the Kazhdan-Lusztig basis coincide for all primes p for $n < 7$. Thus, we will describe the case $n = 7$ where the situation is quite remarkable.

For all primes $p \neq 2$ the p-canonical basis and the Kazhdan-Lusztig basis agree. We have ${}^{2}\underline{H}_x \neq \underline{H}_x$ for exactly 38 out of 40320 elements in S_8 and these examples fall into four classes.

In the following we will denote for a subset $I \subseteq S$ the corresponding parabolic subgroup by $W_I = \langle s \in I \rangle \subseteq W$. If W_I is finite, its longest element will be denoted by w_I. A permutation $\phi \in S_8$ will be displayed as a string $\phi(1)\phi(2)\ldots\phi(8)$.

The Kashiwara-Saito singularity ([**KS97**]): This corresponds to the permutation $w = 62845173$. We have

$$^2\underline{H}_w = \underline{H}_w + \underline{H}_{w_I}$$

with $I = \{1,3,4,5,7\}$. There is a cluster of 16 elements around the Kashiwara-Saito singularity described as follows. If we let $J = \{2,6\}$, then we have

$$^2\underline{H}_{uwv} = \underline{H}_{uwv} + \underline{H}_{uw_Iv}.$$

for all $u, v \in W_J$ unless $u = v = w_J$ in which case

$$^2\underline{H}_{w_Jww_J} = \underline{H}_{w_Jww_J} + \underline{H}_{w_Jw_Iw_J} + \underline{H}_{w_K}$$

where $K = \{1,2,3,5,6,7\}$.

The Hexagon singularity (Braden's example in the appendix of [**WB12**]): Consider the permutation $w = 46718235$. We have

$$^2\underline{H}_w = \underline{H}_w + \underline{H}_{w_I}$$

with $I = \{2,3,5,6\}$. In this case we get a cluster of size 4. For any $u, v \in \langle s_4 \rangle$ we have

$$^2\underline{H}_{uwv} = \underline{H}_{uwv} + \underline{H}_{uw_Iv}$$

unless $u = v = s_4$ in which case

$$^2\underline{H}_{s_4ws_4} = \underline{H}_{s_4ws_4} + \underline{H}_{s_4w_Is_4} + \underline{H}_{w_K}$$

where $K = \{1,3,4,5,7\}$.

For the Kashiwara-Saito singularity and the Hexagon singularity the calculation of the the local intersection form of $\underline{w}$ at w_I using the formula in the nil Hecke ring can be found in [**HW15**, §6.1 and §6.2]. In both cases one can find $\underline{w}$, a reduced expression for w, such that this local intersection form is a 1×1 matrix.

The waterfall: Consider the permutation $w_1 = 67283415$. Then we have

$$^2\underline{H}_{w_1u} = \underline{H}_{w_1u} + \underline{H}_{w_Iu}.$$

for $I = \{1,2,4,5,6\}$ and $w_J \neq u \in W_J$ with $J = \{3,7\}$.

Similarly, if we let $w_2 = 57813462$, then we have

$$^2\underline{H}_{vw_2} = \underline{H}_{vw_2} + \underline{H}_{vw'_I}$$

where $I' = \{2,3,4,6,7\}$ and $w_{J'} \neq v \in W_J$ with $J = \{1,5\}$.

The situation becomes more complicated due to the fact that for $u = w_J$ and $v = w_{J'}$ we have $w_1u = vw_2 =: w$. In this case we get

$$^2\underline{H}_w = \underline{H}_w + \underline{H}_{w_Iu} + \underline{H}_{vw'_I}.$$

Note that, unlike the Kashiwara-Saito and hexagon permutations discussed above, the clusters containing w_1 and w_2 are neither swapped nor fixed by the automorphism $s_i \mapsto s_{8-i}$ and thus by applying the graph automorphism one obtains another seven elements for which $^2\underline{H}_x \neq \underline{H}_x$. Hence the two "waterfall" clusters contain 14 elements in total.

The basket: Consider the permutation

$$w = 84627351.$$

Then for all $u, v \in \langle s_4 \rangle$ one has

$$^2\underline{H}_{uwv} = \underline{H}_{uwv} + \underline{H}_{uw_Iv}$$

where $I = \{1,2,3,5,6,7\}$.

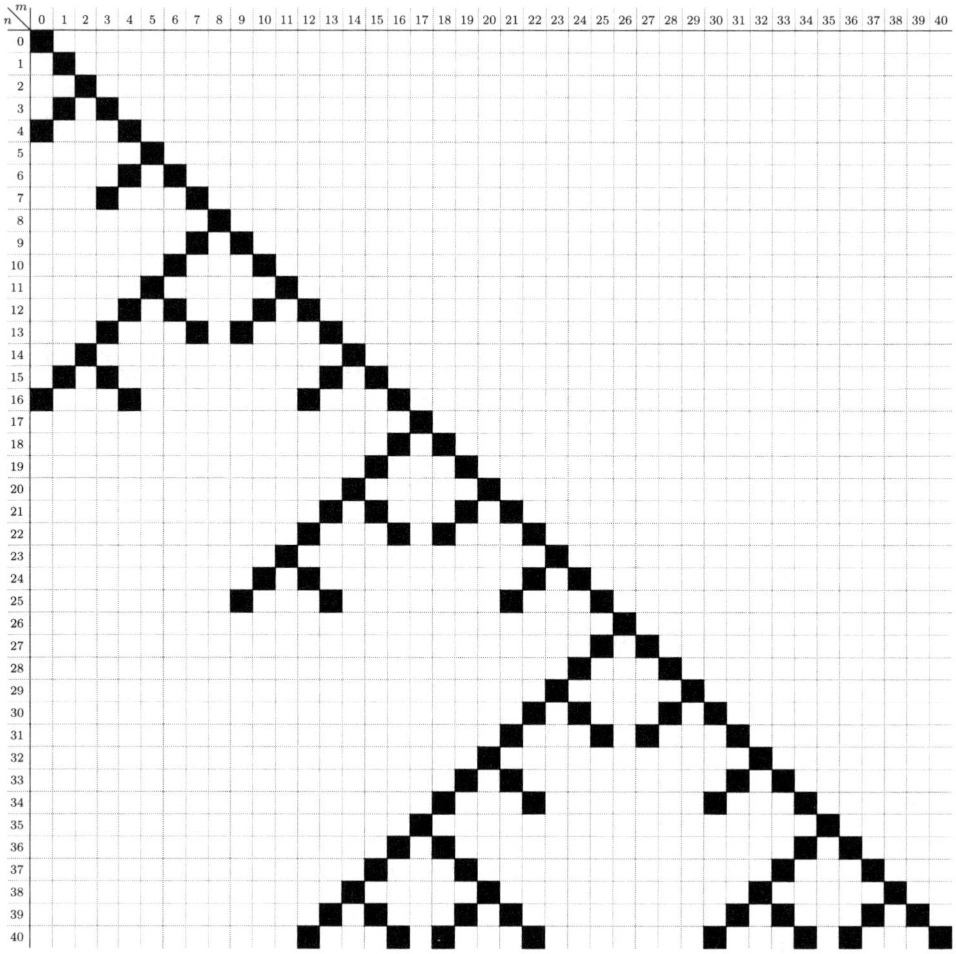

FIGURE 1. The multiplicities of $\Delta(m)$ in $T(n)$ for $p = 3$.

References

[AR13] Pramod N. Achar and Laura Rider, *Parity sheaves on the affine Grassmannian and the Mirković-Vilonen conjecture*, Acta Math. **215** (2015), no. no. 2, 183–216. MR3455233

[AR14] Pramod N. Achar and Simon Riche, *Modular perverse sheaves on flag varieties I: tilting and parity sheaves* (English, with English and French summaries), Ann. Sci.Éc. Norm. Supér. (4) **49** (2016), no. no. 2, 325–370. With a joint appendix with Geordie Williamson. MR3481352

[Ari96] Susumu Ariki, *On the decomposition numbers of the Hecke algebra of $G(m, 1, n)$*, J. Math. Kyoto Univ. **36** (1996), no. 4, 789–808. MR1443748

[BBD82] A. A. Beĭlinson, J. Bernstein, and P. Deligne, *Faisceaux pervers* (French), Analysis and topology on singular spaces, I (Luminy, 1981), Astérisque, vol. 100, Soc. Math. France, Paris, 1982, pp. 5–171. MR751966

[BL94] Joseph Bernstein and Valery Lunts, *Equivariant sheaves and functors*, Lecture Notes in Mathematics, vol. 1578, Springer-Verlag, Berlin, 1994. MR1299527

[dCM09] Mark Andrea A. de Cataldo and Luca Migliorini, *The decomposition theorem, perverse sheaves and the topology of algebraic maps*, Bull. Amer. Math. Soc. (N.S.) **46** (2009), no. 4, 535–633, DOI 10.1090/S0273-0979-09-01260-9. MR2525735

[Del80] Pierre Deligne, *La conjecture de Weil. II* (French), Inst. Hautes Études Sci. Publ. Math. **52** (1980), 137–252. MR601520

[DH05] Stephen Doty and Anne Henke, *Decomposition of tensor products of modular irreducibles for* SL_2, Q. J. Math. **56** (2005), no. 2, 189–207, DOI 10.1093/qmath/hah027. MR2143497

[Don93] Stephen Donkin, *On tilting modules for algebraic groups*, Math. Z. **212** (1993), no. 1, 39–60, DOI 10.1007/BF02571640. MR1200163

[EH02] Karin Erdmann and Anne Henke, *On Ringel duality for Schur algebras*, Math. Proc. Cambridge Philos. Soc. **132** (2002), no. 1, 97–116, DOI 10.1017/S0305004101005485. MR1866327

[EK10] Ben Elias and Mikhail Khovanov, *Diagrammatics for Soergel categories*, Int. J. Math. Math. Sci. (2010), Art. ID 978635, 58. MR3095655

[Eli13] Ben Elias, *The two-color Soergel calculus*, Compos. Math. **152** (2016), no. no. 2, 327–398. MR3462556

[EW12] B. Elias and G. Williamson, *Diagrammatics for Coxeter groups and their braid groups*, 2012, to appear in Quantum Topology, arXiv:1405.4928.

[EW13] Ben Elias and Geordie Williamson, *Soergel calculus*, Represent. Theory **20** (2016), 295–374. MR3555156

[EW14] Ben Elias and Geordie Williamson, *The Hodge theory of Soergel bimodules*, Ann. of Math. (2) **180** (2014), no. 3, 1089–1136, DOI 10.4007/annals.2014.180.3.6. MR3245013

[FW14] Peter Fiebig and Geordie Williamson, *Parity sheaves, moment graphs and the p-smooth locus of Schubert varieties* (English, with English and French summaries), Ann. Inst. Fourier (Grenoble) **64** (2014), no. 2, 489–536. MR3330913

[Gör10] Ulrich Görtz, *Affine Springer fibers and affine Deligne-Lusztig varieties*, Affine flag manifolds and principal bundles, Trends Math., Birkhäuser/Springer Basel AG, Basel, 2010, pp. 1–50.

[Gro99] I. Grojnowski, *Affine $\mathfrak{sl}_p$ controls the representation theory of the symmetric group and related Hecke algebras*, 1999, arXiv:math/9907129v1.

[Hum90] James E. Humphreys, *Reflection groups and Coxeter groups*, Cambridge Studies in Advanced Mathematics, vol. 29, Cambridge University Press, Cambridge, 1990. MR1066460

[HW15] Xuhua He and Geordie Williamson, *Soergel calculus and Schubert calculus*, 2015, arXiv:1502.04914.

[Jam90] Gordon James, *The decomposition matrices of* $\mathrm{GL}_n(q)$ *for* $n \leq 10$, Proc. London Math. Soc. (3) **60** (1990), no. 2, 225–265, DOI 10.1112/plms/s3-60.2.225. MR1031453

[JMW14a] Daniel Juteau, Carl Mautner, and Geordie Williamson, *Parity sheaves*, J. Amer. Math. Soc. **27** (2014), no. 4, 1169–1212, DOI 10.1090/S0894-0347-2014-00804-3. MR3230821

[JMW14b] Daniel Juteau, Carl Mautner, and Geordie Williamson, *Parity sheaves and tilting modules* (English, with English and French summaries), Ann. Sci. Éc. Norm. Supér. (4) **49** (2016), no. 2, 257–275. MR3481350

[KL79] David Kazhdan and George Lusztig, *Representations of Coxeter groups and Hecke algebras*, Invent. Math. **53** (1979), no. 2, 165–184, DOI 10.1007/BF01390031. MR560412

[KL80] David Kazhdan and George Lusztig, *Schubert varieties and Poincaré duality*, Geometry of the Laplace operator (Proc. Sympos. Pure Math., Univ. Hawaii, Honolulu, Hawaii, 1979), Proc. Sympos. Pure Math., XXXVI, Amer. Math. Soc., Providence, R.I., 1980, pp. 185–203. MR573434

[KS97] Masaki Kashiwara and Yoshihisa Saito, *Geometric construction of crystal bases*, Duke Math. J. **89** (1997), no. 1, 9–36, DOI 10.1215/S0012-7094-97-08902-X. MR1458969

[Kum02] Shrawan Kumar, *Kac-Moody groups, their flag varieties and representation theory*, Progress in Mathematics, vol. 204, Birkhäuser Boston, Inc., Boston, MA, 2002. MR1923198

[Lam01] T. Y. Lam, *A first course in noncommutative rings*, 2nd ed., Graduate Texts in Mathematics, vol. 131, Springer-Verlag, New York, 2001. MR1838439

[Lib08] Nicolas Libedinsky, *Sur la catégorie des bimodules de Soergel* (French, with French summary), J. Algebra **320** (2008), no. 7, 2675–2694, DOI 10.1016/j.jalgebra.2008.05.027. MR2441994

[Lib10] Nicolas Libedinsky, *Presentation of right-angled Soergel categories by generators and relations*, J. Pure Appl. Algebra **214** (2010), no. 12, 2265–2278, DOI 10.1016/j.jpaa.2010.02.026. MR2660912

[Lib15] Nicolas Libedinsky, *Light leaves and Lusztig's conjecture*, Adv. Math. **280** (2015), 772–807, DOI 10.1016/j.aim.2015.04.022. MR3350234

[LLT96] Alain Lascoux, Bernard Leclerc, and Jean-Yves Thibon, *Hecke algebras at roots of unity and crystal bases of quantum affine algebras*, Comm. Math. Phys. **181** (1996), no. 1, 205–263. MR1410572

[Lus80] George Lusztig, *Some problems in the representation theory of finite Chevalley groups*, The Santa Cruz Conference on Finite Groups (Univ. California, Santa Cruz, Calif., 1979), Proc. Sympos. Pure Math., vol. 37, Amer. Math. Soc., Providence, R.I., 1980, pp. 313–317. MR604598

[Lus83] George Lusztig, *Singularities, character formulas, and a q-analog of weight multiplicities*, Analysis and topology on singular spaces, II, III (Luminy, 1981), Astérisque, vol. 101, Soc. Math. France, Paris, 1983, pp. 208–229. MR737932

[Mak15] Ruslan Maksimau, *Canonical basis, KLR algebras and parity sheaves*, J. Algebra **422** (2015), 563–610, DOI 10.1016/j.jalgebra.2014.08.052. MR3272091

[MR15] Carl Mautner and Simon Riche, *Exotic tilting sheaves, parity sheaves on affine Grassmannians, and the Mirkovic-Vilonen conjecture*, 2015, arXiv:1501.07369v2.

[MV07] I. Mirković and K. Vilonen, *Geometric Langlands duality and representations of algebraic groups over commutative rings*, Ann. of Math. (2) **166** (2007), no. 1, 95–143, DOI 10.4007/annals.2007.166.95. MR2342692

[Rou12] Raphaël Rouquier, *Quiver Hecke algebras and 2-Lie algebras*, Algebra Colloq. **19** (2012), no. 2, 359–410, DOI 10.1142/S1005386712000247. MR2908731

[RW15] Simon Riche and Geordie Williamson, *Tilting modules and the p-canonical basis*, 2015, arXiv:1512.08296.

[Sai90] Morihiko Saito, *Decomposition theorem for proper Kähler morphisms*, Tohoku Math. J. (2) **42** (1990), no. 2, 127–147, DOI 10.2748/tmj/1178227650. MR1053945

[SGA70] *Schémas en groupes. III: Structure des schémas en groupes réductifs* (French), Séminaire de Géométrie Algébrique du Bois Marie 1962/64 (SGA 3). Dirigé par M. Demazure et A. Grothendieck. Lecture Notes in Mathematics, Vol. 153, Springer-Verlag, Berlin-New York, 1970. MR0274460

[Soe97] Wolfgang Soergel, *Kazhdan-Lusztig-Polynome und eine Kombinatorik für Kipp-Moduln* (German, with English summary), Represent. Theory **1** (1997), 37–68 (electronic), DOI 10.1090/S1088-4165-97-00006-X. MR1445511

[Soe98] Wolfgang Soergel, *Combinatorics of Harish-Chandra modules*, Representation theories and algebraic geometry (Montreal, PQ, 1997), NATO Adv. Sci. Inst. Ser. C Math. Phys. Sci., vol. 514, Kluwer Acad. Publ., Dordrecht, 1998, pp. 401–412. MR1653039

[Soe00] Wolfgang Soergel, *On the relation between intersection cohomology and representation theory in positive characteristic*, J. Pure Appl. Algebra **152** (2000), no. 1-3, 311–335, DOI 10.1016/S0022-4049(99)00138-3. Commutative algebra, homological algebra and representation theory (Catania/Genoa/Rome, 1998). MR1784005

[Soe01] Wolfgang Soergel, *Langlands' philosophy and Koszul duality*, Algebra—representation theory (Constanta, 2000), NATO Sci. Ser. II Math. Phys. Chem., vol. 28, Kluwer Acad. Publ., Dordrecht, 2001, pp. 379–414, DOI 10.1007/978-94-010-0814-3_17. MR1858045

[Soe07] Wolfgang Soergel, *Kazhdan-Lusztig-Polynome und unzerlegbare Bimoduln über Polynomringen* (German, with English and German summaries), J. Inst. Math. Jussieu **6** (2007), no. 3, 501–525, DOI 10.1017/S1474748007000023. MR2329762

[Spr82] T. A. Springer, *Quelques applications de la cohomologie d'intersection* (French), Bourbaki Seminar, Vol. 1981/1982, Astérisque, vol. 92, Soc. Math. France, Paris, 1982, pp. 249–273. MR689533

[Spr89] T. A. Springer, *Linear algebraic groups*, 2nd ed., Modern Birkhäuser Classics, Birkhäuser Boston, Inc., Boston, MA, 2009. MR2458469

[VV11] M. Varagnolo and E. Vasserot, *Canonical bases and KLR-algebras*, J. Reine Angew. Math. **659** (2011), 67–100, DOI 10.1515/CRELLE.2011.068. MR2837011

[VW13] Kari Vilonen and Geordie Williamson, *Characteristic cycles and decomposition numbers*, Math. Res. Lett. **20** (2013), no. 2, 359–366, DOI 10.4310/MRL.2013.v20.n2.a11. MR3151652

[WB12] Geordie Williamson and Tom Braden, *Modular intersection cohomology complexes on flag varieties*, Math. Z. **272** (2012), no. 3-4, 697–727, DOI 10.1007/s00209-011-0955-y. MR2995137

[Wil12] Geordie Williamson, *Some examples of parity sheaves*, March 2012, Oberwolfach report for the meeting "Enveloping Algebras and Geometric Representation Theory", available under: http://people.mpim-bonn.mpg.de/geordie/GWilliamson-EnvAlg.pdf.

[Wil13] Geordie Williamson, *Schubert calculus and torsion explosion*, 2013, arXiv:1309.5055.

[Wil14] Geordie Williamson, *A reducible characteristic variety in type A*, Representations of reductive groups, Prog. Math. Phys., vol. 312, Birkhäuser/Springer, Cham, 2015, pp. 517–532. MR3496286

Max Planck Institute for Mathematics, Vivatsgasse 7, 53111 Bonn
E-mail address: `ltjensen@mpim-bonn.mpg.de`

Max Planck Institute for Mathematics, Vivatsgasse 7, 53111 Bonn
E-mail address: `geordw@gmail.com`
URL: `http://people.mpim-bonn.mpg.de/geordie/`

Selected Published Titles in This Series

683 **Anna Beliakova and Aaron D. Lauda, Editors,** Categorification and Higher Representation Theory, 2017

682 **Gregory Arone, Brenda Johnson, Pascal Lambrechts, Brian A. Munson, and Ismar Volić, Editors,** Manifolds and K-Theory, 2017

681 **Shiferaw Berhanu, Nordine Mir, and Emil J. Straube, Editors,** Analysis and Geometry in Several Complex Variables, 2017

680 **Sergei Gukov, Mikhail Khovanov, and Johannes Walcher, Editors,** Physics and Mathematics of Link Homology, 2016

679 **Catherine Bénéteau, Alberto A. Condori, Constanze Liaw, William T. Ross, and Alan A. Sola, Editors,** Recent Progress on Operator Theory and Approximation in Spaces of Analytic Functions, 2016

678 **Joseph Auslander, Aimee Johnson, and Cesar E. Silva, Editors,** Ergodic Theory, Dynamical Systems, and the Continuing Influence of John C. Oxtoby, 2016

677 **Delaram Kahrobaei, Bren Cavallo, and David Garber, Editors,** Algebra and Computer Science, 2016

676 **Pierre Martinetti and Jean-Christophe Wallet, Editors,** Noncommutative Geometry and Optimal Transport, 2016

675 **Ana Claudia Nabarro, Juan J. Nuño-Ballesteros, Raúl Oset Sinha, and Maria Aparecida Soares Ruas, Editors,** Real and Complex Singularities, 2016

674 **Bogdan D. Suceavă, Alfonso Carriazo, Yun Myung Oh, and Joeri Van der Veken, Editors,** Recent Advances in the Geometry of Submanifolds, 2016

673 **Alex Martsinkovsky, Gordana Todorov, and Kiyoshi Igusa, Editors,** Recent Developments in Representation Theory, 2016

672 **Bernard Russo, Asuman Güven Aksoy, Ravshan Ashurov, and Shavkat Ayupov, Editors,** Topics in Functional Analysis and Algebra, 2016

671 **Robert S. Doran and Efton Park, Editors,** Operator Algebras and Their Applications, 2016

670 **Krishnendu Gongopadhyay and Rama Mishra, Editors,** Knot Theory and Its Applications, 2016

669 **Sergiĭ Kolyada, Martin Möller, Pieter Moree, and Thomas Ward, Editors,** Dynamics and Numbers, 2016

668 **Gregory Budzban, Harry Randolph Hughes, and Henri Schurz, Editors,** Probability on Algebraic and Geometric Structures, 2016

667 **Mark L. Agranovsky, Matania Ben-Artzi, Greg Galloway, Lavi Karp, Dmitry Khavinson, Simeon Reich, Gilbert Weinstein, and Lawrence Zalcman, Editors,** Complex Analysis and Dynamical Systems VI: Part 2: Complex Analysis, Quasiconformal Mappings, Complex Dynamics, 2016

666 **Vicenţiu D. Rădulescu, Adélia Sequeira, and Vsevolod A. Solonnikov, Editors,** Recent Advances in Partial Differential Equations and Applications, 2016

665 **Helge Glöckner, Alain Escassut, and Khodr Shamseddine, Editors,** Advances in Non-Archimedean Analysis, 2016

664 **Dihua Jiang, Freydoon Shahidi, and David Soudry, Editors,** Advances in the Theory of Automorphic Forms and Their L-functions, 2016

663 **David Kohel and Igor Shparlinski, Editors,** Frobenius Distributions: Lang-Trotter and Sato-Tate Conjectures, 2016

662 **Zair Ibragimov, Norman Levenberg, Sergey Pinchuk, and Azimbay Sadullaev, Editors,** Topics in Several Complex Variables, 2016

661 **Douglas P. Hardin, Doron S. Lubinsky, and Brian Z. Simanek, Editors,** Modern Trends in Constructive Function Theory, 2016

For a complete list of titles in this series, visit the
AMS Bookstore at **www.ams.org/bookstore/conmseries/**.